Advances in Experimental Medicine and Biology

For futher volumes:
http://www.springer.com/series/5584

Alexander S. Kauffman • Jeremy T. Smith
Editors

Kisspeptin Signaling in Reproductive Biology

Springer

Editors
Alexander S. Kauffman
Department of Reproductive Medicine
University of California, San Diego
La Jolla, CA, USA

Jeremy T. Smith
School of Anatomy, Physiology
and Human Biology
The University of Western Australia
Nedlands, WA, Australia

ISSN 0065-2598
ISBN 978-1-4614-6198-2 ISBN 978-1-4614-6199-9 (eBook)
DOI 10.1007/978-1-4614-6199-9
Springer New York Heidelberg Dordrecht London

Library of Congress Control Number: 2013932777

Printed on acid-free paper

Springer is part of Springer Science+Business Media (www.springer.com)

Preface

We were lucky to get into the kisspeptin "field" during its early days, though we had different avenues of entry. In early 2004, Sasha was a postdoctoral fellow in Dr. Emilie Rissman's lab in Virginia, where he was studying the effects of various reproductive neuropeptides, such as GALP and GnRH variants, on sex behavior. Dr. Rissman asked whether Sasha would be interested in studying sexual behavior and sexual differentiation in a newly created transgenic mouse that was lacking an orphan receptor called GPR54. This was just months after several high-profile (and soon-to-be seminal) papers had been published linking this receptor—and its ligand kisspeptin—to reproductive status in humans and mice. The prospect of studying this new, highly uncharacterized reproductive factor was too enticing to pass up.

Meanwhile, across the country in Seattle, Washington, Jeremy was also conducting postdoctoral research in the lab of Dr. Robert Steiner, who had also recently stumbled upon the exciting new kisspeptin system and was currently testing its gonadotropin-releasing ability and sex steroid regulation in mice. Such was our initial foray into kisspeptin reproductive biology, and our paths soon crossed when Sasha headed west in 2005 to work in the Steiner lab. Since then, much has been learned—by us and many others—about kisspeptin and GPR54. Satisfyingly, much of what the field has learned in the past decade has come from a growing community of international scientists and labs situated all around the globe, not to mention from a large variety of species and animal models. Indeed, kisspeptin has now been shown, by numerous investigators, to be important for puberty and reproduction in not only humans and mice, as originally demonstrated in 2003, but in a great assemblage of vertebrate species, ranging from mammals of all shapes and sizes down to frogs and fish.

As we rapidly approach the 10-year anniversary of the "discovery" of kisspeptin's role in reproductive biology, we believe this textbook is particularly timely. The goal was to critically highlight—in a single cohesive volume—the functions and regulation of kisspeptin as it relates to reproductive biology. To this end, we present in-depth reviews of a wide range of kisspeptin topics, including (but not limited to) hormone-releasing effects of kisspeptin in a variety of species (including humans); consequences and underlying mechanisms of impaired kisspeptin systems;

utility and benefits of specific kisspeptin experimental "tools" (including a variety of new mouse models and novel chemical analogues); development of the kisspeptin system; the role of kisspeptin in puberty; the regulation of kisspeptin circuits by hormones, photoperiod, circadian signals, metabolic conditions, and stress, as well as by other reproductive neuropeptides; alterations in kisspeptin induced by endocrine-disrupting chemicals; and the newest kid on the block: electrophysiological properties of kisspeptin neurons. In addition, this book emphasizes several larger themes, including the importance and benefits of comparative biology (despite—or perhaps because of—the occasional "species differences"), as well as the value of using both molecular and physiological analyses to advance the field, not to mention the growing importance of bedside-to-benchside (and back) research. Another prominent message throughout the book is that there is still much to learn. While a good deal of the "low-hanging fruit" on the kisspeptin experimental tree has now been plucked, there are still many intriguing questions to be answered and many fundamental gaps in our knowledge that demand filling. These numerous key areas of future direction underscore the notion that, while we have come incredibly far in the past decade, we still have a long way to go to solve some of reproductive biology's greatest puzzles. Thus, our aim for this book is to summarize and celebrate the key findings from the past decade of kisspeptin research, as well as stimulate both future experimentation and further refinement of clinical use of kisspeptin for medical and therapeutic endeavors.

In sum, we hope this cohesive and timely volume of kisspeptin reviews will be a valuable educational and reference tool, useful for both the seasoned kisspeptin researcher and clinician, as well as for new students and doctors wishing to dive into the kisspeptin pool. We believe the comprehensive discussion herein of the numerous kisspeptin topics will provide a fair, yet critical, assessment of the current state of the kisspeptin field, as well as identify key areas of targeted future research and clinical applicability. Indeed, as we pass the first 10-year marker on our kisspeptin journey, it is likely that many of the open questions identified throughout this book will guide the next series of exciting experimental and therapeutic kisspeptin ventures well into the next decade.

La Jolla, CA, USA — Alexander S. Kauffman
Nedlands, WA, Australia — Jeremy T. Smith

Contents

Contributors

Ali Abbara, B.Sc., M.B.B.S. Division of Diabetes, Endocrinology and Metabolism, Department of Investigative Medicine, Imperial College London, London, UK

Meenakshi Alreja, Ph.D. Yale University School of Medicine, New Haven, CT, USA

Suzy Drummond Carvalho Bianco, Pharm.D., Ph.D. Department of Molecular and Cellular Pharmacology and Department of Medicine, University of Miami Miller School of Medicine, Miami, FL, USA

Colin H. Brown, Ph.D. Centre for Neuroendocrinology and Department of Physiology, University of Otago, Dunedin, New Zealand

Alain Caraty, Ph.D. Unité Mixte de Recherche, Physiologie de la Reproduction et des Comportements, Université de Tours/Haras Nationaux, Nouzilly, France

Juan Manuel Castellano, Ph.D. Department of Cell Biology, Physiology and Immunology, University of Cordoba, Cordoba, Spain

Yee-Ming Chan, M.D., Ph.D. Division of Endocrinology, Department of Medicine, Boston Children's Hospital, Boston, MA, USA

Reproductive Endocrine Unit and Harvard Reproductive Sciences Center, Department of Medicine, Massachusetts General Hospital, Boston, MA, USA

Iain J. Clarke, Ph.D. Department of Physiology, School of Biomedical Sciences, Monash University, Clayton, VIC, Australia

William H. Colledge, B.Sc., Ph.D. Department of Physiology, Development and Neuroscience, University of Cambridge, Cambridge, UK

Waljit S. Dhillo, B.Sc., M.B.B.S., F.R.C.P., F.R.C.Path. Division of Diabetes, Endocrinology and Metabolism, Department of Investigative Medicine, Imperial College London, London, UK

Joanne Doran Takeda Cambridge, Cambridge, UK

Robert L. Goodman, Ph.D. Department of Physiology and Pharmacology, West West Virginia University School of Medicine, Morgantown, WV, USA

Pasha Grachev, B.Sc., M.Sc. Division of Women's Health, School of Medicine, King's College London, London, UK

Kathryn A. Guerriero, Ph.D. Wisconsin National Primate Research Center, University of Wisconsin, Madison, WI, USA

Stanley M. Hileman, Ph.D. Department of Physiology and Pharmacology, West Virginia University School of Medicine, Morgantown, WV, USA

Channa N. Jayasena, B.Sc., M.B.B.S., Ph.D. Division of Diabetes, Endocrinology and Metabolism, Department of Investigative Medicine, Imperial College London, London, UK

Ursula B. Kaiser, M.D. Division of Endocrinology, Department of Medicine, Harvard Medical School/Brigham and Women's Hospital, Boston, MA, USA

Shinji Kanda, Ph.D. Department of Biological Sciences, Graduate School of Science, University of Tokyo, Bunkyo, Tokyo, Japan

Alexander S. Kauffman, Ph.D. Department of Reproductive Medicine, University of California, San Diego, La Jolla, CA, USA

Martin J. Kelly, Ph.D. Physiology and Pharmacology, Oregon Health and Science University, Portland, OR, USA

Lance J. Kriegsfeld, Ph.D. Department of Psychology and Helen Wills Neuroscience Institute, Neurobiology Laboratory, University of California, Berkeley, CA, USA

Ana Claudia Latronico Department of Internal Medicine, Division of Endocrinology and Metabolism, Hospital das Clinicas, Sao Paulo University, Sao Paulo, Brazil

Michael N. Lehman, Ph.D. Department of Neurobiology and Anatomical Sciences, University of Mississippi Medical Center, Jackson, MS, USA

Xiao Feng Li, M.D., Ph.D. Division of Women's Health, School of Medicine, King's College London, London, UK

Kei-ichiro Maeda, D.V.M., Ph.D. Animal Breeding, Veterinary Medical Sciences, The University of Tokyo, Tokyo, Japan

Hua Mei, Ph.D. Department of Ophthalmology, Cornea Division, University of California, Los Angeles, CA, USA

Robert P. Millar, M.Sc., Ph.D., M.R.C.Path., F.R.C.P.Path. Department Zoology and Entomology, University of Pretoria, Pretoria, South Africa

Víctor M. Navarro, Ph.D. Division of Endocrinology, Diabetes and Hypertension, Brigham and Women's Hospital and Harvard Medical School, Boston, MA, USA

Kevin O'Byrne, B.Sc., Ph.D. Division of Women's Health, King's College London, London, UK

Satoshi Ohkura, Ph.D. Graduate School of Bioagricultural Sciences, Laboratory of Animal Production, Nagoya University, Nagoya, Japan

Yoshitaka Oka, Ph.D. Department of Biological Sciences, Graduate School of Science, University of Tokyo, Bunkyo, Tokyo, Japan

Hiroaki Okamura, Ph.D., D.V.M. Animal Physiology Research Unit, Division of Animal Sciences, National Institute of Agrobiological Sciences, Tsukuba, Ibaraki, Japan

Heather B. Patisaul, Ph.D. Department of Biology, North Carolina State University, Raleigh, NC, USA

Tony M. Plant, Ph.D. University of Pittsburgh School of Medicine, Pittsburgh, PA, USA

Risheka Ratnasabapathy, B.Sc., M.B.B.S. Division of Diabetes, Endocrinology and Metabolism, Department of Investigative Medicine, Imperial College London, London, UK

Oline K. Rønnekleiv, Ph.D. Physiology and Pharmacology, Oregon Health and Science University, Portland, OR, USA

Antonia Kathryn Roseweir, Ph.D. Institute of Pharmacy and Biomedical Sciences, University of Strathclyde, Glasgow, Scotland, UK

Institute of Cancer Sciences, School of Medicine, Veterinary Medicine and Life Sciences, University of Glasgow, Glasgow, Scotland, UK

Victoria Scott, Ph.D. Centre for Neuroendocrinology and Department of Physiology, University of Otago, Dunedin, New Zealand

Sheila J. Semaan, Ph.D. Department of Reproductive Medicine, University of California, San Diego, La Jolla, CA, USA

Stephanie Beth Seminara, M.D. Department of Internal Medicine, Reproductive Endocrine Unit, Bartlett Hall Extension 504, Endocrine Division, Massachusetts General Hospital, Boston, MA, USA

Letícia Gontijo Silveira, M.D. Department of Internal Medicine, Division of Endocrinology and Metabolism, Hospital das Clinicas, Sao Paulo University, Sao Paulo, Brazil

Jeremy T. Smith Department of Anatomy, Physiology and Human Biology, The University of Western Australia, Nedlands, WA, Australia

Robert A. Steiner Department of Physiology and Biophysics, University of Washington, Seattle, WA, USA

Manuel Tena-Sempere, M.D., Ph.D. Department of Cell Biology, Physiology and Immunology, University of Cordoba, Cordoba, Spain

Ciberobn Fisiopatologia de la Obesidad y Nutricion, Cordoba, Spain

Ei Terasawa, Ph.D. Wisconsin National Primate Research Center, University of Wisconsin, Madison, WI, USA

Department of Pediatrics, University of Wisconsin, Madison, WI, USA

Kristen P. Tolson, Ph.D. Department of Reproductive Medicine, University of California, San Diego, La Jolla, CA, USA

Hiroko Tsukamura, Ph.D. Graduate School of Bioagricultural Sciences, Laboratory of Reproductive Science, Nagoya University, Nagoya, Japan

Yoshihisa Uenoyama, Ph.D. Graduate School of Bioagricultural Sciences, Laboratory of Reproductive Science, Nagoya University, Nagoya, Japan

Yoshihiro Wakabayashi, Ph.D. Animal Physiology Research Unit, Division of Animal Sciences, National Institute of Agrobiological Sciences, Tsukuba, Ibaraki, Japan

Part I
Kisspeptin Signaling In Vivo and In Vitro

Chapter 1
Kisspeptin: Past, Present, and Prologue

Robert A. Steiner

Abstract Research in the nineteenth and early twentieth century established that the brain awakens reproduction, governs reproductive activity in the adult of virtually all vertebrates. By 1950, nearly 100 years later, scientists realized that the hypothalamus and its neurosecretory products play a key role in regulating gonadal function in both males and females. Another 20 years would be required to reveal the chemical identity of GnRH and establish that neurons producing GnRH represent the final common pathway through which the brain regulates gonadotropin secretion. It had also become clear that GnRH neurons behave more like motor neurons—better perhaps at going than stopping—and are themselves regulated by a complex network of afferent inputs, which guide the tempo of sexual maturation, regulate estrous and menstrual cycles, control seasonal breeding, and stop reproduction under adversity. In 2003, the revelation that kisspeptin and its receptor are critical for reproduction opened a floodgate of research documenting the role of kisspeptin neurons as central processors of reproduction. Today, there is wide consensus that kisspeptin signaling in the brain is essential, providing the impetus to GnRH neurons to awaken at puberty and reigning the activity of these neurons when discretion is advised. We celebrate this watershed moment—with full knowledge that time and discovery will provide context and perspective to even these heady days.

In the autumn of 1848, A.A. Berthold fixed his eye upon two sad roosters. The German physiologist had castrated the young cockerels some months earlier, after which their tail feathers were lost, their combs had yellowed and drooped, their voices became monotonic, and the creatures grew disinterested in chasing hens. Berthold had castrated four other roosters at the same time as the first two, but in these animals,

R.A. Steiner (✉)
Department of Physiology and Biophysics, University of Washington,
Box 357290, Seattle, WA, 98195, USA
e-mail: Steiner@uw.edu

A.S. Kauffman and J.T. Smith (eds.), *Kisspeptin Signaling in Reproductive Biology*,
Advances in Experimental Medicine and Biology 784, DOI 10.1007/978-1-4614-6199-9_1,

he had either removed one testis and "relocated" it to within the abdomen or removed both testes and transplanted a spanking new testis into the animal's abdominal cavity. To his astonishment—despite the testicular swapping, the cockerels retained their lusty ways and avian pulchritude. The transplanted testes had become vascularized and produced sperm, but notably lacked nervous innervation. Berthold wrote (in a translation by D.P. Quiring): *... So far as voice, sexual urge, belligerence, and growth of combs and wattles are concerned, such birds remain true cockerels. Since, however, transplanted testes are no longer connected with their original site and severed from their innervations ... it follows that the results in question are determined by the productive function of the testes ... their action on the blood stream ... and then the entire organism, of which it is true, the nervous system represents a considerable part* [1]. Such was the beginning of experimental endocrinology and the first whiff that *hormones act upon the brain to regulate reproduction.*

At the turn of the twentieth century, two neurologists, Joseph Babinski and Alfred Frölich, described patients with a condition called dystrophia adiposogenitalis (later called Frölich's or Babinski–Fröhlich syndrome), whose clinical features include sexual immaturity and which were known to be caused by tumors of the basal forebrain and pituitary. A debate raged for some 20 years thereafter about whether this syndrome was attributable to lesions of the brain or pituitary. In 1912 Bernhard Aschner exhorted: *Not only hypophysectomy—but even a mere wound in the base of the thalamencephalon leads to atrophy of the gonads in male and female dogs* [2]. However, 30 years would pass until seminal studies in rats by Walter Hohlweg and Karl Junkmann demonstrated that ... *internal secretion of the anterior pituitary and the gonads is controlled by a nervous system* [3]. They accomplished this remarkable feat by showing that the capacity of the pituitary to respond to castration required its close proximity to the brain—and the so-called "nervous sex center" (now recognized to include the hypothalamus), which regulates gonadotropin secretion as a feedback control system. Over the next 20 years, scientists, including F.L. Dey [4], C.H. Sawyer, and G.W. Harris, placed lesions and evoked electrical activity within discrete regions of the forebrain and observed their effects on reproduction, and by the early 1950s it became widely accepted that the hypothalamus constitutes the central processor of reproduction—governing the onset of puberty, gonadotropin secretion in both sexes, and timing of ovulation in females. In the 1930s, the anatomy of the hypothalamo-hypophyseal portal system was argued and settled, and the concept of "neurosecretion" gained traction. Belief in the existence of GnRH became widespread by the 1950s, but the final characterization of GnRH would come only after 20 more years of protein chemistry and physiology—in the late 1960s and early 1970s. Thus, the cornerstone of the brain-pituitary-gonadal axis was laid—and acknowledged by the Nobel Committee in 1977. But mysteries remained.

The 20 years that followed awarding the Nobel Prize in Physiology or Medicine to Roger Guilleman and Andrew Schally produced highly reductionist studies of hypothalamic circuitry, which characterized many afferent inputs to GnRH neurons, with the use of pharmacological, neurosurgical, biophysical and, eventually, molecular techniques. This research imparted appreciation to the complexity of the mechanisms that govern GnRH neurons, the final common pathway through which the brain regulates reproduction, and put upfront the classical neurotransmitters (e.g., acetylcholine, norepinephrine, dopamine), certain amino acids (e.g., glutamate and GABA), as well as a constellation of neuropeptides that regulate GnRH neurons. The list of neuropeptides implicated in the regulation of GnRH secretion in mammals boggles the mind—and becomes even more astonishing when we consider those discovered among other nonmammalian vertebrates. Among those with particular prominence include neuropeptide Y, proopiomelanocortin (and its products), galanin, and the RF amide family—not to mention their extended family of receptor subtypes. Yet, by the turn of the twenty-first century, many in the neuroendocrinology community felt that something was missing. Although neuropharmacology and molecular mapping argued persuasively that these neuropeptides play key regulatory roles in the regulation of GnRH secretion, all the hoot 'n holler didn't change the fact that knockouts of *Npy*, *Gal*, *Agrp*, *Pomc*, and the like failed to produce remarkable reproductive phenotypes. The field of dreams had become parched under intense illumination and precision mowing. Although it remains true that redundancy and developmental compensation may excuse the absence of one or another players, it had become clear that we hadn't really discovered *the real* secret how the brain regulates GnRH neurons. Indeed, much of what had been learned in the 1980s and 1990s was on margin—worthwhile, but no Holy Grail. We still didn't know how the brain triggers the onset of puberty; we didn't know the identity of molecular and cellular pathways that allow sex steroids to communicate with GnRH neurons, (since these cells don't express the appropriate receptors); we had nary a clue about the mechanisms that link the circadian clock to GnRH neurons and induce ovulation at a precise time of day, nor how season is perceived by the neuroendocrine reproductive axis—not to mention how a tiny network of GnRH neurons is capable of generating discrete "pulses" that drive LH secretion.

The world as we knew it changed in late 2003, when it was discovered that mutations in an obscure cancer-related gene that encodes the receptor for kisspeptin (aka *KISS1R/kiss1r* or *GPR5/Gpr54*) cause profound hypogonadotropic hypogonadism in humans and mice—a rare finding reported by two separate groups in papers that launched a thousand ships. Thus began the saga of kisspeptin signaling in the neuroendocrine regulation of reproduction. Today, almost 10 years and a 1,000 papers later, kisspeptin has emerged as the *chargé d'affaires*—providing impetus, information, and guidance to GnRH neurons, whose ancient role in reproduction may more akin to that of a motor neuron—an essential servant but no savant. If the GnRH neuron is the prima ballerina, certainly the kisspeptin neuron must be her choreographer. Indeed, we've learned a lot about kisspeptin and reproduction in the past decade!

Scientists enjoy a wide consensus about many aspects of kisspeptin signaling. We know that kisspeptin acts through a $G\alpha_{q/11}$-coupled mechanism to activate GnRH neurons, which express Kiss1r, by inhibiting A-type and inwardly rectifying K^+ currents and activating TRPC currents to induce depolarization and sustained action potentials. We've learned that kisspeptin-expressing neurons reside in many parts of the brain (and in cells elsewhere in the body) and likely serve many different functions in adulthood and development. There are many cellular "phenotypes" of kisspeptin neurons, whose anatomy and physiology varies among species and as a function of age and sex. Some kisspeptin neurons coexpress neurokinin B (NKB) and dynorphin (at the very least!), whereas others express a completely different array of co-transmitters (such as tyrosine hydroxylase). Sex steroids, including estradiol and testosterone, act through the estradiol receptor α and the androgen receptor to regulate the expression of *Kiss1* in kisspeptin neurons. Moreover, the effect of these sex steroids depend on the particular phenotype of their target cells, as well as the age and sex of the animal. Metabolic hormones, such as leptin, act directly (and indirectly) on kisspeptin neurons to regulate their function, and may play a role in inhibiting reproduction during lactation and stress [5]. We also suspect that the *Kiss1* gene is subject to developmental regulation—perhaps through epigenetic mechanisms, which become manifest in adulthood.

Notwithstanding these accomplishments, controversies remain. Debate surrounds the precise role of certain anatomical subsets of kisspeptin neurons in regulating gonadotropin secretion. *Which population of kisspeptin controls the onset of puberty? Which mediates the negative and positive feedback effects of estradiol on GnRH secretion? Do metabolic hormones exert direct or indirect effects on kisspeptin neurons? Are kisspeptin neurons "responsible" for driving pulsatile GnRH (and LH) secretion? How does the functional anatomy of different populations of kisspeptin neurons differ among species?* Controversy also attends the nature and interpretation of studies in transgenic animal models expressing GFP and/or Cre recombinase (Cre) under the *Kiss1* promoter and the phenotype of animals generated by crossing the Cre mice with floxed alleles. Such studies are complicated by the promiscuous nature of Cre expression in *Kiss1* transgenic crosses with reporters and in *Kiss1* knock-in mice, wherein the stochastic properties of Cre expression may (or may not) produce offspring with the expected results of targeted manipulation [6, 7]. This all translates into confusion about (1) interpreting electrophysiological results predicated on identifying kisspeptin neurons in slice preparations based on the presence of GFP—produced by crossing *Kiss1-Cre* mice with transgenic mice reporters (GFP) and (2) creating cell-specific knockouts of genes that are coexpressed in kisspeptin neurons by crossing *Kiss1-Cre* lines with floxed alleles.

So, *what does the future hold? What remains to be learned about kisspeptin signaling?* Certainly, we can dissect more about the details of kisspeptin's molecular action on GnRH neurons—worthy, perhaps, but may find rough sledding in funding agencies. We can (and certainly will) explore the diversity of kisspeptin's action in regions of the brain outside of the hypothalamus (e.g., hippocampus, cortex, amygdala)—also worthwhile, but outside of the context of a physiological problem, such ventures may be viewed as molecular bird-watching. It is essential that we learn more about kisspeptin signaling in the brain of species besides rodents, including humans, to learn more about how kisspeptin neurons are regulated in creatures like us and what it means for puberty and menstrual cycle function. Understanding the role of kisspeptin neurons in the generation of pulsatile GnRH secretion remains one of neuroendocrinology's most fundamental problems—a keystone. Parsing the significance of kisspeptin's co-transmitters in the arcuate nucleus (i.e., NKB and dynorphin) should help to unravel the mystery of pulsatile kisspeptin (and GnRH) secretion. Identifying the afferent inputs to kisspeptin neurons is essential. We need to learn more about the development and lineage of *Kiss1* neurons. *What are major challenges we face*? A lack of understanding of the strengths and limitation of the cellular, molecular, and transgenic tools we currently possess (or would like to develop) stands as an impediment to dialogue, collaboration, and progress. *What's next*? *Is there life after kisspeptin*? History would suggest that neuropeptides enjoy a theater run of about 7 years. *To everything there is a season*. However, important work remains—at least in the short term. *What are the best opportunities ahead*? Collaboration. In this era of economic constriction, groups need to pool resources, combine technologies, and forge interdisciplinary approaches to address the stickiest problems.

References

1. Berthod AA (1849) Archiv für Anatomie, Physiologie und wissenschaftleche Medicin, pp 42–46
2. Aschner B (1912) Zur Physiologie des Zwischenhirns. Wein Klin Wochenschr 27:1042–10043
3. Hohlweg W, Junkmann K (1932) Die hormonal-nervöse Regulierung des Hypophysenvorderlappens und der Keimdrüsen. Klin Wochenschr 11:321–323
4. Dey FL (1943) Evidence of hypothalamic control of hypophyseal gonadotropic functions in the female guinea pig. Endocrinology 33:75–82
5. Pinilla L, Aguilar E, Dieguez C, Millar RP, Tena-Sempere M (2012) Kisspeptins and reproduction: physiological roles and regulatory mechanisms. Physiol Rev 92:1235–1316
6. Padilla SL, Reef D, Zeltser LM (2012) Defining POMC neurons using transgenic reagents: impact of transient *Pomc* expression in diverse immature neuronal populations. Endocrinology 153:1219–1231
7. Morrison CD, Munzberg H (2012) Capricious Cre: the devil is in the details. Endocrinology 153:1005–1007

Chapter 2
Structure, Synthesis, and Phylogeny of Kisspeptin and its Receptor

Shinji Kanda and Yoshitaka Oka

Abstract The kisspeptin system is considered to be essential for successful mammalian reproduction. In addition to the Kiss1 peptide, Kiss2, the product of *kiss2* (the *kiss1* paralogue), has also been shown to activate kisspeptin receptor signaling pathways in nonmammalian species. Furthermore, in nonmammalian species, there are two subtypes of receptors, Gpr54-1 (known as GPR54 or Kiss1R in mammals) and Gpr54-2. Although complete understanding of the two kisspeptin—two kisspeptin receptor systems in vertebrates is not so simple, a careful examination of the phylogeny of their genes may provide insights into the functional generality and differences among the kisspeptin systems in different animal phyla. In this chapter, we first discuss the structure of kisspeptin ligands, Kiss1 and Kiss2, and their characteristics as physiologically active peptides. Then, we discuss the evolutionary traits of *kiss1* and *kiss2* genes and their receptor genes, *gpr54-1* and *gpr54-2*. It appears that each animal species has selected either *kiss1* or *kiss2* rather randomly, leading us to propose that some of the important characteristics of kisspeptin neurons, such as steroid sensitivity and the anatomical relationship with the hypophysiotropic GnRH1 neurons, may be the keys to understanding the general functions of different kisspeptin neuronal populations throughout vertebrates. Species differences in *kiss1/kiss2* may also provide insights into the evolutionary mechanisms of paralogous gene-expressing neuronal systems. Finally, because kisspeptins belong to one of the members of the RFamide peptide families, we discuss the functional divergence of kisspeptins from the other RFamide peptides, which may be explained from phylogenetic viewpoints.

S. Kanda, Ph.D • Y. Oka, Ph.D. (✉)
Department of Biological Sciences, Graduate School of Science, University of Tokyo,
7-3-1 Hongo, Bunkyo, Tokyo 113-0033, Japan
e-mail: okay@biol.s.u-tokyo.ac.jp

A.S. Kauffman and J.T. Smith (eds.), *Kisspeptin Signaling in Reproductive Biology*,
Advances in Experimental Medicine and Biology 784, DOI 10.1007/978-1-4614-6199-9_2,

Introduction

As introduced in Chap. 1, kisspeptin is now considered to be an essential component of the central regulation of reproduction, because the lack of *Kiss1* or kisspeptin receptor (*Kiss1r* or *Gpr54*) genes causes hypogonadotropic hypogonadism in humans and rodents. On the other hand, accumulating evidence suggests that *kiss2*, a paralogous gene for *kiss1*, widely exists in the vast majority of vertebrates, although this gene appears to have been lost in placental mammals. Because the peptide products of these paralogous genes, Kiss1 and Kiss2,[1] show the same extent of receptor activation for Gpr54, these genes, which probably have arisen by genome-wide duplication, should be considered as "kisspeptins" from both receptor affinity and phylogenetic viewpoints.

In this chapter, we will introduce the structure and phylogeny of kisspeptin peptides first. Then, projections and steroid sensitivity of the kisspeptin-expressing neurons will be discussed. Finally, because it is known that kisspeptins, Kiss1 and Kiss2, belong to the RFamide family, we also discuss the characteristics of kisspeptins as members of the RFamide family peptides.

Structure of Kisspeptin

Like other peptide neurotransmitters/neuromodulators, kisspeptin is initially translated into long kisspeptin precursors, which are cleaved or processed to form shorter mature peptides. For instance, the human kisspeptin is first translated to prepro-kisspeptin (Kisspeptin-145, consisting of 145 amino acids), including the signal peptide to be loaded to the peptide vesicles. The peptide is then proteolytically cleaved at the site next to the dibasic residues by subtilisin-like convertase, and the C terminal-RFG is amidated by carboxypeptidase, as in the processing of GnRH peptides [1]. Because several other kisspeptin peptides shorter than 54 amino acid residues have also been found, it is suggested that the peptides are degraded from the N terminus to produce shorter but still active peptides (Fig. 2.1a). From human placental extracts, for example, kisspeptin-54, -14, and -13 have been purified [1, 2]. Although the relative potency for their activation of the receptor Gpr54 varies slightly when the N terminal amino acid length changes, it was suggested that 10 amino acid residues from the C terminal RY-NH_2 (for instance, YNWNSFGLRY-NH_2 for rodent *Kiss1*, and YNWNSFGLRF-NH_2 for primates; see Fig. 2.1b) are essential and sufficient for the activation of Gpr54 signaling pathways. For Kiss2, a kisspeptin-12 isoform (SKFNFNPFGLRF-NH_2) has been isolated from *Xenopus laevis*

[1]According to the Zebrafish Information Network, ZFIN; http://zfin.org/zfinfo/nomen.html, we will italicize gene names, such as *kiss1* and *kiss2*, and romanize protein and peptide name, such as Kiss1 and Kiss2 in this chapter. We will call the receptor for kisspeptins as "GPR54" because of the promiscuous nature of ligands and receptors for RF amide families, including kisspeptin. For details, see Kanda and Oka [37].

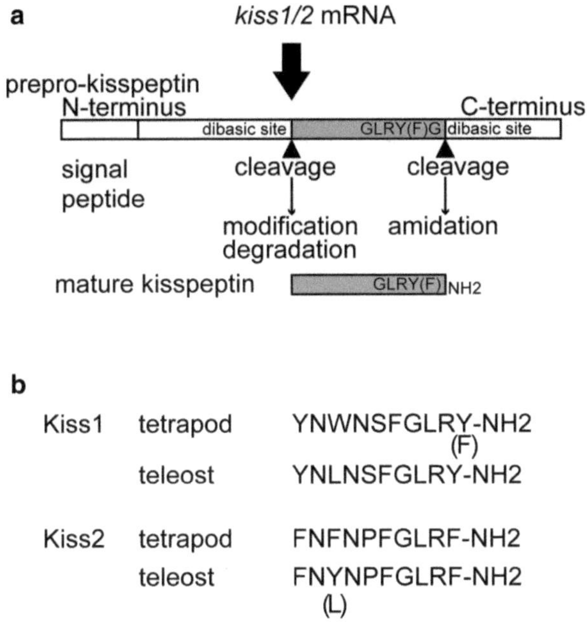

Fig. 2.1 Schematic illustration of Kiss1/Kiss2 peptide maturation process and conserved peptide core sequence in vertebrates. (**a**) Prepro-kisspeptin molecules are cleaved into shorter kisspeptin by the dibasic cleavage sites. The C terminus of kisspeptin is amidated to the characteristic RF or RY motif. After cleavage, amidation occurs at C terminus, and degradation and/or modification such as pyroglutamate formation may occur at N terminus. (**b**) A summary of core sequence of Kiss1 and Kiss2 in vertebrates. Note that tyrosine and tryptophan possess similar side-chain

brains [3]. From the prediction of a cleavage site and subsequent binding assay studies, it has been shown that the 10 amino acid "core sequence" is essential and sufficient for the full activation of Gpr54 by Kiss1 and Kiss2 throughout vertebrates in general [3–5] (Fig. 2.2). Consequently, many researchers refer to the peptides that possess the highly conserved 10 amino acid core sequence as "kisspeptins" and have used kisspeptin-10 as kisspeptin in many studies. However, not many studies have purified "native" forms of kisspeptins in various vertebrate species, and we should therefore be careful about the interpretation of physiological experiments using only the kp-10 as kisspeptin ligands, since there may be some other physiological functions that are slightly different when conveyed by the natural peptides.

Evolution of Kisspeptins and Their Receptors

Phylogeny of Kiss1 and Kiss2 Genes

For proteins that possess longer amino acid residues, sequence similarity of proteins can be used for the construction of phylogenetic trees rather easily. However, shorter

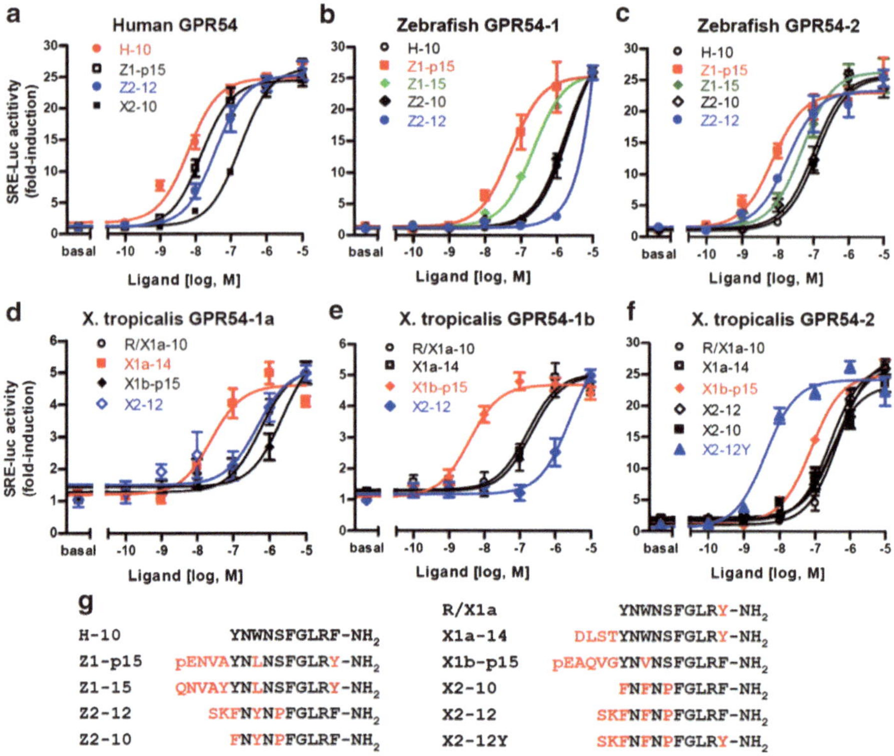

Fig. 2.2 Both Kiss1 and Kiss2 activate both Gpr54-1 and Gpr54-2 in *Xenopus* and zebrafish. Although the relative potency differs among different receptor subtypes, Kiss1 or Kiss2 longer than 10 amino acid residues show activation at physiological concentrations, suggesting that Kiss1 and Kiss2 are the ligands for Gpr54-1 and Gpr54-2. Luciferase assay. Adapted with permission from Lee YR, Tsunekawa K, Moon MJ, Um HN, Hwang JI, et al. (2009) Molecular evolution of multiple forms of kisspeptins and GPR54 receptors in vertebrates. Endocrinology 150: 2837–2846. 2009 © Endocrine Society

peptides, including the kisspeptins, share only a small number of conservative sequences in common. Therefore, to unveil the phylogenetic relationship of certain genes among or within species, especially for shorter peptides, the synteny analysis will often give us powerful evidence for such relationship. It can be used to predict if certain genes in the other species are the homologous gene or just analogous ones by chance.

In genetics, synteny describes the physical co-localization of genes in a certain genetic locus. Recent synteny analysis of kisspeptin genes proposes that *kiss1* and *kiss2* are paralogous to each other and that they have arisen as a result of gene duplication at the locus level. The synteny analysis of *kiss1* and *kiss2* genes in several vertebrate species strongly suggested that *kiss1* and *kiss2* are duplicated together with some surrounding genes such as *goltla/b*, *plekha5/6*, *pik3c2b/cg*, and *etnk1/2*

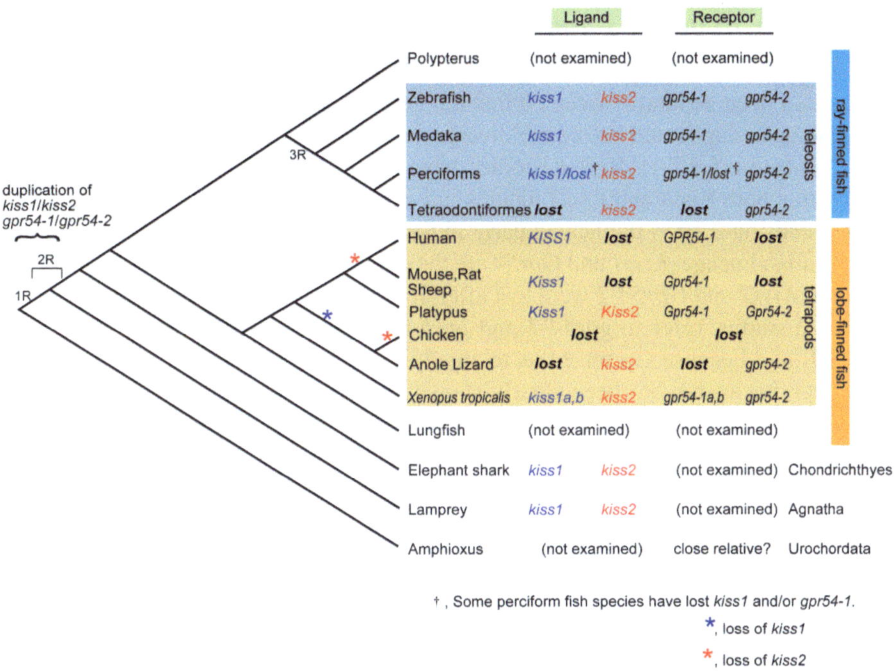

Fig. 2.3 Summary of ligand and receptor genes for kisspeptin systems (*kiss1*/*kiss2*) along the vertebrate lineage. *kiss1* and *kiss2* are suggested to be duplicated before the emergence of lamprey, probably due to the whole genome duplication of the ancestral vertebrate. It is suggested that *kiss2* and *gpr54-2* were lost in marsupial and placental mammals after the divergence from the monotreme during mammalian evolution. Some teleost species have lost *gpr54-1*, but no teleosts have lost *gpr54-2*, suggesting the significance of gpr54-2 in the regulation of teleost reproduction, which is opposite to the case in mammals. It is also consistent with the higher level and wider distribution of expression of *gpr54-2* compared to that of *gpr54-1* in teleost brains. The loss of *kiss1* (*blue*) or *kiss2* (*red*) is indicated by *asterisks* (figure as originally published in Kanda S and Oka Y (2012) Evolutionary insights into the steroid sensitive *kiss1* and *kiss2* neurons in the vertebrate brain. Front. Endocrin. 3: doi: 10.3389/fendo.2012.00028)

[6, 7], because these genes are located in the same locus as their paralogues, suggesting that they have been duplicated at the locus level. Because the phylogenetically old vertebrate, lamprey, has both *kiss1* and *kiss2* (see Fig. 2.3), the duplication of the locus probably occurred in the basal vertebrate. Presumably, the 1R or 2R whole genome duplication[2] produced the paralogous system of *kiss1* and *kiss2*. It is speculated that these peptides shared their receptor Gpr54 in each species in the long evolutionary history, as discussed below.

[2]In the vertebrate lineage, whole genome duplication (WGD) events [8] are considered to have taken place three times in teleosts (1R-3R) and twice (1R and 2R) in tetrapods [9].

Phylogeny of *gpr54-1* and *gpr54-2*

It has also been generally accepted that the gene for the kisspeptin receptors, *gpr54*, has also been duplicated before the divergence of teleosts and tetrapods, as shown in Fig. 2.3 (see ref. [37]). Recent studies have shown by in vitro luciferase reporter assays that both Kiss1 and Kiss2 ligands activate both Gpr54-1 and Gpr54-2, and that these genes have been duplicated early in the vertebrate lineage [7, 8]. Thus, in the mammalian lineage, *Kiss1* and *Gpr54* are the only kisspeptin and kisspeptin receptor, because *Kiss2* and *Gpr54-2* were lost after divergence from the monotremes.

Some studies refer to *gpr54-1* and *gpr54-2* as *kiss1r* and *kiss2r* in accordance with [9] and because Gpr54-2 has relatively higher affinity for Kiss2 in zebrafish. However, because it has been clearly shown that both Kiss1 and Kiss2 bind to and activate both Gpr54-1 and Gpr54-2 [3–5], and, in contrast, the one-to-one relationship between kisspeptin (Kiss1 or Kiss2) and the receptor (Gpr54-1 or Gpr54-2) has not been unequivocally demonstrated by either anatomical or physiological methods, we simply call them *gpr54-1* and *gpr54-2* in this chapter. We are of the opinion that we should wait for the anatomical and/or physiological demonstration of the projection of Kiss1/Kiss2 neurons and the distribution of kisspeptin receptors before we can refer to them as *kiss1r* or *kiss2r*.

Figure 2.3 summarizes, as a phylogenetic tree, the expression of *kiss1/2* and *gpr54-1/-2* in some representative species. Interestingly, *gpr54-2* is suggested to play more critical roles in teleosts, because some species lack *gpr54-1*, and broader distribution of *gpr54-2* expressions in the brain has been reported in some teleosts. In contrast, it is interesting that *gpr54-2* has been lost in mammals during evolution, and *gpr54-1* appears to have taken its place. Although there has been no study to systematically examine the distribution or cellular localization of *gpr54-1* or *gpr54-2* in the nonmammalian tetrapod brain, the inverse situation in teleosts and mammals is intriguing. Likewise, it is interesting that wider variety of species in teleosts appear to possess *kiss2* compared to *kiss1*. Moreover, *kiss1* is lost in reptiles, while *kiss2* remains intact. Thus, except for mammals, *kiss2* appears to be more widely conserved throughout vertebrate species. However, the contribution of *kiss1* or *kiss2* to the central regulation of the hypothalamic–pituitary–gonadal (HPG) axis should not be evaluated only by the existence or absence of the gene(s) in the phylogenetic tree, because the loss of the gene can be functionally compensated for by the other genes, especially by close relative genes.

By taking the phylogenetic tree into account, it is clear that both the genes for ligands and receptors have been duplicated at least before the divergence of teleosts and tetrapods. In other words, the common ancestor of teleosts and tetrapods are considered to have possessed two ligands and two receptors. Except for the complete loss of the kisspeptin system in avian species, reported in chicken and zebrafinch [7], the genes for at least one ligand and one receptor remained in each vertebrate species. The losses of genes seem to have taken place randomly, but, as described above, there appears to be some basic rules for the gene loss by natural selection in each branch. In the long history of vertebrate evolution, it appears that either one of the genes, *kiss1* or *kiss2*, and *gpr54-1* or *gpr54-2*, acquired predominant functions in the HPG axis regulation, which were different among branches. That is because Kiss1 and Kiss2 show similar binding activity to Gpr54-1 and Gpr54-2 immediately after their divergence. Interestingly, there is an observed tendency for the species that lost Gpr54-1 to also have lost Kiss1. Obviously, much more extensive analyses using different species in different phylogenetic branches are needed to verify the general significance of this phenomenon.

Functional Evolution of Kiss1 and Kiss2

As described above, the kisspeptin system is well conserved among vertebrate species, except for the avian species. However, the general physiological functions of kisspeptin in vertebrates still remain to be elucidated. To date, numerous studies have reported on the involvement of kisspeptin(s) in the regulation of the HPG axis. In mice, kisspeptins have been reported to act on GnRH neurons directly, by acting on some intrinsic ion channels to produce strong persistent depolarization [10–15] (see Chap. 6). From the initial reports that loss of *Kiss1* or *GPR54* genes in human and rodents disrupts puberty and leads to hypogonadotropic hypogonadism [16–18], and Kiss1 peptide administration induces LH release by activating Gpr54 localized in GnRH1 neurons [11, 19, 20], it is clear that the kisspeptin-Gpr54 system plays a critical role in the regulation of HPG axis at least in placental mammals (reviewed in ref. [21]). On the other hand, there are fewer studies supporting similar regulatory mechanisms of the kisspeptin system in nonmammalian vertebrates. Moreover, it should be noted that the avian lineage can reproduce in spite of the fact that they lack both the *kiss1* and *kiss2* systems (see Fig. 2.3). Therefore, there are presumably other mechanisms in addition to the kisspeptin system for the central regulation of reproduction, at least in birds.

Studies in teleost kisspeptin systems seem to have yielded more complex situations. There are conflicting results that either support the presence of co-expression of *gpr54* in GnRH1 neurons in European seabass [22] and a tilapia *Oreochromis niloticus* [23] or their absence in another species of tilapia *Astatotilapia burtoni* [24]. Our unpublished results in medaka also showed the absence of *gpr54* mRNA in GnRH1 neurons. Thus, the results of previous reports suggest that the situation is different among different species. Studies of exogenous administration of kisspeptins have been performed both in mammals and teleosts. In mammals, to our knowledge, all the studies to date showed an increase in plasma LH [20, 25–34]. In contrast to the wealth of knowledge in mammals, a much smaller number of studies have been performed in nonmammalian vertebrates. Kiss1 and/or Kiss2 increased LH mRNA or serum LH concentration in zebrafish [35], sea bass [6], and goldfish [5]. The experimental conditions for the occurrence of a rise in LH induced by kisspeptin, and the time of LH/FSH rise after administration of kisspeptin, vary among the different studies. Thus, in teleosts, the situation is somewhat different from those obtained from the placental mammals, where kisspeptin is essential for reproduction.

To our knowledge, there has been no report on the kisspeptin neuronal systems of nonmammalian species other than teleosts and placental mammals, and there is a big gap between these two branches of animals. In fact, during the long history of vertebrate phylogeny, birds are the only species that have lost both *kiss1* and *kiss2*, suggesting that the kisspeptin system may have increased genetic fitness in each species. Thus, the identification of the general functions in vertebrates, including the regulation on HPG axis, is yet to be concluded.

Steroid Sensitivity of Kisspeptin Neurons Is Conserved Among Vertebrates

As discussed above, the kisspeptin system has been suggested to play important roles in the regulation of reproduction in mammals, but not in birds, while it may also be involved in the regulation of some reproductive functions in teleosts. From the survey of literature on the functional aspects of kisspeptin in vertebrates, the apparent conservative nature of the kisspeptin-Gpr54 system undoubtedly suggests considerable contribution of kisspeptin on "evolutionary fitness." Interestingly, the sex steroid sensitivity of kisspeptin-expressing neurons appears to be well conserved among teleosts [36–39] and mammals [40–44], suggesting that this sex steroid sensitive nature of kisspeptin neurons was already present before the divergence of teleosts and tetrapods; the steroid sensitivity is likely to be a general feature of all the kisspeptin systems throughout vertebrates. As described in the previous section, in placental mammals, the *Kiss1* neurons are strongly suggested to mediate sex steroid feedback effects; these neurons receive sex steroid signals from the gonads and directly and/or indirectly modulate the activity of GnRH1 neurons [10–15]. The mediators of sex steroid feedback, i.e., the neurons that directly receive sex steroids and control the release of GnRH, have long been searched for, because GnRH neurons themselves lack estrogen receptor alpha (reviewed in ref. [45]), which is essential for the sex steroid control of reproduction in mammals [46–48]. Because *Kiss1* neurons in mammals express ER alpha, and *Kiss1* mRNA expression in many placental mammalian species is negatively regulated in the arcuate nucleus and positively regulated in the anteroventral periventricular nucleus (AVPV)/preoptic area (POA) [20, 40–44, 49, 50], the *Kiss1* neurons are the most plausible candidate as the "missing link" in the steroid feedback mechanism. From several lines of recent experimental evidence, it is now hypothesized that the arcuate kisspeptin neurons and the AVPV kisspeptin neurons are involved in negative and positive feedback, respectively [40, 51] (see Chap. 13).

In nonmammalian species, the sex steroid sensitivity of kisspeptin neurons has been demonstrated experimentally only in teleosts, medaka, and goldfish [36, 39]. In medaka, among several populations of *kiss1* and *kiss2* neurons in the brain, only the *kiss1* neurons in a hypothalamic nucleus, nucleus ventralis tuberis (NVT), show sex steroid sensitivity; NVT *kiss1* expression is positively regulated by gonadal steroids, probably directly via sex steroid hormone receptors [36, 39]. On the other

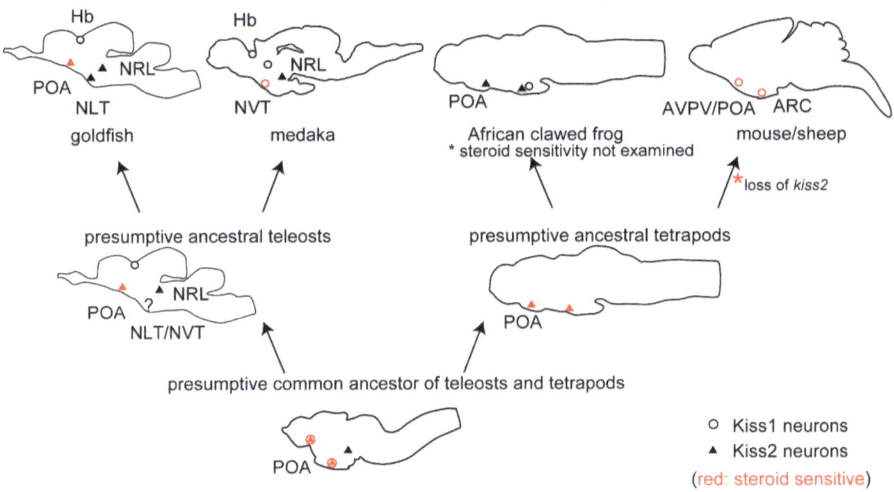

Fig. 2.4 Schematic illustrations for the distribution of *kiss1* and *kiss2* neurons in vertebrate brains, including some hypotheses. Because *kiss1* and *kiss2* are duplicated paralogues, they are considered to have been co-expressed in the same neurons in the common ancestor of teleosts and tetrapods. Given that both amphibians and teleosts express *kiss2* in POA, the ancestral teleosts and ancestral tetrapods should have expressed *kiss2*. Because *Kiss2* was lost in the mammalian lineage, we hypothesize that *Kiss1* began to be expressed where *Kiss2* used to be expressed, to compensate for the loss of *Kiss2* during mammalian evolution. *Open circles* indicate *kiss1*, and *filled triangles* indicate *kiss2* neurons. *Circles/triangles* in *red* are *kiss1/kiss2* neurons that are steroid sensitive (figure as originally published in Kanda S and Oka Y (2012) Evolutionary insights into the steroid sensitive *kiss1* and *kiss2* neurons in the vertebrate brain. Front. Endocrin. 3: doi: 10.3389/fendo.2012.00028)

hand, in goldfish, which lack *kiss1* neurons in NVT, POA *kiss2* neurons are the only population of kisspeptin neurons that shows steroid sensitive kisspeptin mRNA expression in the brain [37]. These POA *kiss2* neurons are also positively regulated by gonadal steroids, as in the NVT *kiss1* neurons in medaka, and thus there has been no report of negative regulation of kisspeptin expression in teleost brain so far. Considering the report that positive or negative steroid feedback regulation can be rather easily changed by the composition of co-expressing transcription factors [52], the important common feature of the vertebrate kisspeptin neurons may be that steroid sensitive kisspeptin neurons are localized in NVT and POA, which are anatomically similar to arcuate and POA/AVPV in mammals, respectively. Although the precise homology of brain nuclei between mammalian and nonmammalian (especially teleost) kisspeptin neurons should be carefully discussed, the presence or absence of sex steroid sensitivity in each nucleus may be one of the strongest pieces of evidence to argue such homology.

An evolutionary working hypothesis of kisspeptin neuronal systems in vertebrates is shown in Fig. 2.4. In this hypothesis, *kiss1*- and *kiss2*-expressing neurons are differentially distributed in the brains of mammal and other vertebrates. It is suggested that the loss of *Kiss2* gene during mammalian evolution has been probably been compensated functionally by the closely related *Kiss1* gene (for detailed discussion, see ref. [38]). From the arguments about the sex steroid sensitivity of kisspeptin neurons thus far, it is suggested that this property of kisspeptin neurons is one of the most general and important properties among the vertebrate kisspeptin systems; it may have been already acquired before the divergence of teleosts and tetrapods.

Axonal Projections of the Kisspeptin Neurons and the Distribution of Kisspeptin Receptors

Projections of Kiss1 Neurons in Mammals

In mammalian species, several studies have analyzed anatomical relationships between the Kiss1 neurons and the other components in the HPG axis, most importantly, the GnRH1 neurons. Recently, Clarkson et al. detailed the projections of Kiss1 neurons in the mouse brain [53]. It was shown that Kiss1-immunoreactive (ir) fibers were abundant in the ventral aspect of the lateral septum and the hypothalamus, running in periventricular and ventral retrochiasmatic pathways, except for the suprachiasmatic and ventromedial nuclei. Moreover, Kiss1-ir fibers were observed in the internal zone of the median eminence, but not in its external layer where GnRH and other hypophysiotropic hormone-containing axons are proposed to terminate. In addition, a small number of kisspeptin fibers were also observed outside the hypothalamus, in the bed nucleus of the stria terminalis, subfornical organ, medial amygdala, paraventricular thalamus, periaqueductal grey and locus coeruleus. These findings are consistent with a study in the rat brain using different antiserum, although there were some discrepancies in the distribution outside the hypothalamus [54]. In mammalian species, heavy projections to some hypothalamic and preoptic nuclei now seem to be the general consensus on the distribution of Kiss1 neurons. Moreover, Kiss1 fibers have been shown to project to the median eminence, and make close contacts with GnRH1 fibers in mammals [55–57]. Because GnRH1 neurons express *gpr54* in mammals, Kiss1 neurons are hypothesized to stimulate GnRH release from GnRH axons at the nerve terminals in or near the median eminence.

Projections of Kiss1/Kiss2 Neurons in Teleosts

There are a limited number of neuroanatomical studies detailing the axonal projections of Kiss1 and Kiss2 neurons and the distribution of Gpr54 in nonmammalian species, and most of the studies have been carried out in teleosts. As described below, the projections of Kiss1 and Kiss2 neurons in nonmammalian species are complicated; the steroid sensitive Kiss1 neurons in medaka project to POA, but only Kiss2 neurons project to POA in zebrafish. On the other hand, the distribution of receptor subtypes show rather consistent results in most teleost studies so far;

gpr54-2 seems to have significant expression in POA in zebrafish and medaka, while *gpr54-1* has been lost in some species.

In zebrafish, Kiss1 neurons, located in the habenula, were demonstrated to project to the interpeduncular and raphe nuclei. On the other hand, Kiss2 neurons, whose cell bodies were shown to be localized in the dorsal and ventral hypothalamus, widely projected to the ventral telencephalon, POA, thalamus, and ventral/caudal hypothalamus, suggesting that Kiss2 neurons mainly function as a homeostatic regulator in this species [58]. *Gpr54-2* was shown to be predominantly expressed in POA, ventral telencephalon, hypothalamus, and several nuclei of the brain. However, *gpr54-1* was only expressed in habenula. Close apposition of Kiss2-ir fibers to GnRH1 neurons was also observed in zebrafish, although the authors did not examine the co-expression of *gpr54* in GnRH neurons.

On the other hand, in medaka, in addition to the habenulo-interpeduncular pathway, the Kiss1 neurons were shown to project many fibers to the POA, ventral telencephalon, and hypothalamus, but not to the rest of the brain [59], which is consistent with our recent results of immunohistochemical demonstration of the projections of Kiss1 neurons in the Kiss1-EGFP medaka established by us (Shimada et al., unpublished observations). This predominant distribution of Kiss1 fibers in the hypothalamus and POA correspond to the dense distribution of *kiss1* mRNA expressing neurons in the hypothalamus of medaka, whereas zebrafish lacks such Kiss1 neurons in the hypothalamus. Because the Kiss1 neurons in the medaka hypothalamic nucleus NVT show high sex steroid sensitivity [36] it is suggested that the release of Kiss1 in the POA, hypothalamus, and ventral telencephalon will also vary according to the breeding state. Furthermore, the receptor distribution was also examined for *gpr54-1*and *-2* by in situ hybridization, and *gpr54-2* was shown to be widely expressed in the medaka brain, especially in regions that are involved in homeostatic regulation, consistent with the Kiss1 neuron projections. On the other hand, as in the case of zebrafish, the expression of *gpr54-1* was practically confined to the habenula and POA (Kanda et al., unpublished data). Recently, indirect effects of kisspeptin on non-hypophysiotropic GnRH3 neurons was reported [60]. Further studies of hypophysiotropic and non-hypophysiotropic function of kisspeptin neurons are important for the understandings of physiological functions of kisspeptin systems.

In a Cichlid fish, *A. burtoni*, the distribution of kisspeptin receptor was examined by in situ hybridization [24]. The authors reported that the fish lacked *gpr54-1*. The cells expressing *gpr54-2* were localized in hypothalamus, POA, and ventral telencephalon, which is similar to the results of zebrafish and medaka. In addition, *gpr54-2* was shown to be expressed in the dorsal telencephalon and some other brain regions, with the most prominent expression in the olfactory bulb. Since *gpr54-2* was not expressed in the GnRH1 neurons, the authors suggested the existence of non-GnRH1 neurons that express *gpr54-2* and may exhibit sensitivity to kisspeptin, according to the social state or sexual maturation of the fish [24].

From these studies, it may be suggested that *gpr54-2* is likely to play a wide variety of roles in teleosts, because *gpr54-1* positive cells are localized mainly in habenula and other restricted regions in the brain in medaka and zebrafish, or even absent in some species [24, 61–63], whereas there is no report of teleost species that lack *gpr54-2*. Therefore, it may be suggested that *gpr54-1* became predominant in the mammalian lineage (probably tetrapod lineage), whereas *gpr54-2* became predominant in the teleost lineage. To date, there is no in situ hybridization study of *gpr54* in amphibians. Such studies may give us clues to understanding the history of selection of *gpr54-1* or *-2* during vertebrate evolution.

Kisspeptin as a Peptide Belonging to the RFamide Family of Peptide

A phylogenic tree for the RFamide family of peptides is shown in Fig. 2.5. As shown in the figure, the kisspeptin genes have already diverged from the other RFamide families and form an independent branch. The first identification of

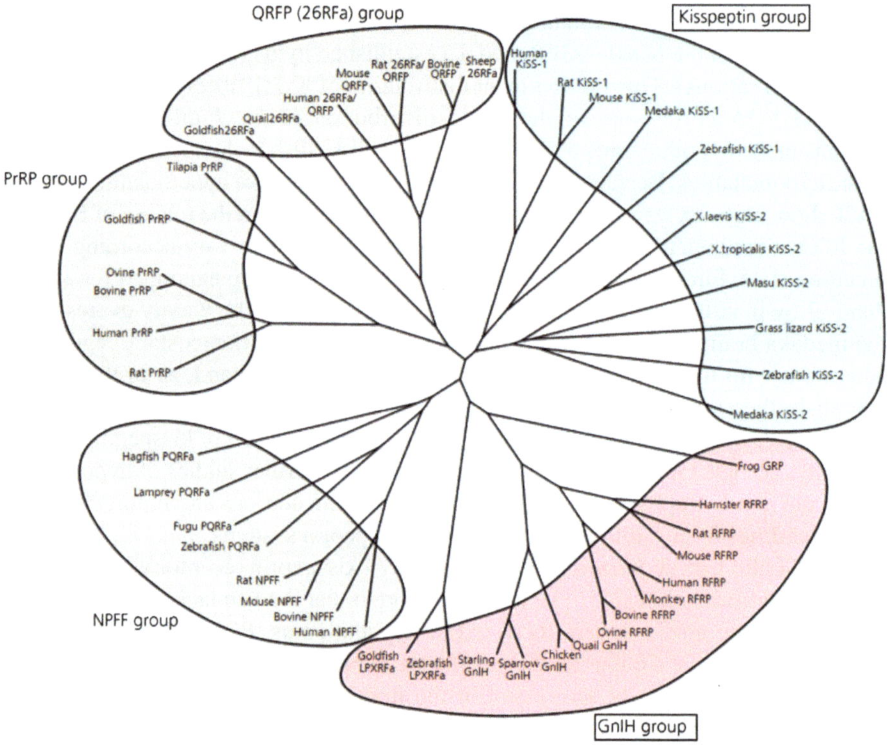

Fig. 2.5 A phylogenetic tree of RFamide peptide family, including *kiss1* and *kiss2*. *npff* group and *gnih* (*rfrp*) group share the same receptor, Gpr147 and 74. In addition, Kiss1 and Kiss2 have been shown to activate Gpr147 and Gpr74 in some species as well. Comprehensive understanding of the promiscuous relationship of ligands and receptors in the RFamide family of peptides is necessary for the future studies. Adapted from Tsutsui et al., Discovery and Evolutionary History of Gonadotrophin-Inhibitory Hormone and Kisspeptin: New Key Neuropeptides Controlling Reproduction. J Neuroendron 716–727. 2010 © The British Society for Neuroendocrinology. With permission from John Wiley and Sons

RFamide peptide in animals goes back to the finding of FMRFamide in sunray Venus clam *Macrocallista nimbosa* [64]. Many RFamides have been identified in invertebrates since then [65], although their evolutionary relationship to the vertebrate RFamide family is yet to be elucidated. Among the vertebrate RFamide family, it is suggested that *kiss1*, *kiss2*, *rfrp/gnih*, and *npff/pqrf* had already diverged in the basal vertebrate, because these four genes have been identified in lamprey [3, 66–68]. Like the paralogous relationship of *kiss1* and *kiss2*, *rfrp* and *npff* are also considered to be duplicated during the whole genome duplication, because they are closely located to the HoxA and HoxC cluster respectively, which are supposed to be duplicated in the whole genome duplication [69]. On the other hand, *prrp* and *qrfp* (*26rf*) have been identified only in teleosts and tetrapods so far, and the presence of these genes in the ancestral vertebrate is still argued. Thus, including *kiss1* and *kiss2*, the RFamide family has already been listed in the early evolution of vertebrate lineage.

It is important to note that their receptors are also close to one another both in sequence and binding capacity for the relative ligands. In fact, human Kiss1 is shown to activate Gpr74 and Gpr147 [70, 71], suggesting the promiscuous relationship between the ligand and receptor among the RFamide group. Interestingly, in spite of this promiscuous relationship, the properties of the receptors are completely different; Gpr54 couples to Gq [2], whereas Gpr147 and Gpr74 both couple to Gi and Gs [72]. In fact, in contrast to kisspeptin, RFRP has been shown to inhibit reproduction [73–79]. Thus, the effect of in vivo or in vitro administration of peptides should be considered with caution, because of the pharmacological side effects via different types of receptors. In the central nervous system, the precise information on the projection of each neuron makes it possible to discriminate the promiscuous ligand–receptor relationship of RFamide and their receptor family.

Conclusions

In this chapter, the structure, function, and phylogeny of kisspeptin and Gpr54, and projection and steroid sensitivity of kisspeptin neurons were discussed. It is a complex but interesting situation that the RFamide family of peptides show promiscuous ligand–receptor relationships, thereby making the results of in vivo pharmacological experiments not always possible to tell the natural or physiological effects of RFamides in the central nervous system. Comprehensive understanding of the morphological/anatomical and physiological characteristics of the neurons, receptor distributions, and ontogenic expression may explain the kisspeptin functions at the organismal level and the mechanisms of kisspeptin actions in the brain. Moreover, the *kiss1* and *kiss2* genes, and the neurons expressing these genes, may be considered as the model for the study of paralogous gene functions, because the genes have avoided strong selection pressure in vertebrates other than mammals, unlike the GnRH system. Consequently, studies of kisspeptin may open a new era of understanding the physiological functions of paralogous genes in addition to a better understanding of neuroendocrine systems.

References

1. Ohtaki T, Shintani Y, Honda S, Matsumoto H, Hori A et al (2001) Metastasis suppressor gene KiSS-1 encodes peptide ligand of a G-protein-coupled receptor. Nature 411:613–617
2. Kotani M, Detheux M, Vandenbogaerde A, Communi D, Vanderwinden JM et al (2001) The metastasis suppressor gene KiSS-1 encodes kisspeptins, the natural ligands of the orphan G protein-coupled receptor GPR54. J Biol Chem 276:34631–34636
3. Lee YR, Tsunekawa K, Moon MJ, Um HN, Hwang JI et al (2009) Molecular evolution of multiple forms of kisspeptins and GPR54 receptors in vertebrates. Endocrinology 150:2837–2846
4. Moon JS, Lee YR, Oh DY, Hwang JI, Lee JY et al (2009) Molecular cloning of the bullfrog kisspeptin receptor GPR54 with high sensitivity to Xenopus kisspeptin. Peptides 30:171–179
5. Li S, Zhang Y, Liu Y, Huang X, Huang W et al (2009) Structural and functional multiplicity of the kisspeptin/GPR54 system in goldfish (Carassius auratus). J Endocrinol 201:407–418
6. Felip A, Zanuy S, Pineda R, Pinilla L, Carrillo M et al (2009) Evidence for two distinct KiSS genes in non-placental vertebrates that encode kisspeptins with different gonadotropin-releasing activities in fish and mammals. Mol Cell Endocrinol 312:61–71
7. Um HN, Han JM, Hwang JI, Hong SI, Vaudry H et al (2010) Molecular coevolution of kisspeptins and their receptors from fish to mammals. Ann N Y Acad Sci 1200:67–74
8. Kim DK, Cho EB, Moon MJ, Park S, Hwang JI et al (2012) Molecular coevolution of neuropeptides gonadotropin-releasing hormone and kisspeptin with their cognate G protein-coupled receptors. Front Neurosci 6:3
9. Gottsch ML, Clifton DK, Steiner RA (2009) From KISS1 to kisspeptins: An historical perspective and suggested nomenclature. Peptides 30(1): 4–9
10. Dumalska I, Wu M, Morozova E, Liu R, van den Pol A et al (2008) Excitatory effects of the puberty-initiating peptide kisspeptin and group I metabotropic glutamate receptor agonists differentiate two distinct subpopulations of gonadotropin-releasing hormone neurons. J Neurosci 28:8003–8013
11. Han SK, Gottsch ML, Lee KJ, Popa SM, Smith JT et al (2005) Activation of gonadotropin-releasing hormone neurons by kisspeptin as a neuroendocrine switch for the onset of puberty. J Neurosci 25:11349–11356
12. Liu X, Lee K, Herbison AE (2008) Kisspeptin excites gonadotropin-releasing hormone neurons through a phospholipase C/calcium-dependent pathway regulating multiple ion channels. Endocrinology 149:4605–4614
13. Pielecka-Fortuna J, Chu Z, Moenter SM (2008) Kisspeptin acts directly and indirectly to increase gonadotropin-releasing hormone neuron activity and its effects are modulated by estradiol. Endocrinology 149:1979–1986
14. Pielecka-Fortuna J, Moenter SM (2010) Kisspeptin increases gamma-aminobutyric acidergic and glutamatergic transmission directly to gonadotropin-releasing hormone neurons in an estradiol-dependent manner. Endocrinology 151:291–300
15. Zhang C, Roepke TA, Kelly MJ, Ronnekleiv OK (2008) Kisspeptin depolarizes gonadotropin-releasing hormone neurons through activation of TRPC-like cationic channels. J Neurosci 28:4423–4434
16. de Roux N, Genin E, Carel JC, Matsuda F, Chaussain JL et al (2003) Hypogonadotropic hypogonadism due to loss of function of the KiSS1-derived peptide receptor GPR54. Proc Natl Acad Sci U S A 100:10972–10976
17. Funes S, Hedrick JA, Vassileva G, Markowitz L, Abbondanzo S et al (2003) The KiSS-1 receptor GPR54 is essential for the development of the murine reproductive system. Biochem Biophys Res Commun 312:1357–1363
18. Seminara SB, Messager S, Chatzidaki EE, Thresher RR, Acierno JS Jr et al (2003) The GPR54 gene as a regulator of puberty. N Engl J Med 349:1614–1627
19. Herbison AE, de Tassigny X, Doran J, Colledge WH (2010) Distribution and postnatal development of Gpr54 gene expression in mouse brain and gonadotropin-releasing hormone neurons. Endocrinology 151:312–321

20. Irwig MS, Fraley GS, Smith JT, Acohido BV, Popa SM et al (2004) Kisspeptin activation of gonadotropin releasing hormone neurons and regulation of KiSS-1 mRNA in the male rat. Neuroendocrinology 80:264–272
21. Oakley AE, Clifton DK, Steiner RA (2009) Kisspeptin signaling in the brain. Endocr Rev 30:713–743
22. Zmora N, Stubblefield J, Zulperi Z, Biran J, Levavi-Sivan B et al (2012) Differential and gonad stage-dependent roles of kisspeptin1 and kisspeptin2 in reproduction in the modern teleosts, Morone species. Biol Reprod 86(6):177
23. Parhar IS, Ogawa S, Sakuma Y (2004) Laser-captured single digoxigenin-labeled neurons of gonadotropin-releasing hormone types reveal a novel G protein-coupled receptor (Gpr54) during maturation in cichlid fish. Endocrinology 145:3613–3618
24. Grone BP, Maruska KP, Korzan WJ, Fernald RD (2010) Social status regulates kisspeptin receptor mRNA in the brain of Astatotilapia burtoni. Gen Comp Endocrinol 169:98–107
25. Gottsch ML, Cunningham MJ, Smith JT, Popa SM, Acohido BV et al (2004) A role for kisspeptins in the regulation of gonadotropin secretion in the mouse. Endocrinology 145:4073–4077
26. Matsui H, Takatsu Y, Kumano S, Matsumoto H, Ohtaki T (2004) Peripheral administration of metastin induces marked gonadotropin release and ovulation in the rat. Biochem Biophys Res Commun 320:383–388
27. Navarro VM, Castellano JM, Fernandez-Fernandez R, Barreiro ML, Roa J et al (2004) Developmental and hormonally regulated messenger ribonucleic acid expression of KiSS-1 and its putative receptor, GPR54, in rat hypothalamus and potent luteinizing hormone-releasing activity of KiSS-1 peptide. Endocrinology 145:4565–4574
28. Navarro VM, Castellano JM, Fernandez-Fernandez R, Tovar S, Roa J et al (2005) Effects of KiSS-1 peptide, the natural ligand of GPR54, on follicle-stimulating hormone secretion in the rat. Endocrinology 146:1689–1697
29. Navarro VM, Castellano JM, Fernandez-Fernandez R, Tovar S, Roa J et al (2005) Characterization of the potent luteinizing hormone-releasing activity of KiSS-1 peptide, the natural ligand of GPR54. Endocrinology 146:156–163
30. Messager S, Chatzidaki EE, Ma D, Hendrick AG, Zahn D et al (2005) Kisspeptin directly stimulates gonadotropin-releasing hormone release via G protein-coupled receptor 54. Proc Natl Acad Sci U S A 102:1761–1766
31. Shahab M, Mastronardi C, Seminara SB, Crowley WF, Ojeda SR et al (2005) Increased hypothalamic GPR54 signaling: a potential mechanism for initiation of puberty in primates. Proc Natl Acad Sci U S A 102:2129–2134
32. Seminara SB, Dipietro MJ, Ramaswamy S, Crowley WF Jr, Plant TM (2006) Continuous human metastin 45-54 infusion desensitizes G protein-coupled receptor 54-induced gonadotropin-releasing hormone release monitored indirectly in the juvenile male Rhesus monkey (Macaca mulatta): a finding with therapeutic implications. Endocrinology 147: 2122–2126
33. Plant TM, Ramaswamy S, Dipietro MJ (2006) Repetitive activation of hypothalamic G protein-coupled receptor 54 with intravenous pulses of kisspeptin in the juvenile monkey (Macaca mulatta) elicits a sustained train of gonadotropin-releasing hormone discharges. Endocrinology 147:1007–1013
34. Dhillo WS, Chaudhri OB, Patterson M, Thompson EL, Murphy KG et al (2005) Kisspeptin-54 stimulates the hypothalamic-pituitary gonadal axis in human males. J Clin Endocrinol Metab 90:6609–6615
35. Kitahashi T, Ogawa S, Parhar IS (2009) Cloning and expression of kiss2 in the zebrafish and medaka. Endocrinology 150:821–831
36. Kanda S, Akazome Y, Matsunaga T, Yamamoto N, Yamada S et al (2008) Identification of KiSS-1 product kisspeptin and steroid-sensitive sexually dimorphic kisspeptin neurons in medaka (Oryzias latipes). Endocrinology 149:2467–2476
37. Kanda S, Karigo T, Oka Y (2012) Steroid sensitive kiss2 neurones in the goldfish: evolutionary insights into the duplicate kisspeptin gene-expressing neurones. J Neuroendocrinol 24: 897–906

38. Kanda S, Oka Y (2012) Evolutionary insights into the steroid sensitive *kiss1* and *kiss2* neurons in the vertebrate brain. Front Endocrinol 3:28
39. Mitani Y, Kanda S, Akazome Y, Zempo B, Oka Y (2010) Hypothalamic Kiss1 but not Kiss2 neurons are involved in estrogen feedback in medaka (Oryzias latipes). Endocrinology 151:1751–1759
40. Smith JT, Cunningham MJ, Rissman EF, Clifton DK, Steiner RA (2005) Regulation of Kiss1 gene expression in the brain of the female mouse. Endocrinology 146:3686–3692
41. Smith JT, Dungan HM, Stoll EA, Gottsch ML, Braun RE et al (2005) Differential regulation of KiSS-1 mRNA expression by sex steroids in the brain of the male mouse. Endocrinology 146:2976–2984
42. Revel FG, Saboureau M, Masson-Pevet M, Pevet P, Mikkelsen JD et al (2006) Kisspeptin mediates the photoperiodic control of reproduction in hamsters. Curr Biol 16:1730–1735
43. Smith JT, Clay CM, Caraty A, Clarke IJ (2007) KiSS-1 messenger ribonucleic acid expression in the hypothalamus of the ewe is regulated by sex steroids and season. Endocrinology 148:1150–1157
44. Smith JT, Coolen LM, Kriegsfeld LJ, Sari IP, Jaafarzadehshirazi MR et al (2008) Variation in kisspeptin and RFamide-related peptide (RFRP) expression and terminal connections to gonadotropin-releasing hormone neurons in the brain: a novel medium for seasonal breeding in the sheep. Endocrinology 149:5770–5782
45. Herbison AE, Pape JR (2001) New evidence for estrogen receptors in gonadotropin-releasing hormone neurons. Front Neuroendocrinol 22:292–308
46. Dorling AA, Todman MG, Korach KS, Herbison AE (2003) Critical role for estrogen receptor alpha in negative feedback regulation of gonadotropin-releasing hormone mRNA expression in the female mouse. Neuroendocrinology 78:204–209
47. Couse JF, Yates MM, Walker VR, Korach KS (2003) Characterization of the hypothalamic-pituitary-gonadal axis in estrogen receptor (ER) null mice reveals hypergonadism and endocrine sex reversal in females lacking ERalpha but not ERbeta. Mol Endocrinol 17:1039–1053
48. Wintermantel TM, Campbell RE, Porteous R, Bock D, Grone HJ et al (2006) Definition of estrogen receptor pathway critical for estrogen positive feedback to gonadotropin-releasing hormone neurons and fertility. Neuron 52:271–280
49. Adachi S, Yamada S, Takatsu Y, Matsui H, Kinoshita M et al (2007) Involvement of anteroventral periventricular metastin/kisspeptin neurons in estrogen positive feedback action on luteinizing hormone release in female rats. J Reprod Dev 53:367–378
50. Ansel L, Bolborea M, Bentsen AH, Klosen P, Mikkelsen JD et al (2010) Differential regulation of kiss1 expression by melatonin and gonadal hormones in male and female Syrian hamsters. J Biol Rhythms 25:81–91
51. Tena-Sempere M (2005) Hypothalamic KiSS-1: the missing link in gonadotropin feedback control? Endocrinology 146:3683–3685
52. Li D, Mitchell D, Luo J, Yi Z, Cho SG et al (2007) Estrogen regulates KiSS1 gene expression through estrogen receptor alpha and SP protein complexes. Endocrinology 148:4821–4828
53. Clarkson J, d'Anglemont de Tassigny X, Colledge WH, Caraty A, Herbison AE (2009) Distribution of kisspeptin neurones in the adult female mouse brain. J Neuroendocrinol 21:673–682
54. Desroziers E, Mikkelsen J, Simonneaux V, Keller M, Tillet Y et al (2010) Mapping of kisspeptin fibres in the brain of the pro-oestrous rat. J Neuroendocrinol 22:1101–1112
55. Clarkson J, Herbison AE (2006) Postnatal development of kisspeptin neurons in mouse hypothalamus; sexual dimorphism and projections to gonadotropin-releasing hormone neurons. Endocrinology 147:5817–5825
56. Ramaswamy S, Guerriero KA, Gibbs RB, Plant TM (2008) Structural interactions between kisspeptin and GnRH neurons in the mediobasal hypothalamus of the male rhesus monkey (Macaca mulatta) as revealed by double immunofluorescence and confocal microscopy. Endocrinology 149:4387–4395
57. Uenoyama Y, Inoue N, Pheng V, Homma T, Takase K et al (2011) Ultrastructural evidence of kisspeptin-gonadotrophin-releasing hormone (GnRH) interaction in the median eminence of

female rats: implication of axo-axonal regulation of GnRH release. J Neuroendocrinol 23:863–870
58. Servili A, Le Page Y, Leprince J, Caraty A, Escobar S et al (2011) Organization of two independent kisspeptin systems derived from evolutionary-ancient kiss genes in the brain of zebrafish. Endocrinology 152:1527–1540
59. Kanda S, Akazome Y, Okubo K, Okamura H, Oka Y (2009) Kisspeptin neurons act closely but indirectly on GnRH 1 neurons via local interneurons but not on GnRH 2 or 3 neurons in medaka. Society for Neuroscience Abstract. 2009. Chicago, IL
60. Zhao Y, Wayne NL (2012) Effects of kisspeptin1 on electrical activity of an extrahypothalamic population of gonadotropin-releasing hormone neurons in medaka (Oryzias latipes). PLoS One 7:e37909
61. Shahjahan M, Motohashi E, Doi H, Ando H (2010) Elevation of Kiss2 and its receptor gene expression in the brain and pituitary of grass puffer during the spawning season. Gen Comp Endocrinol 169:48–57
62. Mechaly AS, Vinas J, Piferrer F (2009) Identification of two isoforms of the Kisspeptin-1 receptor (kiss1r) generated by alternative splicing in a modern teleost, the Senegalese sole (Solea senegalensis). Biol Reprod 80:60–69
63. Biran J, Ben-Dor S, Levavi-Sivan B (2008) Molecular identification and functional characterization of the kisspeptin/kisspeptin receptor system in lower vertebrates. Biol Reprod 79:776–786
64. Price DA, Greenberg MJ (1977) Structure of a molluscan cardioexcitatory neuropeptide. Science 197:670–671
65. Lopez-Vera E, Aguilar MB, Heimer de la Cotera EP (2008) FMRFamide and related peptides in the phylum mollusca. Peptides 29:310–317
66. Osugi T, Daukss D, Gazda K, Ubuka T, Kosugi T et al (2012) Evolutionary origin of the structure and function of gonadotropin-inhibitory hormone: insights from lampreys. Endocrinology 153(5):2362–2374
67. Osugi T, Ukena K, Sower SA, Kawauchi H, Tsutsui K (2006) Evolutionary origin and divergence of PQRFamide peptides and LPXRFamide peptides in the RFamide peptide family. Insights from novel lamprey RFamide peptides. FEBS J 273:1731–1743
68. Osugi T, Uchida K, Nozaki M, Tsutsui K (2011) Characterization of novel RFamide peptides in the central nervous system of the brown hagfish: isolation, localization, and functional analysis. Endocrinology 152:4252–4264
69. Larhammar D, Lundin LG, Hallbook F (2002) The human Hox-bearing chromosome regions did arise by block or chromosome (or even genome) duplications. Genome Res 12:1910–1920
70. Lyubimov Y, Engstrom M, Wurster S, Savola JM, Korpi ER et al (2010) Human kisspeptins activate neuropeptide FF2 receptor. Neuroscience 170:117–122
71. Oishi S, Misu R, Tomita K, Setsuda S, Masuda R et al (2011) Activation of neuropeptide FF receptors by kisspeptin receptor ligands. ACS Med Chem Lett 2:53–57
72. Gouarderes C, Mazarguil H, Mollereau C, Chartrel N, Leprince J et al (2007) Functional differences between NPFF1 and NPFF2 receptor coupling: high intrinsic activities of RFamide-related peptides on stimulation of [35S]GTPgammaS binding. Neuropharmacology 52:376–386
73. Clarke IJ, Qi Y, Puspita Sari I, Smith JT (2009) Evidence that RF-amide related peptides are inhibitors of reproduction in mammals. Front Neuroendocrinol 30(3):371–378
74. Clarke IJ, Sari IP, Qi Y, Smith JT, Parkington HC et al (2008) Potent action of RFamide-related peptide-3 on pituitary gonadotropes indicative of a hypophysiotropic role in the negative regulation of gonadotropin secretion. Endocrinology 149:5811–5821
75. Gibson EM, Humber SA, Jain S, Williams WP III, Zhao S et al (2008) Alterations in RFamide-related peptide expression are coordinated with the preovulatory luteinizing hormone surge. Endocrinology 149:4958–4969
76. Anderson GM, Relf HL, Rizwan MZ, Evans JJ (2009) Central and peripheral effects of RFamide-related peptide-3 on luteinizing hormone and prolactin secretion in rats. Endocrinology 150:1834–1840

77. Kadokawa H, Shibata M, Tanaka Y, Kojima T, Matsumoto K et al (2009) Bovine C-terminal octapeptide of RFamide-related peptide-3 suppresses luteinizing hormone (LH) secretion from the pituitary as well as pulsatile LH secretion in bovines. Domest Anim Endocrinol 36:219–224
78. Kriegsfeld LJ, Mei DF, Bentley GE, Ubuka T, Mason AO et al (2006) Identification and characterization of a gonadotropin-inhibitory system in the brains of mammals. Proc Natl Acad Sci U S A 103:2410–2415
79. Tsutsui K, Ubuka T, Bentley GE, Kriegsfeld LJ (2012) Gonadotropin-inhibitory hormone (GnIH): discovery, progress and prospect. Gen Comp Endocrinol 177(3):305–314

Chapter 3
Neuroanatomy of the Kisspeptin Signaling System in Mammals: Comparative and Developmental Aspects

Michael N. Lehman, Stanley M. Hileman, and Robert L. Goodman

Abstract Our understanding of kisspeptin and its actions depends, in part, on a detailed knowledge of the neuroanatomy of the kisspeptin signaling system in the brain. In this chapter, we will review our current knowledge of the distribution of kisspeptin cells, fibers, and receptors in the mammalian brain, including the development, phenotype, and projections of different kisspeptin subpopulations. A fairly consistent picture emerges from this analysis. There are two major groups of kisspeptin cell bodies: a large number in the arcuate nucleus (ARC) and a smaller collection in the rostral periventricular area of the third ventricle (RP3V) of rodents and preoptic area (POA) of non-rodents. Both sets of neurons project to GnRH cell bodies, which contain *Kiss1r*, and the ARC kisspeptin population also projects to GnRH axons in the median eminence. ARC kisspeptin neurons contain neurokinin B and dynorphin, while a variable percentage of those cells in the RP3V of rodents contain galanin and/or dopamine. Neurokinin B and dynorphin have been postulated to contribute to the control of GnRH pulses and sex steroid negative feedback, while the role of galanin and dopamine in rostral kisspeptin neurons is not entirely clear. Kisspeptin neurons, fibers, and *Kiss1r* are found in other areas, including widespread areas outside the hypothalamus, but their physiological role(s) in these regions remains to be determined.

M.N. Lehman
Department of Neurobiology and Anatomical Sciences, University of Mississippi Medical Center, Jackson, MS, USA
e-mail: mlehman@umc.edu

S.M. Hileman • R.L. Goodman, Ph.D. (✉)
Department of Physiology and Pharmacology, West Virginia University School of Medicine, PO Box 9229, Morgantown, WV 26505-9229, USA
e-mail: rgoodman@hsc.wvu.edu

A.S. Kauffman and J.T. Smith (eds.), *Kisspeptin Signaling in Reproductive Biology*, Advances in Experimental Medicine and Biology 784, DOI 10.1007/978-1-4614-6199-9_3,

Introduction

A clear understanding of the anatomy of the kisspeptin system is a necessary foundation for current and future work on the physiology and pathology of kisspeptin. Indeed, the early recognition of two distinct subpopulations of kisspeptin cell bodies in rodents was critical to the development of current models for the role of kisspeptin in mediating the feedback actions of steroids [1]. As the physiological functions of kisspeptin expand, knowledge of the phenotype, and of the afferent and efferent connections, of different kisspeptin cell populations becomes even more critical. Thus, the primary purpose of this chapter is to review our current understanding of this neural circuitry. The neuroanatomy of this system has been reviewed in the last few years [2–4] so this material should be considered an update of these reviews, with an emphasis on comparative aspects of kisspeptin neuroanatomy in mammals. It is important to recognize that there is also considerable information on this subject in nonmammalian species, particularly in fish, but this is beyond the scope of our chapter. Interested readers are referred to recent reviews of this topic [5, 6] and Chap. 2.

This review will focus on distribution of kisspeptin cell bodies, fibers, and kisspeptin receptors, with some discussion of the overlap of the latter two. We will also consider co-localization of neuropeptides, steroid receptors, and other neurotransmitters in different kisspeptin subpopulations and evidence on where these subpopulations project. Finally, we think it is important to consider developmental neuroanatomy and sexual dimorphism of the kisspeptin system. Although this will be somewhat redundant with other chapters in this book, this information is required for a full understanding of the current neuroanatomical literature on kisspeptin expression in mammals.

Distribution of Kisspeptin and *Kiss1* in the Adult Brain

Since the discovery of its central role in reproduction in 2003, there have been a number of studies documenting the localization of kisspeptin in the brain using either immunocytochemistry (ICC) for cell bodies and fibers or in situ hybridization (ISH) for cell bodies. The original studies using ICC to identify kisspeptin-positive cells and fibers were confounded by cross-reactivity of the antibodies with related peptide members of the RFamide family [7], but more recently, a number of antibodies have been generated which have been shown by the use of careful positive and negative controls to be specific to kisspeptin [8–10]. Using these specific antibodies, and cDNA and RNA probes against kisspeptin sequences, the distribution of kisspeptin cells and fibers has now been mapped out in a variety of mammalian species. Not surprisingly, much of this work has been done in mice [8, 11–17] and rats [14, 18–25], but some data is also available in other rodents (hamster [26–29] and guinea pig [30]). Ruminants, particularly sheep [9, 10, 31–35] and goats [36–38], have also been studied extensively, while there is less information on expression in monkeys [39–43] and humans [41, 44] (Tables 3.1 and 3.2).

Table 3.1 Location of kisspeptin/*Kiss1* cells in the adult brain

Species	ARC	POA	RP3V	ME/Inf. stalk	DMH	VMH	Medial amygdala[a]	BNST	Other regions	ISH reference[e]	ICC reference[e]
Human	+++	+		+						[41]	[44]
Rhesus monkey	+++	++		+				+		[41–43]	[39, 40]
Mouse	++	+	++		+		+	+	+[b]	[12–16]	[8, 11, 17]
Rat	+++		++			+	+	+	+[c]	[14, 18, 20, 21, 23–25]	[19, 22, 24, 25]
Hamster	++		++							[26, 29]	[27, 28]
Guinea pig	+++		+		+				+[d]	[30]	[30]
Sheep	+++	++			+	+				[32–35]	[9, 10, 31, 34, 35]
Goat	+++	+									[36–38]
Horse	+++	+			+						[46, 47]

+++, Many (50–150); ++, moderate (15–50); +, few (<15 or numbers not reported)

[a]Numbers based on diestrus females, ++ in proestrus females and males

[b]Neocortex, insular cortex, piriform cortex, lateral septum, nucleus of the solitary tract (based on [40], Kiss1 gene-driven Cre activity)

[c]Anterior parvocellular portion of the PVN

[d]Intermediate and caudal periventricular area

[e]Reference numbers are from citations at end of chapter

Table 3.2 Distribution of kisspeptin fibers in the adult brain (identified KNDy projections, based on dual immunostaining for peptides and/or tract tracing, in yellow)

Species	ARC	POA	RP3V	ME Inter	ME Exter	PVN	VMH	DMH	LHA	SON	BNST	OVLT	Medial Septum	Lateral Septum	DBB	Other areas	References[d]
Human	++	++	++	++	+	+	+	+	+					+			44
Rhesus Monkey	++	+		++	+												39,40
Mouse	++	+	+	++		+		+	+	+	+	+	+	+		+[a]	8,11,17, 106
Rat	++	+	+	++	+	++		+	+	+	+	+	+	+	+	+[b]	18,19,22, 24,107,109
Hamster	++		+														28
Guinea pig	++	+	+	++	++		+	+			+	+	+	+	+	+[c]	30
Sheep	++	+		++	+		+	+			+						9,10,34,35
Goat	++	+		++	+												36-38
Horse	++	+		++													46

++, Dense fibers; +, moderate or few fibers

[a] Other regions include lateral preoptic area, contralateral ARC, posterior hypothalamic area, medial amygdala, paraventricular thalamic nucleus, periaqueductal gray, locus coeruleus

[b] Other regions include septohypothalamic area, contralateral ARC, retrochiasmatic area, suprachiasmatic nucleus; medial tuberal nucleus; posterior hypothalamus

[c] Other regions include ventral premammillary nucleus

Kisspeptin Cell Bodies

Across the species examined, there are two major populations of kisspeptin cells that have been identified in the diencephalon: a group in the arcuate nucleus (infundibular nucleus in humans) and the other in the preoptic region. The arcuate (ARC) population is the largest group of kisspeptin cells seen in the mammalian hypothalamus [2]. In rodents, kisspeptin cells in this group are present at all rostral-caudal levels [12, 15], while in monkeys and sheep, they are located mostly at middle and caudal levels [10, 39]. In addition to the ARC, a second prominent diencephalic group of kisspeptin cells is seen in the preoptic area (POA). In rodents, the latter group is located in the rostral periventricular of the third ventricle (RP3V), and consists of kisspeptin cells clustered in the anteroventral periventricular nucleus (AVPV) that extend caudally into the adjacent periventricular preoptic zone (PeN). This distribution in rodents is based largely on studies in females, since males have few, if any, kisspeptin cells in this region (see section Sex Differences: Developmental Changes in Males). In contrast to female rodents, other female mammals (primates and ruminants) appear to lack a well-defined RP3V population, and instead kisspeptin cells are scattered slightly more laterally within the medial preoptic region. It seems likely that kisspeptin cells in the RP3V of rodents, and those in the preoptic region in sheep, goats and primates, are homologous, but the precise functional roles of each of the populations may differ between species [45]. The only species in which a distinct preoptic population has yet to be demonstrated is the horse, despite the use of specific antibodies [46, 47]. Since these rostral kisspeptin populations have been implicated in the estrogen-induced preovulatory LH surge in many species [45], the absence of them in the horse correlates with evidence that the preovulatory LH increase in mares is due to withdrawal of steroid negative feedback, rather than the stimulatory actions of estradiol [48].

Identification of precise cell numbers in these populations is somewhat complicated by the fact that kisspeptin mRNA and peptide expression in the preoptic region and ARC are under opposite regulatory control by gonadal steroid hormones. Thus, in general, estradiol in females stimulates kisspeptin expression in the RP3V and POA, while inhibiting it in the ARC [45]. Nonetheless, comparison of cell numbers in the female brain under optimal hormonal conditions (estradiol treatment in the case of the preoptic population, and ovariectomy in the case of the ARC) suggests that the absolute number of kisspeptin cells in the ARC is generally two- to fourfold greater than that in the RP3V or POA [2]. Thus, the ARC kisspeptin cell population is most consistent among mammals in its presence and contains the greatest number of cells.

Another complication that raises both technical and interesting biological issues is the effect of endocrine status on location of kisspeptin-ir within cells. Specifically, estradiol (E_2) may alter the location of kisspeptin protein within parts of the ARC neurons. Thus, in rats [18] and mice [49, 50], intact and E_2-treated OVX animals show a dense network of fibers with few, if any, visible cell bodies in the ARC; in contrast, tissues from OVX females contain kisspeptin-ir cell bodies, but few fibers.

A similar effect of E_2 has recently been reported in guinea pigs [30]. It has been argued that the kisspeptin-ir fibers in the ARC arise from the AVPV because E_2 stimulates kisspeptin expression in that area, but not the ARC [51]. However, data using dual ICC to identify efferents from ARC kisspeptin cells in rats, sheep, and monkeys indicate that many of the kisspeptin fibers in the ARC arise from cell bodies in this area (see below). Moreover, the dissociation of effects of E_2 on kisspeptin expression in AVPV cell bodies and ARC fibers of adult hpg mice is not consistent with these fibers arising from the AVPV in this species [50]. If these kisspeptin fibers do originate from ARC cell bodies, then the increase in immunoreactivity could be due to either (1) increased, or no change, in transport out of the cells, coupled with a decrease in kisspeptin synthesis, or (2) simply a build-up of kisspeptin protein because secretion has been inhibited more than synthesis and transport. Interestingly, this phenomenon is not seen in male rats [18] and is seen 2 days after colchicine administration in females [18], so it may reflect a chronic shift in intracellular location of kisspeptin proteins. Regardless of the mechanisms, the potential for changes in distribution of kisspeptin within a neuron argues for caution in interpreting changes based on ICC analysis.

In addition to the ARC and preoptic region, there are a number of other hypothalamic areas that have been shown to contain populations of kisspeptin cells in the mammalian brain (Table 3.1). In the monkey and human, a small number of kisspeptin cells, likely an extension of the ARC group, are seen in the median eminence (monkey [39]) and infundibular stalk (human [44]). The dorsomedial nucleus of the hypothalamus (DMH) contains a small group of kisspeptin cells, demonstrated by ICC using specific antibodies, in mice, guinea pigs, sheep, and horses, but these cells are not present in the rat or hamster (Table 3.1). Likewise, there is species variation in the presence of cells in the ventromedial hypothalamic (VMH) nucleus, with cells being detected in the sheep (in some reports [32] but not others [33]) and rats [24], but not in other species; as in the case of kisspeptin cells in the median eminence [39], these cells may be an extension of the ARC population, in this case laterally. Regardless, evidence of kisspeptin cells in the DMH and VMH in a number of species rests primarily on ICC data, and additional ISH data would be worthwhile to verify their presence in these species.

There are also a number of kisspeptin populations that reside outside of the classical boundaries of the hypothalamus. These include a cluster of *Kiss1* mRNA-expressing cells in the medial nucleus of the amygdala, seen in both rats and mice [14], and a small number of *Kiss1* cells in the bed nucleus of the stria terminalis (BNST) of mice, rats, and rhesus monkeys (Table 3.1). Like the RP3V population, *Kiss1* expression in the medial amygdala is under the stimulatory influence of gonadal steroid hormones [14]. The localization of these populations in circuitry that mediates pheromonal control of sexual behavior [52, 53] and neuroendocrine function [54] suggests that they may play a role in these functions, but this remains to be explored.

Finally, recent findings suggest that kisspeptin cells may be present in widespread areas of the brain outside the hypothalamus and limbic system. This evidence is based on observations of transgenic mice in which *Kiss1* drives the expression of Cre recombinase and other reporter genes [55]. Using such mice, Cre expression has been detected in widespread cortical areas, including layers 5 and 6 of neocortex, insular cortex, and piriform cortex, as well as in the lateral septum, and in the nucleus of the solitary tract in the brainstem. A potential caveat to these observations is the possibility that Cre reporter expression in some of these regions represents kisspeptin that is transiently expressed during development but not during adulthood [56]. However, at least in the neocortex, *Kiss1*-driven Cre activity is confirmed by the presence of light *Kiss1* mRNA-labeling by ISH, although peptide is not detectable in these cells perhaps because of the low level of mRNA expression. The function(s) of kisspeptin in cortex and these other regions is at present a mystery, but seems likely to portend functions of this peptide that extend far beyond its recognized roles in reproduction and neuroendocrine function.

Kisspeptin Fibers

The overall pattern and distribution of kisspeptin fibers has been analyzed in a number of ICC studies (Table 3.2), representing the same range of species in which cell bodies have been studied. In addition, in the case of the RP3V and ARC populations, the specific projections of each have been analyzed using either tract tracing combined with ICC or multiple-label ICC, and these findings are summarized in section Anatomical Connections of Kisspeptin Cells. In all species examined to date, the densest accumulation of kisspeptin fibers and terminals appears to be within the regions that contain the two major hypothalamic populations, the ARC and RP3V, as well as within the internal zone of the median eminence (Table 3.2). Kisspeptin fibers, albeit fewer in number, have also been reported in the external zone of the median eminence in most species, the site of neurosecretory release of GnRH. As noted previously [2], the paucity and/or lack (e.g., mice) of kisspeptin fibers in the external zone suggests that if kisspeptin is to effect release of GnRH at the level of the median eminence [57], it likely does so via diffusion and/or volume transmission from the internal to external layers.

Thus far, the species where kisspeptin fibers have been thoroughly mapped outside of the ARC, RP3V, and median eminence include human, mouse, rat, guinea pig, and sheep, and in general, the studies have revealed a fairly consistent pattern with a majority of kisspeptin fibers being located predominantly within medially located hypothalamic nuclei and preoptic regions (Table 3.2). For example, in each of these species, kisspeptin-immunoreactive fibers are found in the DMH; in addition, in all species except humans, kisspeptin fibers are seen in the BNST. There are also some species differences: for example, the PVN contains moderate to low numbers of kisspeptin fibers in humans, mice, and rats, but not in guinea pigs or sheep; by contrast, the VMH contains fibers in humans, guinea pigs, and sheep but is devoid of such fibers in rats and mice. Since immunoreactive fibers can represent either terminal boutons or axons of passage, these differences may reflect variation in postsynaptic targets of kisspeptin cells or simply in the route which fibers take to reach those targets.

Table 3.3 Percentage of ARC kisspeptin cells co-localizing other neuropeptides/transmitters

Species	NKB	Dyn	Galanin	Met-Enk	Glutamate	GABA
Human	77 [44]					
Rhesus monkey	40–60[a]					
Mouse	90[b], 94[c]	92[b], 86[c]	12[d], 65[e]	0[e]	90 [55]	50 [55]
Rat	97 [107]					
Sheep	80[f]	94[f]				
Goat	99 [38]	78 [38]				

[a]Castrated males [40]
[b]OVX [63]
[c]Castrated and castrated + T males [61]
[d]OVX [106]
[e]Intact mice treated with colchicines [67]
[f]OVX + E [10]

In addition to medially located nuclei, in the mouse, rat, and human hypothalamus, a few kisspeptin fibers are also seen consistently in the lateral preoptic and lateral hypothalamic areas (Table 3.2). Rostrally, kisspeptin-immunoreactive fibers in these species are present in a variety of forebrain structures, including the medial and lateral septum, the diagonal band of Broca, and organum vasculosum of the lamina terminalis (OVLT). Finally, in mice and rats, kisspeptin fibers have been reported in the caudal hypothalamus, and in mice also in more distant locations: the periaqueductal gray of the midbrain, the locus coeruleus, the paraventricular nucleus of the thalamus, as well as the medial amygdala (Table 3.2). Whether kisspeptin fibers in these more widespread areas are seen in other species is not known. Likewise, the precise origin of these fibers remains to be determined; while some are likely to arise from the ARC or RP3V kisspeptin cells (see section Anatomical Connections of Kisspeptin Cells), it is possible that others originate from populations located outside of the hypothalamus (see above).

Co-localization of Other Peptide/Transmitters

There is increasing evidence of anatomical heterogeneity among kisspeptin cell populations, specifically with respect to their co-expression of other neuropeptides and neurotransmitters. The most consistent example of this to date is in the ARC population, where a majority of kisspeptin neurons express two other neuropeptides, neurokinin B and dynorphin, each of which has been strongly implicated in the physiological control of GnRH secretion [58, 59]. The high degree of co-localization of kisspeptin, NKB, and dynorphin in the ARC population (Table 3.3) and its conservation across species (Fig. 3.1) has led to the acronym "KNDy" neurons [60]. KNDy cells form reciprocal connections with each other, as well as project to GnRH neurons, and the combination of excitatory (kisspeptin, NKB) and inhibitory (dynorphin) actions within this circuitry has provided the foundation for models of how this kisspeptin population may be involved in the generation and control of GnRH pulses [38, 60–62]. Although the percentage of KNDy peptide

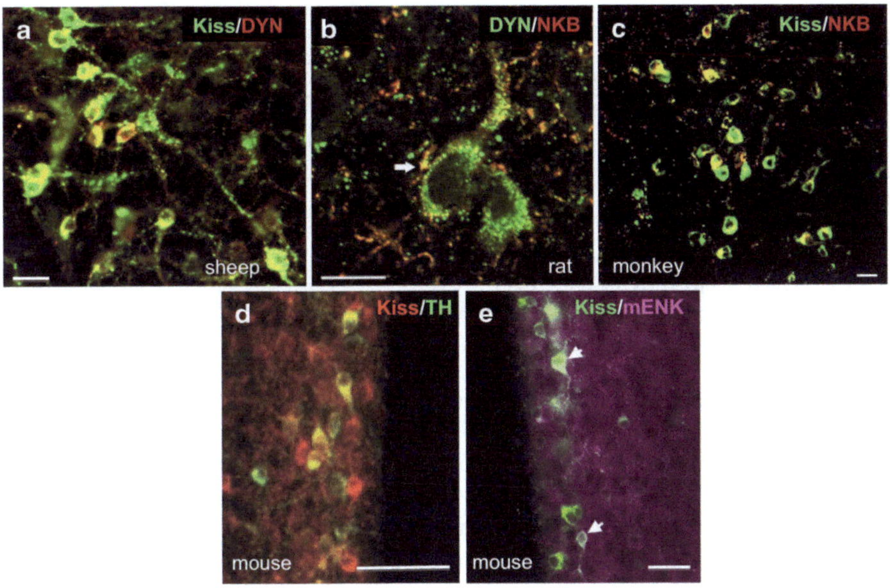

Fig. 3.1 Co-localization of kisspeptin with other neuropeptides/transmitters. Examples of the dual-labeling of ARC kisspeptin cells (*top row*) with dynorphin (DYN) and neurokinin B (NKB) in sheep, rats, and monkeys, and of RP3V kisspeptin cells (*bottom row*) with tyrosine hydroxylase (TH, a marker for dopamine) and met-enkephalin (mENK) in mice. Dual-labeled cells appear *yellow* in (**a–d**), and *white* in (**e**) (indicated by *arrows*). *Arrow* in (**b**) shows an example of a close contact between a DYN/NKB terminal and a DYN/NKB (KNDy) cell in the ARC. Images are modified with permission from [10] (**a**); [105] (**b**); [40] (**c**); [64] (**d**); and [67] (**e**). Scale bars in (**a**) 20 μm, (**b**) 10 μm, (**c**) 20 μm, (**d**) 40 μm, and (**e**) 20 μm. (**a**) Modified from Goodman RL, Lehman MN, Smith JT, Coolen LM, De Oliveira CVR, Jafarzadehshirazi MR et al. Kisspeptin neurons in the arcuate nucleus of the ewe express both dynorphin A and neurokinin B. Endocrinology 2007; 148(12):5752–5760 (with permission from The Endocrine Society). (**b**) Modified from Burke MC, Letts PA, Krajewski SJ, Rance NE. Coexpression of dynorphin and neurokinin B immunoreactivity in the rat hypothalamus: morphologic evidence of interrelated function within the arcuate nucleus. Journal of Comparative Neurology 2006; 498(5):712–726 (with permission from John Wiley & Sons). (**c**) Modified from Ramaswamy S, Seminara SB, Ali B, Ciofi P, Amin NA, Plant TM. Neurokinin B stimulates GnRH release in the male monkey (Macaca mulatta) and is co-localized with kisspeptin in the arcuate nucleus. Endocrinology 2010; 151(9):4494–4503 (with permission from The Endocrine Society). (**d**) Modified from Clarkson J, Herbison AE. Dual phenotype kisspeptin-dopamine neurones of the rostral periventricular area of the third ventricle project to gonadotropin-releasing hormone neurones. Journal of Neuroendocrinology 2011; 23(4):293–301 (with permission from John Wiley & Sons). (**e**) Modified from Porteous R, Petersen SL, Yeo SH, Bhattarai JP, Ciofi P, de Tassigny XD et al. Kisspeptin neurons co-express met-enkephalin and galanin in the rostral periventricular region of the female mouse hypothalamus. Journal of Comparative Neurology 2011; 519(17):3456–3469 (with permission from John Wiley & Sons)

co-localization is uniformly high among female mammals and male rodents and ruminants (Table 3.3), it is slightly lower in male monkeys [40] and humans [44], even in the absence of gonadal steroids that inhibit kisspeptin expression in the ARC. In addition, perturbations in the organizational effects of gonadal steroids during development can lead to an altered ratio of KNDy peptides within the adult ARC population [31].

Table 3.4 Percentage of RP3V or POA kisspeptin cells co-localizing other neuropeptides/transmitters

Species	NKB	Dyn	Galanin	Met-Enk	TH	Glutamate	GABA
Mouse	10 [63]	33 [63]	87[a], 7[b]	28–38[b]	51–68[c], (80) [65]	20 [55]	75 [55]
Rat					5, 20–50[d]		
Sheep	0[e]	0[e]			0[e]		
Goat	0 [36]						

[a]OVX+E [106]
[b]Intact mice treated with colchicine [67]: for Met-Enk, AVPV, 28%; PVpo, 38%
[c]Diestrus 51–55%; proestrus 58–68% [64]
[d]ICC: OVX+E <5%; ISH: OVX+E <20%, intact diestrus 30%, OVX 50% [14]
[e]OVX+E [10]

Preoptic kisspeptin cells in sheep and goats do not appear to co-localize either NKB or dynorphin, although a subset of RP3V neurons does appear to contain both peptides in the mouse [63] (Table 3.4); other species have not as yet been examined for this possible co-localization. However, a sizeable percentage of RP3V cells in both the mouse [64, 65] and rat [21] appear to co-localize tyrosine hydroxylase (TH) (Fig. 3.1d), the rate-limiting enzyme for dopamine biosynthesis [64], although the precise percentage in the rat appears to vary by steroidal status and by whether double-label ICC or ISH was used (Table 3.4). By contrast, kisspeptin cells in the POA or ARC of the sheep do not contain TH [66] even though they are close to adjacent dopaminergic neurons in both regions. In addition, a sizable percentage of RP3V kisspeptin cells in the mouse co-localize met-enkephalin (Fig. 3.1e) and some co-localize galanin [67]; KNDy neurons in the mouse also co-localize galanin but not met-enkephalin (Table 3.3).

In addition to co-localization of other peptides, cells of both the ARC and RP3V kisspeptin populations contain the classical amino acid transmitters, glutamate and GABA (Tables 3.3 and 3.4), as revealed by co-localization of the markers, vesicular glutamate transporter-2 (vGlut2) and gamma amino acid decarboxylase (GAD)-67, respectively. However, the ARC population is predominantly glutamatergic, whereas RP3V cells are mostly GABAergic [55]. Preliminary observations in the sheep [68] suggest that this distinction between ARC and RP3V populations holds in other mammals as well. Finally, it should be noted that kisspeptin cells located in other regions (e.g., medial amygdala) have not yet been examined for possible co-expression of other peptides or transmitters. Given the likely functional heterogeneity of these anatomically distributed populations, it seems probable that their neurochemical phenotypes will be similarly diverse.

Co-localization of Steroid Receptors

The key role that kisspeptin cells play in the steroid feedback control of GnRH neurons is underlined by the high degree of co-localization of nuclear receptors for gonadal sex steroids in these cells. Indeed, in all species examined, a majority of both ARC and RP3V/preoptic kisspeptin cells co-localize estrogen receptor (ER)-alpha (the isoform responsible for estradiol's feedback actions upon GnRH secretion), progesterone receptor (PR), and androgen receptor (AR) [2, 15, 69]. In the ARC population of rats, mice, and sheep, the percentages of co-localization of ER-alpha range from 70 to 99%, whereas in the RP3V and preoptic region they range from 50 to 99% [9, 15, 18, 33, 70]. The percentage of both ARC and RP3V populations that co-localize the beta isoform of the estrogen receptor is considerably less, ranging from 11 to 25% in the ARC and 21–31% in the RP3V of rats and mice [15, 23]. Other kisspeptin populations have not been directly examined for steroid receptor co-localization, but given that sex steroids regulate *Kiss1* expression in the medial amygdala in rats and mice [14], and the presence of ER and AR in this area [71], it seems likely that kisspeptin cells in the amygdala are also a direct target for gonadal steroids. Preliminary evidence suggests that other nuclear steroid receptors are also present in kisspeptin cells: approximately 50% of KNDy cells in the ovine ARC co-localize type II glucocorticoid receptors [72]. Furthermore, there is recent evidence that other types of receptors for circulating hormones are present in ARC kisspeptin cells: receptors for prolactin [73] and insulin [74] have each been co-localized to a subset of KNDy neurons. Taken together, these findings point to a potential convergence of endogenous hormonal cues onto the ARC kisspeptin population, which may place them in a unique position to respond to multiple signals related to stress, nutrition, and the environment, as well as reproductive endocrine status.

Distribution of *Kiss1r*

In contrast to the wealth of information on the neuroanatomical distribution of kisspeptin cells and fibers, there is very limited data available on the location of *Kiss1r* mRNA and no data on Kiss1r protein. The only data in humans is from early studies before the role of *Kiss1r* in reproductive neuroendocrinology was recognized, so they provide almost no information on hypothalamic expression of this receptor [75, 76]. Moreover, most of the studies since then in monkeys and rodents used RT-PCR of mRNA extracted from large tissue blocks or micro-dissected areas (Table 3.5). Consequently, these reports do not provide any information on the location of *Kiss1r* within these relatively large volumes of tissue. Quantitative comparisons between areas using this approach are also problematic because the ratio of *Kiss1r* mRNA to a housekeeping gene is partially dependent on the percentage of *Kiss1r*-containing cells within the block. Thus, variations in the precision of micro-dissection contribute significantly to the values reported. There have been a number of studies using ISH, or related techniques, that can provide cellular resolution, but most of these have been focused on whether GnRH neurons contain *Kiss1r* and do not provide more general neuroanatomical information. Thus, detailed descriptions in this section rely largely on two studies. One of these used ISH in rats, but provided only a few low-power images [77]. The other used transgenic mice in which IRES-LacZ cassettes had been inserted into the *Kiss1r* gene so that β-galactosidase could be identified with Xgal staining as a marker for *Kiss1r*-containing cells [78]. Although this provides the only detailed description of *Kiss1r* expression in the

Table 3.5 Distribution of Kiss1r mRNA based on RT-PCR, in situ hybridization using radioactive probes (ISH) or immunohistochemistry for β-galactosidase (β-gal)

Species	MBH/ARC[a]	POA	RP3V	AHA	DMH	PH	LHA	Medial septum	Hippo/ DG	Medial amygdala	Locus coeruleus	GnRH neurons	Other regions	References
Human (PCR)	++								+++	+++	++		+[b]	[75, 76]
Rhesus monkey (PCR)	+(M) ++(F)	++(M) +(F)												[42, 79, 80]
Mouse (PCR)	++	++						+++	+				+[c]	[85]
Mouse (ISH)												85–95%		[13, 100]
Mouse (β-gal)	–	++	–	–	–	++	–	++	+++	–	+	40–70%	+[d]	[78, 88]
Rat (PCR)	+	++	++						+					[22, 81–84]
Rat (ISH)	+++	++		++	+++	++	+++	p	+++	+++	+++	77%	+[e]	[20, 77]
Sheep (ISH)	p	p			+							78–90%		[35, 86, 87]

p Present but not quantified (from reports of *Kiss1r* in GnRH neurons)

[a]PCR data is from MBH in humans and ARC in rodents and sheep

[b]Other regions include thalamus, substantial nigra, cingulated gyrus

[c]Other regions include cerebellum, striatum

[d]Other regions include supramammillary nucleus, anteroventralthalamus, habenula; periaqueductal gray

[e]Other regions include: zona incerta, olfactory cortex, ventral premammillary, periaqueductal gray

murine brain, the results must be interpreted conservatively until confirmed with conventional ISH or ICC.

Hypothalamic Expression

Even with these caveats in mind, some clear general patterns emerge when *Kiss1r* expression in the hypothalamus is examined across species. In all species studied, there is clear *Kiss1r* expression in the POA (Table 3.5). In monkeys [42, 79, 80], rats [22, 81–84], and mice [85], *Kiss1r* was identified using RT-PCR of tissue block extracts, and many of the latter studies used microdissected POA tissue [81, 82, 84, 85]. Moreover, this conclusion is supported by ISH data in sheep [35, 86, 87], rats [20], and mice [13, 78, 88], which also demonstrated co-localization of *Kiss1r* in GnRH neurons (Table 3.5).

The second relatively consistent location of *Kiss1r* within the hypothalamus is in the ARC (Table 3.5). *Kiss1r* mRNA has been found in the ARC using both RT-PCR in rats [22, 81–84] and mice [85], and ISH in sheep [86] and rats [77]; it is also likely that the *Kiss1r* found in the MBH of monkeys [42, 79, 80] also reflects, in part, expression in the ARC. In contrast, no *Kiss1r* expression was found in mice using Xgal ICC [78], but this observation needs to be confirmed using other approaches since *Kiss1r* was readily detectable in micropunches of the murine ARC [85]. There are two consistent reports of *Kiss1r* in the posterior hypothalamus of rats [77] and mice [78] and in the RP3V of rats [83, 84], although it was not observed in the latter area in mice using Xgal ICC [78]. Similarly, this receptor has been observed in the AHA, DMH, LHA, ventral premammillary nuclei, and zona incerta of rats [77], but not mice [78]; whether this reflects species or technical differences awaits further work.

Extra-hypothalamic Expression

There is even less data on expression of *Kiss1r* in areas outside the hypothalamus, and the only species in which there is information on other neural areas are rats [77], mice [78], and humans [75, 76]. As noted above, the latter comes from two early reports using RT-PCR before the role of *Kiss1r* in reproduction had been discovered. The highest levels of *Kiss1r* are consistently found in the hippocampus, with relatively high expression also observed in the amygdala (humans and rats), periaqueductal gray (rats, mice), and locus coeruleus (humans, rats). Relatively high expression of *Kiss1r* was also reported in the supramammillary nuclei of mice, but data on this region is not available for other species. Similarly, there is evidence for *Kiss1r* expression in the primary olfactory cortex of rats, and the caudate nucleus of humans that is not available in other species. Finally, there are a number of reports that *Kiss1r* is found in the anterior pituitary of humans [75], baboons [89], sheep [90, 91], pigs [92], and rats [93, 94]. One of these studies observed that co-localization in the rat pituitary was limited to gonadotropes using dual ICC [94], but there was no differential expression of *Kiss1r* in cellular fractions of ovine pituitary enriched for gonadotropes, compared to fractions enriched for somatotropes or lactotropes

[91]. The physiological role of *Kiss1r* in the pituitary remains unclear because effects of kisspeptin on LH secretion from pituitary cells in vitro are not consistently observed [95, 96], and when kisspeptin stimulates LH secretion, its effects are generally modest compared to those of GnRH [91, 93, 97, 98]. Moreover, although kisspeptin was detected in hypophysial portal blood, the low concentrations and lack of correlation with LH concentrations in jugular samples led the authors to conclude that hypothalamic release of kisspeptin played no physiological role in regulation of LH in the ewe [91]. It should be noted that kisspeptin mRNA and protein have been observed in the anterior pituitary in species ranging from rats [94] to primates [99], so *Kiss1r* could play a paracrine role in this tissue.

Cellular Expression

Not surprisingly, most studies that have examined cellular *Kiss1r* localization have focused on GnRH neurons. Studies using dual ISH (Table 3.5) consistently report that a high percentage of GnRH cells from the POA contain *Kiss1r*; in female sheep this percentage ranged from 78 to 90% [35, 86, 87], in male rats it averaged 77% [20], and in adult male [13] and female [100] mice greater than 85% co-localization was observed. A slightly lower percentage ranging from 55% [88] to 68% [78] was observed in male mice using Xgal as a marker for *Kiss1r*, and the lowest percentage of co-localization (3/8 cells) was obtained using single cell RT-PCR [101]. The relative low level of expression in the latter study may have been due to the use of fetal cells [78]. A high degree of expression of *Kiss1r* in GnRH neurons is also supported by the 90% of murine GnRH neurons that respond to kisspeptin in slice preparations [13], since this effect is independent of other neural input [13, 102].

Only two other cell types have been examined for possible expression of *Kiss1r*. Based on dual ISH techniques, kisspeptin neurons in the POA and ARC of the sheep do not contain *Kiss1r* [86]. On the other hand, ARC POMC neurons in mice most likely do contain *Kiss1r* because these cells consistently respond to direct application of kisspeptin in slice preparation, even when pharmacologically isolated from other neural inputs [85]. It will be important, however, to confirm this conclusion using anatomical approaches.

Matches/Mismatches of Kiss1r and Kisspeptin Fibers

Under most circumstances, *Kiss1r* receptors are presumably only physiologically important if they are located in neurons innervated by kisspeptin-containing synapses. Thus, the occurrence of receptor-ligand matches and mismatches in the kisspeptin system can provide potentially useful information about physiological significance. On the other hand, the presence of kisspeptin terminals in areas devoid of *Kiss1r* raises the possibility that kisspeptin exerts its effect in these areas via other, as yet unidentified, receptors. Alternatively, since kisspeptin is often co-localized with a number of other important neuropeptides (i.e., NKB and dynorphin) and transmitters (i.e., glutamate) (see section Co-localization of Other Peptide/ Transmitters), such synapses may be providing input from other components within these terminals, as appears to be the case for kisspeptin input to KNDy cells [60]. It should also be noted that because of the paucity of data on *Kiss1r*, only limited conclusions can be drawn on receptor-ligand matches/mismatches for the kisspeptin system at this time, and these are based largely on *Kiss1r* data in rats [77] and mice [78]. For example, we cannot determine whether Kiss1r is matched with the kisspeptin-ir fibers that are observed in the median eminence of all species (Table 3.1), because there is no information on distribution of Kiss1r protein. In light of the expression of *Kiss1r* mRNA in GnRH cell bodies (see above), it is likely that the receptor protein is present in GnRH terminals in the median eminence, but this needs to be directly confirmed.

Within the hypothalamic-POA areas, the most obvious match between *Kiss1r* and kisspeptin-ir fibers is in the POA, where both have been observed in mice, rats, sheep, and monkeys (Tables 3.1 and 3.2), which reflects, in part, kisspeptin innervation of GnRH neurons (Fig. 3.2). The second most consistent area of overlap is the ARC, where kisspeptin fibers have been observed in all species and evidence for *Kiss1r* reported in mice, rats, and sheep, although data in mice are conflicting because *Kiss1r* was not observed in this area using Xgal ICC. The final hypothalamic areas with a consistent match are the medial/lateral septum and the posterior hypothalamus of rats and mice. For both the DMH and RP3V, matches have been observed in rats, while mice apparently have kisspeptin fibers, but not *Kiss1r* (based on Xgal ICC). Other areas of mismatch include the PVN and SON of rats and the LHA of mice, which contains kisspeptin-ir but no detectable *Kiss1r*; conversely, the LHA of rats contains *Kiss1r*, but no corresponding fibers. At a cellular level, there is clear match of *Kiss1r* in GnRH cells and kisspeptin-positive synapses onto GnRH neurons in all species examined to date. However, since not all GnRH cells appear to be innervated by kisspeptin fibers [2], it is unclear whether this correspondence is seen at an individual cell by cell level. In contrast, a majority of ARC KNDy neurons are innervated by other KNDy neurons, but none of these contain *Kiss1r*, at least in the sheep.

Outside the hypothalamus, there is a major mismatch in the dentate gyrus and other areas of the hippocampus in which there is strong evidence for *Kiss1r*, in the absence of kisspeptin-positive fibers in both rats and mice. It should be noted, however, that there is evidence for *Kiss1* mRNA (by PCR) in the dendate gyrus [103, 104] so this possible mismatch should be further investigated. A similar mismatch is evident in the supramammillary nuclei of mice, but there are no data on this area in rats. Conversely, kisspeptin-ir fibers are present in the medial amygdala of mice [8], with no evidence for *Kiss1r* at this time. The clearest areas of matching localization are the periaqueductal gray and locus coeruleus of mice [8]; these areas also contain *Kiss1r* in rats [77], but whether kisspeptin neurons project to these regions in this species remains to be determined.

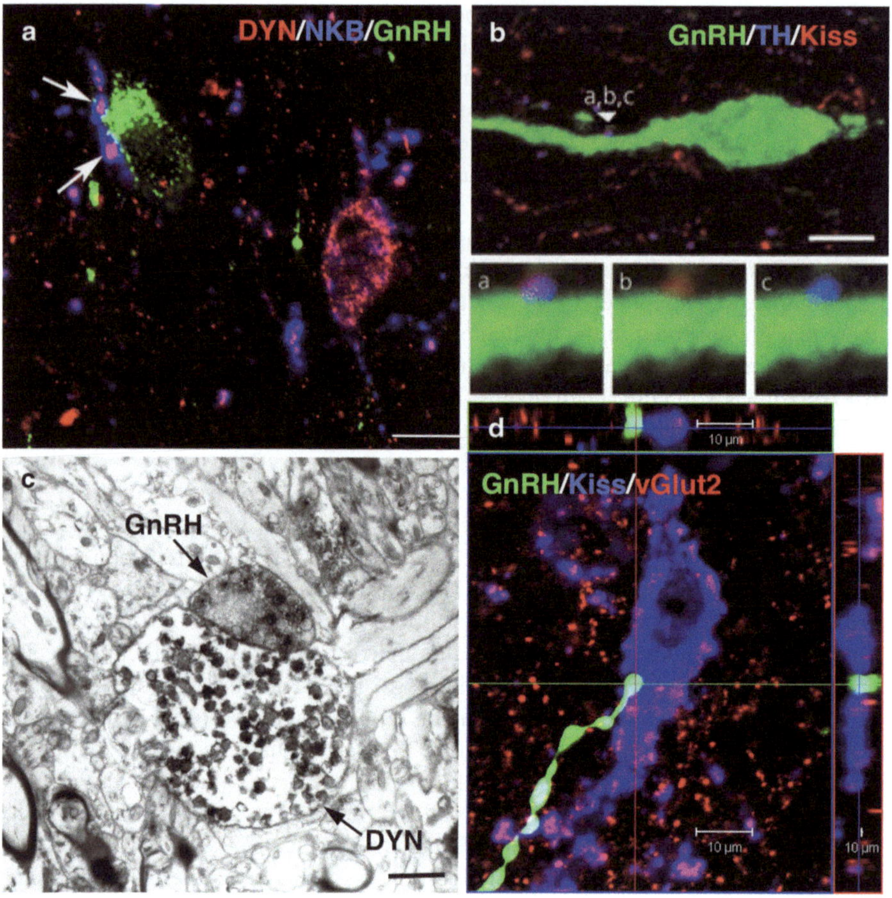

Fig. 3.2 Kisspeptin and KNDy connections with GnRH neurons. (**a**) KNDy inputs (*arrows*, labeled with dynorphin and NKB) to a GnRH cell body in the mediobasal hypothalamus of the sheep; a KNDy cell body (*magenta*) is seen nearby. Bar = 10 µm. Taken from Lehman MN, Coolen LM, Goodman RL. Minireview: kisspeptin/neurokinin B/dynorphin (KNDy) cells of the arcuate nucleus: a central node in the control of gonadotropin-releasing hormone secretion. Endocrinology 2010; 151(8):3479–3489 (with permission from The Endocrine Society). (**b**) RP3V kisspeptin input (*arrowhead*, labeled with kisspeptin and TH) to a preoptic GnRH neuron in the mouse; (*a*) is a higher power view of the contact, (*b*, *c*) show the same image with the TH (*b*) or kisspeptin (*c*) channel removed. Bar = 5 µm. Modified from Clarkson J, Herbison AE. Dual phenotype kisspeptin-dopamine neurones of the rostral periventricular area of the third ventricle project to gonadotropin-releasing hormone neurones. Journal of Neuroendocrinology 2011; 23(4):293–301 (with permission from John Wiley & Sons). (**c**) Electron micrograph showing a dynorphin terminal containing dense-core vesicles (*arrowheads*) in direct contact with a GnRH terminal in the sheep median eminence. Bar = 2 µm. Taken from Lehman MN, Coolen LM, Goodman RL. Minireview: kisspeptin/neurokinin B/dynorphin (KNDy) cells of the arcuate nucleus: a central node in the control of gonadotropin-releasing hormone secretion. Endocrinology 2010; 151(8):3479–3489 (with permission from The Endocrine Society). (**d**) GnRH fiber contacting a kisspeptin (KNDy) cell in the ARC of the sheep; confocal orthogonal views through the close contact are shown above and to the right. Section was also labeled for vGlut-2 showing co-localization in kisspeptin fibers and terminals. Bar = 10 µm. Unpublished data from Lehman, Cernea and Goodman, 2012

Anatomical Connections of Kisspeptin Cells

Efferent Projections

To date, the projections of specific subsets of kisspeptin cells have been investigated using two approaches. The first takes advantage of the co-expression of other neuropeptides and transmitters that are co-expressed uniquely in one population: the most common example of this is the co-localization of KNDy peptides which can be used to define the projections arising from ARC kisspeptin cells [44, 86, 105–107]. The second approach has been to combine stereotaxic injections of anterograde tract tracers (Fluoro-Gold, biotinylated dextran amine) with ICC for kisspeptin, so that fibers dual-labeled for both the tracer and kisspeptin can be inferred to have arisen in the injected region (e.g., ARC) [108, 109].

Dual-label immunostaining of kisspeptin fibers with either NKB and/or dynorphin has been the most common approach to selectively analyze the projections of the ARC kisspeptin population, and has been employed in sheep [60], rats [105], mice [106], and human tissue [44]. In addition, dual-label ICC for NKB and dynorphin, which is unique to the KNDy population, has been used in rats [105], goats [36], and monkeys [40]. The results suggest some common, shared projections of ARC kisspeptin cells across species, but some significant differences as well (Table 3.2). In all species examined, fibers arising from KNDy cells form the densest projections locally, within the ARC and to the internal zone of the median eminence. There are also consistent projections of KNDy fibers to the preoptic region, including the RP3V in rodents, although the density of these fibers is generally less than in the ARC. In the rat, sheep, and goat, a few KNDy fibers also extend into the external zone of the median eminence. In sheep, rats, and mice, KNDy cell efferents appear to be more widespread than in the monkey or human, and include projections in the DMH, BNST, and in rodents, the PVN, LHA, and septal region (Table 3.2). Tracer injections into the ARC of mice combined with kisspeptin ICC revealed a similar distribution of KNDy efferents as did co-localization of the peptides, with projections to the PVN, DMH, and LHA, as well as revealing more distant targets including the posterior hypothalamic region and the periaqueductal gray of the midbrain [6]. In both rats and mice, KNDy cells also project across the midline to the contralateral ARC, perhaps serving to coordinate activity between the KNDy population on both sides of the brain, and their ipsilateral projections to the POA (and presumably GnRH neurons).

In contrast to the ARC population, much less is known about the projections of preoptic and RP3V kisspeptin cells. Dual-labeling of kisspeptin and TH in mice has revealed projections from RP3V kisspeptin cells to the POA, where they contact GnRH neurons [64]. Similarly, anterograde tracer injections into the mouse RP3V, combined with kisspeptin ICC, revealed a set of projections from this population that includes several nuclei of the POA, as well as other areas in the hypothalamus and septal region [108]. The latter projection sites closely overlap those of the ARC population, and include the BNST, PVN, DMH, and periaqueductal gray, as well as projections to the ipsilateral ARC itself. However, there are a few areas that receive inputs from either the

RP3V or ARC populations but not both: the lateral septum receives input from RP3V but not ARC kisspeptin cells, while the lateral preoptic region and lateral hypothalamic area contain projections from KNDy cells but not the RP3V [108].

There are two significant caveats to these observations on the origin of kisspeptin fibers. First, co-localization of kisspeptin and other neuropeptides in individual fibers and boutons does not exclude the possibility that these are fibers of passage en route to another target. Second, as noted above, caution must be used when evaluating projections based on co-localization of KNDy peptides, since the ability to detect these markers is clearly dependent on gonadal hormonal status as well as the dynamics of peptide storage/release at the axon terminal. The recent generation of several strains of *Kiss1*-Cre mice [110] may soon provide another approach that has the potential to circumvent these limitations. Specifically, *Kiss1*-Cre mice can be crossed with transgenic strains bearing Cre-inducible markers such as TdTomato or mCherry to provide complete anterograde filling of axons arising specifically from kisspeptin cells. To achieve selective anterograde labeling of individual *Kiss1* cell populations (e.g., ARC, RP3V, amygdala), Cre-inducible virus lines expressing these markers could be injected into these regions in *Kiss1*-Cre mice. Alternatively, markers for other neuropeptides co-expressed in kisspeptin cells (see section Distribution of Kisspeptin and *Kiss1* in the Adult Brain) could be incorporated into this strategy (e.g., crossing NKB-FLP mice with FLP-inducible Kiss-Cre). In addition, markers could be linked to synaptophysin or other synaptic terminal proteins to allow for identification of boutons at a light microscopic level that represent bona fide synaptic terminals rather than fibers of passage. The use of transgenic/viral vector approaches for neuroanatomical studies of the kisspeptin system holds much promise, and should provide critical information on the anatomy and function of *Kiss1* cells and their connections.

Afferent Inputs

The array of synaptic inputs received by kisspeptin cells has only recently begun to be systematically studied, so there is little data to compare among species or different kisspeptin cell populations. The best-studied kisspeptin cell population with respect to afferents is the ARC subset. One of the major conserved anatomical features of KNDy cells are the reciprocal connections that exist between them, and that have been demonstrated at a light microscopic level using confocal microscopy [105, 111] as well as at an electron microscopic level [112]. These so-called "KNDy–KNDy" connections have been hypothesized to serve as a structural basis for synchronization of activity among KNDy cells, underlying their proposed role as a component of the GnRH pulse generator [38, 60, 62, 63]. It should be noted that it is not known whether these reciprocal connections represent axon collaterals within a single neuron (e.g., autosynapses), connections from neighboring KNDy cells, or inputs from a segregated subset of KNDy cells. Finally, in addition to KNDy–KNDy connections, evidence from tracing studies indicates that there are bilateral connections between KNDy cells on either side of the hypothalamus [109], as well as reciprocal connection between the POA/RP3V and KNDy populations [108]. Thus the anatomical substrate exists for communication, not just among the ARC KNDy cells, but also between the two major hypothalamic kisspeptin populations, as well as between kisspeptin cells on both sides of the brain.

As noted earlier, a majority of KNDy cells are glutamatergic (Table 3.3), and not surprisingly, KNDy terminals in contact with other KNDy cells also contain glutamatergic markers [68]; in addition, KNDy cells receive input from non-KNDy glutamatergic fibers presumably arising from other regions. Preliminary observations in the sheep suggest that KNDy cells also receive GABAergic inputs, although whether these derive in part from neighboring KNDy cells is not known.

Transgenic mice in which the leptin receptor (LepR) drives expression of the anterograde transneuronal tracer, wheat germ agglutinin (WGA), have been used to demonstrate that KNDy cells receive synaptic input from LepR-containing cells [113], consistent with evidence that KNDy cells play a functional role in conveying the influence of leptin on the reproductive axis [114]. Dual-label ICC studies have also shown close contacts between NPY and POMC fibers and KNDy cells in the sheep ARC [115], suggesting that afferents from these metabolic control neurons may be part of the pathways by which leptin modulates the activity of kisspeptin neurons and their control of GnRH [114]. Such input may be particularly important in sheep since KNDy neurons in this species appear to lack LepR [113].

Less is known about the specific afferents that contact RP3V or preoptic kisspeptin cells. In the mouse, there is clear evidence that RP3V cells of the AVPV receive direct input from vasopressin (VP) and vasoactive-intestinal polypeptide (VIP) cells of the suprachiasmatic nucleus (SCN), the site of a central circadian clock that synchronizes the phase of clock cells in other regions of the brain [116]. This input is likely to function as part of the circadian gate regulating the timing of the GnRH/LH surge in rodents [117], and may differ in species in which the surge is not under circadian control [45]. Finally, while there is little anatomical data on other inputs to RP3V neurons, recent electrophysiological studies of mouse AVPV kisspeptin neurons in slice preparations [118] have provided evidence that these cells like KNDy neurons are under the presynaptic influence of glutamate and GABA, as well as the inhibitory RFamide peptide, RFRP-3, the mammalian ortholog of avian gonadotropin-inhibiting hormone [119].

Reciprocal Connections with GnRH Neurons

There is now clear evidence that kisspeptin cells provide direct synaptic input to GnRH neurons (Fig. 3.2), from both light and electron microscopic (EM) studies in rodents, goats, and sheep. Using confocal ICC, kisspeptin-positive close contacts have been observed upon GnRH cell bodies and dendrites in mice [11], sheep [34], horses [46], monkeys [39], and humans [44]. In sheep, these contacts have been colocalized with synaptophysin, providing further evidence as to their identity as bona fide synaptic terminals [120]. More importantly, there is now direct EM evidence in mice of axo-dendritic and axo-somatic contacts between kisspeptin fibers and preoptic GnRH cells [106]. This compelling observation needs to be confirmed in other species, such as monkeys and sheep, where GnRH neurons examined at an EM level are frequently surrounded and separated from nearby presynaptic terminals by astroglial processes [121, 122].

In addition to inputs at the level of GnRH cell bodies, there is also EM evidence that direct membrane appositions exist between kisspeptin and GnRH terminals within the median eminence. In rats, kisspeptin-positive terminals formed direct membrane contacts with GnRH terminals, although these contacts were seen in the internal zone of the median eminence [123]. Similarly, in the goat [36], kisspeptin and GnRH terminals were seen to form direct axo-axonic contacts; however these contacts were seen in the external zone, unlike those observed in rats. In both species, axo-axonic contacts between kisspeptin and GnRH terminals lacked typical synaptic morphological specializations (e.g., synaptic densities, clefts) so that the mechanism of communication remains unclear. Consistent with the fact that in most species, kisspeptin fibers are sparse in the external zone (Table 3.2), kisspeptin cells in mice are not labeled by peripheral injections of tracers, indicating that they do not have access to fenestrated capillaries as GnRH cells do [108]. Thus, even given the presence of direct axo-axonic connections, it seems likely that for kisspeptin to regulate GnRH release within the median eminence, it must act via diffusion either to the external zone or through actions on other local intermediaries (e.g., glial cells). However, there is no evidence for the presence of *Kiss1r* in glial cells at this time.

While the precise origin of all kisspeptin inputs to GnRH cell bodies and terminals has yet to be defined, using the techniques described above ("efferent projections"), there is evidence that at least some of this input arises from both RP3V and ARC populations. In mice, dual-labeled kisspeptin/TH terminals arising from the RP3V population innervate POA GnRH neurons (Fig. 3.2b), although they represent less than 20% of all kisspeptin contacts on those cells [64]. Recent work using galanin as a co-marker for RP3V kisspeptin cells and NKB as a co-marker for ARC kisspeptin (KNDy) cells confirmed inputs to GnRH cell bodies from RP3V kisspeptin cells in mice, as well as showing direct inputs from the ARC population in this species [106]. Once again, the identified inputs represented a small percentage of the total number of kisspeptin inputs to GnRH neurons. In the sheep, kisspeptin terminals arising from the ARC contact GnRH neurons in both the POA and MBH (where they are also located in this species) (Fig. 3.2a), and double-labeled KNDy terminals appear to account for the largest percentage of the total number of kisspeptin afferents. Finally, there is evidence in multiple species that KNDy cells provide input to the median eminence (Table 3.2), and that at least some of this input forms the close contacts observed with GnRH terminals, at least in the internal zone. In rats and sheep, dynorphin fibers of ARC origin (because they co-localize NKB) innervate the median eminence [105, 111], and dynorphin terminals in the sheep make direct contacts with GnRH terminals at an EM level (Fig. 3.2c). Thus, the available evidence to date suggests that while both RP3V and ARC kisspeptin populations contribute direct inputs to GnRH cell bodies, inputs at the level of GnRH terminals in the median eminence arise from the ARC. We would note that these identified inputs represent a small percentage of the total kisspeptin input to GnRH neurons, but it is not clear whether this reflects limitations of the use of peptide co-markers (see above) or the presence of inputs from other kisspeptin cell populations. Again, the use of transgenic approaches where all axonal projections from a given kisspeptin population can be completely labeled should help to resolve this question.

Finally, in addition to kisspeptin inputs to GnRH neurons and terminals, there is also evidence that the reciprocal connection exists, that is, GnRH afferent input to kisspeptin cells. Specifically, confocal studies have shown close contacts between GnRH fibers in the MBH, and kisspeptin cells of the ARC in the rhesus monkey [30] and sheep (Fig. 3.2d). Thus, in these species, kisspeptin (KNDy) neurons, which are in themselves reciprocally interconnected, may comprise part of a larger reciprocal circuitry that includes GnRH neurons and the POA kisspeptin population (see above). One may speculate that if both GnRH and kisspeptin neurons are capable of displaying intrinsic pulsatile activity [37, 124], a reciprocal network involving interconnections at multiple levels may be important in conferring synchronization of phase upon this rhythm in order to generate a coherent GnRH pulse.

Development and Sex Differences in the Kisspeptin Signaling System

In comparing kisspeptin and *Kiss1* expression across mammalian species, it is important to keep in mind developmental changes and sexual differences in expression. Sexually dimorphic expression of at least some kisspeptin populations has been found in all mammals in which it has been examined. Moreover, it is not surprising that the developmental trajectory of both kisspeptin and *Kiss1* varies with the time course of maturation in each species. In this section, we will review the data available on these two issues for the limited number of species in which it is available. In-depth discussion of the development and sexual differentiation of kisspeptin neurons is also available in Chap. 11.

Development of Kisspeptin and Kiss1 Expression in Females

Because kisspeptin was implicated in the onset of puberty at the time that its reproductive function was discovered, it is not surprising that most developmental studies have focused on the pubertal transition. Thus, although the role of kisspeptin in puberty is discussed in detail elsewhere in this book (Chaps. 11 and 12), it will also be considered here, as it represents an integral part of most reports on its developmental expression. Early studies in rats reported an increase in *Kiss1* mRNA expression associated with pubertal development [125]. However, those studies used RT-PCR to measure mRNA levels from whole hypothalami, so individual areas of the hypothalamus were not examined. As detailed previously, kisspeptin cells in the RP3V area and the ARC are differentially regulated. Thus, expression changes in each area must be considered separately.

One technical challenge to interpreting developmental changes in RP3V kisspeptin neurons, particularly in the rat, is that detection of kisspeptin-ir cells within this region seems to depend upon the use of colchicine pretreatment in order to enhance detection of immunoreactive peptide in cell bodies. In studies that have not used colchicine pretreatment, detection of kisspeptin-ir cells in the RP3V is difficult [19, 22, 126–128]. However, kisspeptin-positive cells are readily apparent in studies that have used colchicine [18, 25] or monitored *Kiss1* mRNA [127, 129], perhaps suggesting a very rapid secretion or turnover of the peptide in RP3V neurons of the rat. In sheep, much like non-colchicine treated rats, kisspeptin neurons in the POA (but not those in the ARC) were difficult to detect in young ewes and could not be quantified during the pubertal transition [130]. In contrast, RP3V kisspeptin-ir neurons in the mouse are easily detectable without colchicine treatment. In mice and colchicine-treated rats, RP3V kisspeptin neurons are not usually detected before about PND10 [11, 49, 50, 65, 127, 129]. A pubertal increase in RP3V kisspeptin immunoreactivity and *Kiss1* mRNA expression has been reported in several studies for females of both species [25, 49–51, 127]. In mice, kisspeptin-positive fibers become evident around GnRH neurons beginning at PND25, a time when changes in RP3V kisspeptin cell numbers and *Kiss1* mRNA expression are increasing dramatically [11]. In ovariectomized, estradiol-implanted ewes, POA *Kiss1* mRNA-containing cell numbers increased around the time of puberty [131], but this change was independent of changes in LH pulse frequency associated with puberty in this species.

In the ARC, kisspeptin-positive cells or *Kiss1* mRNA are detectable within the first few days postnatally in the female rat [126, 127, 129]. Subsequently, kisspeptin immunoreactivity and *Kiss1* mRNA levels increase in a puberty-associated manner [25, 126, 127]. More recently, the number of kisspeptin-positive cells in the ARC was found to be higher in young, postpubertal ewes compared to prepubertal animals; these changes mirrored differences in LH pulse frequency [130]. Changes in ARC kisspeptin-ir cell numbers in ewes from that study also paralleled an increase in the percentage of POA GnRH neurons that exhibited close appositions of kisspeptin-immunopositive varicosities, suggesting that potential changes in kisspeptin input to GnRH neurons during puberty in that species arises from the ARC kisspeptin cells. This is further supported by preliminary evidence that 50–70% of GnRH neurons receive close contacts from kisspeptin fibers that also contain dynorphin, indicating that they arise from KNDy neurons (Lehman and Goodman, unpublished data). A role for ovine ARC kisspeptin neurons in puberty is consistent with the positive correlation between the number of *Kiss1* mRNA-containing cells in the middle ARC and an increase in LH pulse frequency during puberty in estradiol-treated ovariectomized ewes [131]. These data fit well with those in primates, where *Kiss1* mRNA expression increased in female monkeys during the midpubertal phase of development [42]. Interestingly, this increase in *Kiss1* mRNA expression is paralleled by an increase in kisspeptin release in the primate median eminence [132] and in the amount of kisspeptin secretion induced by a GABA receptor antagonist [133]. Thus, in these species, kisspeptin input from the ARC may play an important role in puberty onset.

In contrast to the rat, sheep, and primate, the relative importance of ARC kisspeptin neurons in puberty onset of female mice is unclear. One limitation in studying this area in the mouse is that kisspeptin fiber density in the arcuate nucleus is usually so high that assessing numerical changes in kisspeptin-ir cell bodies has not been attempted. However, in one study that used RT-PCR, no changes in kisspeptin mRNA levels were observed in the ARC of female mice from PND10 to PND60 [50]. In contrast, kisspeptin fiber density in the ARC was reported in that study to increase successively from PND10 to PND30 and from PND30 to PND45. This raises the possibility of a mismatch between mRNA levels and protein production if kisspeptin was being produced locally (see above). Alternatively, these kisspeptin-ir fibers could come from the RP3V because changes in RP3V kisspeptin cell numbers often parallel changes in ARC kisspeptin fiber density [18, 25, 49, 84]. It was also reported that at least 40% of AVPV kisspeptin neurons project to the ARC in female mice [108]. Based on these findings, it has been suggested that kisspeptin from the RP3V in the female mouse is more important than that from the ARC for puberty onset in this species [134]. However, changes in kisspeptin expression in the ARC have been noted in the female *hpg* mouse, with a significant increase noted by PND30 [50]. In addition, Kauffman et al. [135] reported increased ARC kisspeptin and NKB cell numbers following gonadectomy in juvenile female mice, and suggested that disinhibition of ARC kisspeptin/NKB neurons in the ARC constitutes a critical element of the puberty triggering mechanism. This hypothesis is supported by the advancement of vaginal opening and increase in ARC *Kiss1* mRNA levels in mice in which ERα was deleted from kisspeptin neurons [51]. Clearly, more work is needed to determine the relative roles of the RP3V and ARC kisspeptin neurons in puberty onset for this species.

In addition to increased expression and release of kisspeptin during pubertal development, there may also be an increase in the ability of GnRH neurons to respond to kisspeptin as well. Very few studies have examined changes in *Kiss1r* over development and none have looked at protein expression of this receptor. Shahab et al. [42] reported a threefold increase in MBH *Kiss1r* mRNA levels during pubertal development in intact female monkeys. Takase et al. [25] observed an increase in *Kiss1r* mRNA around the time of puberty in the OVLT/POA of rats, while *Kiss1r* mRNA expression in the ARC did not change during this period. In mice, the percentage of GnRH neurons expressing *Kiss1r* was about 40% by PND5 and rose to approximately 70% (or adult levels) by PND20 [78]. Given that both kisspeptin expression in the RP3V and input to GnRH neurons increase around PND25 in the female mouse (see above) and that *Kiss1r* expression is maximal by PND20, the level of *Kiss1r* in GnRH neurons would appear not to be a limiting factor in the timing of puberty. This conclusion is consistent with evidence that administration of kisspeptin to pre- or midpubertal animals robustly stimulates GnRH/LH secretion in several species [42, 125, 131, 132, 136] and chronic administration has been shown to advance the timing of puberty in female rats [95, 125]. Interestingly, even though GnRH neurons in mice appear to express *Kiss1r* well before the normal timing of puberty onset, Han et al. [13] reported that the percentage of GnRH neurons in male mice that were activated by kisspeptin, as determined by gramicidin perforated patch clamping, increased during pubertal development. However, more recently, Dumalska et al. [137] reported that the response to kisspeptin in vGlut2-GFP-tagged cells in the medial septum that co-expressed GnRH did not change during pubertal development. More work will be necessary to confirm if increased coupling of *Kiss1r* to the electrical response of GnRH neurons plays a role in timing pubertal onset.

Sex Differences: Developmental Changes in Males

Although a great deal of work has been done in females, there is evidence that changes in kisspeptin occur during development in males as well. While cell numbers in the RP3V of male rodents are much lower than that for females (discussed below), Han et al. [13] showed that the number of RP3V cells expressing kisspeptin mRNA was greater in adults than juveniles and that expression level per cell increased as well. Clarkson and Herbison [11] reported that kisspeptin-positive cell numbers began to increase at PND25 and peaked at PND45, but these changes may reflect increased circulating levels of gonadal steroids since they were abolished by gonadectomy at PND20 and restored to normal with either testosterone or estradiol treatment [138]. In contrast, others have reported very low cell numbers with no change [50, 126] or only small changes [128] in the RP3V.

In the ARC, most changes reported in the male mouse have been fairly unremarkable. Kisspeptin fiber density increased from PND10 to PND45 in male mice [11], but no change in cell number or mRNA expression level has been shown to occur with development [13, 50]. It is important to note that kisspeptin-ir cell numbers increased in male *hpg* mice during development, and that while kisspeptin cell numbers are not increased 4 days after castration at PND14 in juvenile male mice, they are greater in adult male mice castrated at PND14 [135]. Thus steroid-independent changes occur in kisspeptin expression with development in male mice, although the relative importance of those changes remains to be determined. In male rats, ARC kisspeptin cell numbers increase with pubertal development [126–128]. In sheep, ARC kisspeptin-positive cell numbers decline with age in ram lambs between 6 and 12 months of age [130]. Although these ages correspond to puberty onset in female lambs, male sheep develop reproductively at a much earlier age [139]; the significance of decline with age is unclear, but a similar pattern has been reported previously for cells in the ARC expressing *Kiss1* mRNA in male rats after the pubertal increase in cell number [127, 128].

Hormonal Control of Sexual Dimorphism in Development

Because there are many more RP3V kisspeptin neurons in female than male rodents [11, 21], and this area is the site where E_2 acts to induce the preovulatory GnRH surge in these species (reviewed by Herbison [140]), work on the hormonal control of this sexual dimorphism focused on the well-established organizational effects of testosterone during the perinatal period. As described in detail elsewhere in this book (Chap. 11), studies in rats using neonatal gonadectomy and/or steroid treatments [21, 44] support the hypothesis that testicular secretion of testosterone during this period is converted to E_2, which then produces a permanent decrease in kisspeptin-expressing cells in the RP3V. This conclusion is also supported by the effects of genetic manipulations in mice that knocked out the aromatase enzyme [141] or alpha-fetoprotein [142]. A similar sexual dimorphism in POA kisspeptin cell number has been observed in sheep [31] and humans [44], but the factors responsible remain unknown.

The effects of estradiol on kisspeptin cell numbers in the RP3V may not be limited to the perinatal period, since OVX of mice at PND15 reduced *Kiss1* mRNA expression in the adult RP3V to male levels [49]. On the other hand, in hypogonadal (*hpg*) mice [50], the absence of gonadal steroids did not alter the age-related increase in RP3V kisspeptin-ir cell numbers (although the maximal number was suppressed), but male *hpg* mice showed a similar increase in RP3V kisspeptin cells that was not evident in their WT littermates. It should be noted, however, that no age-related increases in *Kiss1* mRNA (qRT-PCR) were observed in either male or female *hpg* mice [50], so there appears to be a mismatch between *Kiss1* mRNA and protein expression in these mice.

In sheep [31] and humans [44], there are also significantly more kisspeptin neurons in the ARC of females than males, while in rodents ARC kisspeptin expression is similar in both sexes [21, 135, 143]. There is no information on factors responsible for this sexual dimorphism in humans, but in sheep, organizational effects of steroids likely play a role because this sex difference is seen in gonadectomized animals [130]. However, prenatal masculinization of females with testosterone treatment during gestation did not influence kisspeptin cell numbers, but it did decrease the number of NKB-ir and dynorphin-ir cells to levels seen in males [31]. Interestingly, the sex differences in ovine ARC kisspeptin expression develop between 6 and 12 months of ages, as cell numbers in males decline, and are independent of gonadal steroids because it is evident in gonadectomized animals [130]. In contrast, mice show a sexual dimorphism in ARC kisspeptin neurons that is lost during development. Specifically, OVX of prepubertal females results in a doubling of kisspeptin cell number, while castration of males prior to puberty has no effect [135], a sex difference not seen in adult mice [135].

Conclusions

The mammalian kisspeptin system is remarkably consistent across all species studied to date, at least that portion of it that has a well-defined role in control of GnRH secretion. Major groups of cell bodies are found in two areas: a large number in the ARC and smaller set in the POA, which is concentrated in the RP3V of rodents (Fig. 3.3). Most neurons in the ARC subpopulation also contain NKB and dynorphin,

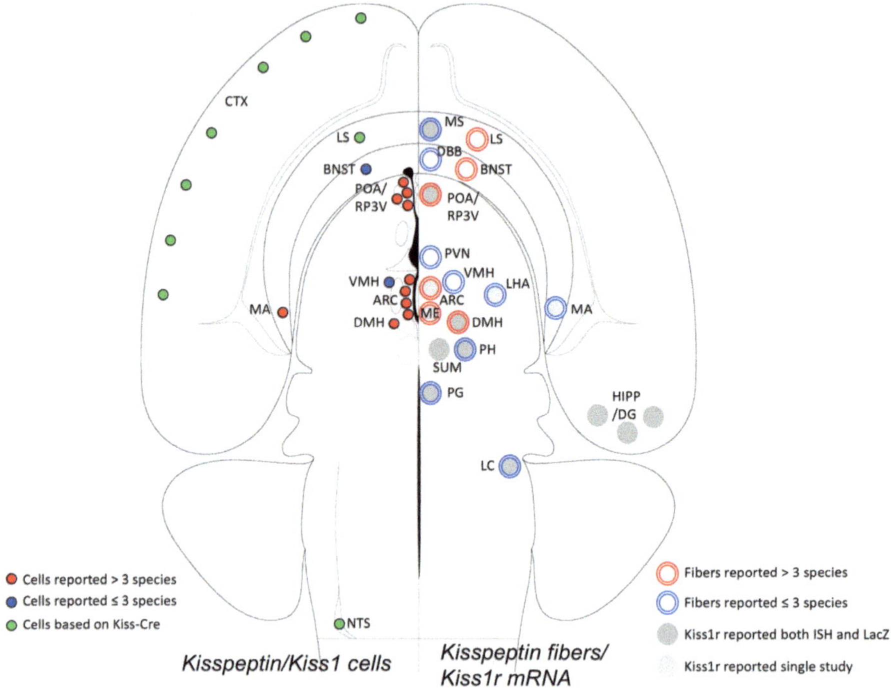

Fig. 3.3 Overall distribution of kisspeptin/Kiss1 cells and kisspeptin fibers/Kiss1r mRNA, in the mammalian brain. The locations of kisspeptin/*Kiss1* cells (*left side*) and kisspeptin fibers/*Kiss1r* mRNA (*right*), based on studies performed to date, are mapped out on a schematic horizontal drawing of the mammalian forebrain, midbrain, and hindbrain. Background is modified from schematic in Swanson, Brain Maps III, Elsevier ©2004, courtesy of Dr. Larry Swanson with permission from Elsevier

while a variable percentage of those in the RP3V of rodents contain dopamine (as indicated by TH) and/or galanin (Tables 3.3 and 3.4); whether these neurotransmitters are found in POA kisspeptin neurons in most other species remains to be determined. Moreover, the POA population is sexually dimorphic in a number of species, with more kisspeptin neurons in females than in males; a similar sexual dimorphism for the ARC population has been reported in sheep and human, but data on this in rodents is conflicting. Additional kisspeptin neurons are found in other hypothalamic and extra-hypothalamic areas in some species, but it is unclear whether or not this is a common distribution and the functional role of these neurons is unknown.

Based on the distribution of kisspeptin fibers, *Kiss1r*, and studies of afferent projections, both populations have functional projections to GnRH cell bodies, which represent an important site at which kisspeptin acts to stimulate GnRH secretion. Kisspeptin-ir fibers are also found in the external zone of the median eminence and there are direct membrane contacts between kisspeptin and GnRH axons in this region. These kisspeptin fibers appear to arise primarily from ARC KNDy neurons, although contacts from more rostral kisspeptin cells cannot be ruled out at this time.

Although Kiss1r has yet to be identified in GnRH terminals, this is likely another site of kisspeptin action because this peptide stimulates GnRH release in vitro from the median eminence of rats [144] and sheep [86]. It should be noted, however, that neither ARC nor RP3V kisspeptin cells in mice [108] or rats [123] have direct access to the portal vessels and, as such, are unlikely to act directly on GnRH terminals at the primary capillary bed of the median eminence.

There is also evidence for reciprocal connections within and between these two kisspeptin populations. The strongest data is for extensive reciprocal connections among KNDy neurons in a number of species, but there is also data that less abundant connections occur within the POA kisspeptin population in sheep. There is also clear evidence that KNDy neurons project to the RP3V of rodents and to some POA kisspeptin neurons in sheep. Conversely, almost half of the RP3V kisspeptin neurons project to the ARC in mice. It is important to note that the functional significance of kisspeptin release at these reciprocal connections is unclear because there is no evidence that kisspeptin neurons in the ARC or RP3V/POA contain *Kiss1r*. The other neurotransmitters within ARC KNDy neurons have been postulated to play a key role in synchronizing this population and driving episodic GnRH secretion. It is thus likely that reciprocal connections within the ARC play an important functional role, but whether that is true for the more rostral kisspeptin populations awaits further study.

Finally, although detailed discussion of the functional roles of these kisspeptin populations is beyond the scope of this chapter, as described elsewhere in this book there also appear to be common functional roles for these two kisspeptin populations. The rostral population is clearly involved in the preovulatory GnRH surge in many species, and the ARC population has been implicated in the negative feedback action of gonadal steroids and in possibly synchronizing GnRH neural activity during episodic secretion. The latter appears to play a key role in the onset of puberty in rodents, sheep, and primates, although the RP3V population likely contributes to this process in mice.

This review also identified several important gaps in our understanding of the neuroanatomy of kisspeptin signaling. The most obvious of these is the paucity of detailed information on the distribution of *Kiss1r* mRNA and the complete lack of any data on location of Kiss1r protein. Until more complete anatomical information is available on the location of these receptors, any conclusions as to physiological significance of most kisspeptin projections must be considered tentative at best. Information is also beginning to be developed on extra-hypothalamic kisspeptin cells and fibers, but this remains limited to just a few species. Thus, these data need to be extended to a broader range of species and studies on their functional significance developed. Another anatomical issue of importance revolves around identifying inputs to kisspeptin cell bodies. While some data is available on afferents to ARC and POA/RP3V kisspeptin cell bodies, this information is fragmentary. A fuller understanding of the complete complement of presynaptic inputs to these, and other, kisspeptin cells will obviously be important for understanding their neural control. Finally, kisspeptin cells in the ARC, POA/RP3V, and elsewhere are likely to play a significant role in the regulation of other physiological systems beyond reproduction. Identifying the full extent of kisspeptin efferents, and the distribution of their postsynaptic targets, is essential as a foundation for future functional dissection of these roles in the mammalian nervous system.

Acknowledgements We thank Dr. Lique M. Coolen for her valuable assistance in the preparation of Fig. 3.3. This work is supported by a grant from NIH (NIH RO1 HD033916)

References

1. Dungan HM, Clifton DK, Steiner RA (2006) Minireview: kisspeptin neurons as central processors in the regulation of gonadotropin-releasing hormone secretion. Endocrinology 147(3):1154–1158
2. Lehman MN, Merkley CM, Coolen LM, Goodman RL (2010) Anatomy of the kisspeptin neural network in mammals. Brain Res 1364:90–102
3. Mikkelsen JD, Simonneaux V (2009) The neuroanatomy of the kisspeptin system in the mammalian brain. Peptides 30(1):26–33
4. Oakley AE, Clifton DK, Steiner RA (2009) Kisspeptin signaling in the brain. Endocr Rev 30(6):713–743
5. Akazome Y, Kanda S, Okubo K, Oka Y (2010) Functional and evolutionary insights into vertebrate kisspeptin systems from studies of fish brain. J Fish Biol 76(1):161–182
6. Tena-Sempere M, Felip A, Gomez A, Zanuy S, Carrillo M (2012) Comparative insights of the kisspeptin/kisspeptin receptor system: lessons from non-mammalian vertebrates. Gen Comp Endocrinol 175(2):234–243
7. Brailoiu GC, Dun SL, Ohsawa M, Yin DL, Yang J, Chang JK et al (2005) KiSS-1 expression and metastin-like immunoreactivity in the rat brain. J Comp Neurol 481(3):314–329
8. Clarkson J, de Tassigny XD, Colledge WH, Caraty A, Herbison AE (2009) Distribution of kisspeptin neurones in the adult female mouse brain. J Neuroendocrinol 21(8):673–682
9. Franceschini I, Lomet D, Cateau M, Delsol G, Tillet Y, Caraty A (2006) Kisspeptin immunoreactive cells of the ovine preoptic area and arcuate nucleus co-express estrogen receptor alpha. Neurosci Lett 401(3):225–230
10. Goodman RL, Lehman MN, Smith JT, Coolen LM, De Oliveira CVR, Jafarzadehshirazi MR et al (2007) Kisspeptin neurons in the arcuate nucleus of the ewe express both dynorphin A and neurokinin B. Endocrinology 148(12):5752–5760
11. Clarkson J, Herbison AE (2006) Postnatal development of kisspeptin neurons in mouse hypothalamus; sexual dimorphism and projections to gonadotropin-releasing hormone neurons. Endocrinology 147(12):5817–5825
12. Gottsch ML, Cunningham MJ, Smith JT, Popa SM, Acohido BV, Crowley WF et al (2004) A role for kisspeptins in the regulation of gonadotropin secretion in the mouse. Endocrinology 145(9):4073–4077
13. Han SK, Gottsch ML, Lee KJ, Popa SM, Smith JT, Jakawich SK et al (2005) Activation of gonadotropin-releasing hormone neurons by kisspeptin as a neuroendocrine switch for the onset of puberty. J Neurosci 25(49):11349–11356
14. Kim J, Semaan SJ, Clifton DK, Steiner RA, Dhamija S, Kauffman AS (2011) Regulation of Kiss1 expression by sex steroids in the amygdala of the rat and mouse. Endocrinology 152(5):2020–2030
15. Smith JT, Cunningham MJ, Rissman EF, Clifton DK, Steiner RA (2005) Regulation of Kiss1 gene expression in the brain of the female mouse. Endocrinology 146(9):3686–3692
16. Smith JT, Dungan HM, Stoll EA, Gottsch ML, Braun RE, Eacker SM et al (2005) Differential regulation of KiSS-1 mRNA expression by sex steroids in the brain of the male mouse. Endocrinology 146(7):2976–2984

17. Hanchate NK, Parkash J, Bellefontaine N, Mazur D, Colledge WH, de Tassigny XD et al (2012) Kisspeptin-GPR54 signaling in mouse NO-synthesizing neurons participates in the hypothalamic control of ovulation. J Neurosci 32(3):932–945
18. Adachi S, Yamada S, Takatsu Y, Matsui H, Kinoshita M, Takase K et al (2007) Involvement of anteroventral periventricular metastin/kisspeptin neurons in estrogen positive feedback action on luteinizing hormone release in female rats. J Reprod Dev 53(2):367–378
19. Desroziers E, Mikkelsen J, Simonneaux V, Keller M, Tillet Y, Caraty A et al (2010) Mapping of kisspeptin fibres in the brain of the pro-oestrous rat. J Neuroendocrinol 22(10):1101–1112
20. Irwig MS, Fraleyb GS, Smith JT, Acohido BV, Popa SM, Cunningham MJ et al (2004) Kisspeptin activation of gonadotropin releasing hormone neurons and regulation of KiSS-1 mRNA in the male rat. Neuroendocrinology 80(4):264–272
21. Kauffman AS, Gottsch ML, Roa J, Byquist AC, Crown A, Clifton DK et al (2007) Sexual differentiation of Kiss1 gene expression in the brain of the rat. Endocrinology 148(4):1774–1783
22. Kinoshita M, Tsukamura H, Adachi S, Matsui H, Uenoyama Y, Iwata K et al (2005) Involvement of central metastin in the regulation of preovulatory luteinizing hormone surge and estrous cyclicity in female rats. Endocrinology 146(10):4431–4436
23. Smith JT, Popa SM, Clifton DK, Hoffman GE, Steiner RA (2006) Kiss1 neurons in the forebrain as central processors for generating the preovulatory luteinizing hormone surge. J Neurosci 26(25):6687–6694
24. Xu ZF, Kaga S, Mochiduki A, Tsubomizu J, Adachi S, Sakai T et al (2012) Immunocytochemical localization of kisspeptin neurons in the rat forebrain with special reference to sexual dimorphism and interaction with GnRH neurons. Endocr J 59(2):161–171
25. Takase K, Uenoyama Y, Inoue N, Matsui H, Yamada S, Shimizu M et al (2009) Possible role of oestrogen in pubertal increase of Kiss1/kisspeptin expression in discrete hypothalamic areas of female rats. J Neuroendocrinol 21(6):527–537
26. Ansel L, Bolborea M, Bentsen AH, Klosen P, Mikkelsen JD, Simonneaux V (2010) Differential regulation of Kiss1 expression by melatonin and gonadal hormones in male and female Syrian hamsters. J Biol Rhythms 25(2):81–91
27. Greives TJ, Mason AO, Scotti MAL, Levine J, Ketterson ED, Kriegsfeld LJ et al (2007) Environmental control of kisspeptin: implications for seasonal reproduction. Endocrinology 148(3):1158–1166
28. Mason AO, Greives TJ, Scotti MAL, Levine J, Frommeyer S, Ketterson ED et al (2007) Suppression of kisspeptin expression and gonadotropic axis sensitivity following exposure to inhibitory day lengths in female Siberian hamsters. Horm Behav 52(4):492–498
29. Revel FG, Saboureau M, Masson-Pevet M, Pevet P, Mikkelsen JD, Simonneaux V (2006) Kisspeptin mediates the photoperiodic control of reproduction in hamsters. Curr Biol 16(17):1730–1735
30. Bosch MA, Xue CH, Ronnekleiv OK (2012) Kisspeptin expression in guinea pig hypothalamus: effects of 17 beta-estradiol. J Comp Neurol 520(10):2143–2162
31. Cheng G, Coolen LM, Padmanabhan V, Goodman RL, Lehman MN (2010) The kisspeptin/neurokinin B/dynorphin (KNDy) cell population of the arcuate nucleus: sex differences and effects of prenatal testosterone in sheep. Endocrinology 151(1):301–311
32. Estrada KM, Clay CM, Pompolo S, Smith JT, Clarke IJ (2006) Elevated KiSS-1 expression in the arcuate nucleus prior to the cyclic preovulatory gonadotrophin-releasing hormone/lutenising hormone surge in the ewe suggests a stimulatory role for kisspeptin in oestrogen-positive feedback. J Neuroendocrinol 18(10):806–809
33. Smith JT, Clay CM, Caraty A, Clarke IJ (2007) KiSS-1 messenger ribonucleic acid expression in the hypothalamus of the ewe is regulated by sex steroids and season. Endocrinology 148(3):1150–1157
34. Smith JT, Coolen LM, Kriegsfeld LJ, Sari IP, Jaafarzadehshirazi MR, Maltby M et al (2008) Variation in kisspeptin and RFamide-related peptide (RFRP) expression and terminal connections to gonadotropin-releasing hormone neurons in the brain: a novel medium for seasonal breeding in the sheep. Endocrinology 149(11):5770–5782
35. Smith JT, Li Q, Pereira A, Clarke IJ (2009) Kisspeptin neurons in the ovine arcuate nucleus and preoptic area are involved in the preovulatory luteinizing hormone surge. Endocrinology 150(12):5530–5538

36. Matsuyama S, Ohkura S, Mogi K, Wakabayashi Y, Mori Y, Tsukamura H et al (2011) Morphological evidence for direct interaction between kisspeptin and gonadotropin-releasing hormone neurons at the median eminence of the male goat: an immunoelectron microscopic study. Neuroendocrinology 94(4):323–332
37. Ohkura S, Wakabayashi Y, Uenoyama Y, Steiner RA, Tsukamura H, Maeda K et al (2010) Gonadotropin-releasing hormone pulse generator activity from the arcuate nucleus kisspeptin neurons in the goat hypothalamus. Endocr J 57:S516
38. Wakabayashi Y, Nakada T, Murata K, Ohkura S, Mogi K, Navarro VM et al (2010) Neurokinin B and dynorphin A in kisspeptin neurons of the arcuate nucleus participate in generation of periodic oscillation of neural activity driving pulsatile gonadotropin-releasing hormone secretion in the goat. J Neurosci 30(8):3124–3132
39. Ramaswamy S, Guerriero KA, Gibbs RB, Plant TM (2008) Structural interactions between kisspeptin and GnRH neurons in the mediobasal hypothalamus of the male rhesus monkey (Macaca mulatta) as revealed by double immunofluorescence and confocal microscopy. Endocrinology 149(9):4387–4395
40. Ramaswamy S, Seminara SB, Ali B, Ciofi P, Amin NA, Plant TM (2010) Neurokinin B stimulates GnRH release in the male monkey (Macaca mulatta) and is colocalized with kisspeptin in the arcuate nucleus. Endocrinology 151(9):4494–4503
41. Rometo AM, Krajewski SJ, Voytko ML, Rance NE (2007) Hypertrophy and increased kisspeptin gene expression in the hypothalamic infundibular nucleus of postmenopausal women and ovariectomized monkeys. J Clin Endocrinol Metab 92(7):2744–2750
42. Shahab M, Mastronardi C, Seminara SB, Crowley WF, Ojeda SR, Plant TM (2005) Increased hypothalamic GPR54 signaling: a potential mechanism for initiation of puberty in primates. Proc Natl Acad Sci U S A 102(6):2129–2134
43. Smith JT, Periera A, Shahab M, Pau KYF, Clarke IJ (2010) Hypothalamic expression of the kisspeptin gene (Kiss1) and the RFamide-related peptide (RFRP) gene during the menstrual cycle of a non-human primate. Endocr J 57:S534
44. Hrabovszky E, Ciofi P, Vida B, Horvath MC, Keller E, Caraty A et al (2010) The kisspeptin system of the human hypothalamus: sexual dimorphism and relationship with gonadotropin-releasing hormone and neurokinin B neurons. Eur J Neurosci 31(11):1984–1998
45. Goodman RL, Lehman MN (2012) Kisspeptin neurons from mice to men: similarities and differences. Endocrinology 153(11):5105–5118
46. Decourt C, Tillet Y, Caraty A, Franceschini I, Briant C (2008) Kisspeptin immunoreactive neurons in the equine hypothalamus interactions with GnRH neuronal system. J Chem Neuroanat 36(3–4):131–137
47. Magee C, Foradori CD, Bruemmer JE, Arreguin-Arevalo JA, Mccue PM, Handa RJ et al (2009) Biological and anatomical evidence for kisspeptin regulation of the hypothalamic-pituitary-gonadal axis of estrous horse mares. Endocrinology 150(6):2813–2821
48. Ginther OJ, Gastal EL, Gastal MO, Beg MA (2005) Regulation of circulating gonadotropins by the negative effects of ovarian hormones in mares. Biol Reprod 73(2):315–323
49. Clarkson J, Boon WC, Simpson ER, Herbison AE (2009) Postnatal development of an estradiol-kisspeptin positive feedback mechanism implicated in puberty onset. Endocrinology 150(7):3214–3220
50. Gill JC, Wang OL, Kakar S, Martinelli E, Carroll RS, Kaiser UB (2010) Reproductive hormone-dependent and -independent contributions to developmental changes in kisspeptin in GnRH-deficient hypogonadal mice. PLoS One 5(7):e11911
51. Mayer C, Acosta-Martinez M, Dubois SL, Wolfe A, Radovick S, Boehm U et al (2010) Timing and completion of puberty in female mice depend on estrogen receptor alpha-signaling in kisspeptin neurons. Proc Natl Acad Sci U S A 107(52):22693–22698
52. Newman SW (1999) The medial extend amygdala in male reproductive behavior. A node in the mammalian social behavior network. Ann N Y Acad Sci 877:242–257
53. Veening JG, Coolen LM (1998) Neural activation following sexual behavior in the male and female rat brain. Behav Neurosci 92:181–193

54. Beltramino C, Taleusbik S (1978) Facilitory and inhibitory effects of electrochemical stimulation of the amygdala on the release of luteinizing hormone. Brain Res 144:95–107
55. Cravo RM, Margatho LO, Osborne-Lawrence S, Donato J, Atkin S, Bookout AL et al (2011) Characterization of Kiss1 neurons using transgenic mouse models. Neuroscience 173:37–56
56. Padilla SL, Carmody JS, Zeltser LM (2010) Pomc-expressing progenitors give rise to antagonistic neuronal populations in hypothalamic feeding circuits. Nat Med 16(4):403–405
57. de Tassigny XD, Fagg LA, Carlton MBL, Colledge WH (2008) Kisspeptin can stimulate gonadotropin-releasing hormone (GnRH) release by a direct action at GnRH nerve terminals. Endocrinology 149(8):3926–3932
58. Goodman RL, Coolen LM, Anderson GM, Hardy SL, Valent M, Connors JM et al (2004) Evidence that dynorphin plays a major role in mediating progesterone negative feedback on gonadotropin-releasing hormone neurons in sheep. Endocrinology 145(6):2959–2967
59. Topaloglu AK, Reimann F, Guclu M, Yalin AS, Kotan LD, Porter KM et al (2009) TAC3 and TACR3 mutations in familial hypogonadotropic hypogonadism reveal a key role for Neurokinin B in the central control of reproduction. Nat Genet 41(3):354–358
60. Lehman MN, Coolen LM, Goodman RL (2010) Minireview: kisspeptin/neurokinin B/dynorphin (KNDy) cells of the arcuate nucleus: a central node in the control of gonadotropin-releasing hormone secretion. Endocrinology 151(8):3479–3489
61. Navarro VM, Gottsch ML, Wu M, Garca-Galiano D, Hobbs SJ, Bosch MA et al (2011) Regulation of NKB pathways and their roles in the control of Kiss1 neurons in the arcuate nucleus of the male mouse. Endocrinology 152(11):4265–4275
62. Rance NE, Krajewski SJ, Smith MA, Cholanian M, Dacks PA (2010) Neurokinin B and the hypothalamic regulation of reproduction. Brain Res 1364:116–128
63. Navarro VM, Gottsch ML, Chavkin C, Okamura H, Clifton DK, Steiner RA (2009) Regulation of gonadotropin-releasing hormone secretion by kisspeptin/dynorphin/neurokinin B neurons in the arcuate nucleus of the mouse. J Neurosci 29(38):11859–11866
64. Clarkson J, Herbison AE (2011) Dual phenotype kisspeptin-dopamine neurones of the rostral periventricular area of the third ventricle project to gonadotrophin-releasing hormone neurones. J Neuroendocrinol 23(4):293–301
65. Semaan SJ, Murray EK, Poling MC, Dhamija S, Forger NG, Kauffman AS (2010) BAX-dependent and BAX-independent regulation of Kiss1 neuron development in mice. Endocrinology 151(12):5807–5817
66. Merkley CM, Coolen LM, Goodman RL, Lehman MN. Colocalization of met-enkephalin, but not galanin or tyrosine hydroxylase, within kisspeptin neurons in the sheep. In: Annual Meeting of the Endocrine Society 2012
67. Porteous R, Petersen SL, Yeo SH, Bhattarai JP, Ciofi P, de Tassigny XD et al (2011) Kisspeptin neurons co-express met-enkephalin and galanin in the rostral periventricular region of the female mouse hypothalamus. J Comp Neurol 519(17):3456–3469
68. Merkley CM, Jackson L, Goodman RL, Lehman MN. Evidence for transcriptional activation of arcuate kisspeptin neurons, and glutamatergic input to kisspeptin during the preovulatory GnRH surge in sheep. In: Annual Meeting of the Endocrine Society 2009, P3–220
69. Wintermantel TM, Campbell RE, Porteous R, Bock D, Grone HJ, Todman MG et al (2006) Definition of estrogen receptor pathway critical for estrogen positive feedback to gonadotropin-releasing hormone neurons and fertility. Neuron 52(2):271–280
70. Clarkson J, d'Anglemont X, Moreno AS, Colledge WH, Herbison AE (2008) Kisspeptin-GPR54 signaling is essential for preovulatory gonadotropin-releasing hormone neuron activation and the luteinizing hormone surge. J Neurosci 28(35):8691–8697
71. Simerly RB, Chang C, Muramatsu M, Swanson LW (1990) Distribution of androgen and estrogen-receptor messenger RNA-containing cells in the rat-brain—an in situ hybridization study. J Comp Neurol 294(1):76–95
72. Oakley AE, Coolen LM, Lehman MN, Wagenmaker ER, Karsch FJ. Are dynorphin neurons in the arcuate nucleus responsive to cortisol and influenced by the combined presence of cortisol and estradiol. In: Annual Meeting of the Endocrine Society 2009

73. Li Q, Rao A, Pereira A, Clarke IJ, Smith JT (2011) Kisspeptin cells in the ovine arcuate nucleus express prolactin receptor but not melatonin receptor. J Neuroendocrinol 23(10):871–882
74. Cernea M, Phillips R, Padmanabhan V, Coolen LM, Lehman MN. Colocalization of insulin receptors in KNDy but not GnRH neurons inthe ewe: effects of excess prenatal testosterone. In: Annual Meeting of the Endocrine Society 2012
75. Kotani M, Detheux M, Vandenbbogaerde A, Communi D, Vanderwinden JM, Le Poul E et al (2001) The metastasis suppressor gene KiSS-1 encodes kisspeptins, the natural ligands of the orphan G protein-coupled receptor GPR54. J Biol Chem 276(37):34631–34636
76. Muir AI, Chamberlain L, Elshourbagy NA, Michalovich D, Moore DJ, Calamari A et al (2001) AXOR12, a novel human G protein-coupled receptor, activated by the peptide KiSS-1. J Biol Chem 276(31):28969–28975
77. Lee DK, Nguyen T, O'Neill GP, Cheng R, Liu Y, Howard AD et al (1999) Discovery of a receptor related to the galanin receptors. FEBS Lett 446(1):103–107
78. Herbison AE, de Tassigny XD, Doran J, Colledge WH (2010) Distribution and postnatal development of Gpr54 gene expression in mouse brain and gonadotropin-releasing hormone neurons. Endocrinology 151(1):312–321
79. Kim W, Jessen HM, Auger AP, Terasawa E (2009) Postmenopausal increase in KiSS-1, GPR54, and luteinizing hormone releasing hormone (LHRH-1) mRNA in the basal hypothalamus of female rhesus monkeys. Peptides 30(1):103–110
80. Shibata M, Friedman RL, Ramaswamy S, Plant TM (2007) Evidence that down regulation of hypothalamic KiSS-1 expression is involved in the negative feedback action of testosterone to regulate luteinising hormone secretion in the adult male rhesus monkey (Macaca mulatta). J Neuroendocrinol 19(6):432–438
81. Kinsey-Jones JS, Li XF, Knox AMI, Wilkinson ES, Zhu XL, Chaudhary AA et al (2009) Down-regulation of hypothalamic kisspeptin and its receptor, Kiss1r, mRNA expression is associated with stress-induced suppression of luteinising hormone secretion in the female rat. J Neuroendocrinol 21(1):20–29
82. Knox AMI, Li XF, Kinsey-Jones JS, Wilkinson ES, Wu XQ, Cheng YS et al (2009) Neonatal lipopolysaccharide exposure delays puberty and alters hypothalamic Kiss1 and Kiss1r mRNA expression in the female rat. J Neuroendocrinol 21(8):683–689
83. Quennell JH, Rizwan MZ, Relf HL, Anderson GM (2010) Developmental and steroidogenic effects on the gene expression of RFamide related peptides and their receptor in the rat brain and pituitary gland. J Neuroendocrinol 22(4):309–316
84. Yamada S, Uenoyama Y, Kinoshita M, Iwata K, Takase K, Matsui H et al (2007) Inhibition of metastin (kisspeptin-54)-GPR54 signaling in the arcuate nucleus-median eminence region during lactation in rats. Endocrinology 148(5):2226–2232
85. Fu LY, van den Pol AN (2010) Kisspeptin directly excites anorexigenic proopiomelanocortin neurons but inhibits orexigenic neuropeptide Y cells by an indirect synaptic mechanism. J Neurosci 30(30):10205–10219
86. Smith JT, Li Q, Yap KS, Shahab M, Roseweir AK, Millar RP et al (2011) Kisspeptin is essential for the full preovulatory LH surge and stimulates GnRH release from the isolated ovine median eminence. Endocrinology 152(3):1001–1012
87. Li Q, Roa J, Clarke IJ, Smith JT (2012) Seasonal variation in the gonadotropin-releasing hormone response to kisspeptin in sheep: possible kisspeptin regulation of the kisspeptin receptor. Neuroendocrinology 96(3):212–221
88. Messager S, Chatzidaki EE, Ma D, Hendrick AG, Zahn D, Dixon J et al (2005) Kisspeptin directly stimulates gonadotropin-releasing hormone release via G protein-coupled receptor 54. Proc Natl Acad Sci U S A 102(5):1761–1766
89. Luque RM, Kineman RD, Tena-Sempere M (2007) Regulation of hypothalamic expression of KiSS-1 and GPR54 genes by metabolic factors: analyses using mouse models and a cell line. Endocrinology 148(10):4601–4611

90. Bellingham M, Fowler PA, Amezaga MR, Rhind SM, Cotinot C, Mandon-Pepin B et al (2009) Exposure to a complex cocktail of environmental endocrine-disrupting compounds disturbs the kisspeptin/GPR54 system in ovine hypothalamus and pituitary gland. Environ Health Perspect 117(10):1556–1562
91. Smith JT, Rao A, Pereira A, Caraty A, Millar RP, Clarke IJ (2008) Kisspeptin is present in ovine hypophysial portal blood but does not increase during the preovulatory luteinizing hormone surge: evidence that gonadotropes are not direct targets of kisspeptin in vivo. Endocrinology 149(4):1951–1959
92. Li S, Ren J, Yang G, Guo Y, Huang L (2008) Characterization of the porcine Kisspeptins receptor gene and evaluation as candidate for timing of puberty in sows. J Anim Breed Genet 125(4):219–227
93. Gutierrez-Pascual E, Martinez-Fuentes AJ, Pinilla L, Tena-Sempere M, Malagon MM, Castano JP (2007) Direct pituitary effects of kisspeptin: activation of gonadotrophs and somatotrophs and stimulation of luteinising hormone and growth hormone secretion. J Neuroendocrinol 19(7):521–530
94. Richard N, Galmiche G, Corvaisier S, Caraty A, Kottler ML (2008) KiSS-1 and GPR54 genes are co-expressed in rat gonadotrophs and differentially regulated in vivo by oestradiol and gonadotrophin-releasing hormone. J Neuroendocrinol 20(3):381–393
95. Matsui H, Takatsu Y, Kumano S, Matsumoto H, Ohtaki T (2004) Peripheral administration of metastin induces marked gonadotropin release and ovulation in the rat. Biochem Biophys Res Commun 320(2):383–388
96. Thompson EL, Patterson M, Murphy KG, Smith KL, Dhillo WS, Todd JF et al (2004) Central and peripheral administration of kisspeptin-10 stimulates the hypothalamic-pituitary-gonadal axis. J Neuroendocrinol 16(10):850–858
97. Navarro VM, Castellano JM, Fernandez-Fernandez R, Tovar S, Roa J, Mayen A et al (2005) Characterization of the potent luteinizing hormone-releasing activity of KiSS-1 peptide, the natural ligand of GPR54. Endocrinology 146(1):156–163
98. Suzuki S, Kadokawa H, Hashizume T (2008) Direct kisspeptin-10 stimulation on luteinizing hormone secretion from bovine and porcine anterior pituitary cells. Anim Reprod Sci 103(3–4):360–365
99. Ramaswamy S, Gibbs RB, Plant TM (2009) Studies of the localisation of kisspeptin within the pituitary of the rhesus monkey (Macaca mulatta) and the effect of kisspeptin on the release of non-gonadotropic pituitary hormones. J Neuroendocrinol 21(10):795–804
100. Poling MC, Kim J, Dhamija S, Kauffman AS (2012) Development, sex steroid regulation, and phenotypic characterization of RFamide-related peptide (Rfrp) gene expression and RFamide receptors in the mouse hypothalamus. Endocrinology 153(4):1827–1840
101. Quaynor S, Hu L, Leung PK, Feng H, Mores N, Krsmanovic LZ et al (2007) Expression of a functional G protein-coupled receptor 54-kisspeptin autoregulatory system in hypothalamic gonadotropin-releasing hormone neurons. Mol Endocrinol 21(12):3062–3070
102. Pielecka-Fortuna J, Chu Z, Moenter SM (2008) Kisspeptin acts directly and indirectly to increase gonadotropin-releasing hormone neuron activity and its effects are modulated by estradiol. Endocrinology 149:1979–1986
103. Arai AC, Xia YF, Suzuki E, Kessler M, Civelli O, Nothacker HP (2005) Cancer metastasis-suppressing peptide metastin upregulates excitatory synaptic transmission in hippocampal dentate granule cells. J Neurophysiol 94(5):3648–3652
104. Arai AC, Orwig N (2008) Factors that regulate KiSS1 gene expression in the hippocampus. Brain Res 1243:10–18
105. Burke MC, Letts PA, Krajewski SJ, Rance NE (2006) Coexpression of dynorphin and neurokinin B immunoreactlivity in the rat hypothalamus: morphologic evidence of interrelated function within the arcuate nucleus. J Comp Neurol 498(5):712–726
106. Kallo I, Vida B, Deli L, Molnar CS, Hrabovszky E, Caraty A et al (2012) Co-localisation of kisspeptin with galanin or neurokinin B in afferents to mouse GnRH neurones. J Neuroendocrinol 24(3):464–476

107. True C, Kirigiti M, Ciofi P, Grove KL, Smith MS (2011) Characterisation of arcuate nucleus kisspeptin/neurokinin B neuronal projections and regulation during lactation in the rat. J Neuroendocrinol 23(1):52–64
108. Yeo SH, Herbison AE (2011) Projections of arcuate nucleus and rostral periventricular kisspeptin neurons in the adult female mouse brain. Endocrinology 152(6):2387–2399
109. Krajewski SJ, Burke MC, Anderson MJ, McMullen NT, Rance NE (2010) Forebrain projections of arcuate neurokinin B neurons demonstrated by anterograde tract-tracing and monosodium glutamate lesions in the rat. Neuroscience 166(2):680–697
110. Dungan Lemko HM, Elias CF (2012) Kiss of the mutant mouse: how genetically altered mice advanced our understanding of kisspeptin's role in reproductive physiology. Endocrinology 153(11):5119–5129
111. Foradori CD, Amstalden M, Goodman RL, Lehman MN (2006) Colocalisation of dynorphin A and neurokinin B immunoreactivity in the arcuate nucleus and median eminence of the sheep. J Neuroendocrinol 18(7):534–541
112. Foradori CD, Coolen LM, Fitzgerald ME, Skinner DC, Goodman RL, Lehman MN (2002) Colocalization of progesterone receptors in parvicellular dynorphin neurons of the ovine preoptic area and hypothalamus. Endocrinology 143(11):4366–4374
113. Louis GW, Greenwald-Yarnell M, Phillips R, Coolen LM, Lehman MN, Myers MG (2011) Molecular mapping of the neural pathways linking leptin to the neuroendocrine reproductive axis. Endocrinology 152(6):2302–2310
114. Tena-Sempere M (2006) KiSS-1 and reproduction: focus on its role in the metabolic regulation of fertility. Neuroendocrinology 83(5–6):275–281
115. Backholer K, Smith JT, Rao A, Pereira A, Iqbal J, Ogawa S et al (2010) Kisspeptin cells in the ewe brain respond to leptin and communicate with neuropeptide Y and proopiomelanocortin cells. Endocrinology 151(5):2233–2243
116. Mohawk JA, Green CB, Takahasi JS (2012) Central and peripheral circadian clocks in mammals. Annu Rev Neurosci 35:445–462
117. Robertson JL, Clifton DK, de la Iglesia HO, Steiner RA, Kauffman AS (2009) Circadian regulation of Kiss1 neurons: implications for timing the preovulatory gonadotropin-releasing hormone/luteinizing hormone surge. Endocrinology 150(8):3664–3671
118. Ducret E, Gaidamaka G, Herbison AE (2010) Electrical and morphological characteristics of anteroventral periventricular nucleus kisspeptin and other neurons in the female mouse. Endocrinology 151(5):2223–2232
119. Bentley GE, Ubuka T, McGuire NL, Calisi R, Perfito N, Kriegsfeld LJ et al (2009) Gonadotrophin-inhibitory hormone: a multifunctional neuropeptide. J Neuroendocrinol 21(4):276–281
120. Smith JT (2008) Kisspeptin signalling in the brain: steroid regulation in the rodent and ewe. Brain Res Rev 57(2):288–298
121. Lehman MN, Karsch FJ, Robinson JE, Silverman AJ (1988) Ultrastructure and synaptic organization of luteinizing-hormone-releasing hormone (Lhrh) neurons in the anestrous ewe. J Comp Neurol 273(4):447–458
122. Witkin JW, O'Sullivan H, Miller R, Ferin M (1997) GnRH perikarya in medial basal hypothalamus of pubertal female rhesus macaque are ensheathed with glia. J Neuroendocrinol 9(12):881–885
123. Uenoyama Y, Inoue N, Pheng V, Homma T, Takase K, Yamada S et al (2011) Ultrastructural evidence of kisspeptin-gonadotrophin-releasing hormone (GnRH) interaction in the median eminence of female rats: implication of axo-axonal regulation of GnRH release. J Neuroendocrinol 23(10):863–870
124. Nunemaker CS, DeFazio RA, Geusz ME, Herzog ED, Pitts GR, Moenter SM (2001) Long-term recordings of networks of immortalized GnRH neurons reveal episodic patterns of electrical activity. J Neurophysiol 86(1):86–93
125. Navarro VM, Fernandez-Fernandez R, Castellano JM, Roa J, Mayen A, Barreiro ML et al (2004) Advanced vaginal opening and precocious activation of the reproductive axis by KiSS-1 peptide, the endogenous ligand of GPR54. J Physiol 561(Pt 2):379–386

126. Desroziers E, Mikkelsen JD, Duittoz A, Franceschini I (2012) Kisspeptin-immunoreactivity changes in a sex- and hypothalamic-region specific manner across rat postnatal development. J Neuroendocrinol 24(8):1154–1165
127. Takumi K, Iijima N, Ozawa H (2011) Developmental changes in the expression of kisspeptin mRNA in rat hypothalamus. J Mol Neurosci 43(2):138–145
128. Bentsen AH, Ansel L, Simonneaux V, Tena-Sempere M, Juul A, Mikkelsen JD (2010) Maturation of kisspeptinergic neurons coincides with puberty onset in male rats. Peptides 31(2):275–283
129. Cao J, Patisaul HB (2011) Sexually dimorphic expression of hypothalamic estrogen receptors alpha and beta and Kiss1 in neonatal male and female rats. J Comp Neurol 519(15):2954–2977
130. Nestor CC, Briscoe AM, Davis SM, Valent M, Goodman RL, Hileman SM (2012) Evidence of a role for kisspeptin and neurokinin B in puberty of female sheep. Endocrinology 153(6):2756–2765
131. Redmond JS, Baez-Sandoval GM, Spell KM, Spencer TE, Lents CA, Williams GL et al (2011) Developmental changes in hypothalamic Kiss1 expression during activation of the pulsatile release of luteinising hormone in maturing ewe lambs. J Neuroendocrinol 23(9):815–822
132. Keen KL, Wegner FH, Bloom SR, Ghatei MA, Terasawa E (2008) An increase in kisspeptin-54 release occurs with the pubertal increase in luteinizing hormone-releasing hormone-1 release in the stalk-median eminence of female rhesus monkeys in vivo. Endocrinology 149(8):4151–4157
133. Kurian JR, Keen KL, Guerriero KA, Terasawa E (2012) Tonic control of kisspeptin release in prepubertal monkeys: implications to the mechanism of puberty onset. Endocrinology 153(7):3331–3336
134. Clarkson J, Han SK, Liu X, Lee K, Herbison AE (2010) Neurobiological mechanisms underlying kisspeptin activation of gonadotropin-releasing hormone (GnRH) neurons at puberty. Mol Cell Endocrinol 324(1–2):45–50
135. Kauffman AS, Navarro VM, Kim J, Clifton DK, Steiner RA (2009) Sex differences in the regulation of Kiss1/NKB neurons in juvenile mice: implications for the timing of puberty. Am J Physiol Endocrinol Metab 297(5):E1212–E1221
136. Plant TM, Ramaswamy S, Dipietro MJ (2006) Repetitive activation of hypothalamic G protein-coupled receptor 54 with intravenous pulses of kisspeptin in the juvenile monkey (Macaca mulatta) elicits a sustained train of gonadotropin-releasing hormone discharges. Endocrinology 147(2):1007–1013
137. Dumalska I, Wu M, Morozova E, Liu R, van den PA, Alreja M (2008) Excitatory effects of the puberty-initiating peptide kisspeptin and group I metabotropic glutamate receptor agonists differentiate two distinct subpopulations of gonadotropin-releasing hormone neurons. J Neurosci 28(32):8003–8013
138. Clarkson J, Shamas S, Mallinson S, Herbison AE (2012) Gonadal steroid induction of kisspeptin peptide expression in the rostral periventricular area of the third ventricle during postnatal development in the male mouse. J Neuroendocrinol 24(6):907–915
139. Olster DH, Foster DL (1986) Control of gonadotropin secretion in the male during puberty: a decrease in response to steroid inhibitory feedback in the absence of an increase in steroid-independent drive in the sheep. Endocrinology 118(6):2225–2234
140. Herbison AE (2008) Estrogen positive feedback to gonadotropin-releasing hormone (GnRH) neurons in the rodent: the case for the rostral periventricular area of the third ventricle (RP3V). Brain Res Rev 57(2):277–287
141. Bakker J, Pierman S, Gonzalez-Martinez D (2010) Effects of aromatase mutation (ArKO) on the sexual differentiation of kisspeptin neuronal numbers and their activation by same versus opposite sex urinary pheromones. Horm Behav 57(4–5):390–395
142. Gonzalez-Martinez D, De Mees C, Douhard Q, Szpirer C, Bakker J (2008) Absence of gonadotropin-releasing hormone 1 and Kiss1 activation in alpha-fetoprotein knockout mice:

prenatal estrogens defeminize the potential to show preovulatory luteinizing hormone surges. Endocrinology 149(5):2333–2340

143. Homma T, Sakakibara M, Yamada S, Kinoshita M, Iwata K, Tomikawa J et al (2009) Significance of neonatal testicular sex steroids to defeminize anteroventral periventricular kisspeptin neurons and the GnRH/LH surge system in male rats. Biol Reprod 81(6):1216–1225
144. D'Anglemont de Tassigny X, Fagg LA, Dixon JPC, Day K, Leitch HG, Hendrick AG et al (2007) Hypogonadotropic hypogonadism in mice lacking a functional Kiss1 gene. Proc Natl Acad Sci U S A 104(25):10714–10719

Chapter 4
The Effects of Kisspeptin on Gonadotropin Release in Non-human Mammals

Ali Abbara, Risheka Ratnasabapathy, Channa N. Jayasena, and Waljit S. Dhillo

Abstract The *Kiss1* gene encodes a 145-amino acid pre-peptide, kisspeptin, which is cleaved into smaller peptides of 54, 14, 13, and 10 amino acids. This chapter reviews in detail the effects of kisspeptin on gonadotropin secretion in non-human mammals. Studies of kisspeptin's effects have included both acute and chronic administration regimens via a number of administration routes. Acute kisspeptin stimulates gonadotropin secretion in a wide range of species of non-human mammals, including rats, mice, hamsters, sheep, pigs, goats, cows, horses, and monkeys. In general, the stimulatory effect of kisspeptin treatment is more pronounced for LH than FSH secretion. Kisspeptin is thought to exert its stimulatory effects on LH and FSH release via stimulation of GnRH release from the hypothalamus, since pre-administration of a GnRH antagonist prevents kisspeptin's stimulation of gonadotropin secretion. Although the kisspeptin receptor is also expressed on anterior pituitary cells of some species, and incubation of anterior pituitary cells with high concentrations of kisspeptin can stimulate in vitro LH release, the contribution of direct effects of kisspeptin on the pituitary is thought to be negligible in vivo. Continuous kisspeptin administration results in reduced sensitivity to the effects of kisspeptin, in some species. This desensitization is thought to occur at the level of the kisspeptin receptor, since the response of the pituitary gland to exogenous GnRH is maintained. Overall, the findings discussed in this chapter are invaluable to the understanding of the reproductive role of kisspeptin and the potential therapeutic uses of kisspeptin for the treatment of fertility disorders.

A. Abbara • R. Ratnasabapathy • C.N. Jayasena • W.S. Dhillo (✉)
Division of Diabetes, Endocrinology and Metabolism, Department of Investigative Medicine, Imperial College London, London, W12 ONN, UK
e-mail: ali.abbara@imperial.ac.uk; w.dhillo@imperial.ac.uk

A.S. Kauffman and J.T. Smith (eds.), *Kisspeptin Signaling in Reproductive Biology*, Advances in Experimental Medicine and Biology 784, DOI 10.1007/978-1-4614-6199-9_4,

Introduction

Administration of kisspeptin has been shown to stimulate gonadotropin release in all non-human mammalian species studied to date. The conserved stimulating effect of kisspeptin on gonadotropin release in all species studied suggests that kisspeptin plays a critical and fundamental role in the control of reproduction. However, there are important differences in the reproductive systems of various mammalian species. In addition, although the amino acid structure of kisspeptin-10 is relatively conserved across species, there may be up to two amino acid differences between species-specific isoforms, which might conceivably result in subtle alterations in kisspeptin activity.

In this chapter, we summarize the effects of different isoforms of kisspeptin (kisspeptin-54/kisspeptin-52, kisspeptin-14, kisspeptin-10 of murine or human origin) on gonadotropin release in various non-human mammals after acute and chronic administration via various routes (intracerebroventricular [ICV], intravenous bolus [IV bolus], intravenous infusion [IVI], intraperitoneal [IP], subcutaneous [SC]). Importantly, the hormone-releasing actions of kisspeptin may be influenced by metabolic status, sex, maturation, and underlying hormonal milieu; these will also be considered in this chapter. Detailed discussion of the effects of kisspeptin in humans is present in Chap. 5.

The Effects of Acute Administration of Kisspeptin on Gonadotropin Secretion

Kisspeptin has been shown to reliably and rapidly stimulate gonadotropin secretion when given acutely by a number of different routes. Whilst kisspeptin is considered to be a neuropeptide, with very low circulating levels in the blood, peripheral administration has been shown to reliably stimulate gonadotropin release in a number of species.

Acute Peripheral Administration of Human Kisspeptin-54 or Rat Kisspeptin-52

In 2004, Matsui et al. performed one of the first studies of kisspeptin administration to any mammalian species. A subcutaneous (SC) bolus of 6.7 nmol of human kisspeptin-54 (called metastin at the time) in pre-pubertal (25 day old) female Wistar rats significantly elevated plasma LH levels tenfold from a baseline of ~3 ng/mL and FSH levels sixfold from a baseline of ~10 ng/mL at 2 h post-injection [1]. Kisspeptin-54 was also able to elicit ovulation in a large proportion of follicles, similar to that seen with human chorionic gonadotropin (hCG) [1].

SC administration of human kisspeptin-54, at a higher dose of 100 nmol/kg, to 10-week-old adult male Wistar rats also led to a rise in LH from a barely detectable baseline to ~32 ng/mL at 2 h, before falling to 10 ng/mL at the end of the 4 h sampling period. In these males, FSH rose threefold from a baseline of 10 ng/mL and was yet to plateaux at the end of the 4 h sampling period [1]. The effect of a single 50 nmol bolus of SC kisspeptin-54 was compared with an equimolar dose of GnRH in rats. Kisspeptin-54 increased plasma LH ~5-fold at 60 min, FSH ~2-fold, and testosterone ~2.5-fold at 60 min. Kisspeptin-54 and GnRH were essentially equally effective in stimulating gonadotropin release at this dose [2].

Acute Central Administration of Kisspeptin

Gottsch et al. were the first authors to study the in vivo effects of exogenous kisspeptin [3]. Central administration of a 1 fmol dose of kisspeptin-54 to adult male C57BL/6 mice stimulated LH secretion threefold at 30 min when compared with vehicle [3]. Although doses of up to 1.2 nmol were examined, maximal LH secretion at 30 min post-administration (~7-fold) was found following a 10 fmol dose [3].

ICV administration of kisspeptin-10 in adult male Wistar rats dose-dependently increased plasma LH levels at 1 h (saline: 0.3 ng/mL; 1 nmol kisspeptin-10: 5 ng/mL; 3 nmol kisspeptin-10: 9.4 ng/mL) [4]. Plasma FSH was only significantly increased at this time following the 1 nmol dose [4]. After a 3 nmol dose of ICV kisspeptin-10 treatment, LH was significantly increased at just 10 min post-injection (saline: 1.0 ng/mL; kisspeptin: 1.7 ng/mL), and continued to rise, peaking at 60 min post-injection (saline: 0.3 ng/mL; kisspeptin: 5.2 ng/mL). Enhancement of FSH secretion was slower, being only significantly increased at 60 min post-injection (saline: 16.9 ng/mL; kisspeptin: 34.4 ng/mL) [4]. In another study, ICV administration of murine kisspeptin-10 (at doses of 10–1,000 pmol) in pubertal (45 day old) male Wistar rats elicited LH rises of 7–8-fold at 15 min post-injection, and dose-dependent rises at 60 min post-injection, when compared with controls (Table 4.1) [5].

ICV murine kisspeptin-10 in pre-pubertal gilts (130 day old female pigs) at doses of 10 μg (7.5 nmol) or 100 μg (75 nmol) increased LH levels from 0.25 ng/mL to peak levels of 2 ng/mL and 3.5 ng/mL, respectively. FSH was also increased ~2-fold at both doses [6]. Hence, ICV kisspeptin was effective from a dose of 10 pmol in the rat and 7.5 nmol in the gilt, raising LH levels from as early as 10 min post-administration and achieving highest levels by 60 min post-injection.

Acute Intravenous Administration

Kisspeptin-10 has a short in vivo plasma half-life of only 4 min in humans [7]; therefore, an intravenous bolus (IV bolus) of kisspeptin-10 would be expected to have a short duration of effects. An IV bolus of murine kisspeptin-10, via intracardiac

Table 4.1 Summary of the effect of acute administration of kisspeptin-10 in different non-human mammals

Species	Route	Dose range	Effect on LH	References
Rats	ICV	0.1 pmol to 100 nmol	2–62.5× inc	[4, 5, 35, 47, 60, 62]
	IP	7.5–300 nmol	>10× inc	[4, 5, 19]
	IC	0.3–30 nmol/kg	2–8× inc	[8]
	IV	0.3 nmol to 75 nmol/kg	9–11× inc	[47]
Mice	ICV	1 fmol to 1 nmol	3-7× inc	[3]
Hamster	IP	1–7.67 nmol	3–7× inc	[61]
Pigs	ICV	7.5–75 nmol	8–14× inc	[6]
	IV	171–854 nmol	6× inc	[6]
Sheep	IV	1.6–100 nmol	2–10× inc	[25]
Goat	IV	0.77–7.67 nmol/kg	4–5× inc	[10]
Cows	IV	3.85 nmol/kg	9.7× inc	[51]
	IM	3.85 nmol/kg	3× inc	[51]
Monkeys	ICV and IV	7.67 nmol	25× inc	[9]

ICV intracerebroventricular; *IP* intra-peritoneal; *IV* intravenous; *IM* intramuscular; *IC* intracardiac; *Inc* increase in plasma levels

cannulae, in adult male Sprague Dawley rats stimulated LH release fourfold following a 0.3 nmol/kg dose and eightfold following 3.0 or 30 nmol/kg doses, when compared with pre-injection values and controls. LH secretion was stimulated within 15 min of injection and the duration of effect of the 0.3 nmol/kg dose of kisspeptin-10 lasted half as long (45 min) as the two higher doses. Based on this data, the ED50 for kisspeptin-10 (the dose which caused a rise in LH in 50% of rats given kisspeptin-10) after IV administration was calculated to be ~0.5 nmol/kg body weight [8].

In 2005, Shahab et al. studied the effects of central and peripheral administration of human kisspeptin-10 in GnRH-primed agonadal male Rhesus monkeys. The researchers observed that 100 μg kisspeptin-10 induced a significant release of gonadotropins, regardless of route of administration. For both ICV and IV treatment, plasma LH levels were increased >25-fold by 30 min after the start of kisspeptin-10 infusion and remained elevated for ~2–3 h (Fig. 4.1) [9]. A smaller ICV dose of 30 μg kisspeptin-10 also stimulated LH secretion in a similar fashion. Moreover, both the ICV and IV bolus effects of kisspeptin in this study were blocked with GnRH antagonist treatment [9].

The effects of IV administration of kisspeptin were also studied in ungulates. An IV bolus of kisspeptin-10 at doses of 1, 2.5, and 5 mg (0.77, 1.54 and 3.84 μmol) to pre-pubertal gilts (female pigs) stimulated LH to a similar degree following all three doses, whereas FSH was only significantly elevated following the highest dose [6]. In another study, human kisspeptin-10 was administered IV to female native Japanese goats (aged 4–5 years old) in the luteal phase, at doses of 1, 5, or 10 μg/kg (0.77, 3.85 or 7.69 nmol/kg). Baseline LH levels, which were between 0.26 and 0.45 ng/mL, rose to maximal levels (1.50–1.93 ng/mL) by 20–30 min post-injection [10]. In adult castrated male Shiba goats, an IV bolus of human kisspeptin-10 at a dose of 0.38 μmol increased pulsatile secretion of LH [11]. Similarly in cows, an IV

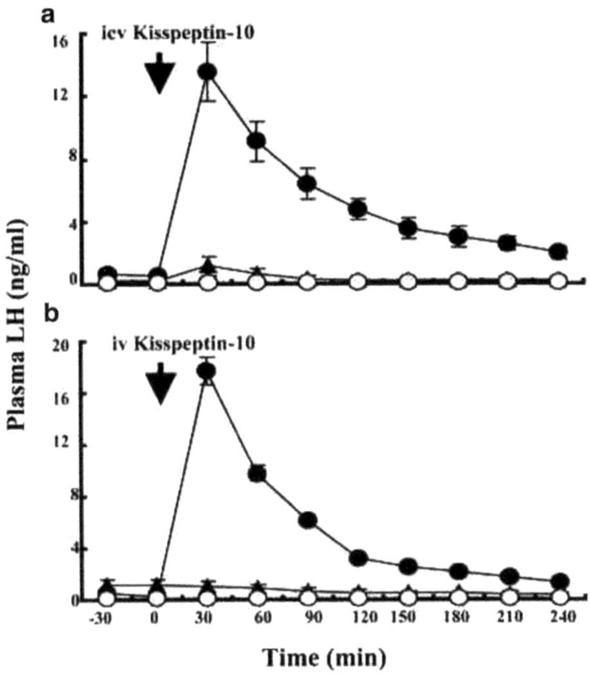

Fig. 4.1 Effect of central (**a**) or intravenous (**b**) administration of 77 nmol of kisspeptin-10 (*filled circle*), vehicle (*filled triangle*), or 77 nmol of kisspeptin-10 with acyline pre-treatment (*open circle*) on plasma LH levels (mean ± SEM) in GnRH primed agonadal juvenile male monkeys. Kisspeptin was injected at time 0 (*arrow*) and elicited robust LH secretion following both central and peripheral administration. The LH-releasing action of ICV or IV kisspeptin was abolished by pre-treatment with acyline. From Shahab et al. [9]. Reprinted with permission from National Academy of Sciences, U.S.A. Copyright (2005)

bolus of kisspeptin-10 at 100 or 200 pmol/kg, with or without additional sex steroids, significantly increases plasma LH levels [12]. The effect of an IV bolus of murine kisspeptin-10 (1 mg; 768 nmol) was also examined in light horse mares (3–15 years of age) during diestrus (when endogenous LH and FSH are low). LH levels increased by ~2-fold within 20–30 min of IV injection, returning to baseline within 4 h [13].

Acute Intraperitoneal Administration

In 2004, Thompson and colleagues observed that intraperitoneal (IP) injections of 10, 30, and 100 nmol of kisspeptin-10 in male Wistar rats dose-dependently increased plasma LH at 20 min, although only significantly following the highest dose (LH 0.6 ng/mL after saline vs. 5.9 ng/mL after 100 nmol kisspeptin) (Table 4.1) [4].

FSH was not significantly increased following IP kisspeptin-10 administration at these doses [4]. Although gonadotropin levels were not increased at 60 min post-injection, testosterone levels were significantly elevated at this time, and a larger 300 nmol dose of IP kisspeptin-10 significantly increased both LH and FSH levels at 60 min post-injection [4]. Navarro et al. found that an IP injection of 7.5 nmol of kisspeptin-10 significantly increased plasma LH levels to a similar extent as central administration of 1 nmol kisspeptin-10 at 15 min post-injection in male Wistar rats [5]. However, the decay in plasma LH levels following IP kisspeptin-10 was more rapid when compared with ICV kisspeptin-10. Therefore, the rapid onset of action of and subsequent rapid decay in LH levels following IP kisspeptin-10 may explain why Thompson et al. [4] were unable to detect an effect on gonadotropin secretion at 60 min post-injection with the lower doses employed. Hence, kisspeptin is capable of robustly stimulating gonadotropin secretion in a number of species, including when given peripherally.

The Effect of Repeated Administration of Kisspeptin on Gonadotropin Secretion

The effects of acute kisspeptin treatment on gonadotropin secretion are relatively short-lived (several hours at most). It is therefore important, from a therapeutic perspective, to ascertain whether kisspeptin is able to stimulate gonadotropin release in chronic dosing regimens.

Four IP injections of 1 nmol of mouse kisspeptin-(105–119) at 30-min intervals in 2–4-month-old male mice resulted in a ~5-fold increase in plasma gonadotropin levels measured 30 min after the last injection [14]. Likewise, four IP injections of 1 nmol kisspeptin-10 every 30 min to adult ovary-intact female Siberian hamsters housed in a summer (breeding) photoperiod significantly increased LH levels at 30 min after the last injection [15]. Hence, repeated IP administration of kisspeptin is effective in stimulating gonadotropin secretion when given at 30 min intervals in mice and hamsters. In addition, four IV boluses of 30 nmol/kg kisspeptin-10, via intracardiac cannulae, to mobile male rats at 75 min intervals elicited LH secretory bursts comparable to a single administration experiment (Table 4.2) [8].

The effects of repetitive hourly kisspeptin administration were studied in primates and in ewes. Hourly IV pulses of 2 μg (1.54 nmol) of kisspeptin-10 for 48 h to GnRH-primed juvenile agonadal male Rhesus monkeys (a model considered to be analogous to a peri-pubertal state in humans) stimulated a pulsatile pattern of LH secretory pulses, similar in magnitude to those elicited by GnRH priming [16]. This kisspeptin-induced LH response was maintained for 48 h, and FSH secretion was qualitatively similar to that of LH, although FSH pulses were less well-defined [16]. In pre-pubertal ewe lambs (28 weeks of age) treated with hourly IV boluses of kisspeptin at 20 μg (15.4 nmol) for 24 h, plasma LH levels were elevated within 15 min of the first injection, as were the frequency and amplitude of LH pulses. Four of six kisspeptin-treated lambs had LH surges 17 h after the start of the hourly

Table 4.2 Summary of the effect of repeated vs. continuous administration of kisspeptin-10 in different non-human mammals

	Repeated administration			Continuous administration			
Species	Route	Dose	Effect on LH	Route	Dose	Effect on LH	References
Rat	ICV	1 nmol	10× inc	ICV	7.5 nmol	2× inc to day 2, baseline day 4	[19, 24, 54]
	ICV	1 nmol	10× inc	ICV	7.5 nmol	2× inc to day 5, baseline day 7	
Hamster	IP	1 nmol	No reversal of seasonal gonadal regression	ICV	0.25 nmol/h	Restored testicular activity	[15, 26]
	IP	1 nmol	3× inc in long day	SC	0.2 nmol/h	No reversal of seasonal gonadal regression	
	SC	1 nmol	6× inc post last injection				
Ewes	IV	12.6 nmol	18× inc (first injection), 10× inc (second), 10× inc(third)	IV	12.4 nmol/h	Sig inc at 2 h, placebo at 48 h	[25, 67]
	IV	15.3 nmol	Inc frequency and amplitude	IV	15.2 nmol/h	9× inc 5 h, 3× inc 10 h	
				IV	15.2 nmol/h	6 h 12.5%, 12 h 50%, 24 h 75% had LH surge	
Goats	IV	4.23 nmol/kg	11× inc (first injection), 22× inc (second), 15× inc (third)				[11]
Cow				IV	500 pmol/kg	Inc for 140 min, then decline	[50]
Monkey	IV	1.54 nmol	2× inc maintained for 48 h	IV	7.7 nmol/h	2× inc to 1–2 h, then decline	[16, 21]

ICV intracerebroventricular; *IP* intra-peritoneal; *IV* intravenous; *SC* subcutaneous; *Inc* increase in plasma levels kisspeptin therapy; however, kisspeptin therapy did not accelerate the onset of puberty in the lambs when compared with placebo [17]. Hence, hourly IV pulses of kisspeptin were effective in stimulating LH secretion for as long as 48 h in primates and at least 24 h in ewes. This hourly administration regime may most accurately reflect the inter-pulse interval of endogenous kisspeptin secretion [18]. Using microdialysis, kisspeptin immunoreactivity (IR) was measured within the stalk median eminence of female Rhesus monkeys. Kisspeptin IR had a pulsatile pattern of secretion, with an inter-pulse interval of ~60 min, and 75% of kisspeptin pulses corresponded to GnRH pulses [18]. Furthermore, kisspeptin-10 infusion into the stalk median eminence stimulates GnRH release in a dose-dependent manner [18]. This data suggests that pulsatile kisspeptin secretion stimulates pulsatile release of GnRH, causing subsequent pulsatile gonadotropin secretion from the pituitary gland.

Repeated IV administration of kisspeptin was also studied in ungulates. Adult female goats in the luteal phase were given three injections of human kisspeptin-10 or GnRH at 5 μg/kg (4.23 nmol/kg) at 2 h intervals. Plasma LH levels were increased from baseline (0.41–0.56 ng/mL) to maximum values of 5.56 ng/mL, 10.72 ng/mL and 7.72 ng/mL for each injection, respectively [10]. Plasma FSH levels increased in a similar time-frame from a baseline of 0.62–0.78 ng/mL to peaks of 1.16 ng/mL, 2.02 ng/mL, and 1.60 ng/mL after the first, second, and third kisspeptin-10 injection, respectively. GnRH elevated gonadotropins more gradually, but to a greater maximum (LH: 34.99 ng/mL; FSH: 3.32 ng/mL) 30 min after the second injection [10]. Hence, kisspeptin was able to stimulate gonadotropin secretion when administered peripherally at 2 h intervals, although to a lesser extent when compared with GnRH.

Repeated ICV administration of 1 nmol of mouse kisspeptin-10 every 12 h to pre-pubertal (day 26) female Wistar rats for 5 days increased LH levels tenfold. While absent in all of the controls, 74% of repeatedly kisspeptin-treated immature rats showed physical evidence of puberty [18]. Hence, repetitive ICV kisspeptin administration was effective in inducing sexual maturation in juvenile female rats. Furthermore, repeated central administration of kisspeptin-10 (1 nmol every 12 h for 1 week) to under-nourished (30% restriction in food intake) pre-pubertal (day 30) female rats induced physical markers of puberty in 60% of kisspeptin-treated animals vs. none in the control group, along with elevated serum LH and FSH levels [19].

Collectively, repeated kisspeptin administration in mice, rats, hamsters, and goats was effective in stimulating gonadotropin levels (Table 4.2). Furthermore, repeated administration of kisspeptin in immature rats was able to induce pubertal maturation (although not in ewe lambs) with the dosing regimen employed. These data suggest that kisspeptin could be used therapeutically to increase gonadotropin and sex hormone levels, although it is not known whether kisspeptin would offer any therapeutic advantage over existing therapies for reproductive disorders.

Desensitization to the Effects of Kisspeptin with Persistent (Chronic) Administration

A number of hormones are subject to tachyphylaxis, where the stimulatory effects are diminished with continued administration. A prime example is GnRH, where initial stimulation is replaced by inhibition after continuous non-pulsatile administration [20]. Similarly, there is evidence that the stimulatory effects of exogenous kisspeptin diminish with persistent administration. An important study by Seminara et al. demonstrated the phenomenon of tachyphylaxis in primates. In the primed male agonadal Rhesus monkey model, human kisspeptin-10 was administered as a 10 μg (7.7 nmol) IV bolus, followed by a continuous 98 h infusion of 100 μg/h (77 nmol/h). The initial IV bolus increased LH levels twofold compared with the preceding GnRH priming dose (0.15 μg/min for 2 min every hour). Plasma LH levels peaked 1–2 h after the start of the constant infusion, reaching a level ~2.5-fold higher than initial LH levels. However, LH levels then declined, despite continued exposure to kisspeptin-10, suggesting that desensitization to the stimulatory effects of kisspeptin-10 had occurred [21]. During the final 3 h of human kisspeptin-10 infusion, an IV bolus of GnRH, *N*-methyl-d-aspartate (NMDA, a glutamate agonist which stimulates GnRH secretion), or human kisspeptin-10 was administered. Although NMDA and GnRH each elicited LH secretion, kisspeptin-10 failed to do so, suggesting that the lack of effectiveness of kisspeptin-10 administration was not due to a failure of GnRH secretion or reduced pituitary sensitivity to GnRH. This indicates that the inability of human kisspeptin-10 to elicit an LH response after continuous infusion may be due to down-regulation at the level of the kisspeptin receptor. The authors reported that responsiveness to kisspeptin-10 was recovered at approximately 21 h after the infusion had concluded [21].

Desensitization to chronic kisspeptin has also been demonstrated in rodents. Kisspeptin-10 administered IP twice daily to 35-day-old male pre-pubertal rats for 12 days, at doses of 15 pmol, 1.5 nmol, and 1.5 μmol, reduced plasma LH and testosterone at the two higher kisspeptin doses, when compared with controls (FSH remained unaltered). These data suggest that twice daily IP kisspeptin-10 at doses of at least 1.5 nmol in pre-pubertal rats resulted in tachyphylaxis. Similarly, a continuous SC infusion of kisspeptin-54, at 50 nmol/day via a subcutaneous mini-pump, for 13 days in adult rats also resulted in inhibition of LH secretion [22, 23]. To ensure that the observed reduction in LH and FSH stimulation with chronic kisspeptin administration was not due to peptide degradation in the mini-pump, bioactivity studies were performed to demonstrate equivalent potency of freshly prepared kisspeptin-54, and of kisspeptin-54 that had been incubated in a mini-pump for 14 days at 37°C [7]. In another study examining the time-course of desensitization, plasma LH levels were increased ~15-fold after 6 h of constant SC kisspeptin-54 administration, ~10-fold after 12 h, but only ~2-fold after 24 h. Plasma FSH showed a similar desensitization pattern, being elevated ~3.5-fold at

6 h, ~2-fold at 12 h, and back to baseline by 24 h. Hence, desensitization to the effects of continuous kisspeptin-54 occurs rapidly, commencing within just 12 h [2].

On the contrary, Roa et al. examined whether nutritional status influences the effect of chronic kisspeptin-10 treatment in rodents. Kisspeptin-10, when administered by continuous ICV infusion at a dose of 7.5 nmol/day to female pubertal (30 day old) rats, significantly elevated LH levels at day 7 of the infusion and modestly elevated FSH levels [24]. The experiment was repeated in pubertal rats exposed to a 30% deficit in calorie intake, and gonadotropin levels were significantly elevated at day 7 to a greater degree than their fed counterparts. It is interesting as to why no desensitization occurred in this experiment. One possibility is that the route of administration (central vs. peripheral) may play a role. Alternatively, a state of energy deficiency seems to enhance the stimulatory effect of kisspeptin-10 and reduces the tachyphylaxis seen with a continuous infusion of kisspeptin-10 [24].

Tachyphylaxis to kisspeptin has also been demonstrated in sheep. A group of acyclic non-breeding Ile de France ewes received a constant IVI of 600 nmol of murine kisspeptin-10 over 48 h (12.6 nmol/h). Plasma LH was significantly elevated at 2 h, yet, by 48 h, plasma LH levels were similar to placebo. Importantly, 80% of the ewes receiving murine kisspeptin-10 by constant infusion ovulated as compared with only 17% of the ewes which had received placebo alone. This data suggests that kisspeptin-10 can induce ovulatory LH surges (which occurred ~30 h after the start of kisspeptin administration), despite desensitization to the effects of kisspeptin on LH secretion by 48 h, perhaps as a result of an initial stimulatory effect on gonadotropin secretion [25].

In the Syrian hamster, reproductive activity and hypothalamic *Kiss1* expression are reduced during short day (SD) winter-like conditions. Furthermore, this effect is reversed by chronic central administration of murine kisspeptin-10 at 6 nmol/day for 4 weeks [26]. In contrast, a chronic SC infusion of 10 nmol/day of human kisspeptin-54 to male SD hamsters for 28 days failed to increase testicular weight and plasma testosterone. However, twice daily IP injection of 10 nmol kisspeptin for 5 weeks increased testicular weight twofold and tended to increase plasma testosterone [27]. Therefore, kisspeptin may be a pivotal regulator of seasonal reproduction, but the regimen of administration is critical for determining whether tachyphylaxis is encountered, at least in SD hamsters.

In summary, desensitization to the effects of kisspeptin has been observed from as early as 3 h after onset of continuous administration and is likely occurring at the level of the kisspeptin receptor. Desensitization does not usually occur with repeated administration of kisspeptin at intervals of 30–75 min, which may possibly mimic the frequency of endogenous kisspeptin pulsatility [18]. Repeated administration at 12 hourly frequencies results in tachyphylaxis in some, but not all, studies, and appears to be associated with higher dosages of kisspeptin. Tachyphylaxis in response to kisspeptin is also recognized during repeated administration of kisspeptin-54 to women with hypothalamic amenorrhoea and is an important consideration for development of therapeutic dosing regimens [28].

Site of Action of Kisspeptin

Kisspeptin Actions in the Brain

Kisspeptin is thought to act in the hypothalamus and results in GnRH release, which in turn acts on the pituitary gland to cause gonadotropin release. This mode of action is supported by several lines of evidence. First, the stimulatory effects of kisspeptin on gonadotropin release are attenuated in the presence of a GnRH antagonist in a number of species, including ewes, rats, mice, and primates [1, 5, 16, 29–31]. Second, peripherally administered kisspeptin-54 elicits c-fos activation in 60% of GnRH neurons in the hypothalamus [1]. Third, kisspeptin treatment leads to a rise in GnRH levels in the CSF of sheep [14, 25]. Fourth, patch clamp studies demonstrate that kisspeptin activates GnRH neuronal firing in brain slices from adult female proestrus mice [32].

Anatomical data also support the notion that kisspeptin acts on GnRH neurons in the brain. In the rat, more than 75% of GnRH neurons co-express kisspeptin receptor mRNA [31], and this has also been shown for several other species. In the rodent brain, there are two high density populations of kisspeptin immunoreactive cells in the hypothalamus; firstly, in the periventricular continuum of cells within the rostral part of the third ventricle (which includes the antero-ventral periventricular nucleus [AVPV] and the periventricular nucleus [PeN]), and secondly, in the arcuate nucleus (ARC). It has been shown in rodents that kisspeptin immunoreactive nerve elements directly adjoin GnRH immunoreactive nerve elements in the median eminence [33]. Hence, kisspeptin may in part stimulate GnRH release from GnRH neurons by non-synaptic axo-axonal interactions at the median eminence [33]. In possible support for this, incubation of rat hypothalamic explants with kisspeptin-10 stimulated the release of GnRH [4].

The brain is protected from circulating substances in the peripheral blood by tight endothelial junctions forming the blood–brain barrier. The circulating blood comes in contact with the brain only at specific areas, including the median eminence and organum vasculosum of the lamina terminalis (OVLT). GnRH neurons project axons to the median eminence, from where they secrete GnRH into the portal vasculature to control pituitary gonadotropin release. A sub-population of GnRH neurons located in the rostral preoptic area possesses complex branching dendritic branches, which fill the OVLT [34]. It was shown that these GnRH neurons in the OVLT are susceptible to stimulation by kisspeptin [34]. Therefore, kisspeptin may also stimulate GnRH release via interaction with GnRH neuronal fibers at the OVLT and/or the median eminence [34].

In the rat, kisspeptin neurons are present in the hypothalamic ARC and the AVPV. To investigate the direct site of action of kisspeptin in the brain, administration of kisspeptin-10 into the region of the ARC or the medial preoptic area (which includes the AVPV) was assessed in estrogen-replaced ovariectomized adult female rats. Both intra-ARC and intra-mPOA administration of kisspeptin-10, at 1, 10, and

100 pmol, over 5 min resulted in a dose-dependent increase in plasma LH levels that lasted ~1 h [35]. Three bolus injections of peptide 234, a kisspeptin antagonist, administered at 30-min intervals into the ARC reduced LH pulse frequency (although not amplitude of LH pulses), suggesting that the ARC plays an important role in the modulation of GnRH pulse frequency by kisspeptin [35].

In the male Rhesus monkey, kisspeptin-immunoreactive cells have been identified in the ARC, but not in the AVPV [36]. Alternatively, female Rhesus monkeys have population of *Kiss1* mRNA expressing neurons in the ARC and the POA (potentially a homologous region to the rodent AVPV) [37]. In the ARC, there are only infrequent contacts made between GnRH neurons and kisspeptin neurons. However, at the median eminence, kisspeptin and GnRH axons are extensively and intimately associated, allowing for potential non-synaptic interaction between these two groups of neurons [36]. Kisspeptin-54 release in the stalk median eminence of monkeys is pulsatile, and ~75% of kisspeptin-54 pulses are correlated with GnRH pulses [18]. Microdialysis infusion of 10 or 100 nM human kisspeptin-10 into the medial basal hypothalamus (MBH; which includes the ARC) and the stalk median eminence for 20 min stimulated GnRH release in a dose-dependent manner in both pre-pubertal (low estradiol) and pubertal (elevated estradiol) ovarian-intact female monkeys. Conversely, infusion of peptide 234, a kisspeptin antagonist, suppressed GnRH release. Interestingly, the stimulation of GnRH release by kisspeptin-10 was absent in ovariectomized post-pubertal, but not pre-pubertal, monkeys; this kisspeptin stimulation of GnRH could be partially restored in post-pubertal monkeys by estradiol replacement. This suggests an important permissive role for estradiol in the neural action of kisspeptin in monkeys, which is apparent only after puberty has occurred; there may be a switch from estradiol-independent to estradiol-dependent kisspeptin signaling within the hypothalamus after puberty [38].

Kisspeptin Actions in the Pituitary

While the predominant effect of kisspeptin is via the hypothalamus, likely via stimulation of GnRH neurons, it has also been enquired whether kisspeptin may also have an additional effect via the pituitary gland. In primary pituitary cell cultures prepared from adult cyclic female baboons (*Papio anubis*), kisspeptin-10 (10^{-14} to 10^{-6} M) can dose-dependently stimulate LH release. Interestingly, estradiol enhanced the relative LH response to kisspeptin-10, both alone and in combination with GnRH [39]. However, while incubation of rat hypothalamic explants with kisspeptin-10 stimulated the release of GnRH at a dose of 10 nM, kisspeptin doses of up to 1,000 nM had no effect on the release of LH or FSH from rat pituitary cultures in vitro [4]. In another study, incubation of pituitary cultures from peri-pubertal male rats with kisspeptin-10 induced a rise in free cytosolic Ca^{2+} concentration in only 10% of rat pituitary cells [40]. Moreover, another study in rats found that incubation of cultured male or female pituitary cells with kisspeptin-54 did not result in in vitro gonadotropin release, even when primed with estradiol [1]. However, Navarro et al. [5] incubated rat pituitary cultures in kisspeptin-10 and reported that

LH was increased in a dose-dependent manner after 60 and 180 min of incubation. Even so, the relative contribution of direct pituitary effects to the potent LH releasing effect of kisspeptin-10 in vivo is likely to be small, as in vitro doses of kisspeptin required to generate LH release from pituitary cells are much higher when compared with in vivo kisspeptin dosages. Taken together, this suggests that the primary effect of kisspeptin on gonadotropin release in vivo occurs in the brain via GnRH release [5].

The sheep is an ideal model to study the effects of kisspeptin on the pituitary, because the pituitary gland is large and a hypothalamo-pituitary disconnection (HPD) model allows for exploration of direct actions of kisspeptin on GnRH-stimulated gonadotropin secretion in vivo. The HPD procedure removes all neural inputs to the median eminence, but gonadotropin secretion can be restored by pulsatile IV administration of GnRH. The kisspeptin receptor is expressed in pituitary cultures from ewes (although only during the follicular phase of the estrous cycle). Incubation of ovine pituitary cells with kisspeptin increased the concentration of LH in culture media by 80%, compared with control treatment. Despite this in vitro finding, kisspeptin was unable to stimulate in vivo LH secretion in GnRH-replaced ovariectomized HPD ewes. Furthermore, while low levels of kisspeptin were detected in the ovine hypophyseal portal circulation, these kisspeptin levels did not change during an estrogen-induced LH surge. Hence, the in vitro stimulatory effect of kisspeptin on isolated pituitary cultures may not relate to kisspeptin's signaling in vivo [41].

In pigs, anterior pituitary cells have been harvested from 6-month-old triple cross bred (Landrace × Large White × Duroc) barrows (castrated male pigs) and incubated in 10 nM GnRH or 10–1,000 nM kisspeptin-10. The 1,000 nM dose of kisspeptin-10 significantly increased in vitro porcine LH levels by 31%, compared with the control. However, the LH-releasing potency of kisspeptin-10 in porcine anterior pituitary cells was far less potent than that of 10 nM of GnRH [42]. Cultured bovine anterior pituitary cells incubated with kisspeptin-10 secreted LH, but only at a high kisspeptin concentration of 1,000 nM [22]. Bovine anterior pituitary cells from 7-week-old male calves incubated with human kisspeptin-10 released LH, when also pre-incubated with estradiol and testosterone, but not progesterone [43].

Overall, despite the presence of kisspeptin receptor in the pituitary, most of the data support the theory that the predominant action of kisspeptin is at the hypothalamus, to increase GnRH release, which subsequently stimulates gonadotropin release. Any direct effects of kisspeptin on the pituitary gland are likely to be negligible in vivo.

Comparison of Different Kisspeptin Isoforms

The 145 amino acid gene product of the *Kiss1* gene is post-translationally cleaved into peptides of varying amino acid lengths, including kisspeptin-54, kisspeptin-14, kisspeptin-13, and kisspeptin-10. Kisspeptin-10 is identical between humans and Macaque monkeys; the sequence of kisspeptin-10 in the rat, mouse, pig, cow, and

Table 4.3 Summary of effects of acute administration of kisspeptin-10, -52, and -54 on LH secretion in the rat

Peptide	Route	Dose range	Effect on LH	References
Kiss 10	ICV	0.1 pmol to 100 nmol	2–62.5× inc	[4, 5, 35, 47, 60, 62]
	IP	7.5–300 nmol	>10× inc	[4, 5, 19]
	IC	0.3–30 nmol/kg	2–8× inc	[8]
	IV	0.3 nmol to 75 nmol/kg	9–11× inc	[47]
Kiss 52	ICV	0.1 nmol	50× inc	[47]
	IV	1–10 nmol/kg	4–14× inc	[8, 47]
Kiss 54	SC	0.1–50 nmol	5–32× inc	[1, 2]

ICV intracerebroventricular; *IP* intra-peritoneal; *IV* intravenous; *SC* subcutaneous; *IC* intracardiac; *Inc* increase in plasma levels

sheep, termed "murine kisspeptin-10," is one amino acid different from the sequence of human kisspeptin-10. Horse kisspeptin-10 is one further amino acid different from rodent, pig, cow, and sheep kisspeptin-10 [44]. Kisspeptin-10 is simpler and cheaper to synthesize when compared with kisspeptin-54; therefore more studies use kisspeptin-10 than kisspeptin-54 (or kisspeptin-52 in rodents). The plasma half-life of kisspeptin-54 is 27.6 min, whereas the in vivo plasma half-life of kisspeptin-10 is much shorter, at ~4 min in humans [7, 45]. Kisspeptin-10 is the minimum sequence required for activation of the kisspeptin receptor and is common to the C-terminus of all kisspeptin isoforms [46]. However, this difference in half-lives may result in a different time-profile for activation of the kisspeptin receptor and hence differences in gonadotropin release (Table 4.3).

Central bolus administration of murine kisspeptin-52 and human kisspeptin-10 was compared in adult male C57BL/6 mice. Murine kisspeptin-52 stimulated LH release from a baseline of 0.2–3.1 ng/mL at 30 min, compared to an LH release of 2.8 ng/mL after human kisspeptin-10 [3]. In another study, an IV dose of kisspeptin-10 was compared with an equimolar dose of full-length kisspeptin (rat kisspeptin-52) in adult Sprague Dawley male rats [8]. Kisspeptin-10 elicited a robust plasma LH burst, with levels peaking at 15–30 min after injection and returning to baseline by 75–90 min; in contrast, the effects of kisspeptin-52 lasted longer, being still detectable at 120 min post-injection. Consequently, the overall magnitude of total LH secretion following kisspeptin-52 was greater when compared with kisspeptin-10 [8]. Another report in male rats compared the effects of an IV bolus of 10 nmol/kg of rat kisspeptin-52, rat kisspeptin-10, or human kisspeptin-10; all three isoforms resulted in peak plasma LH levels at 30 min. However, whereas rat kisspeptin-52 increased plasma LH levels to a peak of ~2.8 ng/mL, returning to baseline by 2 h, rat and human kisspeptin-10 both had shorter-lasting effects, with plasma LH returning to baseline levels by 1 h (both increased plasma LH levels to a similar peak of ~2 ng/mL) [47]. Hence, human and murine kisspeptin-10 have nearly identical effects on LH secretion when given IV, and both elicit a smaller and shorter-lasting LH response than IV kisspeptin-52. Likewise, in Corriedale ewes, murine kisspeptin-10 stimulates gonadotropin secretion to the same degree as human kisspeptin-10 [25].

In one study, ICV administration of 0.1 nmol of murine kisspeptin-52 to male Wistar-Imamichi rats (8–10 weeks of age) stimulated a gradual increase in plasma LH levels from a baseline of 0.2 ng/mL to a peak of ~10 ng/mL at 2 h, which persisted throughout the 3 h sampling period. However, ICV administration of the same dose of murine or human kisspeptin-10 failed to increase plasma LH levels [47]. ICV administration of 1 nmol human kisspeptin-10 increased LH levels to ~2 ng/mL, whereas the same dose of murine kisspeptin-10 increased LH levels to a peak of ~4.5 ng/mL at 2 h [47]. Hence, centrally administered rat kisspeptin-52 stimulates LH to a greater degree and for longer than either human or murine kisspeptin-10. Oddly, murine kisspeptin-10 has a greater effect than human kisspeptin-10 when administered centrally at the same dose to rats, unlike when given IV. It would be interesting to determine whether human or murine kisspeptin-10 have higher potencies in humans or primates.

IP administrations of murine kisspeptin-10, human kisspeptin-10, murine kisspeptin-52, and human kisspeptin-54 were compared in adult male NMRI (Naval Medical Research Unit) mice. Murine kisspeptin-52 was most effective in stimulating release of testosterone, followed by murine kisspeptin-10, then human kisspeptin-54, and finally human kisspeptin-10 [48]. It is interesting that murine kisspeptin was more effective than human kisspeptin in stimulating testosterone release in rodents, although the reasons for this remain unclear. Also of interest, both murine and human kisspeptin-10 had a more rapid onset of action when compared with the longer sequences of kisspeptin. Human kisspeptin-54 appeared to have the longest duration of action, with a peak effect on free testosterone occurring at 60 min [48]. Hence, it appears that shorter sequences may have a more rapid onset of action, whereas longer sequences may have a longer duration of effect.

Recently, an analogue of kisspeptin-10, [dY]1KP-10, was developed. Although [dY]1KP-10 has similar in vitro binding to the kisspeptin receptor, it is more potent in vivo when given to adult male C57Bl/6 mice. IP [dY]1KP-10 increased plasma LH and testosterone levels at 20 min post-injection to a greater degree than kisspeptin-10. Furthermore, at 60 min, a 0.15 nmol dose of [dY]1KP-10 significantly increased total testosterone levels, whereas the same dose of kisspeptin-10 had no significant effect [49]. The reason for the higher potency of the kisspeptin analogue is not immediately clear.

Comparison of Different Routes of Administration

Remarkably, kisspeptin has been shown to be effective in stimulating gonadotropin release when given by a number of routes of administration, including ICV, IP, SC, and intramuscular (IM) injection. However, different routes of administration may result in varying peak concentrations and duration of circulating kisspeptin levels; these may, in turn, result in differential stimulation of the kisspeptin receptor. The volume of distribution of ICV administration is smaller when compared with peripheral administration, and kisspeptin is thought to have a central location of action; it is therefore expected that a centrally administered dose of kisspeptin would be more effective in stimulating gonadotropin secretion when compared with peripheral administration.

In one study in rats, an IV bolus dose of 7.5 nmol of kisspeptin-10 increased plasma LH levels to a similar extent, although for a shorter duration, when compared with ICV administration of the same dose [5]. Furthermore, an IV bolus of 1,000 pmol/kg of human kisspeptin-10 increased LH secretion to a similar degree when compared with ICV administration of a much smaller dose (200 pmol/kg) in ovariectomized adult ewes [50]. A comparison was also made between 150 nmol of murine kisspeptin-10 administered as a SC bolus or as a 6 h constant IV infusion in ewes. After SC bolus administration, plasma LH secretion was increased 3.6-fold for 90 min post-injection (a smaller rise in plasma FSH levels was also noted). On contrast, after an IV infusion, plasma LH levels rose more gradually, with a peak response 2–4 h later. Interestingly, there was a progressive decline in plasma LH levels after this peak, despite continuing IV infusion, which may suggest gradual desensitization to the ongoing presence of kisspeptin [25]. Hence, an IV infusion was able to stimulate LH levels to a greater peak level when compared with SC administration; however, peak levels following IV infusion occur later when compared with SC bolus administration. The effects of IM and IV injection of human kisspeptin-10 were also compared in Japanese pre-pubertal calves (5–6 months of age). A 3.85 nmol/kg dose of human kisspeptin-10 stimulated a ~10-fold increase in plasma LH following IV injection. By comparison, only a threefold increase in plasma LH was observed following IM administration [51].

Collectively, the above results indicate that, although kisspeptin is likely to exert its action centrally, IV bolus administration is associated with a rapid and powerful short-term stimulation of gonadotropin secretion. Moreover, IM or SC kisspeptin administration may be associated with more sustained absorption of kisspeptin; this may cause a lower peak but longer duration of gonadotropin stimulation when compared with IV bolus administration. IV infusion of kisspeptin results in less rapid yet more sustained stimulation when compared with bolus SC administration, but with greater risk of eventual desensitization to the effects of kisspeptin.

Effect of Nutritional State on Gonadotrophic Response to Kisspeptin

Fertility is reduced during undernutrition. Hypothalamic kisspeptin neurons have been proposed as an intermediary in the communication between leptin and GnRH neurons (see Chap. 17). In adult male Rhesus monkeys, 50 μg (38.4 nmol) of human kisspeptin-10 was administered IV in a fed state or following a 48 h fast. Following kisspeptin administration, peak testosterone secretion was significantly reduced and slower to rise in fasted monkeys compared with fed monkeys. However, both fed and fasted monkeys responded similarly to human chorionic gonadotropin (hCG, which has LH-like activity), suggesting that there was no difference in gonadal sensitivity [52]. Furthermore, expression of both kisspeptin and its receptor in the hypothalamus was reduced in fasted monkeys compared with fed animals, suggesting that the reduced response to exogenous kisspeptin-10 in the fasted state may have been due to reduced expression of the kisspeptin receptor [53]. By contrast, following a 48 h fast, immature female rats maintain responsiveness to a 1 nmol dose of ICV kisspeptin-10 when compared with ad libitum-fed females [54]. Similarly, in adult rats fasted for 48 h, a 30 nmol/kg dose of IV kisspeptin-10 stimulates LH secretion to similar level as fed rats, albeit from a lower baseline LH [8]. These data therefore suggest that monkeys and rodents may respond differently to exogenous kisspeptin during undernutrition; this may partially be explained by considering that kisspeptin receptor expression is increased in rodents (rather than decreased in monkeys) under caloric restriction [19].

Central administration of 1 nmol of murine kisspeptin-10 to pre-pubertal (30 day old) male and female Wistar rats stimulated serum LH levels at 15 min post-injection similarly in both sexes (9.0–10-fold increase). Another group of rats was food-deprived for 72 h; interestingly, the ability of kisspeptin to induce LH secretion was notably enhanced in these fasted animals, with 50–60-fold increases in LH in this condition [19]. In this study, food deprivation reduced endogenous kisspeptin expression, but increased kisspeptin receptor expression, perhaps explaining the heightened response to kisspeptin treatment [19]. Furthermore, kisspeptin-10 induces greater GnRH release from in vitro hypothalamic fragments taken from fasted female rats when compared with ad libitum fed rats [19]. Roa et al. [24] centrally administered 7.5 nmol/day of kisspeptin-10 by continuous ICV infusion for 7 days to adult female rats. In ad libitum fed animals, LH levels were elevated for the first 2 days of kisspeptin-10 infusion, but then dropped precipitously, whereas FSH secretion was increased throughout the 7-day infusion [24]. Conversely, in rats exposed to calorie restriction (50% decrease in daily calorie intake), the stimulatory LH response to kisspeptin was maintained until day 5, but FSH secretion was only maintained until day 1 [24]. Hence in female rats, the stimulatory effect of kisspeptin on LH secretion is subject to tachyphylaxis, which becomes less pronounced during undernutrition. Paradoxically, the stimulatory effect of kisspeptin on FSH was not subject to tachyphylaxis, except during a state of undernutrition [24].

To summarize, in monkeys, fasting antagonizes the effects of exogenous kisspeptin on gonadotropin secretion; however, in the rat, kisspeptin effects are enhanced by undernutrition. Interestingly, women with hypothalamic amenorrhoea are more sensitive to acute administration of exogenous kisspeptin when compared with healthy women, but tachyphylaxis is associated with twice daily administration [28]. Further work is needed in order to determine to what extent species, dosing, and treatment duration influence undernutrition's effects on kisspeptin signaling.

Effects of Development and Reproductive Maturation on Response to Kisspeptin

Kisspeptin is thought to play an important role in pubertal development. Indeed, inactivating mutations in the genes encoding kisspeptin or its receptor lead to hypogonadotrophic hypogonadism in humans and mice [55, 56], whereas activating mutations lead to precocious puberty [57]. Recently, circulating plasma kisspeptin levels were shown to be higher in central precocious puberty when compared with normal pre-pubertal girls [58]. The onset of puberty is also known to be linked to metabolic cues, with leptin and adipose tissue thought to be important factors in the onset of puberty [59]. Furthermore, the effects of kisspeptin may be altered by prevailing levels of sex hormones, which change with puberty. Hence, a number of factors may result in differences in the response to exogenous kisspeptin during pubertal maturation.

The effects of IP kisspeptin-10 on LH secretion were assessed in male and female rats at different stages of postnatal maturation, namely the neonatal (5 day old), late infantile (15 day old), and juvenile (25 day old) periods. IP injection of kisspeptin-10 at 0.75 nmol/10 g elicited greater absolute LH responses in rats in the late infantile group (15 day old), when compared with juvenile (25 day old) and neonatal (5 day old) rats [60]. However, LH was stimulated to similar degrees (relative to baseline levels) in all three stages of development. This data suggests that rats are responsive to exogenous kisspeptin throughout development, but the pre-existing level of activity of the reproductive axis influences how much gonadotropin secretion is triggered following kisspeptin administration.

In hamsters, greatest sensitivity to a single IP injection of 1 nmol human kisspeptin-10 was observed in pubertal animals (day 45 old) when compared with pre-pubescent (15 day) or adult (75 day) hamsters [25]. In contrast, in mice, greater response to kisspeptin was seen in adult mice than juvenile mice. Central administration of mouse kisspeptin-52 was carried out at doses of 10 and 100 fmol in juvenile (day 18) and adult male C57BL/6J mice. In adults, both 10 and 100 fmol kisspeptin-54 increased plasma LH levels at 30 min post-injection [32]. However, in juvenile mice, only the 100 fmol dose elicited a rise in LH. Furthermore, the percentage of GnRH neurons responding to kisspeptin increased from ~25% in juveniles to ~50% in pre-pubertal mice and to >90% in adult mice. Therefore, this study suggests that sensitivity to kisspeptin-52 increases with age in male mice [32].

Overall, the sensitivity of the hypothalamo-pituitary–gonadotrophic axis to exogenous kisspeptin appears to increase after the onset of puberty, with greatest responses in peri-pubertal periods and during adulthood; these ages may relate to more mature stages of reproductive development with higher levels of sex hormones, which might influence sensitivity to exogenous kisspeptin administration.

Effect of Sex and Stage of Female Estrus Cycle on Response to Kisspeptin

Interestingly, there may be sex differences in the effects of kisspeptin. Ezzat et al. [51] administered kisspeptin-10 to male and female Japanese pre-pubertal calves at 5–6 months of age. Maximum LH (7.2 ng/mL vs. 17.4 ng/mL) and FSH (0.6 ng/mL vs. 1.5 ng/mL) secretion was lower in females when compared with males [51]. However, it should be noted that the observed sex difference in this study might be due to differing developmental stages, given that males and females go through puberty at different times. Healthy adult men are more responsive to the effects of kisspeptin-10 than adult females in the follicular phase [7]. In keeping with this human data, male hamsters had greater LH responses to IP kisspeptin-10 than females [61]. However, female rats were more sensitive to kisspeptin-10, as assessed by LH stimulation, when compared with male rats [60]. Thus, while some studies have reported sex differences in response to kisspeptin, the direction of the sex difference (female > male or male > female) differs between studies and species, and the exact reasons for the observed sex differences have yet to be teased out.

The stage of adult female reproductive cycle also plays an important role in influencing sensitivity to exogenous kisspeptin. Roa et al. [62] centrally administered 1 nmol kisspeptin-10 to adult Wistar female rats at different stages of the estrus cycle. Plasma LH stimulation was greatest during estrus (sixfold increase); LH increased fourfold during diestrus, and only twofold during proestrus [62]. Gonadotropin responses following GnRH administration are also increased during ovulation, due to increased pituitary sensitivity associated with high circulating estrogen levels [63]. It is therefore possible that kisspeptin induces different levels of LH secretion during the female rodent reproductive cycle, as a consequence of differential sex steroid priming of the pituitary gland. Maximal FSH stimulation following kisspeptin administration was observed during diestrus [62]. It is therefore important to consider that FSH secretion does not always mirror LH secretion. Moreover, it appears clear that levels of circulating sex steroids play an important role in modulating the effects of kisspeptin on gonadotropin secretion. In ovariectomized rats, replacement of estradiol suppressed baseline LH levels, but did not affect LH stimulation by kisspeptin-10 administration. Progesterone replacement did not influence the effects of kisspeptin-10 on LH secretion. However, combined replacement of progesterone and estrogen permitted kisspeptin to elicit a 22-fold increase in LH secretion when compared with animals without progesterone and estrogen replacement. Hence, progesterone replacement seems to sensitize estrogen-replaced ovariectomized rats to the effects of kisspeptin-10 [62].

The reproductive axis clearly plays an important role in determining the effects of kisspeptin in an animal. The question remains, however, if the sensitivity to kisspeptin reflects direct changes in GnRH sensitivity to kisspeptin, or rather, a change in the pituitary sensitivity to GnRH. It is well known that pituitary sensitivity to GnRH is influenced by levels of circulating sex steroids [63] and activity of GnRH neurons also changes during different stages of the female menstrual cycle. Likewise, levels of kisspeptin expression also change during the menstrual cycle and after experimental sex steroid manipulations, such as gonadectomy [64]. Further work is needed to determine the precise extent to which differences in kisspeptin sensitivity are consequences of differential hypothalamic (GnRH) vs. pituitary sensitivity.

Conclusions

Kisspeptin has been shown to induce gonadotropin secretion in all non-human mammalian species studied, and the effects are mediated via GnRH release. However, it is important to note that endogenous secretion of kisspeptin is likely to be pulsatile (~60 min inter-pulse interval) [18] and in much smaller quantities than the pharmacological dosing strategies employed during the studies described in this chapter.

Although most GnRH neurons have kisspeptin receptors, the ability of peripheral kisspeptin to stimulate gonadotropin release is striking, given that kisspeptin is unlikely to be able to cross the blood–brain barrier. Peripherally circulating kisspeptin may interact with GnRH neurons at the OVLT and at the median eminence to cause GnRH release [34, 65]. Another possibility is that kisspeptin may act through intermediary neurons, such as inhibitory GABA neurons located in circumventricular regions of the brain, to modulate GnRH neuronal activity [66]. The ability of kisspeptin to stimulate GnRH release, even when administered peripherally, is important for ease of administration in the development of therapeutic kisspeptin regimens in the future. Further lessons from animal studies include the finding that continuous exposure may result in tachyphylaxis to the effects of kisspeptin. This is an important consideration for any translational usage of kisspeptin during chronic dosing regimens.

There are only a small number of studies which directly compare the effects of different kisspeptin isoforms on gonadotropin release. There are many more studies utilizing kisspeptin-10 than longer isoforms, partly due to the greater availability and easier synthesis of the former. In the in vivo situation, the effects of kisspeptin are not only influenced by dosing regimens, kisspeptin isoforms, and species, but also by nutritional status, age, stage of estrus cycle, and underlying hormonal milieu. It can therefore be difficult to compare findings between various kisspeptin studies, with occasional non-congruent results being observed. However, intriguing differences in the effects of kisspeptin administration between individual mammalian species may reveal important differences in reproductive physiology, which require further investigation.

Acknowledgments The section is funded by grants from the MRC, BBSRC, NIHR, an Integrative Mammalian Biology (IMB) Capacity Building Award, an FP7-HEALTH-2009-241592 EuroCHIP grant and is supported by the NIHR Imperial Biomedical Research Centre Funding Scheme. A.A. is sponsored by a Wellcome Trust Clinical Training Fellowship, R.R. is sponsored by an NIHR Academic Training Fellowship, C.N.J. is supported by an NIHR Clinical Lectureship and Wellcome/AMS Starter Grant for Clinical Lecturers. W.S.D. is supported by an NIHR Career Development Fellowship.

References

1. Matsui H, Takatsu Y, Kumano S, Matsumoto H, Ohtaki T (2004) Peripheral administration of metastin induces marked gonadotropin release and ovulation in the rat. Biochem Biophys Res Commun 320(2):383–388
2. Thompson EL, Amber V, Stamp GW, Patterson M, Curtis AE, Cooke JH, Appleby GF, Dhillo WS, Ghatei MA, Bloom SR, Murphy KG (2009) Kisspeptin-54 at high doses acutely induces testicular degeneration in adult male rats via central mechanisms. Br J Pharmacol 156(4):609–625
3. Gottsch ML, Cunningham MJ, Smith JT, Popa SM, Acohido BV, Crowley WF, Seminara S, Clifton DK, Steiner RA (2004) A role for kisspeptins in the regulation of gonadotropin secretion in the mouse. Endocrinology 145(9):4073–4077
4. Thompson EL, Patterson M, Murphy KG, Smith KL, Dhillo WS, Todd JF, Ghatei MA, Bloom SR (2004) Central and peripheral administration of kisspeptin-10 stimulates the hypothalamic-pituitary-gonadal axis. J Neuroendocrinol 16(10):850–858
5. Navarro VM, Castellano JM, Fernández-Fernández R, Tovar S, Roa J, Mayen A, Nogueiras R, Vazquez MJ, Barreiro ML, Magni P, Aguilar E, Dieguez C, Pinilla L, Tena-Sempere M (2005) Characterization of the potent luteinizing hormone-releasing activity of KiSS-1 peptide, the natural ligand of GPR54. Endocrinology 146(1):156–163
6. Lents CA, Heidorn NL, Barb CR, Ford JJ (2008) Central and peripheral administration of kisspeptin activates gonadotropin but not somatotropin secretion in prepubertal gilts. Reproduction 135(6):879–887
7. Jayasena CN, Nijher GM, Comninos AN, Abbara A, Januszewki A, Vaal ML, Sriskandarajah L, Murphy KG, Farzad Z, Ghatei MA, Bloom SR, Dhillo WS (2011) The effects of kisspeptin-10 on reproductive hormone release show sexual dimorphism in humans. J Clin Endocrinol Metab 96(12):E1963–E1972
8. Tovar S, Vázquez MJ, Navarro VM, Fernández-Fernández R, Castellano JM, Vigo E, Roa J, Casanueva FF, Aguilar E, Pinilla L, Dieguez C, Tena-Sempere M (2006) Effects of single or repeated intravenous administration of kisspeptin upon dynamic LH secretion in conscious male rats. Endocrinology 147(6):2696–2704
9. Shahab M, Mastronardi C, Seminara SB, Crowley WF, Ojeda SR, Plant TM (2005) Increased hypothalamic GPR54 signaling: a potential mechanism for initiation of puberty in primates. Proc Natl Acad Sci U S A 102(6):2129–2134
10. Hashizume T, Saito H, Sawada T, Yaegashi T, Ezzat AA, Sawai K, Yamashita T (2010) Characteristics of stimulation of gonadotropin secretion by kisspeptin-10 in female goats. Anim Reprod Sci 118(1):37–41
11. Ohkura S, Takase K, Matsuyama S, Mogi K, Ichimaru T, Wakabayashi Y, Uenoyama Y, Mori Y, Steiner RA, Tsukamura H, Maeda KI, Okamura H (2009) Gonadotropin-releasing hormone pulse generator activity in the hypothalamus of the goat. J Neuroendocrinol 21(10):813–821
12. Whitlock BK, Daniel JA, Wilborn RR, Rodning SP, Maxwell HS, Steele BP, Sartin JL (2008) Interaction of estrogen and progesterone on kisspeptin-10-stimulated luteinizing hormone and growth hormone in ovariectomized cows. Neuroendocrinology 88(3):212–215

13. Magee C, Foradori CD, Bruemmer JE, Arreguin-Arevalo JA, McCue PM, Handa RJ, Squires EL, Clay CM (2009) Biological and anatomical evidence for kisspeptin regulation of the hypothalamic-pituitary-gonadal axis of estrous horse mares. Endocrinology 150(6): 2813–2821
14. Messager S, Chatzidaki EE, Ma D, Hendrick AG, Zahn D, Dixon J, Thresher RR, Malinge I, Lomet D, Carlton MB, Colledge WH, Caraty A, Aparicio SA (2005) Kisspeptin directly stimulates gonadotropin-releasing hormone release via G protein-coupled receptor 54. Proc Natl Acad Sci U S A 102(5):1761–1766
15. Mason AO, Greives TJ, Scotti MA, Levine J, Frommeyer S, Ketterson ED, Demas GE, Kriegsfeld LJ (2007) Suppression of kisspeptin expression and gonadotropic axis sensitivity following exposure to inhibitory day lengths in female Siberian hamsters. Horm Behav 52(4):492–498
16. Plant TM, Ramaswamy S, Dipietro MJ (2006) Repetitive activation of hypothalamic G protein-coupled receptor 54 with intravenous pulses of kisspeptin in the juvenile monkey (Macaca mulatta) elicits a sustained train of gonadotropin-releasing hormone discharges. Endocrinology 147(2):1007–1013
17. Redmond JS, Macedo GG, Velez IC, Caraty A, Williams GL, Amstalden M (2011) Kisspeptin activates the hypothalamic-adenohypophyseal-gonadal axis in prepubertal ewe lambs. Reproduction 141(4):541–548
18. Keen KL, Wegner FH, Bloom SR, Ghatei MA, Terasawa E (2008) An increase in kisspeptin-54 release occurs with the pubertal increase in LHRH-1 release in the stalk-median eminence of female rhesus monkeys in vivo. Endocrinology 149(8):4151–4157
19. Castellano JM, Navarro VM, Fernández-Fernández R, Nogueiras R, Tovar S, Roa J, Vazquez MJ, Vigo E, Casanueva FF, Aguilar E, Pinilla L, Dieguez C, Tena-Sempere M (2005) Changes in hypothalamic KiSS-1 system and restoration of pubertal activation of the reproductive axis by kisspeptin in undernutrition. Endocrinology 146(9):3917–3925
20. Belchetz PE, Plant TM, Nakai Y, Keogh EJ, Knobil E (1978) Hypophyseal responses to continuous and intermittent delivery of hypothalamic gonadotropin-releasing hormone. Science 202(4368):631–633
21. Seminara SB, Dipietro MJ, Ramaswamy S, Crowley WF Jr, Plant TM (2006) Continuous human metastin 45-54 infusion desensitizes G protein-coupled receptor 54-induced gonadotropin-releasing hormone release monitored indirectly in the juvenile male Rhesus monkey (Macaca mulatta): a finding with therapeutic implications. Endocrinology 147(5):2122–2126
22. Ramzan F, Qureshi IZ (2011) Intraperitoneal kisspeptin-10 administration induces dose-dependent degenerative changes in maturing rat testes. Life Sci 88(5–6):246–256
23. Thompson EL, Murphy KG, Patterson M, Bewick GA, Stamp GW, Curtis AE, Cooke JH, Jethwa PH, Todd JF, Ghatei MA, Bloom SR (2006) Chronic subcutaneous administration of kisspeptin-54 causes testicular degeneration in adult male rats. Am J Physiol Endocrinol Metab 291(5):E1074–E1082
24. Roa J, Vigo E, García-Galiano D, Castellano JM, Navarro VM, Pineda R, Diéguez C, Aguilar E, Pinilla L, Tena-Sempere M (2008) Desensitization of gonadotropin responses to kisspeptin in the female rat: analyses of LH and FSH secretion at different developmental and metabolic states. Am J Physiol Endocrinol Metab 294(6):E1088–E1096
25. Caraty A, Smith JT, Lomet D, Ben Saïd S, Morrissey A, Cognie J, Doughton B, Baril G, Briant C, Clarke IJ (2007) Kisspeptin synchronizes preovulatory surges in cyclical ewes and causes ovulation in seasonally acyclic ewes. Endocrinology 148(11):5258–5267
26. Revel FG, Saboureau M, Masson-Pévet M, Pévet P, Mikkelsen JD, Simonneaux V (2006) Kisspeptin mediates the photoperiodic control of reproduction in hamsters. Curr Biol 16(17):1730–1735
27. Ansel L, Bentsen AH, Ancel C, Bolborea M, Klosen P, Mikkelsen JD, Simonneaux V (2011) Peripheral kisspeptin reverses short photoperiod-induced gonadal regression in Syrian hamsters by promoting GNRH release. Reproduction 142(3):417–425

28. Jayasena CN, Nijher GM, Chaudhri OB, Murphy KG, Ranger A, Lim A, Patel D, Mehta A, Todd C, Ramachandran R, Salem V, Stamp GW, Donaldson M, Ghatei MA, Bloom SR, Dhillo WS (2009) Subcutaneous injection of kisspeptin-54 acutely stimulates gonadotropin secretion in women with hypothalamic amenorrhea, but chronic administration causes tachyphylaxis. J Clin Endocrinol Metab 94(11):4315–4323
29. Smith JT, Li Q, Yap KS, Shahab M, Roseweir AK, Millar RP, Clarke IJ (2011) Kisspeptin is essential for the full preovulatory LH surge and stimulates GnRH release from the isolated ovine median eminence. Endocrinology 152(3):1001–1012
30. Roseweir AK, Kauffman AS, Smith JT, Guerriero KA, Morgan K, Pielecka-Fortuna J, Pineda R, Gottsch ML, Tena-Sempere M, Moenter SM, Terasawa E, Clarke IJ, Steiner RA, Millar RP (2009) Discovery of potent kisspeptin antagonists delineate physiological mechanisms of gonadotropin regulation. J Neurosci 29(12):3920–3929
31. Irwig MS, Fraley GS, Smith JT, Acohido BV, Popa SM, Cunningham MJ, Gottsch ML, Clifton DK, Steiner RA (2004) Kisspeptin activation of gonadotropin releasing hormone neurons and regulation of KiSS-1 mRNA in the male rat. Neuroendocrinology 80(4):264–272
32. Han SK, Gottsch ML, Lee KJ, Popa SM, Smith JT, Jakawich SK, Clifton DK, Steiner RA, Herbison AE (2005) Activation of gonadotropin-releasing hormone neurons by kisspeptin as a neuroendocrine switch for the onset of puberty. J Neurosci 25(49):11349–11356
33. Uenoyama Y, Inoue N, Pheng V, Homma T, Takase K, Yamada S, Ajiki K, Ichikawa M, Okamura H, Maeda KI, Tsukamura H (2011) Ultrastructural evidence of kisspeptin-gonadotropin-releasing hormone (GnRH) interaction in the median eminence of female rats: implication of axo-axonal regulation of GnRH release. J Neuroendocrinol 23(10): 863–870
34. Herde MK, Geist K, Campbell RE, Herbison AE (2011) Gonadotropin-releasing hormone neurons extend complex highly branched dendritic trees outside the blood-brain barrier. Endocrinology 152(10):3832–3841
35. Li XF, Kinsey-Jones JS, Cheng Y, Knox AM, Lin Y, Petrou NA, Roseweir A, Lightman SL, Milligan SR, Millar RP, O'Byrne KT (2009) Kisspeptin signalling in the hypothalamic arcuate nucleus regulates GnRH pulse generator frequency in the rat. PLoS One 4(12):e8334
36. Ramaswamy S, Guerriero KA, Gibbs RB, Plant TM (2008) Structural interactions between kisspeptin and GnRH neurons in the mediobasal hypothalamus of the male rhesus monkey (Macaca mulatta) as revealed by double immunofluorescence and confocal microscopy. Endocrinology 149(9):4387–4395
37. Smith JT, Shahab M, Pereira A, Pau KY, Clarke IJ (2010) Hypothalamic expression of KISS1 and gonadotropin inhibitory hormone genes during the menstrual cycle of a non-human primate. Biol Reprod 83(4):568–577
38. Guerriero KA, Keen KL, Millar RP, Terasawa E (2012) Developmental changes in GnRH release in response to kisspeptin agonist and antagonist in female rhesus monkeys (Macaca mulatta): implication for the mechanism of puberty. Endocrinology 153(2):825–836
39. Luque RM, Córdoba-Chacón J, Gahete MD, Navarro VM, Tena-Sempere M, Kineman RD, Castaño JP (2011) Kisspeptin regulates gonadotroph and somatotroph function in nonhuman primate pituitary via common and distinct signaling mechanisms. Endocrinology 152(3): 957–966
40. Gutiérrez-Pascual E, Martínez-Fuentes AJ, Pinilla L, Tena-Sempere M, Malagón MM, Castaño JP (2007) Direct pituitary effects of kisspeptin: activation of gonadotrophs and somatotrophs and stimulation of luteinising hormone and growth hormone secretion. J Neuroendocrinol 19(7):521–530
41. Smith JT, Clay CM, Caraty A, Clarke IJ (2007) KiSS-1 messenger ribonucleic acid expression in the hypothalamus of the ewe is regulated by sex steroids and season. Endocrinology 148(3):1150–1157
42. Suzuki S, Kadokawa H, Hashizume T (2008) Direct kisspeptin-10 stimulation on luteinizing hormone secretion from bovine and porcine anterior pituitary cells. Anim Reprod Sci 103(3–4):360–365

43. Ezzat AA, Saito H, Sawada T, Yaegashi T, Goto Y, Nakajima Y, Jin J, Yamashita T, Sawai K, Hashizume T (2010) The role of sexual steroid hormones in the direct stimulation by kisspeptin-10 of the secretion of luteinizing hormone, follicle-stimulating hormone and prolactin from bovine anterior pituitary cells. Anim Reprod Sci 121(3–4):267–272
44. Oakley AE, Clifton DK, Steiner RA (2009) Kisspeptin signaling in the brain. Endocr Rev 30(6):713–743
45. Dhillo WS, Chaudhri OB, Patterson M, Thompson EL, Murphy KG, Badman MK, McGowan BM, Amber V, Patel S, Ghatei MA, Bloom SR (2005) Kisspeptin-54 stimulates the hypothalamic-pituitary gonadal axis in human males. J Clin Endocrinol Metab 90(12): 6609–6615
46. Kotani M, Detheux M, Vandenbogaerde A, Communi D, Vanderwinden JM, Le Poul E, Brézillon S, Tyldesley R, Suarez-Huerta N, Vandeput F, Blanpain C, Schiffmann SN, Vassart G, Parmentier M (2001) The metastasis suppressor gene KiSS-1 encodes kisspeptins, the natural ligands of the orphan G protein-coupled receptor GPR54. J Biol Chem 276(37):34631–34636
47. Pheng V, Uenoyama Y, Homma T, Inamoto Y, Takase K, Yoshizawa-Kumagaye K, Isaka S, Watanabe TX, Ohkura S, Tomikawa J, Maeda K, Tsukamura H (2009) Potencies of centrally- or peripherally-injected full-length kisspeptin or its C-terminal decapeptide on LH release in intact male rats. J Reprod Dev 55(4):378–382
48. Mikkelsen JD, Bentsen AH, Ansel L, Simonneaux V, Juul A (2009) Comparison of the effects of peripherally administered kisspeptins. Regul Pept 152(1–3):95–100
49. Curtis AE, Cooke JH, Baxter JE, Parkinson JR, Bataveljic A, Ghatei MA, Bloom SR, Murphy KG (2010) A kisspeptin-10 analog with greater in vivo bioactivity than kisspeptin-10. Am J Physiol Endocrinol Metab 298(2):E296–E303
50. Whitlock BK, Daniel JA, Wilborn RR, Maxwell HS, Steele BP, Sartin JL (2010) Interaction of kisspeptin and the somatotropic axis. Neuroendocrinology 92(3):178–188
51. Ezzat Ahmed A, Saito H, Sawada T, Yaegashi T, Yamashita T, Hirata T, Sawai K, Hashizume T (2009) Characteristics of the stimulatory effect of kisspeptin-10 on the secretion of luteinizing hormone, follicle-stimulating hormone and growth hormone in prepubertal male and female cattle. J Reprod Dev 55(6):650–654
52. Wahab F, Aziz F, Irfan S, Zaman WU, Shahab M (2008) Short-term fasting attenuates the response of the HPG axis to kisspeptin challenge in the adult male rhesus monkey (Macaca mulatta). Life Sci 83(19–20):633–637
53. Wahab F, Ullah F, Chan YM, Seminara SB, Shahab M (2011) Decrease in hypothalamic Kiss1 and Kiss1r expression: a potential mechanism for fasting-induced suppression of the HPG axis in the adult male rhesus monkey (Macaca mulatta). Horm Metab Res 43(2):81–85
54. Navarro VM, Fernández-Fernández R, Castellano JM, Roa J, Mayen A, Barreiro ML, Gaytan F, Aguilar E, Pinilla L, Dieguez C, Tena-Sempere M (2004) Advanced vaginal opening and precocious activation of the reproductive axis by KiSS-1 peptide, the endogenous ligand of GPR54. J Physiol 561(pt 2):379–386
55. De Roux N, Genin E, Carel JC et al (2003) Hypogonadotropic hypogonadism due to loss of function of the KiSS1-derived peptide receptor GPR54. Proc Natl Acad Sci U S A 100: 10972–10976
56. Seminara SB, Messager S, Chatzidaki EE et al (2003) The GPR54 gene as a regulator of puberty. N Engl J Med 349:1614–1627
57. Teles MG, Bianco SD, Brito VN, Trarbach EB, Kuohung W, Xu S, Seminara SB, Mendonca BB, Kaiser UB, Latronico AC (2008) A GPR54-activating mutation in a patient with central precocious puberty. N Engl J Med 358:709–715
58. Pita J, Barrios V, Gavela-Pérez T, Martos-Moreno G, Muñoz-Calvo MT, Pozo J, Rovira A, Argente J, Soriano-Guillén L (2011) Circulating kisspeptin levels exhibit sexual dimorphism in adults, are increased in obese prepubertal girls and do not suffer modifications in girls with idiopathic central precocious puberty. Peptides 32(9):1781–1786
59. Terasawa E, Kurian JR, Keen KL, Shiel NA, Colman RJ, Capuano SV (2012) Body weight impact on puberty: effects of high-calorie diet on puberty onset in female rhesus monkeys. Endocrinology 153(4):1696–1705

60. Castellano JM, Navarro VM, Fernández-Fernández R, Castaño JP, Malagón MM, Aguilar E, Dieguez C, Magni P, Pinilla L, Tena-Sempere M (2006) Ontogeny and mechanisms of action for the stimulatory effect of kisspeptin on gonadotropin-releasing hormone system of the rat. Mol Cell Endocrinol 257–258:75–83
61. Greives TJ, Long KL, Burns CM, Demas GE (2011) Response to exogenous kisspeptin varies according to sex and reproductive condition in Siberian hamsters (Phodopus sungorus). Gen Comp Endocrinol 170(1):172–179
62. Roa J, Vigo E, Castellano JM, Navarro VM, Fernández-Fernández R, Casanueva FF, Dieguez C, Aguilar E, Pinilla L, Tena-Sempere M (2006) Hypothalamic expression of KiSS-1 system and gonadotropin-releasing effects of kisspeptin in different reproductive states of the female Rat. Endocrinology 147(6):2864–2878
63. Knobil E (1980) The neuroendocrine control of the menstrual cycle. Recent Prog Horm Res 36:53–88
64. Smith JT, Dungan HM, Stoll EA, Gottsch ML, Braun RE, Eacker SM, Clifton DK, Steiner RA (2005) Differential regulation of KiSS-1 mRNA expression by sex steroids in the brain of the male mouse. Endocrinology 146(7):2976–2984
65. d'Anglemont de Tassigny X, Fagg LA, Carlton MB, Colledge WH (2008) Kisspeptin can stimulate gonadotropin-releasing hormone (GnRH) release by a direct action at GnRH nerve terminals. Endocrinology 149(8):3926–3932
66. Zhang C, Bosch MA, Ronnekleiv OK, Kelly MJ (2009) Gamma-aminobutyric acid B receptor mediated inhibition of gonadotropin-releasing hormone neurons is suppressed by kisspeptin-G protein-coupled receptor 54 signaling. Endocrinology 150:2388–2394
67. Sébert ME, Lomet D, Saïd SB, Monget P, Briant C, Scaramuzzi RJ, Caraty A (2010) Insights into the mechanism by which kisspeptin stimulates a preovulatory LH surge and ovulation in seasonally acyclic ewes: potential role of estradiol. Domest Anim Endocrinol 38(4):289–298

Chapter 5
Effects of Kisspeptin on Hormone Secretion in Humans

Yee-Ming Chan

Abstract Studies of the actions of kisspeptin in human subjects have examined the effects of different kisspeptin isoforms, doses, and routes of administration on LH secretion, a surrogate measure of GnRH release. These studies, in addition to detailing how these different variables affect LH secretion in response to kisspeptin, have produced new insights into kisspeptin physiology: (1) Brief exposure to kisspeptin results in sustained GnRH release lasting ~17 min in men. (2) Women in different phases of the menstrual cycle have differences in their response to kisspeptin, suggesting that endogenous kisspeptin secretion and GnRH neuronal responsiveness vary in response to the changing sex-steroid environment across the menstrual cycle. (3) Kisspeptin resets the GnRH pulse generator in men, but does not appear to do so in women. (4) Continuous exposure to kisspeptin results in desensitization to kisspeptin, and thus kisspeptin has the potential to either stimulate or suppress reproductive endocrine activity depending on the mode of administration. These findings pave the way for future studies using kisspeptin as a physiologic, diagnostic, and therapeutic tool in both healthy adults and in patients with reproductive disorders.

Introduction

The role of kisspeptin in reproduction was discovered through genetic studies in both humans and mice [1–3]. Since that initial discovery, studies predominantly in animal models have revealed that the hypothalamic neurons that secrete kisspeptin are

Y.-M. Chan, M.D., Ph.D. (✉)
Division of Endocrinology, Department of Medicine, Boston Children's Hospital, Boston, MA, USA

Reproductive Endocrine Unit and Harvard Reproductive Sciences Center, Department of Medicine, Massachusetts General Hospital, Boston, MA, USA
e-mail: ymchan@partners.org

A.S. Kauffman and J.T. Smith (eds.), *Kisspeptin Signaling in Reproductive Biology*, Advances in Experimental Medicine and Biology 784, DOI 10.1007/978-1-4614-6199-9_5,

essential for stimulating normal GnRH secretion, and that these kisspeptin neurons integrate multiple inputs (developmental changes, sex steroids, metabolic status, stress, photoperiod, etc.) and relay these inputs to GnRH neurons. Though there have been only a relatively small number of studies of the effects of kisspeptin in humans, these studies have added to work in animal models to enhance our knowledge of kisspeptin physiology.

Studies in the human can be challenging to execute and are constrained by safety, ethical, and regulatory issues, but these difficulties are offset by distinct advantages of the human model. One obvious advantage is that only human studies can provide definitive information about the human physiology of kisspeptin, as well as the safety of kisspeptin in humans. Another advantage is that human subjects can cooperate with complex dosing regimens, such as repetitive injections and prolonged infusions, and can also self-administer medications. A third advantage is the ability to perform frequent blood sampling for extended periods of time, allowing detailed characterization of endogenous reproductive endocrine activity at baseline, which serves as an important reference point to which responses to exogenous kisspeptin can be compared.

Pharmacokinetics of Kisspeptin in Humans

Human studies involving kisspeptin administration have used different isoforms of kisspeptin (kisspeptin-54 and kisspeptin-10), different routes of administration (intravenous and subcutaneous), and different modes of administration (single boluses, multiple boluses, and infusions of varying duration). These multiple variables complicate comparisons between studies. Fortunately, it appears that differences in the effects of these various kisspeptin isoforms and routes and modes of administration can largely be explained by differences in pharmacokinetics—that is, how quickly kisspeptin accumulates after administration and how quickly it subsequently degrades. Kisspeptin-10 and kisspeptin-54, when applied directly to cultured cells in vitro, have similar activity and potency [4–6]. In contrast, in vivo experiments in rodents have demonstrated that kisspeptin-54 has a slightly longer onset and duration of action than kisspeptin-10 when given at the same molar concentration and by the same route [7, 8]. Because the two isoforms have essentially identical activity in vitro, these in vivo differences are likely to be due to differences in kisspeptin pharmacokinetics.

Thus, understanding the pharmacokinetics of the various kisspeptin isoforms and routes of administration in humans is vital for proper interpretation of the results of studies of kisspeptin administration in humans. However, the seemingly straightforward task of determining the pharmacokinetics of kisspeptin has proved to be challenging, due both to limitations of existing assays for measurement of kisspeptin and to gaps in our knowledge of the metabolism and degradation of kisspeptin. Thus, while several studies have examined the pharmacokinetics of kisspeptin (summarized in Table 5.1), our understanding remains incomplete. Nevertheless, these studies provide an essential backdrop for the interpretation of studies of kisspeptin administration to humans.

Table 5.1 Pharmacokinetic studies of kisspeptin in humans

Isoform	Route of administration	Assay	Results	References
Kisspeptin-10	Intravenous bolus (men, women)	RIA, sheep polyclonal	IR peaked at first time point (15 min), undetectable by 30 min	[14]
	Intravenous infusion (men, women)	RIA, sheep polyclonal	IR peaked 30 min after infusion started; IR half-life 4 min after infusion stopped	[14]
	Subcutaneous bolus (women)	RIA, sheep polyclonal	IR peaked at first time point (15 min), then gradually declined to baseline over 45–120 min, half-life ~20 min	[14]
	In vitro (plasma)	Mass spectrometry	Half-life 55 s	[11]
Kisspeptin-54	Intravenous infusion (men)	RIA, sheep polyclonal	IR peaked 30 min after infusion started; IR half-life 28 min after infusion stopped	[13]
	Intravenous bolus (women)	RIA, sheep polyclonal	Gradual increase in IR, peaking 40 min after bolus and still measurable by the end of the 3-h study	[14]
	Subcutaneous bolus (women)	RIA, sheep polyclonal	IR peaked at first time point (15 min after bolus); at low doses IR disappeared by 3.5 h, at high doses IR essentially undiminished by the end of the 4-h study	[15]

IR immunoreactivity; *RIA* radioimmunoassay

Challenges in Determining the Pharmacokinetics of Kisspeptin

A key component of a pharmacokinetic study is an accurate assay for measuring the concentration of a drug. Ideally, the assay can distinguish the drug and its active metabolites from inactive breakdown products. Two methods have been used to measure kisspeptin: immunoassays and mass spectrometry.

Immunoassays use antibodies that recognize specific epitopes on the target molecule. Assays based on monoclonal antibodies recognize a single epitope; if this epitope is destroyed through proteolysis or altered through some other form of modification, the resulting metabolite cannot be detected by the assay, regardless of whether the metabolite is biologically active or inactive. Assays based on polyclonal antibodies recognize multiple epitopes and are therefore less susceptible to having molecules rendered "invisible" by the loss of a single epitope. However, assays based on polyclonal antibodies also do not distinguish active from inactive metabolites and may actually be more likely to detect biologically inactive metabolites simply by virtue of being able to detect more breakdown products. Furthermore, if a molecule is cleaved into two fragments, each of these fragments may be detected by the assay, resulting in an apparent and paradoxical increase in the concentration of the molecule with ongoing degradation.

Assays based on mass spectrometry can theoretically overcome some of these problems by identifying and quantifying the intact molecule and all of its metabolites. However, interpretation of mass spectrometic results requires knowledge of the degradation pathways for the molecule and the bioactivity of the various breakdown products. For kisspeptin, these pathways have yet to be fully characterized.

Further complicating the measurement of kisspeptin in human blood samples is the fact that kisspeptin undergoes ongoing degradation even after samples are collected. Ramachandran et al. observed that kisspeptin concentrations, measured with a radioimmunoassay based on a polyclonal antibody raised against kisspeptin-54, declined gradually over time in human plasma and very rapidly in serum stored at room temperature after sample collection [9]. Similarly, kisspeptin-10 added to human plasma degrades with a half-life of 55 s in vitro [10, 11]. The principal breakdown product of kisspeptin-10 is a 9-amino acid peptide that results from N-terminal truncation of the kisspeptin decapeptide [10]. While the biological activity of this 9-amino acid metabolite has not been tested formally, it is likely to retain some biological activity because peptides consisting of just eight or even five of the C-terminal amino acids of kisspeptin are capable of stimulating the kisspeptin receptor [5, 12].

Pharmacokinetics of Kisspeptin-10

The pharmacokinetics of kisspeptin-10 have been examined using a radioimmunoassay based on a polyclonal sheep antibody that recognizes both kisspeptin-54 and kisspeptin-10 and has minimal cross-reactivity with other peptides of the RF-amide family [13, 14]. After a single intravenous bolus of kisspeptin-10 in men and women, kisspeptin immunoreactivity peaked at the next measured time point (10 min after injection). By 20 min, kisspeptin immunoreactivity was detectable only with high doses of kisspeptin (3 nmol/kg or higher) [14].

The half-life of kisspeptin-10 was determined more precisely after the end of an infusion of kisspeptin-10. Kisspeptin-10 immunoreactivity decayed with a half-life of 4 min in men, women in the early to mid-follicular phase, and women in the late follicular phase in the days before ovulation [14]. As noted above, a liquid chromatography/mass spectrometry assay that specifically detects kisspeptin-10 has been used to demonstrate that kisspeptin-10 has an in vitro half-life of 55 s in human plasma at 37°C [10, 11]. The most plausible explanation to reconcile these two results is that the radioimmunoassay used in the in vivo studies also detects breakdown products (such as the 9-amino acid metabolite that has been observed), resulting in a longer apparent half-life. Without complete knowledge of the identity and biological activity of these kisspeptin breakdown products, it is difficult to know which one of these results more accurately reflects the half-life of the bioactivity of kisspeptin. Alternatively, there may be factors that prolong the half-life of kisspeptin in vivo, such as binding of kisspeptin to binding proteins or distribution to compartments other than blood. In any case, these studies demonstrate that the half-life of kisspeptin-10 is very short, on the order of 1–4 min.

The kinetics of the subcutaneous route of administration of kisspeptin-10 have also been examined using the same polyclonal sheep immunoassay [14]. After a single subcutaneous bolus of kisspeptin-10, kisspeptin immunoreactivity peaked at 15 min (the first time point after administration) then decayed with a half-life of ~20 min.

Pharmacokinetics of Kisspeptin-54

The half-life of kisspeptin-54 in men has also been examined using the sheep polyclonal radioimmunoassay [13]. Men received an intravenous infusion of kisspeptin-54. At the start of the infusion, kisspeptin immunoreactivity rose quickly and peaked at 30 min after the start of the infusion. After the infusion was stopped, kisspeptin immunoreactivity decayed in a monophasic pattern with a half-life of 28 min.

The same radioimmunoassay was used to examine the pharmacokinetics of a single intravenous bolus of kisspeptin-54 in women [14]. As expected, kisspeptin immunoreactivity was elevated over baseline at the next time point, 10 min after IV bolus injection. Oddly, kisspeptin immunoreactivity continued to rise gradually and did not peak until 40 min after injection. Kisspeptin immunoreactivity then declined gradually, but still remained detectable 3 h after injection. Because most substances are distributed throughout the circulation within 1–2 min after IV bolus administration, at face value this observation suggests complex kinetics of kisspeptin-54. One possibility is that kisspeptin-54 molecules formed aggregates that slowly dissociated after intravenous administration. However, a more likely explanation is that kisspeptin breakdown products interfered with the kisspeptin immunoassay that was used in this study.

This immunoassay has also been used to study the kinetics of the subcutaneous route of administration of kisspeptin-54 in women, with doses ranging from 0.2 to 6.4 nmol/kg [15]. At all doses studied, kisspeptin immunoreactivity peaked at the next time point (15 min) then declined at rates that varied by dose. At the smallest dose (0.2 nmol/kg), kisspeptin immunoreactivity returned to baseline by 3.5 h after injection. In contrast, after subcutaneous injection of kisspeptin-54 6.4 nmol/kg, kisspeptin immunoreactivity was minimally decreased at the end of the 4-h study period. Thus, kisspeptin-54 behaves as a sustained-release formulation when administered subcutaneously at high doses.

In summary, though some details remain to be worked out regarding the pharmacokinetics of the various isoforms of kisspeptin, it is clear that kisspeptin-10 decays very rapidly, kisspeptin-54 less so. Furthermore, as with most drugs, subcutaneous dosing results in more sustained elevation of kisspeptin immunoreactivity than intravenous administration. These pharmacokinetic studies are an important backdrop to the interpretation of studies of human kisspeptin administration. In particular, the finding that subcutaneous administration of kisspeptin-54 at high doses results in sustained elevation of kisspeptin immunoreactivity is important for the interpretation of the studies involving multiple subcutaneous doses of this isoform (described below in section "Effects of Chronic Kisspeptin Administration in Women with Hypothalamic Amenorrhea").

Stimulation of Reproductive Endocrine Activity by Acute Administration of Kisspeptin

Shortly after the finding that mutations in *KISS1R*/*Kiss1r* cause hypogonadotropic hypogonadism in humans and mice, several studies demonstrated that kisspeptin potently and directly stimulates GnRH secretion and, in turn, gonadotropin secretion [16–19]. A number of studies have now shown that kisspeptin similarly stimulates gonadotropin secretion in humans and have further extended these findings to reveal previously unrecognized aspects of kisspeptin physiology. The results of studies of human kisspeptin administration are summarized in Table 5.2.

These human studies all share certain limitations. One is that it is nearly impossible to measure hypothalamic GnRH secretion directly in humans. All studies have therefore used LH as an indicator of GnRH secretion, as LH pulses are a well-validated surrogate measure of GnRH secretion under physiologic conditions [20, 21]. However, LH becomes less reliable under non-physiologic conditions. For example, when exogenous pulsatile GnRH is given at a high pulse frequency (every 30 min or faster), the one-to-one concordance between GnRH pulses and LH pulses begins to break down [22]. In these situations, other markers such as the free alpha subunit more accurately reflect GnRH secretion [23], but to date measurements of the free alpha subunit have not been included in human studies of kisspeptin. Human kisspeptin studies have also measured FSH and testosterone in men or estradiol in women, and these have generally changed in parallel with LH; this chapter will focus on the LH response to kisspeptin, as this most closely correlates with GnRH secretion.

To date, most human studies have been performed in reproductively normal adults. These healthy subjects have normal endogenous secretion of kisspeptin, GnRH, and LH, and this endogenous reproductive endocrine activity complicates the interpretation of studies involving exogenous kisspeptin administration. In particular, the unpredictable timing of endogenous pulses of GnRH and LH secretion results in a constantly moving baseline that can confound the interpretation of LH measurements after kisspeptin administration. This is illustrated by the observation in several studies (e.g., [14, 15, 24]) that mean LH often decreases after subjects

Table 5.2 LH responses to kisspeptin administration

Subjects	Isoform	Dose and route	Results	References
Healthy men	Kisspeptin-10	0.24 nmol/kg IV ×1	Immediate LH pulse in all men, estimated 17 min of GnRH secretion induced, resetting of the GnRH pulse generator	[11]
	Kisspeptin-10	0.0077–2.3 nmol/kg IV ×1	LH increased in a largely dose-dependent fashion, but a smaller LH response was seen at the highest dose of kisspeptin (2.3 nmol/kg)	[24]
	Kisspeptin-10	0.3–10 nmol/kg IV ×1	LH increased, with all doses producing a similar size response	[14]
	Kisspeptin-54	0.125–40 pmol/kg/min IV ×30 min, then half rate ×60 min (cumulative dose 0.0075–2.4 nmol/kg)	LH increased in a dose-dependent fashion	[13]
	Kisspeptin-10	1.1 nmol/kg/h IV infusion ×8 h	LH increased across the course of the study and became erratic towards the end of the infusion	[24]
	Kisspeptin-10	3.1 nmol/kg/h IV infusion ×22.5 h	LH increased across the study and was erratic through most of the infusion	[24]
Healthy women	Kisspeptin-10	0.24 nmol/kg IV ×1	Kisspeptin strongly induced LH secretion in preovulatory and mid-luteal women, but responses in early follicular women were smaller and less consistent	[31]
	Kisspeptin-10	1–10 nmol/kg IV ×1	Kisspeptin strongly induced LH secretion in preovulatory women, but little effect was seen in early to mid-follicular women	[14]
	Kisspeptin-10	2–32 nmol/kg SC ×1	No clear responses to kisspeptin were seen in early to mid-follicular women	[14]

(continued)

Table 5.2 (continued)

Subjects	Isoform	Dose and route	Results	References
	Kisspeptin-10	20–720 pmol/kg/min IV ×30 min, then half rate ×60 min (cumulative dose 1.2–43.2 nmol/kg)	No clear responses to kisspeptin were seen in early to mid-follicular women	[14]
	Kisspeptin-54	1 nmol/kg IV ×1	LH rose gradually across the 3-h study in early to mid-follicular women	[14]
	Kisspeptin-54	0.2–6.4 nmol/kg SC ×1	LH rose across the study in a biphasic pattern, with small responses in early to mid-follicular women, larger responses in mid-luteal women, and very large responses in preovulatory women	[15]
Women with hypothalamic amenorrhea	Kisspeptin-54	6.4 nmol/kg SC twice daily ×2 week	LH rose after initial exposure to kisspeptin, but this response decreased over days	[43, 63]
	Kisspeptin-54	6.4 nmol/kg SC twice weekly ×8 week	LH rose after initial exposure to kisspeptin and continued to respond to kisspeptin (but decreased partially) across the 8-week study period; no folliculogenesis or ovulation was observed	[63]
Men and women with mutations in the neurokinin B pathway	Kisspeptin-10	1.1 nmol/kg/h ×8 h	LH pulse amplitude increased	[45]

IV intravenously; *SC* subcutaneously

receive a control injection of vehicle without kisspeptin. These decreases in LH could lead to the nonsensical conclusion that the pituitary gland is extracting LH from the circulation, but in reality they reflect the ongoing degradation of circulating LH between LH pulses. Alternatively, if an endogenous LH pulse occurs around the same time as kisspeptin administration, it may be incorrectly attributed to kisspeptin. These issues may have resulted in some of the discrepancies seen within and between studies. At the same time, however, studies in reproductively normal subjects provide an important advantage—the ability to study interactions between kisspeptin and the endogenous reproductive endocrine machinery.

Dose-Responsive Stimulation of Gonadotropin Secretion in Healthy Men

The first use of kisspeptin in human studies was reported by Dhillo et al. [13]. In their study, healthy male volunteers received 90-min infusions of kisspeptin-54. Infusion rates in the first 30 min ranged from 0.125 to 40 pmol/kg/min; the rate was then halved for the remaining 60 min of the infusion (for cumulative doses ranging from 0.0075 to 2.4 nmol/kg). These short infusions of kisspeptin-54 resulted in dose-dependent increases in plasma LH, with the dose–response curve reaching a plateau at ~8 pmol/kg/min (cumulative dose 0.48 nmol/kg).

The decapeptide isoform of kisspeptin similarly stimulates gonadotropin secretion in healthy men in a dose-dependent fashion. George et al. [24] administered single intravenous boluses of kisspeptin-10 with doses ranging from 0.0077 to 2.3 nmol/kg. The plateau of the dose–response curve was reached at a dose of 0.23 nmol/kg. Intriguingly, the highest dose tested in this study (2.3 nmol/kg) resulted in a smaller increase in LH than the 0.23 or 0.77 nmol/kg doses. This result raises the possibility that high doses of kisspeptin may induce desensitization of its receptor, a phenomenon that has been observed in vitro [25], ex vivo [26], and in vivo [27–29]. However, these studies and other studies in humans (see section "Effects of Chronic Kisspeptin Administration in Women with Hypothalamic Amenorrhea") suggest that desensitization occurs more slowly, requiring hours to days of exposure. Furthermore, Jayasena et al. [14] found in very similar protocols that single intravenous boluses of kisspeptin-10 at doses of 0.3, 1, 3, and 10 nmol/kg IV ×1 all produced a similar rise in LH, without a decrease in the LH response at higher doses of the decapeptide. These discrepant results may have been due to variations in baseline LH discussed above; thus, further study will be needed to determine whether short-term desensitization of the kisspeptin receptor occurs in humans.

A rough estimate of endogenous kisspeptin secretion can be made by comparing the size of LH pulses induced by exogenous kisspeptin to the size of endogenous LH pulses. This was done in a study that incorporated an extended period of blood sampling to measure endogenous reproductive endocrine activity prior to kisspeptin administration (Fig. 5.1) [11]. The amplitude and area-under-the-curve (AUC) of LH pulses induced by kisspeptin-10 at 0.24 nmol/kg IV ×1 were on average larger

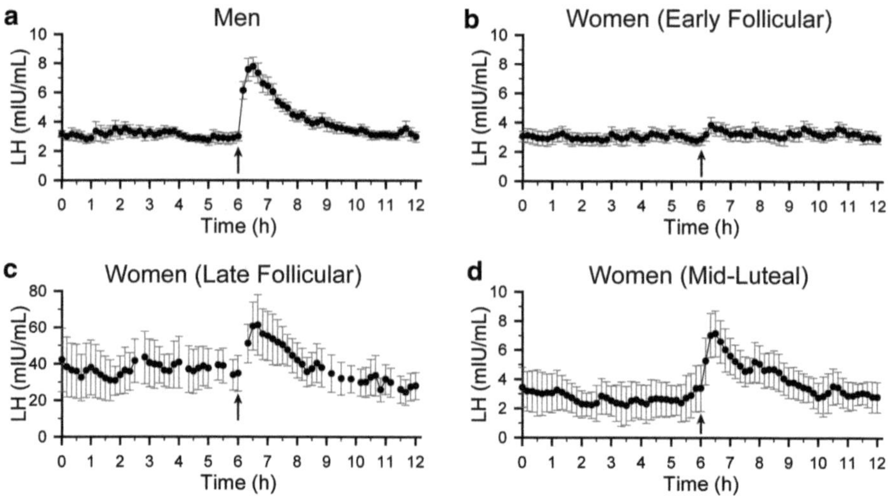

Fig. 5.1 Effects of acute exposure to kisspeptin-10 in healthy men and women. A single intravenous bolus of kisspeptin-10 0.24 nmol/kg was given to (**a**) healthy men ($n = 13$) and healthy women in the (**b**) early follicular ($n = 10$), (**c**) late follicular/preovulatory ($n = 3$; note different *Y*-axis scale), and (**d**) mid-luteal ($n = 14$) phases of the menstrual cycle, with LH measured every 10 min 6 h before and 6 h after kisspeptin administration. Each point indicates the mean LH (±SEM) across all subjects at each time point. *Arrows* indicate the time of kisspeptin administration. Adapted with permission from refs. [11, 31]

than those of endogenous pulses, but there was overlap between the ranges. Thus, the kisspeptin-10 dose of 0.24 nmol/kg produced physiologic to slightly supraphysiologic LH responses. Given the data from dose–response studies described above, which showed that a dose of 0.24 nmol/kg is near the plateau of the dose–response curve, this result suggests that, in men, endogenous kisspeptin "drive" lies only slightly below the threshold required for maximal stimulation of GnRH secretion.

Changes in Kisspeptin Responsiveness Across the Menstrual Cycle in Healthy Women

Studying the effects of kisspeptin in women adds a layer of complexity: the changes in neuroendocrine activity and sex-steroid production across the menstrual cycle [30]. Even more so than in men, various isoforms of kisspeptin and modes of administration have been used in women, resulting in different durations of exposure to kisspeptin. These studies have revealed marked variation in the responses to kisspeptin across the menstrual cycle.

The effects of very brief exposure to kisspeptin have been studied in women in two studies that delivered kisspeptin-10 as single intravenous boluses (Fig. 5.1) [14, 31], which (as noted above) produce a rapid and brief rise in kisspeptin immunoreactivity in the circulation. In both of these studies, kisspeptin-10 induced robust LH secretion in women in the late follicular phase of the menstrual cycle (Fig. 5.1), in the days before ovulation when pituitary sensitivity to GnRH is greatly enhanced due to positive estradiol feedback [30]. One study also showed that women in the mid-luteal phase exhibit robust LH responses to kisspeptin (Fig. 5.1), though not as large as in preovulatory women [31].

In contrast, markedly attenuated responses to kisspeptin were seen in the early to mid-follicular phase (Fig. 5.1), the time in the menstrual cycle when circulating estradiol concentrations are lowest [30]. In one study, no significant increases in LH were seen in women in the early to mid-follicular phase after administration of kisspeptin-10 as single intravenous boluses (1–10 nmol/kg), or even as single subcutaneous boluses (2–32 nmol/kg) or 90-min intravenous infusions (cumulative doses 1.2–43.2 nmol/kg) [14]. In the other study, LH increases could be detected in the early follicular phase, but the responses were both small and inconsistent, with immediate LH pulses seen in only half of subjects [31].

The effects of slightly more sustained exposure to kisspeptin were examined by delivering single intravenous boluses of the 54-amino acid isoform of kisspeptin to women in the early to mid-follicular phase [14]. As described in section "Pharmacokinetics of Kisspeptin-10," this results in sustained elevation of kisspeptin immunoreactivity, and the gonadotropin response to kisspeptin-54 was similarly sustained: LH rose gradually across the study and did not reach a plateau by the end of the 3-h study period.

The effects in women of even more prolonged exposure to kisspeptin have been examined using the subcutaneous route of administration for kisspeptin-54 [15]. Because subcutaneous administration of kisspeptin-54 behaves as a sustained release formulation of kisspeptin, these studies explored the gonadotropin response to prolonged kisspeptin exposure. In women in the early to mid-follicular phase, single subcutaneous doses of kisspeptin-54, ranging from 0.2 to 6.4 nmol/kg, resulted in dose-dependent increases in LH and FSH [15]. The 0.4 nmol/kg dose was further studied in women in different phases of the menstrual cycle: early to mid-follicular, late follicular/preovulatory, and mid-luteal. Mirroring the results with kisspeptin-10, the largest gonadotropin responses were seen in the preovulatory period (LH increase ~20 IU/L peaking 60 min after injection). In the early follicular phase, the gonadotropin responses to kisspeptin-54 were much smaller and were also relatively delayed (LH increase ~1 IU/L peaking at 150 min after injection).

Women in the mid-luteal phase exhibited a biphasic response to subcutaneous kisspeptin-54 0.4 nmol/kg, with an initial peak 30–45 min after injection (~2 IU/L above baseline) and a second, larger peak 180–210 min after injection (~4 IU/L above baseline). An identical biphasic response had been observed with extended infusions of GnRH [32] and administration of long-acting GnRH analogs [33]. Thus, the biphasic LH response to sustained exposure to kisspeptin can be attributed to the behavior of the pituitary gonadotropes in response to sustained activation of the GnRH receptor.

Comparing kisspeptin-induced pulses to endogenous pulses, the size (AUC) of pulses induced by kisspeptin-10 0.24 nmol/kg IV was greater than that of endogenous pulses in preovulatory and mid-luteal women, similar to what had been observed in men [31]. However, in the early follicular phase, kisspeptin-induced pulses were comparable in size to endogenous pulses [31]. These results suggest that the dose of kisspeptin 0.24 nmol/kg was slightly supraphysiologic in preovulatory and mid-luteal women, as it was in men, but may have been more physiologic in women in the early follicular phase.

The differences in the size of the LH response to exogenous kisspeptin in different phases of the menstrual cycle mirrors those seen in response to exogenous GnRH, with modest responses in the early follicular phase, larger responses in the luteal phase, and very large responses in the late follicular/preovulatory phase [34]. Whether changes in pituitary sensitivity to GnRH can fully account for the differences in the size of the LH response to kisspeptin across the cycle, or whether GnRH neuronal responsiveness to kisspeptin also changes across the cycle, remains to be determined. The latter is likely to be true, as estradiol has been shown to enhance the GnRH response to kisspeptin in various models [35–37].

In the early follicular phase, kisspeptin-10 elicited only small and inconsistent responses [14] and kisspeptin-54 appeared to elicit only a late response and not an acute response [15]. These results raise the possibility that GnRH neurons are relatively resistant to kisspeptin in this phase. However, even large doses of kisspeptin-10 failed to elicit a response in early-follicular women [14]. This suggests an alternative possibility: that endogenous kisspeptin secretion already provides near-maximal stimulation of the GnRH neurons in the early follicular phase, such that they have only a limited capacity to mount an acute response to additional exogenous kisspeptin. If so, the early follicular phase would be unique in providing greater kisspeptin "tone" than other phases of the menstrual cycle. Because the early follicular phase is characterized by a relatively low concentration of sex steroids [30], this would be consistent with observations in rhesus monkeys that kisspeptin secretion is negatively regulated by estradiol [38].

In summary, while studies in women have used a wide array of kisspeptin isoforms and doses and routes of administration of kisspeptin, the results of these studies appear to reflect differences in the duration of kisspeptin exposure resulting from these various protocols, with sustained exposure to kisspeptin inducing similarly sustained release of GnRH. Moreover, the response of exogenous GnRH neurons to kisspeptin varies across the menstrual cycle, being large in preovulatory women, intermediate in luteal-phase women, and markedly attenuated in follicular-phase women. It remains to be determined whether differences in the sex-steroid milieu or other factors are responsible for these differences in kisspeptin responsiveness.

Stimulatory Effects of Kisspeptin in Women with Hypothalamic Amenorrhea

In addition to healthy adults, patients with reproductive disorders have been studied in kisspeptin-administration protocols. One disorder that has been explored extensively is functional hypothalamic amenorrhea (HA) in women. HA results from the suppression of reproductive endocrine activity by stress, excessive exercise, and/or negative energy balance [39]. In rodent models, starvation and stress also suppress reproductive endocrine activity, and this is associated with a decrease in kisspeptin expression in the arcuate nucleus of the hypothalamus [40–42]. In stressed/starved rodents, kisspeptin can still elicit LH secretion [40, 41], demonstrating that pathways downstream of the kisspeptin receptor remain intact under these conditions. These results indicate that, at least in some situations, stress inhibits reproductive endocrine activity by suppressing kisspeptin release.

It is therefore logical to postulate that restoring kisspeptin would rescue the reproductive deficits induced by stress/starvation. To test this possibility, Jayasena et al. administered kisspeptin to women with HA [43]. The researchers found that single subcutaneous boluses of kisspeptin-54 at a dose of 6.4 nmol/kg caused significant elevations in LH, FSH, and estradiol. This important proof of principle demonstrated that pathways downstream from kisspeptin remain intact in women with HA. This group's efforts to restore reproductive endocrine activity to women with HA by chronic treatment with kisspeptin are described below in section "Effects of Chronic Kisspeptin Administration in Women with Hypothalamic Amenorrhea."

Effects of Kisspeptin in Patients with GnRH Deficiency

Individuals who are unable to secrete or respond to GnRH have a condition variously termed idiopathic/congenital hypogonadotropic hypogonadism or idiopathic/congenital GnRH deficiency [44]. These individuals fail to go through puberty and are infertile. The past 15 years have seen important advances in our understanding of the genetics of isolated GnRH deficiency [44], and two basic pathophysiological mechanisms for this condition have emerged. "Neurodevelopmental" defects result from a problem in the development and/or migration of the GnRH neurons, such that GnRH neurons are absent or present in reduced number in the central nervous system. In contrast, individuals with a "neuroendocrine" defect have a normal complement of GnRH neurons, but these neurons fail to receive the signals to induce GnRH secretion, they are unable to produce functional GnRH, or the pituitary gland is unable to respond to GnRH [44]. Though these two pathophysiological mechanisms are largely distinct, they can produce virtually identical phenotypes in patients. Because kisspeptin directly stimulates GnRH secretion, it can be used as the first available in vivo probe of GnRH neuronal integrity, just as GnRH has been used for decades to probe pituitary gonadotrope function [34].

To date, one study has examined the effects of kisspeptin-10 in patients with GnRH deficiency [45]. This study administered 8-h infusions of kisspeptin to patients with mutations in *TAC3* or *TACR3*, which encode neurokinin B and its receptor, respectively [45]. These genes fall in the "neuroendocrine" class of genes, as neurokinin B signaling is thought to directly or indirectly modulate the secretion of GnRH neurons [46–48]. When given a control infusion with saline, these patients exhibited low-amplitude but detectable LH pulses. When given an infusion of kisspeptin, the amplitude of LH pulses increased [45]. There was also an apparent increase in the number of detectable pulses during the 8-h infusion period, but it is unclear whether this observation was due to an increase in pulse frequency of the GnRH pulse generator or secondary to the increase in pulse amplitude, such that previously undetectable pulses could now be detected.

These results demonstrated that GnRH neurons are present in patients with mutations affecting neurokinin B signaling and that kisspeptin does not require neurokinin B signaling for its ability to enhance GnRH secretion, consistent with the model that neurokinin B acts through kisspeptin to stimulate GnRH release [49]. However, neurokinin B may also have effects independent of kisspeptin—though kisspeptin enhanced LH pulse amplitude in patients with *TAC3* and *TACR3* mutations, this enhancement was not as pronounced as what had been observed in healthy adults.

This study scratches the surface of potential uses of kisspeptin in patients with isolated GnRH deficiency. Kisspeptin will undoubtedly be used in future studies to further probe the pathophysiology of isolated GnRH deficiency, to establish an earlier diagnosis, and possibly to predict later outcomes such as success at achieving fertility. Kisspeptin also has the potential for treating patients with isolated GnRH deficiency who retain responsiveness to kisspeptin. The recent identification of patients with homozygous deleterious mutations in *KISS1*, the gene that encodes kisspeptin [50], may permit detailed studies of the physiological effects of kisspeptin in a "clean" model of kisspeptin deficiency, much as patients with GnRH deficiency have offered a valuable opportunity to characterize the pituitary response to GnRH [51].

An Extended Duration of GnRH Secretion Induced by Kisspeptin

The above studies have demonstrated the general principle that kisspeptin induces LH secretion, a surrogate marker of GnRH secretion, in humans. Further insight into the effects of kisspeptin on GnRH secretion came from examining the shape of the LH pulses induced by kisspeptin [11]. Unlike endogenous LH pulses, which have a triangular, sawtooth appearance with a rapid rise and subsequent peaking of LH, the LH pulses induced by exogenous kisspeptin-10 were more rounded and slower to reach their peaks [11]. A similar rounded, prolonged morphology of LH pulses was seen in a prior study in which exogenous GnRH infusions were delivered for varying lengths of time (from an IV push to 1-, 5-, and 30-min infusions) [52]. These studies were performed in men with GnRH deficiency to eliminate any confounding effects of endogenous GnRH secretion. As the length of the GnRH infusion increased, the resulting LH pulses appeared more rounded and had a longer time from the nadir to the peak of the pulse (Fig. 5.2). This time-to-peak was directly proportional to the duration of the GnRH infusion (Fig. 5.2). Placing the time-to-peak of LH pulses induced by kisspeptin-10 in the context of this prior study suggested that the morphology of the kisspeptin-induced LH pulses could be mimicked by a 17-min infusion of GnRH (Fig. 5.2).

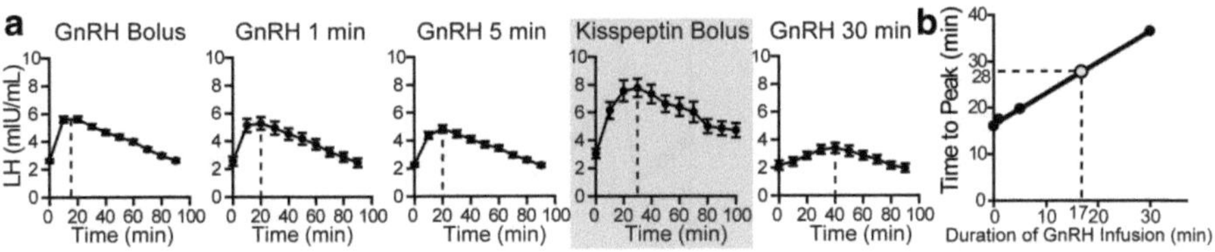

Fig. 5.2 Sustained GnRH secretion induced by kisspeptin-10 in men. (**a**) Men with GnRH deficiency received GnRH as an intravenous bolus or as 1-, 5-, or 30-min intravenous infusions, resulting in increasingly prolonged LH pulses, with a longer time from the nadir to the peak of the pulse. Healthy men received kisspeptin-10 as an intravenous bolus (*shaded graph*). (**b**) The relationship between the length of the GnRH infusion and the time from the nadir to the peak of the resulting LH pulse is linear. When plotted on this graph, the average time from nadir to peak of kisspeptin-induced LH pulses (28 min) corresponds to a 17-min GnRH infusion. Adapted with permission from ref. [11]

Thus, a single IV dose of kisspeptin-10 appears to induce prolonged GnRH secretion. One potential explanation for this is that exogenous kisspeptin-10 could persist for an extended time in the circulation. However, as noted in section "Pharmacokinetics of Kisspeptin-10," the half-life of kisspeptin-10 is very short, making this possibility unlikely. Another possibility is that brief exposure of hypothalamic GnRH neurons to kisspeptin induces prolonged GnRH release. Indeed, in several studies of cultured GnRH neurons in which kisspeptin was applied briefly then washed away, the GnRH neurons exhibited prolonged electrochemical activation and calcium mobilization [53–57]. In one study, the average duration of this activation was 16 min [56], remarkably concordant with the estimate of 17 min of GnRH secretion induced by kisspeptin-10 in humans. Thus, the morphology of the LH pulses induced by kisspeptin in vivo revealed an important in vivo human correlate of a phenomenon observed in ex vivo mouse studies.

Effects of Kisspeptin on GnRH Pulse Generation

The mechanism by which GnRH neurons produce periodic pulses of GnRH secretion remains one of the key mysteries of reproductive neuroendocrinology [58]. How these pulses are generated and how the timing of pulses is determined are largely unknown. Indeed, whether the "pulse generator" physically resides within the GnRH neuronal network or in some other anatomic location has been a topic of lively discussion [59]. One difficulty in studying GnRH pulse generation has been a relative dearth of tools to study this phenomenon. While progesterone, acting through endogenous opioids, is known to cause a chronic slowing of the frequency of GnRH pulses [60], there had not been a factor known to acutely perturb the pattern of GnRH pulse generation. Human studies of kisspeptin have now revealed that kisspeptin may be such a factor.

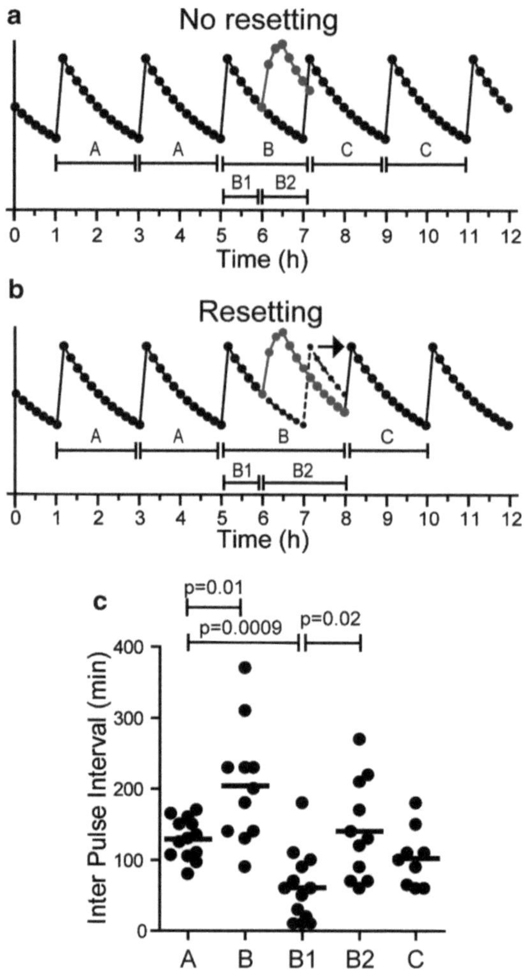

Fig. 5.3 Resetting of the GnRH pulse generator by kisspeptin-10 in men. (**a**) Predicted pulse pattern if kisspeptin were to have no effect on the timing of endogenous pulses. Note that actual pulse profiles show more variability in pulse intervals. (**b**) Predicted pulse pattern if kisspeptin resets the pulse generator, with a delay in the appearance of the next endogenous pulse. (**c**) Observed pulse intervals. Bars indicate means. Reprinted with permission from ref. [11]

Resetting of the GnRH Pulse Generator by Single-Bolus Administration of Kisspeptin

The effects of kisspeptin-10 on GnRH pulse generation were determined in men by examining the timing of endogenous LH pulses before and after kisspeptin administration [11]. Two contrasting models were tested. In one model, the "no resetting model," exogenous kisspeptin has no effect on the underlying GnRH pulse generator (Fig. 5.3). The kisspeptin-induced GnRH pulse would therefore be an extra pulse superimposed on top of the endogenous pulse pattern. One prediction of this model is that the time interval between endogenous pulses should be unaffected by kisspeptin administration. In particular, the interval from the endogenous pulse just before the kisspeptin-induced pulse to the endogenous pulse immediately following the kisspeptin-induced pulse (interval B in Fig. 5.3) should be no different from the interval between other endogenous pulses (interval A in Fig. 5.3). However, in men, this interval was found to be significantly longer than the endogenous pulse interval (Fig. 5.3).

A second prediction of the "no resetting" model stems from the fact that kisspeptin was given without knowledge of the timing of endogenous pulses. If kisspeptin has no effect on the timing of endogenous pulses, then kisspeptin administration could occur at any point within the interval between endogenous pulses, and would be equally likely to occur early in that interval as it would be to occur late in that interval. Thus, given a sufficient number of observations, the average time of kisspeptin administration should fall in the middle of the interval between endogenous pulses, dividing this interval evenly into halves. However, the average time of kisspeptin administration fell closer to the preceding endogenous pulse than to the subsequent endogenous pulse (Fig. 5.3). Thus, neither prediction of the "no resetting model" was upheld by empirical data in men.

An alternative model, the "resetting" model, is that kisspeptin resets the GnRH pulse generator (Fig. 5.3). GnRH pulse generation has been shown to be a renewal process in men, that is, the timing of a GnRH pulse depends on the timing of the previous pulse but not of pulses prior to that [61]. In other words, the process of pulse generation begins anew with each pulse, with each pulse setting a new "time zero" that is used to determine the timing of the next pulse. In the "resetting" model, the kisspeptin-induced pulse replaces the previous endogenous pulse as this time zero. Thus, the next endogenous pulse would be predicted to follow the kisspeptin-induced pulse by an interval that matches the endogenous interpulse interval. This is precisely what was observed (Fig. 5.3). Thus, in men, kisspeptin appeared to have reset the GnRH pulse generator.

A potential alternative explanation for these observations is that kisspeptin may have produced a refractory period during which the reproductive endocrine machinery is unable to produce an LH pulse. This would result in some pulses being "skipped" and thus create an apparent delay in the appearance of the next endogenous pulse. However, in some individuals an endogenous pulse occurred shortly after kisspeptin administration, arguing against a lengthy refractory period [11]. Furthermore, it would be an unlikely coincidence that the duration of this refractory period would result in a delay precisely long enough to match the prediction of the "resetting" model. Resetting therefore remains the most parsimonious explanation for the results observed in men. The phenomenon of resetting appears to be sexually dimorphic, as it was not observed in women [31]. Previous analyses of pulse patterns in healthy adults had also suggested differences in how men and women generate pulses of GnRH secretion [61, 62]. Both the physiologic basis and the teleological explanation for these differences remain obscure, and future studies involving manipulation of the sex-steroid milieu and other factors may elucidate their roles in establishing these differences.

Given our minimal understanding of how GnRH pulses are generated, the mechanisms by which resetting could occur are unclear. Kisspeptin could have a direct effect on the GnRH pulse generator, or it could act indirectly through GnRH, LH, FSH, or other downstream factors. Further exploration of the effects of kisspeptin on GnRH pulsatility may allow investigators to identify the cellular and molecular machinery that generate GnRH pulses.

Erratic LH Secretion Generated by Sustained Infusions of Kisspeptin

Alterations in the pattern of GnRH pulse generation have also been observed in studies of sustained infusions of kisspeptin [24]. In one study, healthy men received 22.5-h infusions of kisspeptin-10 at a rate of 3.1 nmol/kg/h. LH increased across the first several hours of the infusion then remained elevated for the remainder of the infusion. There were several notable features of the pattern of LH secretion during the kisspeptin infusion. One is that desensitization was not observed. In rats, continuous intracerebroventricular administration of kisspeptin has been shown to cause desensitization to kisspeptin [29]. Similarly, continuous intravenous administration of kisspeptin-10 to juvenile and adult male rhesus monkeys at rates of ~25–30 nmol/kg/h causes desensitization, with LH starting to decline within 4 h of the start of the infusion [27, 28]. As detailed in the section below, chronic administration of kisspeptin-54 to women caused desensitization to kisspeptin over several days [43]. Thus, the phenomenon of desensitization has been observed in several mammalian species, including humans. The lack of desensitization in healthy men who received kisspeptin-10 at 3.1 nmol/kg/h ×22.5 h may be due to the lower rate and/or shorter duration of the infusion compared to these other studies.

A second notable feature of the pattern of LH secretion in response to the kisspeptin infusion was that it was erratic, with occasional discrete pulses but mostly apparently chaotic variation. This erratic pattern of LH secretion resembles that observed in studies in which repetitive boluses of GnRH were given to men with GnRH deficiency at high frequencies [22]. While discrete LH pulses were visible when GnRH was given every 1 h, when the frequency of GnRH administration was increased to every 30 min or every 15 min the LH pattern became more chaotic, with peaks and valleys of LH that did not correlate with pulses of exogenous GnRH [22]. Thus, while it is tempting to interpret the presence of LH pulses during a continuous infusion of kisspeptin as demonstrating that pulsatile secretion of GnRH can occur in the absence of pulsatile kisspeptin, because the one-to-one concordance between LH pulses and GnRH pulses is lost under conditions of high GnRH pulse frequency and continuous GnRH infusions, further investigation is required to address this issue conclusively.

Effects of Chronic Kisspeptin Administration in Women with Hypothalamic Amenorrhea

To date, the effects of chronic (>24 h) administration of kisspeptin have been explored only in women with HA. As described above, Jayasena et al. found that acute exposure to kisspeptin caused elevations of FSH, LH, and estradiol in women with HA [43]. Encouraged by this finding, the same group proceeded to study the effects of chronic kisspeptin administration in women with HA by administering kisspeptin-54 6.4 nmol/kg subcutaneously twice daily [43]. Though this dose of kisspeptin caused a rise in LH on the first day, after 2 weeks of treatment kisspeptin

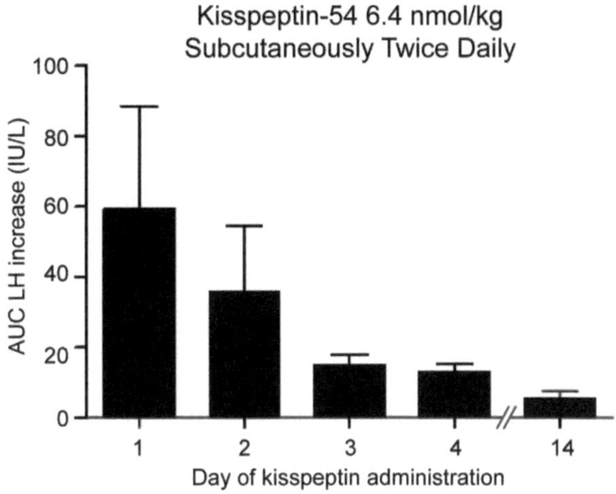

Fig. 5.4 Desensitization induced by chronic administration of kisspeptin to women with hypothalamic amenorrhea. Women with hypothalamic amenorrhea ($n = 10$) exhibited desensitization to kisspeptin when given kisspeptin-54 6.4 nmol/kg subcutaneously twice daily. Error bars indicate SEM. Adapted from Jayasena et al. [63], with permission from Nature Publishing Group

no longer elicited a detectable LH response. The response to exogenously administered GnRH was intact, suggesting that continuous exposure to kisspeptin caused desensitization at the level of the kisspeptin receptor [43]. A subsequent study detailed the time-course of this phenomenon and demonstrated that responses to kisspeptin decreased rapidly during the first few days of treatment (Fig. 5.4) [63].

Because subcutaneous administration of kisspeptin-54 6.4 nmol/kg results in sustained elevation of kisspeptin immunoreactivity lasting at least 4 h [15], it is likely that twice-daily dosing resulted in continuous exposure to kisspeptin. As noted previously, continuous exposure to kisspeptin results in desensitization of the kisspeptin receptor in a variety of experimental models [25–29], and a similar phenomenon appears to have occurred in this study. However, the time course of desensitization in humans was distinct, occurring over days instead of hours [63].

To avoid inducing desensitization, the same group employed twice weekly (instead of twice daily) subcutaneous administration of kisspeptin-54 6.4 nmol/kg [63]. After 8 weeks on this regimen, responses to kisspeptin were still seen, though they were slightly dampened compared to the responses on Day 1. Despite the fact that the gonadotropin responses to kisspeptin remained intact during the 8-week study, no folliculogenesis was observed on ultrasound, and ovulation was not achieved [63]. Because kisspeptin secretion is itself pulsatile [64, 65], rescue of reproductive endocrine activity in women with HA may require pulsatile delivery of kisspeptin, much as exogenous GnRH must be delivered in a pulsatile fashion to stimulate the reproductive endocrine axis [66].

Safety of Kisspeptin in Humans

Collectively, published reports of kisspeptin administration to men and women have reported no adverse events, no subjective complaints, no changes in blood pressure or other vital signs, and no changes in blood cell counts or tests of kidney or liver function [11, 13–15, 24, 31, 43, 63, 67].

Conclusions

Studies of the effects of kisspeptin in human subjects have not only validated findings made in animal models, they have also contributed new insights into the intricate physiology of kisspeptin. Some of these key insights include the findings that brief exposure to kisspeptin results in sustained GnRH release in vivo, that kisspeptin resets the GnRH pulse generator in men, that kisspeptin secretion and GnRH responsiveness to kisspeptin vary across the menstrual cycle, and that desensitization occurs after days of exposure to kisspeptin. The pulsatile pattern of endogenous kisspeptin secretion [64, 65] suggests that modes of kisspeptin administration that result in brief pulses of kisspeptin in the blood are more likely to reflect the normal physiological effects of kisspeptin. Future studies using short-acting boluses of kisspeptin may elucidate the precise effects of sex steroids on the response to exogenous kisspeptin and may thereby result in a better understanding of how changes in both kisspeptin secretory tone and GnRH neuronal responsiveness to kisspeptin contribute to the physiological variation in neuroendocrine activity across the menstrual cycle and across the life cycle. In contrast, protocols that result in prolonged kisspeptin exposure open the door for pharmacologic manipulation of the kisspeptin signaling pathway, and it will be important to determine the precise conditions under which continuous kisspeptin administration can bring about desensitization of the kisspeptin receptor in humans. Importantly, kisspeptin is just starting to be used to probe the pathophysiology of reproductive disorders such as isolated GnRH deficiency, and the excellent safety record for kisspeptin to date opens the door for future studies using kisspeptin as a diagnostic and therapeutic tool for the care of patients with these disorders.

References

1. de Roux N, Genin E, Carel JC, Matsuda F, Chaussain JL, Milgrom E (2003) Hypogonadotropic hypogonadism due to loss of function of the KiSS1-derived peptide receptor GPR54. Proc Natl Acad Sci USA 100:10972–10976
2. Seminara SB, Messager S, Chatzidaki EE, Thresher RR, Acierno JS Jr, Shagoury JK, Bo-Abbas Y, Kuohung W, Schwinof KM, Hendrick AG, Zahn D, Dixon J, Kaiser UB, Slaugenhaupt SA, Gusella JF, O'Rahilly S, Carlton MB, Crowley WF Jr, Aparicio SA, Colledge WH (2003) The GPR54 gene as a regulator of puberty. N Engl J Med 349:1614–1627
3. Funes S, Hedrick JA, Vassileva G, Markowitz L, Abbondanzo S, Golovko A, Yang S, Monsma FJ, Gustafson EL (2003) The KiSS-1 receptor GPR54 is essential for the development of the murine reproductive system. Biochem Biophys Res Commun 312:1357–1363
4. Ohtaki T, Shintani Y, Honda S, Matsumoto H, Hori A, Kanehashi K, Terao Y, Kumano S, Takatsu Y, Masuda Y, Ishibashi Y, Watanabe T, Asada M, Yamada T, Suenaga M, Kitada C, Usuki S, Kurokawa T, Onda H, Nishimura O, Fujino M (2001) Metastasis suppressor gene KiSS-1 encodes peptide ligand of a G-protein-coupled receptor. Nature 411:613–617
5. Muir AI, Chamberlain L, Elshourbagy NA, Michalovich D, Moore DJ, Calamari A, Szekeres PG, Sarau HM, Chambers JK, Murdock P, Steplewski K, Shabon U, Miller JE, Middleton SE, Darker JG, Larminie CG, Wilson S, Bergsma DJ, Emson P, Faull R, Philpott KL, Harrison DC

(2001) AXOR12, a novel human G protein-coupled receptor, activated by the peptide KiSS-1. J Biol Chem 276:28969–28975
6. Kotani M, Detheux M, Vandenbogaerde A, Communi D, Vanderwinden JM, Le Poul E, Brezillon S, Tyldesley R, Suarez-Huerta N, Vandeput F, Blanpain C, Schiffmann SN, Vassart G, Parmentier M (2001) The metastasis suppressor gene KiSS-1 encodes kisspeptins, the natural ligands of the orphan G protein-coupled receptor GPR54. J Biol Chem 276:34631–34636
7. Tovar S, Vazquez MJ, Navarro VM, Fernandez-Fernandez R, Castellano JM, Vigo E, Roa J, Casanueva FF, Aguilar E, Pinilla L, Dieguez C, Tena-Sempere M (2006) Effects of single or repeated intravenous administration of kisspeptin upon dynamic LH secretion in conscious male rats. Endocrinology 147:2696–2704
8. Mikkelsen JD, Bentsen AH, Ansel L, Simonneaux V, Juul A (2009) Comparison of the effects of peripherally administered kisspeptins. Regul Pept 152:95–100
9. Ramachandran R, Patterson M, Murphy KG, Dhillo WS, Patel S, Kazarian A, Ghatei MA, Bloom SR (2008) Preanalytical factors affecting RIA measurement of plasma kisspeptin. Clin Chem 54:615–617
10. Ren C, Liu Z, Jones W, Chen P, Seminara S, Smith N, Covey J, Chan K (2007) An LC-MS/MS method for the quantitation of metastin 45-54 (NSC 741805) and its preclinical pharmacokinetics in rats. AAPS J 9:www.aapsj.org/abstracts/AM_2007/AAPS2007-002199.PDF
11. Chan YM, Butler JP, Pinnell NE, Pralong FP, Crowley WF Jr, Ren C, Chan KK, Seminara SB (2011) Kisspeptin resets the hypothalamic GnRH clock in men. J Clin Endocrinol Metab 96:E908–E915
12. Tomita K, Niida A, Oishi S, Ohno H, Cluzeau J, Navenot JM, Wang ZX, Peiper SC, Fujii N (2006) Structure-activity relationship study on small peptidic GPR54 agonists. Bioorg Med Chem 14:7595–7603
13. Dhillo WS, Chaudhri OB, Patterson M, Thompson EL, Murphy KG, Badman MK, McGowan BM, Amber V, Patel S, Ghatei MA, Bloom SR (2005) Kisspeptin-54 stimulates the hypothalamic-pituitary gonadal axis in human males. J Clin Endocrinol Metab 90:6609–6615
14. Jayasena CN, Nijher GM, Comninos AN, Abbara A, Januszewki A, Vaal ML, Sriskandarajah L, Murphy KG, Farzad Z, Ghatei MA, Bloom SR, Dhillo WS (2011) The effects of kisspeptin-10 on reproductive hormone release show sexual dimorphism in humans. J Clin Endocrinol Metab 96:E1963–E1972
15. Dhillo WS, Chaudhri OB, Thompson EL, Murphy KG, Patterson M, Ramachandran R, Nijher GK, Amber V, Kokkinos A, Donaldson M, Ghatei MA, Bloom SR (2007) Kisspeptin-54 stimulates gonadotropin release most potently during the preovulatory phase of the menstrual cycle in women. J Clin Endocrinol Metab 92:3958–3966
16. Gottsch ML, Cunningham MJ, Smith JT, Popa SM, Acohido BV, Crowley WF, Seminara S, Clifton DK, Steiner RA (2004) A role for kisspeptins in the regulation of gonadotropin secretion in the mouse. Endocrinology 145:4073–4077
17. Navarro VM, Castellano JM, Fernandez-Fernandez R, Barreiro ML, Roa J, Sanchez-Criado JE, Aguilar E, Dieguez C, Pinilla L, Tena-Sempere M (2004) Developmental and hormonally regulated messenger ribonucleic acid expression of KiSS-1 and its putative receptor, GPR54, in rat hypothalamus and potent luteinizing hormone-releasing activity of KiSS-1 peptide. Endocrinology 145:4565–4574
18. Messager S, Chatzidaki EE, Ma D, Hendrick AG, Zahn D, Dixon J, Thresher RR, Malinge I, Lomet D, Carlton MB, Colledge WH, Caraty A, Aparicio SA (2005) Kisspeptin directly stimulates gonadotropin-releasing hormone release via G protein-coupled receptor 54. Proc Natl Acad Sci USA 102:1761–1766
19. Shahab M, Mastronardi C, Seminara SB, Crowley WF, Ojeda SR, Plant TM (2005) Increased hypothalamic GPR54 signaling: a potential mechanism for initiation of puberty in primates. Proc Natl Acad Sci USA 102:2129–2134
20. Levine JE, Pau KY, Ramirez VD, Jackson GL (1982) Simultaneous measurement of luteinizing hormone-releasing hormone and luteinizing hormone release in unanesthetized, ovariectomized sheep. Endocrinology 111:1449–1455

21. Clarke IJ, Cummins JT (1982) The temporal relationship between gonadotropin releasing hormone (GnRH) and luteinizing hormone (LH) secretion in ovariectomized ewes. Endocrinology 111:1737–1739
22. Spratt DI, Finkelstein JS, Butler JP, Badger TM, Crowley WF Jr (1987) Effects of increasing the frequency of low doses of gonadotropin-releasing hormone (GnRH) on gonadotropin secretion in GnRH-deficient men. J Clin Endocrinol Metab 64:1179–1186
23. Hayes FJ, McNicholl DJ, Schoenfeld D, Marsh EE, Hall JE (1999) Free alpha-subunit is superior to luteinizing hormone as a marker of gonadotropin-releasing hormone despite desensitization at fast pulse frequencies. J Clin Endocrinol Metab 84:1028–1036
24. George JT, Veldhuis JD, Roseweir AK, Newton CL, Faccenda E, Millar RP, Anderson RA (2011) Kisspeptin-10 is a potent stimulator of LH and increases pulse frequency in men. J Clin Endocrinol Metab 96:E1228–E1236
25. Teles MG, Bianco SD, Brito VN, Trarbach EB, Kuohung W, Xu S, Seminara SB, Mendonca BB, Kaiser UB, Latronico AC (2008) A GPR54-activating mutation in a patient with central precocious puberty. N Engl J Med 358:709–715
26. d'Anglemont de Tassigny X, Fagg LA, Carlton MB, Colledge WH (2008) Kisspeptin can stimulate gonadotropin-releasing hormone (GnRH) release by a direct action at GnRH nerve terminals. Endocrinology 149:3926–3932
27. Seminara SB, Dipietro MJ, Ramaswamy S, Crowley WF Jr, Plant TM (2006) Continuous human metastin 45-54 infusion desensitizes G protein-coupled receptor 54-induced gonadotropin-releasing hormone release monitored indirectly in the juvenile male rhesus monkey (Macaca mulatta): a finding with therapeutic implications. Endocrinology 147:2122–2126
28. Ramaswamy S, Seminara SB, Pohl CR, DiPietro MJ, Crowley WF Jr, Plant TM (2007) Effect of continuous intravenous administration of human metastin 45-54 on the neuroendocrine activity of the hypothalamic-pituitary-testicular axis in the adult male rhesus monkey (Macaca mulatta). Endocrinology 148:3364–3370
29. Roa J, Vigo E, Garcia-Galiano D, Castellano JM, Navarro VM, Pineda R, Dieguez C, Aguilar E, Pinilla L, Tena-Sempere M (2008) Desensitization of gonadotropin responses to kisspeptin in the female rat: analyses of LH and FSH secretion at different developmental and metabolic states. Am J Physiol Endocrinol Metab 294:E1088–E1096
30. Hall JE (2009) Neuroendocrine control of the menstrual cycle. In: Strauss JF III, Barbieri RL (eds) Yen and Jaffe's reproductive endocrinology, 6th edn. Elsevier, Saunders, pp 139–154
31. Chan YM, Butler JP, Sidhoum VF, Pinnell NE, Seminara SB (2012) Kisspeptin administration to women: a window into endogenous kisspeptin secretion and GnRH responsiveness across the menstrual cycle. J Clin Endocrinol Metab 97(8):E1458–E1467
32. Bremner WJ, Paulsen CA (1974) Two pools of luteinizing hormone in the human pituitary: evidence from constant administration of luteinizing hormone-releasing hormone. J Clin Endocrinol Metab 39:811–815
33. Casper RF, Sheehan KL, Yen SS (1980) Gonadotropin-estradiol responses to a superactive luteinizing hormone-releasing hormone agonist in women. J Clin Endocrinol Metab 50:179–181
34. Yen SS (1975) Gonadotropin-releasing hormone. Annu Rev Med 26:403–417
35. Novaira HJ, Ng Y, Wolfe A, Radovick S (2009) Kisspeptin increases GnRH mRNA expression and secretion in GnRH secreting neuronal cell lines. Mol Cell Endocrinol 311:126–134
36. Tonsfeldt KJ, Goodall CP, Latham KL, Chappell PE (2011) Oestrogen induces rhythmic expression of the Kisspeptin-1 receptor GPR54 in hypothalamic gonadotrophin-releasing hormone-secreting GT1-7 cells. J Neuroendocrinol 23:823–830
37. Guerriero KA, Keen KL, Millar RP, Terasawa E (2012) Developmental changes in GnRH release in response to kisspeptin agonist and antagonist in female rhesus monkeys (Macaca mulatta): implication for the mechanism of puberty. Endocrinology 153:825–836
38. Guerriero KA, Keen KL, Terasawa E (2012) Developmental increase in kisspeptin-54 release in vivo is independent of the pubertal increase in estradiol in female rhesus monkeys (Macaca mulatta). Endocrinology 153:1887–1897

39. Gordon CM (2010) Clinical practice. Functional hypothalamic amenorrhea. N Engl J Med 363:365–371
40. Castellano JM, Navarro VM, Fernandez-Fernandez R, Nogueiras R, Tovar S, Roa J, Vazquez MJ, Vigo E, Casanueva FF, Aguilar E, Pinilla L, Dieguez C, Tena-Sempere M (2005) Changes in hypothalamic KiSS-1 system and restoration of pubertal activation of the reproductive axis by kisspeptin in undernutrition. Endocrinology 146:3917–3925
41. Iwasa T, Matsuzaki T, Murakami M, Shimizu F, Kuwahara A, Yasui T, Irahara M (2008) Decreased expression of kisspeptin mediates acute immune/inflammatory stress-induced suppression of gonadotropin secretion in female rat. J Endocrinol Invest 31:656–659
42. Kinsey-Jones JS, Li XF, Knox AM, Wilkinson ES, Zhu XL, Chaudhary AA, Milligan SR, Lightman SL, O'Byrne KT (2009) Down-regulation of hypothalamic kisspeptin and its receptor, Kiss1r, mRNA expression is associated with stress-induced suppression of luteinising hormone secretion in the female rat. J Neuroendocrinol 21:20–29
43. Jayasena CN, Nijher GM, Chaudhri OB, Murphy KG, Ranger A, Lim A, Patel D, Mehta A, Todd C, Ramachandran R, Salem V, Stamp GW, Donaldson M, Ghatei MA, Bloom SR, Dhillo WS (2009) Subcutaneous injection of kisspeptin-54 acutely stimulates gonadotropin secretion in women with hypothalamic amenorrhea, but chronic administration causes tachyphylaxis. J Clin Endocrinol Metab 94:4315–4323
44. Balasubramanian R, Crowley WF Jr (2011) Isolated GnRH deficiency: a disease model serving as a unique prism into the systems biology of the GnRH neuronal network. Mol Cell Endocrinol 346:4–12
45. Young J, George JT, Tello JA, Francou B, Bouligand J, Guiochon-Mantel A, Brailly-Tabard S, Anderson RA, Millar RP (2012) Kisspeptin restores pulsatile LH secretion in patients with neurokinin B signaling deficiencies: physiological, pathophysiological and therapeutic Implications. Neuroendocrinology 2012 Feb 24 [To appear]
46. Topaloglu AK, Reimann F, Guclu M, Yalin AS, Kotan LD, Porter KM, Serin A, Mungan NO, Cook JR, Ozbek MN, Imamoglu S, Akalin NS, Yuksel B, O'Rahilly S, Semple RK (2009) TAC3 and TACR3 mutations in familial hypogonadotropic hypogonadism reveal a key role for Neurokinin B in the central control of reproduction. Nat Genet 41:354–358
47. Gianetti E, Tusset C, Noel SD, Au MG, Dwyer AA, Hughes VA, Abreu AP, Carroll J, Trarbach E, Silveira LF, Costa EM, de Mendonca BB, de Castro M, Lofrano A, Hall JE, Bolu E, Ozata M, Quinton R, Amory JK, Stewart SE, Arlt W, Cole TR, Crowley WF, Kaiser UB, Latronico AC, Seminara SB (2010) TAC3/TACR3 mutations reveal preferential activation of gonadotropin-releasing hormone release by neurokinin B in neonatal life followed by reversal in adulthood. J Clin Endocrinol Metab 95:2857–2867
48. Young J, Bouligand J, Francou B, Raffin-Sanson ML, Gaillez S, Jeanpierre M, Grynberg M, Kamenicky P, Chanson P, Brailly-Tabard S, Guiochon-Mantel A (2010) TAC3 and TACR3 defects cause hypothalamic congenital hypogonadotropic hypogonadism in humans. J Clin Endocrinol Metab 95:2287–2295
49. Ramaswamy S, Seminara SB, Plant TM (2011) Evidence from the agonadal juvenile male rhesus monkey (Macaca mulatta) for the view that the action of neurokinin B to trigger gonadotropin-releasing hormone release is upstream from the kisspeptin receptor. Neuroendocrinology 94:237–245
50. Topaloglu AK, Tello JA, Kotan LD, Ozbek MN, Yilmaz MB, Erdogan S, Gurbuz F, Temiz F, Millar RP, Yuksel B (2012) Inactivating KISS1 mutation and hypogonadotropic hypogonadism. N Engl J Med 366:629–635
51. Crowley WF Jr, Whitcomb RW, Jameson JL, Weiss J, Finkelstein JS, O'Dea LS (1991) Neuroendocrine control of human reproduction in the male. Recent Prog Horm Res 47:27–62; discussion 62–67
52. Pralong FP, Boepple PA, Conn PM, Whitcomb RW, Butler JP, Schoenfeld D, Crowley WF Jr (1996) Contour of the GnRH pulse independently modulates gonadotropin secretion in the human male. Neuroendocrinology 64:247–256
53. Han SK, Gottsch ML, Lee KJ, Popa SM, Smith JT, Jakawich SK, Clifton DK, Steiner RA, Herbison AE (2005) Activation of gonadotropin-releasing hormone neurons by kisspeptin as a neuroendocrine switch for the onset of puberty. J Neurosci 25:11349–11356

54. Pielecka-Fortuna J, Chu Z, Moenter SM (2008) Kisspeptin acts directly and indirectly to increase gonadotropin-releasing hormone neuron activity and its effects are modulated by estradiol. Endocrinology 149:1979–1986
55. Zhang C, Roepke TA, Kelly MJ, Ronnekleiv OK (2008) Kisspeptin depolarizes gonadotropin-releasing hormone neurons through activation of TRPC-like cationic channels. J Neurosci 28:4423–4434
56. Dumalska I, Wu M, Morozova E, Liu R, van den Pol A, Alreja M (2008) Excitatory effects of the puberty-initiating peptide kisspeptin and group I metabotropic glutamate receptor agonists differentiate two distinct subpopulations of gonadotropin-releasing hormone neurons. J Neurosci 28:8003–8013
57. Constantin S, Caligioni CS, Stojilkovic S, Wray S (2009) Kisspeptin-10 facilitates a plasma membrane-driven calcium oscillator in gonadotropin-releasing hormone-1 neurons. Endocrinology 150:1400–1412
58. Steiner RA, Clifton D (2009) Neuroendocrinology of reproduction. In: Strauss JF III, Barbieri RL (eds) Yen and Jaffe's reproductive endocrinology: physiology, pathophysiology, and clinical management. Saunders, Philadelphia, pp 3–33
59. Maeda K, Ohkura S, Uenoyama Y, Wakabayashi Y, Oka Y, Tsukamura H, Okamura H (2010) Neurobiological mechanisms underlying GnRH pulse generation by the hypothalamus. Brain Res 1364:103–115
60. Marshall JC, Dalkin AC, Haisenleder DJ, Griffin ML, Kelch RP (1993) GnRH pulses—the regulators of human reproduction. Trans Am Clin Climatol Assoc 104:31–46
61. Butler JP, Spratt DI, O'Dea LS, Crowley WF Jr (1986) Interpulse interval sequence of LH in normal men essentially constitutes a renewal process. Am J Physiol 250:E338–E340
62. Santoro N, Butler JP, Filicori M, Crowley WF Jr (1988) Alterations of the hypothalamic GnRH interpulse interval sequence over the normal menstrual cycle. Am J Physiol 255:E696–E701
63. Jayasena CN, Nijher GM, Abbara A, Murphy KG, Lim A, Patel D, Mehta A, Todd C, Donaldson M, Trew GH, Ghatei MA, Bloom SR, Dhillo WS (2010) Twice-weekly administration of kisspeptin-54 for 8 weeks stimulates release of reproductive hormones in women with hypothalamic amenorrhea. Clin Pharmacol Ther 88:840–847
64. Keen KL, Wegner FH, Bloom SR, Ghatei MA, Terasawa E (2008) An increase in kisspeptin-54 release occurs with the pubertal increase in luteinizing hormone-releasing hormone-1 release in the stalk-median eminence of female rhesus monkeys in vivo. Endocrinology 149:4151–4157
65. Smith JT, Rao A, Pereira A, Caraty A, Millar RP, Clarke IJ (2008) Kisspeptin is present in ovine hypophysial portal blood but does not increase during the preovulatory luteinizing hormone surge: evidence that gonadotropes are not direct targets of kisspeptin in vivo. Endocrinology 149:1951–1959
66. Belchetz PE, Plant TM, Nakai Y, Keogh EJ, Knobil E (1978) Hypophysial responses to continuous and intermittent delivery of hypopthalamic gonadotropin-releasing hormone. Science 202:631–633
67. Nijher GM, Chaudhri OB, Ramachandran R, Murphy KG, Zac-Varghese SE, Fowler A, Chinthapalli K, Patterson M, Thompson EL, Williamson C, Kumar S, Ghatei MA, Bloom SR, Dhillo WS (2010) The effects of kisspeptin-54 on blood pressure in humans and plasma kisspeptin concentrations in hypertensive diseases of pregnancy. Br J Clin Pharmacol 70:674–681

Chapter 6
Kisspeptin Excitation of GnRH Neurons

Oline K. Rønnekleiv and Martin J. Kelly

Abstract Kisspeptin binding to its cognate G protein-coupled receptor (GPR54, aka Kiss1R) in gonadotropin-releasing hormone (GnRH) neurons stimulates peptide release and activation of the reproductive axis in mammals. Kisspeptin has pronounced pre- and postsynaptic effects, with the latter dominating the excitability of GnRH neurons. Presynaptically, kisspeptin increases the excitatory drive (both GABA-A and glutamate) to GnRH neurons and postsynaptically, kisspeptin inhibits an A-type and inwardly rectifying K^+ (Kir 6.2 and GIRK) currents and activates nonselective cation (TRPC) currents to cause long-lasting depolarization and increased action potential firing. The signaling cascades and the multiple intracellular targets of kisspeptin actions in native GnRH neurons are continuing to be elucidated. This review summarizes our current state of knowledge about kisspeptin signaling in GnRH neurons.

Relationship Between Kisspeptin and GnRH Secretion

Kisspeptin, encoded by the *Kiss1* gene, is a key factor in the regulation of reproductive development and functions [1–6]. The *Kiss1* gene encodes a 145 amino acid protein, which is proteolytically processed to produce a 54 amino acid peptide, called kisspeptin-54, and several other smaller peptide fragments [7]. Centrally administered kisspeptins stimulate GnRH and gonadotropin secretion in prepubertal and adult animals [1, 8–11]. The central application of kisspeptin induces cFos immunoreactivity within 1–2 h in more than 85% of GnRH neurons, further suggesting that direct activation of the neurons is responsible for the release of luteinizing hormone (LH) and follicle-stimulating hormone (FSH) [8]. As expected,

O.K. Rønnekleiv, Ph.D. (✉) • M.J. Kelly, Ph.D.
Physiology and Pharmacology, Oregon Health and Science University,
L334, 3181 SW Sam Jackson Park Road, Portland, OR 97239, USA
e-mail: ronnekle@ohsu.edu

A.S. Kauffman and J.T. Smith (eds.), *Kisspeptin Signaling in Reproductive Biology*, Advances in Experimental Medicine and Biology 784, DOI 10.1007/978-1-4614-6199-9_6,

kisspeptin is not able to stimulate LH or FSH release in GPR54 knockout animals [11]. Also, the kisspeptin-mediated release of LH is completely inhibited by the GnRH antagonist, acyline [1, 8]. Importantly, CNS administration of kisspeptin in the ewe has conclusively demonstrated a correlation between kisspeptin-induced GnRH and LH release [11]. Therefore, the stimulatory actions of kisspeptin appear to be primarily on GnRH neurons and not the pituitary. Kisspeptin, when applied to GnRH neurons in vitro, potently activates these neurons and causes increased neuronal firing [12–14].

Over the past several years, there have been many publications about the regulation of *Kiss1* gene expression and the role of kisspeptins in regulating GnRH and LH secretion [1, 8, 11, 15–17]. Also, the distribution and regulation of kisspeptin mRNA (*Kiss1*) expression by 17β-estradiol (E_2) has been extensively described in the mouse and rat brain [15, 17, 18]. In these rodent species, it is known that *Kiss1* mRNA is expressed primarily in the anteroventral periventricular nucleus (AVPV) and adjacent periventricular (PeN) areas, as well as in the arcuate nucleus of the hypothalamus [19, 20]. Importantly, E_2 increases the mRNA expression of *Kiss1* in the female AVPV, but decreases the expression in the arcuate nucleus [15, 17]. These findings are consistent with data showing that the AVPV is necessary for E_2 positive feedback on GnRH and LH secretion in these species [21–23]. The number of AVPV kisspeptin neurons is significantly fewer in male rodents than in females, but the numbers are similar in the arcuate nucleus in adults of both sexes [24]. As in females, steroid treatment (testosterone or E_2) increases the number of *Kiss1* neurons in the male AVPV and decreases the number of *Kiss1* expressing cells in the arcuate nucleus [25]. The function of the AVPV kisspeptin neurons in the male rodent is not clear, but these neurons may be involved in generating the basal pulsatile release of LH via a stimulatory action on GnRH neurons.

In other species, such as the guinea pig, sheep, and rhesus monkey, the preoptic area (POA) appears not to be the main region responsible for E_2 positive feedback [26–30]. Thus, it appears that the basal hypothalamus may be sufficient for maintaining steroid-mediated positive feedback regulation of GnRH and LH secretion in these species. Consistent with these findings, kisspeptin neurons within the arcuate nucleus in guinea pig, sheep, and monkey appear to be involved in E_2-mediated positive as well as negative feedback regulation of GnRH neurons [31–34]. The specific kisspeptin neurons within the arcuate nucleus that mediate positive feedback regulation of LH remains to be determined, although evidence suggests that a caudal arcuate population of neurons is involved [31–34]. However, irrespective of the role of the arcuate nucleus in mediating E_2 positive feedback on LH secretion in certain species, evidence suggests that the POA is also involved [31, 34, 35]. Importantly, regardless of differential regulation of *Kiss1* neurons by E_2, in all instances, kisspeptin potently excites GnRH neurons via a phospholipase C (PLC) signaling pathway (see below).

To further study the role of kisspeptin in the regulation of GnRH neurons and LH release, kisspeptin analogs with mixed agonist/antagonist activities have been synthesized [36]. Of these, peptide 234 has primarily antagonist activities in Chinese Hamster Ovary K1 (CHO-K1) cells expressing Kiss1R and inhibits the kisspeptin response by 93%, with an IC50 of 7 nM [36]. This compound also has a binding affinity of 2.7 nM for Kiss1R stably expressed in CHO-K1 cells. Peptide 234 subsequently has been found to inhibit kisspeptin-induced GnRH neuronal firing in vitro, whereas in vivo treatment with the peptide attenuates kisspeptin-induced LH release in intact males [36]. Moreover, peptide 234 attenuates the castration rise in plasma LH levels in mouse and rat, and reduces pulsatile LH release in the ovariectomized ewe and rat [36, 37]. Also, in ovariectomized monkeys, peptide 234 attenuates pulsatile release of GnRH [36]. Collectively, these data support the concept that kisspeptin neurons (in the arcuate nucleus?) are involved in stimulating GnRH and LH release following gonadectomy.

Thus, the estrogen-mediated "negative" feedback inhibition of post-castration GnRH and LH release may be via the differential release of kisspeptin/opioid peptides since this group of arcuate neurons also co-localizes dynorphin [38, 39]. Although there is a robust μ-opioid receptor-mediated inhibition of GnRH neurons in guinea pig [40], a κ-opioid-mediated effect has not been demonstrated.

Kisspeptin Activation of Kiss1R

Kisspeptin-54 has been identified as the endogenous ligand of the orphan G protein-coupled receptor, GPR54 [7, 41], also known as Kiss1R. In addition to kisspeptin-54, the smaller peptide fragments derived from the precursor protein (e.g., kisspeptin 14, 13, and 10) all have biological activity at Kiss1R [7, 42]. These peptides bind with low nanomolar affinities to rat and human Kiss1R expressed in Chinese hamster ovary K1 cells and stimulate PIP2 hydrolysis, Ca^{2+} mobilization, arachidonic acid release, extracellular signal-regulated protein kinase 1 (ERK1), ERK2, and p38 MAP kinase phosphorylation [7]. In mammals, Kiss1R is expressed both in the pituitary and in GnRH neurons [7, 8, 11, 12]. However, as stated above, evidence suggests that the stimulation of gonadotropin secretion by kisspeptin is via direct activation of GnRH neurons and not pituitary gonadotropes [1, 8–10, 43]. Although multiple actions of kisspeptin have been identified (see below), all of the signaling pathways have not been elucidated.

Kisspeptin Activation of Kiss1R in GnRH Neurons: Downstream Signaling Pathways

To date, kisspeptin is the most potent and efficacious neuropeptide/neurotransmitter to excite native GnRH neurons [44–49]. In most studies, kisspeptin is reported to depolarize and excite the vast majority (75–90%) of GnRH neurons (Fig. 6.1), which correlates with the expression of Kiss1R in the majority of GnRH neurons [8, 12, 14, 50]. However, Dumalska et al. found a lower percentage of GnRH

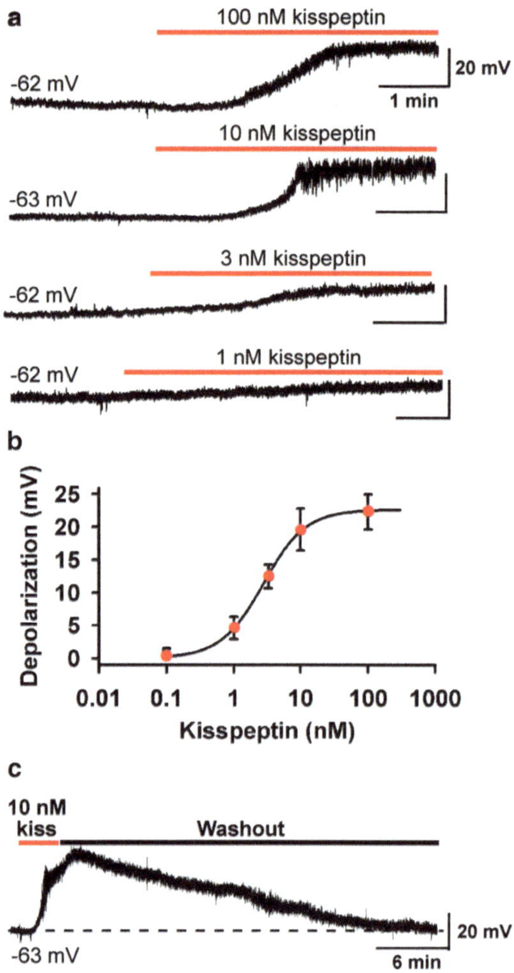

Fig. 6.1 Kisspeptin depolarizes GnRH neurons in a concentration-dependent manner. (**a**) Representative traces showing that kisspeptin (1–100 nM) depolarized GnRH neurons in a concentration-dependent manner. The initial membrane potential for each trace is indicated. Only one cell was recorded from one slice. (**b**) Concentration–response curve of the kisspeptin-induced depolarization. Data are presented as mean ± SEM. The EC_{50} for the kisspeptin-induced depolarization was 2.8 ± 0.2 nM (n = 8–14) based on a logistic equation fit to the data points. (**c**) The kisspeptin (10 nM)-induced depolarization was long lasting and typically took 30 min to recover. From Zhang C, Roepke TA, Kelly MJ, Rønnekleiv OK. Kisspeptin depolarizes gonadotropin-releasing hormone neurons through activation of TRPC-like cationic channels. J Neurosci 2008; 28: 4423–4434. Reprinted with permission from The Society for Neuroscience

neurons responding to kisspeptin and proposed that there are two physiologically distinct populations of GFP-GnRH neurons, one that responds to kisspeptin and the other that responds to the metabotropic glutamate receptor agonist, dihydroxyphenylglycine (DHPG) [51]. One explanation for these differences is that some of the recordings performed by Dumalska and coworkers were made from animals as young as 15 days of age [51], so the reduced response to kisspeptin could be age-related. Although the expression of Kiss1R is similar in juvenile as in adult male mice, and at both age levels Kiss1R can be detected in over 90% of GnRH neurons [12], the percent of GnRH neurons responding to kisspeptin is only about 27% in juvenile vs. 90% in adult males [12]. The reason for the reduced efficacy of kisspeptin in GnRH neurons from juvenile and prepubertal males is not fully understood but could be due to an immature Kiss1R signaling in the younger animals.

In the adult, kisspeptin depolarizes GnRH neurons via the coupling of Kiss1R to a phospholipase Cβ (PLCβ) signaling pathway that activates canonical transient receptor potential (TRPC) channels that allow influx of sodium and to a lesser extent calcium ions (Fig. 6.2) [14]. Besides activating TRPC channels in GnRH neurons, kisspeptin also attenuates resting and ligand-activated inwardly rectifying K^+ (Kir) channels and A-type potassium channels [13, 14, 50, 52]. The inhibition of Kir may be critical because Kir channels (e.g., K_{ATP} and GIRK channels) are highly expressed in GnRH neurons and clamp the cells in a negative resting state of −63 mV [40, 53, 54]. This effect of kisspeptin is also vital for inhibiting GPCR-activated (μ-opioid, $GABA_B$ and perhaps melanin-concentrating hormone, MCH) GIRK (Kir) currents which are prominent in GnRH neurons [40, 54, 55]. Also, A-type K^+ currents are very prominent in GnRH neurons, and E_2 regulation of the A-current may play a role in negative feedback regulation of GnRH neurons [52, 56]. Therefore, kisspeptin inhibition of these K^+ currents would be of high functional significance. Moreover, kisspeptin increases calcium oscillations of mature as well as developing GnRH neurons, and these changes for the most part reflect the coupling of Kiss1R to a PLCβ signaling pathway [14, 50, 57, 58]. Therefore, by inhibiting potassium channels along with the pronounced activation of TRPC channels, kisspeptin depolarizes GnRH neurons to threshold (~−45 mV) and induces sustained firing, which may be accompanied by a sustained calcium ion influx via calcium channels/TRPC channels and augmented GnRH release during positive feedback.

Kisspeptin Activation of TRPC Channels

The mammalian TRPC channel family consists of seven members, TRPC1–7, that appear to function as receptor-operated channels, analogous to the TRP channels involved in *Drosophilia* phototransduction [59]. With the exception of TRPC2, these channels are widely distributed in the mammalian brain [60]. The TRP channels are made of subunits with six membrane-spanning domains that co-assemble as tetrameric complexes similar to what has been described for K^+ channels [61, 62]. TRPC channels appear to co-assemble as heteromeric channels consisting of the TRPC1, 4, and 5 subfamily [63, 64], as well as TRPC3, 6, and 7 subfamily [65, 66]. It is well known that the current–voltage relationship and mechanisms of regulation of TRPC channels depend on the channel subunit composition [59]. However, the

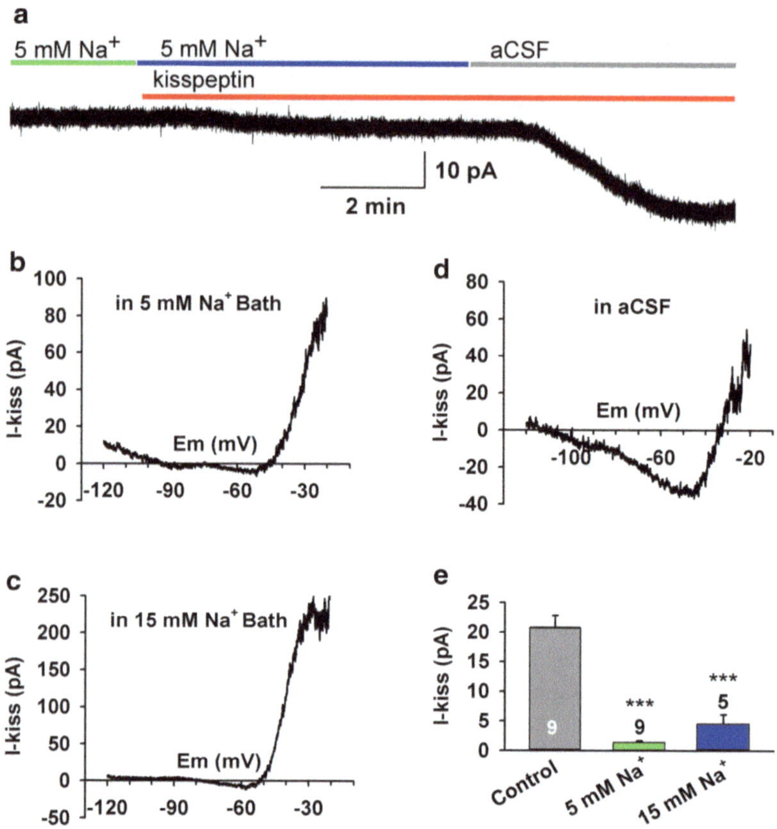

Fig. 6.2 Kisspeptin predominantly activates a sodium-dependent, nonselective cationic (TRPC) channel. (**a**) The kisspeptin-induced inward current (at –60 mV) was greatly reduced in low Na^+ bath solution (5 mM Na^+/140 mM *N*-methyl-d-glucamine [$NMDG^+$]), and switching back to normal aCSF (control) solution revealed a kisspeptin sensitive inward current of 28 pA in this GnRH neuron. (**b**, **c**) The *I–V* relationships of the kisspeptin-evoked current in a low Na^+ bath solution (5 and 15 mM Na^+) between –20 and –120 mV showed a greatly reduced inward current. (**d**) A typical *I–V* relationship of the kisspeptin-induced inward current in normal aCSF (control) solution showed a larger inward current (–30 pA at –60 mV). (**e**) Summary of the effects of the extracellular sodium concentration on the kisspeptin-induced inward current at –60 mV. $^{***}p<0.001$, significantly different from the effects of kisspeptin under control aCSF conditions. Cell numbers tested are indicated for each group. Error bars indicate SEM. From Zhang C, Roepke TA, Kelly MJ, Rønnekleiv OK. Kisspeptin depolarizes gonadotropin-releasing hormone neurons through activation of TRPC-like cationic channels. J Neurosci 2008; 28: 4423–4434. Reprinted with permission from The Society for Neuroscience

functional distinction between these channel subtypes in CNS neurons has been problematic because of a lack of selective pharmacological reagents. The main exception is the discriminatory effects of lathanides to augment TRPC4, 5 channel activity [62]. Whole-cell recording experiments, with K^+ channel blockers on board, have revealed that the current–voltage relationship for the kisspeptin-induced current in GnRH neurons resembles the current–voltage relationship of heteromeric

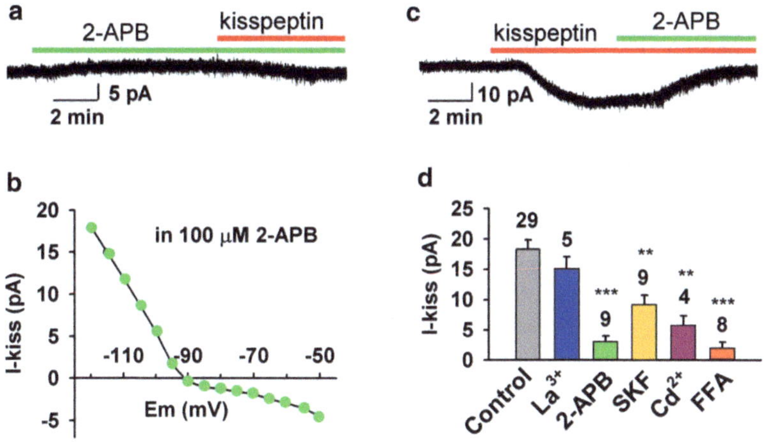

Fig. 6.3 Effects of TRPC channel blockers on the kisspeptin-induced inward currents at –60 mV. (**a**) A representative recording showing that 2-APB (100 μM), which had very little effect on basal holding current, potently blocked the kisspeptin (100 nM)-evoked inward current. (**b**) Mean *I–V* relationship of the kisspeptin-sensitive current in the presence of 2-APB reversed at –90 mV ($n=4$), clearly indicating that a Kir channel was inhibited by kisspeptin. (**c**) A representative recording showing that 2-APB (100 μM) applied after kisspeptin also strongly blocked the kisspeptin-evoked inward current. (**d**) Summary of the effects of different TRPC channel blockers (100 μM La^{3+}, 100 μM 2-APB, 30 μM SKF96365, 250 μM Cd^{2+}, 100 μM flufenamic acid) on the kisspeptin-induced inward currents at –60 mV. Blockers were applied 5–7 min before or after the application of kisspeptin (100 nM). The percent inhibition for the different blockers was as follows: 17.4% for 100 μM La^{3+}, 83.6% for 100 μM 2-APB, 50% for 30 μM SKF, 68.7% for 250 μM Cd^{2+}, and 89.6% for 100 μM FFA. $^{**}p<0.01$ and $^{***}p<0.001$, significantly different from the kisspeptin response under control aCSF conditions. Cell numbers tested are indicated. Error bars indicate SEM. From Zhang C, Roepke TA, Kelly MJ, Rønnekleiv OK. Kisspeptin depolarizes gonadotropin-releasing hormone neurons through activation of TRPC-like cationic channels. J Neurosci 2008; 28: 4423–4434. Reprinted with permission from The Society for Neuroscience

complexes of TRPC 1+4 or TRPC 1+5 subunits expressed in HEK cells with the characteristic negative slope conductance and pronounced outward rectification (Fig. 6.2) [14, 59, 63]. Similar current–voltage relationships have been obtained for the leptin-induced currents in arcuate POMC and kisspeptin neurons and the mGluR1- and CCK2-induced currents in basolateral amygdala neurons [67–70]. All of these neurons have been found to express the same compliment of TRPC channels as GnRH neurons.

Interestingly, GnRH neurons express all of the "brain-type" TRPC channel subunits with the TRPC1, 4, and 5 family being the most prevalent in GnRH neurons [14]. Therefore, based on the current–voltage relationship, pharmacological profile and mRNA expression, TRPC1, 4, and 5 are key players in mediating the excitatory effects of kisspeptin in GnRH neurons (Figs. 6.2 and 6.3) [14]. Traditionally, these channels are known as "store operated calcium channels," but this description is probably the result of poorly understood signaling mechanisms [59, 71]. Therefore, current research has focused on elucidating the signaling pathway(s) by which kisspeptin activates TRPC channels, and the sources of calcium mobilization following kisspeptin activation of GnRH neurons. Although the majority of findings seem to indicate that the initial calcium signal comes via plasma membrane channels [14, 57, 58], there is also evidence that kisspeptin induces the release of calcium from intracellular stores in GnRH neurons via inositol-1,4,5-trisphosphate (IP3) receptors [50, 72]. However, intracellular dialysis with 2-APB, which abrogates the store release of calcium, does not inhibit the effects of kisspeptin [14]. Certainly, sustained calcium release is not required for kisspeptin's actions since calcium mobilization is transient [50].

The mammalian TRPC channels can be activated by G protein-coupled receptors and receptor tyrosine kinases (see refs. [59, 73]). In a heterologous cell expression system (i.e., Chinese hamster ovary K1 cells expressing Kiss1R), kisspeptin is capable of activating multiple signaling pathway resulting in increased IP_3 formation, calcium mobilization, arachidonic acid release, and MAP kinase phosphorylation [7]. Although the kisspeptin induction of GnRH release in hypothalamic explants from immature animals incubated in vitro is reported to involve recruitment of ERK1/2 and p38 kinases, these actions of kisspeptin have not been confirmed in adult GnRH neurons [50, 72]. In native GnRH neurons, the PLC inhibitor U73122 inhibits the effects of kisspeptin [14, 50], and indeed, it is known that all mammalian TRPC channels require PLC for activation [60]. Therefore, it appears that Gq-coupled GPR54 activates PLCβ to signal downstream to open TRPC channels in GnRH neurons, thereby allowing the influx of sodium and calcium. Interestingly, in POMC neurons, PLCγ1 appears to be the isozyme coupled to TRPC channel activation by leptin [69].

Classically, the TRPC3, 6, and 7 subfamily is DAG sensitive [59, 73]. Although TRPC3 and 7, and to a lesser extent TRPC6, transcripts are expressed in GnRH neurons, the surrogate DAG signaling molecule 2-acetyl sn-glycerol (OAG) has only a small effect to activate an inward current (~25% of the kisspeptin-induced current) in GnRH neurons [14]. A potential explanation is that both hydrolysis of PIP_2 by PLCβ and the calcium trigger that facilitates the TRPC channel opening (i.e., the influx of Ca^{2+} through calcium channels [74]) might be missing when applying OAG alone to GnRH neurons. In addition, La^{3+} at a 100 μM concentration, which potentiates TRPC4 and 5 and blocks TRPC3, 6, and 7 channels [75], did not attenuate or augment the kisspeptin-induced current, which indicates that an ensemble of these channel subunits must exist in GnRH neurons as revealed by single-cell RT-PCR [14]. Indeed, extracellular 2-APB (100 μM), which is a potent blocker of TRPC3, 4, 5, and 6 channels, and FFA, which is a potent blocker of TRPC4 and 5 channels, inhibit the effects of kisspeptin in GnRH neurons (Fig. 6.3). These blockers have a similar effects on the leptin activation of TRPC channels in arcuate POMC and kisspeptin neurons, although lanthanum clearly potentiates the leptin-induced activation of TRPC currents in POMC and kisspeptin neurons [69, 70], suggesting subtle differences between the GnRH neurons and the other two cell types. Collectively, these data suggest that, although all of the "brain" TRPC channels are expressed in GnRH neurons, the TRPC1, 4, and 5 family appear to be major (key) players in mediating the effects of kisspeptin in GnRH neurons [14].

Kisspeptin Inhibition of Kir Channels and Their Role in GnRH Neuronal Excitability

Kisspeptin augments the activity of GnRH neurons in part via inhibition of Kir potassium channels [13, 14, 50]. In this respect, the Kir blockers barium (0.3 mM) and tetraethylammonium (20 mM) robustly inhibit the kisspeptin-induced potassium currents in GnRH neurons [13, 14, 50]. The importance of kisspeptin inhibition of Kir is further substantiated by the ability of kisspeptin to attenuate the $GABA_B$-induced hyperpolarization in GnRH neurons [54].

GABA is one of the most important neurotransmitters that regulate the excitability of GnRH neurons. Multiple studies have shown that GABA activates Cl^- currents in GnRH neurons, and these effects are blocked by $GABA_A$ receptor antagonists [44, 46, 47, 76–78]. It is generally accepted that activation of $GABA_A$ receptors depolarizes and excites GnRH neurons [46, 78–80]. Several $GABA_A$ receptor subunits have been identified in GnRH neurons, including $\alpha1$, $\alpha2$, $\alpha3$, $\alpha5$, $\beta1$, $\beta2$, $\beta3$, $\gamma1$, $\gamma2$, and the rho 1 subunits [78, 81, 82]. The $GABA_B$ receptor subunits, R1 and R2, are also found in GnRH neurons [54, 83], and GABA activates $GABA_B$-receptors in GnRH neurons [54, 84]. Moreover, as has been demonstrated in numerous other hypothalamic neurons [85–90], $GABA_B$ receptors are coupled ($G\alpha_{i/o}$) to activation of G protein-coupled inwardly rectifying K^+ (GIRK) channels, resulting in a robust hyperpolarization of GnRH neurons.

The importance of Kir channels in modulating GnRH neuronal excitability has been well documented [13, 14, 40, 53–55, 84]. Female GnRH neurons sit at a relatively negative resting membrane potential (−63 mV) that is due, in part, to the activity of Kir channels including GIRKs and K_{ATP} channels [53, 54]. For example, blocking the K_{ATP} channels with the sulfonylurea tolbutamide significantly depolarizes the cells by 4–6 mV, which puts the membrane potential in the range of most parvocellular hypothalamic neurons [53, 91]. In addition, GABA release is regulated by E_2 through presynaptic mechanisms [85, 87, 92–94] that affect GnRH neuronal activity [40, 95]. Augmented E_2-induced $GABA_B$ receptor activity would further hyperpolarize the membrane through increased GIRK channel activity. However, this inhibitory tone must be attenuated during the excitatory (preovulatory) phase of GnRH neurons. One possible mechanism is that kisspeptin signaling via KissR provides the stimulus to overcome this strong inhibitory tone (Fig. 6.4). Previous investigators have shown that there is a robust kisspeptin drive during E_2 "positive feedback" [13, 15, 96], and kisspeptin counters the hyperpolarizing effects of activation of GIRKs by the $GABA_B$ agonist baclofen, μ-opioid receptor agonists (Zhang et al., unpublished findings), and other Ba^{2+}-sensitive inwardly rectifying K^+ channels in general [13, 14]. Moreover, $G\alpha_{q/11}$-coupled receptors are known to

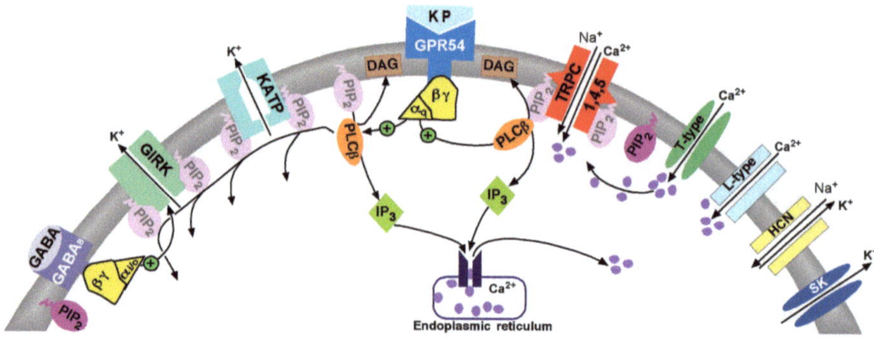

Fig. 6.4 Model of kisspeptin's actions to depolarize GnRH neurons and facilitate burst firing. Kisspeptin binds to its cognate GPR54 receptor, which is Gq-coupled to activate phospholipase Cβ. PLCβ has multiple downstream actions resultant from cleaving phosphatidylinositol 4,5 bisphosphate (PIP_2) to inositol 1,4,5-triphosphate (IP_3) and diacylglycerol (DAG). Since PIP_2 facilitates Kir channel opening, cleavage of this fatty acid attenuates K-ATP and GIRK channel opening. Inhibition of GIRK channels renders a number of $G\alpha_{i/o}$-coupled receptors ineffective to inhibit GnRH neurons ($GABA_B$, μ-opioid, NPY, MCH, etc.). On the other hand, cleavage of PIP_2 facilitates TRPC 4 channel opening [126]. In addition the membrane-associated fatty acid DAG probably activates the TRPC1, 4, 5 channel complex [Note: The OAG (analogue of DAG) was only weakly effective to open the TRPC channels since PIP_2 still exerted a strong inhibition of the TRPC channel complex]. Ca^{2+} potentiates the agonist-activated TRPC1, 4, 5 complex, and plasma membrane calcium channels appear to play a critical role. Intracellular 2-APB dialysis, which effectively blocks IP3 receptor-mediated release of Ca^{2+}, was ineffective, but extracellular Cd^{2+} potently inhibited TRPC1, 4, 5 channel activity. Therefore, we propose that low voltage-activated T-type calcium channels are initially involved in facilitating TRPC channel opening. Once depolarized, Ca^{2+} entry through high voltage-activated Ca^{2+} channels can also contribute to facilitating TRPC channel opening. Also illustrated are other channels contributing to burst firing activity such as the hyperpolarization-activated, cyclic nucleotide-gated (HCN, pacemaker) channel, and the small conductance, Ca^{2+}-activated K^+ channel (SK), which is involved in the repolarization of the membrane following a burst of action potentials

desensitize (i.e., heterologous desensitization) $G\alpha_{i/o}$-coupled receptors through PIP_2 hydrolysis and attenuating the GIRK-mediated hyperpolarization (Fig. 6.4) [97–99]. In addition to attenuating $G\alpha_{i/o}$-coupled receptor-mediated hyperpolarization, kisspeptin activates TRPC channels in GnRH neurons to cause further depolarization [14]. The Kiss1R-$G\alpha_{q/11}$-PLCβ signaling pathway would have a twofold effect to inhibit K^+ channels and activate TRPC channels, which underlies the pronounced excitatory effects of kisspeptin on GnRH neurons [12–14, 50, 52]. Interestingly MCH, although at higher concentrations, can block kisspeptin excitation of septal vesicular glutamate transporter 2 (vGluT2)-GnRH neurons by inhibiting Kir [55], which could be a mechanism by which GnRH neuronal excitability is reduced during certain physiological states.

Presynaptic Effects of Kisspeptin on GnRH Neuronal Excitability

Based on extracellular recording in GnRH neurons in slices obtained from oil- and E_2-treated mice, the kisspeptin-induced neuronal firing rate is potentiated in the E_2-treated females [13, 96]. The E_2-induced potentiation is reduced when GABA and glutamate inputs to GnRH neurons are blocked, suggesting the involvement of these fast synaptic transmitters in the E_2 effect. In addition, it is well known that kisspeptin neurons in the AVPV are positively regulated by E_2 and are believed to contribute to positive feedback input to rodent GnRH neurons [15]. Therefore, E_2 may further augment the effects of kisspeptin in vivo via direct action on AVPV kisspeptin neurons.

The precise localization of the kisspeptin inputs to GnRH neurons has not been identified. The AVPV is a complex nucleus that expresses other neurotransmitters in addition to kisspeptin, such as dopamine, GABA, and glutamate [100, 101]. Although projections from the AVPV to GnRH neurons have been described by a number of investigators, the functional interactions between the AVPV and GnRH neurons are just beginning to be elucidated [102]. Thus, stimulation of the AVPV and recording of responses in GnRH neurons reveals that low stimulation rates (<1 Hz) induce glutamate and GABA synaptic currents in GnRH neurons, whereas higher frequency stimulation (5–10 Hz) induces delayed excitation believed to be kisspeptin mediated since the response is absent in *Kiss1r* knockout animals and antagonized by the kisspeptin antagonist peptide 318 [102]. Therefore, the AVPV kisspeptin neurons may provide a critical excitatory input to GnRH neurons.

Kisspeptin neurons in the arcuate nucleus are negatively regulated by E_2 and are believed to be involved in negative feedback regulation of GnRH secretion [15, 37]. The mechanism by which arcuate neurons negatively regulate GnRH neurons is not completely understood, but has been proposed to also involve arcuate POMC neurons [40]. Interestingly, kisspeptin-immunoreactive fibers in the arcuate nucleus form close contacts onto POMC neurons, and kisspeptin excites POMC neurons via activation of a nonselective cation (TRPC?) channel and activation of a sodium/calcium exchanger [103]. β-endorphin positive fibers, presumably from arcuate POMC neurons, are highly expressed in the POA, and β-endorphin synapses, as well as μ-opioid receptor expression, are found specifically on GnRH neurons [104–107]. Therefore, kisspeptin may influence GnRH neurons indirectly via actions on arcuate POMC neurons. While μ-opioid receptor activation would be inhibitory to GnRH neurons [40], recently it has been shown that an agonist of the melanocortin receptors 3 and 4 excite GnRH neurons [108]. This would suggest that POMC neurons may also excite GnRH neurons via release of αMSH, a POMC product. Clearly, further studies are needed to elucidate the role of arcuate kisspeptin neurons in negative feedback regulation of GnRH neurons, as well as the role of these neurons in GnRH neuronal pulsatility.

Kisspeptin and Burst Firing in GnRH Neurons

It is well known that GnRH is released in a pulsatile manner, and the hypothalamic surge of GnRH and subsequent pituitary release of LH are required for triggering ovulation in the female. Although single action potential-induced calcium influx is enough to spark the release of classical transmitters, burst firing or tetanic stimulation is required for the release of neuropeptides such as vasopressin, oxytocin, substance P, and atrial natriuretic factor [109–111]. Experiments in vitro using perifused hypothalamic tissue, primary hypothalamic cultures, or a GT1 GnRH neuronal cell line have revealed that pulsatile GnRH release is evident in vitro [112–114]. Recordings in slices from genetically modified mice that express the calcium ratiometric indicator Pericam in GnRH neurons have shown that intracellular calcium transients, generated through L-type calcium channels and amplified by calcium release from intracellular stores, are synchronized with burst firing in a subpopulation of GnRH neurons [115]. However, a recent publication suggests that kisspeptin inhibits high voltage-activated (HVA) Ca^{2+} (e.g., L-type) channels, which would attenuate the calcium-activated afterhyperpolarization and thereby promote sustained firing [116]. Regardless of the role of the HVA Ca^{2+} channels in kisspeptin's downstream signaling, T-type calcium channels and hyperpolarization-activated, cyclic nucleotide-gated (HCN) channels and their respective currents are highly expressed in GnRH neurons [117–120]. Both the h-current and T-type calcium current contribute to burst firing [118, 121], and activation of these vital conductances is dependent on membrane (hyper) polarization [121, 122]. In fact, a hyperpolarizing stimulus removes the inactivation of T-type calcium channels and also activates the h-current. Both the GABA (via $GABA_B$) and opioids (via μ-opioid receptors) provide this hyperpolarizing stimulus to GnRH neurons [40, 54, 84]. The membrane hyperpolarization generated by $G\alpha_{i/o}$-coupled receptors during E_2 negative feedback sets the stage for recruiting both the HCN and T-type calcium channels that are critical for phasic burst firing of GnRH neurons [40, 45, 121]. Ultimately, kisspeptin attenuates the hyperpolarized state of "negative feedback" by inhibiting K^+ channel activity and opening up TRPC channels to cause sustained depolarization and firing [13, 14, 50, 52]. Furthermore, the calcium-activated afterhyperpolarizing currents (e.g., small conductance, calcium-activated K^+, SK) would serve to repolarize the cell membrane to allow the continued oscillation and burst firing [115, 123–125].

Summary

It is clear that kisspeptin has pronounced pre- and postsynaptic effects on GnRH neuronal excitability. Presynaptically, kisspeptin increases the excitatory drive (both $GABA_A$ and glutamate) to GnRH neurons, and postsynaptically kisspeptin binds to Kiss1R to activate a PLCβ signaling pathway that has multiple downstream effects to cause a robust and sustained depolarization of GnRH neurons. These downstream effects include inhibition of inwardly rectifying K^+ (K_{ATP} and GIRK) channels and activation of TRPC1, 4, 5 channels (Fig. 6.4). Although all of the intermediary players (signaling molecules) have not been identified, it is clear that the sustained action potential firing in GnRH neurons is due to these membrane circumscribed actions of the Gαq-signaling pathway. In addition, T-type calcium channels are probably involved in the initial facilitation of TRPC channel opening (Fig. 6.4). However, future experiments need to address these nuances of Kiss1R signaling. In addition, the effects of kisspeptin on the presynaptic glutamatergic and GABAergic neurons also need to be elucidated. Regardless, it is clear that the highly potent neuromodulator, kisspeptin, robustly depolarizes GnRH neurons and promotes burst firing via multiple cellular actions.

Acknowledgments The authors thank current and past members of their laboratories who contributed to the work described herein, especially Drs. Chunguang Zhang, Jian Qiu, Yuan Fang, Troy Roepke, and Ms. Martha A. Bosch; also, special thanks to Ms. Martha A. Bosch for her skilled assistance with the illustrations and manuscript preparations. The work from the authors' laboratories was supported by NIH grants NS43330, NS38809, DK68098.

References

1. Gottsch ML, Cunningham MJ, Smith JT, Popa SM, Acohido BV, Crowley WF, Seminara S, Clifton DK, Steiner RA (2004) A role for kisspeptins in the regulation of gonadotropin secretion in the mouse. Endocrinology 145:4073–4077
2. Navarro VM, Castellano JM, Fernández-Fernández R, Barreiro ML, Roa J, Sanchez-Criado JE, Aguilar E, Dieguez C, Pinilla L, Tena-Sempere M (2004) Developmental and hormonally regulated messenger ribonucleic acid expression of KiSS-1 and its putative receptor, GPR54, in rat hypothalamus and potent luteinizing hormone-releasing activity of KiSS-1 peptide. Endocrinology 145:4565–4574
3. Plant TM, Ramaswamy S, DiPietro MJ (2006) Repetitive activation of hypothalamic G protein-coupled receptor 54 with intravenous pulses of kisspeptin in the juvenile monkey (*Macaca mulatta*) elicits a sustained train of gonadotropin-releasing hormone discharges. Endocrinology 147:1007–1013
4. Kuohung W, Kaiser UB (2006) GPR54 and KiSS-1: role in the regulation of puberty and reproduction. Rev Endocr Metab Disord 7:257–263
5. Adachi S, Yamada S, Takatsu Y, Matsui H, Kinoshita M, Takase K, Sugiura H, Ohtaki T, Matsumoto H, Uenoyama Y, Tsukamura H, Inoue K, Maeda K-I (2007) Involvement of anteroventral periventricular metastin/kisspeptin neurons in estrogen positive feedback action on luteinizing hormone release in female rats. J Reprod Dev 53:367–378
6. Kauffman AS, Park JH, McPhie-Lalmansingh AA, Gottsch ML, Bodo C, Hohmann JG, Pavlova MN, Rohde AD, Clifton DK, Steiner RA, Rissman EF (2007) The kisspeptin receptor GPR54 is required for sexual differentiation of the brain and behavior. J Neurosci 27:8826–8835
7. Kotani M, Detheux M, Vandenbogaerde A, Communi D, Vanderwinden J-M, Le Poul E, Brezillon S, Tyldesley R, Suarez-Huerta N, Vandeput F, Blanpain C, Schiffmann SN, Vassart G, Parmentier M (2001) The metastasis suppressor gene KiSS-1 encodes kisspeptins, the natural ligands of the orphan G protein-coupled receptor GPR54. J Biol Chem 276:34631–34636
8. Irwig MS, Fraley GS, Smith JT, Acohido BV, Popa SM, Cunningham MJ, Gottsch ML, Clifton DK, Steiner RA (2004) Kisspeptin activation of gonadotropin releasing hormone neurons and regulation of KiSS-1 mRNA in the male rat. Neuronendocrinology 80:264–272

9. Thompson EL, Patterson M, Murphy KG, Smith KL, Dhillo WS, Todd JF, Ghatei MA, Bloom SR (2004) Central and peripheral administration of kisspeptin-10 stimulates the hypothalamic-pituitary-gonadal axis. J Neuroendocrinol 16:850–858
10. Kinoshita M, Tsukamura H, Adachi S, Matsui H, Uenoyama Y, Iwata K, Yamada S, Inoue K, Ohtaki T, Matsumoto H, Maeda K-I (2005) Involvement of central metastin in the regulation of preovulatory luteinizing hormone surge and estrous cyclicity in female rats. Endocrinology 146:4431–4436
11. Messager S, Chatzidaki EE, Ma D, Hendrick AG, Zahn D, Dixon J, Thresher RR, Malinge I, Lomet D, Carlton MBL, Colledge WH, Caraty A, Aparicio SAJR (2005) Kisspeptin directly stimulates gonadotropin-releasing hormone release via G protein-coupled receptor 54. Proc Natl Acad Sci 102:1761–1766
12. Han S-K, Gottsch ML, Lee KJ, Popa SM, Smith JT, Jakawich SK, Clifton DK, Steiner RA, Herbison AE (2005) Activation of gonadotropin-releasing hormone neurons by kisspeptin as a neuroendocrine switch for the onset of puberty. J Neurosci 25:11349–11356
13. Pielecka-Fortuna J, Chu Z, Moenter SM (2008) Kisspeptin acts directly and indirectly to increase gonadotropin-releasing hormone neuron activity and its effects are modulated by estradiol. Endocrinology 149:1979–1986
14. Zhang C, Roepke TA, Kelly MJ, Rønnekleiv OK (2008) Kisspeptin depolarizes gonadotropin-releasing hormone neurons through activation of TRPC-like cationic channels. J Neurosci 28:4423–4434
15. Smith JT, Cunningham MJ, Rissman EF, Clifton DK, Steiner RA (2005) Regulation of Kiss1 gene expression in the brain of the female mouse. Endocrinology 146:3686–3692
16. Roa J, Vigo E, Castellano JM, Navarro VM, Fernández-Fernández R, Casanueva FF, Dieguez C, Aguilar E, Pinilla L, Tena-Sempere M (2006) Hypothalamic expression of Kiss-1 system and gonadotropin-releasing effects of kisspeptin in different reproductive states of the female rat. Endocrinology 147:2864–2878
17. Smith JT, Popa SM, Clifton DK, Hoffman GE, Steiner RA (2006) Kiss1 neurons in the forebrain as central processors for generating the preovulatory luteinizing hormone surge. J Neurosci 26:6687–6694
18. Oakley AE, Clifton DK, Steiner RA (2009) Kisspeptin signaling in the brain. Endocr Rev 30:713–743
19. Smith JT (2008) Kisspeptin signalling in the brain: steroid regulation in the rodent and ewe. Brain Res Rev 57:288–298
20. Clarkson J, d'Anglemont de Tassigny X, Colledge WH, Caraty A, Herbison AE (2009) Distribution of kisspeptin neurones in the adult female mouse brain. J Endocrinol 21:673–682
21. Wiegand SJ, Terasawa E, Bridson WE, Goy RW (1980) Effects of discrete lesions of preoptic and suprachiasmatic structures in the female rat: alterations in the feedback regulation of gonadotropin secretion. Neuroendocrinology 31:147–157
22. Petersen SL, Barraclough CA (1989) Suppression of spontaneous LH surges in estrogen-treated ovariectomized rats by microimplants of antiestrogens into the preoptic brain. Brain Res 484:279–289
23. Ma YJ, Kelly MJ, Rønnekleiv OK (1990) Pro-gonadotropin-releasing hormone (ProGnRH) and GnRH Content in the preoptic area and the basal hypothalamus of anterior medial preoptic nucleus/suprachiasmatic nucleus-lesioned persistent estrous rats. Endocrinology 127:2654–2664
24. Kauffman AS, Gottsch ML, Roa J, Byquist AC, Crown A, Clifton DK, Hoffman GE, Steiner RA, Tena-Sempere M (2007) Sexual differentiation of kiss1 gene expression in the brain of the rat. Endocrinology 148:1774–1783
25. Smith JT, Dungan HM, Stoll EA, Gottsch ML, Braun RE, Eacker SM, Clifton DK, Steiner RA (2005) Differential regulation of KiSS-1 mRNA expression by sex steroids in the brain of the male mouse. Endocrinology 146:2976–2984
26. Terasawa E, Wiegand SJ (1978) Effects of hypothalamic deafferentation on ovulation and estrous cyclicity in the female guinea pig. Neuroendocrinology 26:229–248

27. King JC, Ronsheim PM, Liu E, Powers L, Slonimski M, Rubin BS (1998) Fos expression in luteinizing hormone-releasing hormone neurons of guinea pigs, with knife cuts separating the preoptic area and the hypothalamus, demonstrating luteinizing hormone surges. Biol Reprod 58:323–329
28. Caraty A, Fabre-Nys C, Delaleu B, Locatelli A, Bruneau G, Karsch FJ, Herbison A (1998) Evidence that the mediobasal hypothalamus is the primary site of action of estradiol in inducing the preovulatory gonadotropin releasing hormone surge in the ewe. Endocrinology 139:1752–1760
29. Plant TM, Krey LC, Moosy J, McCormack JT, Hess DL, Knobil E (1978) The arcuate nucleus and the control of gonadotropin and prolactin secretion in the female Rhesus monkey (*Macaca mulatta*). Endocrinology 102:52–62
30. Weick RF (1981) Induction of the luteinizing hormone surge by intrahypothalamic application of estrogen in the rhesus monkey. Biol Reprod 24:415–422
31. Smith JT, Pereira LA, Clarke IJ (2009) Kisspeptin neurons in the ovine arcuate nucleus and preoptic area are involved in the preovulatory luteinizing hormone surge. Neuroendocrinology 150:5530–5538
32. Clarke IJ, Smith JT, Caraty A, Goodman RL, Lehman MN (2009) Kisspeptin and seasonality in sheep. Peptides 30:154–163
33. Smith JT, Shahab M, Pereira A, Pau K-YF, Clarke IJ (2010) Hypothalamic expression of KISS1 and gonadotropin inhibitory hormone genes during the menstrual cycle of a non-human primate. Biol Reprod 83:568–577
34. Bosch MA, Xue C, Ronnekleiv OK (2012) Kisspeptin expression in guinea pig hypothalamus: effects of 17β-estradiol. J Comp Neurol 520:2143–2162
35. Hoffman GE, Le WW, Franceschini I, Caraty A, Advis JP (2011) Expression of Fos and in Vivo median eminence release of LHRH identifies an active role for preoptic area kisspeptin neurons in synchronized surges of LH and LHRH in the ewe. Neuroendocrinology 152:214–222
36. Roseweir AK, Kauffman AS, Smith JT, Guerriero KA, Morgan K, Pielecka-Fortuna J, Pineda R, Gottsch ML, Tena-Sempere M, Moenter SM, Terasawa E, Clarke IJ, Steiner RA, Millar RP (2009) Discovery of potent kisspeptin antagonists delineate physiological mechanisms of gonadotropin regulation. J Neurosci 29:3920–3929
37. Li X-F, Kinsey-Jones JS, Cheng Y, Knox AMI, Lin Y, Petrou NA, Roseweir A, Lightman SL, Milligan SR, Limmar RP, O'Byrne KT (2009) Kisspeptin signalling in the hypothalamic arcuate nucleus regulates GnRH pulse generator frequency in the rat. PLoS One 4:1–9
38. Goodman RL, Lehman MN, Smith JT, Coolen LM, de Oliveira CVR, Jafarzadehshirazi MR, Pereira A, Iqbal J, Caraty A, Ciofi P, Clarke IJ (2007) Kisspeptin neurons in the arcuate nucleus of the ewe express both dynorphin A and neurokinin B. Endocrinology 148:5752–5760
39. Lehman MN, Coolen LM, Goodman RL (2010) Kisspeptin/neurokinin B/dynorphin (KNDy) cells of the arcuate nucleus: a central node in the control of gonadotropin-releasing hormone secretion. Endocrinology 151:3479–3489
40. Lagrange AH, Rønnekleiv OK, Kelly MJ (1995) Estradiol-17β and μ-opioid peptides rapidly hyperpolarize GnRH neurons: a cellular mechanism of negative feedback? Endocrinology 136:2341–2344
41. Stafford LJ, Xia C, Ma W, Cai Y, Liu M (2002) Identification and characterization of mouse metastasis-suppressor KiSS1 and its G-protein-couple receptor. Cancer Res 62:5399–5404
42. Ohtaki T, Shintani Y, Honda S, Matsumoto H, Hori A, Kanehashi K, Terao Y, Kumano S, Takatsu Y, Masuda Y, Ishibashi Y, Watanabe T, Asada M, Yamada T, Suenaga M, Kitada C, Usuki S, Kurokawa T, Onda H, Nishimura O, Fujino M (2001) Metastasis suppressor gene KiSS-1 encodes peptide ligand of a G-protein-coupled receptor. Nature 411:613–617
43. Clarkson J, d'Anglemont de Tassigny X, Moreno AS, Colledge WH, Herbison AE (2008) Kisspeptin-GPR54 signaling is essential for preovulatory gonadotropin-releasing hormone neuron activation and the luteinizing hormone surge. J Neurosci 28:8691–8697
44. Spergel DJ, Kruth U, Hanley DF, Sprengel R, Seeburg PH (1999) GABA-and glutamate activated channels in green fluorescent protein-tagged gonadotropin-releasing hormone neurons in transgenic mice. J Neurosci 19:2037–2050

45. Kuehl-Kovarik MC, Pouliot WA, Halterman GL, Handa RJ, Dudek FE, Partin KM (2002) Episodic bursting activity and response to excitatory amino acids in acutely dissociated gonadotropin-releasing hormone neurons genetically targeted with green fluorescent protein. J Neurosci 22:2313–2322
46. DeFazio RA, Heger S, Ojeda SR, Moenter SM (2002) Activation of A-type gamma-aminobutyric receptors excites gonadotropin-releasing hormone neurons. Mol Endocrinol 16:2872–2891
47. Han SK, Abraham IM, Herbison AE (2002) Effect of GABA on GnRH neurons switches from depolarization to hyperpolarization at puberty in the female mouse. Endocrinology 143:1459–1466
48. Suter KJ (2004) Control of firing by small (S)-a-amino-3-hydroxy-5methyl-isoxazolepropionic acid-like inputs in hypothalamic gonadotropin releasing-hormone (GnRH) neurons. Neuroscience 128:443–450
49. Iremonger KJ, Constantin S, Liu X, Herbison AE (2010) Glutamate regulation of GnRH neuron excitability. Brain Res 1364:35–43
50. Liu X, Lee K, Herbison AE (2008) Kisspeptin excites gonadotropin-releasing hormone (GnRH) neurons through a phosphilipase C/calcium-dependent pathway regulating multiple ion channels. Endocrinology 149:4605–4614
51. Dumalska I, Wu M, Morozova E, Liu R, Van den Pol AN, Alreja M (2008) Excitatory effects of the puberty-initiating peptide kisspeptin and group I metabotropic glutamate receptor agonists differentiate two distinct subpopulations of gonadotropin-releasing hormone neurons. J Neurosci 28:8003–8013
52. Pielecka-Fortuna J, DeFazio RA, Moenter SM (2011) Voltage-gated potassium currents are targets of diurnal changes in estradiol feedback regulation and kisspeptin action on gonadotropin-releasing hormone neurons in mice. Biol Reprod 85:987–995
53. Zhang C, Bosch MA, Levine JE, Rønnekleiv OK, Kelly MJ (2007) Gonadotropin-releasing hormone neurons express KATP channels that are regulated by estrogen and responsive to glucose and metabolic inhibition. J Neurosci 27:10153–10164
54. Zhang C, Bosch MA, Rønnekleiv OK, Kelly MJ (2009) $GABA_B$ receptor mediated inhibition of GnRH neurons is suppressed by kisspeptin-GPR54 signaling. Endocrinology 150:2388–2394
55. Wu M, Dumalska I, Morozova E, Van den Pol AN, Alreja M (2009) Melanin-concentrating hormone directly inhibits GnRH neurons and blocks kisspeptin activation, linking energy balance to reproduction. Proc Natl Acad Sci U S A 106:17217–17222
56. DeFazio RA, Moenter SM (2002) Estradiol feedback alters potassium currents and firing properties of gonadotropin-releasing hormone neurons. Mol Endocrinol 16:2255–2265
57. Constantin S, Caligioni CS, Stojilkovic S, Wray S (2009) Kisspeptin-10 facilitates a plasma membrane-driven calcium oscillator in gonadotropin-releasing hormone-1 neurons. Endocrinology 150:1400–1412
58. Kroll H, Bolsover S, Hsu J, Kim S-H, Bouloux P-M (2011) Kisspeptin-evoked calcium signals in isolated primary rat gonadotropin-releasing hormone neurones. Neuroendocrinology 93:114–120
59. Clapham DE (2003) TRP channels as cellular sensors. Nature 426:517–524
60. Venkatachalam K, Montell C (2007) TRP channels. Annu Rev Biochem 76:387–417
61. Clapham DE, Runnels LW, Strübing C (2001) The TRP ion channel family. Nat Rev Neurosci 2:387–396
62. Clapham D, Julius D, Montell C, Schultz G (2005) International Union of Pharmacology. XLIX. Nomenclature and structure-function relationships of transient receptor potential channels. Pharmacol Rev 57:427–450
63. Strübing C, Krapivinsky G, Krapivinsky L, Clapham DE (2001) TRPC1 and TRPC5 form a novel cation channel in mammalian brain. Neuron 29:645–655
64. Plant TD, Schaefer M (2003) TRPC4 and TRPC5: receptor-operated Ca^{2+}-permeable nonselective cation channels. Cell Calcium 33:441–450
65. Trebak M, Vazquez G, Bird GS, Putney JW Jr (2003) The TRPC3/6/7 subfamily of cation channels. Cell Calcium 33:451–461

66. Berg AP, Sen N, Bayliss DA (2007) TrpC3/C7 and Slo2.1 are molecular targets for metabotropic glutamate receptor signaling in rat striatal cholinergic interneurons. J Neurosci 27:8845–8856
67. Faber ESL, Sedlak P, Vidovic M, Sah P (2006) Synaptic activation of transient receptor potential channels by metabotropic glutamate receptors in the lateral amygdala. Neuroscience 137:781–794
68. Meis S, Munsch T, Sosulina L, Pape H-C (2007) Postsynaptic mechanisms underlying responsiveness of amygdloid neurons to cholecystrokinin are mediated by a transient receptor potential-like current. Mol Cell Neurosci 35:356–367
69. Qiu J, Fang Y, Rønnekleiv OK, Kelly MJ (2010) Leptin excites proopiomelanocortin neurons via activation of TRPC channels. J Neurosci 30:1560–1565
70. Qiu J, Fang Y, Bosch MA, Rønnekleiv OK, Kelly MJ (2011) Guinea pig kisspeptin neurons are depolarized by leptin via activation of TRPC channels. Endocrinology 152:1503–1514
71. Birnbaumer L (2009) The TRPC class of ion channels: a critical review of their roles in slow, sustained increases in intracellular Ca^{2+} concentrations. Annu Rev Pharmacol Toxicol 49:395–426
72. Castellano JM, Navarro VM, Fernández-Fernández R, Castaño JP, Malagón MM, Aguilar E, Dieguez C, Magni P, Pinilla L, Tena-Sempere M (2006) Ontogeny and mechanisms of action for the stimulatory effect of kisspeptin on gonadotropin-releasing hormone system of the rat. Mol Cell Endocrinol 257–258:75–83
73. Ambudkar IS, Ong HL (2007) Organization and function of TRPC channelsomes. Pflügers Arch Eur J Physiol 455:187–200
74. Blair NT, Kaczmarek JS, Clapham DE (2009) Intracellular calcium strongly potentiates agonist-activated TRPC5 channels. J Gen Physiol 133:525–546
75. Clapham DE (2007) Snapshot: mammalian TRP channels. Cell 129:220
76. Sim JA, Skynner MJ, Pape JR, Herbison AE (2000) Late postnatal reorganization of $GABA_A$ receptor signalling in native GnRH neurons. Eur J Neurosci 12:3497–3504
77. Han SK, Todman MG, Herbison AE (2004) Endogenous GABA release inhibits the firing of adult gonadotropin-releasing hormone neurons. Endocrinology 145:495–499
78. Yin C, Ishii H, Tanaka N, Sakuma Y, Kato M (2008) Activation of A-type gamma-amino butyric acid receptors excites gonadotrophin-releasing hormone neurones isolated from adult rats. J Neuroendocrinol 20:566–575
79. Watanabe M, Sakuma Y, Kato M (2009) GABA receptors mediate excitation in adult rat GnRH neurons. Biol Reprod 81:327–332
80. Herbison AE, Moenter SM (2011) Depolarising and hyperpolarising actions of $GABA_A$ receptor activation on gonadotrophin-releasing hormone neurons: towards an emerging consensus. J Neuroendocrinol 23:557–569
81. Pape JP, Skynner MJ, Sim JA, Herbison AE (2001) Profiling gamma-aminobutyric acid ($GABA_A$) receptor subunit mRNA expression in postnatal gonadotropin-releasing hormone (GnRH) neurons of the male mouse with single cell RT-PCR. Neuroendocrinology 74:300–308
82. Todman MG, Han S-K, Herbison AE (2005) Profiling neurotransmitter receptor expression in mouse gonadotropin-releasing hormone neurons using green fluorescent protein-promoter transgenics and microarrays. Neuroscience 132:703–712
83. Sliwowska JH, Billings HJ, Goodman RL, Lehman MN (2006) Immunocytochemical colocalization of $GABA_B$ receptor subunits in gonadotropin-releasing hormone neurons of the sheep. Neuroscience 141:311–319
84. Liu X, Herbison AE (2011) Estrous cycle- and sex-dependent changes in pre- and postsynaptic $GABA_B$ control of GnRH neuron excitability. Endocrinology 152:1–9
85. Kelly MJ, Loose MD, Rønnekleiv OK (1992) Estrogen suppresses m-opioid and $GABA_B$-mediated hyperpolarization of hypothalamic arcuate neurons. J Neurosci 12:2745–2750
86. Lagrange AH, Wagner EJ, Rønnekleiv OK, Kelly MJ (1996) Estrogen rapidly attenuates a $GABA_B$ response in hypothalamic neurons. Neuroendocrinology 64:114–123
87. Wagner EJ, Rønnekleiv OK, Bosch MA, Kelly MJ (2001) Estrogen biphasically modifies hypothalamic GABAergic function concomitantly with negative and positive control of luteinizing hormone release. J Neurosci 21:2085–2093

88. Stern JE, Li Y, Richards DS (2002) Postsynaptic $GABA_{(B)}$ receptors in supraoptic oxytocin and vasopressin neurons. Prog Brain Res 139:121–125
89. Slugg RM, Zheng SX, Fang Y, Kelly MJ, Rønnekleiv OK (2003) Baclofen inhibits guinea pig magnocellular neurones via activation of an inwardly-rectifying K^+ conductance. J Physiol (Lond) 551:295–308
90. Qiu J, Bosch MA, Tobias SC, Grandy DK, Scanlan TS, Rønnekleiv OK, Kelly MJ (2003) Rapid signaling of estrogen in hypothalamic neurons involves a novel G protein-coupled estrogen receptor that activates protein kinase C. J Neurosci 23:9529–9540
91. Kelly MJ, Rønnekleiv OK (1994) Electrophysiological analysis of neuroendocrine neuronal activity in hypothalamic slices. In: Levine JE (ed) Methods in neurosciences: pulsatility in neuroendocrine systems. Academic, San Diego, pp 47–67
92. Lagrange AH, Rønnekleiv OK, Kelly MJ (1994) The potency of μ-opioid hyperpolarization of hypothalamic arcuate neurons is rapidly attenuated by 17β-estradiol. J Neurosci 14:6196–6204
93. Herbison AE (1997) Estrogen regulation of GABA transmission in rat preoptic area. Brain Res Bull 44:321–326
94. Jackson GL, Kuehl D (2002) Gamma-aminobutyric acid (GABA) regulation of GnRH secretion in sheep. Reproduction 59:15–24
95. Christian CA, Moenter SM (2007) Estradiol induces diurnal shifts in GABA transmission to gonadotropin-releasing hormone neurons to provide a neural signal for ovulation. J Neurosci 27:1913–1921
96. Pielecka-Fortuna J, Moenter SM (2010) Kisspeptin increases γ-aminobutyric acidergic and glutamatergic transmission directly to gonadotropin-releasing hormone neurons in an estradiol-dependent manner. Neuroendocrinology 151:291–300
97. Logothetis DE, Zhang H (1999) Gating of G protein-sensitive inwardly rectifying K^+ channels through phosphatidylinositol 4,5-bisphosphate. J Physiol 520:630
98. Kobrinsky E, Mirshahi T, Zhang H, Jin T, Logothetis DE (2000) Receptor-mediated hydrolysis of plasma membrane messenger PIP2 leads to K^+-current desensitization. Nat Cell Biol 2:507–514
99. Qiu J, Xue C, Bosch MA, Murphy JG, Fan W, Rønnekleiv OK, Kelly MJ (2007) Serotonin 5HT2c receptor signaling in hypothalamic POMC neurons: role in energy homeostasis in females. Mol Pharm 72:885–896
100. Simerly RB (1989) Hormonal control of the development and regulation of tyrosine hydroxylase expression within a sexually dimorphic population of dopaminergic cells in the hypothalamus. Mol Brain Res 6:297–310
101. Ottem EN, Godwin JG, Krishnan S, Petersen SL (2004) Dual-phenotype GABA/glutamate neurons in adult preoptic area: sexual dimorphism and function. J Neurosci 24:8097–8105
102. Liu X, Porteous R, d'Anglemont de Tassigny X, Colledge WH, Millar R, Petersen SL, Herbison AE (2011) Frequency-dependent recruitment of fast amino acid and slow neuropeptide neurotransmitter release controls gonadotropin-releasing hormone neuron excitability. J Neurosci 31:2421–2430
103. Fu L-Y, van den Pol AN (2010) Kisspeptin directly excites anorexigenic proopiomelanocortin neurons but inhibits orexigenic neuropeptide Y cells by an indirect synaptic mechanism. J Neurosci 30:10205–10219
104. Leranth C, MacLusky NJ, Shanabrough M, Naftolin F (1988) Immunohistochemical evidence for synaptic connections between pro-opiomelanocortin-immunoreactive axons and LH-RH neurons in the preoptic area of the rat. Brain Res 449:167–176
105. Chen W-P, Witkin JW, Silverman AJ (1989) β-endorphin and gonadotropin-releasing hormone synaptic input to gonadotropin-releasing hormone neurosecretory cells in the male rat. J Comp Neurol 286:85–95
106. Thornton JE, Loose MD, Kelly MJ, Rønnekleiv OK (1994) Effects of estrogen on the number of neurons expressing β-endorphin in the medial basal hypothalamus of the female guinea pig. J Comp Neurol 341:68–77
107. Zheng SX, Bosch MA, Rønnekleiv OK (2005) Mu-opioid receptor mRNA expression in identified hypothalamic neurons. J Comp Neurol 487:332–344

108. Israel DD, Sheffer-Babila S, de Luca C, Jo Y-H, Lui SM, Xia Q, Spergel DJ, Dun SL, Dun NJ, Chua SC Jr (2012) Effects of leptin and melanocortin signaling interactions on pubertal development and reproduction. Endocrinology 153:1–11
109. Bicknell RJ (1988) Optimizing release from peptide hormone secretory nerve terminals. J Exp Biol 139:51–65
110. Masterson SP, Li J, Bickford ME (2010) Frequency-dependent release of substance P mediates heterosynaptic potentiation of glutamatergic synaptic responses in the rat visual thalamus. J Neurophysiol 104:1758–1767
111. Shakiryanova D, Tully A, Hewes RS, Deitcher DL, Levitan ES (2005) Activity-dependent liberation of synaptic neuropeptide vesicles. Nat Neurosci 8:173–178
112. Kelly MJ, Condon TP, Levine JE, Rønnekleiv OK (1985) Combined electrophysiological, immunocytochemical and peptide release measurements in the hypothalamic slice. Brain Res 345:264–270
113. Krsmanovic LZ, Stojikovic SS, Merelli F, Dufour SM, Virmani MA, Catt KJ (1992) Calcium signaling and episodic secretion of gonadotropin-releasing hormone in hypothalamic neurons. Proc Natl Acad Sci USA 89:8462–8466
114. Constantin S, Caraty A, Wray S, Duittoz AH (2009) Development of gonadotropin-releasing hormone-1 secretion in mouse nasal explants. Endocrinology 150:3221–3227
115. Lee K, Duan W, Sneyd J, Herbison AE (2010) Two slow calcium-activated afterhyperpolarization currents control burst firing dynamics in gonadotropin-releasing hormone neurons. J Neurosci 30:6214–6224
116. Zhang X-B, Spergel DJ (2012) Kisspeptin inhibits high-voltage activated Ca^{2+} channels in GnRH neurons via multiple Ca^{2+} influx and release pathways. Neuroendocrinology 96:68–80
117. Kato M, Ui-Tei K, Watanabe M, Sakuma Y (2003) Characterization of voltage-gated calcium currents in gonadotropin-releasing hormone neurons tagged with green fluorescent protein in rats. Endocrinology 144:5118–5125
118. Zhang C, Bosch MA, Rick EA, Kelly MJ, Rønnekleiv OK (2009) 17β-estradiol regulation of T-type calcium channels in gonadotropin-releasing hormone neurons. J Neurosci 29:10552–10562
119. Chu Z, Takagi H, Moenter SM (2010) Hyperpolarization-activated currents in gonadotropin-releasing hormones (GnRH) neurons contribute to intrinsic excitability and are regulated by gonadal steroid feedback. J Neurosci 30:13373–13383
120. Bosch MA, Tonsfeldt KJ, Ronnekleiv OK (2013) mRNA expression of ion channels in GnRH neurons subtype-specific regulation by 17beta-estradiol. Mol Cell Endocrinol. DOI 10.1016/j.mce.2012.12.021
121. Kelly MJ, Wagner EJ (2002) GnRH neurons and episodic bursting activity. Trends Endocrinol Metab 13:409–410
122. Huguenard JR, McCormick DA (1992) Simulation of the currents involved in rhythmic oscillations in thalamic relay neurons. J Neurophysiol 68:1373–1383
123. Bosch MA, Kelly MJ, Rønnekleiv OK (2002) Distribution, neuronal co-localization and 17β-E2 modulation of small conductance calcium-activated K^+ channel (SK3) mRNA in the guinea pig brain. Endocrinology 143:1097–1107
124. Kato M, Tanaka N, Usui S, Sakuma Y (2006) SK channel blocker apamin inhibits slow afterhyperpolarization currents in rat gonadotropin-releasing hormone neurones. J Physiol 574(2):431–442
125. Liu X, Herbison AE (2008) Small-conductance calcium-activated potassium channels control excitability and firing dynamics in gonadotropin-releasing hormone (GnRH) neurons. Endocrinology 149:3598–3604
126. Otsuguro K-I, Tang J, Tang Y, Xiao R, Freichel M, Tsvilovskyy V, Ito S, Flockerzi V, Zhu MX, Zholos AV (2008) Isoform-specific inhibition of TRPC4 channel by phosphatidylinositol 4,5-bisphosphate. J Biol Chem 283:10026–10036

Chapter 7
Molecular Biology of the Kisspeptin Receptor: Signaling, Function, and Mutations

Suzy Drummond Carvalho Bianco and Ursula B. Kaiser

Abstract Kisspeptin receptor (KISS1R) signaling is essential for the hallmark increase in pulsatile GnRH secretion characteristic of the onset of puberty in humans and experimental animals. Loss-of-function mutations in KISS1R are associated with idiopathic hypogonadotropic hypogonadism in humans. Also, mutations with confirmed association with idiopathic central precocious puberty were identified in kisspeptin and KISS1R. These observations underscore the role of KISS1R signaling for normal pubertal development. Moreover, investigation of the mechanisms underlying the gain-of-function mutation in KISS1R indicates that the duration of KISS1R signaling is critical for the role of this receptor in *timing* the onset of puberty in humans. These findings further endorse the need to uncover the mechanisms, as well as yet-unknown proteins, involved in each step of KISS1R signaling. This knowledge is expected to advance our understanding of normal and abnormal pubertal development, as well as to help uncover the role of KISS1R signaling in non-hypothalamic tissues such as the placenta. This chapter discusses recent advances in the investigation of KISS1R signaling and function, as well as potential pathophysiological implications of naturally occurring mutations in this receptor identified in humans with reproductive disorders.

S.D.C. Bianco
Department of Molecular and Cellular Pharmacology and Department of Medicine, University of Miami Miller School of Medicine, Miami, FL, USA
e-mail: SBianco@med.miami.edu

U.B. Kaiser, M.D. (✉)
Division of Endocrinology, Department of Medicine, Harvard Medical School/Brigham and Women's Hospital, 221 Longwood Avenue, Boston, MA 02115, USA
e-mail: ukaiser@partners.org

A.S. Kauffman and J.T. Smith (eds.), *Kisspeptin Signaling in Reproductive Biology*, Advances in Experimental Medicine and Biology 784, DOI 10.1007/978-1-4614-6199-9_7,

Introduction

The initiation of puberty onset is first detected as an increase in frequency and amplitude of gonadotropin-releasing hormone (GnRH) pulses by the hypothalamus, which is followed by increased secretion of the gonadotropins, luteinizing hormone (LH) and follicle-stimulating hormone (FSH), by the pituitary gland [1]. Failure to increase GnRH or gonadotropin secretion during puberty is the underlying cause of idiopathic hypogonadotropic hypogonadism (IHH) [2]. Conversely, premature activation of GnRH pulsatility leads to idiopathic gonadotropin-dependent (or central) precocious puberty (ICPP). The upstream mechanisms driving the hallmark increase in GnRH pulsatility during puberty, however, are not well defined. Insights into these mechanisms have been provided by the identification and characterization of mutations associated with reproductive disorders in affected patients [3–7]. As a result of these studies, an increasing array of genes has been implicated in the control of pulsatile GnRH secretion and in the etiology of central reproductive disorders. The involvement of kisspeptin and its cognate G protein-coupled receptor (GPCR), kisspeptin receptor (KISS1R), in puberty and reproductive function was not recognized until 2003, when two independent groups identified loss-of-function mutations in *KISS1R* in unrelated patients with a family history of IHH and normal sense of smell (normosmic IHH or nIHH) [8, 9]. Reports of additional loss-of-function mutations in *KISS1R* or its ligand in association with nIHH in affected patients have followed [10–19]. Moreover, targeted disruption of the *Kiss1r*, or of the *Kiss1* gene encoding the ligand, results in a similar phenotype of hypogonadotropic hypogonadism and infertility in mice [9, 20–23], with *Kiss1r* null mice displaying a more severe phenotype than *Kiss1* null mice [24]. Nevertheless, mutations in *Kiss1r* are predicted to account for a small percentage (~5%) of the total human cases of nIHH [7]. This percentage reaches about 20% if only familial cases of normosmic IHH are considered [7].

Five years after the publication of the first cases of *KISS1R* mutants associated with IHH, the critical role of KISS1R on pubertal development was further reinforced when a single amino acid substitution in KISS1R in a girl with CPP was identified. This amino acid substitution was the first genetic mutation reported to be associated with CPP [6]. Two years later, a second mutation was identified in a boy with CPP, this time in the KISS1R ligand, kisspeptin [25]. An additional mutation in kisspeptin was identified in the heterozygous state in two unrelated Brazilian girls with CPP; however, in vitro analyses fail to identify associated alterations in kisspeptin function or activity [25]. Genome-wide association studies have shown significant association of two polymorphisms—one in *KISS1* and the other in *KISS1R*—with CPP in a Chinese population [26, 27]; however, no functional assays have been performed.

To date, kisspeptin/KISS1R is the only ligand/receptor system with confirmed association of identified mutations with both IHH and precocious puberty phenotypes in affected patients, which underscores the role of this ligand/receptor pair in the onset of puberty. This chapter will discuss details of KISS1R trafficking and signaling and their implications for the control of GnRH secretion, as well as the mechanisms by which genetic mutations in this receptor or its ligand kisspeptin may affect these processes, thereby leading to adverse reproductive outcomes.

Kisspeptin Signaling and Onset of Puberty

KISS1R is Expressed in GnRH Neurons

Expression of Kiss1r in hypothalamic GnRH neurons has been reported in cichlid fish [28], rats [29], mice [30], and Rhesus monkeys [31]. Activation of Kiss1r by kisspeptin evokes a powerful and enduring depolarization in more than 90% of GnRH neurons in adult male and female mice [32]. Although the precise mechanisms by which Kiss1r signaling leads to this potent depolarization of GnRH neuron are not clear, this effect is likely associated with the distinctive pulsatile release of GnRH. A significant increase in kisspeptin secretion has been shown in association with the pubertal increase in GnRH release in female Rhesus monkeys [33]. Additionally, GnRH pulses were shown to correlate with kisspeptin pulses in the stalk-median eminence of these females [33].

In mice, essentially all Kiss1r-expressing cells in the rostral preoptic area of the hypothalamus are GnRH neurons [34]. Interestingly, the percentage of GnRH neurons expressing Kiss1r is sharply reduced after birth, before progressively increasing to 70% by postnatal day 20 [34]. While only 27% of the GnRH neurons appear to be activated by kisspeptin in 8–19-day-old mice; this percentage increases to 44% in 26–33-day-old mice, which may be interpreted as a progressive increase in GnRH neuron responsiveness to kisspeptin stimulation at puberty [32]. On the other hand, the release of inhibitory input contributes significantly to the progressive increase in GnRH pulsatility observed at puberty. As an example, a recent study in monkeys shows that infusion of a GABA antagonist (bicuculline) increases GnRH pulsatility and accelerates menarche in female monkeys [35]. Interestingly, this effect was shown to be mediated by kisspeptin secretion and signaling [35]. This suggests that the release of GABAergic inhibition of kisspeptin plays a significant role in the onset of puberty in monkeys.

KISS1R, Puberty, and Pubertal Disorders

The progressive responsiveness of GnRH neurons to the powerful depolarizing effects of KISS1R during postnatal development suggests that signaling by KISS1R is a key event in the process driving the hallmark increase in GnRH pulses during puberty [32]. The essential role of KISS1R signaling for the onset of puberty is confirmed by the phenotype exhibited by humans carrying naturally occurring loss-of-function (or inactivating) mutations in *KISS1R*, who present with IHH [8–11, 13–15], a syndrome characterized by lack of sexual maturation and infertility associated with low levels of circulating gonadotropins and sex steroids [2]. This is further supported by a similar IHH phenotype exhibited by mice lacking Kiss1r [21–23, 36] or kisspeptin [20, 23] expression. Interestingly, mice lacking Kiss1r expression exhibit a reproductive phenotype that is more severe than that of mice lacking kisspeptin. This raises the potential for additional yet-unknown endogenous ligands for Kiss1r other than those expressed by the *Kiss1* gene. However, failed attempts to identify other endogenous ligands [37, 38], as well as the severity of the phenotype of kisspeptin null mice [20, 23], do not support the existence of additional endogenous ligands for Kiss1r. Another possibility would be ligand-independent signaling by the Kiss1r, which is supported by studies showing constitutive activity (in the absence of ligand) of KISS1R in transfected cells [39, 40]. Regardless, the basis for the more severe phenotype in Kiss1r null mice remains to be established. Nonetheless, it is similar to the more severe phenotype exhibited in other mouse models of receptor disruption when compared to that resulting from the disruption of the cognate ligands.

Conversely, the first genetic mutation to be associated with idiopathic gonadotropin-dependent (or central) precocious puberty (CPP), a syndrome characterized by the unexplained premature activation of the hypothalamic-pituitary-gonadal (or reproductive) axis, was a single amino acid substitution in KISS1R (Arg386Pro-KISS1R) identified in an affected girl [6]. Functional characterization indicates that the Arg386Pro substitution impairs KISS1R degradation, which results in gain-of-function [6, 41].

These opposing phenotypes of patients carrying loss- or gain-of-function mutations in *KISS1R* emphasize the critical role of KISS1R signaling for normal reproductive maturation and function, and highlight the potential disease risks of abnormal KISS1R function. In order to prevent abnormal KISS1R activity or function and its associated risks, it is important to understand the precise mechanisms involved. Recent advances in this area include studies of KISS1R intracellular trafficking and binding partners, as well as the use of naturally occurring human mutations, in order to understand the mechanisms by which these mutations affect receptor function. The confirmed or predicted molecular mechanisms underlying loss- or gain-of-function effects of reported KISS1R mutations associated with pubertal disorders will be discussed below.

KISS1R Signaling Pathway

The gene encoding KISS1R was initially named *GPR54* for being the 54th orphan GPCR identified by sequence homology [42], but is now typically referred to as *KISS1R* [43]. GPCRs are plasma membrane receptors that share a seven transmembrane spanning domain organization in which the amino-terminus is extracellular

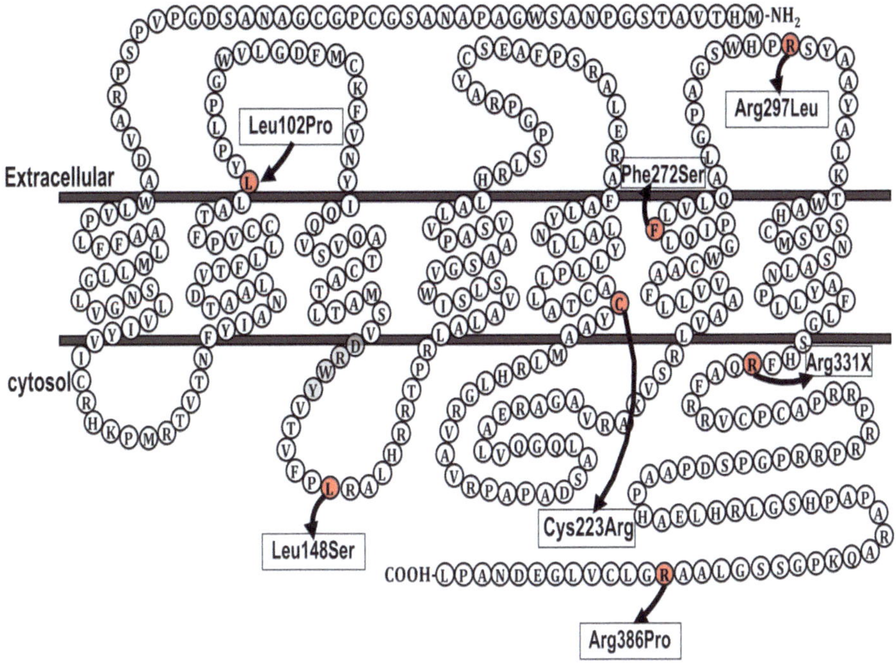

Fig. 7.1 Predicted membrane topology of KISS1R. The typical G protein-coupled receptor topography predicted for KISS1R is depicted with its seven transmembrane domains as well as the three intracellular and three extracellular loops. The location of key genetic mutations identified in humans with reproductive disorders is shown

and the carboxyl-terminus is cytosolic (Fig. 7.1). In addition to the amino- and the carboxyl-terminus, all GPCRs have three extracellular loops and three intracellular (or cytosolic) loops. The predicted seven-transmembrane structure for KISS1R is represented in Fig. 7.1. All GPCRs couple to G proteins, which in the case of KISS1R is primarily $G_{q/11}$. Further analysis of *KISS1R* revealed the presence of consensus sequences encoding residues typical of the Rhodopsin superfamily of GPCRs and a significant degree of identity with the galanin receptor-2. However, no binding to KISS1R was detected for galanin receptor ligands [42]. Kisspeptin (then named metastin) was identified as the endogenous ligand for KISS1R in 2001 [44, 45].

G proteins are composed of three subunits, α, β, and γ, which form a trimeric structure that binds to cytosolic domains of GPCRs. The trimeric structure is inactive, as the inactive, GDP-bound G_α subunit has high affinity for the tightly bound βγ dimer. Activation of the cognate GPCR results in the release of GDP bound to the G_α-subunit, which then binds GTP. The GTP-bound G_α subunit dissociates from the G protein trimer, which thus dissociates from the cognate receptor to activate downstream signaling cascades. Dissociated G protein subunits are active only briefly, as intrinsic GTPase activity within the G_α subunit breaks down the GTP to

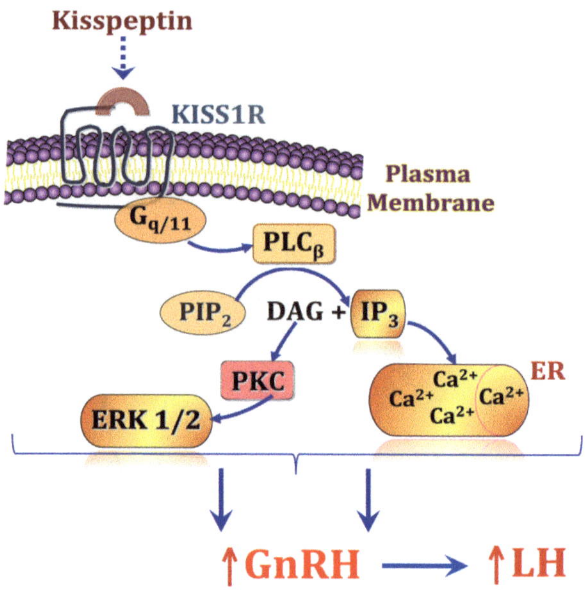

Fig. 7.2 KISS1R signaling in the hypothalamus. Upon kisspeptin binding, G proteins are dissociated and the $G\alpha_{q/11}$ subunit activates phospholipase C beta (PLC_β) to hydrolyze membrane phosphatidyl inositol 4,5-biphosphate (PIP_2), producing inositol triphosphate (IP_3) and diacylglycerol (DAG). IP_3 binds to receptors on the endoplasmic reticulum (ER) to release stored calcium (Ca^{2+}) and DAG contributes to the activation of the calcium-dependent protein kinase (PKC). The extracellular regulated kinase (ERK) is one of the PKC-dependent downstream signals of KISS1R. These signals ultimately lead to increased secretion of GnRH, which then leads to increased LH secretion

GDP, thereby inactivating G_α, which then binds to $G_{\beta\gamma}$ again. Dissociated $G_{\alpha q}$ stimulates phospholipase C_β, a membrane-resident enzyme that hydrolyzes phosphatidylinositol 4,5-bisphosphate (PIP_2) to produce the intracellular messengers diacylglycerol (DAG) and inositol 1,4,5-trisphosphate (IP_3) (Fig. 7.2). Binding of IP_3 to its receptors on the endoplasmic reticulum causes a massive release of calcium from the intracellular stores into the cytosol, with widespread activation of calcium-dependent signaling pathways, whereas DAG contributes to activation of the calcium-dependent protein kinase C (PKC) (Fig. 7.2). Accordingly, stimulation of KISS1R with kisspeptin in Chinese hamster ovary (CHO) cells transfected with the human (or rat) receptor leads to phospholipase C_β activation, PIP_2 hydrolysis and intracellular IP_3 accumulation [37]. Calcium-dependent signals activated by KISS1R include kinases such as PKC, the extracellular signal-regulated kinases 1 and 2 (ERK1/2) and the p38-mitogen-activated protein kinase (p38-MAPK) [44, 45]. The stimulatory effect of kisspeptin on GnRH secretion was also shown to require calcium mobilization, as well as ERK and p38MAPK activation in GnRH neurons of rats [46].

Intracellular Trafficking of KISS1R

KISS1R signaling is essential for the onset and duration of puberty; and abnormal KISS1R signaling may lead to pubertal disorders. As an example, functional assays performed to confirm the association of a gain-of-function mutation in KISS1R with the CPP phenotype of the patient indicated that the mutation affects the duration of KISS1R signaling, suggesting an effect of the mutation on KISS1R desensitization. Thus, investigation of precise mechanisms regulating activation and duration of KISS1R signaling may uncover new proteins or new mechanisms underlying pubertal disorders in patients carrying no mutations in genes currently known to be involved in reproductive disorders.

Recent advances in this area include studies of ligand-induced KISS1R desensitization and internalization as well as KISS1R fate after internalization. As described for other GPCRs, KISS1R was shown to undergo time-dependent ligand-induced receptor desensitization, which occurs in spite of the continuous presence of ligand [41]. This confirms and further explains previous observations of desensitization of biological effects in response to continuous stimulation of KISS1R in vivo [47–49]. These observations include selective desensitization of the LH response to kisspeptin stimulation in female rats [47], as well as in agonadal Rhesus monkeys [48], both of which did not prevent subsequent gonadotropin secretion in response to other secretagogues. Moreover, desensitization of the gonadotropin response to continuous Kiss1r stimulation in male rats is reported to result in testicular degeneration [49]. The testicular degeneration is a consequence of the desensitization of KISS1R signaling, which is typically observed upon continuous stimulation in vitro [6, 41]. Desensitization of biological responses to kisspeptin stimulation such as LH secretion has been described after continuous infusion of kisspeptin in Rhesus monkeys [48, 50] or humans [51]. Surprisingly, continuous infusion of submaximal doses of kisspeptin resulted in a sustained increase in LH pulse frequency in men, suggesting that some responses to kisspeptin elude desensitization [51]. In women treated with a bolus of kisspeptin, serum LH responses varied according to the phase of the menstrual cycle, suggesting variations in the sensitivity to kisspeptin across the menstrual cycle [52].

Acute desensitization of GPCRs is a consequence of uncoupling of the receptor from its signaling pathway, which is typically followed by receptor internalization. This appears to be the case for KISS1R, which has been shown to undergo time-dependent desensitization, despite continuous presence of ligand [41]. Internalization of GPCR is often mediated by arrestin, a membrane-resident protein that binds phosphorylated receptors, prevents further G protein coupling and signaling, and triggers receptor internalization [53–56]. Accordingly, a study using fluorescence microscopy showed that a Flag-tagged KISS1R undergoes rapid ligand-induced internalization in HEK cells [40]. This was confirmed in time-course experiments in which internalization and recycling of KISS1R was measured using ^{125}I-kisspeptin binding [41].

KISS1R internalization may be triggered, at least in part, by G protein-coupled receptor kinase-2 (GRK2)-mediated phosphorylation, as overexpression of GRK2 inhibited G protein-mediated signaling by KISS1R. Conversely, overexpression of a catalytically inactive GRK2 mutant (K220R-GRK2) led to increased G protein signaling by the KISS1R [40]. GRKs are enzymes that selectively phosphorylate activated GPCRs, leading to recruitment and binding of arrestin to the phosphorylated receptor, preventing further G protein coupling or signaling and triggering arrestin-dependent GPCR internalization [56, 57]. Accordingly, KISS1R has been shown to colocalize with β-arrestin-2, which provides further support for arrestin-mediated internalization of KISS1R [58]. Additionally, Saereszewsky et al. showed that overexpression of β-arrestin-2 potentiates the phosphorylation of ERK in fibroblasts lacking arrestin or $G_{q/11}$ and transfected with KISS1R. Conversely, overexpression of β-arrestin-1 in these fibroblasts inhibits KISS1R-dependent ERK phosphorylation, which suggests that this KISS1R effect may specifically require β-arrestin-2 [58].

Following internalization, GPCRs are eventually sorted for recycling or destruction. Destruction of GPCRs is classically mediated by lysosomal degradation, which is responsible for long-term GPCR desensitization [55]. Intriguingly, confocal microscopy of KISS1R failed to detect lysosomal targeting of KISS1R under any experimental conditions studied, including stimulation with a supraphysiological concentration of kisspeptin (10^{-7} M) for up to 5 h [41]. This surprising finding was confirmed by western blot analysis, which showed that levels of KISS1R protein are not affected by treatment with a lysosome inhibitor, which suggests that, unlike most GPCRs, lysosomal degradation of KISS1R is low or absent [41]. Conversely, treatment of the same cells with an inhibitor of proteasomal degradation resulted in massive increases in KISS1R protein when compared to untreated cells, which suggests that KISS1R may undergo proteasomal (rather than lysosomal) degradation [41]. Although unusual, proteasomal degradation has been reported for some GPCRs [59, 60].

Further investigation using confocal microscopy revealed persistent membrane localization of KISS1R for up to 5 h of stimulation with supraphysiological concentrations of kisspeptin [41]. This suggests a dynamic pattern of recycling of internalized KISS1R back to the cell surface, which was confirmed in studies measuring the rate of membrane trafficking of KISS1R using ^{125}I-kisspeptin binding [41]. This behavior indicates that a substantial amount of kisspeptin-dependent KISS1R signaling is, in fact, due to the binding of kisspeptin to newly recycled KISS1R, which demonstrates that KISS1R recycling is relevant for its function [41].

Inactivating Mutations in KISS1R and Normosmic IHH

Central hypogonadotropic hypogonadism refers to delayed or absent sexual maturation due to a central nervous system defect that results in low circulating gonadotropins and sex steroids [2]. These cases are further classified as idiopathic when the source of the hypogonadotropic hypogonadism cannot be identified. IHH affects 1–10 individuals per 100,000 births and its incidence is sexually dimorphic, with males exhibiting a fivefold higher incidence compared to females [2, 61, 62]. Approximately 60% of affected patients have associated anosmia—decreased or absent sense of smell [63]. The underlying cause of the IHH in these patients is defective developmental migration of GnRH neurons from the nasal placode to their final destination in the pre-optic area of the hypothalamus during embryonic development [64]. Mutations identified in patients with IHH associated with anosmia have been traced to genes encoding proteins that regulate GnRH and olfactory neuronal migration during development [65]. In contrast to anosmic patients, the majority of mutations identified in the remaining 40% of patients with a normal sense of smell (nIHH) encode proteins that regulate GnRH secretion or action, such as the KISS1R [66].

The analysis of pedigrees in families with a history of nIHH often indicates an *autosomal recessive* inheritance pattern (Table 7.1). This pattern has been confirmed in families carrying inactivating mutations in KISS1R [8, 9, 11, 12]. Affected members of these families carry the associated loss-of-function mutation in the *homozygous* (or compound heterozygous) state, whereas *heterozygous* parents and siblings have no apparent reproductive phenotype. This suggests that one wild type allele is sufficient for the effect of KISS1R on GnRH secretion in heterozygous patients. On the other hand, heterozygous mutations in *KISS1R* may contribute to milder phenotypes [16, 67].

All inactivating mutations in *KISS1R* identified to date have been shown or predicted to impair G protein signaling by the receptor [8, 9, 11, 12, 14], indicating a vital role of G protein signaling for GnRH secretion. Since affected patients have normal sense of smell, inactivating mutations in this receptor do not appear to affect GnRH neuronal migration during development. Please refer to Table 7.1 for the list of inactivating mutations discussed below.

The Leu148Ser-KISS1R Mutation

A point mutation in the coding sequence of the *KISS1R* gene was identified in the homozygous state in all affected members of a consanguineous Saudi-Arabian family with history of nIHH. This highly consanguineous family includes 3 marriages of first-degree cousins, who gave birth to a total of 16 children, 6 of them presenting with symptoms compatible with complete IHH [9].

This point mutation resulted in the replacement of a leucine residue at position 148 of the KISS1R with a serine (Leu148Ser-KISS1R). This leucine is located within the second intracellular loop of the receptor (Fig. 7.1) and is highly conserved among the members of class A GPCRs, strongly suggestive that this leucine is essential for receptor function [68]. Nonetheless, heterozygous parents and siblings of affected patients in this family have no apparent phenotype [9], in agreement with the predicted autosomal recessive mode of inheritance (Table 7.1).

Functional assays showed that the Leu148Ser-KISS1R mutant has impaired G protein signaling, as indicated by the absence of increased inositol phosphate

Table 7.1 KISS1 and KISS1R mutations associated with idiopathic hypogonadotropic hypogonadism or idiopathic central precocious puberty

Disorder	DNA mutation	Protein mutation	Location	Inheritance mode	Status	Affected	KISS1R Signaling	Reference
IHH	443T>C	Leu148Ser (highly conserved)	ICL-2	Autosomal recessive	Familial	4 males + 2 females	Impaired G protein signaling (inositol phosphate + pERK); ↓membrane localization	[9, 66]
	305T>C	Leu102Pro	ECL-1 (beginning)	Autosomal recessive	Familial	1 males + 2 females (less severe in sisters)	Impaired G protein signaling	[14]
	305T>C	Leu102Pro	ECL-1 (beginning)	Autosomal recessive	Familial	1 female	Impaired G protein signaling	[14]
	T815>C	Phe272Ser	TM-6 (beginning)	Autosomal recessive	Familial	5 males + 1 female (less severe in sister)	Impaired G protein signaling; no membrane localization	[11]
	345C>G	N115K-KiSS-1 (highly conserved)	within the decapeptide	Autosomal recessive	Familial	4 females	Impaired G protein signaling (↓inositol phosphate production)	[18]
	IVS4-13-142del15	truncated protein (presumed)	ICL-3	Autosomal recessive	Familial	4 males + 1 female	Not tested	[8]
	IVS2-4-2delGC insACCG-GCT	aberrant protein (presumed)	3′-end splice site - intron2	Autosomal recessive	Familial	2 males	Not tested	[15]

	991C>T	Arg331Stop	C-tail (start)	Compound heterozygous	Unknown	1 male	Impaired G protein signaling (inositol phosphate production); ↓expression	[9, 54]
	1195T>A	Stop399Arg	C-tail (end)	Compound heterozygous	Unknown	1 male	Decreased G protein signaling (inositol phosphate production); ↓expression	[9]
	667T>C	Cys223Arg	TM-5 (end)	Compound heterozygous	Unknown	1 male	Impaired signaling (Ca^{2+} mobilization)	[13]
	891G>T	Arg297Leu	ECL-3 (middle)	Compound heterozygous	Familial	1 male	Slightly impaired signaling (Ca^{2+} mobilization)	[13]
	1101-1102_insC	Frameshift	C-tail(middle)	Autosomal recessive	Sporadic	1 male	Not tested	[10]
ICPP	1157G>C	Arg386Pro (conserved)	C-tail (middle)	Autosomal dominant	Unknown	1 female	Prolonged G protein signaling (inositol phosphate + pERK); ↓degradation of KISS1R	[6]
	369C>T	Pro74Ser-KISS1 (highly conserved)	PEST site (predicted)	Autosomal dominant	Sporadic	1 male	↑ G protein signaling (inositol phosphate); ↓degradation of kisspeptin	[17]

ICL intracellular loop; *ECL* extracellular loop; *TM* transmembrane domain; *C-tail* carboxyl terminal tail; *pERK* phosphorylated ERK (extracellular-regulated kinase) production in response to kisspeptin stimulation in cells transfected with this mutant receptor [9]. On the other hand, biochemical and immunocytochemical assays demonstrated that this mutation did not affect expression levels, ligand binding, or interaction of KISS1R with arrestin [58, 68]. However, studies using fluorescence resonance energy transfer (FRET assay) indicate that this mutation prevents kisspeptin-induced activation of $G\alpha_q$ [68]. Accordingly, substitution of the corresponding leucine in the second intracellular loop of the α_{1A}-adrenergic receptor, another member of the class A GPCR family, mimics the effects of the Leu148Ser mutation on KISS1R function, suggesting that leucine in this position may belong to a functional motif within the second intracellular loop of class A GPCRs with a role in G protein coupling and/or activation of G protein by ligand [68].

The Leu148Ser-KISS1R Mutant and Fertility After Therapy

Treatment of four male members of the Saudi-Arabian family carrying the homozygous Leu148Ser-KISS1R mutation with exogenous gonadotropins resulted in testicular maturation, spermatogenesis, ejaculation, and subsequent fertility [12]. Similarly, an affected female from this family was able to ovulate, and later become pregnant upon pulsatile GnRH treatment followed by exogenous gonadotropin therapy. She had two unsuccessful pregnancies before carrying a pregnancy to term and delivering a healthy baby. The first was an ectopic pregnancy and the second gestation was lost spontaneously at 6 months [12], raising the possibility that some of her reproductive defects may not have been fully reversed by her treatments. Nonetheless, she was able to conceive a second child years later, which was delivered by cesarean section [12].

The Leu102Pro KISS1R Mutation

A single amino acid substitution in KISS1R (Leu102Pro) identified in patients with normosmic IHH and born to consanguineous marriages was also shown to result in loss of function (Table 7.1) [12, 14]. Located on the first extracellular loop of KISS1R (Fig. 7.1; Table 7.1), this mutation decreases membrane expression of the KISS1R and impairs receptor signaling [14]. As observed for other loss-of-function mutations in KISS1R, heterozygous carriers of Leu102Pro KISS1R have no apparent reproductive phenotype, whereas homozygous carriers present with partial or complete gonadotropin deficiency [14].

A female carrying the Leu102Pro KISS1R mutation in the homozygous state was born to first-degree cousins and presented with primary amenorrhea and incomplete pubertal development. After steroid supplementation for 6 months and two cycles of GnRH stimulation, this patient was able to ovulate and become pregnant. She was able to carry two pregnancies to term and deliver healthy babies [14].

A male patient born to an unrelated consanguineous family and also carrying the Leu102Pro mutation in the homozygous state presented with congenital hypogonadism, as indicated by micropenis and bilateral cryptorchidism at birth [14]. Testosterone treatment did not produce significant changes in testicular volume or facial hair growth in this patient, despite improving his pubic hair growth. He had two sisters with primary amenorrhea, one with incomplete pubertal development and the other with a history of infertility for 5 years [14]. All affected siblings carried the Leu102Pro KISS1R mutation in the homozygous state. Interestingly, the females in this family appear to have a milder phenotype when compared to their brother.

The Phe272Ser-KISS1R Mutation

A recent study identified a novel loss-of-function single amino acid substitution in KISS1R in six patients from two related and highly consanguineous families (Table 7.1) [11]. All five affected males and the one affected female were shown to carry the Phe272Ser KISS1R mutation in the homozygous state, whereas heterozygous parents and siblings had no apparent reproductive phenotype [11]. The replacement of phenylalanine at position 272 of KISS1R with serine (Fig. 7.1) resulted in cryptorchidism, short penis, azoospermia, gonadotropin deficiency, and no spontaneous pubertal development in all affected males, whereas the affected female exhibited primary amenorrhea with low basal LH and undetectable estradiol at age 18 [11]. Functional assays showed impaired G_q signaling by the Phe272Ser-KISS1R mutant, which also lacked membrane localization as indicated by immunofluorescence imaging [11]. Despite the severity of the phenotype in homozygous carriers, pubertal development of a heterozygous male sibling was not affected, again confirming the predicted autosomal recessive inheritance mode.

Compound Heterozygous KISS1R Mutations and IHH

The combination of two distinct mutations carried by separate alleles may also result in loss-of-function. Called *compound heterozygous*, such mutations have been identified in *KISS1R* in association with the IHH phenotype. An example is the insertion of a stop codon in exchange for the normal arginine at position 331 of the KISS1R on one allele (Arg331X) (Fig. 7.1), whereas the other allele had the normal stop codon replaced with an arginine (X399Arg) in a patient with IHH (Table 7.1) [9]. The introduction of the stop codon at position 331 generates a truncated receptor, whereas the replacement of the normal stop codon could result in receptor misfolding, as functional assays revealed impaired signaling by both mutants [9]. The affected male patient presented with azoospermia with germinal hypoplasia and had no LH pulses [12]. Nonetheless, a single intravenous injection of GnRH resulted in a robust LH and FSH response in this patient. Despite the positive response and the signs of sexual maturation, this patient was able to achieve fertility only after a year and a half of GnRH therapy [12].

Another pair of compound heterozygous mutations was identified in a boy with hypogonadism [13]. One allele had leucine replacing the normal arginine at position 297 (Arg297Leu) whereas the other allele had arginine replacing the normal cysteine at position 223 of the KISS1R (Cys223Arg) (Fig. 7.1). The affected boy was born with microphallus and cryptorchidism, and had undetectable serum gonadotropins at 2 months of age [13]. Evaluation at age 10 years showed poor gonadotropin and testosterone response to GnRH or hCG stimulation. The boy's mother and a younger brother, both heterozygous for the Arg297Leu mutation, had no signs of hypogonadism [13].

Calcium release by the Cys223Arg-KISS1R mutant was shown to be impaired. The mutant receptor required 40 times more kisspeptin to reach the response level of the wild-type KISS1R (i.e., EC_{50} was 40-fold lower for the Cys223Arg KISS1R mutant) [13]. Although the mutation in the second allele—Arg297Leu—only slightly impaired KISS1R signaling, the combination with the Cys223Arg KISS1R was associated with hypogonadism (Table 7.1).

Loss-of-Function due to Frame-Shift in the KISS1R Gene

The deletion or insertion of nucleotides within the coding region of the *KISS1R* has also been associated with the nIHH phenotype in patients due to loss-of-function of the receptor protein [8, 10]. The first description of a deletion in the coding sequence of the *KISS1R* gene was published in 2003 [8]. The deletion was identified in the homozygous state in all five affected children of a consanguineous family of eight children.

The four affected males of this family had a phenotype compatible with complete IHH with a normal sense of smell as well as a blunted response to GnRH stimulation [8]. Despite the manifestations of hypogonadism, the only affected female showed signs of partial sexual maturation, such as an episode of uterine bleeding and partial breast development [8]. Her serum estradiol was above prepubertal levels and her LH response to an intravenous injection of GnRH was unusually robust [8, 67]. Interestingly, the heterozygous mother of these children had delayed puberty (menarche at 16 years of age), whereas the heterozygous father reported normal pubertal development [8].

The *KISS1R* deletion in this family eliminated 155 nucleotides of the *KISS1R* gene, starting within intron 4 and including the splice acceptor site between intron 4 and exon 5, as well as part of the coding sequence of exon 5 (Table 7.1) [8]. Although functional assays have not been performed for this mutant, the deletion of the splice acceptor site between intron 4 and exon 5 would be expected to eliminate all amino acids encoded by exon 5 from the protein. If expressed, this mutant may be truncated at the end of the fourth exon, which corresponds to the glutamic acid at position 246 of the KISS1R (Fig. 7.1). Such a truncated receptor would be missing half of the third intracellular loop, the sixth and seventh transmembrane domains and the carboxyl terminal tail of the KISS1R, which would likely result in receptor misfolding and degradation, or impaired activity.

A homozygous insertion of a cytosine after nucleotide 1001 (1001_1002insC) was identified in the *KISS1R* gene of a patient born to a family with no history of IHH (Table 7.1) [10]. However, the parents of this patient were second-degree cousins, which suggests that they may be asymptomatic heterozygous carriers of the mutation. If expressed, this mutant would result in a frame-shift that could potentially increase the size of the KISS1R protein from 398 to 441 amino acids [10]. The affected patient presented with congenital hypogonadotropic hypogonadism as indicated by the undescended testes and mild hypospadias, with subsequent delayed puberty as well as impaired fertility. Semen analysis revealed oligoasthenozoospermia (low concentration and reduced motility of spermatozoids) [10]. The patient's phenotype indicates that the nucleotide insertion results in profound impairment of KISS1R function when present in the homozygous state, despite the absence of hypogonadism in the presumably heterozygous parents. Nonetheless, this patient was able to father a child by in vitro fertilization after 2 years of pulsatile GnRH therapy [10].

An insertion/deletion within the splice acceptor site of the 3′-end of intron 2 of *KISS1R* was identified in two brothers presenting with absent sexual maturation as well as prepubertal serum testosterone levels and low gonadotropins at ages 14 and 20 years [15]. In this family, the nucleotides at position −2 to −4 (GCA) within the 3'-splice acceptor site of intron 2 were missing. In their place was the insertion of a 5-nucleotide sequence (ACCGGT) (Table 7.1). This mutation is predicted to disrupt normal splicing, resulting in the use of alternative acceptor sites and the generation of aberrant KISS1R proteins. The mother was heterozygous for the same deletion/insertion but reported normal sexual maturation [15].

Mutations in *KISS1R* not Associated with Altered Function

Some *KISS1R* mutations identified in IHH patients have not yet been associated with altered receptor function, such as Glu252Gln KISS1R. This mutation was identified in the heterozygous state in a patient with sporadic IHH [15]. However, functional assays performed for this mutant did not detect any changes in activity. Furthermore, the presence of the mutation in the heterozygous state, with the other allele being normal, is in disagreement with the predicted autosomal recessive inheritance mode. This suggests that this patient may carry additional yet-to-be-identified mutation(s) or polymorphism(s) associated with the IHH phenotype.

Other amino acid substitutions identified in patients with reproductive disorders have also been found in control populations and thus are considered to be normal variants, such as the His364Leu KISS1R variant [13, 15].

Mouse Models of Kiss1r Disruption

The phenotype of male and female mice with disruption of the *Kiss1r* gene is consistent with that exhibited by patients carrying biallelic inactivating mutations in *KISS1R*. Male and female mice with congenital disruption of the *Kiss1r* gene or the gene encoding its ligand, *Kiss1*, fail to undergo sexual maturation, have very small gonads, low serum gonadotropins and sex steroid levels, and are infertile [9, 21–23, 36, 69]. These models provide evidence that the absence of expression of either KISS1R or kisspeptin cannot be effectively compensated, which suggests a lack of redundancy for at least some important biological effects of the kisspeptin/KISS1R system on reproductive function and puberty [69]. Despite severe impairment of spermatogenesis in males and of estrous cyclicity and ovulation in females, *Kiss1r* null mice can respond to exogenous stimulation with GnRH (9), consistent with an innate failure to secrete GnRH.

Interestingly, mice with disruption of the Kiss1r ligand, kisspeptin, exhibit a less severe phenotype when compared to *Kiss1r* null mice; nevertheless, *Kiss1* null mice also have impaired spontaneous puberty and are infertile [23, 24]. Males and females null for *Kiss1* have larger gonads and males have a less severe defect in spermatogenesis when compared to *Kiss1r* null mice. Additionally, LH and FSH responsiveness to kisspeptin is preserved in males and females null for *Kiss1*. Conversely, LH and FSH responsiveness to kisspeptin is lost in *Kiss1r* null males and females, which emphasizes the requirement of *Kiss1r* expression for the effect of kisspeptin on gonadotropin secretion [23]. Although males from both lineages exhibited low basal testosterone, the decrease in *Kiss1r* null males was 87%, whereas *Kiss1* null males were less affected, exhibiting only a 41% decrease when compared to the wild type basal testosterone levels. The response of testosterone to stimulation was below normal in males from both lineages; however, the average values reached in *Kiss1* null males were over 12-fold higher than those reached in *Kiss1r* null males [23].

Nonetheless, some phenotypic differences are noticeable among the reported *Kiss1r* mutant lineages, which may be related to the gene disruption strategy used in each case. The strategies used to disrupt *Kiss1r* in mice have been nicely summarized in a review by Colledge [69] (also see Chap. 22). Three of the *Kiss1r* null lineages were generated by gene targeting and deletion of part of the coding sequence of the *Kiss1r* gene [9, 22, 23]. The lineage generated by *Paradigm Therapeutics* has a deletion of 702 base pairs, which includes 92 base pairs from exon 1, the entire intron 1 and 101 base pairs of exon 2 [9]. The lineage generated by *Schering Plough* is missing 52 base pairs within exon 2 of the *Kiss1r* coding sequence [22]; the lineage generated by *Harvard Reproductive Endocrine Sciences Center* has the entire exon 2 of the *Kiss1r* gene deleted from the sequence [23]. The disruption of the *Kiss1r* gene in an additional lineage was created without deletions in the coding sequence, by inserting a retrovirus within intron-2 of the *Kiss1r* gene [21, 36]. The insertion includes two polyadenylation sequences designed to terminate transcription from the *Kiss1r* promoter, which should greatly impair Kiss1r expression [21, 36]. Male mice from this lineage are reported to lack male mounting behavior, which can be rescued by adulthood testosterone supplementation [36]. Normal copulatory behavior was also restored in adult females supplemented with estrogen [36]. On the other hand, other abnormalities exhibited by *Kiss1r* null males could not be rescued by testosterone supplementation, such as the absence of the normal female preference when mounting, the female-like (low) number of motoneurons in the spino-bulbocavernosus nucleus, as well as female-like tyrosine hydroxylase fibers and kisspeptin expression in the AVPV, which suggests that some of the abnormalities caused by disruption of *Kiss1r* gene may be permanent once established [36].

As opposed to female mice from all other lineages carrying *Kiss1r* gene deletions, *Kiss1r* null females from the viral targeting lineage were reported to respond to sex steroid treatment with activation of GnRH neurons (as indicated by cFos induction), as well as a significant increase in serum LH [21]. Similar experiments failed to reproduce these results in females from the *Paradigm Therapeutics* lineage, which showed no cFos activation in GnRH neurons, and none of the *Kiss1r* null females exhibited an LH response to the sex steroid regimens [70]. Incidentally, females that responded to the sex steroid regimen belong to the only lineage of *Kiss1r* gene disruption without any deletions in the coding sequence. Although in situ hybridization did not detect significant *Kiss1r* expression in mice from this lineage, residual gene expression in cases of retroviral insertions has been reported, which could be achieved by splicing out the inserted gene from the RNA [69].

Gain-of-Function of *Kiss1r* and Central Precocious Puberty

Gonadotropin-dependent or CPP is characterized by advanced sexual maturation that is originated at the central nervous system level. These cases are further classified as idiopathic after tumors or other anatomical abnormalities have been excluded [71]. The etiology in idiopathic cases is thought to relate primarily to genetic factors, with additional influences of environmental and dietary factors [72–76].

Prevalence of CPP is substantially higher than that of IHH in any given population, and girls are at least ten times more likely to develop CPP than boys [74, 76–78]. This likelihood is further increased when only idiopathic cases of the disorder are considered [79, 80]. The genetic nature of this disorder is emphasized by the fact that about a third of affected children have a family history of the disorder [77]. Moreover, at least 95% of girls with CPP have the idiopathic form of the disorder [71]. On the other hand, the majority of CPP cases in boys are a consequence of anatomical defects [71]. These differences reveal a sexually dimorphic pattern of incidence of pubertal disorders in children.

The analysis of families with a history of idiopathic CPP suggests an autosomal-dominant inheritance mode of this disorder [77]. Thus, one mutant allele would be sufficient to result in the manisfestation of the CPP phenotype. Despite the higher incidence of CPP, most of the mutations affecting fertility have been identified in patients with IHH. This could suggest that idiopathic CPP would be a polygenic disorder; however, the autosomal-dominant transmission observed in families with a history of idiopathic CPP disagrees with a polygenic origin for this disorder. The first report of a naturally occurring genetic mutation associated with CPP was published quite recently. This mutation is a single amino acid substitution in the carboxyl terminal tail of KISS1R (Arg386Pro) identified in an affected girl [6]. One additional single amino acid substitution in the KISS1R ligand, kisspeptin, has also been associated with the CPP phenotype in an unrelated boy [17]. Both mutations have had their association with the CPP phenotype confirmed in functional assays and were identified in the heterozygous state in the affected children, consistent with the autosomal-dominant inheritance mode of CPP predicted in familial cases of the disorder.

The Arg386Pro-KISS1R, the First Mutation to be Associated with CPP

The Arg386Pro-KISS1R mutation was identified in a girl who exhibited slowly progressing thelarche since birth [6]. At age 8.5 years, the affected girl had advanced breast development and pubic hair as well as additional signs of precocious exposure to estrogen, such as enlarged uterine and ovarian volumes and a 3-year advanced bone age for her chronological age. Accordingly, her serum estradiol was twice the value expected for her age [6]. Despite the elevated estradiol, her basal and stimulated LH responses were within the expected range for her age, suggesting that the underlying defect led to stimulation of the central reproductive axis.

Arginine at position 386 is a conserved amino acid in an otherwise poorly conserved domain of the KISS1R, which suggests that this amino acid is relevant for receptor function. Accordingly, functional assays showed that this mutation resulted in prolonged responsiveness of the mutant receptor to kisspeptin stimulation, as indicated by inositol phosphate production and phosphorylation of ERK1/2, which were used as markers of KISS1R-activated G protein signaling in cells transfected with the Arg386Pro mutant or the wild-type KISS1R [6].

Basal KISS1R activity (i.e., activity in the absence of agonist stimulation) was not affected by the Arg386Pro mutation. This important detail indicates that, unlike the well-established effects of naturally occurring genetic mutations identified in patients with gonadotropin-independent precocious puberty, the Arg386Pro KISS1R mutation does not generate a constitutively active receptor [6]. Instead, the Arg386Pro mutation was associated with a mild gain-of-function effect that was only detected after prolonged stimulation with kisspeptin. This observation points to a slower rate of desensitization of the Arg386Pro KISS1R mutant, consistent with a mutation in the carboxyl terminal tail—a well-known target of intracellular regulators of GPCR activity. In fact, the effect of this mutation could only be detected when kisspeptin signaling was declining.

A Unique Mechanism of Prolonged KISS1R Signaling with Pathophysiological Implications

The investigation of potential mechanisms underlying the slower rate of desensitization of the Arg386Pro KISS1R eliminated the possibility that receptor internalization was affected by the Arg386Pro mutation, which suggested that this mutation affected an event downstream of receptor internalization [41]. Interestingly, the Arg386Pro mutation also did not affect the dynamic recycling of KISS1R following receptor internalization, which was preserved for this mutant. These findings suggest that binding of kisspeptin to newly recycled receptors plays a role in the prolonged responsiveness of the Arg386Pro KISS1R mutant to kisspeptin. Thus, a slower rate of degradation of the mutant receptor combined with dynamic recycling of internalized Arg386Pro KISS1R may account for the increased responsiveness of the Arg386Pro KISS1R to kisspeptin in the absence of changes in KISS1R internalization [41].

The proposed mechanism underlying the effect of the Arg386Pro KISS1R mutation on receptor activity is depicted in Fig. 7.3. Here it is shown that, at baseline, the absolute number and relative proportions of wild type (WT) and Arg386Pro KISS1R receptors would be similar in both the plasma membrane and internalized compartments of a given cell. The model in Fig. 7.3 depicts 70% of WT (top left) and of Arg386Pro (bottom left) KISS1R receptors are in the plasma membrane at baseline; whereas 30% of the receptors in both cases are intracellular. After ligand stimulation, the relative proportions of receptors remain similar in the membrane and intracellular compartments for WT and Arg386Pro KISS1R. This is consistent with the unaltered rate of receptor internalization for the Arg386Pro KISS1R mutant. However, the absolute amount of receptors in each cell compartment after ligand stimulation is no longer similar: a slower rate of degradation would lead to greater accumulation of the Arg386Pro KISS1R mutant in the intracellular compartment compared to WT receptor, which would in turn result in an increase of Arg386Pro mutant KISS1R in the plasma membrane [41].

Decreased receptor degradation has been described as the mechanism underlying other disease-associated mutations. In one example, a mutation associated with gonadotropin-independent precocious puberty prevents lysosomal degradation of the LH receptor, which is also a GPCR [81]. In this case, however, the effect of the mutation is not mild and leads to a marked increase in receptor signaling at baseline, which characterizes constitutive activation of the receptor.

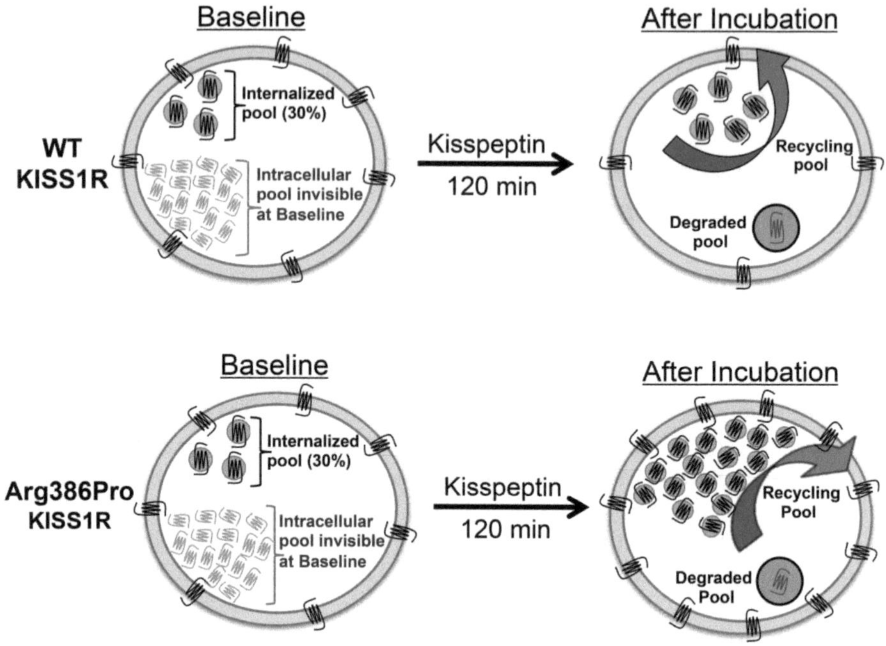

Fig. 7.3 Model of KISS1R desensitization. At baseline, absolute numbers and relative proportions of wild type (WT) and Arg386Pro KISS1R would be similar in both the plasma membrane and internalized compartments of a given cell. In this model, 70% of WT (*top left*) and of Arg386Pro (*bottom left*) receptors are in the plasma membrane whereas the remaining 30% of the receptors in both cases are intracellular. After 120 min of ligand stimulation, relative proportions of receptors remain similar in the membrane and intracellular compartments for WT and Arg386Pro KISS1R, whereas absolute number is no longer similar, with the Arg386Pro KISS1R mutant exhibiting higher absolute number of receptors in both the membrane and intracellular compartments compared to WT KISS1R. From Bianco SD, Vandepas L, Correa-Medina M, Gereben B, Mukherjee A, Kuohung W, Carroll R, Teles MG, Latronico AC, Kaiser UB (2011) KISS1R intracellular trafficking and degradation: effect of the Arg386Pro disease-associated mutation. Endocrinology 152:1616–1626. (Reprinted with permission from The Endocrine Society)

A Little Gain-of-Function Goes a Long Way

The mild gain of function conferred by the Arg386Pro KISS1R mutation, combined with the requirement for ligand activation for the appearance of the effect of this mutation, are predicted to be key details determining the manifestation of the precocious puberty phenotype in the affected patient. This is based on the kinetics of KISS1R activity, which shows that continuous stimulation leads to a peak of activity followed by complete desensitization of KISS1R signaling [41]. This desensitization can be detected in vivo as suppression of gonadotropin secretion in primates [48] or testicular degeneration in rats [49]. An implication of this behavior is that constitutive activation of KISS1R is expected to lead to an IHH phenotype rather than precocious puberty, as the expression of a constitutively active KISS1R would lead to permanent desensitization of KISS1R signaling, regardless of the presence of kisspeptin.

Further support for the prediction that mild gain-of-function is key for the manifestation of a precocious puberty phenotype came from the functional characterization of another naturally occurring mutation associated with idiopathic CPP in a boy. The mutation in this boy is in kisspeptin (Pro74Ser) [25]. Similarly to the Arg386Pro-KISS1R, the Pro74Ser kisspeptin mutant also results in a relatively mild gain-of-function characterized by prolonged KISS1R signaling. Functional assays indicated that the Pro74Ser kisspeptin mutant was degraded more slowly than the wild type kisspeptin [25]. Thus, the mild gain-of-function effects of both the Arg386Pro KISS1R and the Pro74Ser kisspeptin mutations lead to prolonged kisspeptin signaling, which contributes to the precocious puberty phenotype in the patients carrying these mutations.

Final Considerations

Additional mutations in *KISS1R* or *KISS1* not yet characterized in functional assays have been identified in a screen for polymorphisms in Chinese girls with CPP. A Pro196His KISS1R mutation that is located in the second extracellular loop of the receptor and a Pro110Thr kisspeptin mutant were identified in these studies [26, 27]. Whether these mutations are associated with the precocious puberty phenotype in the affected girls has not been established.

The characterization of mutations in *KISS1R* or *KISS1*, as well as the disruption of the genes encoding KISS1R or kisspeptin, have established the relevance of KISS1R signaling for triggering pubertal maturation and maintaining reproductive competence in humans and animal models. Further characterization has revealed a role for the duration of KISS1R signaling in the onset and timing of pubertal maturation. These findings underscore the importance of identification of additional intracellular pathways or networks controlling KISS1R signaling, as mutations in proteins potentially involved in these networks and signaling pathways may be responsible for additional yet-unidentified cases of pubertal and reproductive disorders.

In addition, investigation of the effect of loss- and gain-of-function mutations in *KISS1R* in sites outside of the hypothalamus where the expression of this receptor may contribute to overall reproductive function and/or fertility may help to uncover new functions for the KISS1R.

Acknowledgments This work was supported by the *Eunice Kennedy Shriver* National Institute of Child Health and Human Development, National Institutes of Health (NIH), through cooperative agreement U54 HD28138 as part of the Specialized Cooperative Centers Program in Reproduction and Infertility Research, by R01 HD61577, and by R21 HD66495 (to UBK) as well as by the R21 HD059015 and ARRA R21 HD059015-02S1 (SDCB) and by the 2008 Child Health Research Award from the Charles H. Hood Foundation, Boston, MA (SDCB).

References

1. Plant TM (2002) Neurophysiology of puberty. J Adolesc Health 31:185–191
2. Seminara SB, Hayes FJ, Crowley WF Jr (1998) Gonadotropin-releasing hormone deficiency in the human (idiopathic hypogonadotropic hypogonadism and Kallmann's syndrome): pathophysiological and genetic considerations. Endocr Rev 19:521–539
3. Cadman SM, Kim SH, Hu Y, Gonzalez-Martinez D, Bouloux PM (2007) Molecular pathogenesis of Kallmann's syndrome. Horm Res 67:231–242
4. Chan YM, Broder-Fingert S, Seminara SB (2009) Reproductive functions of kisspeptin and Gpr54 across the life cycle of mice and men. Peptides 30:42–48
5. Iovane A, Aumas C, de Roux N (2004) New insights in the genetics of isolated hypogonadotropic hypogonadism. Eur J Endocrinol 151(Suppl 3):U83–88
6. Teles MG, Bianco SD, Brito VN, Trarbach EB, Kuohung W, Xu S, Seminara SB, Mendonca BB, Kaiser UB, Latronico AC (2008) A GPR54-activating mutation in a patient with central precocious puberty. N Engl J Med 358:709–715
7. Trarbach EB, Silveira LG, Latronico AC (2007) Genetic insights into human isolated gonadotropin deficiency. Pituitary 10:381–391
8. de Roux N, Genin E, Carel JC, Matsuda F, Chaussain JL, Milgrom E (2003) Hypogonadotropic hypogonadism due to loss of function of the KiSS1-derived peptide receptor GPR54. Proc Natl Acad Sci USA 100:10972–10976
9. Seminara SB, Messager S, Chatzidaki EE, Thresher RR, Acierno JS Jr, Shagoury JK, Bo-Abbas Y, Kuohung W, Schwinof KM, Hendrick AG, Zahn D, Dixon J, Kaiser UB, Slaugenhaupt SA, Gusella JF, O'Rahilly S, Carlton MB, Crowley WF Jr, Aparicio SA, Colledge WH (2003) The GPR54 gene as a regulator of puberty. N Engl J Med 349:1614–1627
10. Lanfranco F, Gromoll J, von Eckardstein S, Herding EM, Nieschlag E, Simoni M (2005) Role of sequence variations of the GnRH receptor and G protein-coupled receptor 54 gene in male idiopathic hypogonadotropic hypogonadism. Eur J Endocrinol 153:845–852
11. Nimri R, Lebenthal Y, Lazar L, Chevrier L, Phillip M, Bar M, Hernandez-Mora E, de Roux N, Gat-Yablonski G (2011) A novel loss-of-function mutation in GPR54/KISS1R leads to hypogonadotropic hypogonadism in a highly consanguineous family. J Clin Endocrinol Metab 96:E536–545
12. Pallais JC, Bo-Abbas Y, Pitteloud N, Crowley WF Jr, Seminara SB (2006) Neuroendocrine, gonadal, placental, and obstetric phenotypes in patients with IHH and mutations in the G-protein coupled receptor, GPR54. Mol Cell Endocrinol 254–255:70–77
13. Semple RK, Achermann JC, Ellery J, Farooqi IS, Karet FE, Stanhope RG, O'Rahilly S, Aparicio SA (2005) Two novel missense mutations in g protein-coupled receptor 54 in a patient with hypogonadotropic hypogonadism. J Clin Endocrinol Metab 90:1849–1855
14. Tenenbaum-Rakover Y, Commenges-Ducos M, Iovane A, Aumas C, Admoni O, de Roux N (2007) Neuroendocrine phenotype analysis in five patients with isolated hypogonadotropic hypogonadism due to a L102P inactivating mutation of GPR54. J Clin Endocrinol Metab 92:1137–1144
15. Teles MG, Trarbach EB, Noel SD, Guerra-Junior G, Jorge A, Beneduzzi D, Bianco SD, Mukherjee A, Baptista MT, Costa EM, De Castro M, Mendonca BB, Kaiser UB, Latronico AC (2010) A novel homozygous splice acceptor site mutation of KISS1R in two siblings with normosmic isolated hypogonadotropic hypogonadism. Eur J Endocrinol 163:29–34
16. Chan YM, Broder-Fingert S, Paraschos S, Lapatto R, Au M, Hughes V, Bianco SD, Min L, Plummer L, Cerrato F, De Guillebon A, Wu IH, Wahab F, Dwyer A, Kirsch S, Quinton R, Cheetham T, Ozata M, Ten S, Chanoine JP, Pitteloud N, Martin KA, Schiffmann R, Van der Kamp HJ, Nader S, Hall JE, Kaiser UB, Seminara SB (2011) GnRH-deficient phenotypes in humans and mice with heterozygous variants in KISS1/Kiss1. J Clin Endocrinol Metab 96:E1771–1781
17. Silveria LGNS, Silveira-Neto AP, Abreu AP, Brito VN, Santos MGBS, Kuohung W, Xu S, Gryngarten M, Escobar MEAI, Mendonca BB, Kaiser UB, Latronico AC (2010) Mutations of the KISS1 Gene in Disorders of Puberty. J Clin Endocrinol Metab 95:2276–2280

18. Topaloglu AK, Tello JA, Kotan LD, Ozbek MN, Yilmaz MB, Erdogan S, Gurbuz F, Temiz F, Millar RP, Yuksel B (2012) Inactivating KISS1 mutation and hypogonadotropic hypogonadism. N Engl J Med 366:629–635
19. Breuer O, Abdulhadi-Atwan M, Zeligson S, Fridman H, Renbaum P, Levy-Lahad E, Zangen DH (2012) A novel severe N-terminal splice site KISS1R gene mutation causes hypogonadotropic hypogonadism but enables a normal development of neonatal external genitalia. Eur J Endocrinol 167(2):209–16
20. d'Anglemont de Tassigny X, Fagg LA, Dixon JP, Day K, Leitch HG, Hendrick AG, Zahn D, Franceschini I, Caraty A, Carlton MB, Aparicio SA, Colledge WH (2007) Hypogonadotropic hypogonadism in mice lacking a functional Kiss1 gene. Proc Natl Acad Sci USA 104:10714–10719
21. Dungan HM, Gottsch ML, Zeng H, Gragerov A, Bergmann JE, Vassilatis DK, Clifton DK, Steiner RA (2007) The role of kisspeptin-GPR54 signaling in the tonic regulation and surge release of gonadotropin-releasing hormone/luteinizing hormone. J Neurosci 27:12088–12095
22. Funes S, Hedrick JA, Vassileva G, Markowitz L, Abbondanzo S, Golovko A, Yang S, Monsma FJ, Gustafson EL (2003) The KiSS-1 receptor GPR54 is essential for the development of the murine reproductive system. Biochem Biophys Res Commun 312:1357–1363
23. Lapatto R, Pallais JC, Zhang D, Chan YM, Mahan A, Cerrato F, Le WW, Hoffman GE, Seminara SB (2007) Kiss1-/- mice exhibit more variable hypogonadism than Gpr54-/- mice. Endocrinology 148:4927–4936
24. Mei H, Walters C, Carter R, Colledge WH (2011) Gpr54-/- mice show more pronounced defects in spermatogenesis than Kiss1-/- mice and improved spermatogenesis with age when exposed to dietary phytoestrogens. Reproduction 141:357–366
25. Silveira LG, Noel SD, Silveira-Neto AP, Abreu AP, Brito VN, Santos MG, Bianco SD, Kuohung W, Xu S, Gryngarten M, Escobar ME, Arnhold IJ, Mendonca BB, Kaiser UB, Latronico AC (2010) Mutations of the KISS1 gene in disorders of puberty. J Clin Endocrinol Metab 95:2276–2280
26. Luan X, Yu H, Wei X, Zhou Y, Wang W, Li P, Gan X, Wei D, Xiao J (2007) GPR54 polymorphisms in Chinese girls with central precocious puberty. Neuroendocrinology 86:77–83
27. Luan X, Zhou Y, Wang W, Yu H, Li P, Gan X, Wei D, Xiao J (2007) Association study of the polymorphisms in the KISS1 gene with central precocious puberty in Chinese girls. Eur J Endocrinol 157:113–118
28. Parhar IS, Ogawa S, Sakuma Y (2004) Laser-captured single digoxigenin-labeled neurons of gonadotropin-releasing hormone types reveal a novel G protein-coupled receptor (Gpr54) during maturation in cichlid fish. Endocrinology 145:3613–3618
29. Irwig MS, Fraley GS, Smith JT, Acohido BV, Popa SM, Cunningham MJ, Gottsch ML, Clifton DK, Steiner RA (2004) Kisspeptin activation of gonadotropin releasing hormone neurons and regulation of KiSS-1 mRNA in the male rat. Neuroendocrinology 80:264–272
30. Messager S, Chatzidaki EE, Ma D, Hendrick AG, Zahn D, Dixon J, Thresher RR, Malinge I, Lomet D, Carlton MB, Colledge WH, Caraty A, Aparicio SA (2005) Kisspeptin directly stimulates gonadotropin-releasing hormone release via G protein-coupled receptor 54. Proc Natl Acad Sci USA 102:1761–1766
31. Shibata M GR, Shahab M, Plant TM (2005) GnRH neurons in the peripubertal male rhesus monkey (Macaca mulatta) express GPR54: implication for the control of primate puberty. In: 87th Annual Meeting of the Endocrine Society. San Diego, CA
32. Han SK, Gottsch ML, Lee KJ, Popa SM, Smith JT, Jakawich SK, Clifton DK, Steiner RA, Herbison AE (2005) Activation of gonadotropin-releasing hormone neurons by kisspeptin as a neuroendocrine switch for the onset of puberty. J Neurosci 25:11349–11356
33. Keen KL, Wegner FH, Bloom SR, Ghatei MA, Terasawa E (2008) An increase in kisspeptin-54 release occurs with the pubertal increase in luteinizing hormone-releasing hormone-1 release in the stalk-median eminence of female rhesus monkeys in vivo. Endocrinology 149:4151–4157
34. Herbison AE, de Tassigny X, Doran J, Colledge WH (2010) Distribution and postnatal development of Gpr54 gene expression in mouse brain and gonadotropin-releasing hormone neurons. Endocrinology 151:312–321

35. Kurian JR, Keen KL, Guerriero KA, Terasawa E (2012) Tonic control of kisspeptin release in prepubertal monkeys: implications to the mechanism of puberty onset. Endocrinology 153:3331–3336
36. Kauffman AS, Park JH, McPhie-Lalmansingh AA, Gottsch ML, Bodo C, Hohmann JG, Pavlova MN, Rohde AD, Clifton DK, Steiner RA, Rissman EF (2007) The kisspeptin receptor GPR54 is required for sexual differentiation of the brain and behavior. J Neurosci 27:8826–8835
37. Kotani M, Detheux M, Vandenbogaerde A, Communi D, Vanderwinden JM, Le Poul E, Brezillon S, Tyldesley R, Suarez-Huerta N, Vandeput F, Blanpain C, Schiffmann SN, Vassart G, Parmentier M (2001) The metastasis suppressor gene KiSS-1 encodes kisspeptins, the natural ligands of the orphan G protein-coupled receptor GPR54. J Biol Chem 276:34631–34636
38. Oishi SMR, Tomita K, Setsuda S, Masuda R, Ohno H, Naniwa Y, Ieda N, Inoue N, Ohkura S, Uenoyama Y, Tsukamura H, Maeda K, Hirasawa A, Tsujimoto G, Fijii N (2011) Activation of neuropeptide FF receptors by kisspeptin receptor ligands. ACS Med Chem Lett 2:5
39. Colledge WH, d'Anglemont de Tassigny X (2010) The role of kisspeptin signalling in the regulation of the GnRH-gonadotrophin ovarian axis in mice. Ann Endocrinol (Paris) 71:198–200
40. Pampillo M, Camuso N, Taylor JE, Szereszewski JM, Ahow MR, Zajac M, Millar RP, Bhattacharya M, Babwah AV (2009) Regulation of GPR54 signaling by GRK2 and {beta}-arrestin. Mol Endocrinol 23:2060–2074
41. Bianco SD, Vandepas L, Correa-Medina M, Gereben B, Mukherjee A, Kuohung W, Carroll R, Teles MG, Latronico AC, Kaiser UB (2011) KISS1R intracellular trafficking and degradation: effect of the Arg386Pro disease-associated mutation. Endocrinology 152:1616–1626
42. Lee DK, Nguyen T, O'Neill GP, Cheng R, Liu Y, Howard AD, Coulombe N, Tan CP, Tang-Nguyen AT, George SR, O'Dowd BF (1999) Discovery of a receptor related to the galanin receptors. FEBS Lett 446:103–107
43. Gottsch ML, Clifton DK, Steiner RA (2009) From KISS1 to kisspeptins: An historical perspective and suggested nomenclature. Peptides 30:4–9
44. Ohtaki T, Shintani Y, Honda S, Matsumoto H, Hori A, Kanehashi K, Terao Y, Kumano S, Takatsu Y, Masuda Y, Ishibashi Y, Watanabe T, Asada M, Yamada T, Suenaga M, Kitada C, Usuki S, Kurokawa T, Onda H, Nishimura O, Fujino M (2001) Metastasis suppressor gene KiSS-1 encodes peptide ligand of a G-protein-coupled receptor. Nature 411:613–617
45. Muir AI, Chamberlain L, Elshourbagy NA, Michalovich D, Moore DJ, Calamari A, Szekeres PG, Sarau HM, Chambers JK, Murdock P, Steplewski K, Shabon U, Miller JE, Middleton SE, Darker JG, Larminie CG, Wilson S, Bergsma DJ, Emson P, Faull R, Philpott KL, Harrison DC (2001) AXOR12, a novel human G protein-coupled receptor, activated by the peptide KiSS-1. J Biol Chem 276:28969–28975
46. Castellano JM, Navarro VM, Fernandez-Fernandez R, Castano JP, Malagon MM, Aguilar E, Dieguez C, Magni P, Pinilla L, Tena-Sempere M (2006) Ontogeny and mechanisms of action for the stimulatory effect of kisspeptin on gonadotropin-releasing hormone system of the rat. Mol Cell Endocrinol 257–258:75–83
47. Roa J, Vigo E, Garcia-Galiano D, Castellano JM, Navarro VM, Pineda R, Dieguez C, Aguilar E, Pinilla L, Tena-Sempere M (2008) Desensitization of gonadotropin responses to kisspeptin in the female rat: analyses of LH and FSH secretion at different developmental and metabolic states. Am J Physiol Endocrinol Metab 294:E1088–1096
48. Seminara SB, Dipietro MJ, Ramaswamy S, Crowley WF Jr, Plant TM (2006) Continuous human metastin 45-54 infusion desensitizes G protein-coupled receptor 54-induced gonadotropin-releasing hormone release monitored indirectly in the juvenile male Rhesus monkey (Macaca mulatta): a finding with therapeutic implications. Endocrinology 147:2122–2126
49. Thompson EL, Murphy KG, Patterson M, Bewick GA, Stamp GW, Curtis AE, Cooke JH, Jethwa PH, Todd JF, Ghatei MA, Bloom SR (2006) Chronic subcutaneous administration of

kisspeptin-54 causes testicular degeneration in adult male rats. Am J Physiol Endocrinol Metab 291:E1074–1082

50. Ramaswamy S, Seminara SB, Pohl CR, DiPietro MJ, Crowley WF Jr, Plant TM (2007) Effect of continuous intravenous administration of human metastin 45-54 on the neuroendocrine activity of the hypothalamic-pituitary-testicular axis in the adult male rhesus monkey (Macaca mulatta). Endocrinology 148:3364–3370
51. George JT, Veldhuis JD, Roseweir AK, Newton CL, Faccenda E, Millar RP, Anderson RA (2011) Kisspeptin-10 is a potent stimulator of LH and increases pulse frequency in men. J Clin Endocrinol Metab 96:E1228–1236
52. Chan YM, Butler JP, Sidhoum VF, Pinnell NE, Seminara SB (2012) Kisspeptin administration to women: A window into endogenous kisspeptin secretion and GnRH responsiveness across the menstrual cycle. J Clin Endocrinol Metab 97(8):E1458–67
53. Palczewski K, Buczylko J, Kaplan MW, Polans AS, Crabb JW (1991) Mechanism of rhodopsin kinase activation. J Biol Chem 266:12949–12955
54. Gurevich VV, Benovic JL (1993) Visual arrestin interaction with rhodopsin. Sequential multisite binding ensures strict selectivity toward light-activated phosphorylated rhodopsin. J Biol Chem 268:11628–11638
55. Ferguson SS (2001) Evolving concepts in G protein-coupled receptor endocytosis: the role in receptor desensitization and signaling. Pharmacol Rev 53:1–24
56. Drake MT, Shenoy SK, Lefkowitz RJ (2006) Trafficking of G protein-coupled receptors. Circ Res 99:570–582
57. Gurevich VV, Gurevich EV (2006) The structural basis of arrestin-mediated regulation of G-protein-coupled receptors. Pharmacol Ther 110:465–502
58. Szereszewski JM, Pampillo M, Ahow MR, Offermanns S, Bhattacharya M, Babwah AV (2010) GPR54 regulates ERK1/2 activity and hypothalamic gene expression in a Galpha(q/11) and beta-arrestin-dependent manner. PLoS One 5:e12964
59. Chaturvedi K, Bandari P, Chinen N, Howells RD (2001) Proteasome involvement in agonist-induced down-regulation of mu and delta opioid receptors. J Biol Chem 276:12345–12355
60. Shenoy SK, McDonald PH, Kohout TA, Lefkowitz RJ (2001) Regulation of receptor fate by ubiquitination of activated beta 2-adrenergic receptor and beta-arrestin. Science 294: 1307–1313
61. Semple RK, Topaloglu AK (2010) The recent genetics of hypogonadotrophic hypogonadism - novel insights and new questions. Clin Endocrinol (Oxf) 72:427–435
62. Sykiotis GP, Hoang XH, Avbelj M, Hayes FJ, Thambundit A, Dwyer A, Au M, Plummer L, Crowley WF Jr, Pitteloud N (2010) Congenital idiopathic hypogonadotropic hypogonadism: evidence of defects in the hypothalamus, pituitary, and testes. J Clin Endocrinol Metab 95:3019–3027
63. Pallais JC, Au M, Pitteloud N, Seminara S, Crowley WF (1993) Isolated gonadotropin-releasing hormone (GnRH) deficiency overview. In: Bird TD, Dolan CR, Stephens K, Adam MP, Pagon RA (eds) GeneReviews™. University of Washington, Seattle
64. Cariboni A, Maggi R (2006) Kallmann's syndrome, a neuronal migration defect. Cell Mol Life Sci 63:2512–2526
65. Mitchell AL, Dwyer A, Pitteloud N, Quinton R (2011) Genetic basis and variable phenotypic expression of Kallmann syndrome: towards a unifying theory. Trends Endocrinol Metab 22:249–258
66. Bianco SD, Kaiser UB (2009) The genetic and molecular basis of idiopathic hypogonadotropic hypogonadism. Nat Rev Endocrinol 5:569–576
67. Wahab F, Quinton R, Seminara SB (2011) The kisspeptin signaling pathway and its role in human isolated GnRH deficiency. Mol Cell Endocrinol 346:29–36
68. Wacker JL, Feller DB, Tang XB, Defino MC, Namkung Y, Lyssand JS, Mhyre AJ, Tan X, Jensen JB, Hague C (2008) Disease-causing mutation in GPR54 reveals the importance of the second intracellular loop for class A G-protein-coupled receptor function. J Biol Chem 283:31068–31078

69. Colledge WH (2009) Transgenic mouse models to study Gpr54/kisspeptin physiology. Peptides 30:34–41
70. Clarkson J, d'Anglemont de Tassigny X, Moreno AS, Colledge WH, Herbison AE (2008) Kisspeptin-GPR54 signaling is essential for preovulatory gonadotropin-releasing hormone neuron activation and the luteinizing hormone surge. J Neurosci 28:8691–8697
71. Klein KO (1999) Precocious puberty: who has it? Who should be treated? J Clin Endocrinol Metab 84:411–414
72. Palmert MR, Boepple PA (2001) Variation in the timing of puberty: clinical spectrum and genetic investigation. J Clin Endocrinol Metab 86:2364–2368
73. Palmert MR, Hirschhorn JN (2003) Genetic approaches to stature, pubertal timing, and other complex traits. Mol Genet Metab 80:1–10
74. Parent AS, Teilmann G, Juul A, Skakkebaek NE, Toppari J, Bourguignon JP (2003) The timing of normal puberty and the age limits of sexual precocity: variations around the world, secular trends, and changes after migration. Endocr Rev 24:668–693
75. Aksglaede L, Sorensen K, Petersen JH, Skakkebaek NE, Juul A (2009) Recent decline in age at breast development: the Copenhagen Puberty Study. Pediatrics 123:e932–939
76. Biro FM, Galvez MP, Greenspan LC, Succop PA, Vangeepuram N, Pinney SM, Teitelbaum S, Windham GC, Kushi LH, Wolff MS (2010) Pubertal Assessment Method and Baseline Characteristics in a Mixed Longitudinal Study of Girls. Pediatrics 126:E583–E590
77. de Vries L, Kauschansky A, Shohat M, Phillip M (2004) Familial central precocious puberty suggests autosomal dominant inheritance. J Clin Endocrinol Metab 89:1794–1800
78. Prete G, Couto-Silva AC, Trivin C, Brauner R (2008) Idiopathic central precocious puberty in girls: presentation factors. BMC Pediatr 8:27
79. Teilmann G, Pedersen CB, Jensen TK, Skakkebaek NE, Juul A (2005) Prevalence and incidence of precocious pubertal development in Denmark: an epidemiologic study based on national registries. Pediatrics 116:1323–1328
80. Teles MG, Silveira LF, Tusset C, Latronico AC (2011) New genetic factors implicated in human GnRH-dependent precocious puberty: the role of kisspeptin system. Mol Cell Endocrinol 346(1–2):84–90
81. Galet C, Ascoli M (2006) A constitutively active mutant of the human lutropin receptor (hLHR-L457R) escapes lysosomal targeting and degradation. Mol Endocrinol 20:2931–2945

Chapter 8
Kisspeptin Antagonists

Antonia Kathryn Roseweir and Robert P. Millar

Abstract Kisspeptin is now known to be an important regulator of the hypothalamic-pituitary-gonadal axis and is the target of a range of regulators, such as steroid hormone feedback, nutritional and metabolic regulation. Kisspeptin binds to its cognate receptor, KISS1R (also called GPR54), on GnRH neurons and stimulates their activity, which in turn provides an obligatory signal for GnRH secretion—thus gating down-stream events supporting reproduction. The development of peripherally active kisspeptin antagonists could offer a unique therapeutic agent for treating hormone-dependent disorders of reproduction, including precocious puberty, endometriosis, and metastatic prostate cancer. The following chapter discusses the advances made in the search for both peptide and small molecule kisspeptin antagonists and their use in delineating the role of kisspeptin within the reproductive system. To date, four peptide antagonists and one small molecule antagonist have been designed.

Introduction

The HPG axis regulates reproduction via the modulation of GnRH production as a consequence of multiple pathways acting within the hypothalamus [1–3]. Kisspeptin (KP), acting via its cognate receptor, KISS1R (also called GPR54), has been shown

A.K. Roseweir
Institute of Pharmacy and Biomedical Sciences, University of Strathclyde, Glasgow, Scotland, UK

Institute of Cancer Sciences, School of Medicine, Veterinary Medicine and Life Sciences, University of Glasgow, Glasgow, Scotland, UK
e-mail: toni@roseweir.co.uk

R.P. Millar, M.Sc., Ph.D., M.R.C.Path., F.R.C.P.Path. (✉)
Department Zoology and Entomology, University of Pretoria, Private Bag X 20, Hatfield 0028, South Africa
e-mail: robertpetermillar@gmail.com

A.S. Kauffman and J.T. Smith (eds.), *Kisspeptin Signaling in Reproductive Biology*, Advances in Experimental Medicine and Biology 784, DOI 10.1007/978-1-4614-6199-9_8,

to be one of the most potent regulators of this system known to date, with KP doses as low as 1 fmol (i.c.v.) capable of stimulating GnRH secretion in vivo. KP and KISS1R have been shown to have a role in puberty onset, regulation of the pre-ovulatory LH surge, and integration of other regulatory pathways that affect the reproductive axis, such as steroid hormone feedback and seasonal breeding stimuli [4, 5]. The discovery of the KP system arose from the observation that mutations in *KISS1R* caused delayed or absent puberty, with similar results evident in knockout mice [6, 7]. Sex steroids negatively regulate *Kiss1* mRNA levels within the ARC in various species [8–12]. Oestrogen can also positively regulate *Kiss1* mRNA within either the ARC in sheep and primates or the AVPV in rodents at the time of the pre-ovulatory LH surge, which initiates ovulation [13–15]. *Kiss1* gene expression is also regulated by metabolic and nutritional status, as well as photoperiod. This regulation of *Kiss1* is covered in depth in other chapters in this volume.

Most of the data concerning the physiological role of the KP system to date is derived from the measurement of changes in mRNA levels or altered patterns of immunohistochemical staining of KP peptide. To directly elucidate the role of KP and KISS1R in physiological processes, the development of KP antagonists is needed. To develop these antagonists, the structure of kisspeptin needs to be assessed. The *KiSS1* gene encodes a polypeptide consisting of 145 amino acids, known as the precursor peptide. This precursor peptide gives rise to a secretory precursor protein of 126 amino acids that is proteolytically cleaved and modified to form a C-terminal amide moiety, KP-54 [16]. KP-54 is then cleaved further to smaller fragments of 14, 13, and 10 amino acids in length. These fragments have subsequently been shown to bind to and activate the KISS1R receptor with equal potency [17]. Therefore, the final ten amino acid fragments of the C-terminus [18] have been used to analyse the structure of KP and to design KP antagonists. KP-10 contains the following ten amino acids: Tyr^{1}, Asn^{2}, Trp^{3}, Asn^{4}, Ser^{5}, Phe^{6}, Gly^{7}, Leu^{8}, Arg^{9}, and Phe^{10}. It is thought that the activation domain is within the five most N-terminal residues, where there are many hydrophobic residues, and that the binding domain is within the five most C-terminal residues, which contain both hydrophobic and charged residues to form bonds with the receptor (Fig. 8.2a).

These ten amino acids provide multiple targets for amino acid substitution when designing antagonists. Several strategies for peptide analogue development have been utilised. Alanine screening involves each residue being systematically changed to alanine, which has no functional side chain. This was reported by two groups and both revealed that alanine substitutions at Phe^{6} and Phe^{10} of KP-10 significantly reduced binding to and activation of KISS1R in vitro and in vivo, making them critical to KP peptide activity [19, 20]. Orsini et al. also showed, via alanine screening, that substitution of Arg^{9} also significantly decreased binding and activation, as did substitutions of Leu^{8}. They concluded that Phe^{6}, Arg^{9}, and Phe^{10} create a binding pharmacophore with the two phenyl rings of Phe^{6} and Phe^{10} on top of each other, flanked by Arg^{9} with Leu^{8} on the opposite side of the peptide [19]. The second approach is to make more intuitive amino acid changes based on knowledge of the peptide structure of KP-10. This involves making either conservative changes at positions of interest, such as changing the side chain but not the overall charge of a residue, or changing L-amino acids to D-amino acids to assess the positioning of side chains. Finally, radical structural changes can also be made to change the charge or flexibility of the analogue. This approach is described in detail in this chapter, both in vitro and in vivo. In addition to peptide antagonists, some small molecule non-peptide antagonists have also been designed for KP, as they have more potential for being taken forward into human trials, and these results are also discussed later in this chapter. Any antagonists identified could be very useful tools to investigate whether blockade of the system directly affects GnRH/LH secretion, puberty onset, steroid feedback, and the pre-ovulatory LH surge. Antagonists may also be useful medicines, with potential application as novel contraceptives, and potential for the management of many reproductive disorders which are often associated with irregular LH levels and pulse frequency.

Development of KP Peptide Antagonists and Receptor Binding

KP analogues were synthesised based on the structure of human KP-10, as this is the smallest fragment needed to bind to and activate the receptor [21, 22]. Analogues were tested using CHO cells stably expressing human KISS1R for their ability to bind to the receptor.

Experiments tested whether all ten amino acids of KP-10 were required for binding to the receptor using N-terminally truncated peptides possessing five or seven amino acid residues. These analogues were named peptides 188 (acFGLRF) and 186 (acNSFGLRF), respectively (for continuity, the amino acid residues are numbered to correspond to positions in the full length KP-10). Each of these truncated peptides had a reduced binding affinity for the receptor compared to KP-10 (Fig. 8.1). The reduction in binding affinity seen for the 5aa- and 7aa-truncated analogues suggests that more than 7aa are needed to bind effectively to KISS1R. Therefore, this implies that all 10aa are needed for binding to human KISS1R and that at least one of the N-terminal residues must be involved in this process.

Nevertheless, amino acids within the five amino acid-truncated peptides were substituted to see if the binding affinities could be increased. These are peptides 189, 190, 191, and 200, 201, 202, 203, and 206, 207. All of these analogues had reduced affinity for human KISS1R (Fig. 8.1); however, these analogues did highlight some residues important for binding interactions with the receptor.

As the RFamide motif at the C-terminus is hypothesised to be critical for receptor binding, we firstly investigated substitutions at these two residues. Trp^{10} was introduced into peptide 189 to assess the effects of introducing a bulkier and polar side chain into the RFamide motif. The use of Trp^{10} also introduces rigidity to the phenyl ring and some steric hindrance, thus reducing flexibility of the C-terminus. Peptide 189 (acFGLRW) had an IC_{50} of 9.5×10^{-7} M for human KISS1R (Fig. 8.1). The results imply that introducing a more rigid polar residue into the RFamide motif decreases the ability to bind the receptor. This contrasts with a study by Niida et al. that showed that small peptide analogues with FW-amide at the c-terminus were potent agonists. However, these analogues had Guanidino or bis[(2-pyridinyl) methyl] added at the n-terminus, and receptor binding was not investigated in this

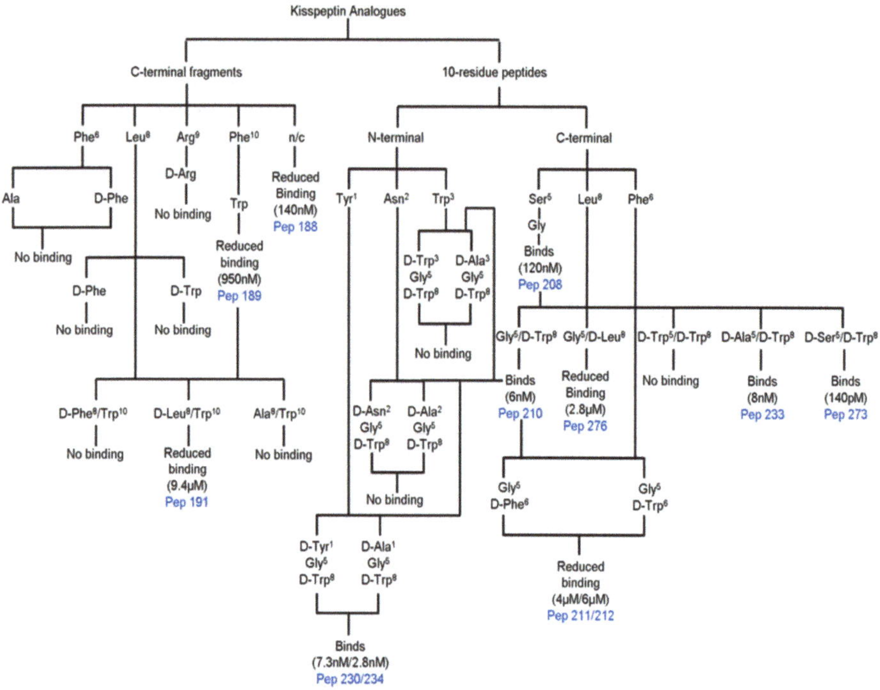

Fig. 8.1 Diagram of amino acid substitution effects on receptor binding. Flowchart for amino acid substitutions within truncated and full-length peptide analogues showing the effects of each substitution on binding. The IC_{50} for analogues that exhibit binding is shown in brackets and peptide number in *blue*

study [23]. Next, a D-Arg^9 was substituted in place of the normal L-Arg^9 in conjunction with the Trp^{10} substitution (peptide 200). This substitution of both residues of the RFamide motif ablated kisspeptin's ability to bind to the receptor. These results confirm the importance of both the residues within this motif for receptor binding interactions and that it is part of an essential pharmacophore (Fig. 8.1) [19].

Thus, Arg^9 and Phe^{10} are important for receptor binding in truncated peptides, and one further residue was also shown to be involved in this interaction within the N-terminus (peptides 202, 203). The first residue of the truncated five amino acid analogues is a bulky aromatic and hydrophobic Phe^6. This is usually within the core of the peptide structure and may evade exposure to an aqueous environment. Substitutions of Phe^6 with Ala (peptide 202) and D-Phe (peptide 203) were made but both had poor binding (Fig. 8.1), suggesting that the side chain of phenylalanine is important for KP-10 to interact with KISS1R, along with Arg^9 and Phe^{10}, possibly via hydrophobic interactions.

As none of the above truncated peptides could bind KISS1R with the same or higher affinity than KP-10, this confirms that more than five residues are needed to fully bind to the receptor. Therefore, further studies with truncated peptides were abandoned and full-length ten-residue analogues were designed and tested. For the first full-length substitutions, the six most C-terminal residues were substituted, as it

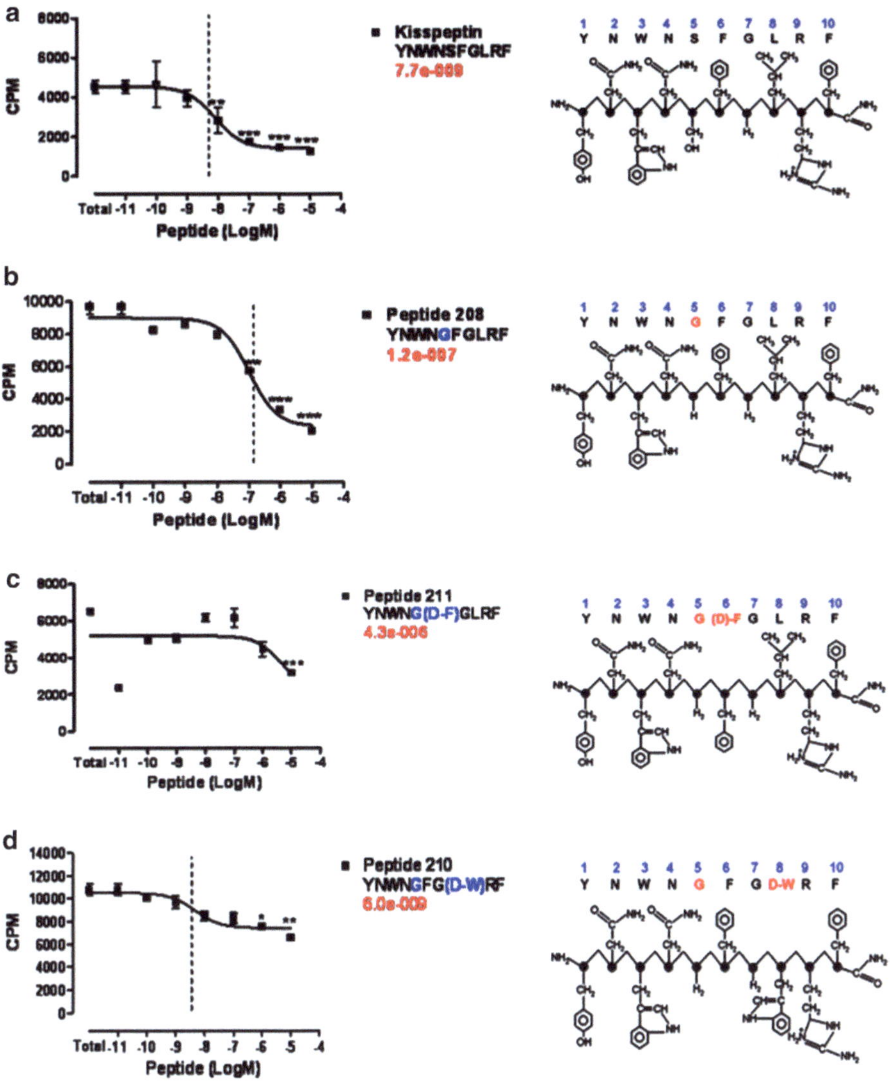

Fig. 8.2 Amino acid changes to KP-10 at the C-terminus affect receptor binding. (**a**) KP-10 binds strongly to GPR54 with an IC_{50} of 7.7 nM. (**b**) Peptide 208, with a glycine at position 5 binds the receptor with an IC_{50} of 120 nM and effectively displaces the labelled ligand by 75% at 10 μM. (**c**) Replacement of Phe^6 with D-Phe^6 significantly reduces the IC_{50} to 4.3 μM (**d**) D-Trp^8 retains a binding affinity with an IC_{50} of 6 nM similar to KP-10. Amino acid structure of the peptides is also shown

had been hypothesised that these residues were involved in binding to the receptor due to the high level of conservation of these residues between vertebrate species and the RFamide motif within this family; peptides 208, 209, 210, 211, 212, 213 and 288 (Figs. 8.2 and 8.3).

In order to increase the flexibility of the peptide and make substitutions potentially more tolerable, Ser^5 was substituted with the achiral glycine in peptide 208.

Peptide	Species	Sequence
FMRFamide	Clam	F M R F NH2
LPLRFamide	Chicken	L P L R F NH2
GnIH	Quail	S I K P S A Y L P L R F NH2
NPFF	Human	F L F Q P Q R F NH2
NPAF	Human	A G E G L S S P F W S L A A P Q R F NH2
PrRP20	Human	T P D I N P A W Y A S P G I R P V G R F NH2
RFRP-1	Bovine	S L T F E E V K D W A P K I K M N K P W N K M P P S A A N L P L R F NH2
RFRP-3	Bovine	A M A H L P L R L G K N R E D S L S R W V P N L P Q R F NH2
Kisspeptin-1 mammalian	Human	G T S L S P P P E S S G S P Q Q P G L S A P H S R Q I P A P Q G A L V Q R E K D L P N Y N W N S F G L R F NH2
	Bovine	G A A L C P P E S S A G P Q R L G P C A P R S R L I P S P R G A V L V Q R E K D V S A Y N W N S F G L R Y NH2
	Rat	R T S P C P P V E N P T G H Q R P P C A T R S R L I P A P R G S V L V Q R E K D M S A Y N W N S F G L R Y NH2
	Mouse	R S S P C P P V E G P A G R Q R P L C A S R S E L I P A P R G A V L V Q R E K D L S T Y N W N S F G L R Y NH2
	Opossum	L A M L C P S D E A S D P L W P G L C P T R S R L I T A P Q G A L L V E R E K D M S T Y N W N S F G L R Y NH2
	Xenopus	L L C R R K K S L S T G H P W S T D S L L P S R S I S A P E G E F L V Q R E K D L S T Y N W N S F G L R Y NH2
fish	Zebrafish	P T D G S P P S K L S A L F S M G A G P Q K N T W W W S P E S P Y T K R R Q N V A Y Y N L N S F G L R Y NH2
	Medaka	K E W P K D R S S D G G T P M V G C W M V K A L H P V A I K K R Q D L S S Y N L N S F G L R Y NH2
	Lamphrey	Y D F P G S G G S V D R A F M S P L H F Y P M L R A R M R S L P A S D A D E K K G S T Y N L N S F G L R F NH2
Kisspeptin-2 fish	Zebrafish	M E R R Q F E E P S A S D D A S L C F F I Q E K D E T S Q I S C K H R L A R S K F N Y N P F G L R F NH2
	Medaka	I L R R S E D D S A A G G A G L C S S L R E D D E Q L L C A D R R S K F N Y N P F G L R F NH2
	Lamphrey	V C R H A A E T P R L L R L R A L R G G H D L D A G L T D G E A L P R S A E Q D V T E F N Y N P F G L R F NH2

Fig. 8.3 Primary sequences for RFamides, including kisspeptin. Primary sequences for RFamides showing species were discovered. Sequences are also shown for kisspeptin-1/2 in a variety of mammalian and fish species. *Boxes* indicate conserved residues

Due to the small hydrogen atom in place of a side chain, this allows glycine to create flexibility within the peptide, to promote conformational changes. This substitution retained binding to the receptor with an IC_{50} of 1.2×10^{-7} M (Fig. 8.1). As this replacement was well-tolerated, it was included in the majority of subsequent analogues. Next, amino acid changes at Leu^8 were attempted. Leu^8 is a hydrophobic and aliphatic residue found in the peptide core, likely to be involved in interactions with other hydrophobic residues. Peptide 209 had a D-Leu^8 substitution, to check if the orientation was important for binding. Peptide 210 and 228 had D-Trp^8 in place of Leu^8 to test the effect of orientation and steric hindrance due to a bulky side chain on binding. In combination with Gly^5, this potentially allows flexibility within the central region. Peptide 228 also had an acetyl group added to the N-terminus to try to convey resistance to amino peptidases and extend the half-life of the peptide. Peptide 209 had decreased ability to bind the receptor, with an IC_{50} of 2.8×10^{-6} M, but exhibited similar capacity to displace radiolabelled ligand compared to KP-10. Peptide 210 was, however, able to bind the receptor with an IC_{50} close to that of KP-10 at 6×10^{-9} M. The results were similar to peptide 228, although the IC_{50} was higher at 1×10^{-11} M (Fig. 8.1). Therefore, it appears that changes within residue 8 that change side chain orientation and increase steric hindrance do not significantly affect the binding properties of KP-10.

The contribution of Phe^6 was next investigated in the full-length analogues, since changes to this residue had ablated binding in the truncated peptides. Two substitutions were made at this position in conjunction with Gly^5. In peptide 211, replacement with D-Phe^6 as in the truncated peptides reduced binding affinity, and the same was true with D-Trp substitution in peptide 212. This suggests that Phe^6 is critical for receptor binding via interactions between its side chain and KISS1R (Fig. 8.1).

The above results examining the effects of substitutions of C-terminal residues suggest that binding involves Phe^6, Arg^9, and Phe^{10}. This is in line with data published describing the pharmacophore created by these residues for binding to KISS1R [19].

Following the demonstration that the C-terminal residues of KP are involved in binding to receptor, attention was turned to the five N-terminal residues to examine if receptor binding also involved N-terminal interactions (Fig. 8.4). As peptide 210 could bind with a similar affinity to KP-10, the D-Trp^8 substitution was incorporated into all further analogues. Because position 5 could tolerate a glycine substitution to promote flexibility, the next changes examined if other residues would also be acceptable. Three more substitutions were studied. These were D-Trp^5 (Peptide 229), D-Ala^5 (peptide 233), and D-Ser^5 (Peptide 273). Peptide 229 was unable to bind to KISS1R; suggesting flexibility is needed in this region. For peptides 233 and 273, the displacement of radiolabelled ligand is lower (Fig. 8.1), implying that only glycine at this position allows high affinity binding of KISS1R and that side chain interactions are not important within this residue. Therefore, flexibility is more important at this position than side chain interactions, and Gly^5 was incorporated into all subsequent analogues.

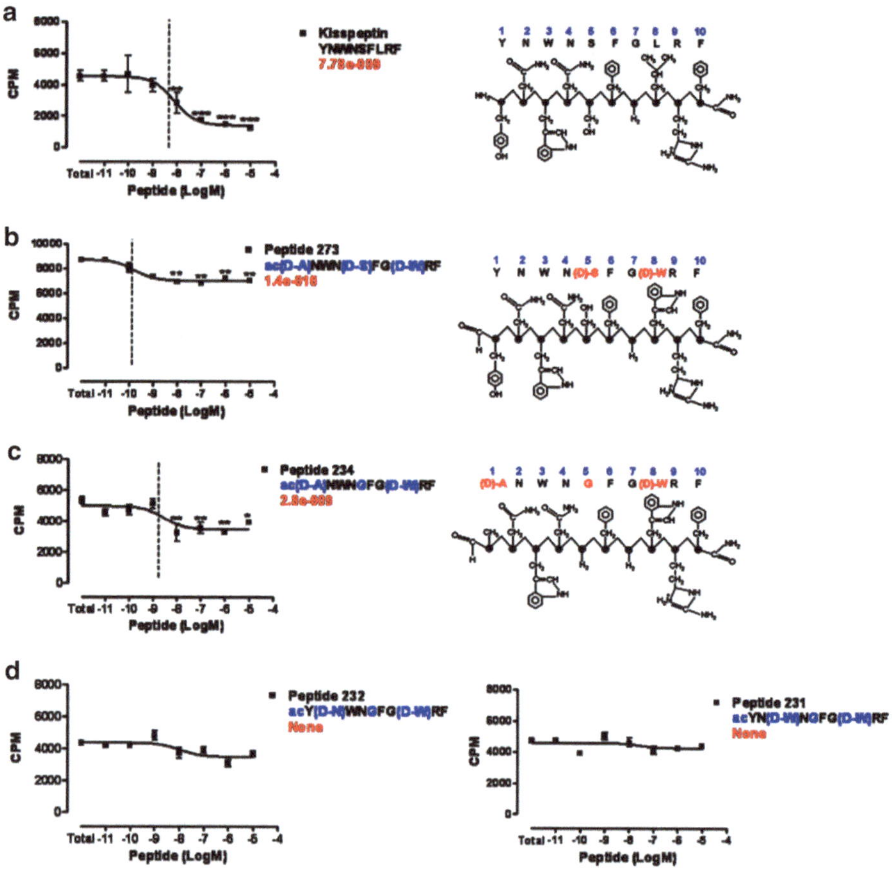

Fig. 8.4 Amino acid changes to KP-10 at the C-terminus and receptor binding. (a) KP-10 binds strongly to GPR54 with an IC_{50} of 7.7 nM. (b) D-Ser[5] retains binding with an IC_{50} of 140 pM (c) Peptide 234 with a D-Ala[1] binds with a similar IC_{50} value to KP-10 of 2.8 nM. (d) D-Asn at position 2 and D-Trp at position 3 cannot bind the receptor. Amino acid composition is also shown for each analogue

Substitutions were then made for the three most N-terminal residues in combination with Gly[5] and D-Trp[8]. Asn[4] was not substituted in these studies, as little effect was seen for this residue with alanine screening [19, 20]. Two changes were made at Tyr[1]; firstly, this residue was replaced with D-Tyr[1] (Peptide 230) to test side chain positioning and then with D-Ala[1] (Peptide 234) to test the importance of side chain interactions for binding. Peptide 230 was still able to bind to the receptor, but the displacement of radiolabelled ligand was decreased, whereas peptide 234 could bind the receptor with similar binding affinity to KP-10 and exhibited a greater amount of displacement compared to peptide 230 (Fig. 8.1). Therefore, we can assume that Tyr[1] is not very important for binding to the receptor. Two similar substitutions were made in place of Asn[2]. Peptide 232 incorporated a D-Asn[2] and peptide 236 had a D-Ala[2] in place of the asparagine removing the charge at this position. Peptide 232 and peptide 236 could not bind to the receptor (Fig. 8.1).

These results imply that Asn^2 interacts with the receptor to assist binding, probably via hydrogen bonding, and that the presence and positioning of the side chain is critical to this interaction. The results also suggest that a charged residue is needed at this position for receptor binding to take place. Finally, the same substitutions were made for Trp^3 (peptide 231 with D-Trp^3 and peptide 235 with D-Ala^3). Peptide 231 and peptide 235 could not bind to KISS1R (Fig. 8.1). This suggests that Trp^3 may also form interactions with the receptor and that the presence and positioning of a side chain is again important for this interaction and binding to occur.

These results from examining the effects of residue substitutions within full-length KP-10 suggest that the main binding residues lie within the C-terminal domain via Phe^6, Arg^9, and Phe^{10}, as was proposed within truncated analogues, and within the N-terminal domain via Asn^2 and Trp^3. This would account for the reduced binding in truncated peptides where Asn^2 and Trp^3 were removed. The study of the full-length analogues also suggests that the functional peptide conformation requires flexibility in the centre of the peptide, as bulky residues at position 5 ablate receptor binding. The introduction of rigidity via D-Trp^8 within the C-terminus is tolerated well, suggesting that this residue does not contribute to receptor binding interactions or that the D-Trp makes new interactions with the receptor to compensate for a loss of interaction from L-Leu. Similar results were seen with Tyr^1 substitutions, again suggesting this residue has no involvement in receptor binding or that the new substitutions make new contacts.

Although further analogues were made, these were used to investigate receptor activation and to screen for KP antagonists, since most residues within KP-10 have been tested for effects on binding to human KISS1R in the above research.

Functional Tests of KP-10 Peptide Analogues In Vitro

Active KP-10 Peptide Analogues Inhibit KP-10-Induced Inositol Phosphate Production and Intracellular Calcium Release

Analogues were tested for receptor activation via phospholipase C-mediated inositol phosphate (IP) production and antagonistic properties by their ability to inhibit kisspeptin-stimulated IP production and intracellular calcium release. A full antagonist binds to the receptor, but does not activate it, and inhibits kisspeptin-stimulated IP production or intracellular calcium release by >80%. KP activates the receptor with a mean EC_{50} of 3.9×10^{-9} M; therefore a dose of 10nM was used for antagonist studies.

Effects on Inositol Phosphate Production

Truncated peptides all activated KISS1R at high doses except peptides 201 and 206, 7, which had substitutions of Leu^8. Truncating the ligand to five amino acids retained its agonistic properties (peptide 188–9). Changes to position 6 (peptide 202, 3) had no antagonistic effects and changes to position 9/10 (peptide 200) could not bind, activate, or antagonise the receptor. However, position 8 changes did not intrinsically activate IP production and had weak antagonistic properties. Peptide 201, possessing Ala^8, could antagonise the KP-10 stimulation by 67%, with an IC_{50} of 7×10^{-9} M. Peptide 206, with a D-Trp^8, could also antagonise KP-10 stimulation by 71%, with an IC_{50} of 5×10^{-9} M, but peptide 207 with a D-Phe^8 could only antagonise by 53%, with an IC_{50} of 3×10^{-9} M (Fig. 8.5). These results suggest that Leu^8 is important for activation of KISS1R and that residues with bulky side chains are the most effective substitutions for antagonism. However, as none of these truncated KP-10 analogues could antagonise by more than 70%, they were not tested further and attentions were turned to the full-length analogues.

Amino acid changes within the C-terminal region were focused on, to determine residues within this region involved in receptor activation or creation of antagonism (Fig. 8.7a–d). Ser^5 to Gly^5 substitution in KP-10 (peptide 208) inhibited KP-10 stimulation of IP by 54% with an IC_{50} of 1×10^{-8} M (Fig. 8.6). Therefore, this change was incorporated into subsequent analogues. However, as this analogue could still stimulate IP release with an EC_{50} of 4.5×10^{-8} M, it implies that this substitution alone is insufficient for antagonism. Analogues comprising Gly^5 and substitutions at Phe^6 exhibited some antagonism but could still activate the receptor at high concentrations. Peptide 211 containing D-Phe^6 and peptide 212 containing D-Trp^6 could not antagonise any further than peptide 208 (Fig. 8.6). This implies that a Phe^6 substitution is not useful for antagonist activity.

As Leu^8 had been shown to create antagonism in the truncated peptides, changes at this residue were tested next in combination with Gly^5. D-Leu^8 (peptide 209) did not activate the receptor but could not antagonise any further than Gly^5 alone, whereas D-Trp^8 (peptides 210 and 228) again elicited no intrinsic IP release but caused further antagonism to 64%, with an apparent IC_{50} of $<1\times10^{-10}$ M. The acetylated version of this peptide could also inhibit KP-10 stimulation of IP by 69%. Combination of changes at positions 6 and 8 with Gly^5 (peptide 213) appeared to disrupt the structural conformation of the peptide, as it could no longer bind, activate, or antagonise the receptor (Fig. 8.6). This suggests that Leu^8 is important for receptor activation and that appropriate substitutions in the C-terminal region promote antagonism. Therefore, D-Trp^8 was incorporated into further analogues along with Gly^5 substitution. To test if glycine was the best residue for antagonism at position 5, changes were made in combination with D-Trp^8. D-Trp^5 (peptide 229) slightly decreased the antagonism, however, D-Ala^5 (peptide 233) increased the antagonism to 71% but the IC_{50} was reduced to 7×10^{-7} M. Therefore, Gly^5 is the best substitution for this position (Fig. 8.6).

The findings above indicated that Gly^5 in combination with D-Trp^8 could antagonise receptor activation and that Leu^8 was involved in receptor activation. In an attempt to enhance this antagonism, substitutions within the N-terminal region were tested in combination with these C-terminal region modifications (Fig. 8.7e–h). Changes were made at positions 1, 2, and 3. All of these analogues (peptides 230, 231, 232, 233, 234, 235, 236 and 243, 4) failed to stimulate IP release and did not have intrinsic agonist activity. Position 1 substitution to D-Tyr^1 (peptide 230)

Kisspeptin Analogues (CHO cells)												Dose Response		Antagonistic IP Inhibition		Possible
No.	Peptide Sequence											Bind(IC50)	IP(EC50)	IC50	% inh max	Antagonists
Kiss	Y	N	W	N	S	F	G	L	R	F	NH_2	7.77E-09	3.90E-09	n/a	n/a	n/a
186			ac	N	S	F	G	L	R	F	NH_2	1.84E-06	5.90E-07	n/a	agonist	poor
187	Y	N	W	N	S	F	G	L	R	W	NH_2	1.50E-07	4.00E-08	n/a	36%	poor
188					ac	F	G	L	R	F	NH_2	1.43E-07	8.70E-08	n/a	agonist	poor
189					ac	F	G	L	R	W	NH_2	9.46E-07	3.60E-08	n/a	agonist	poor
190					ac	F	G	(D)-F	R	W	NH_2	No displacement	2.00E-06	n/a	agonist	poor
191					ac	F	G	(D)-L	R	W	NH_2	9.39E-06	1.80E-06	n/a	agonist	poor
200					ac	F	G	L	(D)-R	W	NH_2	No displacement	No IP	n/a	0%	none
201					ac	F	G	A	R	W	NH_2	4.02E-03	No IP	7.00E-09	67%	poor
202					ac	A	G	L	R	W	NH_2	No displacement	3.40E-05	n/a	0%	none
203					ac	(D)-F	G	L	R	W	NH_2	No displacement	9.10E-06	n/a	40%	poor
206					ac	F	G	(D)-W	R	F	NH_2	No displacement	No IP	5.00E-09	71%	good
207					ac	F	G	(D)-F	R	F	NH_2	No displacement	No IP	3.00E-09	52%	poor

Fig. 8.5 Truncated KP-10 analogues. Table showing IC_{50} values for binding and antagonism, and EC_{50} values for intrinsic IP release. The percentage of maximal inhibition at micromolar concentration is given along with an indication for possible antagonists. *Blue* indicates an analogue with antagonistic potential and *no displacement* no displacement of KP-10

Kisspeptin Analogues (CHO cells)													Dose Response			Antagonistic IP Inhibition		Antagonistic Ca^{2+} Inhibition		Possible
No.	Peptide Sequence												Bind(Kd)	IP(EC50)	Ca^{2+}(EC50)	IC50	% inh max	IC50	% inh max	Antagonists
Kiss		Y	N	W	N	S	F	G	L	R	F	NH$_2$	7.70E-09	3.90E-09	n/a	n/a	n/a	n/a	n/a	n/a
186				ac	N	S	F	G	L	R	F	NH$_2$	1.85E-06	5.90E-07	n/a	n/a	agonist	n/a	n/a	poor
187		Y	N	W	N	S	F	G	L	R	W	NH$_2$	1.50E-07	4.00E-08	n/a	n/a	36%	n/a	n/a	poor
188						ac	F	G	L	R	F	NH$_2$	1.43E-07	8.70E-08	n/a	n/a	agonist	n/a	n/a	poor
189						ac	F	G	L	R	W	NH$_2$	9.46E-07	3.60E-08	n/a	n/a	agonist	n/a	n/a	poor
190						ac	F	G	(D)-F	R	W	NH$_2$	No displacement	2.00E-06	n/a	n/a	agonist	n/a	n/a	poor
191						ac	F	G	(D)-L	R	W	NH$_2$	9.39E-06	1.80E-06	n/a	n/a	agonist	n/a	n/a	poor
200						ac	F	G	L	(D)-R	W	NH$_2$	No displacement	No IP	n/a	n/a	0%	n/a	n/a	none
201						ac	F	G	A	R	W	NH$_2$	4.02E-03	No IP	n/a	7.00E-09	67%	n/a	n/a	poor
202						ac	A	G	L	R	W	NH$_2$	No displacement	3.40E-05	n/a	n/a	0%	n/a	n/a	none
203						ac	(D)-F	G	L	R	W	NH$_2$	No displacement	9.10E-06	n/a	n/a	40%	n/a	n/a	poor
206						ac	F	G	(D)-W	R	F	NH$_2$	No displacement	No IP	n/a	5.00E-09	71%	n/a	n/a	good
207						ac	F	G	(D)-F	R	F	NH$_2$	No displacement	No IP	n/a	3.00E-09	52%	n/a	n/a	poor
208		Y	N	W	N	G	F	G	L	R	F	NH$_2$	1.17E-07	4.56E-08	n/a	1.00E-07	54%	n/a	n/a	poor
209		Y	N	W	N	G	F	G	(D)-L	R	F	NH$_2$	2.80E-06	No IP	n/a	5.00E-08	56%	n/a	n/a	poor
210		Y	N	W	N	G	F	G	(D)-W	R	F	NH$_2$	5.97E-09	No IP	n/a	3.00E-09	64%	n/a	n/a	poor
211		Y	N	W	N	G	(D)-F	G	L	R	F	NH2	4.32E-06	2.57E-06	n/a	4.00E-08	30%	n/a	n/a	poor
212		Y	N	W	N	G	(D)-W	G	L	R	F	NH$_2$	6.30E-06	6.09E-05	n/a	2.00E-08	52%	n/a	n/a	poor
213		Y	N	W	N	G	(D)-L	G	(D)-L	R	F	NH$_2$	1.30E-06	No IP	n/a	n/a	0%	n/a	n/a	none
228	ac	Y	N	W	N	G	F	G	(D)-W	R	F	NH$_2$	1.03E-11	No IP	n/a	4.00E-08	69%	n/a	n/a	poor
229	ac	Y	N	W	N	(D)-W	F	G	(D)-W	R	F	NH$_2$	No displacement	No IP	n/a	5.00E-07	50%	n/a	n/a	poor
230	**ac**	**(D)-Y**	**N**	**W**	**N**	**G**	**F**	**G**	**(D)-W**	**R**	**F**	**NH$_2$**	**7.29E-09**	**No IP**	**n/a**	**3.00E-08**	**81%**	**n/a**	**n/a**	**good**
231	ac	Y	N	(D)-W	N	G	F	G	(D)-W	R	F	NH$_2$	No displacement	No IP	n/a	2.00E-08	60%	n/a	n/a	poor
232	ac	Y	(D)-N	W	N	G	F	G	(D)-W	R	F	NH$_2$	No displacement	No IP	n/a	3.00E-08	67%	n/a	n/a	poor
233	ac	Y	N	W	N	(D)-A	F	G	(D)-W	R	F	NH$_2$	7.90E-09	No IP	None	7.00E-07	71%	5.00E-12	83%	good
234	**ac**	**(D)-A**	**N**	**W**	**N**	**G**	**F**	**G**	**(D)-W**	**R**	**F**	**NH$_2$**	**2.79E-09**	**No IP**	**None**	**7.00E-08**	**93%**	**1.00E-11**	**89%**	**good**
235	ac	Y	N	(D)-A	N	G	F	G	(D)-W	R	F	NH$_2$	No displacement	No IP	n/a	1.00E-07	73%	n/a	n/a	good
236	ac	Y	(D)-A	W	N	G	F	G	(D)-W	R	F	NH$_2$	No displacement	No IP	n/a	2.00E-08	71%	n/a	n/a	good
237	ac	(D)-Y	N	W	N	G	F	G	W	R	F	NH$_2$	n/a	6.40E-06	n/a	None	0%	n/a	n/a	none
238	ac	(D)-Y	N	W	N	S	F	G	(D)-W	R	F	NH$_2$	n/a	2.80E-06	n/a	None	0%	n/a	n/a	none
239	ac	(D)-Y	(D)-N	W	N	S	F	G	W	R	F	NH$_2$	n/a	1.60E-08	n/a	None	0%	n/a	n/a	none
240	ac	(D)-Y	(D)-N	W	N	G	F	G	W	R	F	NH$_2$	n/a	No IP	n/a	None	47%	n/a	n/a	poor
241	ac	(D)-Y	(D)-N	W	N	S	F	G	(D)-W	R	F	NH$_2$	n/a	No IP	n/a	None	32%	n/a	n/a	poor
242	ac	(D)-Y	(D)-N	W	N	G	F	G	(D)-W	R	F	NH$_2$	n/a	No IP	n/a	None	44%	n/a	n/a	poor
243	ac	(D)-W	N	W	N	G	F	G	(D)-W	R	F	NH2	n/a	No IP	n/a	1.00E-10	51%	n/a	n/a	poor
244	ac	(D)-F	N	W	N	G	F	G	(D)-W	R	F	NH2	n/a	No IP	n/a	5.00E-07	59%	n/a	n/a	poor
245	ac	(D)-Y	N	W	N	G	(D)-W	G	(D)-W	R	F	NH2	n/a	No IP	n/a	3.00E-07	67%	n/a	n/a	poor
246	ac	(D)-A	N	W	N	G	(D)-W	G	(D)-W	R	F	NH2	n/a	No IP	n/a	1.00E-07	64%	n/a	n/a	poor
247	ac	(D)-A	N	W	N	S	F	G	(D)-W	R	F	NH2	n/a	No IP	n/a	5.00E-09	78%	n/a	n/a	good
248	ac	(D)-A	N	W	N	G	F	G	W	R	F	NH2	n/a	No IP	n/a	5.00E-06	65%	n/a	n/a	poor
271	P	(D)-A	N	W	N	G	F	G	(D)-W	R	F	NH$_2$	1.56E-08	No IP	n/a	1.00E-06	62%	1.00E-11	100%	poor
273	**ac**	**(D)-A**	**N**	**W**	**N**	**(D)-S**	**F**	**G**	**(D)-W**	**R**	**F**	**NH2**	**1.46E-10**	**No IP**	**None**	**1.00E-10**	**80%**	**1.00E-08**	**63%**	**good**
274	ac	(D)-A	N	W	N	P	F	G	(D)-W	R	F	NH2	n/a	5.30E-04	7.30E-09	n/a	24%	n/a	28%	poor
275	ac	(D)-A	N	W	N	(D)-P	F	G	(D)-W	R	F	NH2	n/a	No IP	None	n/a	45%	n/a	27%	poor
276	**ac**	**(D)-A**	**N**	**W**	**N**	**G**	**F**	**G**	**(D)-L**	**R**	**F**	**NH2**	**n/a**	**No IP**	**None**	**1.00E-12**	**85%**	**1.00E-11**	**100%**	**good**
277	ac	(D)-A	N	W	N	(D)-A	F	G	L	R	F	NH2	n/a	1.50E-06	9.40E-08	1.00E-07	24%	n/a	22%	poor
278	ac	(D)-A	N	W	N	G	F	G	L	(D)-R	F	NH2	n/a	No IP	None	1.00E-06	50%	n/a	21%	poor

Fig. 8.6 Table detailing all KP-10 peptide analogues. Table showing the peptide analogues tested with results for binding and stimulation of IP and calcium release. Antagonistic results for IP and calcium are also shown. The four antagonists found are highlighted in *yellow* and other promising analogues in *blue*

significantly improved antagonism to 81%; and D-Ala[1] substitution (peptide 234) significantly increased antagonism further to 93%, with an IC_{50} of 7×10^{-8} M. However, D-Trp[1] (peptide 243) or D-Phe[1] (peptide 244) substitution at this position did not increase the antagonism (Fig. 8.6). This suggests that removing the charge and hydrogen bonding from this position is important for antagonism and that steric hindrance at this position is detrimental to antagonistic activity. Amino acid

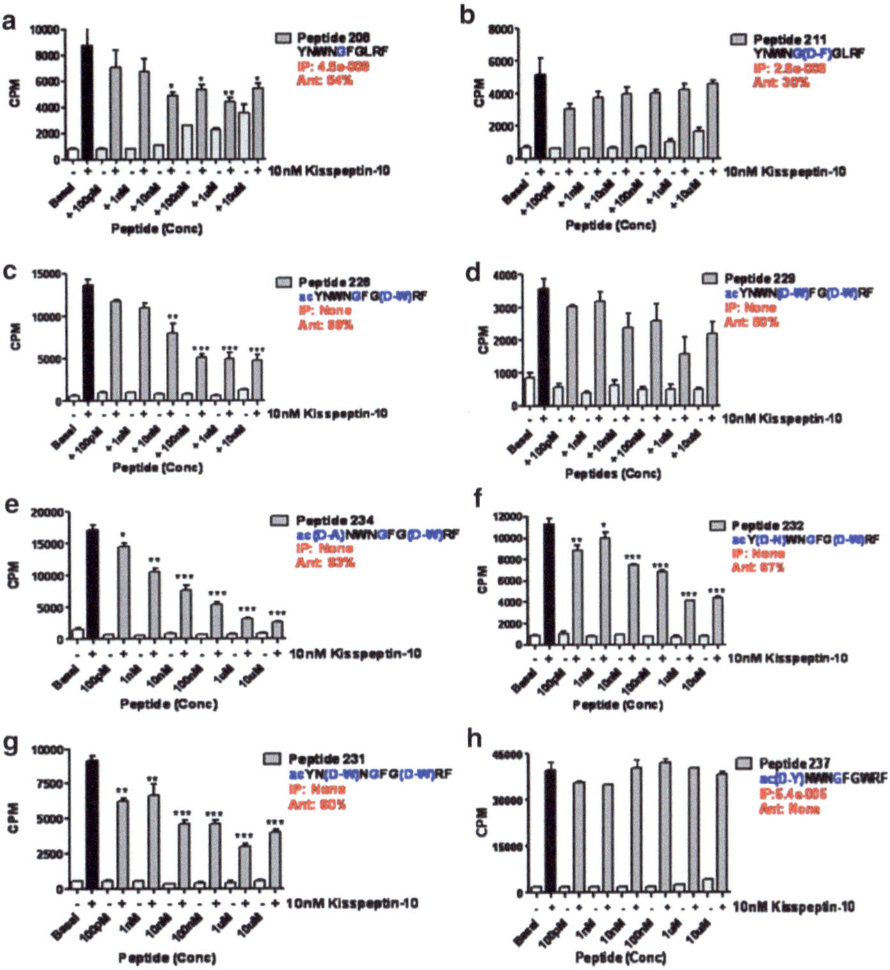

Fig. 8.7 Specific amino acid changes create antagonism. Graphs showing both intrinsic IP release (*light grey*) and antagonism of KP-10 stimulated IP release (*dark grey*). IP is the concentration at which stimulation reached 50% of the maximal (EC_{50}) and Ant is the percentage for maximal antagonism at micromolar doses. (**a**) Substitution of Ser^5 with Gly enhances antagonism at high doses but still activates the receptor at 45 nM. (**b**) Additional substitution at position 6 also stimulates IP release but D-Phe^6 does not increase the antagonism over peptide 208. (**c**) Addition of D-Trp8 increases antagonism to 69% with an IC_{50} of 1 nM. (**d**) Substitution of Gly^5 with a bulky D-Trp^5 reduces this antagonism to 50% suggesting a small flexible amino acid is needed at position 5. (**e**) Adding D-Ala at position 1 of 228 increases antagonism to 93% with and IC_{50} of 70 nM and no intrinsic IP production. (**f**) Adding D-Asn at position 2 of 228 decreases antagonism as does (**g**) adding D-Trp at position 3. (**h**) Removal of D-Trp^8 from peptide 234 completely ablates antagonism

changes at position 2; D-Asn^2 (peptide 232) and D-Ala^2 (peptide 236) did not increase the antagonistic effects over peptide 228. Therefore substitution of this residue does not appear important for antagonism of the receptor. Finally, substitutions of position 3 to D-Trp^3 (peptide 231) or D-Ala^3 (peptide 235) also had no further antagonistic effect on the receptor (Fig. 8.6). Therefore, Tyr^1 is the only residue within the N-terminal region that appeared important for receptor activation and antagonism.

Next, to test whether antagonism was due to a combination of all three substituted residues or whether one was more critical, we took the two position 1 analogues and removed or changed the Gly^5 or D-Trp^8 substitution to examine the effect on the antagonistic properties of the peptide. For analogues with D-Tyr^1, modification of either of these residues completely ablated any antagonistic effect (peptides 237, 8), indicating that the D-Tyr^1 has relatively little effect on the receptor antagonism alone. However, for D-Ala^1 analogues, conversion of Gly^5 back to Ser^5 (peptide 247) slightly reduced the antagonism to 78% and changing of D-Trp^8 to L-Trp^8 (peptide 248) reduced antagonism further to 65%. However, D-Leu^8 at this position (peptide 276) increased antagonism to 85%, suggesting a D-amino acid is needed at this position for antagonism. However, peptide 276 does have intrinsic agonist activity (Fig. 8.6). The results indicate that D-Ala^1 and D-Trp^8 are the most important residues for antagonism. However, Gly^5 must also play a role by keeping the peptide flexible, since other substitutions at this position within peptide 234 have drastic effects except when D-Ser^5 (peptide 273) is substituted here, keeping antagonism high at 80% with an IC_{50} of 1×10^{-10} M. Replacement with Pro^5 (peptide 274), D-Pro^5 (peptide 275), or D-Ala^5 (peptide 277) reduces antagonism to 24%, 45%, and 24%, respectively (Fig. 8.6). Therefore, flexibility, possibly to allow receptor engagement appears to be the key factor in this position. Since substitution of position 1, 5, and 8 had been shown to produce antagonists, combinations of these with position 2 substitutions were tested (peptides 239, 240, 241, 242). However, all of these reduced antagonism. Also, introduction of D-Trp^6 into peptides 230 and 234 to increase steric hindrance at the C-terminal region (peptides 245 and 246) also reduced antagonism to around 60% (Fig. 8.6). The above data shows that amino acid changes to D-Ala^1, Gly^5, and D-Trp^8 (peptide 234) are the most critical for antagonism. However, other residues are tolerated at these positions such as D-Trp^1 in peptide 230, D-Ser^5 in peptide 273, or D-Leu^8 in peptide 276.

Effects of Peptide Analogues on Mobilisation of Intracellular Calcium

To further characterise inhibition of KP-10 signalling by these antagonists, stimulation of intracellular calcium release in model cells was measured. Peptide 234 did not stimulate calcium release alone, but did antagonise KP-stimulated calcium release by 89%, with an IC_{50} of 1×10^{-10} M, further confirming its potency. Peptide 273 could also antagonise intracellular calcium release to 63%, with no intrinsic stimulation, and peptide 276 completely inhibited intracellular calcium secretion (Figs. 8.6 and 8.8). To confirm these were specific effects, three analogues that had reduced antagonism of IP were also tested. Peptides 274, 275, and 277 only antagonised by 28%, 27%, and 22%, respectively, confirming the effects above are specific (Fig. 8.6). Overall, four antagonists have been created (peptide 230, 234, 273, and 276) from this research and identified the residues important for receptor binding

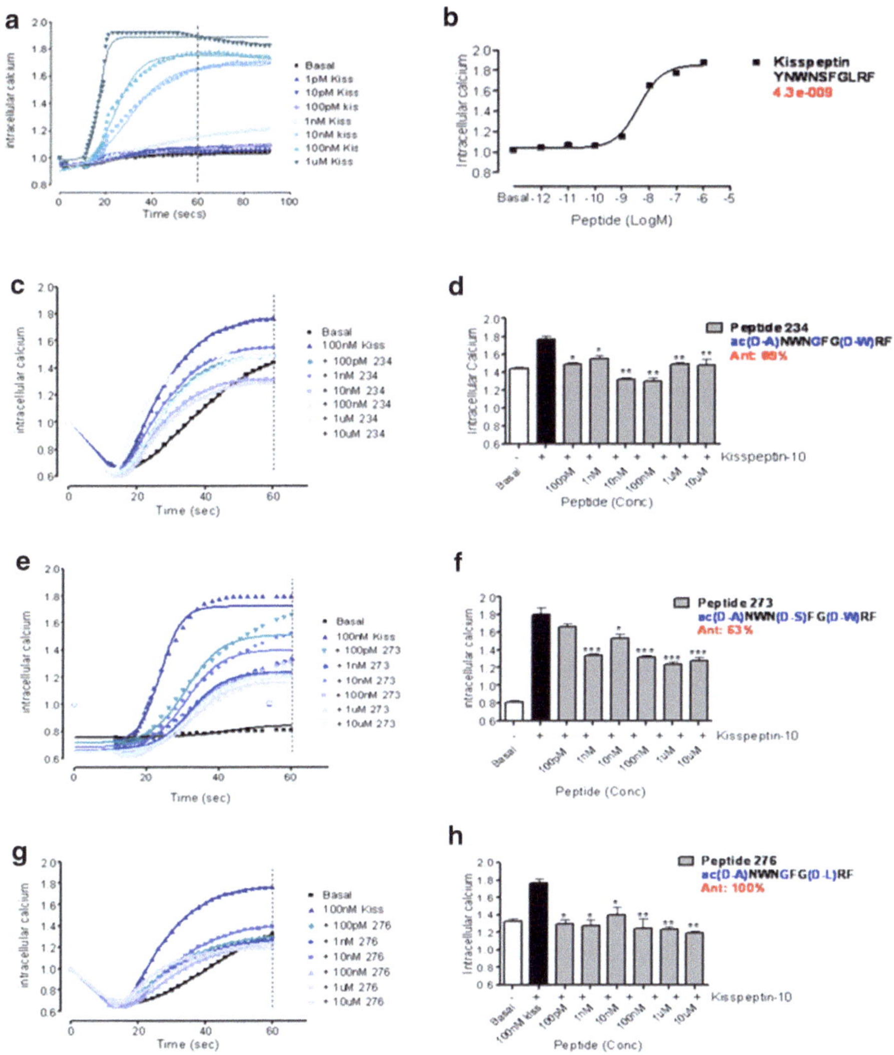

Fig. 8.8 Effects of selected peptides on the elevation of intracellular calcium by KP-10 in CHO/GPR54 cells. Graphs show antagonism of KP-10 stimulated calcium at 60 s. *Ant* is the maximal antagonism achieved at micromolar concentrations. (**a**) Raw calcium transients for KP-10 showing 60 sec time point (*dotted line*). (**b**) Quantification of KP-10 calcium transients. (**c**) Peptide 234 calcium transients showing antagonism of KP-10 stimulated calcium (**d**) Quantification of peptide 234 transients showing 89% antagonism. (**e**) Peptide 273 calcium transient showing partial antagonism of KP-10 stimulated calcium by 63%. (**f**) Quantification of peptide 273 calcium transients. (**g**) Peptide 276 calcium transients showing complete antagonism of KP-10 stimulated calcium. (**h**) Quantification of peptide 276 calcium transients showing 100% antagonism

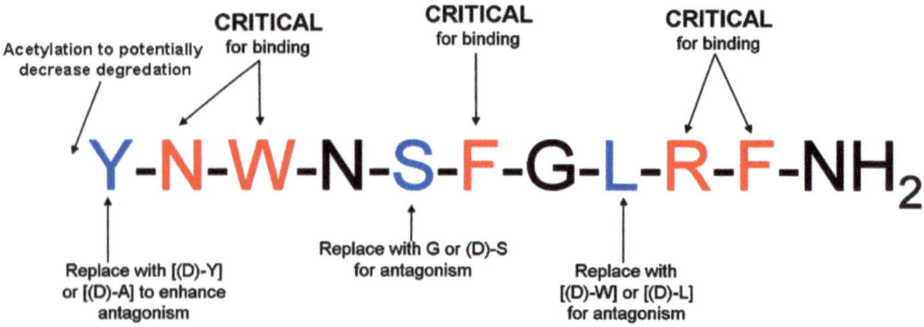

Fig. 8.9 Schematic diagram of important KP-10 residues. Residues that are important for receptor binding (*red*) and receptor activation (*blue*) within KP-10 are highlighted

(Asn^2, Trp^3, Phe^6, Arg^9, and Phe^{10}) and for receptor activation (Tyr^1 and Leu^8). Peptide 234 was the most efficacious and potent antagonist in vitro.

Besides producing four antagonists, the structure–activity relationship has also provided us with a consensus model for antagonism of KP-10 (Fig. 8.9). The model highlights five residues involved in receptor binding (Asn^2, Trp^3, Phe^6, Arg^9, Phe^{10}) and three residues involved in receptor activation (Tyr^1, Ser^5, Leu^8), which concurs with previous alanine scanning results [19, 20].

Design and In Vitro Effects of Small Molecule Antagonists

Small molecule non-peptide KP-10 antagonists containing a 2-acylamino-4,6-diphenylpyridine scaffold have also been designed. Kobayashi et al. adopted a combinatorial chemistry approach to identify a 2-furoyl group to be the most active antagonist of all tested 2-acylamino-4,6-diphenylpyridine derivatives [24]. They identified compound 9l (Fig. 8.10a) with an IC_{50} value of 3.7 nM for receptor binding to KISS1R and with high antagonistic activity against KP-10 stimulated intracellular calcium release [24]. Then, with further optimisation, compound 15a (Fig. 8.10b) was designed containing a piperazine ring and exhibited high affinity binding to both human and rat KISS1R, with IC_{50} values of 3.6 and 15 nM, respectively. Compound 15a also showed high antagonistic activity against KP-10 stimulated calcium release in CHO cells stably expressing KISS1R (Fig. 8.10c) [25].

Compound 15a has a similar receptor binding affinity to peptide 234 above for human KISS1R; however, it does not antagonise calcium release to the same extent. Peptide 234 inhibits calcium with an IC50 of 1nM and a maximal inhibition of 89%, whereas compound 15a has an IC50 of 1 μM with a similar maximal inhibition to peptide 234. However, as compound 15a is a small molecule antagonist, it is likely to be orally active, unlike peptide 234, which needs to be given by injection; therefore compound 15a may be more suitable to being taken forward into clinical trials.

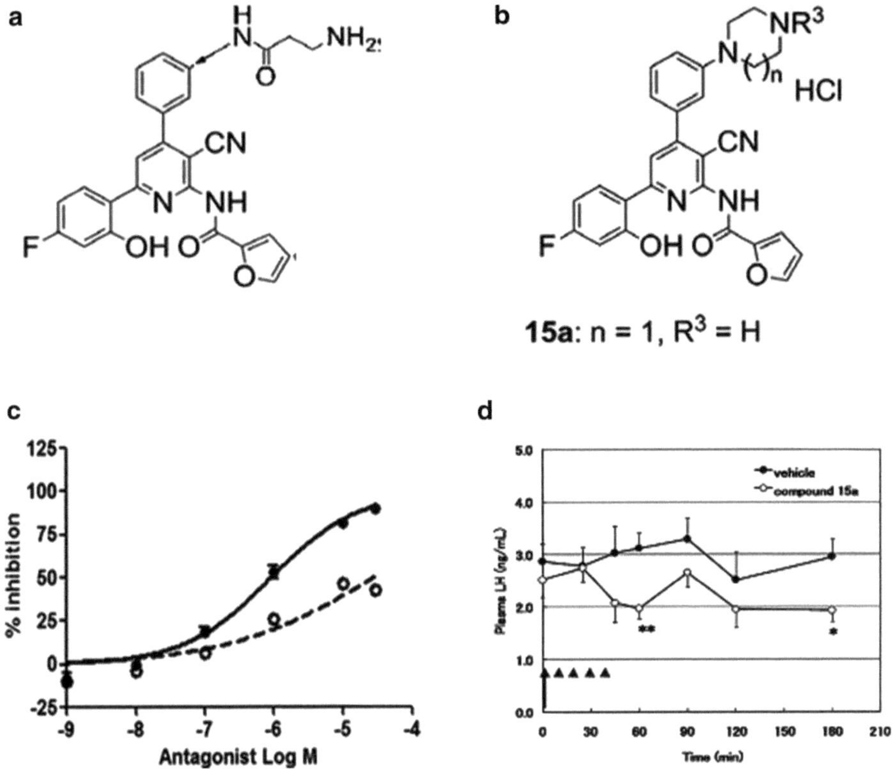

Fig. 8.10 Small molecule KP antagonist. (**a**) Structural model for compound 9l. (**b**) Structural model for compound 15a bearing a piperazine ring. (**c**) The antagonistic activities on calcium mobilisation of 15a (*solid line*) and 15d (*broken line*) against KP-10 in CHO cells stably expressing human GPR54. (**d**) Effect of compound 15a on plasma LH level in castrated male rats. Values are means ± SE, determined from eight experiments. $*P<0.05$ and $**P<0.01$ compared with the vehicle control. (From Kobayashi, T., et al., Synthesis and structure–activity relationships of 2-acylamino-4,6-diphenylpyridine derivatives as novel antagonists of GPR54. Bioorg Med Chem. 18(11):3841–59 and from Kobayashi, T., et al., 2-acylamino-4,6-diphenylpyridine derivatives as novel GPR54 antagonists with good brain exposure and in vivo efficacy for plasma LH level in male rats. Bioorg Med Chem. 18(14):5157–71. Reprinted with permission from Elsevier Limited)

Testing KP's Role in GnRH Neuron Function

Since the primary action of KP is the stimulation of GnRH neurons, we determined whether KP antagonist, peptide 234, could inhibit KP action on mouse GnRH neurons in brain slices. 1nM KP markedly increased GnRH neuron firing activity [26], as previously described [27]. Peptide 234 alone had no effect on GnRH neuron firing, but pretreatment with this peptide at 1, 10, and 100 nM strongly inhibited 10 nM KP-10 stimulation of GnRH firing activity (Fig. 8.11a). Thus, KP antagonist, peptide 234, is a potent inhibitor of KP-stimulated GnRH neuron firing.

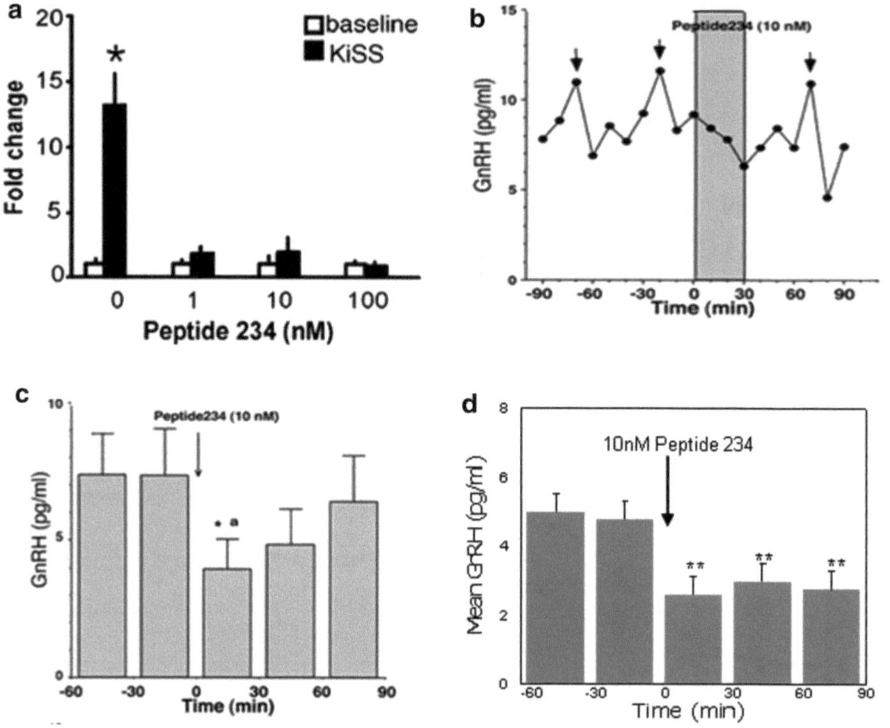

Fig. 8.11 Peptides 234 inhibits KP actions in vivo. (**a**) Summary bar graph showing mean ± SEM fold change in firing rate during baseline (*white bars*) and KP-10 (*black bars*). KP-10 significantly increased firing activity of GnRH neurons ($n=7$, $^*P<0.002$). Response to KP-10 was significantly reduced with the presence of 1, 10, 100 nM peptide 234 (1 nM $n=5$, 10 nM $n=6$, 100 nM $n=7$, $P<0.001$ all groups). (**b**) Representative case from the effects of Peptide 234 on GnRH release and group mean (±SEM, $n=6$) is shown. Pulsatile GnRH release in the hypothalamus was suppressed by 10 nM peptide 234 infusion to the stalk-median eminence regions of female rhesus monkeys (*dark shaded bar*). Short arrows indicate GnRH peaks identified by PULSAR. (**c**) Data analysis indicated that peptide 234 significantly ($P<0.05$) suppressed GnRH release as compared to levels prior to peptide 234 as well as to the vehicle control ($^*P<0.05$ vs. before peptide 234; a $P<0.05$ vs. control at corresponding time period). (**d**) In pre-pubertal female rhesus monkeys, peptide 234 at 10 nM significantly decreased GnRH release compared with baseline levels ($P<0.01$). (From Roseweir, A.K., et al., Discovery of potent kisspeptin antagonists delineate physiological mechanisms of gonadotropin regulation. J Neurosci. 2009. 29(12):3920–9, with permission from The Society for Neuroscience and from Guerriero, K.A., et al., Developmental changes in GnRH release in response to kisspeptin agonist and antagonist in female rhesus monkeys (Macaca mulatta): implication for the mechanism of puberty. Endocrinology. 2012. 153(2):825–36. Reprinted with permission from The Endocrine Society)

Testing KP's Role in GnRH Pulsatility

Microdialysis of 10 nM peptide 234 into the stalk/median eminence of pubertal female rhesus monkeys over 30 min suppressed GnRH pulses (Fig. 8.11b) and mean GnRH levels (Fig. 8.11c) measured in median eminence microdialysate

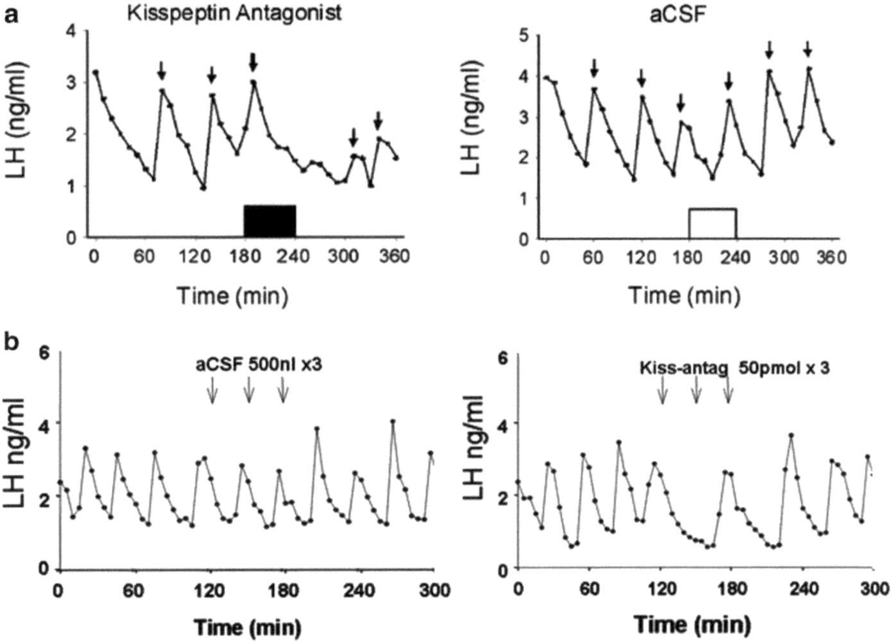

Fig. 8.12 Peptide 234 inhibits LH secretion and pulsatility. (**a**) Concentrations of LH are shown in an oestrogen-replaced ovariectomised ewe treated with peptide 234 (*black bar, left*) or control (*grey bar, right*). Arrows indicate LH pulses. Analysis revealed a significant reduction in the mean LH concentration and pulse amplitude after peptide 234 infusion. (**b**) Representative examples illustrating the effects of intra-ARC injection of 500 nL aCSF (*left hand*; 3 injections at 30 min intervals) or 50 pmol peptide 234 (*right hand*; 3 injections at 30 min intervals) in ovariectomised 17β-estradiol-replaced rats (n=5–6). *Arrows* represent intra-ARC injections. (From Roseweir, A.K., et al., Discovery of potent kisspeptin antagonists delineate physiological mechanisms of gonadotropin regulation. J Neurosci. 2009. 29(12):3920–9, with permission from The Society for Neuroscience and from Li, X.F., et al., Kisspeptin signalling in the hypothalamic arcuate nucleus regulates GnRH pulse generator frequency in the rat. PLoS One. 2009. 4(12):e8334. Open Access)

samples. GnRH pulses were restored when the antagonist infusion was terminated. This provides the first direct evidence that KP is required for GnRH secretion and is responsible for GnRH pulses at puberty [26]. Importantly, the antagonist did not affect basal GnRH levels. Peptide 234 also suppressed GnRH pulses in another study on pre-pubertal female rhesus monkeys during a 30 min infusion, and the suppression continued for 90 min post-infusion (Fig. 8.11d) [28], which may reflect lower levels of KP in pre-pubertal animals or differences in pharmacokinetics.

The role of KP in GnRH pulsatility was also investigated indirectly in ovariectomised ewes by monitoring peptide 234 effects on LH pulses when the antagonist was infused i.c.v. as a 40 μg bolus followed by 40 μg over the ensuing hour. Peptide 234 reduced both LH pulse amplitude and pulse frequency during infusion, and this inhibition continued for 2 h after termination of infusion (Fig. 8.12a) [26]. Since a reduction in pulse amplitude reduces the ability to detect pulses, the effects on pulse frequency are uncertain. However, a reduction in LH pulse frequency is clearly evident in the female rat after peptide 234 administration into the ARC (see below). The effect on LH appears to be specific, as prolactin and cortisol secretion were unaffected in the ovariectomised ewes [26].

In the rat, the KP neurons critical for gonadotropin secretion are located in the hypothalamic ARC and AVPV nuclei. As the ARC is known to be the site of the GnRH pulse generator [29], we explored the effects of KP-10 or peptide 234 administered intra-ARC or intra-medial preoptic area, which includes the AVPV, on pulsatile LH secretion [30]. Intra-ARC administration of peptide 234 profoundly suppressed LH pulse frequency (Fig. 8.12b), but intra-mPOA administration of peptide 234 had no effect [30] (data not shown). These data are the first to identify the ARC as a key site for KP modulation of LH pulse frequency, supporting the notion that KP/KISS1R signalling in this region of the mediobasal hypothalamus is a component of the hypothalamic GnRH pulse generator.

Testing KP's Role in Puberty

Hypogonadotropic hypogonadism and a failure to progress through puberty in mice and humans with inactivating mutations of *KISS1R* is the phenotype which led to the discovery of the importance of the KP/KISS1R system in regulating reproduction. These findings indicated an essential role for the KP system in translating neuroendocrine regulators of puberty into increased GnRH output. Puberty was thus a relevant model to examine the effects of a KP antagonist. To study KP's role in prolonged physiological processes such as puberty, protracted blockade of KP action was required. Central infusion of KP antagonist, peptide 234 (10 nmol/24 h), by osmotic minipump attached to an i.c.v. catheter for 7 days to pubertal (d30) female rats delayed vaginal opening and decreased uterine and ovarian weights (data not shown) at the expected time of puberty, without affecting body weight (data not shown) (Fig. 8.13a) [31]. Eighty percent of animals infused with vehicle displayed complete canalisation of the vagina by d36, while only 13% of the females treated with the antagonist showed complete vaginal opening (Fig. 8.13a). In the study on pubertal female rhesus monkeys described earlier, peptide 234 inhibited GnRH pulses [26]. These findings provided direct evidence for a role of KP in initiating puberty, in support of indirect data showing that exogenous KP administration advances puberty [32], *Kiss1* gene expression increases at puberty in rats, and the puberty is absent in *KiSS1* and *KISS1R* gene inactivation in humans. Thus, we are able to conclude that KP secretion is essential for increasing GnRH secretion and progression through puberty.

Testing if KP Mediates Gonadal Steroid Negative Feedback

The mechanism by which gonadal steroids mediate feedback has been a lively area of research and debate over many years. The discovery of the KP/KISS1R system suggested that it may play a part in negative steroid feedback. Gonadectomy of mice

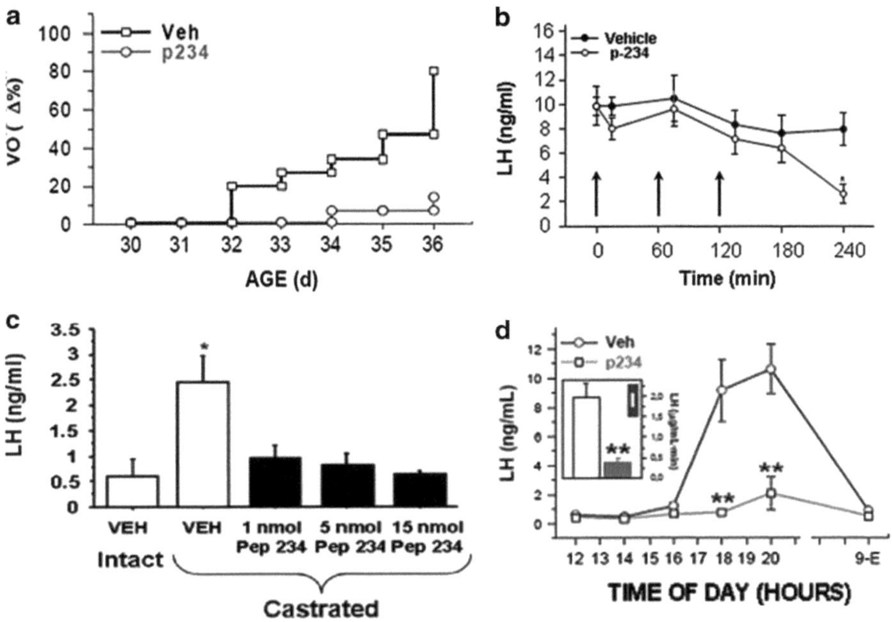

Fig. 8.13 Peptide 234 delays puberty and inhibits castration rise in LH and the pre-ovulatory LH surge. (**a**) Effects of continuous infusion of peptide 234 on puberty onset in female rats. The impact of chronic i.c.v. infusion of the peptide 234, into pubertal female rats (d30–d36) on puberty onset shows a delay. *VO* vaginal opening. (**b**) Castrated male rats were given three infusions (indicated by *arrows*) of 1 nmol peptide 234 at 0, 60, and 120 min, which significantly inhibited LH secretion after 240 min ($n=10$; $*P<0.05$). (**c**) Dose–response graph for castrated male mice given two i.c.v. infusions of peptide 234, 60 min apart, indicating that all three doses tested are able to inhibit the post-castration rise in LH. *Bars* show mean ± SEM ($*P<0.05$ compared with other groups). (**d**) Effects of continuous infusion of peptide 234 on pre-ovulatory surge of LH in rats. Mean serum LH levels in both groups, as well as integrated LH secretion between 14:00 h and 20:00 h of proestrus (AUC; bar graph) in (**d**), are shown. Numeric values on *x*-axis represent daytime (hour) during the afternoon and evening of proestrus. Hormonal levels at the morning of oestrus (*9-E*) are also shown. $**P<0.01$ vs. corresponding control group. (From Roseweir, A.K., et al., Discovery of potent kisspeptin antagonists delineate physiological mechanisms of gonadotropin regulation. J Neurosci. 2009. 29(12):3920–9, with permission from The Society for Neuroscience and from Pineda, R., et al., Critical roles of kisspeptins in female puberty and preovulatory gonadotropin surges as revealed by a novel antagonist. Endocrinology. 151(2):722–30. Reprinted with permission from The Endocrine Society)

and rats induced an increase in *Kiss1* gene expression in the ARC nucleus which is reversed by gonadal steroid replacement [8, 10, 13]. However, there is no direct evidence that gonadectomy results in increased KP secretion. To address this issue, KP antagonist, peptide 234, was infused i.c.v. into castrated male rats and mice [26]. Two or three i.c.v. injections of Peptide 234 inhibited the castration-induced rise in LH in both rats (Fig. 8.13b) and mice (Fig. 8.13c). This is consistent with data that KP neurons express sex steroid receptors [14, 33]. It therefore appears that a substantial part of sex steroid feedback on GnRH secretion is mediated via the KP system (Fig. 8.14).

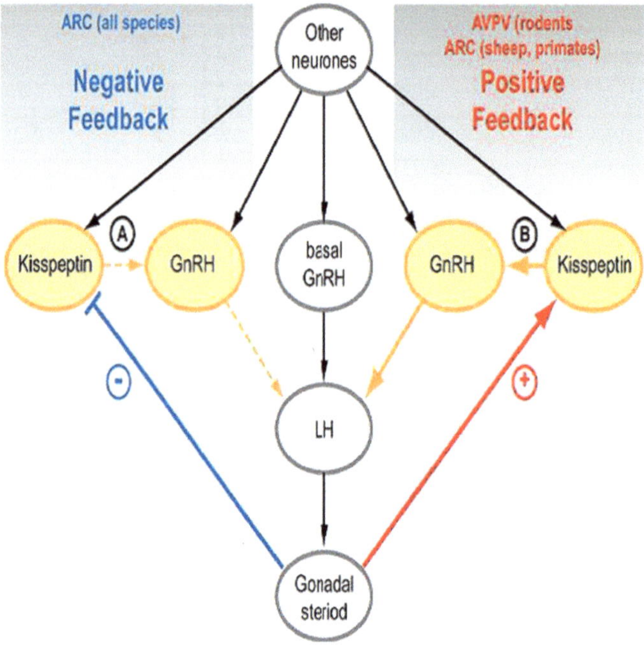

Fig. 8.14 Physiological roles of KP interpreted from KP antagonist studies. Gonadectomy leads to an increase in GnRH and LH pulsatility due to removal of negative feedback in the arcuate nucleus (ARC) which is inhibited by KP antagonist. The positive feedback effects of gonadal steroids in the anteroventral paraventricular nucleus (AVPV) in rodents which induce the LH surge are inhibited by KP antagonist (indicated by "B"). KP antagonist does not inhibit basal GnRH or LH in all animal models studied, suggesting that this is independent of KP ("other neurons"). "A" indicates that KP in the ARC also plays a role in influencing GnRH pulse frequency. (From Millar, R.P., et al., Kisspeptin antagonists: unraveling the role of kisspeptin in reproductive physiology. Brain Res. 1364:81–9. Reprinted with permission from Elsevier Limited)

Testing a Role for KP in Positive Feedback and LH Ovulatory Surge

The switch in the hypothalamus which determines the change in oestrogen feedback from negative to positive to generate the pre-ovulatory LH surge is central to understanding of the female reproductive system. However the pathways and mediators involved have remained elusive. To examine a putative role of KP in the LH surge, regularly cycling female rats were implanted in the morning of oestrus with osmotic mini-pumps to allow i.c.v. delivery of KP antagonist, peptide 234, at a constant rate of 10 nmol/24 h. The infusion was continued until the afternoon of the following proestrus, when the animals were subjected to serial blood sampling during the afternoon/evening of proestrus and the morning of oestrus. In controls, 10 out of 11 animals displayed the expected pre-ovulatory surge of LH during the afternoon of proestrus, with a progressive rise of serum concentrations between 16:00 and 20:00, followed by a decrease in LH levels on the morning of oestrus. However, when peptide 234 was infused, 7 out of 9 females failed to display the prototypical surge of LH at proestrus (Fig. 8.13d) [31]. This experiment provides direct evidence for the indispensable role of KP signalling in mediating the pre-ovulatory surge of gonadotropins (Fig. 8.14). The finding supports a previous demonstration that administration of KP antiserum abolishes the ovulatory LH surge [34].

Design of KP Antagonists to Penetrate the Blood–Brain Barrier

KP-10 and its peptide antagonists can be modified at the NH2 terminus without loss of activity. We have therefore exploited this property by adding the penetrating peptide sequence (which allows passage across the cell membrane) to the NH2 terminus of KP antagonist, peptide 234. This potentially facilitates transfer across the blood–brain barrier. When administered systemically, this antagonist (Peptide 271) inhibited KP-10 stimulation of LH in rats following both i.c.v. and systemic administrations of KP-10 [31]. Peptide 271 has also been shown to decrease both amplitude and frequency of LH pulses in OVX ewes. In these animals, peptide 271 was also shown to inhibit EB-induced LH surge [35], providing further evidence for the role of KP in positive steroid feedback (Fig. 8.15).

Kobayashi et al. also modified the small molecule antagonist, compound 91, to increase penetration of the blood–brain barrier, as compound 91 had relatively low brain exposure when given by i.v. administration to male castrated rats. The modified compound 15a was shown to have high brain exposure levels in castrated male rats and also suppressed plasma LH levels in these animals (Fig. 8.10d) [25].

KP-Independent GnRH Secretion

The majority of the studies described above indicated that KP antagonists did not affect basal LH secretion in intact and castrated male mice and rats, in ovariectomised oestrogen treated ewes, in studies on the ovulatory LH surge in rats and ewes, and when administered into the ARC in ovariectomised, oestrogen-replaced rats. It cannot be ruled out that this is due to the dose of antagonist used in these studies, as higher doses have not yet been tested. However, if this is shown not to be due to the dosage, then these findings suggest that, although KP is clearly involved in regulating LH pulses and the increase in LH secretion after orchidectomy and during the LH surge, the maintenance of basal LH appears to be KP-independent. This proposal is supported by the demonstration that peptide 234 decreased KP-stimulated GnRH neuron firing rate in mouse brain but not the non-stimulated firing rate, and also decreased GnRH pulses, but not basal GnRH secretion, in the rhesus monkey. Moreover, peptide 234 does not lower LH to the same extent as

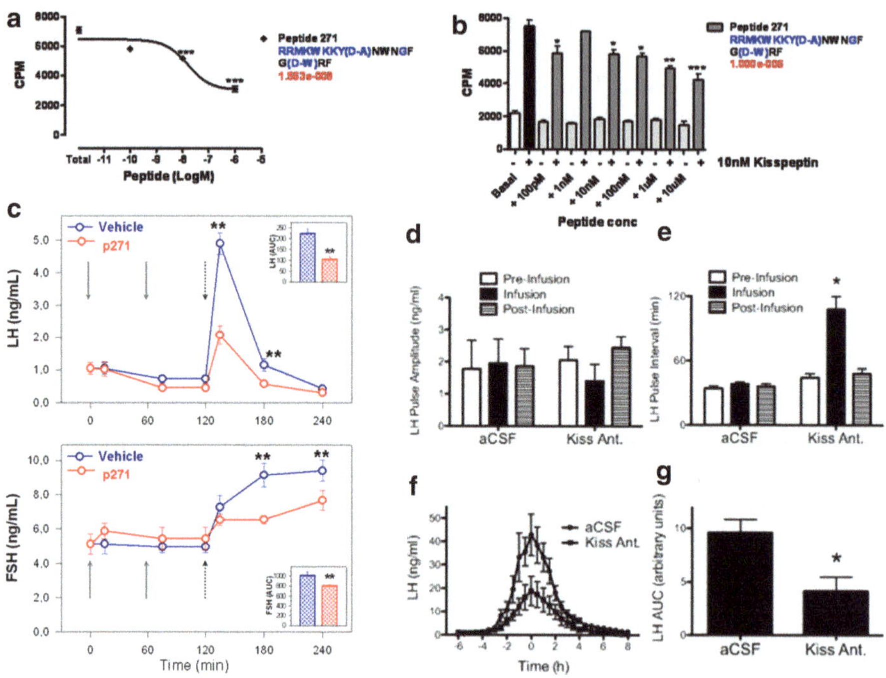

Fig. 8.15 Effects of Peptide 271 in vitro and in vivo. (**a**) Peptide 271 can still bind with an IC_{50} of 15 nM. (**b**) Peptide 271 has no intrinsic IP stimulation but can antagonise KP-10 stimulated IP production by 62% with an IC50 of 1 μM. (**c**) LH and FSH secretory profiles are shown from adult male rats receiving three consecutive i.p. injections (5 nmol/each) of the antagonist of KP, peptide 271 (*denoted by arrows*); the last injection was associated with an i.c.v. bolus of KP-10 (100 pmol). Integrated secretory responses following KP-10 administration, calculated as area under the curve (AUC), are also depicted as bar graphs. (**d**, **e**) Bar graphs show AUC for LH pulse amplitude (**d**) and pulse interval (**e**) during 2-h time periods before infusion, during infusion, and after infusion in OVX ewes ($n=5$ per group). A significantly prolonged pulse interval during peptide 271 infusion was detected. (**f**) Mean LH surges in OVX ewes treated with Peptide 271 or aCSF. The *x*-axis is time (hours) from the peak of the LH surge. (**g**) AUC analysis revealed a significant decrease in the LH surge in ewes treated with Peptide 271. Data are the mean ± SEM. $^{*}P<0.05$, $^{**}P<0.01$ vs. corresponding control group. (From Pineda, R., et al., Critical roles of kisspeptins in female puberty and preovulatory gonadotropin surges as revealed by a novel antagonist. Endocrinology. 151(2):722–30. Reprinted with permission from The Endocrine Society. And from Smith, J.T., et al., Kisspeptin is essential for the full preovulatory LH surge and stimulates GnRH release from the isolated ovine median eminence. Endocrinology. 2011. 152(3):1001–12. Reprinted with permission from The Endocrine Society)

GnRH antagonist. Collectively, the studies indicate that there is an element of GnRH secretion which is independent of KP. These findings have important therapeutic implications, as they indicate that KP antagonists may be utilised to partially suppress gonadal function and steroid hormone secretion. However, as GnRH and LH are not known to have any continuous basal mode of secretion, this may be an artefact of the assays utilised.

Future Applications in Physiological Regulation of Gonadotropins

To date, few studies have examined the effects of KP antagonists on FSH secretion. Although both LH and FSH secretion are dependent on GnRH, low GnRH pulse frequencies favour FSH secretion while high pulse frequencies favour LH secretion [36]. Since KP antagonists slow LH pulse frequencies (and by inference, GnRH pulse frequency) in a number of the models studied, it may be instructive to study the effects on FSH secretion.

Leptin stimulates *Kiss1* gene expression and increases gonadotropins in leptin-deficient and nutritionally deprived mice [37]. Studies on the ability of KP antagonists to inhibit leptin stimulation of gonadotropin in these models would provide direct evidence for leptin stimulation of gonadotropin through stimulation of KP secretion. KP antagonists could also be employed to interrogate other metabolic stimulators of gonadotropins. Another area which has yet to be investigated is the effect of gonadal peptides (inhibin, activin, and follistatin) on the KP system. Conversely, KP antagonists can be used to investigate the potential role of basal gonadotropin secretion on the production of these gonadal peptides.

KP, neurokinin B (NKB), and dynorphin (Dyn) are co-expressed by a population of neurons in the ARC of sheep [38, 39], goats [40], rodents, and humans [41], and a model is emerging that involves a complex dialogue between these peptide hormones and their effects on the GnRH neuron. Putative paracrine and autocrine feedback on the KP/NKB/Dyn neurons may be gainfully investigated by the combined use of KISS1R, TAC3R, and μ-opioid receptor agonists and antagonists. KP antagonists may also be valuable reagents to determine whether other neurotransmitters and neuropeptides which stimulate the GnRH neuron do so directly or by stimulating KP secretion.

Potential Clinical Applications of KP Antagonists

Since KP is a major regulator of GnRH but a degree of GnRH secretion is KP-independent, KP antagonists might be employed as partial inhibitors of gonadotropin and sex steroids. GnRH analogues have found extensive therapeutic applications in hormone-dependent diseases and in IVF. GnRH agonists induce gonadotrope desensitisation, while antagonists prevent receptor activation by endogenous GnRH. Both treatments result in suppression of gonadotropin secretion with consequent reduction in circulating gonadal steroid hormones. In these treatments, steroid hormones are lowered to castrate levels, resulting in side effects such as hot flushes, reduced lean body mass, loss of libido, and bone loss. Our demonstration that KP antagonists reduce LH pulsatility and inhibit the ovulatory LH surge, but do not appear to affect basal LH, suggests that they may find clinical utility in conditions where maximal suppression of sex steroids is contraindicated. Partial sex steroid suppression with KP antagonists may find utility in benign prostatic hyperplasia, endometriosis, and uterine fibroids, where lowering of gonadal steroids could improve the conditions without the side effects of reducing steroid hormones to castrate levels. Polycystic ovarian syndrome (PCOS), which is characterised by increased LH pulse frequency and increased ratio to FSH, is another potential target of KP therapy. Although unproven, lowering of GnRH pulse frequency by KP antagonists should maintain FSH (favoured by low pulse frequency) while lowering LH, as desired in PCOS. As this would allow follicle development and oestrogen production, but inhibit ovulation, it may be an attractive possibility as a female contraceptive. KP antagonists may also find application in IVF to prevent premature luteinisation while maintaining basal LH, which may be of value in some women during superovulation [42, 43].

Conclusion

The discovery of KP antagonists provides a new avenue for investigating the role of KP in the normal physiology of gonadotropin regulation (Fig. 8.14) and in pathological conditions. They also offer possibilities in new therapeutic approaches in the treatment of hormone-dependent diseases without generating side effects of total ablation of gonadal steroids seen in current therapies.

References

1. Centeno ML et al (2007) Hypothalamic gonadotrophin-releasing hormone expression in female monkeys with different sensitivity to stress. J Neuroendocrinol 19(8):594–604
2. Porkka-Heiskanen T et al (1997) Rapid photoperiod-induced increase in detectable GnRH mRNA-containing cells in Siberian hamster. Am J Physiol 273(6 Pt 2):R2032–R2039
3. Millar RP (2005) GnRHs and GnRH receptors. Anim Reprod Sci 88(1–2):5–28
4. Popa SM, Clifton DK, Steiner RA (2008) The role of kisspeptins and GPR54 in the neuroendocrine regulation of reproduction. Annu Rev Physiol 70:213–238
5. Roseweir AK, Millar RP (2009) The role of kisspeptin in the control of gonadotrophin secretion. Hum Reprod Update 15(2):203–212
6. Seminara SB et al (2003) The GPR54 gene as a regulator of puberty. N Engl J Med 349(17):1614–1627
7. de Roux N et al (2003) Hypogonadotropic hypogonadism due to loss of function of the KiSS1-derived peptide receptor GPR54. Proc Natl Acad Sci U S A 100(19):10972–10976
8. Smith JT et al (2005) Differential regulation of KiSS-1 mRNA expression by sex steroids in the brain of the male mouse. Endocrinology 146(7):2976–2984
9. Smith JT et al (2007) KiSS-1 messenger ribonucleic acid expression in the hypothalamus of the ewe is regulated by sex steroids and season. Endocrinology 148(3):1150–1157
10. Pompolo S et al (2006) Colocalization of kisspeptin and gonadotropin-releasing hormone in the ovine brain. Endocrinology 147(2):804–810
11. Shibata M et al (2007) Evidence that down regulation of hypothalamic KiSS-1 expression is involved in the negative feedback action of testosterone to regulate luteinising hormone secretion in the adult male rhesus monkey (Macaca mulatta). J Neuroendocrinol 19(6):432–438

12. Rance NE (2009) Menopause and the human hypothalamus: evidence for the role of kisspeptin/neurokinin B neurons in the regulation of estrogen negative feedback. Peptides 30(1):111–122
13. Smith JT et al (2005) Regulation of Kiss1 gene expression in the brain of the female mouse. Endocrinology 146(9):3686–3692
14. Adachi S et al (2007) Involvement of anteroventral periventricular metastin/kisspeptin neurons in estrogen positive feedback action on luteinizing hormone release in female rats. J Reprod Dev 53(2):367–378
15. Estrada KM et al (2006) Elevated KiSS-1 expression in the arcuate nucleus prior to the cyclic preovulatory gonadotrophin-releasing hormone/lutenising hormone surge in the ewe suggests a stimulatory role for kisspeptin in oestrogen-positive feedback. J Neuroendocrinol 18(10):806–809
16. Takino T et al (2003) Cleavage of metastasis suppressor gene product KiSS-1 protein/metastin by matrix metalloproteinases. Oncogene 22(30):4617–4626
17. Kotani M et al (2001) The metastasis suppressor gene KiSS-1 encodes kisspeptins, the natural ligands of the orphan G protein-coupled receptor GPR54. J Biol Chem 276(37): 34631–34636
18. Palczewski K, Kumasaka T, Hori T, Behnke CA, Motoshima H, Fox BA, Le Trong I, Teller DC, Okada T, Stenkamp RE, Yamamoto M, Miyano M (2000) Crystal structure of rhodopsin: A G protein-coupled receptor. Science 289(5480):739–745; Am J Ophthalmol 130(6):865.
19. Orsini MJ et al (2007) Metastin (KiSS-1) mimetics identified from peptide structure-activity relationship-derived pharmacophores and directed small molecule database screening. J Med Chem 50(3):462–471
20. Gutierrez-Pascual E et al (2009) In vivo and in vitro structure-activity relationships and structural conformation of Kisspeptin-10-related peptides. Mol Pharmacol 76(1):58–67
21. Ohtaki T et al (2001) Metastasis suppressor gene KiSS-1 encodes peptide ligand of a G-protein-coupled receptor. Nature 411(6837):613–617
22. Muir AI et al (2001) AXOR12, a novel human G protein-coupled receptor, activated by the peptide KiSS-1. J Biol Chem 276(31):28969–28975
23. Niida A et al (2006) Design and synthesis of downsized metastin (45-54) analogs with maintenance of high GPR54 agonistic activity. Bioorg Med Chem Lett 16(1):134–137
24. Kobayashi T et al (2010) Synthesis and structure-activity relationships of 2-acylamino-4,6-diphenylpyridine derivatives as novel antagonists of GPR54. Bioorg Med Chem 18(11):3841–3859
25. Kobayashi T et al (2010) 2-Acylamino-4,6-diphenylpyridine derivatives as novel GPR54 antagonists with good brain exposure and in vivo efficacy for plasma LH level in male rats. Bioorg Med Chem 18(14):5157–5171
26. Roseweir AK et al (2009) Discovery of potent kisspeptin antagonists delineate physiological mechanisms of gonadotropin regulation. J Neurosci 29(12):3920–3929
27. Han SK et al (2005) Activation of gonadotropin-releasing hormone neurons by kisspeptin as a neuroendocrine switch for the onset of puberty. J Neurosci 25(49):11349–11356
28. Guerriero KA et al (2012) Developmental changes in GnRH release in response to kisspeptin agonist and antagonist in female rhesus monkeys (Macaca mulatta): implication for the mechanism of puberty. Endocrinology 153(2):825–836
29. Plant TM et al (1978) The arcuate nucleus and the control of gonadotropin and prolactin secretion in the female rhesus monkey (Macaca mulatta). Endocrinology 102(1):52–62
30. Li XF et al (2009) Kisspeptin signalling in the hypothalamic arcuate nucleus regulates GnRH pulse generator frequency in the rat. PLoS One 4(12):e8334
31. Pineda R et al (2010) Critical roles of kisspeptins in female puberty and preovulatory gonadotropin surges as revealed by a novel antagonist. Endocrinology 151(2):722–730
32. Navarro VM et al (2004) Advanced vaginal opening and precocious activation of the reproductive axis by KiSS-1 peptide, the endogenous ligand of GPR54. J Physiol 561(Pt 2):379–386
33. Franceschini I et al (2006) Kisspeptin immunoreactive cells of the ovine preoptic area and arcuate nucleus co-express estrogen receptor alpha. Neurosci Lett 401(3):225–230

34. Kinoshita M et al (2005) Involvement of central metastin in the regulation of preovulatory luteinizing hormone surge and estrous cyclicity in female rats. Endocrinology 146(10): 4431–4436
35. Smith JT et al (2011) Kisspeptin is essential for the full preovulatory LH surge and stimulates GnRH release from the isolated ovine median eminence. Endocrinology 152(3):1001–1012
36. Dalkin AC et al (1989) The frequency of gonadotropin-releasing-hormone stimulation differentially regulates gonadotropin subunit messenger ribonucleic acid expression. Endocrinology 125(2):917–924
37. Smith JT et al (2006) KiSS-1 neurones are direct targets for leptin in the ob/ob mouse. J Neuroendocrinol 18(4):298–303
38. Cheng G et al (2010) The kisspeptin/neurokinin B/dynorphin (KNDy) cell population of the arcuate nucleus: sex differences and effects of prenatal testosterone in sheep. Endocrinology 151(1):301–311
39. Goodman RL et al (2007) Kisspeptin neurons in the arcuate nucleus of the ewe express both dynorphin A and neurokinin B. Endocrinology 148(12):5752–5760
40. Wakabayashi Y et al (2010) Neurokinin B and dynorphin A in kisspeptin neurons of the arcuate nucleus participate in generation of periodic oscillation of neural activity driving pulsatile gonadotropin-releasing hormone secretion in the goat. J Neurosci 30(8):3124–3132
41. Cejudo Roman A et al (2012) Analysis of the expression of neurokinin B, kisspeptin, and their cognate receptors NK(3)R and KISS1R in the human female genital tract. Fertil Steril 97(5):1213–1219
42. Durnerin CI et al (2008) Effects of recombinant LH treatment on folliculogenesis and responsiveness to FSH stimulation. Hum Reprod 23(2):421–426
43. Drakakis P et al (2009) Early hCG addition to rFSH for ovarian stimulation in IVF provides better results and the cDNA copies of the hCG receptor may be an indicator of successful stimulation. Reprod Biol Endocrinol 7:110
44. Millar RP et al (2010) Kisspeptin antagonists: unraveling the role of kisspeptin in reproductive physiology. Brain Res 1364:81–89

Chapter 9
Kisspeptin and Clinical Disorders

Letícia Gontijo Silveira, Ana Claudia Latronico, and Stephanie Beth Seminara

Abstract The hypothalamic hormone GnRH has traditionally been viewed as a central driver of the hypothalamic-pituitary-gonadal axis. Pulsatile GnRH release is required for pulsatile gonadotropin secretion, which then modulates gonadal steroid feedback and brings about full fertility in the adult. Pathways governing GnRH ontogeny and physiology have been discovered by studying humans with disorders of GnRH secretion. In this chapter, the human genetics of the kisspeptin signaling pathway in patients with diverse reproductive phenotypes will be explored. The discovery of defects in the kisspeptin system in several reproductive disorders has shed light on the mechanisms involved in regulating GnRH secretion, revealing the critical role played by the kisspeptin signaling pathway in pubertal initiation and reproductive function.

GnRH Deficiency

GnRH deficiency is a condition characterized by abnormal pubertal development and low gonadotropins and sex steroids. Administration of exogenous pulsatile GnRH long term can restore normal levels of gonadotropins and sex steroids in patients with this disorder, demonstrating the hypothalamic nature of the defect in the vast majority of patients [1, 2]. GnRH deficiency is heterogenous in its clinical

L.G. Silveira • A.C. Latronico
Department of Internal Medicine, Division of Endocrinology and Metabolism, Hospital das Clinicas, Sao Paulo University, Sao Paulo, Brazil
e-mail: leticia_gontijo@yahoo.com.br

S.B. Seminara, M.D. (✉)
Department of Internal Medicine, Reproductive Endocrine Unit, Bartlett Hall Extension 504, Endocrine Division, Massachusetts General Hospital, 55 Fruit Street, Boston, MA 02114, USA
e-mail: sseminara@partners.org

A.S. Kauffman and J.T. Smith (eds.), *Kisspeptin Signaling in Reproductive Biology*, Advances in Experimental Medicine and Biology 784, DOI 10.1007/978-1-4614-6199-9_9,

presentation, with variation in the presence or absence of olfactory defects (and other somatic anomalies), severity of the hypogonadism, and neuroendocrine patterns. Thus, patients with isolated GnRH deficiency represent a unique opportunity to identify genes that awaken the reproductive cascade at the time of sexual maturation and maintain normal reproductive function throughout life.

In 2003, homozygosity mapping and candidate gene analysis of two large consanguineous pedigrees with isolated GnRH deficiency led to the identification of loss-of-function mutations in a then little-known G protein-coupled receptor, *GPR54* (later to be renamed *KISS1R*=kisspeptin receptor), by two investigative groups [3, 4]. The identification of mutations in multiple families, as well as unrelated probands, coupled with a parallel reproductive phenotype in a *Kiss1r* mutant mouse, catapulted kisspeptin into the spotlight as a key regulator of GnRH secretion.

Initial Reports

In searching for novel gene defects associated with isolated GnRH deficiency, one group employed homozygosity mapping of a large consanguineous family with five affected siblings. Chromosome localization and candidate gene sequence analysis led to the identification of a homozygous deletion of 155 nucleotides in the *KISS1R* [3]. This deletion encompassed the splicing acceptor site of intron 4-exon 5 junction and part of exon 5. In the unlikely event that this abnormal transcript was translated, the deleted receptor would be truncated within the third intracellular loop, lacking transmembrane domains 6 and 7. The proband of the index family was a 20-year-old male who presented with abnormal pubertal development, 4 mL testes, and a normal sense of smell. While his affected brothers all had similar clinical features, his affected sister had partial breast development and had experienced a single episode of uterine bleeding. His mother, a heterozygote carrier of the deletion, was noted to have experienced menarche at age 16.

Homozygosity mapping of a different consanguineous family with GnRH deficiency, this time from the Middle East, was the focus of a second investigative group [4]. Candidate gene sequencing of *KISS1R* led to the discovery of a homozygous missense mutation, p.L148S, within the second intracellular loop of the receptor. Transfection of COS-7 cells with a mutant construct representing L148S revealed significantly decreased accumulation of inositol phosphate in vitro compared to wild type [4]. Further sequencing of *KISS1R* led to the identification of compound heterozygote mutations, p.R331X and p.X399R, in an African American male [4]. As mRNAs with premature termination codons are known to be subject to nonsense-mediated decay [5], and mRNAs without an in-frame termination codon had recently been appreciated to be subject to nonstop decay [6, 7], it was hypothesized that the combination of nonstop and nonsense mutations in a single individual would result in the absence of a functional receptor. Quantitative RT-PCR confirmed a significant reduction of *KISS1R* mRNA in immortalized white blood cells from the p.R331X/p.X399R proband. Should a protein have been produced by either the p.R331X or p.X399R transcripts, in vitro studies suggested that each of the mutant proteins would have functioned poorly.

Understanding the functional consequences of the p.R331X/p.X399R mutations was an important backdrop for the interpretation of the clinical presentation of the p.R331X/p.X399R proband [4]. On frequent blood sampling, this individual had low-amplitude pulses of luteinizing hormone (LH), suggesting present, but enfeebled, secretion of GnRH. He also had a left-shifted dose–response curve compared to other patients with GnRH deficiency undergoing the same therapy, suggesting that he may be more sensitive to exogenous GnRH. Thus, despite the clear importance of the kisspeptin pathway in modulating GnRH release, the clinical data from this patient suggested the possibility of residual GnRH activity, a foreshadowing of subsequent studies that were performed in the rodent demonstrating the existence of kisspeptin-independent GnRH secretion [8].

The clinical presentations of the p.L148S and p.R331X/p.X399R patients were also juxtaposed against the phenotype of *Kiss1r*-deficient mice; the mutant mice were striking phenocopies of the GnRH-deficient patients, including lack of sexual maturation associated with low levels of gonadotropins [4]. The strong parallels in presentation between the GnRH-deficient patients and the mutant mice established a central role for *KISS1R*/*Kiss1r* across mammalian species. Moreover, the use of a mouse model allowed quantification of hypothalamic GnRH content, which was found to be normal in the *Kiss1r*-deficient mice, suggesting that *KISS1R*/*Kiss1r* influences the timing of sexual maturation by affecting the processing or secretion of GnRH [4, 9].

Thus, the identification of mutations in *KISS1R* by multiple groups thrust the kisspeptin pathway into the spotlight, and laboratories around the world began assembling expression, physiologic, transgenic, knock out/down, and electrophysiologic data to tell the biologic story of kisspeptin and its important role in regulating GnRH secretion. The hypothesis that kisspeptin does not affect GnRH neuronal migration, but rather GnRH biosynthesis and/or release, was supported shortly thereafter by in vivo studies demonstrating that kisspeptin administration, either centrally or peripherally, triggers robust GnRH-induced LH and FSH secretion [10, 11]. Kisspeptin expression in the hypothalamus of rodents and nonhuman primates was also found to increase at the time of sexual maturation [11, 12], an important finding since kisspeptin is thought to mediate sex steroid feedback [12–17], estrous cycle regulation [17, 18], seasonal breeding [19–25], and to convey information about the energy status of the organism [26, 27]. Thus, relatively quickly, kisspeptin was found to be a key gatekeeper for the activation of the GnRH axis.

Central Themes for Patients Carrying Mutations in the Kisspeptin Signaling Pathway

Once *KISS1R* was added to the roster of genes for GnRH deficiency, several groups began to search for mutations in the kisspeptin signaling pathway in patients with hypogonadotropic hypogonadism. However, mutations in the coding sequence of *KISS1R* have proven to be relatively rare, particularly in comparison with many other genes for GnRH deficiency; for example, *KAL1* has been reported to harbor mutations in 5–14% of GnRH-deficient patients [28–30], considerably higher than the 1% frequency of mutations in *KISS1R* [28]. It is unclear why the prevalence of mutations in the kisspeptin signaling pathway is relatively low, but it is possible that kisspeptin's role in trophoblast invasion [31, 32] or metastasis suppression [33, 34] creates negative selection pressure against the promulgation of mutations in families.

Reported biallelic mutations of *KISS1R* include a homozygous 155 base pair deletion [3], a homozygous frameshift (c.1001_1002insC) [35], a homozygous splice acceptor site mutation [36], homozygous p.L102P [37], p.L148S [4, 38, 39] and p.F272S missense mutations [40], and the compound heterozygous mutations p.R331X/p.X399R [4, 39] and p.R297L/p.C223R [41]. Thus, mutations in this G protein-coupled receptor are variable in type (large deletion, frameshift, splice site, nonsense, nonstop, and missense) and occur throughout the receptor.

Typically, patients carrying biallelic complete loss-of-function mutations serve as the "phenotypic bookends" of the most extreme clinical presentation that can be associated with loss of a particular gene. In general, complete loss of kisspeptin signaling is associated with normosmic GnRH deficiency, but, as noted earlier, the abnormalities in GnRH secretion may be partial. For example, an affected female carrying a homozygous 155 base pair deletion in *KISS1R* (that if translated, would lead to a truncated G protein-coupled receptor unable to stimulate the transduction pathway) presented with some partial breast development and one episode of uterine bleeding [3]. Although her sexual maturation was clearly abnormal, her breast development and uterine bleeding suggest her endogenous estradiol levels were above pre-pubertal values. Another female, this time carrying a homozygous L102P mutation, demonstrated a robust LH level in response to a 100 μg GnRH stimulation test (peak approximately 32 IU/L) at initial evaluation [37]. During 6 h of frequent blood sampling, she manifested low-amplitude LH pulses occurring approximately once per hour. These clinical clues all suggest the presence of some degree of enfeebled endogenous GnRH secretion.

To date, patients with biallelic mutations in *KISS1R* lack the syndromic features that can be found in anosmic forms of this disorder (Kallmann syndrome), including cleft lip and palate, synkinesia, and renal agenesis. This is likely due to the fact that mutations in the kisspeptin signaling pathway do not affect olfactory bulb development and by extension, GnRH neuronal migration, a hypothesis supported by studies of *Kiss1r*$^{-/-}$ mice, which have normal GnRH hypothalamic content and preserved anatomy [4, 55, 56]. However, because GnRH deficiency can be an oligogenic disease [28], it is possible that patients with Kallmann syndrome carry mutations in genes that affect not only GnRH neuronal migration but also kisspeptin signaling [42].

Discovery of Mutations in Closely Related Pathways

Kisspeptin is now appreciated to be co-expressed with other neuropeptides that are likely to work in a cooperative fashion to regulate the hypothalamic control of reproduction. Kisspeptin neurons in the ARC co-express the neuropeptides neurokinin B

(NKB) and dynorphin, giving rise to the term KNDY neurons (*K*isspeptin-*N*eurokinin B-*Dy*norphin); this co-localization has been observed in several mammalian species, including humans [43–45]. About 5 years after the discovery of mutations in the kisspeptin pathway, loss-of-function mutations in the genes encoding neurokinin B (*TAC3*) and its receptor (*TACR3*) in patients with normosmic isolated hypogonadotropic hypogonadism (IHH) and pubertal failure were discovered [46]. Patients with mutations in *KISS1R* have a relatively straightforward phenotype without syndromic features, although there can be evidence of residual GnRH activity [3, 4, 37]. Patients bearing mutations in the neurokinin B signaling pathway also lack anosmia, renal agenesis, and bony abnormalities, but their neuroendocrine phenotype is more complex. A large proportion of patients with mutations in either *TAC3* or *TACR3* have undergone reversal of their hypogonadotropism [47], a phenomenon in which patients undergo spontaneous recovery of their hypothalamic-pituitary-gonadal cascade. *Tacr3*$^{-/-}$ mice have numerous reproductive defects (including abnormal estrous cycles, reduced corpora lutea, decreased uterine weights), but mutant female mice are able to achieve fertility when mated, demonstrating that these mice are more similar to the patients with *TAC3*/*TACR3* mutations than previously appreciated [48]. While reversible GnRH deficiency is not exclusively associated with *TAC3*/*TACR3* [49–54], none of the patients bearing biallelic mutations in the kisspeptin signaling pathway have yet to be reported with this clinical sub-phenotype. While the triggers to reversal remain poorly understood, these clinical observations appear to be providing important clues as to the physiologic hierarchy and relative influence of neurokinin B and kisspeptin in modulating GnRH release. In fact, the possibility that the actions of neurokinin B are proximal to those of kisspeptin is supported by the observation of increased LH pulse frequency during kisspeptin infusion to patients with neurokinin B signaling deficiencies [55].

Fertility Phenotypes

Although fertility data is available for only a subset of patients, mutations in *KISS1R* do not appear to impact fertility potential. Despite bilateral cryptorchidism and mild hypospadius, a male patient carrying a homozygous *KISS1R* c.1001_1002insC mutation responded to exogenous pulsatile GnRH, normalizing testosterone levels and inducing spermatogenesis [35]. Because his semen analysis showed oligoasthenozoospermia, pregnancy, albeit achieved with assisted reproduction, was possible. The male proband harboring p.R331X/p.X399R also received pulsatile GnRH and experienced steady increases in testicular volume and the appearance of normal spermatogenesis, resulting in fertility [39]. A homozygous p.L148S female had (1) intact responses to exogenous GnRH and gonadotropins, (2) multiple conceptions using the aforementioned therapies as well as IVF, and (3) two uncomplicated pregnancies [39]. While details are not available regarding the quality of her follicular response, mutations in *KISS1R* do not appear to preclude steroidogenesis and gametogenesis.

The Missing Link of KISS1

In GnRH deficiency, the discovery of loss-of-function mutations in genes encoding cell membrane-associated receptors has always preceded or accompanied the discovery of disabling mutations in the genes encoding their ligands. Initially, it appeared that *KISS1* was going to escape this genetic tradition, as no mutations in *KISS1* were reported for 8 years after the discovery of mutations in *KISS1R*. However, mutations in *KISS1* are now clearly associated with GnRH deficiency.

In 2011, 15 probands with GnRH deficiency were found to harbor 10 heterozygous rare sequence variants in *KISS1* [56]; in silico, in vitro, and in vivo studies were performed to explore the functional consequences of these variants. p.F117L was found to reduce inositol phosphate generation in vitro. p.G35S and p.C53R were predicted in silico to be deleterious. Lying outside the coding region, the variant g.1-3659C→T was found to impair transcription in vitro while another variant, c.1-7C→T, was noted to sit within the consensus Kozak sequence. Because these variants were monoallelic, and not biallelic, an examination of reproductive phenotypes in heterozygous and double-heterozygous *Kiss1* and *Kiss1r* mice was also performed. Heterozygous *Kiss1* mutations produced reproductive phenotypes in mutant mice and these phenotypes were further accentuated when accompanied by heterozygous mutations in *Kiss1r* [56].

As over 1,000 probands were screened to identify these nucleotide changes, disabling genetic variation in *KISS1* is clearly as rare, if not more so, than that of *KISS1R*. However, a homozygous loss-of-function mutation in *KISS1* was eventually discovered in affected siblings from a consanguineous Kurdish pedigree with normosmic GnRH deficiency (c.345C→G; p.N115K) [57]. The mutant kisspeptin was significantly less potent than wild type kisspeptin in activating GnRH neurons. Thus, both *KISS1R* and *KISS1* are clearly genetic determinants of the timing of sexual maturation in the human.

Insights Garnered from Kiss1$^{-/-}$ and Kiss1r$^{-/-}$ Mice

In general, *Kiss1*$^{-/-}$ and *Kiss1r*$^{-/-}$ mice are phenocopies of humans bearing *KISS1R* mutations. Both *Kiss1*$^{-/-}$ and *Kiss1r*$^{-/-}$ mice have small gonads, low gonadotropins, and abnormal gametogenesis, and infertility [4, 9, 58, 59]. GnRH neuronal migration into the hypothalamus is normal in *Kiss1*$^{-/-}$ animals, along with appropriate axonal connections to the median eminence and total GnRH content [58].

Despite their infertility, *Kiss1*$^{-/-}$ female mice can develop follicles, up to the preovulatory level, although no spontaneous ovulations are observed [8]. Both *Kiss1*$^{-/-}$ and *Kiss1r*$^{-/-}$ females alternate between periods of prolonged diestrus and prolonged estrus. These transitions increase in frequency with increasing age and are not associated with changes in hypothalamic *Gnrh1* mRNA expression. Administration of the competitive GnRH antagonist acyline disrupts the estrus exhibited by *Kiss1*$^{-/-}$

and *Kiss1r*$^{-/-}$ female mice, demonstrating that this estrus is due to GnRH activity. The low-amplitude LH pulsations observed in multiple patients with *KISS1R* mutations appears to be echoed in the persistent GnRH activity documented in *Kiss1*$^{-/-}$ and *Kiss1r*$^{-/-}$ mice. Kisspeptin-independent GnRH activity, whether in mice or men, could be due to low level constitutive activity of GnRH neurons or could be induced by other neuroendocrine pathways that modulate GnRH neuronal secretion.

Surprisingly, mice with targeted ablation of kisspeptin or kisspeptin receptor expressing *cells* (as opposed to deletion of the *Kiss1* and *Kiss1r genes*) are almost entirely reproductively normal suggesting, at initial interpretation, that kisspeptin signaling is not required for puberty and fertility [60]. However, the residual kisspeptin and GnRH neurons present in each of these cellular ablation mouse models, while quite small in number, may in fact be sufficient for sexual maturation and fertility, as initially suggested by preoptic area brain grafts in hypogonadal (*hpg*) mice [61, 62] and a more recent mouse model (*Kiss*$^{Cre/Cre}$) with markedly reduced expression of *Kiss1* [63].

From Loss-of-Function to Gain-of-Function in the Kisspeptin Signaling Pathway

While GnRH deficiency presents with delayed pubertal development, central precocious puberty (CPP) results from early activation of hypothalamic GnRH secreting neurons resulting in precocious pubertal development in childhood [64, 65]. Affected children present with premature development of secondary sexual characteristics, acceleration of linear growth, and progressive skeletal maturation, resulting in premature epiphyseal closure and, consequently, short adult height in untreated cases [66]. CPP has remarkable female gender predominance and most cases are considered idiopathic, with normal central nervous system (CNS) magnetic resonance imaging (MRI) [64–68]. However, up to 75% of boys with CPP have a detectable CNS lesion, mainly hypothalamic hamartomas [64–67]. Familial occurrence has been reported in 20–25% of CPP cases, suggesting a role for genetic factors in its pathogenesis [69, 70]. Segregation analysis of these families suggested an autosomal dominant transmission with incomplete sex-dependent penetrance [69].

In 2008, the kisspeptin signaling pathway was implicated in the pathogenesis of CPP. The first heterozygous activating mutation of the KISS1R (p.R386P) was described in an adopted Brazilian girl with CPP [71]. She presented with slowly progressive thelarche from birth; accelerated growth, skeletal maturation, and progression of breast development were noticed at 7 years of age. She had pubertal estradiol levels and borderline-pubertal LH stimulated levels [71]. In vitro studies demonstrated that the R386P mutation, located in the carboxy terminal tail of the receptor, led to prolonged activation of intracellular signaling pathways in response to kisspeptin [71]. Therefore, in contrast to gain-of-function mutations in many G protein-coupled receptors, which cause constitutive receptor activation, the p.R386P mutation appeared to reduce the rate of desensitization of the mutant KISS1R at the cell surface after ligand-binding and signaling. Indeed, the p.R386P mutation appears to decrease KISS1R degradation, resulting in a net increase of the receptor on the plasma membrane [72].

Given the description of an activating mutation in *KISS1R* causing premature activation of the gonadotropic axis, *KISS1* was another obvious natural candidate gene for precocious puberty. One rare kisspeptin variant, p.P74S, was identified in one child with sporadic CPP [70]. The p.P74S mutation was identified in the heterozygous state in a boy who developed CPP at 1 year of age with remarkably high levels of basal LH and testosterone [70]. Although the majority of boys with CPP, especially younger than 4 years old, have an underlying CNS abnormality [66–68], this boy had no CNS lesions. His mother and maternal grandmother, who had normal pubertal development, also carried the p.P74S mutation in heterozygous state, suggesting incomplete sex-dependent penetrance. After pre-incubating the mutant kisspeptin in human serum to more closely mimic in vivo conditions, the capacity to stimulate signal transduction was significantly greater for p.P74S compared to the wild type, suggesting that this variant might be more resistant to degradation, resulting in greater kisspeptin bioavailability [70].

The patients with *KISS1R* or *KISS1* mutations described above demonstrated adequate response to conventional treatment with GnRH agonists [70, 71]. Depot GnRH agonist treatment resulted in the regression or stabilization of pubertal symptoms in the two patients with activating mutations of *KISS1R* or *KISS1* genes. As expected, a decrease in the release of LH, FSH, and consequently normal pre-pubertal sexual steroids was achieved in these cases. In addition, the discontinuation of depot GnRH agonists treatment at the age higher than 11 years was associated with the reactivation of the reproductive axis in both cases, suggesting that the clinical and hormonal features of patients with activating mutations of the *KISS1R* and *KISS1* gene were not different from children with idiopathic or organic causes of CPP.

Although these case reports expand the genotype–phenotype correlations for the kisspeptin pathway, no other CPP cases with activating *KISS1R* or *KISS1* mutations have been reported, suggesting that these genetic abnormalities are very rare. Other cohorts have been screened, but no mutations have been identified in the *KISS1* gene [73, 74]. Considering the low incidence of mutations in these genes in relation to the frequency of familial CPP, it is possible that other genes involved in GnRH regulation also will bear activating or inactivating mutations. Indeed, loss-of-function mutations affecting repressor genes of the GnRH gene might play a role in the pathogenesis of nonorganic CPP in the future.

Kisspeptin Expression in Organic Central Precocious Puberty

Hypothalamic hamartomas are the most common identifiable cause of CPP. Certain characteristics of anatomy and neuropeptide expression have been proposed to be associated with CPP. The expression of GnRH, GnRH receptor, TGFalpha, KISS1, KISS1R, and GRM1A was investigated in hamartomas associated with or without precocious sexual development [75]. Hypothalamic hamartomas associated with CPP were larger than those not associated with CPP and were more likely to contact the infundibulum or tuber cinereum. However, the expression of *KISS1* and *KISS1R* was similar in both groups, demonstrating that expression of this signaling pathway does not differentiate between hamartomas associated with precocity vs. hamartomas that are identified in the setting of normal pubertal development [75].

Conclusions

Mutations in genes associated with IHH have been identified in approximately 30–40% of the patients with GnRH deficiency. Although mutations in *KISS1/KISS1R* are not a common cause of hypogonadotropism or CPP, the discovery of defects in this pathway in IHH as well as in CPP has shed some light on the mechanisms involved in GnRH secretion regulation, revealing the critical role played by the kisspeptin signaling pathway in pubertal initiation and reproductive function.

Considering the low incidence of mutations in these genes in precocious puberty so far, other genes involved in the HPG axis modulation, particularly factors upstream from, the *KISS1/KiSS1R* systems, might be involved in CPP pathogenesis. New methodologies, such as next generation sequencing and comparative genomic hybridization (CGH), will provide a more comprehensive assessment of genomic abnormalities and allow new genes to be uncovered.

Acknowledgments Dr. Seminara and some of the data referenced in this article were supported by the Eunice Kennedy Shriver NICHD/NIH through cooperative agreement [U54 HD028138] as part of the Specialized Cooperative Centers Program in Reproduction and Infertility Research, as well as R01 HD043341 and K24 HD067388.

References

1. Crowley WF Jr, McArthur JW (1980) Simulation of the normal menstrual cycle in Kallman's syndrome by pulsatile administration of luteinizing hormone-releasing hormone (LHRH). J Clin Endocrinol Metab 51:173–175
2. Hoffman AR, Crowley WF Jr (1982) Induction of puberty in men by long-term pulsatile administration of low-dose gonadotropin-releasing hormone. N Engl J Med 307:1237–1241
3. de Roux N, Genin E, Carel JC, Matsuda F, Chaussain JL, Milgrom E (2003) Hypogonadotropic hypogonadism due to loss of function of the KiSS1-derived peptide receptor GPR54. Proc Natl Acad Sci U S A 100:10972–10976
4. Seminara SB, Messager S, Chatzidaki EE, Thresher RR, Acierno JS Jr, Shagoury JK, Bo-Abbas Y, Kuohung W, Schwinof KM, Hendrick AG, Zahn D, Dixon J, Kaiser UB, Slaugenhaupt SA, Gusella JF, O'Rahilly S, Carlton MB, Crowley WF Jr, Aparicio SA, Colledge WH (2003) The GPR54 gene as a regulator of puberty. N Engl J Med 349:1614–1627
5. Maquat LE, Carmichael GG (2001) Quality control of mRNA function. Cell 104:173–176
6. Frischmeyer PA, van Hoof A, O'Donnell K, Guerrerio AL, Parker R, Dietz HC (2002) An mRNA surveillance mechanism that eliminates transcripts lacking termination codons. Science 295:2258–2261

7. van Hoof A, Frischmeyer PA, Dietz HC, Parker R (2002) Exosome-mediated recognition and degradation of mRNAs lacking a termination codon. Science 295:2262–2264
8. Chan YM, Broder-Fingert S, Wong KM, Seminara SB (2009) Kisspeptin/Gpr54-independent gonadotrophin-releasing hormone activity in Kiss1 and Gpr54 mutant mice. J Neuroendocrinol 21:1015–1023
9. Funes S, Hedrick JA, Vassileva G, Markowitz L, Abbondanzo S, Golovko A, Yang S, Monsma FJ, Gustafson EL (2003) The KiSS-1 receptor GPR54 is essential for the development of the murine reproductive system. Biochem Biophys Res Commun 312:1357–1363
10. Gottsch ML, Cunningham MJ, Smith JT, Popa SM, Acohido BV, Crowley WF, Seminara S, Clifton DK, Steiner RA (2004) A role for kisspeptins in the regulation of gonadotropin secretion in the mouse. Endocrinology 145:4073–4077
11. Shahab M, Mastronardi C, Seminara SB, Crowley WF, Ojeda SR, Plant TM (2005) Increased hypothalamic GPR54 signaling: a potential mechanism for initiation of puberty in primates. Proc Natl Acad Sci U S A 102:2129–2134
12. Navarro VM, Castellano JM, Fernandez-Fernandez R, Barreiro ML, Roa J, Sanchez-Criado JE, Aguilar E, Dieguez C, Pinilla L, Tena-Sempere M (2004) Developmental and hormonally regulated messenger ribonucleic acid expression of KiSS-1 and its putative receptor, GPR54, in rat hypothalamus and potent luteinizing hormone-releasing activity of KiSS-1 peptide. Endocrinology 145:4565–4574
13. Smith JT, Dungan HM, Stoll EA, Gottsch ML, Braun RE, Eacker SM, Clifton DK, Steiner RA (2005) Differential regulation of KiSS-1 mRNA expression by sex steroids in the brain of the male mouse. Endocrinology 146:2976–2984
14. Smith JT, Cunningham MJ, Rissman EF, Clifton DK, Steiner RA (2005) Regulation of Kiss1 gene expression in the brain of the female mouse. Endocrinology 146:3686–3692
15. Irwig MS, Fraley GS, Smith JT, Acohido BV, Popa SM, Cunningham MJ, Gottsch ML, Clifton DK, Steiner RA (2004) Kisspeptin activation of gonadotropin releasing hormone neurons and regulation of KiSS-1 mRNA in the male rat. Neuroendocrinology 80:264–272
16. Shibata M, Friedman RL, Ramaswamy S, Plant TM (2007) Evidence that down regulation of hypothalamic KiSS-1 expression is involved in the negative feedback action of testosterone to regulate luteinising hormone secretion in the adult male rhesus monkey (Macaca mulatta). J Neuroendocrinol 19:432–438
17. Smith JT, Popa SM, Clifton DK, Hoffman GE, Steiner RA (2006) Kiss1 neurons in the forebrain as central processors for generating the preovulatory luteinizing hormone surge. J Neurosci 26:6687–6694
18. Kinoshita M, Tsukamura H, Adachi S, Matsui H, Uenoyama Y, Iwata K, Yamada S, Inoue K, Ohtaki T, Matsumoto H, Maeda K (2005) Involvement of central metastin in the regulation of preovulatory luteinizing hormone surge and estrous cyclicity in female rats. Endocrinology 146:4431–4436
19. Greives TJ, Mason AO, Scotti MA, Levine J, Ketterson ED, Kriegsfeld LJ, Demas GE (2007) Environmental control of kisspeptin: implications for seasonal reproduction. Endocrinology 148:1158–1166
20. Greives TJ, Kriegsfeld LJ, Demas GE (2008) Exogenous kisspeptin does not alter photoperiod-induced gonadal regression in Siberian hamsters (Phodopus sungorus). Gen Comp Endocrinol 156:552–558
21. Smith JT, Coolen LM, Kriegsfeld LJ, Sari IP, Jaafarzadehshirazi MR, Maltby M, Bateman K, Goodman RL, Tilbrook AJ, Ubuka T, Bentley GE, Clarke IJ, Lehman MN (2008) Variation in kisspeptin and RFamide-related peptide (RFRP) expression and terminal connections to gonadotropin-releasing hormone neurons in the brain: a novel medium for seasonal breeding in the sheep. Endocrinology 149:5770–5782
22. Greives TJ, Humber SA, Goldstein AN, Scotti MA, Demas GE, Kriegsfeld LJ (2008) Photoperiod and testosterone interact to drive seasonal changes in kisspeptin expression in Siberian hamsters (Phodopus sungorus). J Neuroendocrinol 20:1339–1347
23. Smith JT, Saleh SN, Clarke IJ (2009) Seasonal and cyclical change in the luteinizing hormone response to kisspeptin in the ewe. Neuroendocrinology 90:283–291

24. Paul MJ, Pyter LM, Freeman DA, Galang J, Prendergast BJ (2009) Photic and nonphotic seasonal cues differentially engage hypothalamic kisspeptin and RFamide-related peptide mRNA expression in Siberian hamsters. J Neuroendocrinol 21:1007–1014
25. Chalivoix S, Bagnolini A, Caraty A, Cognie J, Malpaux B, Dufourny L (2010) Effects of photoperiod on kisspeptin neuronal populations of the ewe diencephalon in connection with reproductive function. J Neuroendocrinol 22(2):110–118
26. Smith JT, Acohido BV, Clifton DK, Steiner RA (2006) KiSS-1 neurones are direct targets for leptin in the ob/ob mouse. J Neuroendocrinol 18:298–303
27. Castellano JM, Navarro VM, Fernandez-Fernandez R, Roa J, Vigo E, Pineda R, Dieguez C, Aguilar E, Pinilla L, Tena-Sempere M (2006) Expression of hypothalamic KiSS-1 system and rescue of defective gonadotropic responses by kisspeptin in streptozotocin-induced diabetic male rats. Diabetes 55:2602–2610
28. Sykiotis GP, Plummer L, Hughes VA, Au M, Durrani S, Nayak-Young S, Dwyer AA, Quinton R, Hall JE, Gusella JF, Seminara SB, Crowley WF Jr, Pitteloud N (2010) Oligogenic basis of isolated gonadotropin-releasing hormone deficiency. Proc Natl Acad Sci U S A 107:15140–15144
29. Bhagavath B, Xu N, Ozata M, Rosenfield RL, Bick DP, Sherins RJ, Layman LC (2007) KAL1 mutations are not a common cause of idiopathic hypogonadotrophic hypogonadism in humans. Mol Hum Reprod 13:165–170
30. Albuisson J, Pecheux C, Carel JC, Lacombe D, Leheup B, Lapuzina P, Bouchard P, Legius E, Matthijs G, Wasniewska M, Delpech M, Young J, Hardelin JP, Dode C (2005) Kallmann syndrome: 14 novel mutations in KAL1 and FGFR1 (KAL2). Hum Mutat 25:98–99
31. Bilban M, Ghaffari-Tabrizi N, Hintermann E, Bauer S, Molzer S, Zoratti C, Malli R, Sharabi A, Hiden U, Graier W, Knofler M, Andreae F, Wagner O, Quaranta V, Desoye G (2004) Kisspeptin-10, a KiSS-1/metastin-derived decapeptide, is a physiological invasion inhibitor of primary human trophoblasts. J Cell Sci 117:1319–1328
32. Roseweir AK, Katz AA, Millar RP (2012) Kisspeptin-10 inhibits cell migration in vitro via a receptor-GSK3 beta-FAK feedback loop in HTR8SVneo cells. Placenta 33:408–415
33. Lee JH, Welch DR (1997) Suppression of metastasis in human breast carcinoma MDA-MB-435 cells after transfection with the metastasis suppressor gene, KiSS-1. Cancer Res 57:2384–2387
34. Miele ME, Robertson G, Lee JH, Coleman A, McGary CT, Fisher PB, Lugo TG, Welch DR (1996) Metastasis suppressed, but tumorigenicity and local invasiveness unaffected, in the human melanoma cell line MelJuSo after introduction of human chromosomes 1 or 6. Mol Carcinog 15:284–299
35. Lanfranco F, Gromoll J, von Eckardstein S, Herding EM, Nieschlag E, Simoni M (2005) Role of sequence variations of the GnRH receptor and G protein-coupled receptor 54 gene in male idiopathic hypogonadotropic hypogonadism. Eur J Endocrinol 153:845–852
36. Teles MG, Trarbach EB, Noel SD, Guerra-Junior G, Jorge A, Beneduzzi D, Bianco SD, Mukherjee A, Baptista MT, Costa EM, De Castro M, Mendonca BB, Kaiser UB, Latronico AC (2010) A novel homozygous splice acceptor site mutation of KISS1R in two siblings with normosmic isolated hypogonadotropic hypogonadism. Eur J Endocrinol 163:29–34
37. Tenenbaum-Rakover Y, Commenges-Ducos M, Iovane A, Aumas C, Admoni O, de Roux N (2007) Neuroendocrine phenotype analysis in five patients with isolated hypogonadotropic hypogonadism due to a L102P inactivating mutation of GPR54. J Clin Endocrinol Metab 92: 1137–1144
38. Bo-Abbas Y, Acierno JS Jr, Shagoury JK, Crowley WF Jr, Seminara SB (2003) Autosomal recessive idiopathic hypogonadotropic hypogonadism: genetic analysis excludes mutations in the gonadotropin-releasing hormone (GnRH) and GnRH receptor genes. J Clin Endocrinol Metab 88:2730–2737
39. Pallais JC, Bo-Abbas Y, Pitteloud N, Crowley WF Jr, Seminara SB (2006) Neuroendocrine, gonadal, placental, and obstetric phenotypes in patients with IHH and mutations in the G-protein coupled receptor, GPR54. Mol Cell Endocrinol 254–255:70–77
40. Nimri R, Lebenthal Y, Lazar L, Chevrier L, Phillip M, Bar M, Hernandez-Mora E, de Roux N, Gat-Yablonski G (2011) A novel loss-of-function mutation in GPR54/KISS1R leads to

hypogonadotropic hypogonadism in a highly consanguineous family. J Clin Endocrinol Metab 96(3):E536–E545
41. Semple RK, Achermann JC, Ellery J, Farooqi IS, Karet FE, Stanhope RG, O'Rahilly S, Aparicio SA (2005) Two novel missense mutations in g protein-coupled receptor 54 in a patient with hypogonadotropic hypogonadism. J Clin Endocrinol Metab 90:1849–1855
42. Wahab F, Quinton R, Seminara SB (2011) The kisspeptin signaling pathway and its role in human isolated GnRH deficiency. Mol Cell Endocrinol 346:29–36
43. Goodman RL, Lehman MN, Smith JT, Coolen LM, de Oliveira CV, Jafarzadehshirazi MR, Pereira A, Iqbal J, Caraty A, Ciofi P, Clarke IJ (2007) Kisspeptin neurons in the arcuate nucleus of the ewe express both dynorphin A and neurokinin B. Endocrinology 148:5752–5760
44. Rance NE (2009) Menopause and the human hypothalamus: evidence for the role of kisspeptin/neurokinin B neurons in the regulation of estrogen negative feedback. Peptides 30:111–122
45. Hrabovszky E, Ciofi P, Vida B, Horvath MC, Keller E, Caraty A, Bloom SR, Ghatei MA, Dhillo WS, Liposits Z, Kallo I (2010) The kisspeptin system of the human hypothalamus: sexual dimorphism and relationship with gonadotropin-releasing hormone and neurokinin B neurons. Eur J Neurosci 31:1984–1998
46. Topaloglu AK, Reimann F, Guclu M, Yalin AS, Kotan LD, Porter KM, Serin A, Mungan NO, Cook JR, Ozbek MN, Imamoglu S, Akalin NS, Yuksel B, O'Rahilly S, Semple RK (2009) TAC3 and TACR3 mutations in familial hypogonadotropic hypogonadism reveal a key role for neurokinin B in the central control of reproduction. Nat Genet 41:354–358
47. Gianetti E, Tusset C, Noel SD, Au MG, Dwyer AA, Hughes VA, Abreu AP, Carroll J, Trarbach E, Silveira LF, Costa EM, de Mendonca BB, de Castro M, Lofrano A, Hall JE, Bolu E, Ozata M, Quinton R, Amory JK, Stewart SE, Arlt W, Cole TR, Crowley WF, Kaiser UB, Latronico AC, Seminara SB (2010) TAC3/TACR3 mutations reveal preferential activation of gonadotropin-releasing hormone release by neurokinin B in neonatal life followed by reversal in adulthood. J Clin Endocrinol Metab 95(6):2857–2867
48. Yang JJ, Caligioni CS, Chan YM, Seminara SB (2012) Uncovering novel reproductive defects in neurokinin B receptor null mice: closing the gap between mice and men. Endocrinology 153:1498–1508
49. Pitteloud N, Boepple PA, DeCruz S, Valkenburgh SB, Crowley WF Jr, Hayes FJ (2001) The fertile eunuch variant of idiopathic hypogonadotropic hypogonadism: spontaneous reversal associated with a homozygous mutation in the gonadotropin-releasing hormone receptor. J Clin Endocrinol Metab 86:2470–2475
50. Dewailly D, Boucher A, Decanter C, Lagarde JP, Counis R, Kottler ML (2002) Spontaneous pregnancy in a patient who was homozygous for the Q106R mutation in the gonadotropin-releasing hormone receptor gene. Fertil Steril 77:1288–1291
51. Pitteloud N, Acierno JS Jr, Meysing AU, Dwyer AA, Hayes FJ, Crowley WF Jr (2005) Reversible Kallmann syndrome, delayed puberty, and isolated anosmia occurring in a single family with a mutation in the fibroblast growth factor receptor 1 gene. J Clin Endocrinol Metab 90:1317–1322
52. Raivio T, Falardeau J, Dwyer A, Quinton R, Hayes FJ, Hughes VA, Cole LW, Pearce SH, Lee H, Boepple P, Crowley WF Jr, Pitteloud N (2007) Reversal of idiopathic hypogonadotropic hypogonadism. N Engl J Med 357:863–873
53. Ribeiro RS, Vieira TC, Abucham J (2007) Reversible Kallmann syndrome: report of the first case with a KAL1 mutation and literature review. Eur J Endocrinol 156:285–290
54. Sinisi AA, Asci R, Bellastella G, Maione L, Esposito D, Elefante A, De Bellis A, Bellastella A, Iolascon A (2008) Homozygous mutation in the prokineticin-receptor2 gene (Val274Asp) presenting as reversible Kallmann syndrome and persistent oligozoospermia: case report. Hum Reprod 23:2380–2384
55. Young J, George JT, Tello JA, Francou B, Bouligand J, Guiochon-Mantel A, Brailly-Tabard S, Anderson RA, Millar RP (2012) Kisspeptin restores pulsatile LH secretion in patients with neurokinin B signaling deficiencies: physiological, pathophysiological and therapeutic implications. Neuroendocrinology. doi:10.1159/000336376
56. Chan YM, Broder-Fingert S, Paraschos S, Lapatto R, Au M, Hughes V, Bianco SD, Min L, Plummer L, Cerrato F, De Guillebon A, Wu IH, Wahab F, Dwyer A, Kirsch S, Quinton R,

Cheetham T, Ozata M, Ten S, Chanoine JP, Pitteloud N, Martin KA, Schiffmann R, Van der Kamp HJ, Nader S, Hall JE, Kaiser UB, Seminara SB (2011) GnRH-deficient phenotypes in humans and mice with heterozygous variants in KISS1/Kiss1. J Clin Endocrinol Metab 96:E1771–E1781
57. Topaloglu AK, Tello JA, Kotan LD, Ozbek MN, Yilmaz MB, Erdogan S, Gurbuz F, Temiz F, Millar RP, Yuksel B (2012) Inactivating KISS1 mutation and hypogonadotropic hypogonadism. N Engl J Med 366:629–635
58. d'Anglemont de Tassigny X, Fagg LA, Dixon JP, Day K, Leitch HG, Hendrick AG, Zahn D, Franceschini I, Caraty A, Carlton MB, Aparicio SA, Colledge WH (2007) Hypogonadotropic hypogonadism in mice lacking a functional Kiss1 gene. Proc Natl Acad Sci U S A 104:10714–10719
59. Lapatto R, Pallais JC, Zhang D, Chan YM, Mahan A, Cerrato F, Le WW, Hoffman GE, Seminara SB (2007) Kiss1$^{-/-}$ mice exhibit more variable hypogonadism than Gpr54$^{-/-}$ mice. Endocrinology 148:4927–4936
60. Mayer C, Boehm U (2011) Female reproductive maturation in the absence of kisspeptin/GPR54 signaling. Nat Neurosci 14:704–710
61. Gibson MJ, Krieger DT, Charlton HM, Zimmerman EA, Silverman AJ, Perlow MJ (1984) Mating and pregnancy can occur in genetically hypogonadal mice with preoptic area brain grafts. Science 225:949–951
62. Gibson MJ, Charlton HM, Perlow MJ, Zimmerman EA, Davies TF, Krieger DT (1984) Preoptic area brain grafts in hypogonadal (hpg) female mice abolish effects of congenital hypothalamic gonadotropin-releasing hormone (GnRH) deficiency. Endocrinology 114:1938–1940
63. Steiner RA (2012) How many kisses are required to change a frog into a princess? In: Presented at the 94th annual meeting of the endocrine society, United States, Houston, TX. Symposium S38-2 "New insights into the regulation of puberty"
64. Palmert MR, Boepple PA (2001) Variation in the timing of puberty: clinical spectrum and genetic investigation. J Clin Endocrinol Metab 86:2364–2368
65. Partsch CJ, Heger S, Sippell WG (2002) Management and outcome of central precocious puberty. Clin Endocrinol (Oxf) 56:129–148
66. Brito VN, Latronico AC, Arnhold IJ, Mendonca BB (2008) Update on the etiology, diagnosis and therapeutic management of sexual precocity. Arq Bras Endocrinol Metabol 52:18–31
67. Kakarla N, Bradshaw KD (2003) Disorders of pubertal development: precocious puberty. Semin Reprod Med 21:339–351
68. Brito VN, Latronico AC, Arnhold IJ, Lo LS, Domenice S, Albano MC, Fragoso MC, Mendonca BB (1999) Treatment of gonadotropin dependent precocious puberty due to hypothalamic hamartoma with gonadotropin releasing hormone agonist depot. Arch Dis Child 80:231–234
69. de Vries L, Kauschansky A, Shohat M, Phillip M (2004) Familial central precocious puberty suggests autosomal dominant inheritance. J Clin Endocrinol Metab 89:1794–1800
70. Silveira LG, Noel SD, Silveira-Neto AP, Abreu AP, Brito VN, Santos MG, Bianco SD, Kuohung W, Xu S, Gryngarten M, Escobar ME, Arnhold IJ, Mendonca BB, Kaiser UB, Latronico AC (2010) Mutations of the KISS1 gene in disorders of puberty. J Clin Endocrinol Metab 95:2276–2280
71. Teles MG, Bianco SD, Brito VN, Trarbach EB, Kuohung W, Xu S, Seminara SB, Mendonca BB, Kaiser UB, Latronico AC (2008) A GPR54-activating mutation in a patient with central precocious puberty. N Engl J Med 358:709–715
72. Bianco SD, Vandepas L, Correa-Medina M, Gereben B, Mukherjee A, Kuohung W, Carroll R, Teles MG, Latronico AC, Kaiser UB (2011) KISS1R intracellular trafficking and degradation: effect of the Arg386Pro disease-associated mutation. Endocrinology 152:1616–1626
73. Ko JM, Lee HS, Hwang JS (2010) KISS1 gene analysis in Korean girls with central precocious puberty: a polymorphism, p.P110T, suggested to exert a protective effect. Endocr J 57:701–709
74. Tommiska J, Sorensen K, Aksglaede L, Koivu R, Puhakka L, Juul A, Raivio T (2011) LIN28B, LIN28A, KISS1, and KISS1R in idiopathic central precocious puberty. BMC Res Notes 4:363
75. Chan YM, Fenoglio-Simeone KA, Paraschos S, Muhammad L, Troester MM, Ng YT, Johnsonbaugh RE, Coons SW, Prenger EC, Kerrigan JF Jr, Seminara SB (2010) Central precocious puberty due to hypothalamic hamartomas correlates with anatomic features but not with expression of GnRH, TGFalpha, or KISS1. Horm Res Paediatr 73:312–319

Chapter 10
Beyond the GnRH Axis: Kisspeptin Regulation of the Oxytocin System in Pregnancy and Lactation

Victoria Scott and Colin H. Brown

Abstract Circulating oxytocin is critical for normal birth and lactation. Oxytocin is synthesised by hypothalamic supraoptic and paraventricular neurons and is released from the posterior pituitary gland into the circulation. Oxytocin secretion depends on action potentials initiated at the cell body, and we have shown that intravenous (IV) administration of kisspeptin-10 transiently increases the firing rate of supraoptic nucleus oxytocin neurons in anaesthetised, non-pregnant, pregnant and lactating rats. This peripheral effect is likely via vagal afferent input, because disruption of vagal afferents prevented the excitation. In our initial studies, intracerebroventricular (icv) administration of kisspeptin-10 did not alter the firing rate of oxytocin neurons in non-pregnant rats. Remarkably, we have now gathered unpublished observations showing that icv kisspeptin-10 transiently excites oxytocin neurons in late pregnancy and during lactation, suggesting that a central kisspeptin excitation of oxytocin neurons emerges at the end of pregnancy, when increased oxytocin secretion is required for delivery of the fetus and for milk letdown after delivery.

Introduction

This chapter considers a newly emerging area of investigation in the kisspeptin field: the effects of kisspeptin on reproduction outside the hypothalamic-pituitary-gonadal axis. When the endogenous ligand for the orphan GPR54 receptor (now commonly known as Kiss1r in rodents and KISS1R in humans) was identified as kisspeptin, its possible role in modulating secretion of the hormone oxytocin was

V. Scott (✉) • C.H. Brown
Centre for Neuroendocrinology and Department of Physiology,
University of Otago, PO Box 913, Dunedin, 9054, New Zealand
e-mail: victoria.scott@otago.ac.nz

A.S. Kauffman and J.T. Smith (eds.), *Kisspeptin Signaling in Reproductive Biology*, Advances in Experimental Medicine and Biology 784, DOI 10.1007/978-1-4614-6199-9_10,

highlighted because intravenous (IV) administration of kisspeptin-10 increased plasma levels of oxytocin in non-pregnant female rats [1]. However, this result was quickly overshadowed by the discovery of the role of Kiss1r/KISS1R and kisspeptin neurons in fertility through actions on GnRH neurons.

In 2011, the possibility of an important role for kisspeptin in the regulation of the oxytocin system was again brought to light by our finding that intravenous kisspeptin-10 increases the firing rate of oxytocin-secreting neurons [2]. Here, we will give an overview of that work and our more recent findings that suggest that central kisspeptin excitation of oxytocin neurons emerges over the course of pregnancy, which might have important implications for successful delivery of the offspring at birth and successful delivery of milk during lactation.

Pregnancy and Lactation

Reproduction is the process by which new offspring are produced and is a fundamental process of life. The hypothalamic-pituitary-gonadal axis, including the kisspeptin neurons, is a critical part of reproduction, but fertility is only the first step in successful reproduction. Reproduction also requires successful pregnancy, parturition and, in mammals, lactation. In all mammals, there are complex interactions of neural, endocrine and behavioural processes that enable a female to successfully meet the challenge of pregnancy and lactation. Elucidating these different mechanisms and pathways is important for understanding the transient physiological states of pregnancy and lactation. The mechanisms that are recruited during these states reflect profound changes in brain neurochemistry and morphology, and this includes the oxytocin system.

The hormone oxytocin has recently come to prominence as an important modulator of various behaviours, including social recognition, pair bonding, anxiety and maternal behaviours [3]. However, oxytocin is best known for its role in parturition and lactation [4]. The word oxytocin comes from the Greek 'quick birth', and oxytocin facilitates mammalian reproduction through uterine contractions during labour (parturition) and milk ejection in response to suckling during lactation [5]. Circulating oxytocin concentrations rise progressively over the course of pregnancy [6, 7], but only induce uterine contraction when specific oxytocin receptors (OTRs) are up-regulated in the uterus immediately prior to birth [8, 9]. The contracting uterus provides positive feedback to induce further oxytocin secretion ('the Ferguson reflex', Fig. 10.1) [10].

Oxytocin is the strongest uterotonic substance known and is widely used to induce labour in humans, yet oxytocin-deficient mice successfully complete parturition [11]. This suggests that oxytocin secretion is not essential for birth. However, the importance of oxytocin for the initiation and maintenance of labour, and for delivery, remains controversial. Oxytocin plays a vital role in timing of delivery, because oxytocin-deficient mice give birth at random times following a circadian clock reset [12]. Furthermore, OTR antagonists administered prior to labour delay

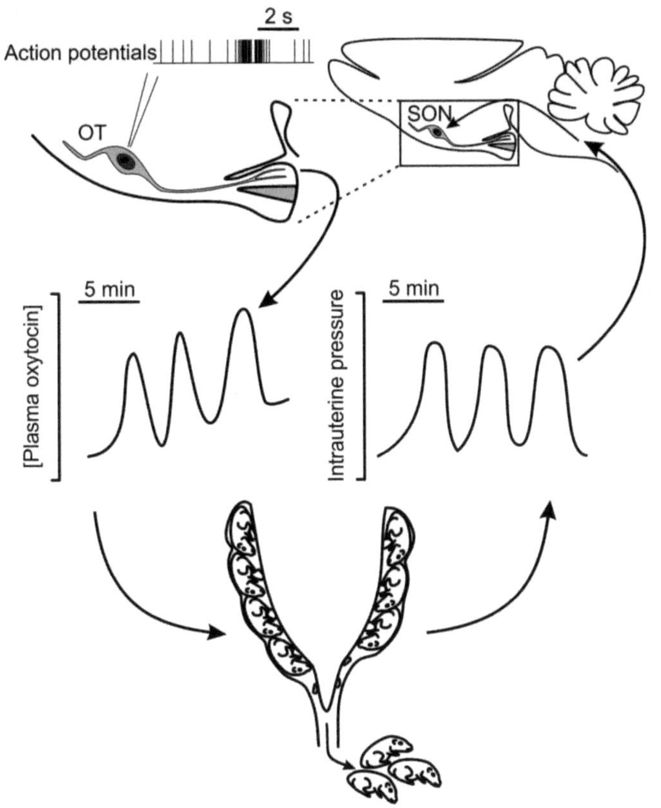

Fig. 10.1 A schematic representation of oxytocin regulation of parturition. Oxytocin cells fire in bursts similar to those seen during lactation but which are superimposed upon a higher baseline activity (*left hand side*). This pattern of activity results in pulsatile secretion of oxytocin from the posterior pituitary gland which acts on myometrial oxytocin receptors to induce uterine contractions and cause fetal expulsion. The uterine contractions feedback to the supraoptic nucleus (SON) via the nucleus tractus solitarii (NTS) in the brainstem (*right hand side*) to further enhance oxytocin cell activity and thus maintain parturition. Each burst of activity precedes the birth of a pup

the onset of delivery, and when given early in delivery, they increase the time between the deliveries of each pup [13]. Additionally, selective oxytocin agonists given centrally accelerate birth, as well as the onset of maternal behaviour [14, 15]. This suggests that, at least in rats, oxytocin is important for the initiation and the maintenance of parturition [13].

While it might be controversial as to whether oxytocin plays an indispensable role in parturition, the critical role that oxytocin plays in milk let-down during lactation is not disputed. The release of milk is mediated by secretion of oxytocin from the posterior pituitary gland, and oxytocin's action at OTR in the mammary gland induces a rise in intra-mammary pressure and release of milk: an oxytocin-mediated reflex upon suckling [16]. The oxytocin knockout mice fail to deliver milk to their offspring, resulting in the death of the pups. These mice have normal milk production, as well as normal ductal/glandular epithelium in the mammary gland, and the pups latch and suckle the nipples but, unlike wild type animals, this fails to induce milk ejection. Exogenous oxytocin administration in these mice can produce sufficient milk let-down to release milk and keep pups alive [11], highlighting the essential role for oxytocin in the milk-ejection reflex.

The Magnocellular Neurosecretory System

Oxytocin is principally synthesised in magnocellular neurons of the supraoptic nucleus (SON) and paraventricular nucleus (PVN) of the hypothalamus and released from the posterior pituitary gland to act in the periphery. The PVN contains magnocellular neurons that synthesise either oxytocin, or the closely related peptide vasopressin, and parvocellular neurons that contain a range of other peptide hormones and project to the median eminence to control anterior pituitary hormone secretion, as well as to other brain regions. By contrast to the PVN, the SON contains only magnocellular oxytocin and vasopressin neurons, which project a single axon caudally and medially to collect in the hypothalamo-neurohypophysial tract. These axons travel through the internal zone of the median eminence to the posterior pituitary gland (the neurohypophysis), where they end in several thousand neurosecretory axon swellings and terminals, filled with dense-core granules (neurosecretory vesicles) containing oxytocin (or vasopressin) [17]. Exocytosis of the neurosecretory vesicles occurs in response to invasion from action potentials, and once released into the extracellular space, oxytocin enters the general circulation by diffusion through fenestrated capillaries in the posterior pituitary gland [18]. The axon terminals in the posterior pituitary gland cannot maintain intrinsic repetitive firing [19], and so hormone secretion is principally determined by the frequency and pattern of action potentials initiated at the cell bodies. Nevertheless, various factors can modulate the release from magnocellular neuron terminals, including ionic conditions [20], purines [21] and neuropeptides [22].

Firing Patterns of Oxytocin Neurons

The profile of oxytocin secretion from the posterior pituitary gland is co-ordinated by the pattern of action potential discharge at the oxytocin cell bodies [23]. Nevertheless, the axon terminals in the posterior pituitary gland actively modulate the secretory response to action potential invasion by increasing the efficiency of stimulus-secretion coupling at higher firing rates so that each action potential releases more oxytocin per action potential at high action potential frequencies than at low frequencies [24]. As a result, when oxytocin neurons respond to stimulation with a linear increase in firing rate, the hormone output from the posterior pituitary gland is facilitated as the frequency of stimulation increases [24]. This frequency-facilitation results, in part, from increased calcium entry into the axon terminals through voltage-gated channels to enhance exocytosis. Frequency-facilitation of oxytocin release is most marked between ~5 and 25 Hz, but continues to increase (albeit at a slower rate) beyond ~50 Hz. By contrast, as a consequence of frequency-facilitation, oxytocin neurons firing at less than ~4 Hz release little or no oxytocin to the posterior pituitary gland in vivo [25]. Hormone secretion is not sustained upon continuous stimulation and there is 'fatigue' in facilitation over time (although this is more pronounced in vasopressin release), which is rapidly reversed when stimulation is stopped for a few tens of seconds [26]. This frequency-facilitation of oxytocin release is of particular importance during parturition and lactation.

Under normal physiological conditions, oxytocin neuron firing rate is highly variable between neurons, and the neurons fire action potentials in a slow continuous or irregular pattern. Generally, the mean firing rate of oxytocin neurons is approximately 3–5 Hz, with intervals of at least 30 ms between consecutive action potentials [27, 28]. However, during parturition and suckling, co-ordinated intermittent high-frequency bursts are superimposed upon this slow/irregular firing (Fig. 10.1) [29, 30]. Each burst lasts just 1 or 2 s but the action potential activity is intense, often with 100 action potentials per burst in each neuron [31–33]. During one of these milk-ejection bursts, the intervals between consecutive action potentials are between 6 and 10 ms [34]. After a burst, each oxytocin neuron typically falls silent for a few tens of seconds (allowing the recovery from frequency-facilitation fatigue).

Remarkably, these milk-ejection bursts are co-ordinated across the population of oxytocin neurons [30]. Because of the co-ordination of bursts between oxytocin neurons and frequency-facilitation of secretion in each oxytocin neuron, oxytocin is released into the circulation in high concentrations for short periods. This pattern of secretion underpins the episodic contraction of the milk ducts for milk ejection, and similar pattern of activity causes rhythmic contraction of the uterus during parturition [35, 36]. Because these bursts occur every few minutes and last for only a couple of seconds, the net increase in firing rate only averages about one to two action potentials per minute for each oxytocin neuron.

Bursting activity occurs in oxytocin neurons only during parturition and lactation. Other stimuli that excite oxytocin neurons simply increase the continuous firing rate of the neurons, which rarely exceeds about 15 action potentials per second [27, 28]. Additionally, during lactation, oxytocin neurons will only fire in bursts in response to suckling and continue to respond to other stimuli in a similar way to virgin rats [37]; indeed stimuli that increase the background firing rate can actually inhibit burst firing [38]. Therefore, the emergence of bursting behaviour at the end of pregnancy does not result simply from changes in the structure or intrinsic properties of oxytocin neurons that fundamentally alter the way in which they respond to excitatory inputs. Rather, it is the nature of the stimulus that determines whether bursts will be triggered, and our recent work suggests the possibility that the emergence of a central kisspeptin projection to oxytocin neurons over the course of pregnancy might fulfil this role. Uniquely, milk-ejection bursts, and bursts during parturition, are synchronised throughout all the magnocellular oxytocin neurons, with neurons in both the SON and PVN firing at the same time [39, 40].

Central Effects of Oxytocin

In addition to release into the circulation, oxytocin is also released into the brain from centrally projecting neurons, as well as from the dendrites of magnocellular oxytocin neurons [41]. While centrally projecting oxytocin neurons have important functions, the oxytocin release into the brain from magnocellular neuron dendrites appears most important for regulating peripheral secretion, particularly secretion underpinned by bursts of action potentials [42]; blocking the effects of dendritically released oxytocin by administration of an OTR antagonist SON delays birth [14]. Therefore, oxytocin release from dendrites might contribute to pulsatile oxytocin secretion needed for parturition and lactation by facilitating the occurrence and synchronisation of action potential bursts in oxytocin neurons [43].

Afferent Inputs to Oxytocin Neurons

Like all other neurons, oxytocin neurons receive afferent inputs from many different peripheral and central sources. The relay of sensory information from the uterus and nipples is, of course, particularly important in allowing for the Ferguson reflex (during parturition) and milk-ejection reflex (during lactation). The vagus nerve provides a major route of sensory input from the periphery to the brainstem and in particular the nucleus of the solitary tract (NTS) [44]. In turn, the NTS and the ventrolateral medulla (VLM) within the medulla oblongata of the brainstem provide afferent input to the SON and PVN [45]. Therefore, vagal afferent fibres provide a pathway by which factors in the periphery can relay information to the SON to modulate the secretion of oxytocin. Electrical stimulation of vagal afferents increases the firing rate of some magnocellular neurons [46] and disrupting the vagal pathway reduces oxytocin secretion in response to peripheral modulators [47]. Brainstem projections are mainly noradrenergic from the A1 group of the VLM and the A2 group of noradrenergic neurons in the NTS to the SON [48], and excitation of A2 noradrenergic neurons excites oxytocin neurons in the SON [49]. Over half the projections from the brainstem neurons to the SON are activated during parturition [50], and many of these are noradrenergic. During parturition, the excited A2 neurons release noradrenaline in the SON [51], contributing to the excitation of oxytocin neurons. This is unlikely to be the only link in the Ferguson reflex afferent pathway, but to date the A2 neurons are the only identified population with a demonstrated role in directly exciting oxytocin neurons during parturition.

Peptides secreted in the periphery also excite oxytocin neurons indirectly, via vagal afferents to the NTS and subsequent excitatory noradrenergic input to the SON. Cholecystokinin (CCK) is a peptide hormone that is released from the stomach following food intake and, in addition to inhibiting feeding [47], CCK increases the firing rate of oxytocin neurons [52], resulting in the secretion of oxytocin into the circulation [47]. The effects of CCK on food intake and oxytocin secretion are mediated though CCKA receptors located on peripheral vagal endings [53], which signal through the NTS in the brainstem to the oxytocin neurons in the hypothalamus [28]. Recently, another gastric peptide hormone, secretin, has also been shown to increase the activity and secretion of oxytocin neurons. This effect appears to also be mediated by the noradrenergic pathway, because intracerebroventricular (icv) injection of the $\alpha 1$ adrenergic antagonist, benoxathian, blocked the increase in firing rate of oxytocin (and vasopressin) neurons [54]. Hence, visceral inputs to oxytocin neurons (including those active in parturition and lactation) converge on the noradrenergic inputs via the vagus nerve.

The SON and PVN also receive prominent inputs from several forebrain areas, including areas in the anteroventral region of the third ventricle (AV3V): the subfornical organ (SFO), the organum vasculosum of the lamina terminalis (OVLT) and the median preoptic nucleus (MnPO) [45]. The SFO and OVLT are located outside the blood brain barrier and so peripheral stimuli can stimulate oxytocin secretion by acting on AV3V inputs. These projections from the AV3V are best characterised as mediating osmoregulation, and lesions of this region reduces oxytocin secretion and consequently impairs sodium excretion, but does not affect the milk-ejection reflex or parturition [55]. In addition to the above afferent inputs that have been extensively studied, the SON and PVN also receive numerous afferent inputs that have been less well characterised and whose physiological function is not well-established. These inputs include (but are not limited to): arcuate nucleus, bed nucleus of the stria terminalis, diagonal band of Broca, raphe nuclei, tuberomammillary nucleus and suprachiasmatic nucleus.

Up to this point, we have provided a brief background of the oxytocin system and its importance in pregnancy, parturition and lactation. There is remarkable plasticity in the activity of oxytocin neurons that emerges at the end of pregnancy to facilitate delivery of the offspring and delivery of milk to the offspring. This emergent behaviour requires afferent input for its expression, and so any afferent input that exhibits plasticity over the course of pregnancy is a likely candidate to be involved in the processes that underpin this change in behaviour at a cellular level. Our recent results suggest that kisspeptin might be an important newly discovered player in the regulation of oxytocin neurons in pregnancy and lactation. The remainder of this chapter focuses on how kisspeptin affects the oxytocin neurons and highlights new data that indicates that central kisspeptin regulation of oxytocin neurons emerges towards the end of pregnancy.

Circulating Kisspeptin Concentrations in Pregnancy and Lactation

Investigations of kisspeptin regulation of fertility have focussed on central interactions with GnRH neurons [56]. However, our initial interest was in the effects of circulating kisspeptin on oxytocin neurons because IV kisspeptin administration increases plasma oxytocin levels in non-pregnant female rats [1], and the plasma concentration of kisspeptin in woman has been reported to increase hugely over the course of pregnancy, with a 900-fold rise in the first trimester that further increases to over 7,000-fold in the third trimester. While the circulating kisspeptin that is found at low concentrations in non-pregnant women (and in men) is likely to be from either vascular endothelial cells [57] or adipose tissue [58], the main source of circulating kisspeptin during pregnancy is likely to be the placenta, because *KISS1* mRNA and kisspeptin protein are detectable in syncytiotrophoblasts in the placental wall [59]. The physiological function of elevated kisspeptin during human pregnancy is unknown; while it is possible that kisspeptin regulates trophoblast invasion, a process that is important for embryonic development [60], we hypothesised that such elevations in circulating kisspeptin might signal the oxytocin neurons to prepare for birth and lactation.

Activation of Oxytocin Neurons by Peripheral Kisspeptin in Non-pregnant Rats

To determine whether circulating kisspeptin might signal the oxytocin neurons to prepare for birth and lactation, we used an in vivo electrophysiology preparation to record spontaneous activity from the neurons in the SON of anaesthetised rats. This preparation causes the least disruption to whole animal physiology, allowing us to record near-normal neuron activity that can be altered by their intact peripheral and central inputs. Our work using this preparation was the first to show that neurons in the SON increase their firing rate following IV administration of kisspeptin-10 [2].

First, extracellular single-unit recordings of action potential firing rate were made from oxytocin neurons in non-pregnant rats across all stages of the estrous cycle. Intravenous administration of 25 μg kisspeptin-10 (which should achieve similar circulating concentrations of kisspeptin reported at the end of human pregnancy [57]) increases plasma oxytocin levels in the blood of virgin female rats [1], and so we administered 25 μg IV kisspeptin-10 to investigate whether changes in oxytocin neuron activity underpin the increased plasma hormone concentration; we found that this IV dose of kisspeptin-10 caused a robust, short-lived (~5 min) increase in the firing rate of every single oxytocin neuron tested (Fig. 10.2). The increase in firing rate following the peripheral kisspeptin injection was rapid, with a clear increase in firing rate within 30 s of administration and the peak rate within 60 s. The majority of the neurons showed a return to basal levels between 5 and 10 min following the injection. Hence, we were able to show that oxytocin neurons are able to rapidly respond to fluctuations in circulating kisspeptin, at least in non-pregnant rats. Importantly, repeated IV injections of kisspeptin-10 every few minutes caused a similar increase in firing rate, showing that the oxytocin neuron response does not desensitise over the time course expected for milk-ejection bursts. Given the short duration of excitation, and the lack of down-regulation of the response, it is likely that breakdown of kisspeptin (rather than a long deactivation or

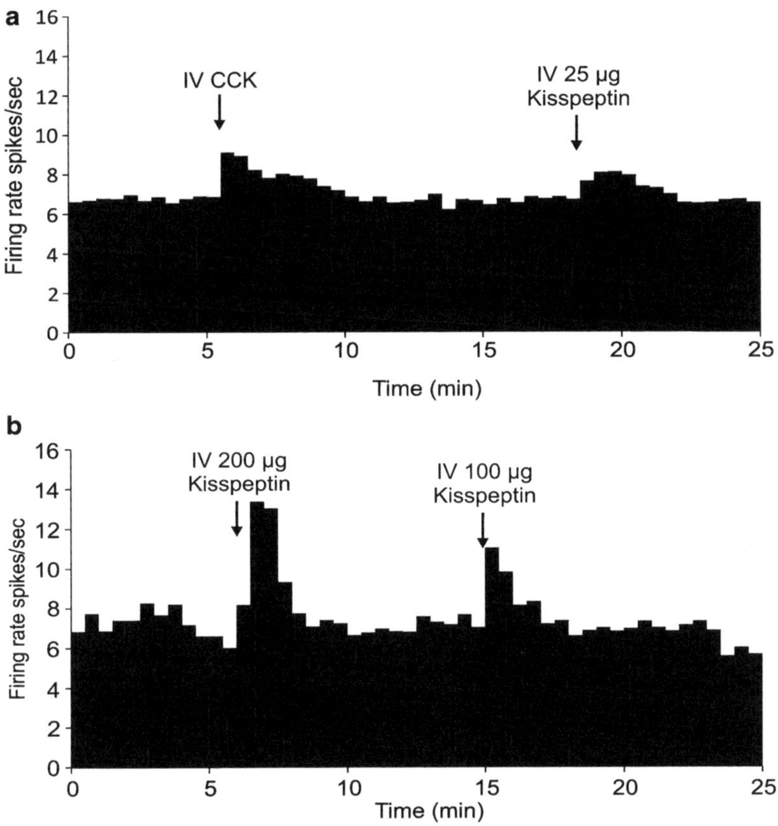

Fig. 10.2 (**a**) Representative ratemeter record (in 30 s bins) showing an increase in firing rate of an oxytocin neuron (identified by a transient excitation following IV CCK) in response to 25 μg IV kisspeptin-10 in a urethane-anaesthetised virgin female rat. (**b**) Representative ratemeter record (in 30 s bins) showing a dose-dependent increase in firing rate of an oxytocin neuron (identified by a transient excitation following IV CCK, not shown) in response to IV kisspeptin-10 in a urethane-anaesthetised virgin female rat

down-regulation of Kiss1r) is the limiting factor in the duration of the oxytocin neuron response to a bolus IV injection of kisspeptin-10.

To further characterise the response of oxytocin neurons to IV kisspeptin, we completed dose-ranging studies that showed a clear dose-dependent increase in oxytocin neuron firing rate to IV kisspeptin. Even at the lowest dose that we tested (5 μg), IV kisspeptin elicited a small but consistent increase in firing rate of oxytocin neurons.

Kisspeptin-10 is known to have comparable, if not greater, biological potency to that of the full length peptide [61, 62], but to eliminate the possibility that our observed responses were specific to kisspeptin-10, we also completed a small number of experiments where we used kisspeptin-54 (the full length peptide). As expected, IV kisspeptin-54 caused a similar increase in oxytocin neuron firing rate to that seen with kisspeptin-10.

As explained earlier, the basal firing rates of individual oxytocin neurons are highly variable, but the responses of oxytocin neurons to IV kisspeptin were completely independent of basal firing rate. This observation, combined with the consistent increase in firing rate evident in every oxytocin neuron challenged with IV kisspeptin-10, makes it probable that, when circulating kisspeptin rises, every single oxytocin neuron increases its firing rate. Hence, circulating kisspeptin is a potential mechanism that could co-ordinate activity across the population of oxytocin neurons; such co-ordinated oxytocin neuron activity is a pre-requisite for successful milk ejection during lactation.

Vagal Mediation of Oxytocin Neuron Activation by Peripheral Kisspeptin

As described earlier, information is relayed to the oxytocin neurons via various afferent input pathways, including noradrenergic pathways from the brainstem that relay vagal signals. This pathway is implicated in the co-ordination of neuroendocrine changes occurring at birth and is well-established as completing the positive feedback loop from the uterus to the oxytocin system in birth and lactation. Hence, we hypothesised that circulating kisspeptin-10 might also converge on this pathway to act on the SON, because Kiss1r are expressed in peripheral tissues, including the stomach and small intestine [1], that are known to be innervated by the vagus nerve.

To test this hypothesis, we repeated our experiments after desensitisation of vagal sensory fibres using the sensory neurotoxin, capsaicin (8-methyl-*N*-vanillyl-6-nonenamide). Capsaicin is a compound found in capsicums and hot chilli peppers that creates a perception of burning in mammals via activation of the transient receptor potential channel-vanilloid receptor subtype 1 (TRPV-1; capsaicin receptor). While capsaicin initially excites thin primary afferent c-fibres expressing TRPV-1, when it is administered in large and/or repeated doses it results in the desensitisation and defunctionalisation of the neurons [63]. TRPV-1 is expressed on gastric primary vagal afferent fibres [64], and capsaicin can act on these fibres to effectively block the vagal pathway into the brain with the advantage that efferent vagal pathways are left intact.

We found that pre-treatment with intraperitoneal (IP) capsaicin completely eliminated the increase in firing rate of oxytocin neurons induced by IV kisspeptin-10, as well as that induced by IV CCK (which is known to be mediated by vagal inputs to the NTS [47]). Thus, it appears that in non-pregnant rats, IV kisspeptin-10 does not directly excite oxytocin neurons, but acts as a hormone on peripheral targets with projections to the SON relayed by vagal afferent fibres. While we do not yet have direct evidence to support the involvement of the NTS, or of noradrenergic neurons, it is likely that NTS relays the vagal input to the SON through noradrenergic input. Of course, further work is still required to fully establish the pathway from the vagus to the SON, which will involve retrograde-labelling from the SON combined with immunohistochemistry for neuronal activation in the retrogradely labelled neurons after IV administration of kisspeptin.

While we have clearly established that peripheral administration of exogenous kisspeptin-10 consistently increases the firing rate of oxytocin neurons in non-pregnant rats [2], the physiological significance of this excitation remains to be established and this has been a focus of our more recent unpublished work.

Activation of Oxytocin Neurons by Peripheral Kisspeptin in Pregnancy and Lactation

We have begun to repeat our experiments by administering IV kisspeptin while recording oxytocin neuron firing rate in rats anaesthetised at various times over pregnancy and lactation. While still preliminary, our results to date show that the excitation of oxytocin neurons by IV kisspeptin is evident throughout pregnancy and into lactation. Superficially, these observations might seem straightforward, but they are difficult to reconcile with the published observations of markedly increased kisspeptin levels over the course of pregnancy, particularly in the third trimester [59]. In the face of hugely increased endogenous kisspeptin levels, one might expect occlusion, or desensitisation to the effects of exogenous kisspeptin. Furthermore, one would expect the endogenous kisspeptin to drive a steady increase in firing rate of oxytocin neurons over the course of pregnancy. We have seen no diminution in the effectiveness of exogenous kisspeptin in pregnant rats, but the progressive rise in circulating oxytocin concentrations over the course of human pregnancy [6, 7] is very much more modest than the increase reported for kisspeptin in humans [59].

However, the published data on circulating kisspeptin levels in pregnancy and lactation are from humans and our experimental model is rats. It appears likely that humans (and possibly higher primates) are the only species in which plasma kisspeptin levels might increase during pregnancy, because kisspeptin does not appear to increase during pregnancy across many non-primate species, including rodents, sheep and horses (Alain Caraty, personal communication). Thus, if the published data on humans genuinely reflect a species difference rather than a lack of specificity of kisspeptin antibodies used in the early studies, the rat might not be the model of choice for kisspeptin regulation of human pregnancy. Notwithstanding any species differences, it appears that kisspeptin probably excites oxytocin neurons in pregnant and lactating rats, and if the placental secretion of kisspeptin is pulsatile at parturition, this might add another level of control for the co-ordination of oxytocin neuron bursts during parturition. Of course, this idea is highly speculative and requires further investigation using peripheral administration of kisspeptin receptor antagonists to determine whether these can disrupt delivery of the offspring (and/or delivery of milk to the new-born).

Activation of Oxytocin Neurons by Central Kisspeptin in Pregnancy and Lactation

Several electrophysiology studies have shown that kisspeptin directly excites various neuronal populations: GnRH neurons, arcuate nucleus neurons, pro-opiomelanocortin neurons and hippocampal neurons are all excited by direct application of kisspeptin to brain slices [65–68]. So, we administered icv kisspeptin-10 in the expectation that we would localise the site of action of IV kisspeptin-10 to the brain. However, to our surprise, we did not find any effect of icv kisspeptin on the firing rate of oxytocin neurons in non-pregnant rats. In fact, in marked contrast to the robust and repeatable excitation following IV kisspeptin, there was no change in firing rate following icv injection at 2 μg or at 40 μg (both in a 2 μL volume) in any oxytocin neuron tested (Fig. 10.3), even in those neurons that were excited by IV kisspeptin-10. The higher dose of icv kisspeptin-10 that we used (40 μg) has been published as inducing a very robust response in the GnRH system to markedly increase circulating luteinising hormone levels [69], so it is very unlikely that the failure of icv kisspeptin-10 to excite oxytocin neurons in non-pregnant rats was due to a failure to deliver sufficient kisspeptin. The unexpected lack of response of oxytocin neurons to icv kisspeptin-10, combined with a robust response to IV kisspeptin, suggests that kisspeptin-10 might not cross the blood brain barrier to act directly on oxytocin neurons, or on Kiss1r that might be expressed by any central inputs to the SON. Consistent with our observed lack of effect of icv kisspeptin-10 on oxytocin neurons in non-pregnant rats, a recent mapping study has shown the presence of only a few kisspeptin fibres in the locality of the SON and it is not known whether they terminate in or form synapses in the area [70]. Additionally, there is no clear indication as to whether the Kiss1r is expressed in the SON [71].

Notwithstanding the lack of effect of icv kisspeptin on oxytocin neurons in non-pregnant rats, we continued to administer icv kisspeptin during our more recent studies into the effects of IV kisspeptin in pregnancy and lactation. The preliminary data that we have generated make the lack of effect of icv kisspeptin on oxytocin neurons in non-pregnant animals even more intriguing because our unpublished observations suggest that a central kisspeptin excitation of oxytocin neurons might emerge over the course of pregnancy in rats and that this is sustained during lactation.

In our latest experiments, we made in vivo extracellular single-unit recordings from oxytocin neurons in rats on different days of pregnancy (day 15–21). As described above, the response of oxytocin neurons to peripheral kisspeptin-10 was maintained throughout pregnancy, with IV injections of kisspeptin-10 continuing to cause a short (~5 min) increase in the firing rate of oxytocin neurons recorded from pregnant rats. This excitation was indistinguishable from the response seen in non-pregnant rats (Fig. 10.2), with the increase in firing rate following the peripheral injection occurring within 30 s and returning to basal levels 5–10 min later.

By contrast to the consistent responses to IV kisspeptin, there appears to be a marked change in the response of oxytocin neurons to icv kisspeptin-10 that arises over the course of pregnancy. In all oxytocin neurons recorded from animals on day

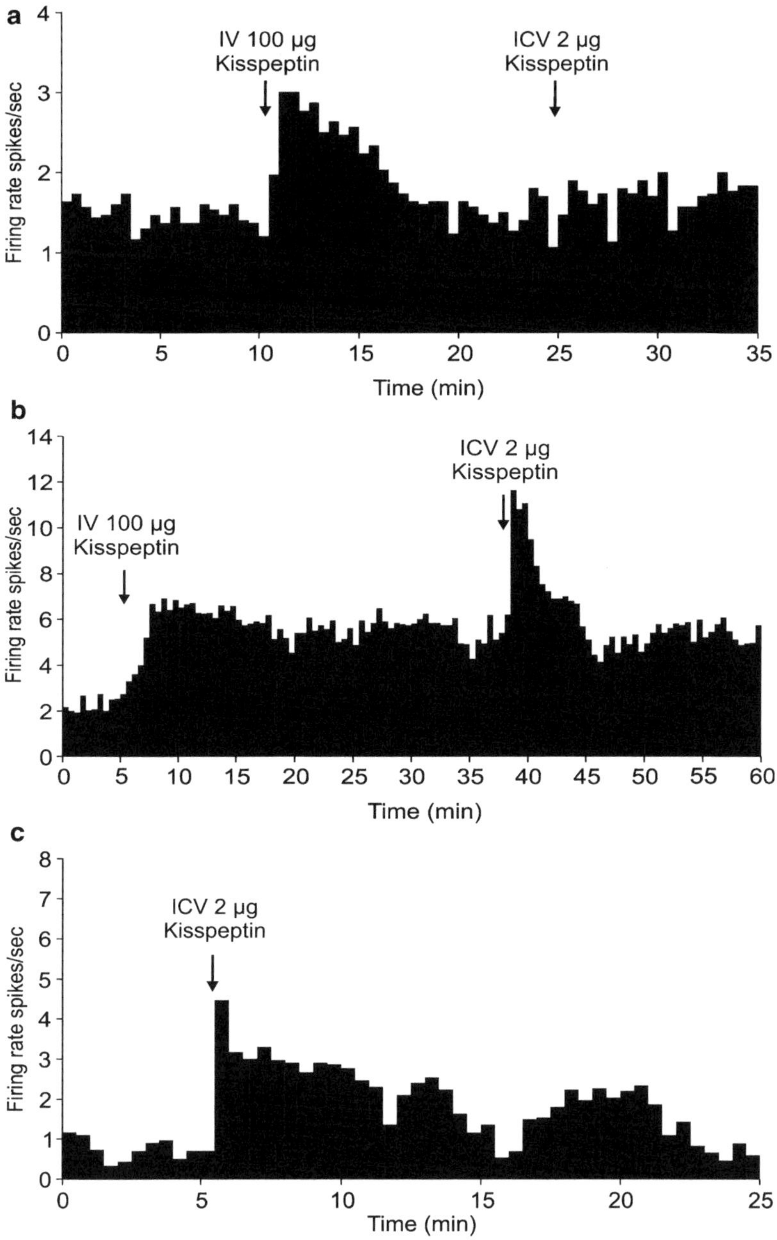

Fig. 10.3 (**a**) Representative ratemeter record (in 30 s bins) of an oxytocin neuron, showing an increase in firing rate after IV kisspeptin-10 but no response to icv kisspeptin-10, in a urethane-anaesthetised virgin female rat. (**b**) Representative ratemeter record (in 30 s bins) of an oxytocin neuron, showing a prolonged increase in firing rate after IV kisspeptin-10 and a rapid ~5 min excitation after icv kisspeptin-10, in a urethane-anaesthetised day 20 pregnant rat. (**c**) Representative ratemeter record (in 30 s bins) of an oxytocin neuron, showing an increase in firing rate after icv kisspeptin-10 in a urethane-anaesthetised day 7 lactating rat

18–21 of pregnancy, icv kisspeptin-10 caused an immediate, robust increase in oxytocin neuron firing rate following the injection. Unlike the response to peripheral kisspeptin-10, the increase in firing rate seen with central administration appeared substantially different between neurons, with some neurons showing a return to basal firing within 10 min, while others showed a sustained shift to a higher firing rate that lasted tens of minutes, somewhat reminiscent of the initial observations of kisspeptin excitation of GnRH neurons in vitro [65]. The most remarkable aspect of this response to icv kisspeptin-10 is that it only becomes apparent during pregnancy.

We have no knowledge of the functional consequences of the emergence of this sensitivity of oxytocin neurons to central kisspeptin during pregnancy, but it is tempting to speculate that it might be involved in driving, or facilitating, bursting behaviour of oxytocin neurons during parturition. Consistent with this speculation, the excitation of oxytocin neurons seen in response to icv kisspeptin-10 in late-pregnant rats was also evident in the one oxytocin neuron that we have recorded from a lactating rat, the only other time in a mammal's life that oxytocin neurons are known to exhibit high-frequency co-ordinated bursts of action potentials to release a bolus of oxytocin into the bloodstream.

Not only do we not know the function of central kisspeptin excitation of oxytocin neurons in late pregnancy, we also do not know the mechanisms that underpin the emergence of this excitation. One possibility is an increased accessibility of kisspeptin to Kiss1r that leads to the excitation of oxytocin neurons by icv kisspeptin. In addition, there might be up-regulation of Kiss1r expression in oxytocin neurons (or their afferent inputs), the laying down of a new kisspeptin projection to oxytocin neurons, and/or the up-regulation of kisspeptin expression in an existing projection to oxytocin neurons. These possibilities are a focus of current work in our laboratory.

Conclusion

Kisspeptin and its receptor, Kiss1r (in rodents), have been described as an essential gatekeeper of reproductive function [72]. Our recent work has expanded our knowledge of the critical role that kisspeptin plays in reproductive function via excitation of oxytocin neurons. During pregnancy and lactation, dynamic and transient changes occur within the oxytocin system. The synthesis, dendritic release and peripheral secretion are all modified in order to facilitate and synchronise the function of oxytocin neurons in labour, birth and lactation, as well as in maternal behaviour. We have shown that circulating kisspeptin excites oxytocin neurons throughout life, which might become important when the placenta secretes kisspeptin during pregnancy. Possibly of more importance for successful reproduction, we have also shown that the excitation of oxytocin neurons by central administration of kisspeptin appears to be dramatically changed over the course of pregnancy, emerging only in late pregnancy. The mechanisms behind this change have yet to be established but may underlie an important new role for kisspeptin in reproduction, the activation and modification of the oxytocin system during pregnancy, parturition and lactation.

Acknowledgments Funded by a New Zealand Health Research Council Project Grant (VS and CHB) and a University of Otago Health Sciences Fellowship (VS). We also thank Dr Rebecca Campbell for constructive criticism of an earlier version of the manuscript.

References

1. Kotani M, Detheux M, Vandenbogaerde A, Communi D, Vanderwinden JM, Le Poul E, Brezillon S, Tyldesley R, Suarez-Huerta N, Vandeput F, Blanpain C, Schiffmann SN, Vassart G, Parmentier M (2001) The metastasis suppressor gene KiSS-1 encodes kisspeptins, the natural ligands of the orphan G protein-coupled receptor GPR54. J Biol Chem 276:34631–34636
2. Scott V, Brown CH (2011) Kisspeptin activation of supraoptic nucleus neurons in vivo. Endocrinology 152:3862–3870
3. Neumann ID (2008) Brain oxytocin: a key regulator of emotional and social behaviours in both females and males. J Neuroendocrinol 20:858–865
4. Russell JA, Leng G, Douglas AJ (2003) The magnocellular oxytocin system, the fount of maternity: adaptations in pregnancy. Front Neuroendocrinol 24:27–61
5. Mitchell BF, Fang X, Wong S (1998) Oxytocin: a paracrine hormone in the regulation of parturition? Rev Reprod 3:113–122
6. Otsuki Y, Yamaji K, Fujita M, Takagi T, Tanizawa O (1983) Serial plasma oxytocin levels during pregnancy and labor. Acta Obstet Gynecol Scand 62:15–18
7. de Geest K, Thiery M, Piron-Possuyt G, Vanden Driessche R (1985) Plasma oxytocin in human pregnancy and parturition. J Perinat Med 13:3–13
8. Kimura T, Takemura M, Nomura S, Nobunaga T, Kubota Y, Inoue T, Hashimoto K, Kumazawa I, Ito Y, Ohashi K, Koyama M, Azuma C, Kitamura Y, Saji F (1996) Expression of oxytocin receptor in human pregnant myometrium. Endocrinology 137:780–785
9. Renthal NE, Chen CC, Williams KC, Gerard RD, Prange-Kiel J, Mendelson CR (2010) miR-200 family and targets, ZEB1 and ZEB2, modulate uterine quiescence and contractility during pregnancy and labor. Proc Natl Acad Sci U S A 107:20828–20833
10. Ferguson JKW (1941) A study of the motility of the intact uterus at term. Surg Gynecol Obstet 73:7
11. Nishimori K, Young LJ, Guo Q, Wang Z, Insel TR, Matzuk MM (1996) Oxytocin is required for nursing but is not essential for parturition or reproductive behavior. Proc Natl Acad Sci U S A 93:11699–11704
12. Roizen J, Luedke CE, Herzog ED, Muglia LJ (2007) Oxytocin in the circadian timing of birth. PLoS One 2:e922
13. Antonijevic IA, Douglas AJ, Dye S, Bicknell RJ, Leng G, Russell JA (1995) Oxytocin antagonists delay the initiation of parturition and prolong its active phase in rats. J Endocrinol 145:97–103
14. Neumann I, Douglas AJ, Pittman QJ, Russell JA, Landgraf R (1996) Oxytocin released within the supraoptic nucleus of the rat brain by positive feedback action is involved in parturition-related events. J Neuroendocrinol 8:227–233
15. Kendrick KM (2000) Oxytocin, motherhood and bonding. Exp Physiol 85 Spec No:111S–124S
16. Lincoln DW, Paisley AC (1982) Neuroendocrine control of milk ejection. J Reprod Fertil 65:571–586

17. Brownstein MJ, Russell JT, Gainer H (1980) Synthesis, transport, and release of posterior pituitary hormones. Science 207:373–378
18. Hatton GI, Yang QZ, Smithson KG (1988) Synaptic inputs and electrical coupling among magnocellular neuroendocrine cells. Brain Res Bull 20:751–755
19. Bourque CW (1990) Intraterminal recordings from the rat neurohypophysis in vitro. J Physiol 421:247–262
20. Marrero HG, Lemos JR (2010) Ionic conditions modulate stimulus-induced capacitance changes in isolated neurohypophysial terminals of the rat. J Physiol 588:287–300
21. Lemos JR, Wang G (2000) Excitatory versus inhibitory modulation by ATP of neurohypophysial terminal activity in the rat. Exp Physiol 85 Spec No:67S–74S
22. Ortiz-Miranda S, Dayanithi G, Custer E, Treistman SN, Lemos JR (2005) Micro-opioid receptor preferentially inhibits oxytocin release from neurohypophysial terminals by blocking R-type Ca2+ channels. J Neuroendocrinol 17:583–590
23. Brown CH, Bourque CW (2006) Mechanisms of rhythmogenesis: insights from hypothalamic vasopressin neurons. Trends Neurosci 29:108–115
24. Bicknell RJ (1988) Optimizing release from peptide hormone secretory nerve terminals. J Exp Biol 139:51–65
25. Dutton A, Dyball RE (1979) Phasic firing enhances vasopressin release from the rat neurohypophysis. J Physiol 290:433–440
26. Bicknell RJ, Brown D, Chapman C, Hancock PD, Leng G (1984) Reversible fatigue of stimulus-secretion coupling in the rat neurohypophysis. J Physiol 348:601–613
27. Leng G, Brown D (1997) The origins and significance of pulsatility in hormone secretion from the pituitary. J Neuroendocrinol 9:493–513
28. Leng G, Brown CH, Russell JA (1999) Physiological pathways regulating the activity of magnocellular neurosecretory cells. Prog Neurobiol 57:625–655
29. Lincoln DW, Wakerley JB (1975) Neurosecretory activation in the rat: correlation of the suckling stimulus with the pulsatile release of oxytocin. J Physiol 245:42P–43P
30. Belin V, Moos F (1986) Paired recordings from supraoptic and paraventricular oxytocin cells in suckled rats: recruitment and synchronization. J Physiol 377:369–390
31. Wakerley JB, Lincoln DW (1973) Proceedings: unit activity in the supra-optic nucleus during reflex milk ejection. J Endocrinol 59:xlvi–xlvii
32. Lincoln DW, Wakerley JB (1974) Electrophysiological evidence for the activation of supraoptic neurones during the release of oxytocin. J Physiol 242:533–554
33. Ingram CD, Wakerley JB (1993) Post-partum increase in oxytocin-induced excitation of neurones in the bed nuclei of the stria terminalis in vitro. Brain Res 602:325–330
34. Dyball RE, Leng G (1986) Regulation of the milk ejection reflex in the rat. J Physiol 380:239–256
35. Summerlee AJ (1981) Extracellular recordings from oxytocin neurones during the expulsive phase of birth in unanaesthetized rats. J Physiol 321:1–9
36. Summerlee AJ, Lincoln DW (1981) Electrophysiological recordings from oxytocinergic neurones during suckling in the unanaesthetized lactating rat. J Endocrinol 90:255–265
37. Brown D, Fontanaud P, Moos FC (2000) The variability of basal action potential firing is positively correlated with bursting in hypothalamic oxytocin neurones. J Neuroendocrinol 12:506–520
38. Moos F, Poulain DA, Rodriguez F, Guerne Y, Vincent JD, Richard P (1989) Release of oxytocin within the supraoptic nucleus during the milk ejection reflex in rats. Exp Brain Res 76:593–602
39. Moos F, Richard P (1988) Characteristics of early- and late-recruited oxytocin bursting cells at the beginning of suckling in rats. J Physiol 399:1–12
40. Moos F, Richard P (1989) Paraventricular and supraoptic bursting oxytocin cells in rat are locally regulated by oxytocin and functionally related. J Physiol 408:1–18
41. Ludwig M, Leng G (2006) Dendritic peptide release and peptide-dependent behaviours. Nat Rev Neurosci 7:126–136
42. Ludwig M, Sabatier N, Bull PM, Landgraf R, Dayanithi G, Leng G (2002) Intracellular calcium stores regulate activity-dependent neuropeptide release from dendrites. Nature 418:85–89

43. Russell JA, Leng G (2000) Veni, vidi, vici: the neurohypophysis in the twentieth century. Exp Physiol 85 Spec No:1S–6S
44. Kalia M, Sullivan JM (1982) Brainstem projections of sensory and motor components of the vagus nerve in the rat. J Comp Neurol 211:248–265
45. Weiss ML, Hatton GI (1990) Collateral input to the paraventricular and supraoptic nuclei in rat. II. Afferents from the ventral lateral medulla and nucleus tractus solitarius. Brain Res Bull 25:561–567
46. Ueta Y, Kannan H, Yamashita H (1991) Gastric afferents to the paraventricular nucleus in the rat. Exp Brain Res 84:487–494
47. Verbalis JG, McCann MJ, McHale CM, Stricker EM (1986) Oxytocin secretion in response to cholecystokinin and food: differentiation of nausea from satiety. Science 232:1417–1419
48. Day TA, Sibbald JR (1988) Direct catecholaminergic projection from nucleus tractus solitarii to supraoptic nucleus. Brain Res 454:387–392
49. Day TA, Sibbald JR (1988) Solitary nucleus excitation of supraoptic vasopressin cells via adrenergic afferents. Am J Physiol 254:R711–R716
50. Meddle SL, Leng G, Selvarajah JR, Bicknell RJ, Russell JA (2000) Direct pathways to the supraoptic nucleus from the brainstem and the main olfactory bulb are activated at parturition in the rat. Neuroscience 101:1013–1021
51. Herbison AE, Voisin DL, Douglas AJ, Chapman C (1997) Profile of monoamine and excitatory amino acid release in rat supraoptic nucleus over parturition. Endocrinology 138:33–40
52. Renaud LP, Tang M, McCann MJ, Stricker EM, Verbalis JG (1987) Cholecystokinin and gastric distension activate oxytocinergic cells in rat hypothalamus. Am J Physiol 253:R661–R665
53. Moriarty P, Dimaline R, Thompson DG, Dockray GJ (1997) Characterization of cholecystokinin A and cholecystokinin B receptors expressed by vagal afferent neurons. Neuroscience 79:905–913
54. Velmurugan S, Brunton PJ, Leng G, Russell JA (2010) Circulating secretin activates supraoptic nucleus oxytocin and vasopressin neurons via noradrenergic pathways in the rat. Endocrinology 151:2681–2688
55. Russell JA, Blackburn RE, Leng G (1988) The role of the AV3V region in the control of magnocellular oxytocin neurons. Brain Res Bull 20:803–810
56. Clarkson J, Han SK, Liu X, Lee K, Herbison AE (2010) Neurobiological mechanisms underlying kisspeptin activation of gonadotropin-releasing hormone (GnRH) neurons at puberty. Mol Cell Endocrinol 324:45–50
57. Dhillo WS, Murphy KG, Bloom SR (2007) The neuroendocrine physiology of kisspeptin in the human. Rev Endocr Metab Disord 8:41–46
58. Brown RE, Imran SA, Ur E, Wilkinson M (2008) KiSS-1 mRNA in adipose tissue is regulated by sex hormones and food intake. Mol Cell Endocrinol 281:64–72
59. Horikoshi Y, Matsumoto H, Takatsu Y, Ohtaki T, Kitada C, Usuki S, Fujino M (2003) Dramatic elevation of plasma metastin concentrations in human pregnancy: metastin as a novel placenta-derived hormone in humans. J Clin Endocrinol Metab 88:914–919
60. Bilban M, Ghaffari-Tabrizi N, Hintermann E, Bauer S, Molzer S, Zoratti C, Malli R, Sharabi A, Hiden U, Graier W, Knofler M, Andreae F, Wagner O, Quaranta V, Desoye G (2004) Kisspeptin-10, a KiSS-1/metastin-derived decapeptide, is a physiological invasion inhibitor of primary human trophoblasts. J Cell Sci 117:1319–1328
61. Muir AI, Chamberlain L, Elshourbagy NA, Michalovich D, Moore DJ, Calamari A, Szekeres PG, Sarau HM, Chambers JK, Murdock P, Steplewski K, Shabon U, Miller JE, Middleton SE, Darker JG, Larminie CG, Wilson S, Bergsma DJ, Emson P, Faull R, Philpott KL, Harrison DC (2001) AXOR12, a novel human G protein-coupled receptor, activated by the peptide KiSS-1. J Biol Chem 276:28969–28975
62. Ohtaki T, Shintani Y, Honda S, Matsumoto H, Hori A, Kanehashi K, Terao Y, Kumano S, Takatsu Y, Masuda Y, Ishibashi Y, Watanabe T, Asada M, Yamada T, Suenaga M, Kitada C, Usuki S, Kurokawa T, Onda H, Nishimura O, Fujino M (2001) Metastasis suppressor gene KiSS-1 encodes peptide ligand of a G-protein-coupled receptor. Nature 411:613–617

63. Holzer P (1991) Capsaicin as a tool for studying sensory neuron functions. Adv Exp Med Biol 298:3–16
64. Patterson LM, Zheng H, Ward SM, Berthoud HR (2003) Vanilloid receptor (VR1) expression in vagal afferent neurons innervating the gastrointestinal tract. Cell Tissue Res 311:277–287
65. Han SK, Gottsch ML, Lee KJ, Popa SM, Smith JT, Jakawich SK, Clifton DK, Steiner RA, Herbison AE (2005) Activation of gonadotropin-releasing hormone neurons by kisspeptin as a neuroendocrine switch for the onset of puberty. J Neurosci 25:11349–11356
66. Liu X, Lee K, Herbison AE (2008) Kisspeptin excites gonadotropin-releasing hormone neurons through a phospholipase C/calcium-dependent pathway regulating multiple ion channels. Endocrinology 149:4605–4614
67. Fu LY, van den Pol AN (2010) Kisspeptin directly excites anorexigenic proopiomelanocortin neurons but inhibits orexigenic neuropeptide Y cells by an indirect synaptic mechanism. J Neurosci 30:10205–10219
68. Arai AC, Xia YF, Suzuki E, Kessler M, Civelli O, Nothacker HP (2005) Cancer metastasis-suppressing peptide metastin upregulates excitatory synaptic transmission in hippocampal dentate granule cells. J Neurophysiol 94:3648–3652
69. Roa J, Vigo E, Castellano JM, Navarro VM, Fernandez-Fernandez R, Casanueva FF, Dieguez C, Aguilar E, Pinilla L, Tena-Sempere M (2006) Hypothalamic expression of KiSS-1 system and gonadotropin-releasing effects of kisspeptin in different reproductive states of the female rat. Endocrinology 147:2864–2878
70. Desroziers E, Mikkelsen J, Simonneaux V, Keller M, Tillet Y, Caraty A, Franceschini I (2010) Mapping of kisspeptin fibres in the brain of the pro-oestrous rat. J Neuroendocrinol 22:1101–1112
71. Herbison AE, de Tassigny X, Doran J, Colledge WH (2010) Distribution and postnatal development of Gpr54 gene expression in mouse brain and gonadotropin-releasing hormone neurons. Endocrinology 151:312–321
72. Reynolds RM, Logie JJ, Roseweir AK, McKnight AJ, Millar RP (2009) A role for kisspeptins in pregnancy: facts and speculations. Reproduction 138:1–7

Part II
Development and Regulation of Kisspeptin Neurons

Chapter 11
The Development of Kisspeptin Circuits in the Mammalian Brain

Sheila J. Semaan, Kristen P. Tolson, and Alexander S. Kauffman

Abstract The neuropeptide kisspeptin, encoded by the *Kiss1* gene, is required for mammalian puberty and fertility. Examining the development of the kisspeptin system contributes to our understanding of pubertal progression and adult reproduction and sheds light on possible mechanisms underlying the development of reproductive disorders, such as precocious puberty or hypogonadotropic hypogonadism. Recent work, primarily in rodent models, has begun to study the development of kisspeptin neurons and their regulation by sex steroids and other factors at early life stages. In the brain, kisspeptin is predominantly expressed in two areas of the hypothalamus, the anteroventral periventricular nucleus and neighboring periventricular nucleus (pre-optic area in some species) and the arcuate nucleus. Kisspeptin neurons in these two hypothalamic regions are differentially regulated by testosterone and estradiol, both in development and in adulthood, and also display differences in their degree of sexual dimorphism. In this chapter, we discuss what is currently known and not known about the ontogeny, maturation, and sexual differentiation of kisspeptin neurons, as well as their regulation by sex steroids and other factors during development.

Introduction

The status of the neuroendocrine reproductive axis is in flux during various stages of perinatal and pubertal development, ranging from being entirely quiescent to fully active. Additionally, the neuroendocrine reproductive system is anatomically and physiologically differentiated between males and females, and these sex differences

S.J. Semaan • K.P. Tolson • A.S. Kauffman, Ph.D. (✉)
Department of Reproductive Medicine, University of California, San Diego, 9500 Gilman Drive, 0674, Leichtag Building, Room 3A-15, La Jolla, CA 92093, USA
e-mail: ssemaan@ucsd.edu; kptolson@ucsd.edu; akauffman@ucsd.edu

A.S. Kauffman and J.T. Smith (eds.), *Kisspeptin Signaling in Reproductive Biology*, Advances in Experimental Medicine and Biology 784, DOI 10.1007/978-1-4614-6199-9_11,

originate during key stages of development. Sex differences are also seen in several reproductive health disorders, such as idiopathic hypogonadotropic hypogonadism, constitutional delayed puberty, and precocious puberty [1–5]. Many of these reproductive health disorders have been attributed to known, or in many cases, unknown developmental defects in the brain. Specific neuronal circuits located in the forebrain and hypothalamus have been implicated as control centers responsible for proper development of reproductive physiology, converging on neurons that release gonadotropin releasing hormone (GnRH). GnRH stimulates the release of luteinizing hormone (LH) and follicle stimulating hormone (FSH) from the pituitary, thereby driving the maturation and activation of the gonads. How and when the developmental changes in the neuroendocrine axis are induced is not completely known, but gaining a clearer understanding may help pinpoint the cause and timing of defects in reproductive maturation.

One of the key upstream hypothalamic circuits involved in the control of GnRH secretion consists of neurons expressing the *Kiss1* gene and its protein product, kisspeptin. Kisspeptin signaling has been implicated as an essential regulator of fertility and puberty in numerous mammalian species, including humans [6–8]. Alterations in the expression of *Kiss1* or kisspeptin over development, along with differences in expression between the sexes, especially during key developmental periods, may be a critical driving force in the maturation of the neuroendocrine reproductive system. Indeed, changes in the *Kiss1* system likely contribute to the timing of puberty onset, sex differences in LH secretion, and other facets of reproductive physiology. It is therefore essential to understand how kisspeptin neurons develop, when and how *Kiss1* gene and protein expression are modified during development, and what possible regulatory mechanisms govern the development of the kisspeptin system. This chapter discusses the current knowledge for these topics in mammals and also pinpoints several important unanswered questions involving the development of kisspeptin neuronal circuits.

Kiss1 and Kisspeptin Expression in the Adult Brain

In order to study the development of the kisspeptin system, it is essential to first understand the localization and phenotype of kisspeptin neurons in the adult state. Until the recent generation of transgenic mice, which label kisspeptin cells with markers such as GFP [9–11], three techniques were used to examine the localization of kisspeptin neurons in the brain: reverse transcriptase PCR (RT-PCR)/quantitative PCR (qPCR), in situ hybridization (ISH), and immunohistochemistry (IHC). RT-PCR was used in early studies to identify high *Kiss1* mRNA expression in large regions of the brain, such as the hypothalamus [12]. However, this technique was weakened by an inability to specifically visualize and separately analyze discrete *Kiss1* populations within these large brain areas. This issue was resolved through the subsequent use of ISH and IHC, which allow for precise neuroanatomical mapping of *Kiss1*- and kisspeptin-synthesizing neurons, respectively. However, in earlier

IHC studies, some kisspeptin antibodies were not very specific, as they were shown to cross react with other RFamide family members [13]. The recent use of more specific kisspeptin antibodies has allowed for more precise detection of kisspeptin immunoreactivity in the brain [14–19]. Utilizing these various methods, many studies have confirmed that in adult rodents, *Kiss1* (or kisspeptin) is expressed in just a few discrete brain regions, including a small population in the medial amygdala (MeA) [20] and two larger hypothalamic populations in the anteroventral periventricular nucleus and neighboring periventricular nucleus (AVPV/PeN) and the arcuate nucleus (ARC) [21–23]. In non-rodent species, such as sheep and non-human primates, *Kiss1* gene expression and kisspeptin immunoreactivity have, for the most part, a similar distribution as in rodents, with expression localized to the pre-optic area (POA) and the ARC/infundibular nucleus (INF) [19, 24–27]. Expression of *Kiss1* or kisspeptin in the MeA of non-rodent species has not yet been examined.

In contrast to *Kiss1*/kisspeptin cell bodies, which are found in just a few discrete brain regions, kisspeptin-immunoreactive (ir) fibers are scattered throughout the brain (discussed in detail in Chap. 3). In adult rodents and sheep, terminals of kisspeptin fibers are found within the POA (and regions containing GnRH neurons), the ARC and medial basal hypothalamus, the paraventricular nucleus, and the median eminence (perhaps targeting GnRH axons/terminals) [14, 15, 28]. In mice, additional regions have been identified containing kisspeptin fibers, including the lateral septum, dorsal-medial nucleus of the hypothalamus, bed nucleus of the stria terminalis (BNST), and the MeA [14]. It also appears that ARC and AVPV/PeN *Kiss1* neurons send a number of projections to one another, perhaps allowing these two populations to directly communicate [29], although currently there is no evidence that *Kiss1* neurons themselves express *Kiss1r* [30].

In adult rodents, the AVPV/PeN region displays sex differences in various morphological parameters [31, 32] and is considered the main anatomical site that drives the sexually dimorphic preovulatory luteinizing hormone (LH) surge that occurs in adult females [33]. Mounting evidence supports a critical involvement of AVPV/PeN kisspeptin neurons in the sexually differentiated LH surge. For example, Estradiol (E_2) dramatically stimulates *Kiss1* expression in the AVPV/PeN, and *Kiss1* neurons in this region co-express sex steroid receptors, including ERα [34]. Moreover, *Kiss1* neuronal activity (as measured by *cfos* induction) in the AVPV/PeN is upregulated in a circadian pattern in complete synchrony with the circadian timing of the LH surge [35]. Additionally, as will be discussed in more detail later, the AVPV/PeN *Kiss1* population itself is sexually differentiated, just like the preovulatory LH surge, with females expressing greater *Kiss1* and kisspeptin expression in this region than males [36]. It is also worth noting that the AVPV/PeN region contains several other sexually dimorphic subpopulations that have been implicated in regulating reproduction, such as dopaminergic neurons, which express the tyrosine hydroxylase (TH) enzyme [37, 38], and cells expressing both GABA and glutamate [39]. Interestingly, most *Kiss1* neurons in the AVPV/PeN co-express TH [40, 41], though the functional significance of such co-expression has yet to be determined.

Kisspeptin neurons in the ARC comprise the largest kisspeptin population in the brain [15, 24, 25, 34, 36, 42, 43]. In contrast to the AVPV/PeN population, *Kiss1* cells in the ARC are not highly sexually dimorphic in adult rodents, especially when the adulthood sex steroid milieu is similar between the sexes [36, 44]. However, there are sex differences in the regulation of *Kiss1* expression in the ARC that are present in mice during the prepubertal period, an intriguing finding that will be discussed more below [44]. Moreover, some species, such as sheep, exhibit sex differences in ARC *Kiss1* neurons in adulthood, which may reflect slightly different roles of these ARC kisspeptin neurons between species. While *Kiss1* expression is upregulated in the AVPV/PeN by E_2 and testosterone (T), the opposite is true in the ARC: sex steroids potently inhibit ARC *Kiss1* expression [22, 34, 36, 45], and *Kiss1* neurons in this region have a high degree of colocalization with ERα, androgen receptor, progesterone receptor, and to a lesser degree, ERβ [15, 22, 24, 46–49]. It is suggested that the inhibition of *Kiss1* expression in the ARC by sex steroids reflects the involvement of ARC kisspeptin neurons in the negative feedback effects of sex steroids on pulsatile GnRH secretion in both sexes [6, 50, 51]. Additionally, ARC *Kiss1* cells co-express several other regulatory neuropeptides, neurokinin B (NKB) and dynorphin (DYN), which appear to also play a role in regulating GnRH/LH secretion [52–57]. Moreover, ARC kisspeptin neurons have been shown to have abundant reciprocal connections with each other [58–60] and to highly express the NKB receptor (NK3R) [16, 55, 58, 60], which may allow these neurons to communicate and synchronize with each other [55, 60] (discussed more in Chap. 15).

The Development of the AVPV/PeN *Kiss1* Population

Ontogeny of Kiss1 Expression in the AVPV/PeN

In adulthood, *Kiss1* is highly expressed in the hypothalamic AVPV/PeN and ARC regions, but this is not always the case earlier in development. In rodents, *Kiss1* expression is present in the embryonic brain but is limited to the ARC region [61], with no detectable AVPV/PeN *Kiss1* expression at this age. Rather, *Kiss1* mRNA and kisspeptin protein expression in the AVPV/PeN of rodents first occurs later in postnatal life. The developmental timing of AVPV/PeN *Kiss1* expression has been studied in both mice and rats. However, different studies describing the development of *Kiss1* (or kisspeptin) expression sometimes use different nomenclature to describe age of birth: some studies define the day of birth as postnatal day 1 (PND 1), others as PND 0 or P0. For consistency, and to permit for direct comparison between studies, in this chapter the day of birth will be denoted as PND 1; if the day of birth used in the original paper was noted as P0 or PND 0, it will be changed to PND 1 and subsequent ages modified accordingly.

Neither *Kiss1* mRNA nor kisspeptin protein expression has been detected in the AVPV/PeN of mice or rats on the day of birth. The earliest documented AVPV/PeN *Kiss1* expression in mice was recently shown, using ISH, to be ~PND 10 [40], though this expression was of very low magnitude compared to the next several days examined, PND 12, and PND 14, when moderately higher levels were observed (Fig. 11.1).

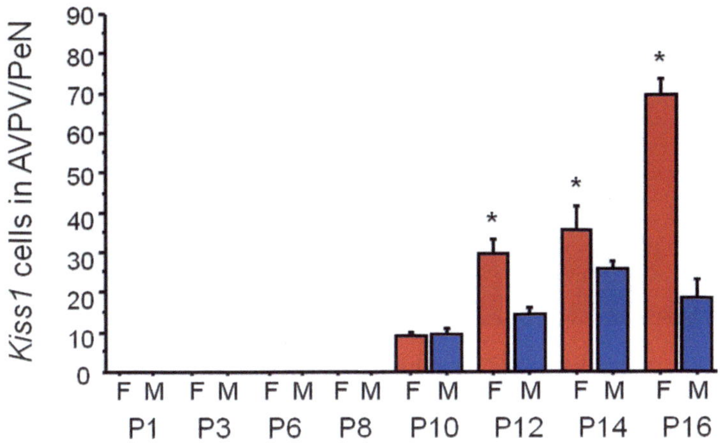

Fig. 11.1 Development of the *Kiss1* sex difference in the anteroventral periventricular nucleus and neighboring periventricular nucleus (AVPV/PeN) of mice. Mean number of *Kiss1* neurons in the AVPV/PeN of female and male mice over the course of early postnatal development. *Kiss1* cells were first detected in each sex on PND 10. The number of *Kiss1* neurons was significantly higher in females than males on PND 12 and later. *Significantly different from males of same age. Modified from Semaan SJ, Murray EK, Poling MC, Dhamija S, Forger NG, Kauffman AS 2010 BAX-dependent and BAX-independent regulation of Kiss1 neuron development in mice. Endocrinology 151:5807–5817

Another developmental study of kisspeptin protein expression in female mice found no kisspeptin-immunoreactive (ir) cells in the AVPV/PeN at PND 10 and only a small number of kisspeptin cells at PND 15, which was the next chronological age examined [62]. Whether the absence of detectable kisspeptin-ir cells at PND 10 reflects differences in the sensitivities of the techniques used (IHC vs. ISH) or a developmental difference at the level of either post-transcriptional processing or translation is currently unknown.

Looking beyond the first 2 weeks of life, kisspeptin-ir cell number was found to steadily increase in the mouse AVPV/PeN from PND 15 to adulthood (assessed every 5 days of age) [62]. Extending this protein data, our lab recently performed a detailed, day-by-day analysis of *Kiss1* mRNA expression in prepubertal and pubertal female mice and found that *Kiss1* cell number in the AVPV/PeN steadily and continually increases from PND 15 through PND 28, at which point it resembles adulthood levels [63]. In rats, like mice, AVPV/PeN kisspeptin expression is not detectable on PND 4 or PND 8, but is noticeably present by the next age examined, PND 22 [64]. Developmental changes in *Kiss1* mRNA expression have also been documented in the POA of ewes [65] at 25, 30, and 35 weeks of age, and an increase in POA *Kiss1* cell number was observed at 30 weeks, corresponding to the time of puberty, with no further increase detectable at 35 weeks [65]. In non-human primates, *KISS1* expression also increases in the hypothalamus during puberty, but this was attributed to kisspeptin neurons in the medial basal hypothalamic region (i.e., the ARC) rather than the POA population [25].

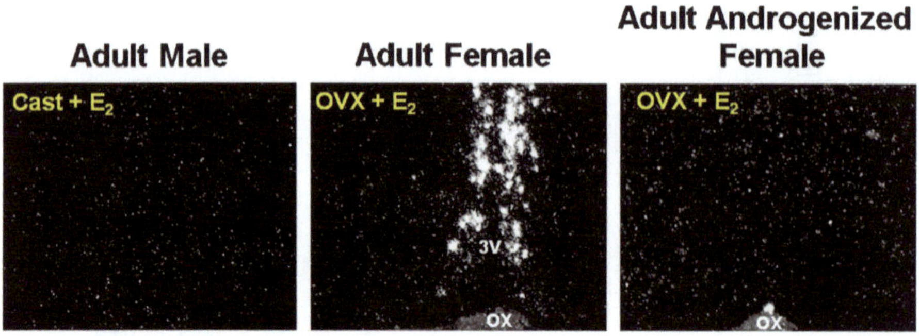

Fig. 11.2 Sex differences in AVPV *Kiss1* cells are organized early in development by perinatal hormones and are unaffected by the activational effects of adult hormones. Representative photomicrographs showing *Kiss1* mRNA-expressing cells in the AVPV of adult male, female, and neonatally androgenized female rats. All animals were treated in adulthood with E_2. *3V* third ventricle; *Cast* castrated; *OVX* ovariectomized. Modified from Kauffman AS, Gottsch ML, Roa J, Byquist AC, Crown A, Clifton DK, Hoffman GE, Steiner RA, Tena-Sempere M 2007 Sexual differentiation of Kiss1 gene expression in the brain of the rat. Endocrinology 148:1774–1783. With permission from The Endocrine Society

In addition to developmental changes in AVPV/PeN kisspeptin cell number, the development of kisspeptin fibers apposing GnRH neurons has been reported to change over development [28]. Detectable kisspeptin fiber appositions to GnRH neurons were absent in female mice younger than PND 25, but such appositions were present on PND 25 and further increased in numbers by PND 31 [28]. The authors postulated that many of these GnRH-apposing kisspeptin fibers were derived from the AVPV/PeN region, though this was not experimentally determined. However, if this assumption is true, then the appearance and subsequent pubertal increase in kisspeptin-GnRH appositions may be a function of the known increase in kisspeptin synthesis in the AVPV/PeN during this age, increasing the visibility and detection of the fibers. Alternatively, it is also possible that a physical increase in the degree of kisspeptin fiber innervation of GnRH neurons occurs peripubertally, perhaps affecting GnRH activation around puberty onset.

Sexual Differentiation of AVPV/PeN Kiss1 Expression During Development

A few years ago, it was discovered that *Kiss1*-expressing neurons in the adult rat AVPV/PeN are sexually differentiated, with adult females possessing more *Kiss1* mRNA (and detectable *Kiss1* cells) than males (Fig. 11.2) [36]. Similar observations have now been reported for *Kiss1* mRNA levels in mice [40] and kisspeptin protein levels in mice and rats [28, 66]. No documented *Kiss1* sex differences have yet been reported in the POA of the ewe or monkey, although a sex difference is observed in the ARC *Kiss1* population of sheep [67] and is discussed later in further detail.

Although activational effects of sex steroids play a role in transiently increasing AVPV/PeN *Kiss1* expression levels in adulthood [22], the adult sex steroid milieu does not account for the observed sex differences in AVPV/PeN *Kiss1* expression [36]. This is evidenced by the fact that male and female rats that are gonadectomized as adults and treated with identical E_2 levels still display sexually dimorphic *Kiss1* expression in the AVPV/PeN [36, 42]. In fact, the AVPV/PeN *Kiss1* sex difference appears to be permanently organized by sex steroid signaling early in postnatal development. In the postnatal "critical period," which is typically the first week of postnatal life in rodents, males normally secrete elevated gonadal T, whereas females secrete little sex steroids at this time. Experiments manipulating the postnatal sex steroid milieu of rodents support the model that the presence of elevated levels of postnatal sex steroids determines whether many sexually dimorphic traits develop to be male-like in adulthood. Thus, in newborn males, elevated sex steroids act to organize neural circuits to permanently develop a male-like phenotype [68]. In contrast, newborn females are not exposed to sufficient levels of sex steroids, and therefore their brains permanently develop to be female-like [32, 37]. Supporting this "organizational" model of sexual differentiation, castration of newborn males, to remove high postnatal T, results in the permanent development of feminized neural populations. Conversely, sex steroid treatment to newborn females, mimicking elevated T secretion in postnatal males, results in the permanent development of masculinized brain circuitry (reviewed in ref. [6]).

A number of studies have determined that the AVPV/PeN *Kiss1* system is organized postnatally by sex steroids. For example, castrating male rats at birth causes a permanent feminization of the developing AVPV/PeN *Kiss1* system (Fig. 11.3) [66]. Conversely, neonatal female rats treated once with T or E_2 exhibit a permanent reduction of *Kiss1*- or kisspeptin-expressing cells in the AVPV/PeN in adulthood, similar to what is exhibited in normal males (Figs. 11.2 and 11.3) [36, 66, 69]. The fact that postnatal E_2 treatment can, like T, permanently alter the development of the AVPV/PeN *Kiss1* system suggests that postnatal masculinization of this system is likely mediated via aromatization of T to E_2. In rats, the effects of postnatal E_2 on *Kiss1* sexual differentiation are likely mediated by ERα and not ERβ, because neonatal treatment with the ERα agonist, PPT, caused a reduction in female AVPV/PeN kisspeptin levels [70], while neonatal administration of the ERβ agonist, DPN, had no significant effect on adulthood *Kiss1* levels [71]. In these experiments, however, males and females were not compared, and further studies are needed to determine if PPT can completely masculinize the female *Kiss1* AVPV/PeN population to male levels. Additionally, the reduction of AVPV/PeN *Kiss1* expression in female rats that were treated neonatally with sex steroids correlates with the inability of these females to generate an E_2-mediated LH surge as adults [66], linking the sexually dimorphic AVPV/PeN *Kiss1* system and the sexually dimorphic LH surge event.

As discussed earlier, *Kiss1* mRNA is expressed in the AVPV/PeN as early as PND 10 in mice of both sexes. However, there are no sex differences in AVPV/PeN *Kiss1* neuron number or *Kiss1* mRNA levels/neuron at this age, even though both of these parameters are well-established sex differences in adulthood [40]. The sex difference in *Kiss1* cell number, however, is evident by PND 12 and becomes even

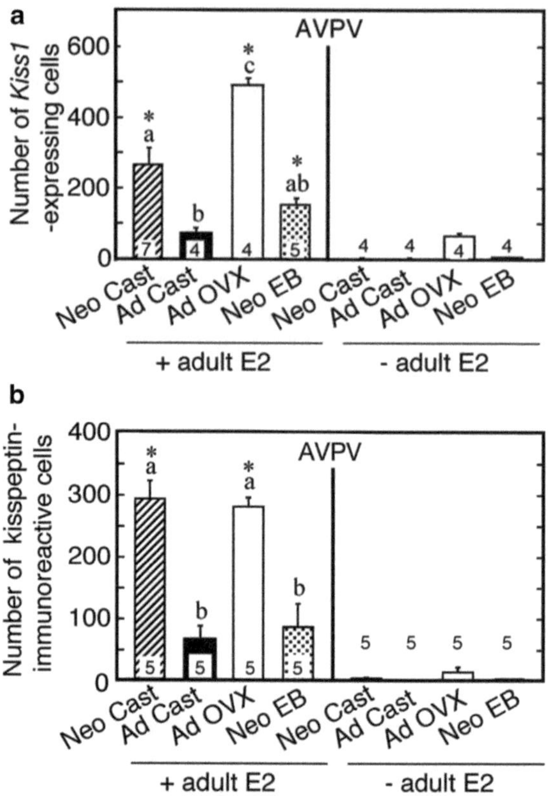

Fig. 11.3 *Kiss1* mRNA and kisspeptin expression in the AVPV of adult rats. Mean level of *Kiss1* mRNA expression (**a**) and kisspeptin immunoreactivity (**b**) in the AVPV of male and female rats that were either neonatally castrated (Neo Cast), neonatally treated with estradiol benzoate (Neo EB), or untreated. All animals were gonadectomized and treated with E_2 or without E_2 as adults to equalize hormone levels. Values with same letters are not significantly different within the group with the same adult E_2 treatment. Values with asterisks are significantly different from corresponding animals without adult E_2 treatment. The number in or on each column indicates the number of animals used. Values are means ± SEM. Modified from Homma T, Sakakibara M, Yamada S, Kinoshita M, Iwata K, Tomikawa J, Kanazawa T, Matsui H, Takatsu Y, Ohtaki T, Matsumoto H, Uenoyama Y, Maeda K-i, Tsukamura H 2009 Significance of Neonatal Testicular Sex Steroids to Defeminize Anteroventral Periventricular Kisspeptin Neurons and the GnRH/LH Surge System in Male Rats. Biology of Reproduction 81:1216–1225. With permission from The Society for the Study of Reproduction

more robust on PND 14 and 16 (Fig. 11.1) [40]. Semi-quantitative analysis of *Kiss1* mRNA levels/cell revealed a significant sex difference only beginning around PND 16 [40]. Thus, AVPV/PeN *Kiss1* mRNA expression first arises at the same time in both sexes (~PND 10), but the sex difference in both cell number and mRNA expression/cell takes several more days to develop, at least as assessed via ISH in mice. As mentioned above, in other rodent studies utilizing IHC, neither sex displays

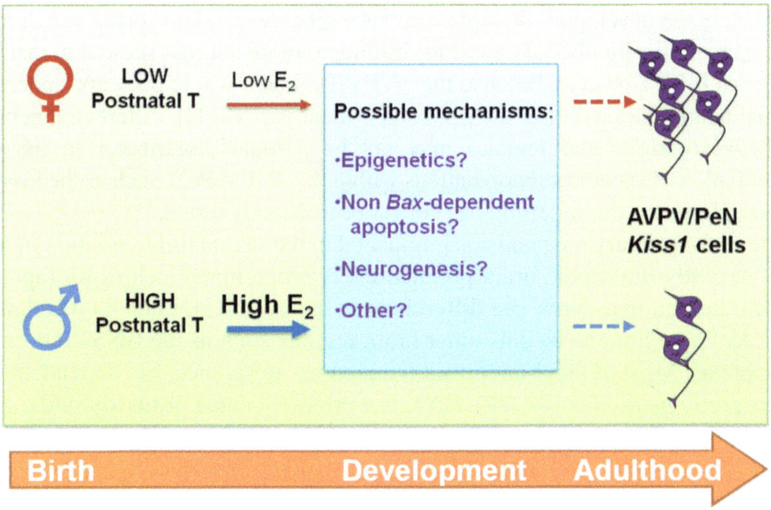

Fig. 11.4 Schematic of the development of sexually dimorphic AVPV/PeN *Kiss1* neurons. Males secrete elevated testosterone (T) at birth, while newborn females secrete negligible levels of sex steroids. Postnatal T converted to estradiol influences the development of neural circuits, leading to their masculinization in adulthood. AVPV/PeN *Kiss1* neurons are more abundant in adult females than males. It is currently unknown how the perinatal sex steroid milieu organizes the sexual differentiation of *Kiss1* neurons, although the AVPV as a whole is differentiated through *Bax*-dependent apoptotic mechanisms. However, the sexual differentiation of *Kiss1* neurons is likely not due to *Bax*-dependent apoptosis [40], though other apoptotic mechanisms have not been ruled out. Current data indicate that the *Kiss1* gene may be more transcriptionally active in females, pointing to epigenetic alterations as a putative mechanism

detectable kisspeptin-ir cells in the AVPV/PeN on or before PND 10 [28, 64]. Currently, kisspeptin protein levels have not been directly compared between sexes from PND 11 through PND 21, but the sex difference in kisspeptin-ir cell number was readily apparent in mice on PND 25 [28] and in rats on PND 21 [64].

Possible Mechanisms of Steroid-Mediated Sexual Differentiation of AVPV/PeN Kiss1 Neurons

Sexual differentiation of the AVPV/Pen *Kiss1* system is dependent on the postnatal sex steroid milieu, but it is unclear exactly how E_2 (aromatized from T) directs this developmental process. Several sex steroid-dependent mechanisms, such as differential neurogenesis, migration, epigenetics, and apoptosis, have been implicated in the sexual differentiation and development of other neuronal populations (Fig. 11.4) [32, 72–74]. E_2, for example, can promote neurogenesis in the olfactory bulb and dentate gyrus of the adult rat hippocampus, leading to more newly formed neurons in females [32].

Likewise, in the developing rat hippocampus, higher levels of postnatal sex steroids in males increase the number of new cells, leading to more neurons present in males than in females [75]. However, because the AVPV/PeN as a whole does not undergo differential neurogenesis during the postnatal critical period [32], differential neurogenesis between males and females may not be a major contributor to the sexual differentiation of specific subpopulations within the AVPV/PeN, such as the kisspeptin neurons. However, this assertion has not yet been directly tested.

One of the primary mechanisms implicated in the sexual differentiation of a number of sexually dimorphic brain populations is programmed cell death (apoptosis) [76–78]. In fact, in rodents, sex differences in the overall size and total cell number of the AVPV region, as well as other brain regions such as the BNST, are induced by apoptosis. Most of these apoptosis-induced sex differences are dependent on the pro-apoptotic gene, *Bax* [77–79]. BAX is a protein located primarily in the cytosol in a healthy cell. In response to cell death signals, BAX translocates to the mitochondria where it precipitates the release of cytochrome c, thereby triggering caspase pathways that culminate in cell death [80]. Interestingly, in the developing rat AVPV, postnatal males have higher *Bax* expression than postnatal females, which possibly initiates more cell death in the former sex [81]. Higher *Bax* expression in postnatal males than females therefore correlates with the presence of fewer AVPV cells in adult males. The sexually dimorphic postnatal *Bax* levels also coincide with higher sex steroid levels in postnatal males than females, suggesting that sex steroids might affect postnatal *Bax* expression [81]. Supporting this prediction, E_2 treatment of neonatal female rats increases the number of apoptotic AVPV neurons [82]. Importantly, a recent study determined that the sex difference in total number of AVPV neurons is eliminated in *Bax* knockout mice [76]. Thus, total cell number in the AVPV is sexually differentiated via *Bax*-dependent apoptotic mechanisms. Despite these findings, the sexual differentiation of *Kiss1* neurons in the AVPV/PeN is surprisingly unaltered in *Bax* knockout mice [40]. When *Kiss1* cell numbers are compared between adult male and female *Bax* KO mice, the *Kiss1* sex difference was still incredibly robust [40]. Thus, the *Kiss1* population is sexually differentiated either by other apoptotic pathways, such as tumor necrosis α-dependent or -independent mechanisms (as may be the case for AVPV GABA-ergic neurons [78]) or, by non-apoptosis-related mechanisms (Fig. 11.4).

It is likely that the sexual differentiation of AVPV/PeN *Kiss1* cells is not induced by mechanisms that affect the physical existence of cells (like apoptosis). Rather, sex differences in *Kiss1* cell number may be induced by developmental mechanisms affecting transcriptional activity of the *Kiss1* gene (Fig. 11.4) [5, 83]. In fact, epigenetic changes, such as histone modifications and DNA methylation, precipitated by postnatal sex steroids are emerging as critical contributors to alterations in neuronal cell number and gene expression between the sexes [73, 84–88]. Histones are proteins that allow for the packaging of DNA into chromatin. When histones are modified, such as by acetylation, transcriptional activity is altered [89]. Histone acetylation, which is generally associated with increased transcriptional activity, has recently been implicated in the sex steroid-induced sexual differentiation of the BNST [73]. For example, inhibiting histone deacetylase (HDAC) during the early postnatal period in mice blocks the sexual differentiation of the size of the BNST and also alters sexually dimorphic vasopressin fiber projections [73, 88]. Postnatal HDAC inhibition also alters the sexual differentiation of male sexual behavior in rats [87]. It is likely that these postnatal alterations in HDAC activity directly or indirectly affected apoptosis in the BNST, as sex differences in this region are known to be governed specifically by *Bax*-dependent apoptosis [77]. Other sexual differentiation studies have addressed the role of DNA methylation, which occurs at CpG sites or on CpG islands in a gene or its promoter, and which is generally associated with the repression of gene expression. Studies investigating the involvement of DNA methylation in neural sex differences have found that DNA methylation levels in the hypothalamus correlate with sexually differentiated expression of sex steroid receptor genes in this region [85, 86]. In addition, the expression of DNA methyl transferase 3a (DNMT3a) in newborn rats was found to be sexually dimorphic in the amygdala, which is a known sexually dimorphic brain region [90].

With all the recent evidence implicating the involvement of epigenetics in the sexual differentiation of various brain parameters, we investigated whether either histone deacetylation and/or DNA methylation contributes to the AVPV/PeN *Kiss1* sex difference [83]. We pharmacologically blocked histone deacetylation during the postnatal period by administering an HDAC inhibitor, valproic acid (VPA), or vehicle to mice on PND 1 and PND 2 and then analyzed AVPV/PeN *Kiss1* expression in adulthood. This postnatal HDAC inhibitor treatment significantly increased the number of detectable *Kiss1* cells in the adult AVPV in each sex. However, the sex difference in *Kiss1* expression was not eliminated, indicating that histone acetylation is not a key process for inducing the *Kiss1* sex difference [83]. Although the sex difference was still robust, the fact that overall *Kiss1* levels were higher in mice treated with HDAC inhibitor suggests that the level of histone H3 acetylation during the critical period may be involved in modulating the development of *Kiss1* neurons in the AVPV.

Interestingly, we found significant sex differences in the CpG methylation status of the AVPV/PeN *Kiss1* gene, predominantly in the putative promoter region (Fig. 11.5). In all cases, these sexually dimorphic *Kiss1* CpG sites were more methylated in females than males. Methylation of CpG sites can have multiple modes of affecting gene activity. We tested if methyl-CpG binding protein-2 (MeCP2) was involved in the sex difference by assessing AVPV/PeN *Kiss1* levels in male and female *Mecp2* mutant mice. The AVPV/PeN sex difference was not eliminated in *Mecp2* mutant mice, suggesting that if DNA methylation influences the AVPV/PeN *Kiss1* sex difference, it likely does so via non-Mecp2 mechanisms, possibly by blocking the binding of transcriptional repressors [83]. Although these experiments increase our knowledge about the involvement (or lack of involvement) of several epigenetic processes in the development of the AVPV/PeN *Kiss1* sex difference, more work is needed to elucidate the exact extent that certain processes, like DNA methylation, are involved. In fact, the AVPV/PeN *Kiss1* sex difference may be induced by several epigenetic processes affected by the postnatal sex steroid milieu, causing a silencing of the *Kiss1* gene in males while simultaneously allowing increased transcriptional activity in females.

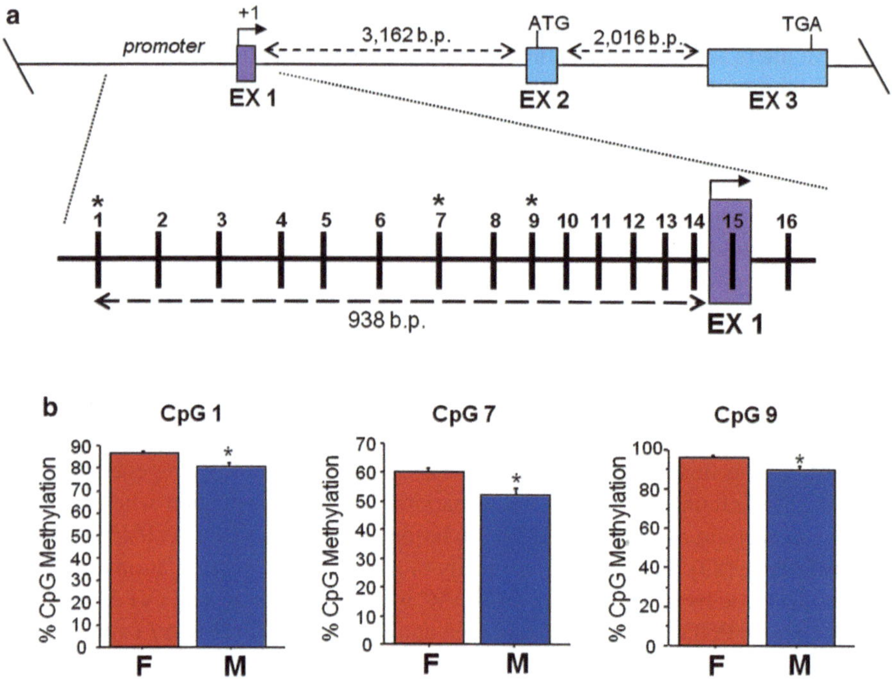

Fig. 11.5 CpG methylation analysis of the putative murine *Kiss1* promoter region in the AVPV/PeN. (**a**) Map of pyrosequenced CpG sites in the putative *Kiss1* promoter region using bisulfite-treated DNA derived from AVPV/PeN micropunches of adult males and females that were E_2-treated for 1 week before sacrifice. (**b**) Mean percentage of methylation of sexually dimorphic CpGs (CpG 1, 7, 9) in the *Kiss1* promoter. *Significantly different than females ($P<0.05$). Modified from Semaan SJ, Dhamija S, Kim J, Ku EC, Kauffman AS 2012 Assessment of epigenetic contributions to sexually dimorphic kiss1 expression in the anteroventral periventricular nucleus of mice. Endocrinology 153:1875–1886. With permission from The Endocrine Society

Regulation of AVPV/PeN Kiss1 *Neurons by Gonadal and Non-gonadal Factors During Peripubertal Development*

The steady developmental increase in kisspeptin cell number in the female AVPV/PeN is likely dependent on the presence of ovarian steroids. As mentioned previously, *Kiss1* gene expression in the AVPV/PeN increases markedly over the pubertal transition [12, 28, 62, 63, 91]. In adult animals, *Kiss1* expression levels in the AVPV/PeN are transiently increased by elevations in the sex steroid milieu and, conversely, decreased by removal of sex steroids [22, 36]. Therefore, it is likely that the observed developmental increase in *Kiss1* gene expression during puberty is caused by increased ovarian sex steroid secretion at this time.

Several studies have addressed the effects of sex steroids or their receptors on the developmental increase of kisspeptin cell number in the AVPV/PeN. In one study, female mice were either ovariectomized (OVX) or sham treated on PND 15 and killed on either PND 30 or PND 60. Mice that were OVX on PND 15 had dramatically reduced levels of kisspeptin in the AVPV/PeN later in adulthood, suggesting that the primary cause of the developmental kisspeptin increase is due to ovarian sex steroid secretion [62]. This was supported by the observation that estrogen replacement in OVX animals from either PND 15–30 or PND 22–30 rescues kisspeptin expression when examined at PND 30 [62]. Thus, kisspeptin neurons are sensitive to ovarian steroids peri-pubertally, which is not surprising given their robust regulation by sex steroids in adulthood. It is currently unknown if kisspeptin neurons, however, are responsive to E_2 at the first time of visible protein expression (PND 15) or mRNA expression (PND 10), or if OVX at earlier time periods has a more permanent effect on the development of kisspeptin expression.

Several studies have used aromatase knockout (ArKO) mice to examine the effects of E_2 signaling on kisspeptin neuron development. One study reported a complete elimination of kisspeptin expression in the AVPV/PeN of adult female ArKO mice [62]. However, E_2 was not replaced in these mice prior to sacrifice, and it was therefore unclear if the absence of AVPV/PeN kisspeptin cells in ArKO females mirrored a chronically OVX condition (since removal of E_2 via OVX in adulthood reduces AVPV/PeN kisspeptin synthesis). More recently, another study looked at kisspeptin cells in ArKO mice that were given E_2 in adulthood. This study found that AVPV/PeN kisspeptin cells were in fact present in adult ArKO mice after E_2-treatment. However, surprisingly, the sex difference in kisspeptin cell number was eliminated in these E_2-treated ArKO mice [92]. But, instead of ArKO males exhibiting high kisspeptin levels similar to that of wild-type (WT) females, as would be predicted due to the lack of E_2-signaling in these males during the postnatal critical period, AVPV/PeN kisspeptin cell number was instead intermediate in level in ArKOs of both sexes, being significantly lower than in normal WT females and significantly higher than in normal WT males [92]. This finding suggests that E_2 during development may normally actively contribute to complete feminization of the AVPV/PeN kisspeptin system in females, although when and how this would occur is unknown.

A similar story has emerged concerning the development of the AVPV/PeN kisspeptin system in hypogonadal (*hpg*) mice. Hypogonadal mice possess a deletion in the *Gnrh* gene and therefore do not secrete GnRH or gonadal sex steroids [93]. In female *hpg* mice, AVPV/PeN kisspeptin-ir cell number during development never reaches the level of WT females and is similar to that of *hpg* males (i.e., the normal kisspeptin sex difference is absent in *hpg* mice) [94]. Similarly, *Kiss1* mRNA expression in pubertal *hpg* females is decreased compared to that of WT females, and sexually dimorphic *Kiss1* expression is eliminated in *hpg* mice [94]. Hormone replacement was not compared in pubertal *hpg* females and males in this particular study and may be critical to fully interpret the results. Interestingly though, 1 week of E_2 replacement in adult *hpg* females did *not* increase AVPV/PeN kisspeptin protein levels to WT female levels, suggesting that gonadal sex steroids may be required at some time during development in order for AVPV/PeN kisspeptin expression to fully mature [94].

Finally, in a different study, ERα was specifically ablated in kisspeptin neurons using cre-lox technology (generating "KERKO" mice) in order to test the role of ERα signaling in kisspeptin neurons [10]. In this model, AVPV/PeN kisspeptin-ir was significantly diminished in adult KERKO females compared to WT females, further implicating the involvement of E_2 and ERα in the development of the AVPV/PeN kisspeptin population in females [10]. Likewise, a number of recent studies have implicated endocrine disruptors (which often mimic the effects of E_2) in affecting the development of AVPV/PeN *Kiss1* gene expression [69, 95, 96], a topic discussed in detail in Chap. 21.

The Development of the ARC *Kiss1* Population

Ontogeny of Kiss1 Expression in the ARC

The second major population of kisspeptin neurons in the hypothalamus is located in the ARC, which is equivalent to the INF in humans [7, 14, 21, 62]. The ARC is the most consistently detected kisspeptin population in mammalian species, although its precise distribution and role may vary from species to species [15, 16, 19, 21, 22]. In general, the developmental pattern of *Kiss1* expression in the ARC, and the specific factors regulating the development of this neuronal population, are not completely understood and are less well-characterized than the AVPV/PeN.

Unlike in the AVPV/PeN, *Kiss1* is expressed in the rodent ARC during embryonic development. Analysis of both kisspeptin-ir and *Kiss1* mRNA (via qPCR) in embryonic rats determined that expression in both sexes begins around E14.5 and increases by E18.5, with a sharp drop in levels just prior to birth [61]. Additionally, double-labeling of kisspeptin-ir/BrdU-ir demonstrated that kisspeptin cell neurogenesis in the ARC begins around E12.5 and peaks several days later [97]. In other recent studies, when crossed to reporter mice that permanently mark all cells that ever expressed *Kiss1* at any point in development, two *Kiss1*-Cre transgenic mouse lines were reported to have "extra" cells in the greater ARC region [9, 11]. One of these reports quantified these "extra" *Kiss1* cells to comprise ~25 % of the total cells that have ever expressed *Kiss1* in the ARC region at some point during development or adulthood [11]. This suggests that the pattern of ARC kisspeptin expression may differ significantly between early developmental and older ages, but it is not clear at what developmental age these "extra" kisspeptin cells become undetectable. Additionally, both male and female adult mice with impaired BAX-mediated apoptosis display "extra" *Kiss1* cells in the ARC compared to adult WT mice, particularly in the male, suggesting that the number of ARC *Kiss1* neurons is also regulated in early development by apoptosis [40]. The extra *Kiss1* cells in the ARC of both the *Bax* KO and the *Kiss1*-Cre mice show a similar expression pattern that is located dorsally in relation to the normal ARC *Kiss1* population (Fig. 11.6). The ontogeny, developmental pattern, and role, if any, of these extra kisspeptin neurons remain unknown.

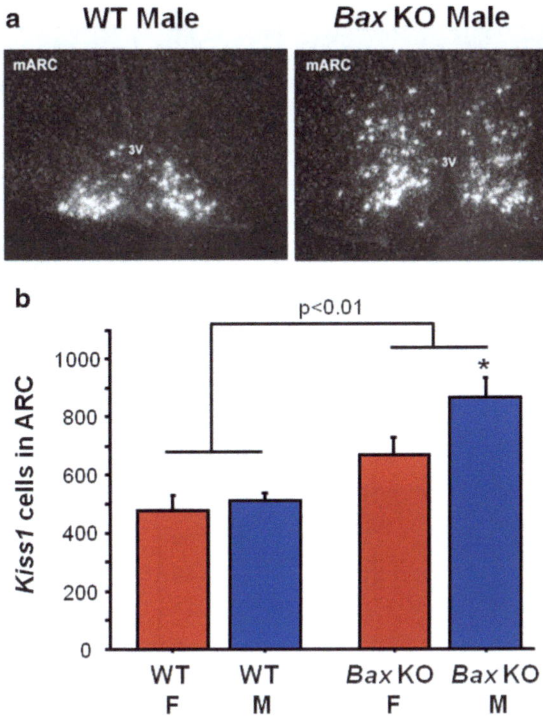

Fig. 11.6 Regulation of *Kiss1* expression in the ARC by apoptosis. (**a**) *Kiss1* expression in the ARC of GDX adult WT and *Bax* KO male mice show extra *Kiss1* cells in the ARC, suggesting that ARC *Kiss1* neurons are regulated in early development by apoptosis. The "extra" *Kiss1* cells in the ARC show an expression pattern that is located dorsally in relation to the normal ARC *Kiss1* population typically found in WT mice. *3V* third ventricle; *mARC* medial ARC. (**b**) Mean number of *Kiss1* cells in the ARC of adult female (F) and male (M) *Bax* KO and WT mice that were GDX for 9 day before sacrifice. *Kiss1* gene expression in the ARC was significantly higher in *Bax* KO mice than WT mice ($P<0.01$), especially in males. *Significantly different from female mice of same genotype. (**a, b**) Modified from Semaan SJ, Murray EK, Poling MC, Dhamija S, Forger NG, Kauffman AS 2010 BAX-dependent and BAX-independent regulation of Kiss1 neuron development in mice. Endocrinology 151:5807–5817. With permission from The Endocrine Society

In rats, initial studies detected *Kiss1* expression in the whole hypothalamus on PND 1 using RT-PCR, but it was not determined which specific nuclei (ARC, AVPV/PeN, etc.) were responsible for this neonatal *Kiss1* expression [12]. Based on the AVPV/PeN developmental time-course discussed earlier, it is unlikely that the source of this PND 1 *Kiss1* expression is the AVPV/PeN. Rather, most, if not all, of the hypothalamic *Kiss1* detected in PND 1 rats was probably from the ARC population. Several recent studies have now provided supporting evidence for this likelihood. Using ISH, *Kiss1* expression in the ARC was readily detected at PND 1 in mice [98] and rats [47, 99], as well as in rats in another study at the first collected

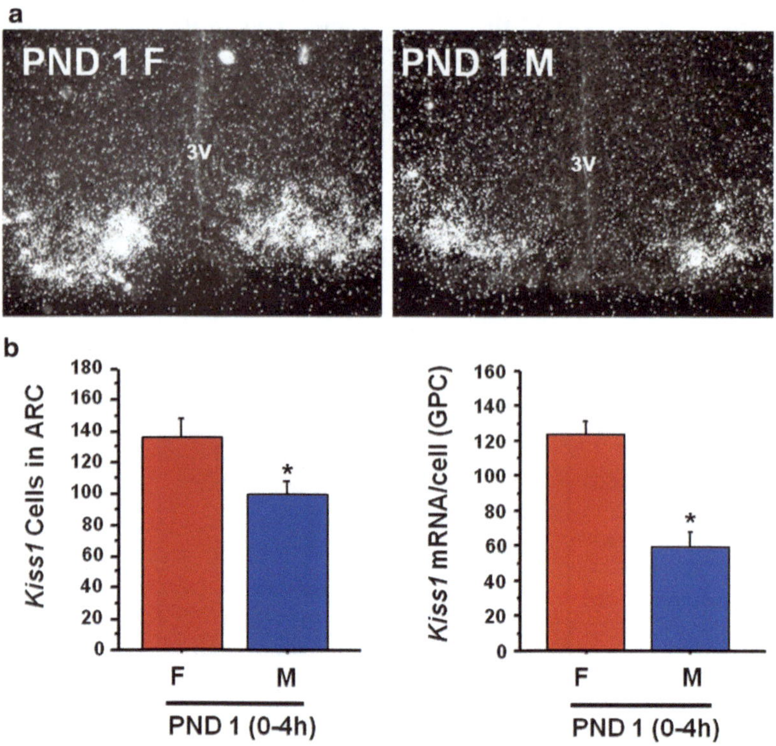

Fig. 11.7 *Kiss1* gene expression in the murine ARC on PND 1. (**a**) Representative photomicrographs for *Kiss1* expression in newborn females (F) and males (M) that were 0–4 h old. (**b**) The mean number of *Kiss1* neurons in the ARC, as well as the relative level of *Kiss1* mRNA/cell, for PND 1 male and female mice that were 0–4 h old. Newborn males had significantly fewer ARC *Kiss1* cells and lower mRNA per cell than newborn females ($P<0.05$). *3V* third ventricle; *GPC* grains per cell. Modified from Poling MC, Kauffman AS 2012 Sexually dimorphic testosterone secretion in prenatal and neonatal mice is independent of kisspeptin-kiss1r and GnRH signaling. Endocrinology 153:782–793. With permission from The Endocrine Society

age of PND 4 [64]. Poling and Kauffman provided greater temporal resolution of expression in PND 1 mice: they found notable *Kiss1* expression in both sexes at both 0–4 h and 16–20 h after birth (Fig. 11.7), without a significant change in *Kiss1* levels between these time-points [98]. Parenthetically, NKB, which is co-expressed in ARC *Kiss1* neurons in adulthood, was also readily expressed in the ARC of both sexes on PND 1 [98]. One rat study noted an increase in *Kiss1* expression in the ARC between PND 1 and 3 in both sexes [47], with an additional increase at PND 8 in females only. Another study noted similar but non-significant trends in both males and females [64].

The role of *Kiss1* in the ARC during early neonatal life is unknown. Because kisspeptin signaling regulates the pubertal and adult reproductive axis, it is possible that kisspeptin plays a similar role earlier in neonatal development. However, since the reproductive axis is typically quiescent at this time, it is not clear what functional significance neonatal kisspeptin signaling would have. The kisspeptin receptor (Kiss1r), also known as Gpr54, is already present in some GnRH neurons at birth and begins appearing in other areas of the brain by PND 6 [100], suggesting the possibility of a functional kisspeptin neuronal network at early postnatal ages. It is possible that neonatal kisspeptin regulates gonadal T secretion, causing the elevation in T that is seen in newborn males but not females [98, 101–105]. Recent evidence has shown that, unlike their WT littermates, newborn *Kiss1r* KO and *hpg* mice both have undetectable FSH and LH, demonstrating that kisspeptin and GnRH signaling are each essential for neonatal gonadotropin secretion [98]. However, the same study found that newborn male *Kiss1r* KO and *hpg* mice both had elevated serum T levels at birth that were similar to their WT male littermates, indicating that neither GnRH nor kisspeptin signaling is required for neonatal T secretion. These results suggest that, while kisspeptin regulates gonadotropin secretion in neonatal rodents, the neonatal T surge in males is independent of regulation by gonadotropins and kisspeptin.

Prepubertal and Pubertal Changes in Kiss1 Expression in the ARC

Changes in peripubertal *Kiss1* expression have been examined in the ARC of rodents, but not as thoroughly as in the AVPV/PeN, and the results are not entirely consistent between studies. Early studies in the whole hypothalamus of the rat showed that *Kiss1* levels increased between PND 1–20 and PND 30 in females, and between PND 1–30 and PND 45 in males, followed by decreased expression in adulthood. These studies demonstrated that increases in *Kiss1* levels correlate with pubertal timing in males and females [12]. However, since the whole hypothalamus was analyzed, these pubertal changes in *Kiss1* expression could be due to changes in AVPV/PeN and/or ARC *Kiss1* levels. To get better spatial resolution of the developmental changes in kisspeptin/*Kiss1*, more recent studies in rodents have used ISH or IHC. Studies using IHC detected kisspeptin-ir in the ARC of juvenile, prepubertal, and peripubertal mice of both sexes, although this kisspeptin-ir was not quantified [28, 94]. An early study using ISH found similar numbers of ARC *Kiss1* cells in PND 18 vs. adult male mice [106], suggesting no difference in ARC *Kiss1* levels between juvenile and adult stages; however, peripubertal ages in between PND 18 and adulthood were not analyzed. This finding was mirrored by another recent ISH study in female rats that reported no difference in ARC *Kiss1* levels between juvenile and late-pubertal animals [107]. Likewise, in male and female mice, *Kiss1* expression in the ARC, determined with qPCR, remained virtually unchanged during several stages of postnatal and pubertal development (PND 11–61), and no significant sex differences were noted at any of the ages studied [94].

In contrast to the above studies showing no major peripubertal changes in ARC *Kiss1* levels, additional experiments demonstrated increases in ARC *Kiss1* expression around the time of early puberty in both female [91] and male rats [108], suggesting that increased ARC *Kiss1* expression may be involved in puberty onset. More recently, initial elevations in ARC *Kiss1* expression in neonatal/juvenile female rats were reportedly followed by a decrease in *Kiss1* levels by 3 weeks of age [47, 64], perhaps concurrent with increased sex steroid production (and negative feedback) at this time. One study also detected another increase in ARC *Kiss1* of females between weeks 5 and 8 [64]. Several studies reported that ARC *Kiss1* levels in male rats remained essentially unchanged during the first several weeks of life (up to around 4 weeks of age) [47, 64], followed in one case by a small temporary increase in ARC *Kiss1* levels around week 5 [64]. Overall, the current data on rodent ARC *Kiss1* expression during development, especially at prepubertal and pubertal stages, is incomplete and fairly inconsistent. Thus, full characterization of *Kiss1* expression in the ARC during development and puberty is still needed in rodents, as well as other species (see below).

Very few studies on developmental changes in ARC kisspeptin neurons have been conducted in non-rodent species. One study using agonadal male and intact female monkeys described an increase in *Kiss1* expression in the ARC during puberty [25]. In peripubertal female lambs with controlled sex steroid levels, a non-significant increase in ARC *Kiss1* cell number was seen over time when comparing weeks 25, 30, and 35. When analyzed specifically in lambs that showed increased LH pulse frequency, the number of *Kiss1* cells in the middle ARC was found to have a significant linear increase between weeks 25, 30, and 35, possibly implicating the middle ARC in puberty onset [65]. However, another recent report in sheep found that kisspeptin-ir cell number increased in the caudal ARC of females between the prepubertal and postpubertal periods [109]. Thus, the specific location(s) within the ovine ARC where *Kiss1* might increase during puberty occurs still needs to be resolved. Collectively, these initial studies in monkeys and sheep suggest that increased levels of *Kiss1* in the ARC of pubertal animals correlate with increased activation of the reproductive axis, but these studies surveyed a low resolution of ages over development, necessitating additional data on this subject.

Sex Differences in ARC Kisspeptin Neurons During Development

Most data agree that, unlike the sexually dimorphic AVPV/PeN [31, 47, 70], the number of ARC *Kiss1* or kisspeptin cells and the quantity of *Kiss1* mRNA per cell are not majorly different between sexes in adult rodents, especially when circulating sex steroid levels are equalized between the sexes [28, 36, 40, 42, 44, 66, 94]. In the few cases where an ARC *Kiss1* sex difference was observed, it was typically in gonadally intact adult animals that have unequal sex steroid levels between the sexes. For example, Adachi et al. [42] reported that intact adult male and female rats had equivalent ARC *Kiss1* expression at all stages of the female estrous cycle except for at diestrous 2. Moreover, adult males and females exhibit similar increases in *Kiss1* levels in the ARC after gonadectomy (GDX) and similar inhibition of ARC *Kiss1* expression when treated equivalently with sex steroids [36, 40, 42, 44]. It should be noted that sex differences exist in the ARC in several other parameters, such as the morphology of astroglia, the number of spine and somatic synapses [110–112], the expression of growth hormone-releasing hormone, and the axonal projections of NKB/Dyn neurons [113–115]. Additionally, the kisspeptin fibers visualized in the ARC by IHC are often denser in adult females, but it is likely that some of these fibers are derived from the sexually dimorphic AVPV/PeN kisspeptin population [28, 29, 62, 116]. The ARC population of *Kiss1* neurons in rodents is hypothesized to provide tonic stimulatory input to GnRH neurons and to relay negative feedback effects of sex steroids to the GnRH axis [36, 47, 117–119]. These processes occur in both sexes, which is consistent with the lack of a major sexual dimorphism in these neurons in adult rodents.

In contrast to adulthood, recent analysis of *Kiss1* expression in the ARC of rodents during development has revealed a different story regarding sexual dimorphism. Work in the *Bax* KO mouse suggests that male mice may initially have more ARC *Kiss1* neurons (or neurons that have the potential to express *Kiss1* later in development) than females prior to birth, which is later offset by a higher rate of perinatal apoptosis [40]. However, a recent report in the rat did not find sex differences in the number of kisspeptin-ir cells during prenatal development, but did not look at sex differences in *Kiss1* expression levels [97]. A recent study in mice found that newborn PND 1 pups exhibit sexual dimorphism in ARC *Kiss1* expression, with newborn females having significantly more *Kiss1* and *NKB* cells in the ARC, as well as higher cellular expression of these mRNAs, than newborn males [98]. Similarly, male rats also have lower ARC *Kiss1* levels than females at PND 3 [47] and PND 4 [64]. Studies using both ISH and IHC found that the number of *Kiss1* neurons in the ARC of rats is still sexually dimorphic between PND 5 and PND 11 [47, 64, 99, 120], with juvenile females having approximately 2–4 times as many ARC *Kiss1* (or kisspeptin) neurons than juvenile males of the same age. This ARC *Kiss1* sex difference begins to diminish with the approach of puberty; while males maintain a slight non-significant trend in increasing ARC *Kiss1* cell numbers from juvenile life through puberty, *Kiss1* expression in the ARC of females decreases around 3 weeks of age to reach levels similar to males [44, 47, 63].

The neonatal and juvenile sex differences in *Kiss1* expression in the rodent ARC may be a temporary sex difference due to sexual dimorphism in the circulating sex steroid milieu: higher levels of circulating sex steroids in young males may provide more negative feedback inhibition than in young females, resulting in lower ARC *Kiss1* expression in the males. This could certainly be the case in newborn animals, in which males secrete elevated T (to drive sexual differentiation) whereas females do not [98]. However, it is not entirely clear what normal sex steroid levels are in male and female rodents, especially mice, during each phase of neonatal and juvenile development, and most studies that have examined *Kiss1* expression early in development have not measured serum sex steroid levels. Moreover, it is also possible that non-steroidal mechanisms also influence the ARC *Kiss1* sexual dimorphism, as PND 1 male mice that are 16–20 h old (when circulating T levels are no longer different between the sexes) still display sex differences in ARC *Kiss1* levels [98]. Other findings support the role of sex steroid-independent factors in regulating *Kiss1*/*NKB* expression in the ARC of juvenile mice in a sexually dimorphic manner

[44] (discussed more later). Additionally, the functional significance, if any, of the sexual dimorphism in the ARC *Kiss1* neuronal population during neonatal and juvenile development is completely unknown, especially since the reproductive axis is essentially quiescent at this time.

Some non-rodent species, such as the sheep, display clear adulthood sex differences in kisspeptin cells in the ARC. Ewes display a greater number of *Kiss1* cells in the ARC than adult males [67]; this may be due to differences in the functional roles of *Kiss1* cells in the ARC/INF of rodents vs. sheep. However, as of yet, no direct comparison has been made between adult ewes and rams with equivalent sex steroid levels. Additionally, while prenatal androgen treatment reduced NKB and DYN expression in the ARC in adult sheep, this prenatal treatment did not alter ARC *Kiss1* expression, which remained sexually dimorphic [67]. Thus, there may be different critical periods for the sexual differentiation of these ARC genes, but it is also possible that the ovine ARC *Kiss1* system is not affected by prenatal sex steroids. Dissimilarities between species may be due to the difference in the role of the ARC; while ARC kisspeptin cells in rodents may mediate negative feedback regulation of GnRH secretion, which is not sexually dimorphic [6, 22, 46, 51, 118], the same kisspeptin population in sheep appears to be key for *both* negative feedback *and* positive feedback (i.e., the preovulatory GnRH surge) [51, 121–124]. Similar sex differences in kisspeptin expression have been recently reported in humans in the INF, as well as the POA, with females showing greater number of cells than males in each region [125]. However, note that in these human studies, regulatory factors that are known to alter kisspeptin levels, such as circulating sex steroids, metabolic status, stress hormones, and circadian status, were not controlled for and could differ dramatically between subjects, thereby confounding the results. Whether non-human primates display adult sex differences in ARC/INF kisspeptin levels is currently unexamined [126].

Regulation of ARC Kiss1 Neurons by Gonadal and Non-gonadal Factors During Development

As noted earlier, under normal developmental conditions, the rodent ARC displays a dense network of kisspeptin-ir fibers that likely originate from both the ARC and the AVPV/PeN. Supporting the hypothesis that some of these fibers are projections from the AVPV, data has shown that when gonadal hormones are removed during development, whether by GDX or genetically, as in the case of the *hpg* and KERKO mice, the number of kisspeptin fibers detected in the ARC are reduced in adulthood [10, 62, 94]. These data also suggest that gonadal steroids during development are necessary for either the physical development of kisspeptin fibers (organizational effects) or for the transient expression of kisspeptin protein within the cells and fibers (activational effects). Considering that the number of kisspeptin fibers in the ARC was restored in adult *hpg* mice treated with E_2, it seems likely that these kisspeptin fibers develop normally but only contain detectable levels of kisspeptin under the correct adult sex steroidal milieu [94]. Thus, the reduction of these fibers in non-E_2 treated *hpg* (and KERKO) mice is likely due to an activational defect (i.e., absence of circulating sex steroids). However, caution must be used in interpreting fiber data since the origin of the fibers cannot always be determined.

The *hpg* and KERKO mouse models have recently been utilized to shed more light on the role of gonadal sex steroids in the development of ARC *Kiss1* neurons. First, kisspeptin cell bodies are more visible in the ARC of adult *hpg* mice than WT mice [94]. This difference in detectable kisspeptin cells is likely due to the lack of obstruction by dense kisspeptin fibers in *hpg* mice, as well as a possible increase in ARC kisspeptin synthesis in *hpg* mice lacking steroid negative feedback (Fig. 11.8). Since differences in ARC kisspeptin-ir expression and cell number were not easily measurable by IHC due to the obstructive overlaying fiber network, *Kiss1* mRNA levels were examined instead using ISH. Both the number of *Kiss1* neurons and the density of silver grains were increased in the ARC of adult *hpg* mice compared to WT littermates, suggesting an increased level of kisspeptin synthesis in *hpg* mice, perhaps due to less steroid negative feedback. Interestingly, at PND 11, only a small difference in *Kiss1* levels was present between WT and *hpg* littermates (as measured by RT-PCR), but by PND 31 this difference was much larger, with dramatically higher ARC *Kiss1* expression in *hpg* mice. This indicates that, at PND 11, there may be non-gonadal hormone factors inhibiting *Kiss1* in the ARC, but by PND 31 this gonadal-independent mechanism is reduced (or absent). These *hpg* data therefore suggest that changes in gonad-independent factors may induce ARC *Kiss1* expression around puberty.

The robust ARC *Kiss1* expression seen in cell bodies of adult *hpg* and ArKO mice may be similar to the hypertrophied *Kiss1* neurons seen in postmenopausal women [43, 127]; in all cases, the elevated *Kiss1* levels may reflect a lack of negative feedback of sex steroids. Finally, female KERKO mice show decreased kisspeptin-ir fibers in the ARC at PND 16, 26, and 35, similar to *hpg* mice [10]. Surprisingly, unlike *hpg* mice, adult KERKO mice did not display the high-expressing kisspeptin-ir cell bodies in the ARC, even though the same kisspeptin antibody was used in both studies. However, when mRNA was analyzed by qPCR in the mediobasal hypothalamus (which includes the ARC), *Kiss1* levels in KERKO mice were in fact increased relative to WT levels, similar to data from *hpg* mice. This unexpected difference between protein and mRNA levels in the KERKO mice has not yet been explained.

Recent work has shown that during the prepubertal period of mice, the *Kiss1* response to GDX is sexually dimorphic. While prepubertal females display elevated *Kiss1* expression in the ARC, along with increased LH levels, several days after GDX, prepubertal male mice do not (Fig. 11.9). Instead, if prepubertal male mice are GDX at PND 14, no increase is seen in ARC *Kiss1* cell number or serum LH levels when measured 2–4 days later (PND 16–18) [44]. Because this prepubertal sex difference is only observable after the removal of gonadal hormones (there is no sex differences in ARC *Kiss1* cell number at this age in gonadally intact mice), these results suggest a sexually dimorphic difference in the *regulation* of ARC *Kiss1* cells at this point in development: *Kiss1*/LH levels in PND 16–18 females appear to

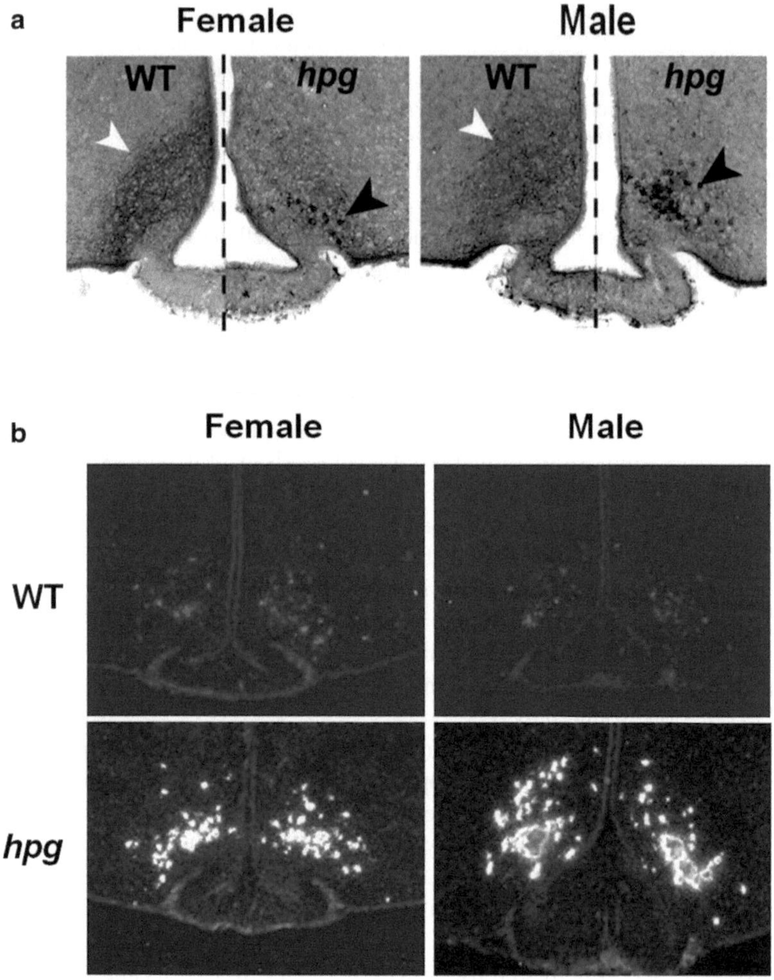

Fig. 11.8 Kisspeptin and *Kiss1* expression are increased in the ARC of female and male *hpg* mice. (**a**) Kisspeptin staining pattern in the ARC of both female and male WT mice (*left* of *dashed line*) is dominated by densely stained fibers that obscure kisspeptin-positive cell bodies (*white arrowheads*). In *hpg* mice (*right* of *dashed line*), kisspeptin immunoreactivity shows reduced fiber staining and increased clusters of large, darkly stained cell bodies (*black arrowheads*). (**b**) Representative dark-field images of silver grain *Kiss1* mRNA signal as measured by ISH in the ARC of pubertal-aged WT and *hpg* females (30 days) and males (45 days). Scale bar = 200 mm. Figure modified from Gill JC, Wang O, Kakar S, Martinelli E, Carroll RS, Kaiser UB 2010 Reproductive hormone-dependent and -independent contributions to developmental changes in kisspeptin in GnRH-deficient hypogonadal mice. PLoS One 5:e11911

be controlled primarily by gonadal sex steroid feedback, whereas PND 16–18 males have an additional mechanism of regulation of their ARC *Kiss1* neurons by unknown non-gonadal factors. Whether this divergence between the sexes is due to a unique, male-specific mechanism not present in similarly aged females or to sex differences

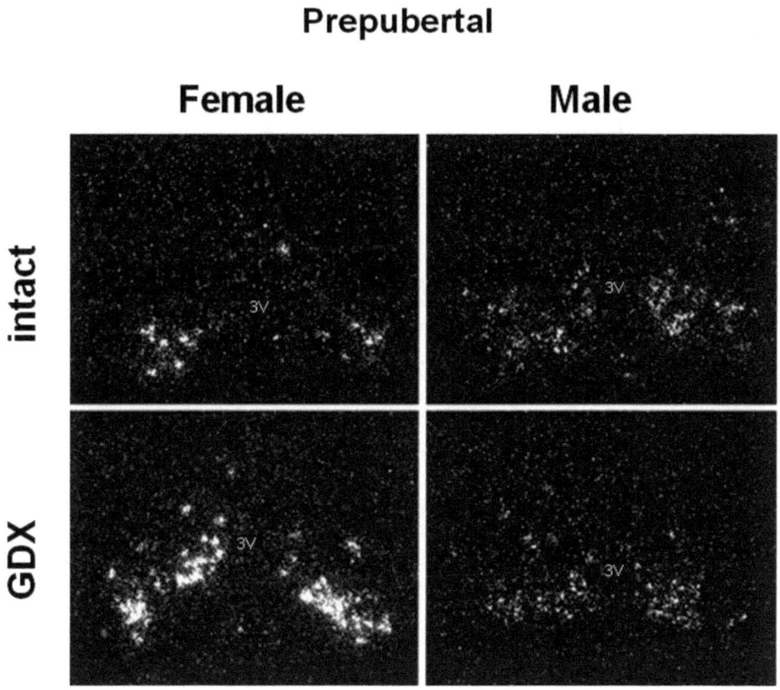

Fig. 11.9 *Kiss1* expression in the ARC of prepubertal mice under different sex steroid conditions. Gonadectomized prepubertal female mice have higher ARC *Kiss1* expression than similarly aged gonadectomized males. The absence of elevated *Kiss1* expression in gonadectomized prepubertal males suggests that some non-gonadal factor(s) acts to suppress the ARC *Kiss1* system in males at this developmental stage. *3V* third ventricle. Figure modified from Kauffman AS 2010 Gonadal and non-gonadal regulation of sex differences in hypothalamic Kiss1 neurons. J Neuroendocrinol 22:682–691 with permission from John Wiley and Sons

in the rate of pubertal maturation (females initiate puberty earlier than males) remains to be seen. This gonadal hormone-independent regulation of *Kiss1* neurons in developing males appears to disappear sometime between PND 18 and 45, as male mice GDX at PND 14 exhibit the expected elevation of ARC *Kiss1* and serum LH levels when analyzed at PND 45. The exact function of this sex difference still needs to be elucidated, though it is possible it relates to the sexually dimorphic timing of pubertal progression.

Conclusions and Perspectives

Kiss1 neurons are critical for puberty and reproduction, and this chapter is focused on the development of these neurons in the hypothalamic AVPV/PeN and ARC (Fig. 11.10). In terms of the AVPV/PeN *Kiss1* population, which emerges in the

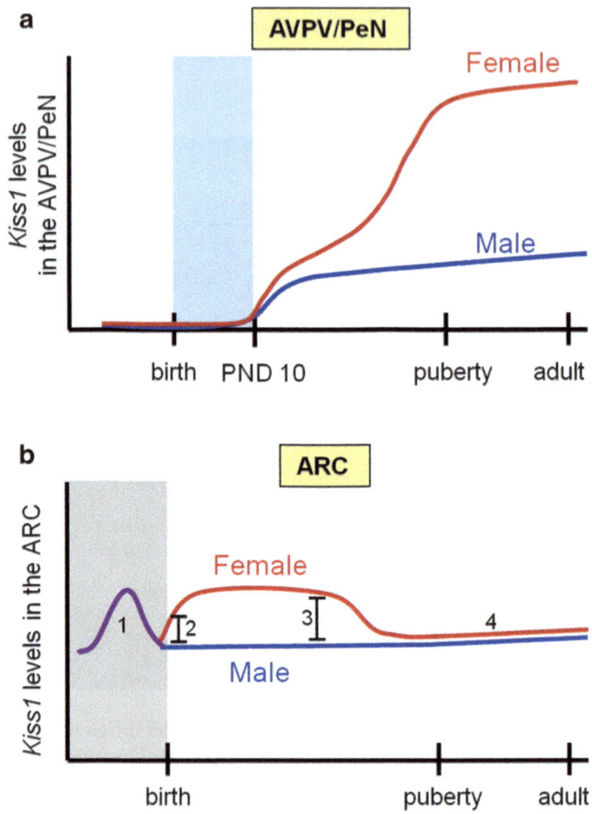

Fig. 11.10 Schematic diagram developmental changes in *Kiss1* expression in the AVPV/PeN (**a**) and ARC (**b**) of rodents. (**a**) *Kiss1* expression in the AVPV/PeN has not been detected on the day of birth or during other days of the "critical period" of perinatal development (*light blue shading*). *Kiss1* mRNA expression is first detected in the AVPV/PeN on postnatal day 10 (PND 10) in both males and females. The sex difference emerges by PND 12. Male *Kiss1* gene expression in the AVPV/PeN only slightly increases after that time to adulthood, whereas females have a steady increase in *Kiss1* expression that reaches adulthood levels around the time of puberty. (**b**) *1 Kiss1* is expressed in the ARC before birth. Analysis of combined male and female embryonic rat brains demonstrates increasing levels of *Kiss1* in the ARC throughout prenatal development (*gray shading*), with a slight drop before birth. Comparative analysis of embryonic *Kiss1* mRNA levels between sexes has not yet been determined. *2* During the neonatal period, *Kiss1* expression in the rodent ARC is sexually dimorphic, with females expressing more *Kiss1* than males. This may correlate with differences in the circulating sex steroid milieu, but requires further investigation. *3* Peripubertal *Kiss1* levels in the ARC decrease significantly in females to levels similar to males. *4* There may be a slight increase in ARC *Kiss1* expression during puberty leading to adulthood levels, though such an increase is controversial at present

second postnatal week of life and is sexually dimorphic, it remains to be answered how this population becomes sexually differentiated. Recent data suggest that epigenetic processes, precipitated by postnatal sex steroid signaling, may induce the AVPV/PeN *Kiss1* sex difference [5], especially because a role for the major neuronal pathway of apoptosis was ruled out [40]. This model predicts that the differences in sex steroid exposure between male and female rodent brains early in neonatal life eventually result in permanent changes in the transcriptional activity of the *Kiss1* gene in the AVPV/PeN, resulting in greater *Kiss1* silencing in males and greater transcriptional activation in females. Supporting this, DNA methylation of the putative promoter of the *Kiss1* gene in the AVPV/PeN is sexually differentiated [83], though we still do not know when or how in development this *Kiss1* methylation difference first occurs.

Although E_2 signaling in males masculinizes the AVPV/PeN *Kiss1* system during the neonatal "critical period," it now appears that exposure to E_2 in developing females also shapes and promotes the complete feminization of AVPV/PeN *Kiss1* expression [10, 62, 94]. Indeed, several studies in mice have suggested that E_2 is required at *some time* during development in order for AVPV/PeN kisspeptin expression to fully develop a female phenotype [10, 62, 94]. However, the mechanism by which this occurs currently remains a mystery and may also involve epigenetics, neurogenesis, or a combination of these and other processes.

The role of the ARC/INF population of kisspeptin neurons in mediating sex steroid negative feedback in rodents appears to be different than some other mammals, such as sheep, in which ARC kisspeptin cells also play a role in positive feedback. Recent studies suggest that *Kiss1* is expressed in the ARC of developing rodents long before the AVPV/PeN, including prenatally (Fig. 11.10). The role of ARC *Kiss1* expression during prenatal/neonatal development is currently unknown and requires further investigation. In many species, the regulation of ARC kisspeptin fibers and their projections in development remains under-studied, in part because, until recently, there has not been a good way to distinguish the anatomical origin of kisspeptin-ir fibers. The recent findings that the ARC *Kiss1* population co-expresses *NKB* and *DYN* (in all species examined to date), whereas AVPV/PeN *Kiss1* neurons do not, should facilitate future studies characterizing the development of kisspeptin neuron projections, as should the use of several newly created transgenic *Kiss1* mouse models. Additionally, some studies suggest that there may be an increase in *Kiss1* expression in the ARC around puberty (Fig. 11.10), but the data is incomplete and inconsistent between studies, and more experiments looking at pubertal *Kiss1*/kisspeptin expression with higher temporal resolution are needed, including in non-rodent species.

While sex differences in ARC *Kiss1* cell number have not been observed by most studies examining adult rodents, such sex differences do exist in the ARC of neonatal and juvenile rodents (Fig. 11.10), as do sex differences in the gonad hormone-independent regulation of ARC *Kiss1* neurons in prepubertal life, at least in mice. The functional significance of these early ARC *Kiss1* sex differences is unknown, as are the specific factors controlling *Kiss1* expression in the ARC at early developmental ages. For example, sex steroids may transiently inhibit ARC *Kiss1* neurons (activational effects) during prenatal and postnatal development, as occurs in adulthood, but this has not been directly tested. Indeed, there is a need for examination of normal sex steroid levels in developing rodents, especially mice, during neonatal, juvenile, and prepubertal stages. Interestingly, it appears that in the complete absence of gonadal sex steroids during development, ARC *Kiss1* neurons still develop normally and have intact hormone responsiveness later in adulthood, suggesting that this *Kiss1* population may only be affected by activational effects of sex steroids, rather than by permanent organizational effects that occur in the AVPV/PeN.

Overall, the development of kisspeptin circuits appears to be complex, region-specific, sex-specific, and influenced by multiple factors. The upstream mechanism(s) driving developmental and pubertal changes in kisspeptin neurons in each hypothalamic region, and how gonadal and non-gonadal factors work together to control development of the *Kiss1* system, remain important questions whose answers will provide a better understanding of the regulation of the reproductive neuroendocrine axis at multiple life stages.

References

1. Cesario SK, Hughes LA (2007) Precocious puberty: a comprehensive review of literature. J Obstet Gynecol Neonatal Nurs 36:263–274
2. Fechner A, Fong S, McGovern P (2008) A review of Kallmann syndrome: genetics, pathophysiology, and clinical management. Obstet Gynecol Surv 63:189–194
3. Bianco SD, Kaiser UB (2009) The genetic and molecular basis of idiopathic hypogonadotropic hypogonadism. Nat Rev Endocrinol 5:569–576
4. Fechner PY (2002) Gender differences in puberty. J Adolesc Health 30:44–48
5. Semaan SJ, Kauffman AS (2010) Sexual differentiation and development of forebrain reproductive circuits. Curr Opin Neurobiol 20:424–431
6. Kauffman AS (2010) Coming of age in the kisspeptin era: sex differences, development, and puberty. Mol Cell Endocrinol 324:51–63
7. Clarkson J, d'Anglemont de Tassigny X, Moreno AS, Colledge WH, Herbison AE (2008) Kisspeptin-GPR54 signaling is essential for preovulatory gonadotropin-releasing hormone neuron activation and the luteinizing hormone surge. J Neurosci 28:8691–8697
8. Pineda R, Garcia-Galiano D, Roseweir A, Romero M, Sanchez-Garrido MA, Ruiz-Pino F, Morgan K, Pinilla L, Millar RP, Tena-Sempere M (2009) Critical roles of kisspeptins in female puberty and preovulatory gonadotropin surges as revealed by a novel antagonist. Endocrinology 151(2):722–730
9. Cravo RM, Margatho LO, Osborne-Lawrence S, Donato J Jr, Atkin S, Bookout AL, Rovinsky S, Frazao R, Lee CE, Gautron L, Zigman JM, Elias CF (2011) Characterization of Kiss1 neurons using transgenic mouse models. Neuroscience 173:37–56
10. Mayer C, Acosta-Martinez M, Dubois SL, Wolfe A, Radovick S, Boehm U, Levine JE (2010) Timing and completion of puberty in female mice depend on estrogen receptor alpha-signaling in kisspeptin neurons. Proc Natl Acad Sci USA 107:22693–22698
11. Gottsch ML, Popa SM, Lawhorn JK, Qiu J, Tonsfeldt KJ, Bosch MA, Kelly MJ, Ronnekleiv OK, Sanz E, McKnight GS, Clifton DK, Palmiter RD, Steiner RA (2011) Molecular properties of Kiss1 neurons in the arcuate nucleus of the mouse. Endocrinology 152:4298–4309
12. Navarro VM, Castellano JM, Fernández-Fernández R, Barreiro ML, Roa J, Sanchez-Criado JE, Aguilar E, Dieguez C, Pinilla L, Tena-Sempere M (2004) Developmental and hormonally regulated messenger ribonucleic acid expression of KiSS-1 and its putative receptor, GPR54, in rat hypothalamus and potent luteinizing hormone-releasing activity of KiSS-1 peptide. Endocrinology 145:4565–4574
13. Brailoiu GC, Dun SL, Ohsawa M, Yin D, Yang J, Chang JK, Brailoiu E, Dun NJ (2005) KiSS-1 expression and metastin-like immunoreactivity in the rat brain. J Comp Neurol 481:314–329

14. Clarkson J, D'Anglemont de Tassigny X, Colledge WH, Caraty A, Herbison AE (2009) Distribution of kisspeptin neurones in the adult female mouse brain. J Neuroendocrinol 21:673–682
15. Franceschini I, Lomet D, Cateau M, Delsol G, Tillet Y, Caraty A (2006) Kisspeptin immunoreactive cells of the ovine preoptic area and arcuate nucleus co-express estrogen receptor alpha. Neurosci Lett 401:225–230
16. Goodman RL, Lehman MN, Smith JT, Coolen LM, de Oliveira CV, Jafarzadehshirazi MR, Pereira A, Iqbal J, Caraty A, Ciofi P, Clarke IJ (2007) Kisspeptin neurons in the arcuate nucleus of the ewe express both dynorphin A and neurokinin B. Endocrinology 148:5752–5760
17. Greives TJ, Mason AO, Scotti MA, Levine J, Ketterson ED, Kriegsfeld LJ, Demas GE (2007) Environmental control of kisspeptin: implications for seasonal reproduction. Endocrinology 148:1158–1166
18. Ohkura S, Takase K, Matsuyama S, Mogi K, Ichimaru T, Wakabayashi Y, Uenoyama Y, Mori Y, Steiner RA, Tsukamura H, Maeda KI, Okamura H (2009) Gonadotrophin-releasing hormone pulse generator activity in the hypothalamus of the goat. J Neuroendocrinol 21:813–821
19. Ramaswamy S, Guerriero KA, Gibbs RB, Plant TM (2008) Structural interactions between kisspeptin and GnRH neurons in the mediobasal hypothalamus of the male rhesus monkey (Macaca mulatta) as revealed by double immunofluorescence and confocal microscopy. Endocrinology 149:4387–4395
20. Kim J, Semaan SJ, Clifton DK, Steiner RA, Dhamija S, Kauffman AS (2011) Regulation of Kiss1 expression by sex steroids in the amygdala of the rat and mouse. Endocrinology 152:2020–2030
21. Gottsch ML, Cunningham MJ, Smith JT, Popa SM, Acohido BV, Crowley WF, Seminara S, Clifton DK, Steiner RA (2004) A role for kisspeptins in the regulation of gonadotropin secretion in the mouse. Endocrinology 145:4073–4077
22. Smith JT, Cunningham MJ, Rissman EF, Clifton DK, Steiner RA (2005) Regulation of Kiss1 gene expression in the brain of the female mouse. Endocrinology 146:3686–3692
23. Kauffman AS, Park JH, McPhie-Lalmansingh AA, Gottsch ML, Bodo C, Hohmann JG, Pavlova MN, Rohde AD, Clifton DK, Steiner RA, Rissman EF (2007) The kisspeptin receptor GPR54 is required for sexual differentiation of the brain and behavior. J Neurosci 27:8826–8835
24. Smith JT, Clay CM, Caraty A, Clarke IJ (2007) KiSS-1 messenger ribonucleic acid expression in the hypothalamus of the ewe is regulated by sex steroids and season. Endocrinology 148:1150–1157
25. Shahab M, Mastronardi C, Seminara SB, Crowley WF, Ojeda SR, Plant TM (2005) Increased hypothalamic GPR54 signaling: a potential mechanism for initiation of puberty in primates. Proc Natl Acad Sci USA 102:2129–2134
26. Kim W, Jessen HM, Auger AP, Terasawa E (2009) Postmenopausal increase in KiSS-1, GPR54, and luteinizing hormone releasing hormone (LHRH-1) mRNA in the basal hypothalamus of female rhesus monkeys. Peptides 30:103–110
27. Smith JT, Shahab M, Pereira A, Pau KY, Clarke IJ (2010) Hypothalamic expression of KISS1 and gonadotropin inhibitory hormone genes during the menstrual cycle of a non-human primate. Biol Reprod 83:568–577
28. Clarkson J, Herbison AE (2006) Postnatal development of kisspeptin neurons in mouse hypothalamus; sexual dimorphism and projections to gonadotropin-releasing hormone neurons. Endocrinology 147:5817–5825
29. Yeo SH, Herbison AE (2011) Projections of arcuate nucleus and rostral periventricular kisspeptin neurons in the adult female mouse brain. Endocrinology 152:2387–2399
30. Smith JT, Li Q, Yap KS, Shahab M, Roseweir AK, Millar RP, Clarke IJ (2011) Kisspeptin is essential for the full preovulatory LH surge and stimulates GnRH release from the isolated ovine median eminence. Endocrinology 152:1001–1012
31. Kauffman AS (2009) Sexual differentiation and the Kiss1 system: hormonal and developmental considerations. Peptides 30:83–93

32. Simerly RB (2002) Wired for reproduction: organization and development of sexually dimorphic circuits in the mammalian forebrain. Annu Rev Neurosci 25:507–536
33. Herbison AE (2008) Estrogen positive feedback to gonadotropin-releasing hormone (GnRH) neurons in the rodent: the case for the rostral periventricular area of the third ventricle (RP3V). Brain Res Rev 57:277–287
34. Smith JT, Popa SM, Clifton DK, Hoffman GE, Steiner RA (2006) Kiss1 neurons in the forebrain as central processors for generating the preovulatory luteinizing hormone surge. J Neurosci 26:6687–6694
35. Robertson JL, Clifton DK, de la Iglesia HO, Steiner RA, Kauffman AS (2009) Circadian regulation of Kiss1 neurons: implications for timing the preovulatory gonadotropin-releasing hormone/luteinizing hormone surge. Endocrinology 150:3664–3671
36. Kauffman AS, Gottsch ML, Roa J, Byquist AC, Crown A, Clifton DK, Hoffman GE, Steiner RA, Tena-Sempere M (2007) Sexual differentiation of Kiss1 gene expression in the brain of the rat. Endocrinology 148:1774–1783
37. Simerly RB (1998) Organization and regulation of sexually dimorphic neuroendocrine pathways. Behav Brain Res 92:195–203
38. Simerly RB, Zee MC, Pendleton JW, Lubahn DB, Korach KS (1997) Estrogen receptor-dependent sexual differentiation of dopaminergic neurons in the preoptic region of the mouse. Proc Natl Acad Sci USA 94:14077–14082
39. Ottem EN, Godwin JG, Krishnan S, Petersen SL (2004) Dual-phenotype GABA/glutamate neurons in adult preoptic area: sexual dimorphism and function. J Neurosci 24:8097–8105
40. Semaan SJ, Murray EK, Poling MC, Dhamija S, Forger NG, Kauffman AS (2010) BAX-dependent and BAX-independent regulation of Kiss1 neuron development in mice. Endocrinology 151:5807–5817
41. Clarkson J, Herbison AE (2011) Dual phenotype kisspeptin-dopamine neurones of the rostral periventricular area of the third ventricle project to gonadotrophin-releasing hormone neurones. J Neuroendocrinol 23:293–301
42. Adachi S, Yamada S, Takatsu Y, Matsui H, Kinoshita M, Takase K, Sugiura H, Ohtaki T, Matsumoto H, Uenoyama Y, Tsukamura H, Inoue K, Maeda K (2007) Involvement of anteroventral periventricular metastin/kisspeptin neurons in estrogen positive feedback action on luteinizing hormone release in female rats. J Reprod Dev 53:367–378
43. Rometo AM, Krajewski SJ, Voytko ML, Rance NE (2007) Hypertrophy and increased kisspeptin gene expression in the hypothalamic infundibular nucleus of postmenopausal women and ovariectomized monkeys. J Clin Endocrinol Metab 92:2744–2750
44. Kauffman AS, Navarro VM, Kim J, Clifton DK, Steiner RA (2009) Sex differences in the regulation of Kiss1/NKB neurons in juvenile mice: implications for the timing of puberty. Am J Physiol Endocrinol Metab 297:E1212–E1221
45. Smith JT (2008) Kisspeptin signalling in the brain: steroid regulation in the rodent and ewe. Brain Res Rev 57:288–298
46. Smith JT, Dungan HM, Stoll EA, Gottsch ML, Braun RE, Eacker SM, Clifton DK, Steiner RA (2005) Differential regulation of KiSS-1 mRNA expression by sex steroids in the brain of the male mouse. Endocrinology 146:2976–2984
47. Cao J, Patisaul HB (2011) Sexually dimorphic expression of hypothalamic estrogen receptors alpha and beta and Kiss1 in neonatal male and female rats. J Comp Neurol 519:2954–2977
48. Shughrue PJ, Merchenthaler I (2001) Distribution of estrogen receptor beta immunoreactivity in the rat central nervous system. J Comp Neurol 436:64–81
49. Simerly RB, Chang C, Muramatsu M, Swanson LW (1990) Distribution of androgen and estrogen receptor mRNA-containing cells in the rat brain: an in situ hybridization study. J Comp Neurol 294:76–95
50. Oakley AE, Clifton DK, Steiner RA (2009) Kisspeptin signaling in the brain. Endocr Rev 30:713–743
51. Smith JT (2009) Sex steroid control of hypothalamic Kiss1 expression in sheep and rodents: comparative aspects. Peptides 30:94–102

52. Topaloglu AK, Reimann F, Guclu M, Yalin AS, Kotan LD, Porter KM, Serin A, Mungan NO, Cook JR, Ozbek MN, Imamoglu S, Akalin NS, Yuksel B, O'Rahilly S, Semple RK (2009) TAC3 and TACR3 mutations in familial hypogonadotropic hypogonadism reveal a key role for neurokinin B in the central control of reproduction. Nat Genet 41:354–358
53. Goodman RL, Coolen LM, Anderson GM, Hardy SL, Valent M, Connors JM, Fitzgerald ME, Lehman MN (2004) Evidence that dynorphin plays a major role in mediating progesterone negative feedback on gonadotropin-releasing hormone neurons in sheep. Endocrinology 145:2959–2967
54. Kinoshita M, Tsukamura H, Adachi S, Matsui H, Uenoyama Y, Iwata K, Yamada S, Inoue K, Ohtaki T, Matsumoto H, Maeda K (2005) Involvement of central metastin in the regulation of preovulatory luteinizing hormone surge and estrous cyclicity in female rats. Endocrinology 146:4431–4436
55. Navarro VM, Gottsch ML, Chavkin C, Okamura H, Clifton DK, Steiner RA (2009) Regulation of gonadotropin-releasing hormone secretion by kisspeptin/dynorphin/neurokinin B neurons in the arcuate nucleus of the mouse. J Neurosci 29:11859–11866
56. Sandoval-Guzman T, Rance NE (2004) Central injection of senktide, an NK3 receptor agonist, or neuropeptide Y inhibits LH secretion and induces different patterns of Fos expression in the rat hypothalamus. Brain Res 1026:307–312
57. Schulz R, Wilhelm A, Pirke KM, Gramsch C, Herz A (1981) Beta-endorphin and dynorphin control serum luteinizing hormone level in immature female rats. Nature 294:757–759
58. Burke MC, Letts PA, Krajewski SJ, Rance NE (2006) Coexpression of dynorphin and neurokinin B immunoreactivity in the rat hypothalamus: morphologic evidence of interrelated function within the arcuate nucleus. J Comp Neurol 498:712–726
59. Foradori CD, Amstalden M, Goodman RL, Lehman MN (2006) Colocalisation of dynorphin a and neurokinin B immunoreactivity in the arcuate nucleus and median eminence of the sheep. J Neuroendocrinol 18:534–541
60. Krajewski SJ, Burke MC, Anderson MJ, McMullen NT, Rance NE (2010) Forebrain projections of arcuate neurokinin B neurons demonstrated by anterograde tract-tracing and monosodium glutamate lesions in the rat. Neuroscience 166:680–697
61. Desroziers E, Droguerre M, Bentsen AH, Robert V, Mikkelsen JD, Caraty A, Tillet Y, Duittoz A, Franceschini I (2012) Embryonic development of kisspeptin neurones in rat. J Neuroendocrinol 24(10):1284–1295
62. Clarkson J, Boon WC, Simpson ER, Herbison AE (2009) Postnatal development of an estradiol-kisspeptin positive feedback mechanism implicated in puberty onset. Endocrinology 150:3214–3220
63. Kauffman AS, Semaan SJ (2011) Pubertal changes in gene expression in reproductive neural circuits of female mice. Society for Neuroscience: Abstract #713
64. Takumi K, Iijima N, Ozawa H (2011) Developmental changes in the expression of kisspeptin mRNA in rat hypothalamus. J Mol Neurosci 43:138–145
65. Redmond JS, Baez-Sandoval GM, Spell KM, Spencer TE, Lents CA, Williams GL, Amstalden M (2011) Developmental changes in hypothalamic Kiss1 expression during activation of the pulsatile release of luteinising hormone in maturing ewe lambs. J Neuroendocrinol 23:815–822
66. Homma T, Sakakibara M, Yamada S, Kinoshita M, Iwata K, Tomikawa J, Kanazawa T, Matsui H, Takatsu Y, Ohtaki T, Matsumoto H, Uenoyama Y, K-I M, Tsukamura H (2009) Significance of neonatal testicular sex steroids to defeminize anteroventral periventricular kisspeptin neurons and the GnRH/LH surge system in male rats. Biol Reprod 81:1216–1225
67. Cheng G, Coolen LM, Padmanabhan V, Goodman RL, Lehman MN (2010) The kisspeptin/neurokinin B/dynorphin (KNDy) cell population of the arcuate nucleus: sex differences and effects of prenatal testosterone in sheep. Endocrinology 151:301–311
68. Morris JA, Jordan CL, Breedlove SM (2004) Sexual differentiation of the vertebrate nervous system. Nat Neurosci 7:1034–1039

69. Bateman HL, Patisaul HB (2008) Disrupted female reproductive physiology following neonatal exposure to phytoestrogens or estrogen specific ligands is associated with decreased GnRH activation and kisspeptin fiber density in the hypothalamus. Neurotoxicology 29:988–997
70. Patisaul HB, Todd KL, Mickens JA, Adewale HB (2009) Impact of neonatal exposure to the ERalpha agonist PPT, bisphenol-A or phytoestrogens on hypothalamic kisspeptin fiber density in male and female rats. Neurotoxicology 30:350–357
71. Patisaul HB, Losa-Ward SM, Todd KL, McCaffrey KA, Mickens JA (2012) Influence of ERbeta selective agonism during the neonatal period on the sexual differentiation of the rat hypothalamic-pituitary-gonadal (HPG) axis. Biol Sex Differ 3:2
72. Forger NG (2009) Control of cell number in the sexually dimorphic brain and spinal cord. J Neuroendocrinol 21:393–399
73. McCarthy MM, Auger AP, Bale TL, De Vries GJ, Dunn GA, Forger NG, Murray EK, Nugent BM, Schwarz JM, Wilson ME (2009) The epigenetics of sex differences in the brain. J Neurosci 29:12815–12823
74. Murray EK, Hien A, de Vries GJ, Forger NG (2009) Epigenetic control of sexual differentiation of the bed nucleus of the stria terminalis. Endocrinology 150:4241–4247
75. Zhang JM, Konkle AT, Zup SL, McCarthy MM (2008) Impact of sex and hormones on new cells in the developing rat hippocampus: a novel source of sex dimorphism? Eur J Neurosci 27:791–800
76. Forger NG, Rosen GJ, Waters EM, Jacob D, Simerly RB, de Vries GJ (2004) Deletion of Bax eliminates sex differences in the mouse forebrain. Proc Natl Acad Sci USA 101:13666–13671
77. Gotsiridze T, Kang N, Jacob D, Forger NG (2007) Development of sex differences in the principal nucleus of the bed nucleus of the stria terminalis of mice: role of Bax-dependent cell death. Dev Neurobiol 67:355–362
78. Krishnan S, Intlekofer KA, Aggison LK, Petersen SL (2009) Central role of TRAF-interacting protein in a new model of brain sexual differentiation. Proc Natl Acad Sci USA 106:16692–16697
79. Holmes MM, McCutcheon J, Forger NG (2009) Sex differences in NeuN- and androgen receptor-positive cells in the bed nucleus of the stria terminalis are due to Bax-dependent cell death. Neuroscience 158:1251–1256
80. Adams JM, Cory S (2007) The Bcl-2 apoptotic switch in cancer development and therapy. Oncogene 26:1324–1337
81. Tsukahara S, Kakeyama M, Toyofuku Y (2006) Sex differences in the level of Bcl-2 family proteins and caspase-3 activation in the sexually dimorphic nuclei of the preoptic area in postnatal rats. J Neurobiol 66:1411–1419
82. Waters EM, Simerly RB (2009) Estrogen induces caspase-dependent cell death during hypothalamic development. J Neurosci 29:9714–9718
83. Semaan SJ, Dhamija S, Kim J, Ku EC, Kauffman AS (2012) Assessment of epigenetic contributions to sexually-dimorphic kiss1 expression in the anteroventral periventricular nucleus of mice. Endocrinology 153:1875–1886
84. Nugent BM, McCarthy MM (2010) Epigenetic underpinnings of developmental sex differences in the brain. Neuroendocrinology 93:150–158
85. Kurian JR, Olesen KM, Auger AP (2010) Sex differences in epigenetic regulation of the estrogen receptor-alpha promoter within the developing preoptic area. Endocrinology 151:2297–2305
86. Schwarz JM, Nugent BM, McCarthy MM (2010) Developmental and hormone-induced epigenetic changes to estrogen and progesterone receptor genes in brain are dynamic across the life span. Endocrinology 151:4871–4881
87. Matsuda KI, Mori H, Nugent BM, Pfaff DW, McCarthy MM, Kawata M (2011) Histone deacetylation during brain development is essential for permanent masculinization of sexual behavior. Endocrinology 152:2760–2767
88. Murray EK, Varnum MM, Fernandez JL, de Vries GJ, Forger NG (2011) Effects of neonatal treatment with valproic acid on vasopressin immunoreactivity and olfactory behaviour in mice. J Neuroendocrinol 23(10):906–914

89. Goldberg AD, Allis CD, Bernstein E (2007) Epigenetics: a landscape takes shape. Cell 128:635–638
90. Kolodkin MH, Auger AP (2011) Sex difference in the expression of DNA methyltransferase 3a in the rat amygdala during development. J Neuroendocrinol 23:577–583
91. Takase K, Uenoyama Y, Inoue N, Matsui H, Yamada S, Shimizu M, Homma T, Tomikawa J, Kanda S, Matsumoto H, Oka Y, Tsukamura H, Maeda KI (2009) Possible role of oestrogen in pubertal increase of Kiss1/kisspeptin expression in discrete hypothalamic areas of female rats. J Neuroendocrinol 21:527–537
92. Bakker J, Pierman S, Gonzalez-Martinez D (2010) Effects of aromatase mutation (ArKO) on the sexual differentiation of kisspeptin neuronal numbers and their activation by same versus opposite sex urinary pheromones. Horm Behav 57:390–395
93. Cattanach BM, Iddon CA, Charlton HM, Chiappa SA, Fink G (1977) Gonadotrophin-releasing hormone deficiency in a mutant mouse with hypogonadism. Nature 269:338–340
94. Gill JC, Wang O, Kakar S, Martinelli E, Carroll RS, Kaiser UB (2010) Reproductive hormone-dependent and -independent contributions to developmental changes in kisspeptin in GnRH-deficient hypogonadal mice. PLoS One 5:e11911
95. Dickerson SM, Cunningham SL, Patisaul HB, Woller MJ, Gore AC (2010) Endocrine disruption of brain sexual differentiation by developmental PCB exposure. Endocrinology 152:581–594
96. Tena-Sempere M (2010) Kisspeptin/GPR54 system as potential target for endocrine disruption of reproductive development and function. Int J Androl 33:360–368
97. Desroziers E, Mikkelsen JD, Duittoz A, Franceschini I (2012) Kisspeptin-immunoreactivity changes in a sex- and hypothalamic-region-specific manner across rat postnatal development. J Neuroendocrinol 24:1154–1165
98. Poling MC, Kauffman AS (2012) Sexually dimorphic testosterone secretion in prenatal and neonatal mice is independent of kisspeptin-kiss1r and GnRH signaling. Endocrinology 153:782–793
99. Cao J, Mickens JA, McCaffrey KA, Leyrer SM, Patisaul HB (2012) Neonatal bisphenol A exposure alters sexually dimorphic gene expression in the postnatal rat hypothalamus. Neurotoxicology 33:23–36
100. Herbison AE, de Tassigny X, Doran J, Colledge WH (2010) Distribution and postnatal development of Gpr54 gene expression in mouse brain and gonadotropin-releasing hormone neurons. Endocrinology 151:312–321
101. Corbier P (1985) Sexual differentiation of positive feedback: effect of hour of castration at birth on estradiol-induced luteinizing hormone secretion in immature male rats. Endocrinology 116:142–147
102. Corbier P, Edwards DA, Roffi J (1992) The neonatal testosterone surge: a comparative study. Arch Int Physiol Biochim Biophys 100:127–131
103. Corbier P, Kerdelhue B, Picon R, Roffi J (1978) Changes in testicular weight and serum gonadotropin and testosterone levels before, during, and after birth in the perinatal rat. Endocrinology 103:1985–1991
104. Motelica-Heino I, Castanier M, Corbier P, Edwards DA, Roffi J (1988) Testosterone levels in plasma and testes of neonatal mice. J Steroid Biochem 31:283–286
105. Baum MJ, Brand T, Ooms M, Vreeburg JT, Slob AK (1988) Immediate postnatal rise in whole body androgen content in male rats: correlation with increased testicular content and reduced body clearance of testosterone. Biol Reprod 38:980–986
106. Han SK, Gottsch ML, Lee KJ, Popa SM, Smith JT, Jakawich SK, Clifton DK, Steiner RA, Herbison AE (2005) Activation of gonadotropin-releasing hormone neurons by kisspeptin as a neuroendocrine switch for the onset of puberty. J Neurosci 25:11349–11356
107. Navarro VM, Ruiz-Pino F, Sanchez-Garrido MA, Garcia-Galiano D, Hobbs SJ, Manfredi-Lozano M, Leon S, Sangiao-Alvarellos S, Castellano JM, Clifton DK, Pinilla L, Steiner RA, Tena-Sempere M (2012) Role of neurokinin B in the control of female puberty and its modulation by metabolic status. J Neurosci 32:2388–2397
108. Bentsen AH, Ansel L, Simonneaux V, Tena-Sempere M, Juul A, Mikkelsen JD (2010) Maturation of kisspeptinergic neurons coincides with puberty onset in male rats. Peptides 31:275–283

109. Nestor CC, Briscoe AM, Davis SM, Valent M, Goodman RL, Hileman SM (2012) Evidence of a role for kisspeptin and neurokinin B in puberty of female sheep. Endocrinology 153:2756–2765
110. Mong JA, Glaser E, McCarthy MM (1999) Gonadal steroids promote glial differentiation and alter neuronal morphology in the developing hypothalamus in a regionally specific manner. J Neurosci 19:1464–1472
111. Mong JA, McCarthy MM (2002) Ontogeny of sexually dimorphic astrocytes in the neonatal rat arcuate. Brain Res Dev Brain Res 139:151–158
112. Mong JA, Roberts RC, Kelly JJ, McCarthy MM (2001) Gonadal steroids reduce the density of axospinous synapses in the developing rat arcuate nucleus: an electron microscopy analysis. J Comp Neurol 432:259–267
113. Ciofi P, Leroy D, Tramu G (2006) Sexual dimorphism in the organization of the rat hypothalamic infundibular area. Neuroscience 141:1731–1745
114. Mong JA, Kurzweil RL, Davis AM, Rocca MS, McCarthy MM (1996) Evidence for sexual differentiation of glia in rat brain. Horm Behav 30:553–562
115. Nurhidayat, Tsukamoto Y, Sasaki F (2001) Role of the gonads in sex differentiation of growth hormone-releasing hormone and somatostatin neurons in the mouse hypothalamus during postnatal development. Brain Res 890:154–161
116. Losa SM, Todd KL, Sullivan AW, Cao J, Mickens JA, Patisaul HB (2011) Neonatal exposure to genistein adversely impacts the ontogeny of hypothalamic kisspeptin signaling pathways and ovarian development in the peripubertal female rat. Reprod Toxicol 31:280–289
117. Castellano JM, Bentsen AH, Mikkelsen JD, Tena-Sempere M (2010) Kisspeptins: bridging energy homeostasis and reproduction. Brain Res 1364:129–138
118. Kauffman AS, Clifton DK, Steiner RA (2007) Emerging ideas about kisspeptin- GPR54 signaling in the neuroendocrine regulation of reproduction. Trends Neurosci 30:504–511
119. Dungan HM, Clifton DK, Steiner RA (2006) Minireview: kisspeptin neurons as central processors in the regulation of gonadotropin-releasing hormone secretion. Endocrinology 147:1154–1158
120. Iijima N, Takumi K, Sawai N, Ozawa H (2011) An immunohistochemical study on the expressional dynamics of kisspeptin neurons relevant to GnRH neurons using a newly developed anti-kisspeptin antibody. J Mol Neurosci 43:146–154
121. Caraty A, Fabre-Nys C, Delaleu B, Locatelli A, Bruneau G, Karsch FJ, Herbison A (1998) Evidence that the mediobasal hypothalamus is the primary site of action of estradiol in inducing the preovulatory gonadotropin releasing hormone surge in the ewe. Endocrinology 139:1752–1760
122. Hess DL, Wilkins RH, Moossy J, Chang JL, Plant TM, McCormack JT, Nakai Y, Knobil E (1977) Estrogen-induced gonadotropin surges in decerebrated female rhesus monkeys with medial basal hypothalamic peninsulae. Endocrinology 101:1264–1271
123. Krey LC, Butler WR, Knobil E (1975) Surgical disconnection of the medial basal hypothalamus and pituitary function in the rhesus monkey. I. Gonadotropin secretion. Endocrinology 96:1073–1087
124. Foster DL, Jackson LM, Padmanabhan V (2006) Programming of GnRH feedback controls timing puberty and adult reproductive activity. Mol Cell Endocrinol 254–255:109–119
125. Hrabovszky E, Ciofi P, Vida B, Horvath MC, Keller E, Caraty A, Bloom SR, Ghatei MA, Dhillo WS, Liposits Z, Kallo I (2010) The kisspeptin system of the human hypothalamus: sexual dimorphism and relationship with gonadotropin-releasing hormone and neurokinin B neurons. Eur J Neurosci 31:1984–1998
126. Kauffman AS (2010) Gonadal and nongonadal regulation of sex differences in hypothalamic Kiss1 neurones. J Neuroendocrinol 22:682–691
127. Rance NE (2009) Menopause and the human hypothalamus: evidence for the role of kisspeptin/neurokinin B neurons in the regulation of estrogen negative feedback. Peptides 30:111–122

Chapter 12
Kisspeptin and Puberty in Mammals

Ei Terasawa, Kathryn A. Guerriero, and Tony M. Plant

Abstract Since the discovery of the G-protein coupled receptor 54 (kisspeptin receptor) and its ligand, kisspeptin, our understanding of the neurobiological mechanisms that govern the pituitary-gonadal axis has evolved dramatically. In this chapter, we have reviewed progress regarding the relationship between kisspeptin and puberty, and have proposed a novel hypothesis for the role of kisspeptin signaling in the onset of this crucial developmental event. According to this hypothesis, although kisspeptin neurons in the arcuate nucleus (ARC) are critical for puberty, this is simply because these cells are an integral component of the hypothalamic GnRH pulse generating mechanism that drives intermittent release of the decapeptide, as an increase in GnRH is obligatory for the onset of puberty. In our model, ARC kisspeptin neurons play no "regulatory" role in controlling the timing of puberty. Rather, as a component of the neural network responsible for GnRH pulse generation, they subserve upstream regulatory mechanisms that are responsible for the timing of puberty.

Abbreviations

ARC	Arcuate nucleus
AVPV	Anteroventral periventricular nucleus
E_2	Estradiol
ERα	Estrogen receptor alpha
GABA	γ-Aminobutyric acid

E. Terasawa, Ph.D. (✉) • K.A. Guerriero
Wisconsin National Primate Research Center, University of Wisconsin,
1223 Capitol Court, Madison, WI 53715-1299, USA
e-mail: terasawa@primate.wisc.edu

T.M. Plant
University of Pittsburgh School of Medicine, Pittsburgh, PA, USA

A.S. Kauffman and J.T. Smith (eds.), *Kisspeptin Signaling in Reproductive Biology*, Advances in Experimental Medicine and Biology 784, DOI 10.1007/978-1-4614-6199-9_12,

IPI	Inter-pulse interval
KISS1	Kisspeptin gene (primates)
Kiss1	Kisspeptin gene (non-primates)
KISS1R	Kisspeptin-1 receptor (primates)
Kiss1r	Kisspeptin-1 receptor (non-primates)
KP	Kisspeptin
MBH	Medial basal hypothalamus
ME	Median eminence
S-ME	Stalk-median eminence
NPY	Neuropeptide Y
POA	Preoptic area

Introduction

The discovery, nearly a decade ago, that the G-protein coupled receptor, KISS1R (aka GPR54), and its ligand, kisspeptin, encoded by the genes *KISS1R* and *KISS1*, respectively, play a major role in regulating the hypothalamic-pituitary-gonadal axis has provided a new perspective on the mystery of puberty. As discussed in Chap. 9, Seminara et al. [1] and de Roux et al. [2] first described amino acid mutations of KISS1R in human patients with a delay in puberty onset or an abnormality in pubertal development. Subsequently, several reports also described mutations at different sites of the *KISS1R* gene in patients with either an absence of or a delay in puberty [3–7] or with precocious puberty [8]. Moreover, it has been reported that a genetically targeted deletion of either *Kiss1r* or *Kiss1* in mice results in hypogonadotropic hypogonadism, including delayed pubertal maturation [1, 9–11]. Most recently, impairment of pubertal progression in a human family with a mutation of *KISS1* was described [12].

Despite a plethora of reports on kisspeptin and its receptor in relation to puberty over the last 10 years, a critical evaluation of the role of kisspeptin signaling in the timing of puberty onset is missing. In this review, we will discuss (1) postnatal development of kisspeptin neurons and the kisspeptin receptor in relation to parallel changes in activity of the GnRH neuronal network, an increase in which is obligatory for puberty onset, (2) recent findings on development of kisspeptin signaling in the rhesus monkey, and (3) our conceptualization of the role played by kisspeptin signaling in the mechanism that controls the onset and progression of puberty.

Developmental Changes in GnRH Release

An increase in GnRH release from the hypothalamus triggers puberty. Pulsatile infusion of GnRH induces precocious puberty in sexually immature female and male monkeys and female guinea pigs [13–15] and increased pubertal release of

GnRH and/or gonadotropin has been described in many mammalian species, including humans (see [16–20]). In males, an increase in pulsatile GnRH release at puberty activates tonic gonadotropin secretion that, in turn, results in the onset of elevated levels of testicular testosterone secretion, which in combination with FSH, initiates spermatogenesis. Tonic LH secretion is composed of intermittent secretory episodes of the hormone, which reflect a corresponding pattern of pulsatile GnRH release by the hypothalamus [21]. In females, an increase in pulsatile GnRH release also drives tonic gonadotropin secretion, which is responsible for folliculogenesis and estradiol (E_2) secretion. Ovulation in most mammalian species, however, also requires development of the capacity to induce a large surge of GnRH in response to the positive feedback action of the rising circulating E_2 levels secreted by the follicle(s) destined to ovulate at mid cycle [22]. Currently, the mechanism for these two modes of GnRH release (pulsatile vs. surge) is unclear.

There are two basic developmental patterns of pulsatile GnRH release from birth until the onset of puberty. In highly evolved primates, such as man and macaques, GnRH pulsatility is robust during the infantile period after birth, but is subsequently dampened during juvenile development (and childhood in humans), resulting in a hypogonadotropic state and relative quiescence of the gonad [18, 20]. The hiatus in pulsatile GnRH release during the juvenile period may be viewed as a consequence of a neurobiological "brake" that holds GnRH release in check until the initiation of the onset of puberty [15]. It is important to note that this is a conceptual brake and may be accounted for by either the imposition of an inhibitory input and/or the loss of a stimulatory input to GnRH neurons [18]. Our current viewpoint is that this conceptual brake is an inhibitory neurocircuit in the brain [17]. The juvenile phase of primate development is terminated by release from the brake, leading to a *reactivation* of robust GnRH pulsatility [15]. Because this juvenile restraint on pulsatile GnRH release is observed in neonatally castrated monkeys [23, 24] and in agonadal humans [25, 26], and because low levels of LH and GnRH release during the juvenile period in ovariectomized female monkeys are not further suppressed by ovarian steroids [27], the hiatus of pulsatile GnRH release during the juvenile period of primate development is independent of ovarian or testicular steroids.

This control system may be contrasted to that in non-primate species, in which LH release (and presumably GnRH release) immediately after birth is minimal but increases before the onset of puberty, with the prepubertal gonad playing a critical role in restraining GnRH release prior to puberty. For example, in sheep and rodents, gonadotropin secretion (and presumably GnRH release) is suppressed by small amounts of gonadal steroid after birth through the juvenile period, but at a time prior to puberty, low levels of steroids are no longer inhibitory [19, 28]. Moreover, neonatal gonadectomy in sheep, rats, and guinea pigs increases LH levels, and in sheep and rats, administration of gonadal steroids suppresses LH levels [29, 30]. Therefore, the control system governing reactivation of GnRH release at puberty in primates is different from that regulating the postnatal development of pulsatile GnRH release in non-primates.

Developmental Changes in the Kisspeptin Neuronal System

Kisspeptin neurons in the adult hypothalamus are typically found in both the medial basal hypothalamus (MBH) and the preoptic area (POA) (see Chap. 3). In the MBH, kisspeptin neurons are localized in the arcuate nucleus (ARC, synonymous with infundibular nucleus in humans), and in the POA, these cells are found in the anteroventral periventricular nucleus (AVPV) in rodents and in similar areas in other species. Kisspeptin neurons in the ARC are considered to be an important component of the hypothalamic control of tonic gonadotropin secretion in all species, while kisspeptin neurons in the AVPV of rodents are critical for surge secretion of GnRH and LH, and therefore for ovulation [31].

Overall expression of *Kiss1* mRNA in the hypothalamus (AVPV and ARC combined) is significantly elevated around the time of puberty in both male and female rats [32]. In the ARC, *Kiss1* mRNA levels in female rats at postnatal day 26 (P26), i.e., 3–4 days before vaginal opening, are over fourfold higher than those at P21 [33], although changes in the number of *Kiss1*-expressing neurons from P3 to adulthood are unremarkable [34]. In male rats, *Kiss1* mRNA levels in the ARC at P45 are significantly higher than those at P15 [35], and the number of *Kiss1* neurons increases progressively throughout postnatal development [34]. In male mice, however, a developmental increase in *Kiss1* mRNA levels in the ARC has not been observed [36–38]. Moreover, whereas ovariectomy in female mice at P14 dramatically increases expression of ARC *Kiss1* mRNA by P16-P18, i.e., well before puberty onset, castration at P14 in male mice does not result in increased ARC *Kiss1* mRNA or LH release at P18 [39]. However, expression of both the ARC *Kiss1* mRNA and secretion of the gonadotropin were elevated at P45 in male mice castrated at P14 [39]. Interestingly, in both male and female mice, *Kiss1* expression is detected in the ARC on P1, and at least in females, kisspeptin receptor signaling appears to be driving gonadotropin release at this early stage of development [40]. The absence of a post-castration LH response in prepubertal male mice has been previously reported [41], and differs from the situation in rats and guinea pigs where prepubertal orchidectomy elicits a robust increase in LH secretion [29, 30, 42]. In an alternative paradigm to eliminate the confounding effect of testicular steroid feedback on the development of *Kiss1* expression in mice, Gill et al. [38] studied the *hpg* mouse, a GnRH deficient hypogonadal animal, and found that ARC *Kiss1* expression increases dramatically between P10 and P30 (as it also did in the *hpg* female). Clearly, the developmental pattern in *Kiss1* expression in the ARC of the male mouse requires further study.

In the rhesus monkey, pubertal increases in *KISS1* mRNA in the MBH (presumably in the ARC) in ovarian intact female and agonadal male monkeys have been observed [43]. Although a gonadal steroid-independent pubertal increase in *KISS1* mRNA expression in female monkeys has not been examined, an ovarian steroid-independent increase in kisspeptin release in the region of the ARC-median eminence (ARC-ME) has been observed (see next section).

The developmental pattern of ARC kisspeptin expression as assessed by immunohistochemistry is less clear. Studies in mice describe an increase in intensity of kisspeptin fibers in the ARC during postnatal development in both males and females, but developmental changes in kisspeptin cell number have not been reported [38, 44, 45]. It is possible that the pubertal increase in kisspeptin fibers in the ARC may reflect an increased kisspeptin output from kisspeptin cell bodies in the AVPV (see below), as direct innervation of the ARC by AVPV kisspeptin neurons has been reported [46]. In the ewe, the number of kisspeptin neurons in the ARC is significantly greater in postpubertal animals compared to prepubertal lambs [47]. In the agonadal male monkey, developmental changes in the number of immunopositive kisspeptin neurons in the ARC parallel changes in pulsatile GnRH release, with both infant and pubertal animals exhibiting numerous and intensely stained ARC perikarya [48]. The importance of ARC kisspeptin neuronal network for generating pulsatile GnRH release in the infant monkey is consistent with the observation that circulating gonadotropin levels were undetectable in a 2-month old infantile boy bearing a loss-of-function mutation of *KISS1R* [3].

In the case of AVPV kisspeptin neurons, it has been clearly shown that the cell number in female mice progressively increases until the age of puberty [38, 39, 44, 49]. Moreover, the developmental increase in the number of kisspeptin neurons in the AVPV in females is dependent on the presence of circulating E_2, as ovariectomy of prepubertal mice reduces and/or masks this developmental change [49]. Similarly, in the *hpg* mouse, the prepubertal increase in expression of both kisspeptin and *Kiss1* is blunted [38], and in aromatase knockout mice there is virtually no kisspeptin expression [49]. This action of E_2 appears to be exerted directly on the AVPV kisspeptin neurons, as conditional knockout of estrogen receptor alpha (ERα) resulted in a marked decrease in the number of kisspeptin immunopositive neurons in this nucleus [50]. Kisspeptin- or *KISS1*-expressing neurons have also been described in the POA of women and female monkeys [51–53], but developmental changes in this particular population of neurons have not been studied in primates.

Hypothalamic (POA and ARC combined) levels of *Kiss1r* mRNA increase at the age of puberty in both male and female rats [32]. Specifically, in the female, *Kiss1r* expression in the AVPV increased at the age of puberty [33]. However, neither the neuronal phenotype in the POA/AVPV exhibiting this pubertal increase in *Kiss1r* expression, nor the gonadal steroid dependency of this phenomenon in rodents, has been studied. In ovarian intact female rhesus monkeys, *KISS1R* mRNA in the MBH also increases across puberty onset [43], and functionally, developmental changes in GnRH response to KP-10 depend on the pubertal increase in E_2 (see next section). Thus, it is possible that the pubertal increase in *Kiss1r*/*KISS1R* mRNA in females is due to the increase in estrogens at this stage of development. However, this view needs further examination, as *KISS1R* mRNA expression does not change across puberty in agonadal male monkeys [43].

Kiss1r is expressed in approximately 80% of GnRH neurons in cichlid fish, and in adult mice and rats [54–56]. During the first few days of postnatal life in mice, only ~40% of GnRH neurons express *Kiss1r* but this increases to adult levels by P20 [57]. Although expression of *KISS1R* in primate GnRH neurons has not been reported, GnRH neurons in both male and female prepubertal monkeys respond to exogenous kisspeptin [43, 58, 59]. Kiss1r may also be present in embryonic mouse GnRH neurons, as they respond to exogenous kisspeptin in vitro [60]. Embryonic primate GnRH neurons, however, do not respond to kisspeptin (Keen and Terasawa, unpublished observation), suggesting that GnRH neurons in rhesus monkeys may not acquire KISS1R until later in gestation.

Taking the foregoing considerations together, it seems reasonable to propose that an increase in expression of both kisspeptin mRNA and peptide in the ARC occurs in association with the onset of puberty in both sexes of most mammalian species, and this is likely correlated with an increase in kisspeptin release in the ARC-ME region, as demonstrated for the monkey (see next section). In primates, the postnatal pattern in ARC kisspeptin expression is fundamentally dictated by a central inhibition that is independent of gonadal steroids, rather than by ovarian and testicular feedback, as is the case in rats [19, 42]. Additionally, it appears that in female rodents an estrogen-dependent developmental increase in kisspeptin peptide and mRNA in the AVPV occurs, leading presumably to an increase in the secretory activity of this rostral population of kisspeptin neurons.

Changes in Kisspeptin Release and KISS1R Responsiveness to Kisspeptin During the Pubertal Process

As discussed in Chap. 2, human preprokisspeptin is cleaved to form kisspeptin-54 and further cleaved to kisspeptin-14, -13, or -10, which are all biologically active [61, 62]. To determine the role kisspeptin plays in the pubertal increase in GnRH release, it is important to understand (1) the developmental pattern of kisspeptin-54 release and (2) developmental changes in the function of KISS1R expressed by GnRH neurons and/or afferent neurons to the GnRH network. The maturational changes in the responsiveness of the GnRH neurosecretory system can be tested by the kisspeptin agonist, human kisspeptin-10 (hKP-10), and the synthetic kisspeptin antagonist, peptide 234, as described by Roseweir et al. [63]. Accordingly, Terasawa and colleagues conducted a series of studies using a microdialysis method, which allows for (1) in vivo measurements of kisspeptin-54 and GnRH release in serially collected dialysate samples from the stalk-median eminence (S-ME) of monkeys, and (2) for infusion of hKP-10 and peptide 234 through the microdialysis probe [64].

In an initial series of studies, the developmental pattern of kisspeptin-54 release was examined in both intact and ovariectomized monkeys. Kissspeptin-54 release is pulsatile, and mean kisspeptin-54 release increases along with the pubertal increase in mean GnRH release [65]. Moreover, kisspeptin-54 pulses during the prepubertal period are of low amplitude with a long inter-pulse interval (IPI), whereas kisspeptin-54 pulses during the pubertal period have a higher amplitude with a shorter IPI [66]. This pubertal modulation of pulsatile kisspeptin-54 release leads to higher mean levels of the peptide in the S-ME of pubertal animals, which is parallel to those seen in GnRH release during the course of puberty in female monkeys [67, 68].

As discussed in the previous section, developmental changes in GnRH release in the rhesus monkey are independent of circulating gonadal steroids. Similar to ovarian intact females [67], developmental increases in the pulse frequency and pulse amplitude of GnRH release do not occur until the age of puberty in ovariectomized monkeys [68]. Likewise, the pulse frequency and pulse amplitude of kisspeptin-54 release in ovariectomized monkeys do not increase until the age that puberty would have been anticipated had the animals remained intact (Fig. 12.1) [66]. Importantly, the IPI of kisspeptin-54 release in ovarian intact and ovariectomized females at the prepubertal stage is ~80 min, which is very similar to that of GnRH release [67, 68], whereas the IPI of kisspeptin-54 release at the pubertal stage is ~50 min, regardless of the presence or absence of the ovary, which is, again, similar to the IPI of GnRH release in animals at the same developmental stage (Fig. 12.1) [67, 68]. (The role of kisspeptin in GnRH pulse generation will be further discussed in a later section.)

An impact of the ovary on kisspeptin-54 release is only observed in the pubertal monkey, where both the pulse amplitude and mean release of kisspeptin-54 is markedly increased by ovariectomy, presumably due to loss of negative feedback from the ovarian steroid E_2 (Fig. 12.1). In fact, administration of E_2 can suppress kisspeptin-54 release in pubertal monkeys, whereas kisspeptin-54 release in prepubertal monkeys is insensitive to E_2 [66]. This developmental change in ovarian steroid regulation of kisspeptin-54 release is similar to that seen with GnRH release [27]. Collectively, these observations indicate that the pubertal increase in kisspeptin-54 release occurs independently from an ovarian steroid hormone feedback mechanism. Rather, the pubertal increase in pulsatile release of kisspeptin-54 in female rhesus monkeys (and presumably male primates) requires a developmental change in an upstream neuronal signal to the kisspeptin neuronal network.

Because developmental changes in KISS1R may also contribute to the pubertal increase in GnRH release, in a second series of studies, Terasawa and colleagues examined the developmental changes in GnRH release in response to the kisspeptin agonist, hKP-10, and antagonist, peptide 234, administered directly into the S-ME. While the GnRH response to hKP-10 is dose dependent in both ovarian intact prepubertal and pubertal monkeys, a smaller response to a 10 nM dose of hKP-10 is consistently observed in prepubertal monkeys as compared to pubertal monkeys [58]. Release of GnRH in both prepubertal and pubertal monkeys is also suppressed by peptide 234. These results suggest that the pubertal increase in pulsatile GnRH release is, in part, due to an increased responsiveness of KISS1R in GnRH neurons during the progression of puberty. This view is consistent with studies in transgenic mice expressing GFP in GnRH neurons, in which electrical firing activity of GnRH neurons stimulated by KP-10 increases across male puberty [37].

To further determine whether the enhanced responses of GnRH neurons to hKP-10 in pubertal monkeys are due to higher levels of circulating E_2 at puberty, a similar experiment examining the GnRH responsiveness to hKP-10 in ovariectomized monkeys was conducted. While ovariectomy in prepubertal monkeys did not

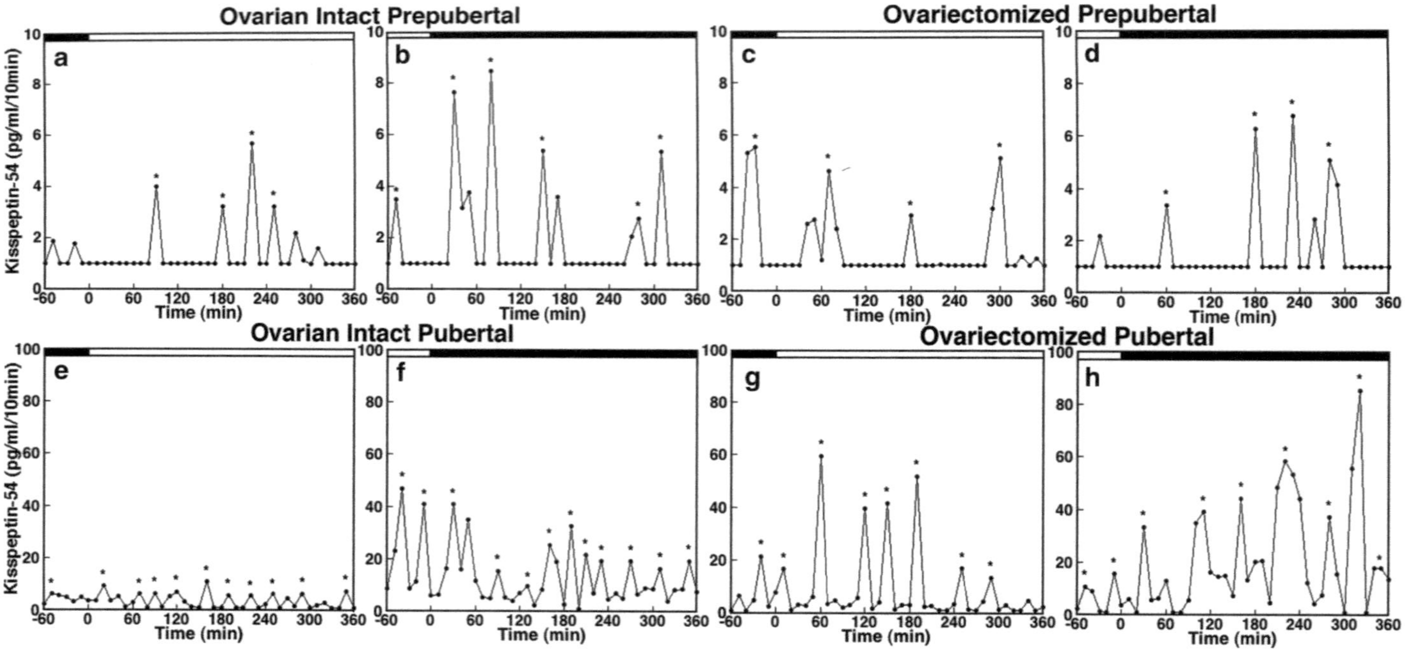

Fig. 12.1 Developmental increases in kisspeptin-54 (KP-54) release are independent of the presence or absence of the ovary in female monkeys. In vivo KP-54 release from the S-ME of ovarian intact prepubertal (**a**, **b**) and pubertal (**e**, **f**) monkeys as well as ovariectomized prepubertal (**c**, **d**) and pubertal (**g**, **h**) monkeys are shown. Samples were obtained during the morning period (**a**, **c**, **e**, **g**) and during the evening period (**b**, **d**, **f**, **h**) as indicated by the *open* and *closed bars*, respectively, at the *top* of each graph. Both pulse frequency and amplitude of KP-54 release in ovarian intact pubertal monkeys (**e**, **f**) are higher than those in ovarian intact prepubertal monkeys (**a**, **b**). Similarly, pulse frequency and amplitude of KP-54 release in ovariectomized pubertal monkeys (**g**, **h**) are higher than those in ovariectomized prepubertal monkeys (**a**, **b**). Importantly, ovariectomy does not cause any change in KP-54 release ((**a**, **b**) vs. (**c**, **d**)) in prepubertal monkeys, whereas ovariectomy increases the pulse amplitude of KP-54 release in pubertal monkeys ((**e**, **f**) vs. (**g**, **h**)). *Asterisks* indicate peaks as determined by PULSAR. Note that the scale of the *y*-axis in (**e**–**h**) (pubertal monkeys) is tenfold higher than that in (**a**–**d**) (prepubertal monkeys). From Guerriero KA, Keen KL, Terasawa E. Developmental increase in kisspeptin-54 in vivo is independent of the pubertal increase in estradiol in female rhesus monkeys (*Macaca mulatta*). Endocrinology. 2012 153:1887–97. Modified with permission from The Endocrine Society

modify the GnRH response to hKP-10 nor peptide 234, it completely eliminated both the hKP-10-induced stimulation and peptide 234-induced GnRH suppression of GnRH release in pubertal monkeys ([58], also Guerriero and Terasawa, unpublished observation). Moreover, replacement of E_2 in OVX pubertal monkeys only partially restored the hKP-10-induced GnRH release that was absent in OVX pubertal monkeys [58]. These observations suggest that while, in prepubertal monkeys, the response of KISS1R on GnRH neurons is independent of E_2, in pubertal monkeys, functional changes in KISS1R occur as a consequence of the exposure to increased circulating E_2 after puberty onset, such that KISS1R responsiveness is enhanced by E_2. Collectively, once the pubertal increase in E_2 occurs in the female monkey, as a consequence of pubertal activation of the GnRH pulse generating mechanism, the presence of E_2 appears to enhance the response of GnRH neurons to kisspeptin [58]. Although to date, developmental changes in *KISS1R* mRNA in ovariectomized monkeys have not been examined, it will be important to address this issue further.

Kisspeptin Signaling and GnRH Pulse Generation

The hypothesis that kisspeptin neurons are a part of the neurocircuitry underlying the GnRH pulse generating mechanism has been proposed by Goodman and colleagues, Maeda and colleagues, and Steiner and colleagues [69, 70] (see also Chap. 14). It is posited that pulsatility originates in ARC kisspeptin neurons containing neurokinin B and dynorphin (called KNDy neurons) by reciprocal interactions of neurokinin B (stimulatory) and dynorphin (inhibitory), and that an intermittent output to the GnRH neuronal network is mediated by kisspeptin. This hypothesis is based on several observations. First, periodic increases in multiunit activity obtained from electrodes in the MBH are associated with LH pulses in several species [71, 72], and specifically in the ARC, as shown in the goat [73]. Second, the neurokinin B receptor agonist, senktide, is a potent stimulator of ARC kisspeptin neurons (presumably KNDy neurons) in the mouse [74], and the site of the stimulatory action of neurokinin B on GnRH-dependent LH release in the monkey appears to be upstream of kisspeptin [75]. Third, in pubertal monkeys, pulses of kisspeptin-54 released in the ARC-ME correlate to GnRH pulses 75% of the time [65]. Fourth, repetitive iv injections of hKP-10 induce trains of GnRH-dependent LH pulses in juvenile male monkeys, in which endogenous GnRH pulsatility is minimal [59], presumably by activating KISS1R on GnRH terminals in the ME, as kisspeptin and GnRH fibers are found in extensive and intimate association in the ME (Fig. 12.2) [76]. Fifth, intra-ARC, not intra-POA, administration of the kisspeptin antagonist, peptide 234, profoundly suppressed LH pulse frequency [77], although again the site of action of the antagonist is likely to be at the ME, as recent electrophysiological studies by Alreja and Steiner indicate that kisspeptin is unable to stimulate KNDy neurons in the mouse (see Chap. 16). The contemporary notion regarding the integral role played by KNDy neurons in GnRH pulse generation

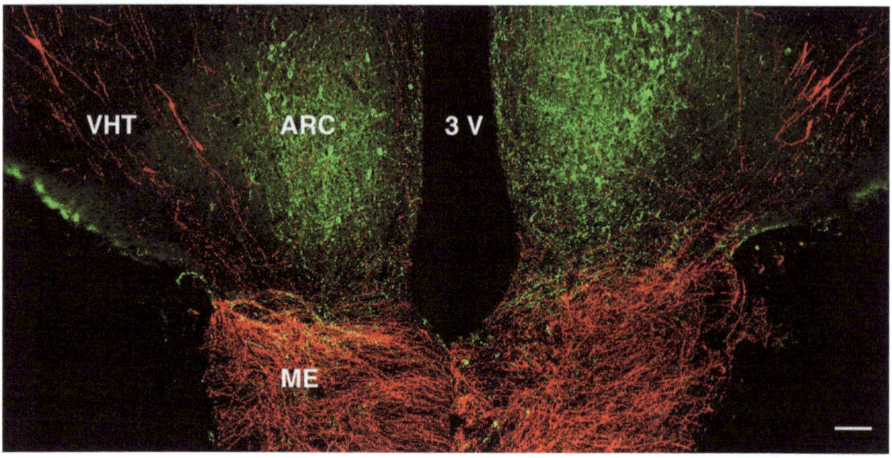

Fig. 12.2 A confocal projection illustrating the relationship between kisspeptin neurons (*green*) in the arcuate nucleus (ARC) and GnRH cell bodies and projections (*red*) to the median eminence in a coronal section of the mediobasal hypothalamus of a castrated adult male rhesus monkey. *VHT* ventral hypothalamic tract; *3V* third ventricle; *ME* median eminence. Scale bar, 100 μm. From Ramaswamy S, Guerriero KA, Gibbs RB, Plant TM. Structural interactions between kisspeptin and GnRH neurons in the mediobasal hypothalamus of the male rhesus monkey (*Macaca mulatta*) as revealed by double immunofluorescence and confocal microscopy. Endocrinology. 2008 149:4387–95. Reprinted with permission from The Endocrine Society

is consistent with the classical findings that complete surgical deafferentation of the rat and monkey MBH does not eliminate pulsatile LH release [78, 79], and that selective lesions of the ARC in female monkeys abolishes pulsatile LH release [80]. It is also consistent with the recent finding that selective ablation of KNDy neurons in the rat dramatically truncates the ovariectomy-induced increase in LH release [81].

A Novel View on the Role of Kisspeptin in Puberty Onset

As discussed above, the genetic evidence for the view that kisspeptin neurons are critical for the onset of puberty is overwhelming. Together with results from compelling physiological and pharmacological studies indicating that kisspeptin is the most potent GnRH secretagogue [82], a dogma has emerged that the genes encoding kisspeptin and its receptor regulate puberty, which in turn has led to the perception that kisspeptin signaling represents the key neural substrate that controls the timing of the onset of puberty. Here, we offer an alternative possibility. Namely, while kisspeptin-expressing neurons in ARC are critical for puberty, this is simply because these cells comprise an integral component of the hypothalamic GnRH pulse generating mechanism that generates intermittent release of the decapeptide,

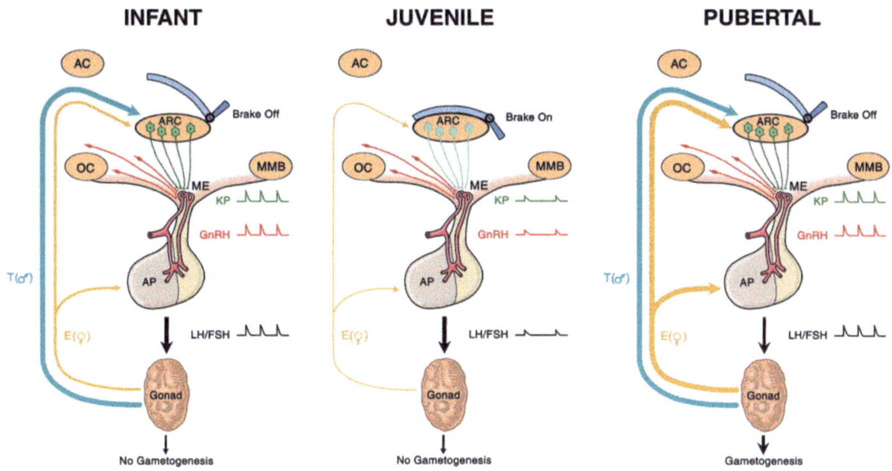

Fig. 12.3 A model for the control of the timing of puberty in primates, in which the role of kisspeptin (KP, *green*) signaling is posited to be a critical component of the neural machinery essential for generation of pulsatile GnRH (*red*) release in the hypothalamus. In this model, the GnRH pulse generating mechanism resides in the arcuate nucleus (ARC) and the output of this signaling is relayed to GnRH terminals in the median eminence (ME) by KP projections arising from perikarya in the ARC. During infancy (*left panel*), ARC GnRH pulse generating activity is robust leading to intermittent release of KP in the ME, resulting in a corresponding pattern of GnRH release into the portal circulation. This, in turn, drives pulsatile gonadotropin (LH and FSH) secretion. In the transition from infancy to the juvenile phase of development (*middle panel*), a neurobiological brake (central inhibition) holds the ARC GnRH pulse generating mechanism in check and pulsatile release of KP in the ME is markedly suppressed. This leads to reduced GnRH release and to a hypogonadotropic state in the juvenile period. Puberty is triggered when the brake is removed and GnRH pulse generation with robust intermittent release of KP in the ME is reactivated (*right panel*). According to this model, the mystery of primate puberty lies in the nature of the neurobiological brake, i.e., the mechanism that times its application during infancy and its release at the end of the juvenile phase of development. The thickness of the *blue* (T, testosterone) and *gold* (E, estradiol) *arrows* indicating negative feedback by the testis and ovary, respectively, reflect the degree of gonadal steroid inhibition exerted on LH secretion at these three stages of primate development. *AC* anterior commissure; *AP* anterior pituitary gland; *ARC* arcuate nucleus; *OC* optic chiasm; *ME* median eminence; *MMB* mammillary body

an increase of which is obligatory for the onset of puberty. According to this model, kisspeptin neurons in the ARC play no regulatory role in controlling the timing of puberty. Rather, as a component of hypothalamic GnRH pulse generation, they subserve upstream regulatory mechanisms determining the timing of puberty onset. In the case of primates, the upstream control system(s), which is independent of gonadal steroids, first suppress pulsatile GnRH release in infancy and, subsequently, a reduction in this suppression reactivates pulsatility of GnRH release at the end of juvenile development (Fig. 12.3). In rodents, the early postnatal ontogeny of pulsatile GnRH release is less clear, but later in prepubertal development, steroid-dependent mechanisms dictate the timing of puberty by suppressing GnRH pulse generation. This being the case, loss-of-function mutations in *KISS1/Kiss1* or *KISS1R/Kiss1r*, or ablation of neurons expressing either kisspeptin or its receptor, would likely lead to a loss or impairment in GnRH pulsatility that secondarily results in delayed or absent puberty and infertility, regardless of species. While this is indeed the case in situations where the genes have been manipulated either spontaneously or experimentally [1, 9–11, 50, 83], interestingly, embryonic ablation of kisspeptin cells in mice did not dramatically influence the timing of puberty or prevent fertility [45]. It should be noted that failure to change the timing of puberty in this study may be due to the 5% of kisspeptin neurons in the AVPV that escaped ablation [45]. In the context of the results of the study employing kisspeptin neuron ablation, *Kiss1* or *Kiss1r* null mice exhibit some degree of GnRH release as they age [83]. Therefore, the difference in the phenotypes between these two models may be quantitative, and perhaps explained by differences in the extent to which the GnRH neuronal network is intrinsically able to generate intermittent GnRH release following a genetic or ablative insult to the GnRH pulse generating mechanism that normally drives gonadotropin secretion in the adult.

The notion that kisspeptin signaling is necessary for the onset of puberty only because of its critical role in GnRH pulse generation may be most readily appreciated when the concept is applied to puberty in the male, where initiation of this developmental event requires only robust pulsatile GnRH release to drive tonic LH and FSH secretion. In the case of puberty onset in the human female, the validity of the idea that *KISS1* may simply be regarded as a "pulse generating" gene is tenable, because the preovulatory LH surge is triggered by E_2 positive feedback action within the MBH-pituitary unit to amplify pulsatile GnRH release and/or the response of the pituitary gonadotrophs to pulsatile GnRH stimulation [22]. The situation in the female rodent is more complex because the positive feedback action of E_2 is exerted, at least in part, on kisspeptin neurons in the AVPV [31]. Nevertheless, as discussed above, the development of kisspeptin neurons in the AVPV in female mice is dependent on ovarian E_2 secretion, which, in turn, is dependent on tonic gonadotropin secretion that is driven by pulsatile GnRH release. Thus, it seems reasonable to propose that (1) the primary role of kisspeptin signaling in the control of puberty across species may be restricted to its crucial role in GnRH pulse generation, (2) the time of puberty onset is dictated by kisspeptin-independent mechanisms that control the ontogeny of GnRH pulse generation, and (3) *Kiss1* in the rodent may be viewed as a "surge generating gene," as well as a pulse generating gene (see below for further discussion).

Neuronal Substrates of Central Inhibition on GnRH in Juvenile Primates

According to the model proposed above, the key to the mystery of puberty in primates is to understand (1) the neural substrate that underlies the gonadal steroid-independent reduction in GnRH pulse generation from infancy to puberty, and (2) the signals responsible for timing the application and removal of this central neurobiological brake. In this section, we discuss possible neuronal substrates responsible for "central inhibition."

Two laboratories have each proposed a different neuronal subtype. First, Terasawa and her colleagues have proposed the hypothesis that tonic inhibition by γ-amino butyric acid (GABA) neurotransmission is responsible for this central inhibition in female rhesus monkeys [17]. This hypothesis is based on the observations that (1) GABA levels are higher when GnRH release is low in prepubertal monkeys, whereas GABA levels are lower after the onset of puberty when GnRH release is elevated [84], (2) infusion of the $GABA_A$ receptor antagonist, bicuculline, into the S-ME stimulates GnRH release to a much greater extent in prepubertal, than in pubertal, monkeys, whereas infusion of GABA is effective in suppressing GnRH release in pubertal, but not prepubertal, monkeys, presumably because of the reduction in tonic GABA inhibition at the onset of puberty [84], and (3) a long-term infusion of bicuculline into the S-ME of juvenile female primates results in precocious puberty and first ovulation [85]. Second, Plant and his colleagues have proposed the hypothesis that neuropeptide Y (NPY) neurons are responsible for the central inhibition of pulsatile GnRH release during juvenile development in male monkeys. This hypothesis is based on the finding that mRNA and peptide levels of NPY in the MBH are significantly lower during the neonatal period compared to those during the juvenile period, whereas mRNA and peptide levels of NPY in the MBH decrease, while GnRH mRNA levels increase, across puberty in male monkeys [86]. Presently, whether the sex-differences noted in the "juvenile hiatus" in gonadotropin secretion are attributable to central inhibition mediated by GABA neurons in females vs. NPY neurons in males is unclear. Nonetheless, it is possible that the same population of neurons in the MBH is responsible for gonadal steroid-independent central inhibition, as a large number of GABA neurons in the rat ARC express NPY [87, 88].

A recent study from Terasawa's group indicates that bicuculline infusion into the S-ME of prepubertal female monkeys stimulates kisspeptin-54 release (Fig. 12.4a) [89], similar to the bicuculline-induced stimulation of GnRH release observed in prepubertal monkeys [84]. Moreover, the bicuculline-induced GnRH release was blocked by simultaneous infusion of the kisspeptin antagonist, peptide 234 (Fig. 12.4b) [89]. These latter results are consistent with the view that inhibitory GABA neurotransmission is an important component in the upstream suppression of the GnRH pulse generating mechanism during juvenile development in primates. It is, however, unclear what reduces GABA inhibition prior to puberty and whether additional (or alternative) neuronal substrates and somatic cues [90–93] are involved in the upstream control of GnRH pulse generation. Thus, the most important question of exactly what triggers the onset of puberty in primates remains a mystery.

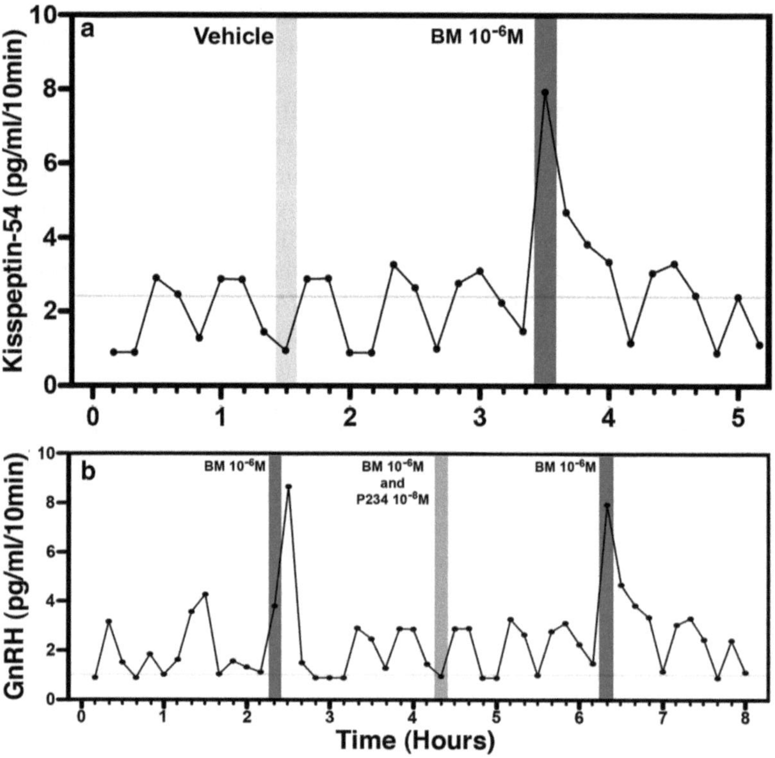

Fig. 12.4 (**a**) $GABA_A$ antagonist bicuculline (BM) stimulates KP-54 release in prepubertal female monkeys (but not pubertal monkeys, data not shown). An example showing that bicuculline infusion the S-ME (*dark shaded bar*) for 10 min induces an increase KP-54 release, whereas vehicle (*light shaded bar*) infusion does not. (**b**) Blockade of the bicuculline-induced GnRH release by the kisspeptin receptor antagonist, peptide 234 (P234), in a prepubertal monkey. The stimulated GnRH release by bicuculline infusion in the S-ME (*dark shaded bars*) is not seen in the presence of P234 (*light shaded bar*). From Kurian JR, Keen KL, Guerriero KA, Terasawa E. Tonic control of kisspeptin release in prepubertal monkeys: Implications to the mechanism of puberty onset. Endocrinology. 2012 153:3331–6. Modified with permission from The Endocrine Society

Neural Substrate for Steroid Inhibition of GnRH Release in Juvenile Rodents

As discussed above, in contrast to primates, the prepubertal restraint on the GnRH pulse generating mechanism in rodents is gonadal steroid dependent. In this regard, studies in sheep and mice indicate that the majority of kisspeptin neurons express ERα [94–96], and it is well established in the adult rodent that ovariectomy increases, and E_2 replacement decreases, *Kiss1* expression in ARC kisspeptin neurons [31]. As might be expected, therefore, transgenic mice with a conditional knockout of *ERα* in kisspeptin neurons exhibit elevated *Kiss1* mRNA levels in ARC at a prepubertal age, and this is associated with high circulating concentrations of LH (and presumably E_2) and a dramatic advancement of the age of vaginal opening [50]. Interestingly, in contrast to the mRNA data, kisspeptin immunoreactivity in the ARC was greatly reduced in the conditional knockout, suggesting perhaps enhanced release of kisspeptin. Thus, in the case of the female mouse, it seems reasonable to conclude that the site of the prepubertal ovarian steroid suppression on pulsatile GnRH release is on the GnRH pulse generating mechanism itself, and specifically on kisspeptin (KNDy) neurons in the ARC. This view is consistent with the long-standing "differential sensitivity to E_2" theory, which has been proposed in female rats and sheep [97–100]. During the postnatal period through the juvenile period, the hypothalamus (presumably the GnRH neurosecretory system) is inhibited by E_2, and, sometime prior to puberty, the GnRH pulse generating mechanism in the ARC escapes from suppression by E_2. It has been proposed that this escape is the result of an E_2-induced increase in activity of kisspeptin neurons in the AVPV, which in turn amplifies GnRH neuronal activity, leading to puberty onset [36, 49, 50]. The precise mechanism by which the initial prepubertal elevation of E_2 is triggered in non-primate species, however, is unknown. (Note that, in women and female rhesus monkeys, a similar escape of E_2-dependent inhibition of GnRH release occurs well after the initiation of puberty onset, between menarche and first ovulation [101–103]).

Summary

In this chapter, we have reviewed progress regarding the relationship between kisspeptin and puberty onset, and have proposed a novel hypothesis for the role of kisspeptin signaling in controlling the timing of this major event in postnatal development. We posit that the profound impact of loss-of-function mutations in the genes encoding either kisspeptin or its receptor on the onset and progression of puberty in all species can be attributed primarily to the critical role of ARC kisspeptin neurons in the generation of pulsatile GnRH release, which is obligatory for pubertal activation of the pituitary-gonadal axis. According to this hypothesis, kisspeptin neurons do not determine the timing of puberty (see Fig. 12.3). Rather, this important developmental event is achieved by upstream neuronal mechanisms that govern the timing of the pubertal activation (rodents) or reactivation (primates) of robust pulsatile GnRH release at the end of the juvenile phase of development. Validation of this hypothesis requires future studies.

Acknowledgements Supported by grants R01 HD15433 and R01 HD11355 for ET, R01 HD 013254 and U54 HD 08160 for TMP, T32 HD041921 for KAG, and P51 0D011106 for WNPRC.

References

1. Seminara SB, Messager S, Chatzidaki EE, Thresher RR, Acierno JS Jr, Shagoury JK, Bo-Abbas Y, Kuohung W, Schwinof KM, Hendrick AG, Zahn D, Dixon J, Kaiser UB, Slaugenhaupt SA, Gusella JF, O'Rahilly S, Carlton MB, Crowley WF Jr, Aparicio SA, Colledge WH (2003) The GPR54 gene as a regulator of puberty. N Engl J Med 349:1614–1627
2. de Roux N, Genin E, Carel JC, Matsuda F, Chaussain JL, Milgrom E (2003) Hypogonadotropic hypogonadism due to loss of function of the KiSS1-derived peptide receptor GPR54. Proc Natl Acad Sci USA 100:10972–10976
3. Semple RK, Achermann JC, Ellery J, Farooqi IS, Karet FE, Stanhope RG, O'rahilly S, Aparicio SA (2005) Novel missense mutations in g protein-coupled receptor 54 in a patient with hypogonadotropic hypogonadism. J Clin Endocrinol Metab 90:1849–1855
4. Cerrato F, Shagoury J, Kralickova M, Dwyer A, Falardeau J, Ozata M, Van Vliet G, Bouloux P, Hall JE, Hayes FJ, Pitteloud N, Martin KA, Welt C, Seminara SB (2006) Coding sequence analysis of GnRHR and GPR54 in patients with congenital and adult onset forms of hypogonadotropic hypogonadism. Eur J Endocrinol 155:S3–S10
5. Tenenbaum-Rakover Y, Commenges-Ducos M, Iovane A, Aumas C, Admoni O, de Roux N (2007) Neuroendocrine phenotype analysis in five patients with isolated hypogonadotropic hypogonadism due to a L102P inactivating mutation of GPR54. J Clin Endocrinol Metab 92:1137–1144
6. Nimri R, Lebenthal Y, Lazar L, Chevrier L, Phillip M, Bar M, Hernandez-Mora E, de Roux N, Gat-Yablonski G (2011) A novel loss-of-function mutation in GPR54/KISS1R leads to hypogonadotropic hypogonadism in a highly consanguineous family. J Clin Endocrinol Metab 96:E536–E545
7. Lanfranco F, Gromoll J, von Eckardstein S, Herding EM, Nieschlag E, Simoni M (2005) Role of sequence variations of the GnRH receptor and G protein-coupled receptor 54 gene in male idiopathic hypogonadotropic hypogonadism. Eur J Endocrinol 153:845–852
8. Teles MG, Bianco SD, Brito VN, Trarbach EB, Kuohung W, Xu S, Seminara SB, Mendonca BB, Kaiser UB, Latronico AC (2008) A GPR54-activating mutation in a patient with central precocious puberty. N Engl J Med 358:709–715
9. Funes S, Hedrick JA, Vassileva G, Markowitz L, Abbondanzo S, Golovko A, Yang S, Monsma FJ, Gustafson EL (2003) The KiSS-1 receptor GPR54 is essential for the development of the murine reproductive system. Biochem Biophys Res Commun 312:1357–1363
10. d'Anglemont de Tassigny X, Fagg LA, Dixon JPC, Day K, Leitch HG, Hendrick AG, Zahn D, Franceschini I, Caraty A, Carlton MBL, Aparicio SAJR, Colledge WH (2007) Hypogonadotropic hypogonadism in mice lacking a functional Kiss1 gene. Proc Natl Acad Sci USA 104:10714–10719
11. Lapatto R, Pallais JC, Zhang D, Chan YM, Mahan A, Cerrato F, Le WW, Hoffman GE, Seminara SB (2007) Kiss1–/– mice exhibit more variable hypogonadism than Gpr54–/– mice. Endocrinology 148:4927–4936
12. Topaloglu AK, Tello JA, Kotan LD, Ozbek MN, Yilmaz MB, Erdogan S, Gurbuz F, Temiz F, Millar RP, Yuksel B (2012) Inactivating KISS1 mutation and hypogonadotropic hypogonadism. N Engl J Med 366:629–635
13. Wildt L, Marshall G, Knobil E (1980) Experimental induction of puberty in the infantile female rhesus monkey. Science 207:1373–1375
14. Loose MD, Terasawa E (1985) Pulsatile infusion of luteinizing hormone-releasing hormone induces precocious puberty (vaginal opening and first ovulation) in the immature female guinea pig. Biol Reprod 33:1084–1093
15. Plant TM (2001) Neurobiological bases underlying the control of the onset of puberty in the rhesus monkey: a representative higher primate. Front Neuroendocrinol 22:107–139
16. Grumbach MM, Styne DM (1998) Puberty, ontogeny, neuroendocrinology, physiology, and disorders. In: Williams RH, Foster DW, Kroenenberg H, Larsen PR, Zorab R (eds) Williams textbook of endocrinology, 9th edn. W.B. Saunders, Philadelphia, PA, pp 1509–1625

17. Terasawa E, Fernandez DL (2001) Neurobiological mechanisms of the onset of puberty in primates. Endocr Rev 22:111–151
18. Plant TM, Witchel SF (2006) Puberty in nonhuman primates and primates. In: Neill J (ed) The physiology of reproduction, vol 2, 3rd edn. Academic, San Diego, CA, pp 2177–2230
19. Ojeda SR, Skinner MK (2006) Puberty in the rat. In: Neill J (ed) The physiology of reproduction, vol 2, 3rd edn. Academic, Elsevier, San Diego, CA, pp 2061–2126
20. Terasawa E, Kurian JR (2012) Neuroendocrine mechanism of puberty. In: Fink G, Pfaff DW, Levine JE (eds) Handbook of neuroendocrinology. Academic, Elsevier, London, pp 433–484
21. Plant TM (1986) Gonadal regulation of hypothalamic gonadotropin-releasing hormone release in primates. Endocr Rev 7:75–88
22. Plant TM (2012) A comparison of the neuroendocrine mechanisms underlying the initiation of the preovulatory LH surge in the human, Old World monkey and rodent. Front Neuroendocrinol 33:160–168
23. Plant TM (1985) A study of the role of the postnatal testes in determining the ontogeny of gonadotropin secretion in the male rhesus monkey (*Macaca mulatta*). Endocrinology 116:1341–1350
24. Pohl CR, de Ridder CM, Plant TM (1995) Gonadal and nongonadal mechanisms contribute to the prepubertal hiatus in gonadotropin secretion in the female rhesus monkey (*Macaca mulatta*). J Clin Endocrinol Metab 80:2094–2101
25. Conte FA, Grumbach MM, Kaplan SL (1975) A diphasic pattern of gonadotropin secretion in patients with the syndrome of gonadal dysgenesis. J Clin Endocrinol Metab 40:670–674
26. Ross JL, Loriaux DL, Cutler GB Jr (1983) Developmental changes in neuroendocrine regulation of gonadotropin secretion in gonadal dysgenesis. J Clin Endocrinol Metab 57:288–293
27. Chongthammakun S, Terasawa E (1993) Negative feedback effects of estrogen on luteinizing hormone-releasing hormone release occur in pubertal, but not prepubertal, ovariectomized female rhesus monkeys. Endocrinology 132:735–743
28. Foster DL, Jackson LM (2006) Puberty in the sheep. In: Neill J (ed) The physiology of reproduction, vol 2, 3rd edn. Academic, Elsevier, San Diego, CA, pp 1415–1482
29. Goldman BD, Gorski RA (1971) Effects of gonadal steroids on the secretion of LH and FSH in neonatal rats. Endocrinology 89:112–115
30. Fraser MO, Plant TM (1989) Further studies of the role of the gonads in determining the ontogeny of gonadotropin secretion in the guinea pig (*Cavia porcelus*). Endocrinology 125:906–911
31. Oakley AE, Clifton DK, Steiner RA (2009) Kisspeptin signaling in the brain. Endocr Rev 30:713–743
32. Navarro VM, Castellano JM, Fernandez-Fernandez R, Barreiro ML, Roa J, Sanchez-Criado JE, Aguilar E, Dieguez C, Pinilla L, Tena-Sempere M (2004) Developmental and hormonally regulated messenger ribonucleic acid expression of KiSS-1 and its putative receptor, GPR54, in rat hypothalamus and potent luteinizing hormone-releasing activity of KiSS-1 peptide. Endocrinology 145:4565–4574
33. Takase K, Uenoyama Y, Inoue N, Matsui H, Yamada S, Shimizu M, Homma T, Tomikawa J, Kanda S, Matsumoto H, Oka Y, Tsukamura H, Maeda KI (2009) Possible role of oestrogen in pubertal increase of Kiss1/kisspeptin expression in discrete hypothalamic areas of female rats. J Neuroendocrinol 21:527–537
34. Takumi K, Iijima N, Ozawa H (2011) Developmental changes in the expression of kisspeptin mRNA in rat hypothalamus. J Mol Neurosci 43:138–145
35. Bentsen AH, Ansel L, Simonneaux V, Tena-Sempere M, Juul A, Mikkelsen JD (2010) Maturation of kisspeptinergic neurons coincides with puberty onset in male rats. Peptides 31:275–283
36. Clarkson J, Herbison AE (2011) Dual phenotype kisspeptin-dopamine neurones of the rostral periventricular area of the third ventricle project to gonadotrophin-releasing hormone neurones. J Neuroendocrinol 23:293–301
37. Han SK, Gottsch ML, Lee KJ, Popa SM, Smith JT, Jakawich SK, Clifton DK, Steiner RA, Herbison AE (2005) Activation of gonadotropin-releasing hormone neurons by kisspeptin as a neuroendocrine switch for the onset of puberty. J Neurosci 25:11349–11356

38. Gill JC, Wang O, Kakar S, Martinelli E, Carroll RS, Kaiser UB (2010) Reproductive hormone-dependent and -independent contributions to developmental changes in kisspeptin in GnRH-deficient hypogonadal mice. PLoS One 5:e11911
39. Kauffman AS, Navarro VM, Kim J, Clifton DK, Steiner RA (2009) Sex differences in the regulation of Kiss1/NKB neurons in juvenile mice: implications for the timing of puberty. Am J Physiol Endocrinol Metab 297:E1212–E1221
40. Poling MC, Kauffman AS (2012) Sexually dimorphic testosterone secretion in prenatal and neonatal mice is independent of kisspeptin-Kiss1r and GnRH signaling. Endocrinology 153:782–793
41. Jean-Faucher C, el Watik N, Berger M, De Turckheim M, Veyssiere G, Jean C (1985) Regulation of gonadotrophin secretion in male mice from birth to adulthood. Response to LHR injection, castration, and testosterone replacement therapy. Acta Endocrinol (Copenh) 110:193–199
42. Ojeda SR, Ramirez VD (1973–1974) Short-term steroid treatment on plasma LH and FSH in castrated rats from birth to puberty. Neuroendocrinology 13:100–114
43. Shahab M, Mastronardi C, Seminara SB, Crowley WF, Ojeda SR, Plant TM (2005) Increased hypothalamic GPR54 signaling: a potential mechanism for initiation of puberty in primates. Proc Natl Acad Sci USA 102:2129–2134
44. Clarkson J, Herbison AE (2006) Postnatal development of kisspeptin neurons in mouse hypothalamus; sexual dimorphism and projections to gonadotropin-releasing hormone (GnRH) neurons. Endocrinology 147:5817–5825
45. Mayer C, Boehm U (2011) Female reproductive maturation in the absence of kisspeptin/GPR54 signaling. Nat Neurosci 14:704–710
46. Yeo SH, Herbison AE (2011) Projections of arcuate nucleus and rostral periventricular kisspeptin neurons in the adult female mouse brain. Endocrinology 152:2387–2399
47. Nestor CC, Briscoe AM, Davis SM, Valent M, Goodman RL, Hileman SM (2012) Evidence of a role for kisspeptin and neurokinin B in puberty of female sheep. Endocrinology 153:2756–2765
48. Dwarki K, Ramaswamy S, Gibbs R and Plant TM. The arrest of GnRH pulsatility during infancy that guarantees the quiescence of the primate gonad during juvenile development is correlated with a reduction in immunopositive kisspeptin neurons in the arcuate nucleus of the male rhesus monkey (*Macaca mulatta*). In: 93rd Annual Meeting of The Endocrine Society, Boston, June 2011, Abstract #P2-262
49. Clarkson J, Boon WC, Simpson ER, Herbison AE (2009) Postnatal development of an estradiol-kisspeptin positive feedback mechanism implicated in puberty onset. Endocrinology 150:3214–3220
50. Mayer C, Acosta-Martinez M, Dubois SL, Wolfe A, Radovick S, Boehm U, Levine JE (2010) Timing and completion of puberty in female mice depend on estrogen receptor alpha-signaling in kisspeptin neurons. Proc Natl Acad Sci USA 107:22693–22698
51. Rometo AM, Krajewski SJ, Voytko ML, Rance NE (2007) Hypertrophy and increased kisspeptin gene expression in the hypothalamic infundibular nucleus of postmenopausal women and ovariectomized monkeys. J Clin Endocrinol Metab 92:2744–2750
52. Smith JT, Shahab M, Pereira A, Pau KY, Clarke IJ (2010) Hypothalamic expression of KISS1 and gonadotropin inhibitory hormone genes during the menstrual cycle of a non-human primate. Biol Reprod 83:568–577
53. Hrabovszky E, Ciofi P, Vida B, Horvath MC, Keller E, Caraty A, Bloom SR, Ghatei MA, Dhillo WS, Liposits Z, Kallo I (2010) The kisspeptin system of the human hypothalamus: sexual dimorphism and relationship with gonadotropin-releasing hormone and neurokinin B neurons. Eur J Neurosci 31:1984–1998
54. Parhar IS, Ogawa S, Sakuma Y (2004) Laser-captured single digoxigenin-labeled neurons of gonadotropin-releasing hormone types reveal a novel G protein-coupled receptor (Gpr54) during maturation in cichlid fish. Endocrinology 145:3613–3618
55. Messager S, Chatzidaki EE, Ma D, Hendrick AG, Zahn D, Dixon J, Thresher RR, Malinge I, Lomet D, Carlton MB, Colledge WH, Caraty A, Aparicio SA (2005) Kisspeptin directly

stimulates gonadotropin-releasing hormone release via G protein-coupled receptor 54. Proc Natl Acad Sci USA 102:1761–1766
56. Irwig MS, Fraley GS, Smith JT, Acohido BV, Popa SM, Cunningham MJ, Gottsch ML, Clifton DK, Steiner RA (2004) Kisspeptin activation of gonadotropin releasing hormone neurons and regulation of KiSS-1 mRNA in the male rat. Neuroendocrinology 80:264–272
57. Herbison AE, de Tassigny X, Doran J, Colledge WH (2010) Distribution and postnatal development of Gpr54 gene expression in mouse brain and gonadotropin-releasing hormone neurons. Endocrinology 151:312–321
58. Guerriero KA, Keen KL, Millar RP, Terasawa E (2012) Developmental changes in GnRH release in response to kisspeptin agonist and antagonist in female Rhesus monkeys (*Macaca mulatta*): Implication for the mechanism of puberty. Endocrinology 153:825–836
59. Plant TM, Ramaswamy S, DiPietro MJ (2006) Repetitive activation of hypothalamic G protein coupled receptor 54 with intravenous pulses of kisspeptin in the juvenile monkey (*Macaca mulatta*) elicits a sustained train of gonadotropin-releasing hormone discharges. Endocrinology 147:1007–1013
60. Constantin S, Caligioni CS, Stojilkovic S, Wray S (2009) Kisspeptin-10 facilitates a plasma membrane-driven calcium oscillator in gonadotropin-releasing hormone-1 neurons. Endocrinology 150:1400–1412
61. Ohtaki T, Shintani Y, Honda S, Matsumoto H, Hori A, Kanehashi K, Terao Y, Kumano S, Takatsu Y, Masuda Y, Ishibashi Y, Watanabe T, Asada M, Yamada T, Suenaga M, Kitada C, Usuki S, Kurokawa T, Onda H, Nishimura O, Fujino M (2001) Metastasis suppressor gene KiSS-1 encodes peptide ligand of a G-protein-coupled receptor. Nature 411:613–617
62. Kotani M, Detheux M, Vandenbogaerde A, Communi D, Vanderwinden JM, Le Poul E, Brezillon S, Tyldesley R, Suarez-Huerta N, Vandeput F, Blanpain C, Schiffman SN, Vassart G, Parmentier M (2001) The metastasis suppressor gene KiSS-1 encodes kisspeptins, the natural ligands of the orphan G protein-coupled receptor GPR54. J Biol Chem 276:34631–34636
63. Roseweir AK, Kauffman AS, Smith JT, Guerriero KA, Morgan K, Pielecka-Fortuna J, Pineda R, Gottsch ML, Tena-Sempere M, Moenter SM, Terasawa E, Clarke IJ, Steiner RA, Miller RP (2009) Discovery of potent kisspeptin antagonists delineate physiological mechanisms of gonadotropin regulation. J Neurosci 29:3920–3929
64. Frost SI, Keen KL, Levine JE, Terasawa E (2008) Microdialysis methods for *in vivo* neuropeptide measurement in the stalk-median eminence in the rhesus monkey. J Neurosci Methods 168:26–34
65. Keen KL, Wegner FH, Bloom SR, Ghatei MA, Terasawa E (2008) An increase in kisspeptin-54 release occurs with the pubertal increase in luteinizing hormone-releasing hormone-1 release in the stalk-median eminence of female rhesus monkeys *in vivo*. Endocrinology 149:4151–4157
66. Guerriero KA, Keen KL, Terasawa E (2012) Developmental increase in kisspeptin-54 *in vivo* is independent of the pubertal increase in estradiol in female rhesus monkeys (*Macaca mulatta*). Endocrinology 153:1887–1897
67. Watanabe G, Terasawa E (1989) *In vivo* release of luteinizing hormone releasing hormone (LHRH) increases with puberty in the female rhesus monkey. Endocrinology 125:92–99
68. Chongthammakun S, Claypool LE, Terasawa E (1993) Ovariectomy increases *in vivo* LHRH release in pubertal, but not prepubertal, female rhesus monkeys. J Neuroendocrinol 5:41–50
69. Goodman RL, Lehman MN, Smith JT, Coolen LM, de Oliveira CV, Jafarzadehshirazi MR, Pereira A, Iqbal J, Caraty A, Ciofi P, Clarke IJ (2007) Kisspeptin neurons in the arcuate nucleus of the ewe express both dynorphin A and neurokinin B. Endocrinology 148:5752–5760
70. Wakabayashi Y, Nakada T, Murata K, Ohkura S, Mogi K, Navarro VM, Clifton DK, Mori Y, Tsukamura H, Maeda K, Steiner RA, Okamura H (2010) Neurokinin B and dynorphin A in kisspeptin neurons of the arcuate nucleus participate in generation of periodic oscillation of neural activity driving pulsatile gonadotropin-releasing hormone secretion in the goat. J Neurosci 30:3124–3132

71. Kawakami M, Uemura T, Hayashi R (1982) Electrophysiological correlates of pulsatile gonadotropin release in rats. Neuroendocrinology 35:63–67
72. Wilson RC, Kesner JS, Kaufman JM, Uemura T, Akema T, Knobil E (1984) Central electrophysiologic correlates of pulsatile luteinizing hormone secretion in the rhesus monkey. Neuroendocrinology 39:256–260
73. Ohkura S, Takase K, Matsuyama S, Mogi K, Ichimaru T, Wakabayashi Y, Uenoyama Y, Mori Y, Steiner RA, Tsukamura H, Maeda KI, Okamura H (2009) Gonadotrophin-releasing hormone pulse generator activity in the hypothalamus of the goat. J Neuroendocrinol 21:813–821
74. Navarro VM, Gottsch ML, Wu M, Garcia-Galiano D, Hobbs SJ, Bosch MA, Pinilla L, Clifton DK, Dearth A, Ronnekleiv OK, Braun RE, Palmiter RD, Tena-Sempere M, Alreja M, Steiner RA (2011) Regulation of NKB pathways and their roles in the control of Kiss1 neurons in the arcuate nucleus of the male mouse. Endocrinology 152:4265–4275
75. Ramaswamy S, Seminara SB, Plant TM (2011) Evidence from the agonadal juvenile male rhesus monkey (*Macaca mulatta*) for the view that the action of neurokinin B to trigger gonadotropin-releasing hormone release is upstream from the kisspeptin receptor. Neuroendocrinology 94:237–245
76. Ramaswamy S, Guerriero KA, Gibbs RB, Plant TM (2008) Structural interactions between kisspeptin and GnRH neurons in the mediobasal hypothalamus of the male rhesus monkey (*Macaca mulatta*) as revealed by double immunofluorescence and confocal microscopy. Endocrinology 149:4387–4395
77. Li XF, Kinsey-Jones JS, Cheng Y, Knox AM, Lin Y, Petrou NA, Roseweir A, Lightman SL, Milligan SR, Millar RP, O'Byrne KT (2009) Kisspeptin signalling in the hypothalamic arcuate nucleus regulates GnRH pulse generator frequency in the rat. PLoS One 4:e8334
78. Blake CA, Sawyer CH (1974) Effects of hypothalamic deafferentation on the pulsatile rhythm in plasma concentrations of luteinizing hormone in ovariectomized rats. Endocrinology 94:730–736
79. Krey LC, Hess DL, Butler WR, Espinosa-Campos J, Lu KH, Piva F, Plant TM, Knobil E (1981) Medial basal hypothalamic disconnection and the onset of puberty in the female rhesus monkey. Endocrinology 108:1944–1948
80. Plant TM, Moossy J, Hess DL, Nakai Y, McCormack JT, Knobil E (1979) Further studies on the effects of lesions in the rostral hypothalamus on gonadotropin secretion in the female rhesus monkey (*Macaca mulatta*). Endocrinology 105:465–473
81. Mittelman-Smith MA, Williams H, Krajewski-Hall SJ, Lai J, Ciofi P, McMullen NT, Rance NE (2012) Arcuate kisspeptin/neurokinin B/dynorphin (KNDy) neurons mediate the estrogen suppression of gonadotropin secretion and body weight. Endocrinology 153:2800–2812
82. Clarkson J, Han SK, Liu X, Lee K, Herbison AE (2010) Neurobiological mechanisms underlying kisspeptin activation of gonadotropin-releasing hormone (GnRH) neurons at puberty. Mol Cell Endocrinol 324:45–50
83. Chan YM, Broder-Fingert S, Wong KM, Seminara SB (2009) Kisspeptin/Gpr54-independent gonadotrophin-releasing hormone activity in Kiss1 and Gpr54 mutant mice. J Neuroendocrinol 21:1015–1023
84. Mitsushima D, Hei DL, Terasawa E (1994) GABA is an inhibitory neurotransmitter restricting the release of luteinizing hormone-releasing hormone before the onset of puberty. Proc Natl Acad Sci USA 91:395–399
85. Keen KL, Burich AJ, Mitsushima D, Kasuya E, Terasawa E (1999) Effects of pulsatile infusion of the GABAA receptor blocker bicuculline on the onset of puberty in female rhesus monkeys. Endocrinology 140:5257–5266
86. El Majdoubi M, Sahu A, Ramaswamy S, Plant TM (2000) Neuropeptide Y: a hypothalamic brake restraining the onset of puberty in primates. Proc Natl Acad Sci USA 97:6179–6184
87. Horvath TL, Bechmann I, Naftolin F, Kalra SP, Leranth C (1997) Heterogeneity in the neuropeptide Y-containing neurons of the rat arcuate nucleus: GABAergic and non GABAergic subpopulations. Brain Res 756:283–286

88. Vong L, Ye C, Yang Z, Choi B, Chua S Jr, Lowell BB (2011) Leptin action on GABAergic neurons prevents obesity and reduces inhibitory tone to POMC neurons. Neuron 71:142–154
89. Kurian JR, Keen KL, Guerriero KA, Terasawa E (2012) Tonic control of kisspeptin release in prepubertal monkeys: implications to the mechanism of puberty onset. Endocrinology 153:3331–3336
90. Plant TM, Fraser MO, Medhamurthy R, Gay VL (1989) Somatogenic control of GnRH neuronal synchronization during development in primates: a speculation. In: Plant TM, van Rees GP, Schoemaker J, Delamarre-van de Waal HA (eds) Control of the onset of puberty III. Excerpta Medica, Amsterdam, pp 111–121
91. Wilson ME, Gordon TP, Rudman CG, Tanner JM (1989) Effects of growth hormone on the tempo of sexual maturation in female rhesus monkeys. J Clin Endocrinol Metab J68:29–38
92. Mann DR, Plant TM (2010) The role and potential sites of action of thyroid hormone in timing the onset of puberty in male primates. Brain Res 1364:175–185
93. Terasawa E, Kurian JR, Keen KL, Shiel NA, Colman RJ, Capuano SV (2012) Body weight impact on puberty: effects of high-calorie diet on puberty onset in female rhesus monkeys. Endocrinology 153:1696–1705
94. Smith JT, Popa SM, Clifton DK, Hoffman GE, Steiner RA (2006) Kiss1 neurons in the forebrain as central processors for generating the preovulatory luteinizing hormone surge. J Neurosci 26:6687–6694
95. Cravo RM, Margatho LO, Osborne-Lawrence S, Donato J Jr, Atkin S, Bookout AL, Rovinsky S, Frazão R, Lee CE, Gautron L, Zigman JM, Elias CF (2011) Characterization of Kiss1 neurons using transgenic mouse models. Neuroscience 173:37–56
96. Franceschini I, Lomet D, Cateau M, Delsol G, Tillet Y, Caraty A (2006) Kisspeptin immunoreactive cells of the ovine preoptic area and arcuate nucleus co-express estrogen receptor alpha. Neurosci Lett 401:225–230
97. Foster DL, Ryan KD (1979) Endocrine mechanisms governing transition into adulthood: a marked decrease in inhibitory feedback action of estradiol on tonic secretion of luteinizing hormone in the lamb during puberty. Endocrinology 105:896–904
98. Dohrn M, Hohlweg W (1931) Hormonale beziehungen zwischen hypohysenvorderlappen und keimdrusen. In: Proceedings of the second International Congress on Sex Research. Oliver and Boyd, Edinburgh, pp 436–442
99. Steele RE, Weisz J (1974) Changes in sensitivity of estradiol-LH feedback system with puberty in the female rat. Endocrinology 95:513–520
100. Andrews WW, Advis JP, Ojeda SR (1981) The maturation of estradiol-negative feedback in female rats: evidence that the resetting of the hypothalamic "gonadostat" does not precede the first preovulatory surge of gonadotropins. Endocrinology 109:2022–2031
101. Kulin HE, Grumbach MM, Kaplan SL (1969) Changing sensitivity of the pubertal gonadal hypothalamic feedback mechanism in man. Science 166:1012–1013
102. Rapisarda JJ, Bergman KS, Steiner RA, Foster DL (1983) Response to estradiol inhibition of tonic luteinizing hormone secretion decreases during the final stage of puberty in the rhesus monkey. Endocrinology 112:1172–1179
103. Wilson ME (1995) IGF-1 administration advances the decrease in hypersensitivity to oestradiol negative feedback inhibition of serum LH in adolescent female monkeys. J Endocrinol 145:121–130

Chapter 13
Sex Steroid Regulation of Kisspeptin Circuits

Jeremy T. Smith

Abstract Kisspeptin cells appear to be the "missing link," bridging the divide between levels of gonadal steroids and feedback control of gonadotropin-releasing hormone (GnRH) secretion. Kisspeptin neurons are important in the generation of both sex steroid negative and estrogen positive feedback signals to GnRH neurons, the former being involved in the tonic regulation of GnRH secretion in males and females and the latter governing the preovulatory GnRH/luteinizing hormone (LH) surge in females. In rodents, kisspeptin-producing cells populate the anteroventral periventricular nucleus (AVPV) and the arcuate nucleus (ARC), and estrogen regulation of kisspeptin has been extensively studied in these regions. Kisspeptin cells in the ARC appear to receive and forward signals applicable to negative feedback regulation of GnRH. In the female rodent AVPV, kisspeptin cells are important for positive feedback regulation of GnRH and the preovulatory LH surge. In sheep and primates, a rostral population of kisspeptin cells is located in the dorsolateral preoptic area (POA) as well as the ARC. Initial studies showed kisspeptin cells in the latter were involved in both the positive and negative feedback regulation of GnRH. Interestingly, further studies now suggest that kisspeptin cells in the ovine POA may also play an important role in generating estrogen positive feedback. This chapter discusses the current consensus knowledge regarding the interaction between sex steroids and kisspeptin neurons in mammals.

J.T. Smith (✉)
Department of Anatomy, Physiology and Human Biology, The University of Western Australia, M309, Hackett Entrance 2, Hackett Drive, Nedlands, Perth, WA 6009, Australia
e-mail: jeremy.smith@uwa.edu.au

A.S. Kauffman and J.T. Smith (eds.), *Kisspeptin Signaling in Reproductive Biology*, Advances in Experimental Medicine and Biology 784, DOI 10.1007/978-1-4614-6199-9_13,

Introduction: First Kiss

Not since the discovery of gonadotropin-releasing hormone (GnRH) has a neuropeptide created such a stir of activity within the field of reproduction. Kisspeptin research has been rapid and momentous, and it is undeniable that this peptide is critical for many neuroendocrine aspects of fertility (as indicated in this textbook). The neuroendocrine reproductive axis is governed through meticulously controlled neural and hormonal interactions between the brain, pituitary, and gonads. In the brain, the final common factor in the control of this hierarchical system is the release of GnRH from neurons extending to the median eminence. GnRH stimulates the pituitary gonadotropes and leads the production of gonadal sex steroids (androgens, estrogens, and progestins), which then "feedback" to GnRH neurons to control their activity [1]. In males and females, testosterone, estradiol, and progesterone act via tonic negative feedback to suppress GnRH secretion, which is necessary for optimal steroidogenesis and gametogenesis. In females, there is an additional switch from negative feedback to acute estrogen positive feedback, which precedes and causes the GnRH/LH surge, necessary for ovulation. Although GnRH was discovered more than 30 years ago, the precise cellular and molecular mechanisms governing sex steroid negative and positive feedback on GnRH secretion have proved far more challenging to uncover. GnRH neurons do not express androgen receptor, estrogen receptor α (ERα), or progesterone receptor (PR) [2–4]; thus, the feedback effects of sex steroids on GnRH secretion must be transferred by other steroid-sensitive neurons. Enter kisspeptins.

Produced from the *Kiss1* gene, kisspeptin peptides are robust stimulators of GnRH secretion [5]. Evidence for this now includes (1) the stimulatory effect of kisspeptin on gonadotropin secretion is blocked by GnRH antagonists [5–8]; (2) injections of kisspeptin directly in to the preoptic area (POA), where GnRH neurons reside, stimulate LH secretion [9]; (3) kisspeptin activates (as determined by fos induction) GnRH neurons [6, 7]; (4) kisspeptin directly stimulates the electrophysiological properties of GnRH neurons [10, 11]; (5) kisspeptin-immunoreactive fibers appose GnRH neurons [12–14]; (6) kisspeptin stimulates GnRH release into the portal circulation of sheep [15]; and (7) almost all GnRH neurons express the kisspeptin receptor (Kiss1r) [6, 10, 16]. Importantly, the effects of kisspeptin are absent in Kiss1r knockout mice, showing specificity to this receptor [17–19].

First Clues: *Kiss1* Distribution and Kisspeptin Projections

Discrete localization of *Kiss1* mRNA in the brain was first determined in the mouse by in situ hybridization and immediately provided clues to the physiological role of kisspeptin. Cellular populations were identified primarily in the anteroventral periventricular nucleus (AVPV), extending to periventricular nucleus (PEN), and the arcuate nucleus (ARC) [5]. Smaller, less dense populations of *Kiss1*-expressing

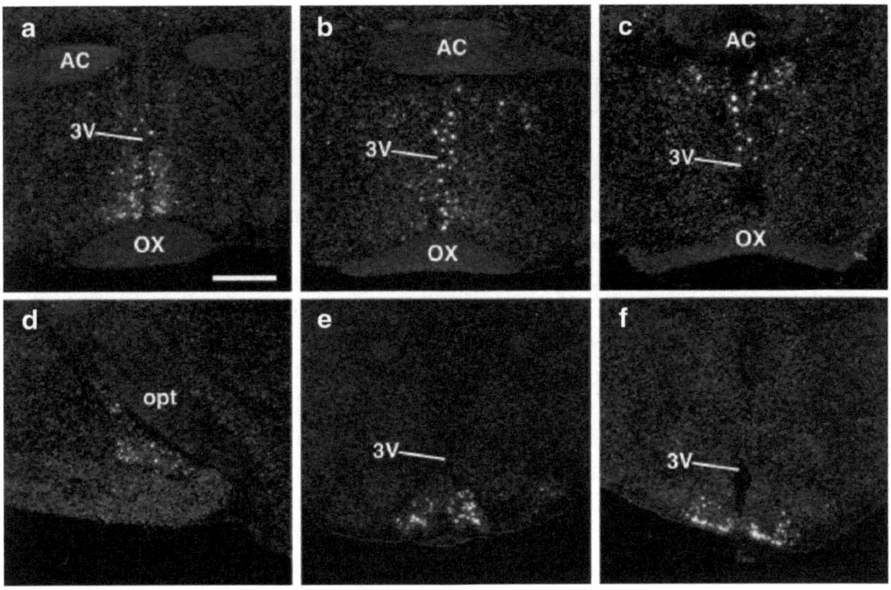

Fig. 13.1 The original photomicrographs detailing *Kiss1* mRNA distribution in the hypothalamus. Silver grain clusters indicate areas where the labeled riboprobe is concentrated revealing *Kiss1* mRNA-expressing neurons. Cells were observed in the AVPV (**a**), periventricular nucleus (**b**), anterodorsal preoptic nucleus (**c**), the medial amygdala (**d**), and the ARC (**e**, **f**). *3V* third ventricle; *AC* anterior commissure; *opt* optic tract; *OX* optic chiasm. Scale bar = 500 μm. Data taken from Smith JT, Dungan HM, Stoll EA, Gottsch ML, Braun RE, Eacker SM, Clifton DK, Steiner RA 2005 Differential regulation of KiSS-1 mRNA expression by sex steroids in the brain of the male mouse. Endocrinology 146:2976–2984 with permission from The Endocrine Society

cells were also noted in the anterodorsal POA, the medial amygdala, and the bed nucleus of the stria terminalis (Fig. 13.1). *Kiss1* mRNA localization in the AVPV and ARC was also apparent in rats [6, 20, 21] and hamsters [22, 23]. Confirmation of kisspeptin production in the AVPV and ARC was later provided with immunohistochemistry studies in mice [12, 24], rats [25], and hamsters [22]. In sheep, *Kiss1* mRNA-expressing cells were shown to be located in the dorsolateral region of the POA (perhaps a homologous population to the rodent AVPV/PEN) and the ARC [26, 27], with a similar distribution of kisspeptin-immunoreactive cells [14, 28]. Populations of kisspeptin cells (mRNA and protein) in the POA and ARC have also been reported in both human and nonhuman primates [8, 29, 30].

Kisspeptin Neurons Finding Their Way

If kisspeptin cells were critical in the feedback control of GnRH secretion, it would be expected that kisspeptin neuronal fibers make putative contacts with GnRH neuronal cell bodies. In mice, kisspeptin-immunoreactive fibers are located within the ventral aspect of the lateral septum and the hypothalamus, running in periventricular and ventral retrochiasmatic pathways [24], with a similar distribution in sheep [28] and monkeys [29]. Kisspeptin fibers make connections and close appositions to GnRH neurons in mice [12], rats [13], sheep [14], and monkeys [29]. Despite this, the anatomical origins of kisspeptinergic inputs to GnRH neurons are yet to be fully determined in any species. In mice, evidence suggests that direct input to GnRH cell bodies is from AVPV kisspeptin cells, but unlikely to be from kisspeptin cells originating in the ARC [31]. However, more recent studies from this group indicate ARC kisspeptin cells do project to the POA [32], and a small percentage (<20%) do appose GnRH neurons [33]. A similar phenomenon may exist in sheep, because ARC kisspeptin cells do not provide substantial input to GnRH cells in the ventromedial POA, whereas POA kisspeptin cells do [34]. It was suggested (at least in sheep) that kisspeptin cells in the POA could form an interneuronal bridge linking ARC kisspeptin cells to GnRH cell bodies; this may be particularly relevant in terms of estrogen positive feedback regulation, but now appears unlikely because no kisspeptin neurons express Kiss1r [15]. Alternatively, it is now proposed that the majority of ARC kisspeptin neurons find their way to GnRH neurons through terminal-to-terminal communication at the median eminence. Evidence for this stems from in vitro cultures of mediobasal hypothalamic explants challenged with kisspeptin [35]. Similar data also show direct stimulation of GnRH release from the isolated ovine median eminence [15]. Kisspeptin fibers are abundant in the external zone of the median eminence of sheep [36] and monkeys [29] where they appear to appose GnRH terminals. However, the distribution of kisspeptin fibers in the median eminence of mice (12, 24) and rats (12, 24, 25) appears to be less abundant. Whether these kisspeptin terminals are apposed to GnRH terminals remains to be determined.

Kisspeptin, the "Missing Link" in Sex Steroid Control of GnRH Secretion

In rodent species, sex steroid-sensitive neurons projecting from the ARC have been implicated in the negative feedback control of GnRH secretion by estrogen [37–41]. Alternatively, the rodent AVPV/PEN region is a sexually dimorphic nucleus and is recognized to play an important role in mediating positive feedback effects of estrogen to induce the preovulatory surge of GnRH and LH [31, 42–44]. Kisspeptin cells in the ARC and AVPV are ideally placed in rodents to mediate these feedback effects. Moreover, like the LH surge itself, *Kiss1* mRNA expression in the AVPV is sexually dimorphic, with the female AVPV harboring far more *Kiss1*-positive cells than the male [12, 20].

In sheep, key differences are apparent in feedback regulation of GnRH compared to rodents. Estrogen-sensitive cells in the ARC appear central for *both* the negative feedback regulation of GnRH and the positive feedback response to estradiol that causes the preovulatory GnRH/LH surge [45, 46]. Again, kisspeptin cells located in the ARC are prime candidates to mediate these feedback effects. Importantly, in the ovine species, kisspeptin-expressing cells in the ARC are more abundant in females than males [47], which likely relates to a role in positive feedback. In primates, classical studies in rhesus monkeys where the mediobasal hypothalamus was surgically isolated from the rest of the brain show no interference in either estrogen negative or positive feedback [48–50]. While such data convey the likely importance of ARC kisspeptin cells and potentially eliminate a critical role for the POA kisspeptin cells, in sex steroid feedback, an active role for the POA cannot be ruled out. Bilateral lesions to the POA in monkeys compromised positive feedback [51]. Similar results were also noted after separation of the anterior hypothalamus and mediobasal hypothalamus [52]. Thus, in the primate, the anterior hypothalamic nuclei (potentially kisspeptin cells in the POA) may play a role in mediating some aspects of sex steroid feedback control. Whether there is a sex difference in kisspeptin expression in primates has yet to be determined.

Sex Steroid Regulation of *Kiss1* in Females

Estradiol robustly and differentially regulates the expression of *Kiss1* mRNA in a site-specific manner. These data were uncovered using in situ hybridization, which allows both quantification and histological mapping of *Kiss1* mRNA. In the ARC, ovariectomized (OVX) female mice show a pronounced upregulation in the number of *Kiss1* mRNA-expressing cells, which is prevented by estradiol replacement (Fig. 13.2) [53]. In striking contrast, in the AVPV, OVX mice show reduced expression of *Kiss1* mRNA, and estradiol replacement stimulated its expression (Fig. 13.2) [53]. Similar data were subsequently forthcoming in female rats [21]. These findings led to the hypothesis that kisspeptin neurons in the ARC transmit signals to GnRH neurons pertinent to estrogen negative feedback, while kisspeptin cells in the AVPV relay estradiol signals for positive feedback.

In rodents, virtually all kisspeptin cells in the ARC and AVPV express ERα [21, 53], and the importance of ERα expression on kisspeptin neurons in females is unquestionable. Unlike female wild-type mice, female ERα knockout mice are unable to regulate AVPV or ARC *Kiss1* mRNA expression in the face of OVX and estradiol replacement (Fig. 13.3) [53]. Similarly, in female rats, the selective ERα agonist, propyl pyrazole triol, is able to suppress hypothalamic *Kiss1* mRNA expression, as determined by RTPCR (however, this data lack anatomical specificity) [54]. Interestingly, the selective ERβ agonist, diarylpropionitrile, was unable to inhibit the post-OVX rise in *Kiss1* mRNA [54], and so the role of this estrogen receptor subtype for kisspeptin regulation is questionable. Despite this, a surprisingly significant proportion, but not the majority, of ARC and AVPV kisspeptin neurons express ERβ [53]. However, in ERβ knockout mice, the ability of OVX and estradiol replacement to regulate *Kiss1* mRNA expression is maintained [53]. These data reinforce the notion that ERβ does not play a significant role in estrogen feedback

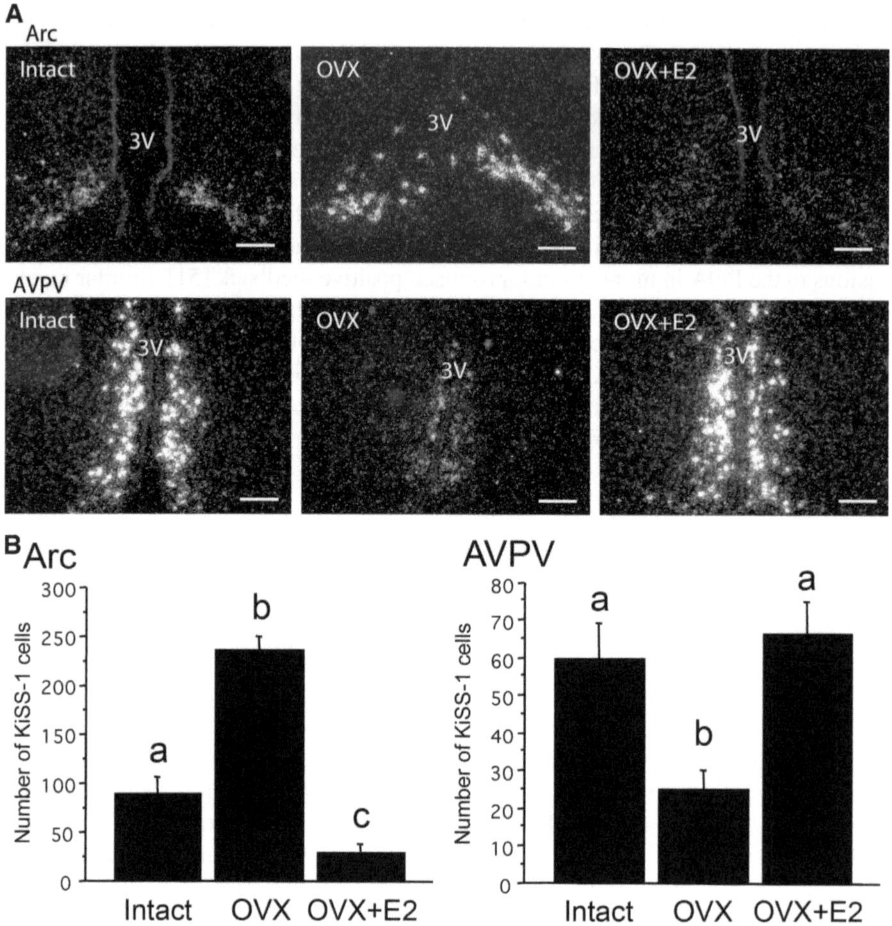

Fig. 13.2 Differential estradiol regulation of *Kiss1* mRNA in the anteroventral periventricular nucleus (AVPV) and arcuate nucleus (ARC) of the female mouse. (**a**) Photomicrographs show *Kiss1* mRNA-expressing cells in representative sections of the AVPV and ARC from ovary-intact, ovariectomized (OVX), and OVX+estradiol-treated (E2) mice. *3V* third ventricle. Scale bars = 100 μm. (**b**) Quantification of the data appears below. Values (mean ± SEM) without common notations (a, b, c) differ significantly ($P<0.01$). Data taken from Smith JT, Cunningham MJ, Rissman EF, Clifton DK, Steiner RA 2005 Regulation of Kiss1 gene expression in the brain of the female mouse. Endocrinology 146:3686–3692 with permission from The Endocrine Society

regulation of GnRH neurons. Alternatively, ERβ may play some other role in kisspeptin neurons, perhaps independent of *Kiss1* gene regulation.

In sheep, OVX stimulates *Kiss1* mRNA expression in the ARC, and estradiol replacement prevents this effect [14, 27], similar to that seen in rodents. In sheep, progesterone replacement has also been shown to inhibit *Kiss1* expression in the ARC, albeit to a lesser extent than estradiol [27]. Given the pronounced role for progesterone in negative feedback in this species, this result is somewhat surprising and may indicate a lesser role for progesterone in *Kiss1* regulation (few studies have

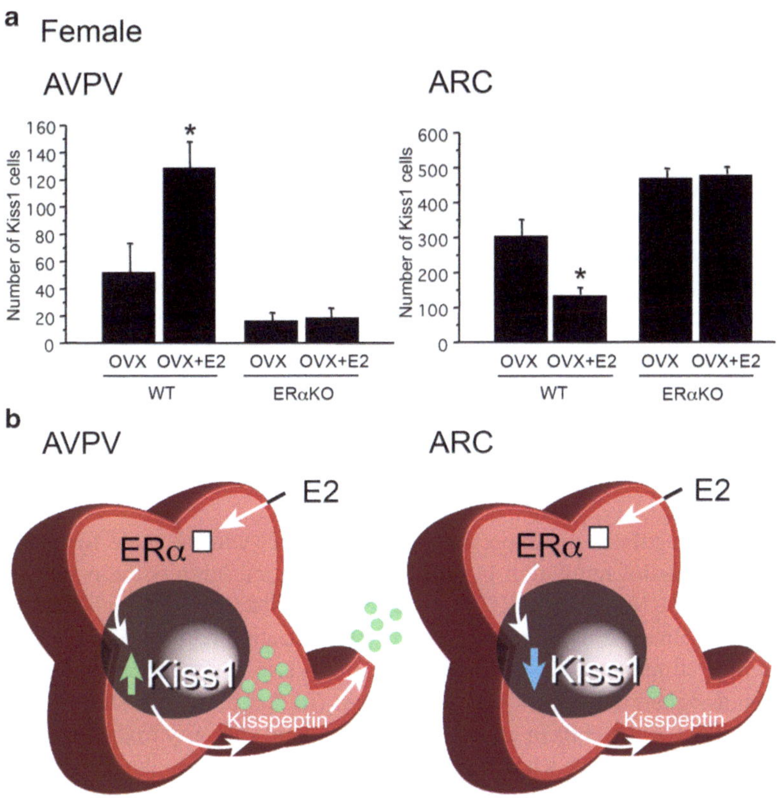

Fig. 13.3 In females, estradiol (E2) regulation of *Kiss1* in the anteroventral periventricular nucleus (AVPV) and arcuate nucleus (ARC) is mediated via estrogen receptor α (ERα) signaling. (**a**) Data show the effects of ovariectomy (OVX) with or without estradiol (E2) replacement on *Kiss1* mRNA in the AVPV and ARC of wild-type (WT) and ERα knockout (ERαKO) mice. Note the lack of effect in ERαKO mice. *Asterisk*, $P<0.05$. Data taken from Smith JT, Cunningham MJ, Rissman EF, Clifton DK, Steiner RA 2005 Regulation of Kiss1 gene expression in the brain of the female mouse. Endocrinology 146:3686–3692 with permission from The Endocrine Society. (**b**) Schematic diagram demonstrating the likely pathways leading to *Kiss1* regulation by E2 in the female. *ERα* estrogen receptor α

sought to determine the independent role of progesterone on kisspeptin neurons), or be more likely due to reduced progesterone receptor expression in the absence of estradiol in OVX ewes [55]. Not surprisingly, all kisspeptin neurons in the ovine ARC express ERα [28], as well as progesterone receptor [27] (determined in ovary-intact animals during the luteal phase), but a much smaller proportion of kisspeptin cells in the POA appears to do so [28]. In the ovine POA, estradiol treatment increases *Kiss1* mRNA expression [27], as it does in the rodent AVPV, and so the differential regulation of kisspeptin persists in this animal model, again suggesting different regions controlling negative (ARC) and positive feedback (POA). In this species, however, an additional layer of complexity exists (as described earlier) in that estradiol-induced positive feedback is generated from the mediobasal hypothalamus (containing the ARC), not the POA [45, 46]. Thus, it appears in this species that kisspeptin neurons in the ARC are additionally able to respond in a manner appropriate for positive feedback (see below).

In humans, *KISS1* mRNA expression is increased in the ARC of postmenopausal women to a similar degree to that seen in OVX cynomolgus macaques [30]. Given that menopause results from ovarian failure, the model may be regarded as similar to chronic OVX. These data then suggest that kisspeptin neurons in the ARC mediate estrogen negative feedback in humans, as they appear to do in rodents and sheep. Interestingly, kisspeptin neurons also populate the POA in humans [30]; however there have been no reports of this population's regulation via estradiol or change in postmenopausal women.

Sex Steroid Regulation of *Kiss1* in Males

Testosterone regulates the expression of *Kiss1* mRNA in the male hypothalamus. Similar to estradiol regulation in females, testosterone inhibits the post-castration rise of *Kiss1* mRNA in the ARC of mice [56]. Consistent data are also apparent in male rats [6]. Interestingly, in the male AVPV, castration reduces *Kiss1* mRNA and testosterone stimulates its expression [56]. Thus, kisspeptin neurons in the ARC can again be assigned the task of relaying steroid (in this case, testosterone) negative feedback. Kisspeptin cells in the male AVPV respond similarly to the female AVPV, but we are left to ponder what physiological role they play (males normally do not generate sex steroid positive feedback). There is a significant sex difference in the abundance of kisspeptin neurons in the AVPV, with females harboring far more *Kiss1*-positive cells in the AVPV than males [12, 20] (discussed in greater detail in Chap. 11). Nevertheless, kisspeptin cells in the male AVPV are clearly different than those in the ARC, especially in their regulation by sex steroids, and may play a role in other testosterone-mediated processes.

The majority of kisspeptin neurons in the male AVPV and ARC co-express ERα and AR [56, 57]. Unlike females, the signaling mechanisms mediating the actions of testosterone appear to be different in the two populations of kisspeptin cells. In the AVPV, the effects of testosterone appear to be mediated by ERα or ERβ, because estradiol treatment is able to fully mimic the effect of testosterone (Fig. 13.4) [56]. Treatment with the non-aromatizable dihydrotestosterone (DHT) had no effect, indicating no action via the androgen receptor [56]. Using ERα knockout mice, the effect of testosterone on kisspeptin regulation in the AVPV was maintained [56], indicating ERα did not mediate this effect and it was therefore possibly due to ERβ. In the male ARC, both DHT and estradiol were able to mimic the inhibitory effect of testosterone, indicating the regulation of *Kiss1* here is mediated by both estrogen receptor (most likely ERα) and androgen receptor (Fig. 13.4) [56]. Not surprising then was that in male mice lacking ERα or complete functional androgen receptor, there was no disturbance in testosterone-mediated inhibition of *Kiss1* in the ARC (with one receptor compensating for the absence of the other) [56].

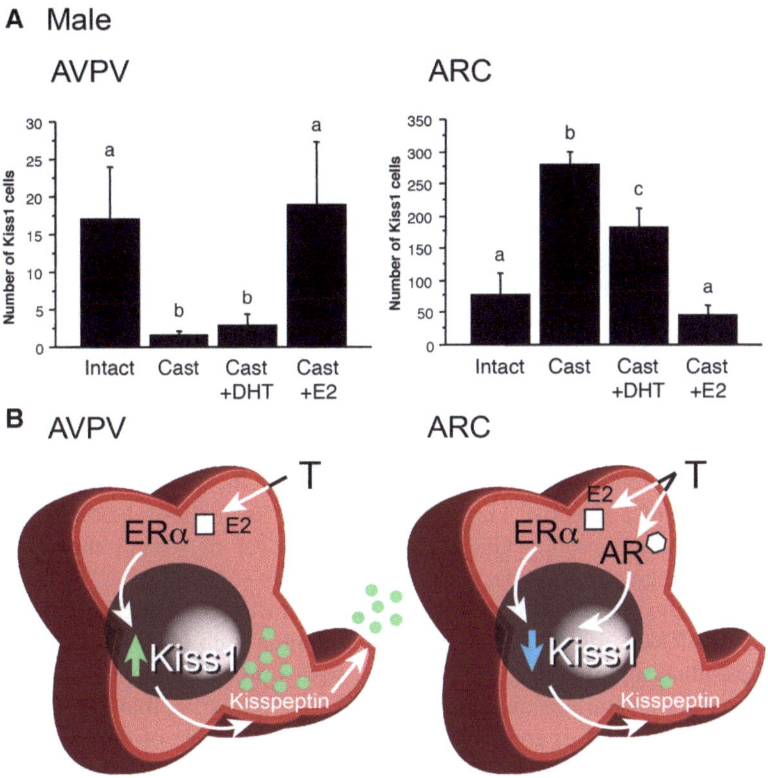

Fig. 13.4 In male mice, testosterone regulation of *Kiss1* in the anteroventral periventricular nucleus (AVPV) is mediated via estrogen signaling while in the arcuate nucleus (ARC), testosterone acts through both androgen and estrogen signaling. (**a**) Data show the effects of castration (Cast) and dyhydrotestosterone (DHT) or estradiol (E2) treatment on *Kiss1* mRNA in the AVPV and ARC. Values (mean ± SEM) without common notation (a, b, c) differ significantly ($P<0.05$). Data taken from Smith JT, Dungan HM, Stoll EA, Gottsch ML, Braun RE, Eacker SM, Clifton DK, Steiner RA 2005 Differential regulation of KiSS-1 mRNA expression by sex steroids in the brain of the male mouse. Endocrinology 146:2976–2984 with permission from The Endocrine Society. (**b**) Schematic diagram demonstrating the likely pathways leading to *Kiss1* regulation by testosterone (T) in the male. *AR* androgen receptor; *ERα* estrogen receptor α

Recent data show sex steroids also regulate kisspeptin expression outside the hypothalamus. *Kiss1* mRNA expression in the medial nucleus of the amygdala, a region implicated in various aspects of reproduction, including social and emotional behaviors, appears greater in males than in females [58]. Moreover, the expression appears to be upregulated by testosterone and/or estradiol. Testosterone's inductive effect on *Kiss1* expression in the male amygdala most likely occurs through estrogen receptor-dependent pathways, not through the androgen receptor, because DHT treatment had no affect on amygdala *Kiss1* levels. The precise role for kisspeptin in the amygdala is not yet known. Sexual behavior is an obvious candidate, but it should be noted that male Kiss1r knockout mice display normal sexual behaviors when adequate sex steroid levels are provided [18].

Kisspeptin Controls Negative Feedback Regulation of GnRH Secretion

Sex steroids clearly inhibit kisspeptin cells in the ARC in numerous species. Paired with the aforementioned data linking cells in the ARC to negative feedback, kisspeptin cells are well placed to be central to this process. Consistent with negative feedback, the post-gonadectomy rise in *Kiss1* expression is paired with a corresponding rise in LH secretion, in an attempt to increase sex steroid production [53, 56]. In Kiss1r knockout mice, gonadectomy and sex steroid replacement are able to regulate *Kiss1* mRNA expression appropriately, but in these mice, the subsequent stimulation of LH release was not evident [17], even after the withdrawal of sex steroid treatment [59] (although a modest response in FSH secretion was detectable), indicating the importance of kisspeptin signaling in relaying negative feedback. Consistent with this, LH secretion in gonadectomized rats and mice, and pulsatile LH secretion in female monkeys and OVX sheep, is inhibited by central administration of a selective kisspeptin antagonist [60]. These data again demonstrate that the stimulation of GnRH/LH after gonadectomy—via removal of negative feedback regulation—is dependent on kisspeptin signaling.

Alternatively, in mice with a targeted deletion of ERα only in kisspeptin neurons (termed "KERKO" mice), gonadotropin secretion is high [61]. Indeed, in female KERKOs, abnormally high gonadotropins cause a precocious puberty-like model where the external markers of puberty arise 15–16 days earlier than wild-type counterparts. In this model, it appears there is a chronic absence of negative feedback causing unrestrained gonadotropin release. Importantly, in these animals the number of identifiable kisspeptin cells in the AVPV is low, but in the ARC, the number of identifiable kisspeptin cells and the expression of *Kiss1* mRNA are elevated [61], reflecting the absence of sex steroid feedback to these populations. These mice illustrate both the critical role of ARC kisspeptin cells in negative feedback and the importance of ERα signaling on kisspeptin neurons for this process.

In goats, evidence suggests that the intrinsic GnRH pulse generator (controlled by negative feedback) is located in the caudal ARC, where kisspeptin cell bodies are located [62] (discussed in greater detail within this textbook). Growing evidence also suggests ARC kisspeptin cells drive pulsatile GnRH secretion via the autosynaptic activity of the neuropeptides neurokinin B (NKB) and dynorphin, which are co-expressed in virtually all ARC kisspeptin expression neurons [63]. Neurokinin B (stimulatory) and dynorphin (inhibitory) are proposed to coordinate the pulsatile discharge/release of kisspeptin from the ARC [47, 64, 65] (discussed in greater detail within this textbook). Thus, kisspeptin cells in the ARC are geared to receive estrogen negative feedback signals and modulate their activity appropriately, which in turn regulates the stimulatory control of GnRH neurons to maintain the tonic/pulsatile secretion of GnRH and LH.

Kisspeptins Are Critical for the Positive Feedback Preovulatory GnRH/LH Surge

A Case for the AVPV in Rodent Species

In mice and rats, estradiol stimulates *Kiss1* expression in the AVPV [21, 25, 53, 56]. In these species, the GnRH/LH surge is generated by estrogen-sensitive cells in the AVPV [31, 42–44]. Moreover, lesions of the AVPV can block the surge in rats [66]. This and other evidence strongly suggest the sexually differentiated population of *Kiss1* cells in the AVPV provides the conduit for positive feedback signals to reach GnRH neurons in rodent species. At the time of the preovulatory GnRH/LH surge (the evening of proestrus), *Kiss1* expression in the AVPV of mice and rats is upregulated, and these cells also become transcriptionally activated (showing an induction of Fos) (Fig. 13.5) [21, 25, 67]. Moreover, recent data show that the GnRH/LH surge can be blocked in rats with a central infusion of a kisspeptin antagonist [68], and Kiss1R knockout mice are unable to mount an LH surge in response to estradiol treatment designed to induce positive feedback [67]. Similarly, in mice with targeted deletions of ERα only in kisspeptin cells (KERKO mice), no LH surges are apparent in "postpubertal" females [61], showing kisspeptin cells, and specifically ERα signaling in these cells is critical for the surge.

It appears that a rise in *Kiss1* expression in the AVPV in response to estradiol does not always predicate a GnRH/LH surge. OVX mice treated with chronic estradiol have elevated *Kiss1* mRNA in the AVPV, but "for the most part" have suppressed levels of LH [53]. Given that a rise in AVPV *Kiss1* expression from the morning to the evening of proestrus in rats is seen in the face of relatively constant levels of estradiol [21], it is hypothesized that other inputs, namely, circadian signals, also regulate AVPV *Kiss1* expression (discussed in greater detail in Chap. 18). Projections from the suprachiasmatic nucleus (SCN) to the AVPV are known to constrain the GnRH/LH surge to a circadian onset [69, 70]. Not surprising then, *Kiss1* expression and activation in the AVPV of estradiol treated mice also show a circadian pattern in synchrony to the timing of the GnRH/LH surge [71]. In hamsters, it has been shown that the SCN targets AVPV kisspeptin neurons via vasopressinergic projections [72], and recent data in rats show it is the dorsomedial SCN that times *Kiss1* expression in the AVPV to be in phase with the LH surge [73, 74]. This circadian activation of *Kiss1* in the AVPV also appears to be estrogen dependent [71]. Overall, it is clear that kisspeptin neurons within the AVPV region of the mouse and rat are stimulated by estradiol and well placed to provide critical inputs to GnRH neurons to stimulate the preovulatory GnRH/LH surge.

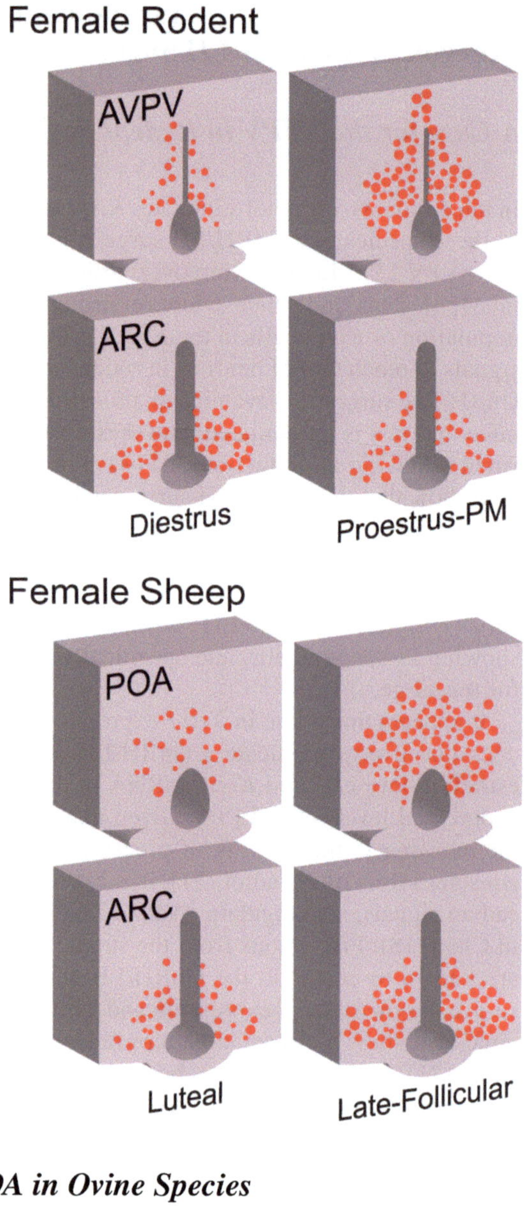

Fig. 13.5 Schematic diagram demonstrating *Kiss1* upregulation during the GnRH/LH surge in female rodents and sheep. In rodents, *Kiss1* mRNA (*red dots*) increases in the anteroventral periventricular nucleus (AVPV) at the time of the preovulatory GnRH/LH surge (Proestrus PM). In the arcuate nucleus (ARC), *Kiss1* mRNA is reduced at this time. In sheep, *Kiss1* mRNA is stimulated in the caudal ARC and preoptic area (POA) immediately prior to the GnRH/LH surge (late follicular)

A Case for the ARC and POA in Ovine Species

In sheep, kisspeptin administration is able to synchronize preovulatory LH surges in ovary-intact cycling ewes and also stimulate the surge in seasonally acyclic ewes [75, 76]. Conversely, kisspeptin antagonist administration is able to inhibit the estradiol-induced LH surge [15]. In this species, there are three types of estrogen feedback important for the control of GnRH secretion: short-term negative feedback, long-term negative feedback, and acute positive feedback [77]. As is the case for negative feedback, cells within the mediobasal hypothalamus are thought to be crucial in mediating estrogen positive feedback effects on GnRH secretion [45, 46]. Consistent with this, *Kiss1* expression is greatest in the caudal region of the ovine ARC immediately prior to the GnRH/LH surge [16, 26]. Moreover, transcriptional activation of kisspeptin cells increases in the mid- and caudal ARC of OVX ewes when treated with a positive feedback-inducing estradiol stimulus [16]. Thus, it appears kisspeptin cells in the ARC are important for both estrogen negative (as discussed above) and positive feedback effects (Fig. 13.5). The discriminating factor here being kisspeptin cells in the mid- to caudal region of the ARC are involved in the estrogen positive feedback GnRH/LH surge, while kisspeptin cells across the whole ARC appear responsive to chronic estrogen negative feedback regulation. It is possible that discrete regions of the ARC distinguish different estrogen stimuli (acute rise in estradiol vs. chronic levels) and transmit negative or positive regulation of GnRH secretion. Alternatively, the same kisspeptin cells in the caudal ARC may respond to both negative and positive feedback stimuli, possibly involving classical and nonclassical estrogen signaling pathways, as discussed below [78]. In this regard, it is pertinent to note that acute (surge-inducing) estradiol treatment induced Fos expression in the vast majority of caudal ARC kisspeptin cells in sheep, while after OVX, approximately half of the kisspeptin cells in the same region had induced Fos expression [16]. Thus, it is probable that some kisspeptin cells are responding to both positive and negative estrogen feedback signals, at least in sheep.

Similar to data in the caudal ARC, *Kiss1* expression in the ovine POA was also greater just prior to the preovulatory GnRH/LH surge (Fig. 13.5) [16]. More recent data show transcriptional activation of kisspeptin cells in the POA during the surge [79]. Because estradiol treatment appears to increase *Kiss1* expression in the ovine POA [14], with similar reports in the pig [80], a parallel is drawn between the ovine POA and the rodent AVPV. Moreover, these data implicate kisspeptin cells in both the caudal ARC and the POA as central processors of the feedback effects of estradiol that cause the GnRH/LH surge in the ovine species (Fig. 13.5). Data also show upregulation of *Kiss1* in the rhesus monkey ARC and POA at the time of the GnRH/LH surge, suggesting both populations are involved in the preovulatory GnRH/LH surge of nonhuman primates [81]. It is unclear exactly how POA kisspeptin neurons in the sheep participate in the estradiol-induced GnRH/LH surge. Unlike the ARC or the rodent AVPV, fewer (~50%) *Kiss1* cells in the ovine POA express ERα [28]. It is entirely possible that estrogen can activate these neurons directly, but one alternative possibility is that kisspeptin cells in the POA are activated indirectly by estrogen positive feedback.

Mechanism for Differential Regulation

The mechanism for differential regulation of *Kiss1* expression in the ARC and AVPV is not entirely known. Clues may arise through the search for phenotypical differences between kisspeptin cells in the ARC and AVPV. In mice, between 50 and

80% of the kisspeptin neurons in the AVPV also express tyrosine hydroxylase (indicating the sexually differentiated population of dopaminergic neurons), whereas no tyrosine hydroxylase co-expression is noted in the ARC [33, 74]. However, lower *Kiss1*-TH co-expression is seen in the AVPV of other species. In the female rat, kisspeptin neurons in the AVPV and ARC are virtually distinct from TH neurons in these regions (5–30% of *Kiss1* co-express tyrosine hydroxylase in the rat AVPV) [20]. Similar data are also apparent in the POA and ARC of sheep (*Smith JT, unpublished observation*). The distribution of galanin, neurotensin, met-enkephalin (mENK), and cholecystokinin (CCK)-immunoreactive cells was recently determined within the AVPV of female mice [82]. A small proportion (<10%) of kisspeptin cells in the AVPV express galanin whereas a larger percentage (30–40%) expressed mENK. Interestingly, kisspeptin cells in the ARC had a similar degree of galanin co-expression but did not co-express mENK [82]. Virtually all kisspeptin neurons in the ARC co-express the NKB and dynorphin, but no kisspeptin neurons in the AVPV (or POA) do so [63, 64, 83]. Importantly, ARC kisspeptin cells also express genes for the NKB receptor (NK3) and the dynorphin receptor [the kappa opioid receptor (KOR)]. Expression of dynorphin, NKB, KOR, and NK3 is inhibited by estradiol in the ARC [64]. From this, a model whereby NKB and dynorphin act autosynaptically on kisspeptin neurons in the ARC to potentially shape pulsatile expression and secretion of kisspeptin has been proposed. This is then thought to drive the pulsatile release of GnRH from fibers in the median eminence.

Within kisspeptin neurons, the response to ERα may be regulated by the actions of SP transcription factor protein complexes. In GT1-7 cells, estradiol increases *Kiss1* promoter activity through the Sp1 binding site at the proximal promoter region [84]. Sp1 and ERα form a complex resulting in estrogen-induced activation of the *Kiss1* promoter; however, decreasing the Sp1 to Sp3 ratio negatively regulates *Kiss1* promoter activity [84]. A differing Sp1 to Sp3 ratio between the ARC and the AVPV could potentially explain the differential regulation of *Kiss1* in response to estrogen, but this remains to be determined. On the other hand, GT1-7 cells are dedifferentiated GnRH cells, so one should be cautious interpreting the relevance of this data. The presence of *Kiss1* expression in these cells does not match animal data (GnRH cells do not express *Kiss1* in vivo). Thus, it is possible that aspects of *Kiss1* expression and regulation in GT1-7 cells do not relate to normal in vivo physiology.

In vivo data suggest it may be differences in estrogen signaling pathways that defines the regulation of kisspeptin. Estrogen can act via ERα through multiple signaling pathways, the major pathways being grouped into classical vs. nonclassical estrogen signaling. Classical signaling involves the translocation of ERα into the nucleus, where it recruits cofactors to the estrogen response element (ERE) regulatory sites to alter gene transcription, while for nonclassical signaling, the ERα employ ERE-independent genomic pathways that entail interactions with transcription factors. Data from a mouse model, permitting only nonclassical estrogen signaling, demonstrate that the inhibition of *Kiss1* mRNA in the ARC is mediated by an ERE-independent ERα signaling mechanism—nonclassical estrogen signaling (Fig. 13.6) [78]. Conversely, the stimulation of *Kiss1* in the AVPV is likely to occur through classical, but not nonclassical, signaling pathways (Fig. 13.6) [78].

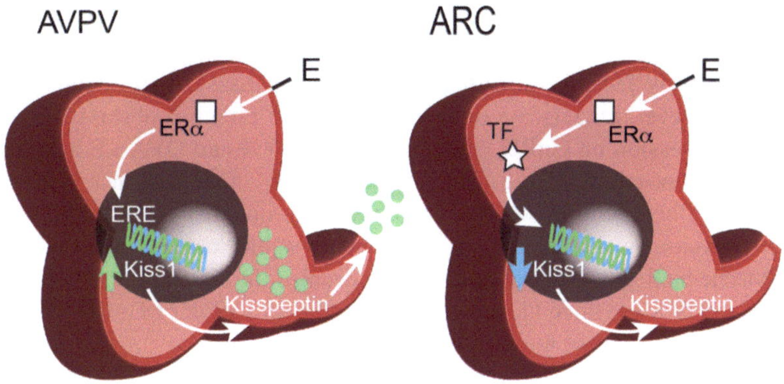

Fig. 13.6 Schematic diagram demonstrating the likely pathways resulting in differential regulation of *Kiss1* in the anteroventral periventricular nucleus (AVPV) and arcuate nucleus (ARC) of female mice. In the AVPV, the stimulation of *Kiss1* is likely to occur through classical and not nonclassical signaling pathways. Conversely, in the ARC, the inhibition of *Kiss1* mRNA is mediated by a nonclassical ERE-independent ERα signaling mechanism possibly involving interactions with transcription factors (TF)—nonclassical estrogen signaling [5]

This is consistent with a report showing the positive feedback effects of estradiol require classical signaling, whereas negative feedback involves nonclassical signaling mechanisms [85].

Sex Steroid Regulation of Kisspeptin Sensitivity

It is clear that kisspeptin "output" from specific hypothalamic nuclei is increased at the time of the preovulatory GnRH/LH surge. In addition to this, the sensitivity of the hypothalamic–pituitary–gonadal axis to kisspeptin varies across different reproductive states. In humans, the LH response to kisspeptin treatment appears to be highest during the preovulatory phase of the menstrual cycle in women [86], with similar data in sheep [15, 87] and rats [88]. At face value, these data suggest that the GnRH/LH surge is stimulated by kisspeptin in a "two-step mechanism" consisting of increased kisspeptin output from the hypothalamus and increased kisspeptin sensitivity at the GnRH neuron. Moreover, these data support a role for sex steroids in the modulation of LH responses to kisspeptin. In rats, maximal LH and FSH responses to kisspeptin require replacement of estradiol and progesterone in an OVX model [88]. In terms of estrogen action, subsequent studies indicate the LH response to kisspeptin is modulated by ERα, with blockade in female rats blunting the LH response, which could alternatively be increased after selective activation of ERα [89]. In contrast, antagonism of ERβ augmented acute LH responses to kisspeptin [89]. Alternatively, antagonism of ERα or ERβ equally blunted the FSH response to kisspeptin [90].

Any effects of sex steroids on the gonadotropin response to kisspeptin are likely to be mediated by effects on Kiss1r expression on GnRH neurons. Data demonstrating a link between estrogen and Kiss1r remain mixed, with studies showing estrogen inhibition [54] and others showing no effect [91]. Both these studies utilized RTPCR and subsequently lack the specificity of observing Kiss1r expression on GnRH neurons. However, recent data in sheep to this end failed to observe any change in Kiss1r expression on GnRH neurons between OVX and OVX+E animals [92]. An alternative possibility for estrogen regulation of GnRH response may relate to the indirect effects of afferent inputs to GnRH neurons that are estrogen sensitive. In electrophysiological recordings of GnRH neurons, blockade of ionotropic gamma-aminobutyric acid and glutamate receptors prevented the estrogen-potentiated GnRH response in OVX+E mice [11]. Thus, kisspeptin activation of GnRH neurons can be gated by transsynaptic mechanisms.

Conclusions

Since 2003, kisspeptin has been cast to the forefront of neuroendocrine research, and remarkable advances in our understanding of the reproductive axis have been made. It is clear that kisspeptin signaling is fundamental to the reproductive system. Specifically, kisspeptin cells in the hypothalamus are strong candidates to act as key conduits providing the "missing link" in the sex steroid feedback control of GnRH secretion. We now know that kisspeptin cells in the ARC are well placed to drive the estrogen negative feedback signals that control the tonic pulsatile release of GnRH. This appears to be true for multiple species. Concerning estrogen positive feedback, which drives the preovulatory GnRH/LH surge, more rostral hypothalamic populations of kisspeptin cells in the AVPV are implicated in rodent species. In sheep and primates, kisspeptin cells in the ARC are also poised to play key roles in the positive feedback regulation of GnRH. Importantly, it remains to be determined if it is the same kisspeptin cells that show both estrogen negative feedback and estrogen positive feedback responses. Growing evidence suggests the rostral POA populations in these species may also play a role in the estrogen positive feedback process (drawing comparisons to rodent species). It is still unclear exactly how kisspeptin neurons appear to be differentially regulated by essentially the same hormonal stimulus (estradiol). Evidence shows that classical and nonclassical estrogen signaling pathways are involved, but it is still unknown how the divergent pathways arise. These and other challenges in the kisspeptin field remain for future studies.

References

1. Moore CR, Price D (1932) Gonadal hormone functions and the reciprocal influence between gonads and hypophysis, with its bearing on sex hormone antagonism. Am J Anat 50:13–71
2. Herbison AE, Theodosis DT (1992) Localization of oestrogen receptors in preoptic neurons containing neurotensin but not tyrosine hydroxylase, cholecystokinin or luteinizing hormone-releasing hormone in the male and female rat. Neuroscience 50:283–298

3. Shivers BD, Harlan RE, Morrell JI, Pfaff DW (1983) Absence of oestradiol concentration in cell nuclei of LHRH-immunoreactive neurones. Nature 304:345–347
4. Skinner DC, Caraty A, Allingham R (2001) Unmasking the progesterone receptor in the preoptic area and hypothalamus of the ewe: no colocalization with gonadotropin-releasing neurons. Endocrinology 142:573–579
5. Gottsch ML, Cunningham MJ, Smith JT, Popa SM, Acohido BV, Crowley WF, Seminara S, Clifton DK, Steiner RA (2004) A role for kisspeptins in the regulation of gonadotropin secretion in the mouse. Endocrinology 145:4073–4077
6. Irwig MS, Fraley GS, Smith JT, Acohido BV, Popa SM, Cunningham MJ, Gottsch ML, Clifton DK, Steiner RA (2004) Kisspeptin activation of gonadotropin releasing hormone neurons and regulation of KiSS-1 mRNA in the male rat. Neuroendocrinology 80:264–272
7. Matsui H, Takatsu Y, Kumano S, Matsumoto H, Ohtaki T (2004) Peripheral administration of metastin induces marked gonadotropin release and ovulation in the rat. Biochem Biophys Res Commun 320:383–388
8. Shahab M, Mastronardi C, Seminara SB, Crowley WF, Ojeda SR, Plant TM (2005) Increased hypothalamic GPR54 signaling: a potential mechanism for initiation of puberty in primates. Proc Natl Acad Sci USA 102:2129–2134
9. Patterson M, Murphy KG, Thompson EL, Patel S, Ghatei MA, Bloom SR (2006) Administration of kisspeptin-54 into discrete regions of the hypothalamus potently increases plasma luteinising hormone and testosterone in male adult rats. J Neuroendocrinol 18:349–354
10. Han SK, Gottsch ML, Lee KJ, Popa SM, Smith JT, Jakawich SK, Clifton DK, Steiner RA, Herbison AE (2005) Activation of gonadotropin-releasing hormone neurons by kisspeptin as a neuroendocrine switch for the onset of puberty. J Neurosci 25:11349–11356
11. Pielecka-Fortuna J, Chu Z, Moenter SM (2008) Kisspeptin acts directly and indirectly to increase gonadotropin-releasing hormone neuron activity and its effects are modulated by estradiol. Endocrinology 149:1979–1986
12. Clarkson J, Herbison AE (2006) Postnatal development of kisspeptin neurons in mouse hypothalamus; sexual dimorphism and projections to gonadotropin-releasing hormone neurons. Endocrinology 147:5817–5825
13. Kinoshita M, Tsukamura H, Adachi S, Matsui H, Uenoyama Y, Iwata K, Yamada S, Inoue K, Ohtaki T, Matsumoto H, Maeda K (2005) Involvement of central metastin in the regulation of preovulatory luteinizing hormone surge and estrous cyclicity in female rats. Endocrinology 146:4431–4436
14. Smith JT, Coolen LM, Kriegsfeld LJ, Sari IP, Jafarzadehshirazi MR, Maltby M, Bateman K, Goodman RL, Tilbrook AJ, Ubuka T, Bentley GE, Clarke IJ, Lehman MN (2008) Variation in kisspeptin and RFamide-related peptide (RFRP) expression and terminal connections to gonadotropin-releasing hormone neurons in the brain: a novel medium for seasonal breeding in the sheep. Endocrinology 149:5770–5782
15. Smith JT, Li Q, Yap KS, Shahab M, Roseweir AK, Millar RP, Clarke IJ (2011) Kisspeptin is essential for the full preovulatory LH surge and stimulates GnRH release from the isolated ovine median eminence. Endocrinology 152:1001–1012
16. Smith JT, Li Q, Pereira A, Clarke IJ (2009) Kisspeptin neurons in the ovine arcuate nucleus and preoptic area are involved in the preovulatory luteinizing hormone surge. Endocrinology 150:5530–5538
17. Dungan HM, Gottsch ML, Zeng H, Gragerov A, Bergmann JE, Vassilatis DK, Clifton DK, Steiner RA (2007) The role of kisspeptin-GPR54 signaling in the tonic regulation and surge release of gonadotropin-releasing hormone/luteinizing hormone. J Neurosci 27:12088–12095
18. Kauffman AS, Park JH, McPhie-Lalmansingh AA, Gottsch ML, Bodo C, Hohmann JG, Pavlova MN, Rohde AD, Clifton DK, Steiner RA, Rissman EF (2007) The kisspeptin receptor GPR54 is required for sexual differentiation of the brain and behavior. J Neurosci 27:8826–8835
19. Messager S, Chatzidaki EE, Ma D, Hendrick AG, Zahn D, Dixon J, Thresher RR, Malinge I, Lomet D, Carlton MB, Colledge WH, Caraty A, Aparicio SA (2005) Kisspeptin directly stimulates gonadotropin-releasing hormone release via G protein-coupled receptor 54. Proc Natl Acad Sci USA 102:1761–1766

20. Kauffman AS, Gottsch ML, Roa J, Byquist AC, Crown A, Clifton DK, Hoffman GE, Steiner RA, Tena-Sempere M (2007) Sexual differentiation of Kiss1 gene expression in the brain of the rat. Endocrinology 148:1774–1783
21. Smith JT, Popa SM, Clifton DK, Hoffman GE, Steiner RA (2006) Kiss1 neurons in the forebrain as central processors for generating the preovulatory luteinizing hormone surge. J Neurosci 26:6687–6694
22. Greives TJ, Mason AO, Scotti MA, Levine J, Ketterson ED, Kriegsfeld LJ, Demas GE (2007) Environmental control of kisspeptin: implications for seasonal reproduction. Endocrinology 148:1158–1166
23. Revel FG, Ansel L, Klosen P, Saboureau M, Pevet P, Mikkelsen JD, Simonneaux V (2007) Kisspeptin: a key link to seasonal breeding. Rev Endocr Metab Disord 8:57–65
24. Clarkson J, d'Anglemont de Tassigny X, Colledge WH, Caraty A, Herbison AE (2009) Distribution of kisspeptin neurones in the adult female mouse brain. J Neuroendocrinol 21:673–682
25. Adachi S, Yamada S, Takatsu Y, Matsui H, Kinoshita M, Takase K, Sugiura H, Ohtaki T, Matsumoto H, Uenoyama Y, Tsukamura H, Inoue K, Maeda KI (2007) Involvement of anteroventral periventricular metastin/kisspeptin neurons in estrogen positive feedback action on luteinizing hormone release in female rats. J Reprod Dev 53:367–378
26. Estrada KM, Clay CM, Pompolo S, Smith JT, Clarke IJ (2006) Elevated KiSS-1 expression in the arcuate nucleus prior to the cyclic preovulatory gonadotrophin-releasing hormone/luteinising hormone surge in the ewe suggests a stimulatory role for kisspeptin in oestrogen-positive feedback. J Neuroendocrinol 18:806–809
27. Smith JT, Clay CM, Caraty A, Clarke IJ (2007) KiSS-1 messenger ribonucleic acid expression in the hypothalamus of the ewe is regulated by sex steroids and season. Endocrinology 148:1150–1157
28. Franceschini I, Lomet D, Cateau M, Delsol G, Tillet Y, Caraty A (2006) Kisspeptin immunoreactive cells of the ovine preoptic area and arcuate nucleus co-express estrogen receptor alpha. Neurosci Lett 401:225–230
29. Ramaswamy S, Guerriero KA, Gibbs RB, Plant TM (2008) Structural interactions between kisspeptin and GnRH neurons in the mediobasal hypothalamus of the male rhesus monkey (Macaca mulatta) as revealed by double immunofluorescence and confocal microscopy. Endocrinology 149:4387–4395
30. Rometo AM, Krajewski SJ, Voytko ML, Rance NE (2007) Hypertrophy and increased kisspeptin gene expression in the hypothalamic infundibular nucleus of postmenopausal women and ovariectomized monkeys. J Clin Endocrinol Metab 92:2744–2750
31. Wintermantel TM, Campbell RE, Porteous R, Bock D, Grone HJ, Todman MG, Korach KS, Greiner E, Perez CA, Schutz G, Herbison AE (2006) Definition of estrogen receptor pathway critical for estrogen positive feedback to gonadotropin-releasing hormone neurons and fertility. Neuron 52:271–280
32. Yeo SH, Herbison AE (2011) Projections of arcuate nucleus and rostral periventricular kisspeptin neurons in the adult female mouse brain. Endocrinology 152:2387–2399
33. Clarkson J, Herbison AE (2011) Dual phenotype kisspeptin-dopamine neurones of the rostral periventricular area of the third ventricle project to gonadotrophin-releasing hormone neurones. J Neuroendocrinol 23:293–301
34. Backholer K, Smith JT, Clarke IJ (2009) Melanocortins may stimulate reproduction by activating orexin neurons in the dorsomedial hypothalamus and kisspeptin neurons in the preoptic area of the ewe. Endocrinology 150:5488–5497
35. d'Anglemont de Tassigny X, Fagg LA, Carlton MB, Colledge WH (2008) Kisspeptin can stimulate gonadotropin-releasing hormone (GnRH) release by a direct action at GnRH nerve terminals. Endocrinology 149:3926–3932
36. Smith JT (2009) Sex steroid control of hypothalamic Kiss1 expression in sheep and rodents: comparative aspects. Peptides 30:94–102
37. Shughrue PJ, Lane MV, Merchenthaler I (1997) Comparative distribution of estrogen receptor-alpha and -beta mRNA in the rat central nervous system. J Comp Neurol 388:507–525
38. Simerly RB (2002) Wired for reproduction: organization and development of sexually dimorphic circuits in the mammalian forebrain. Annu Rev Neurosci 25:507–536

39. Simerly RB, Chang C, Muramatsu M, Swanson LW (1990) Distribution of androgen and estrogen receptor mRNA-containing cells in the rat brain: an in situ hybridization study. J Comp Neurol 294:76–95
40. Simonian SX, Spratt DP, Herbison AE (1999) Identification and characterization of estrogen receptor alpha-containing neurons projecting to the vicinity of the gonadotropin-releasing hormone perikarya in the rostral preoptic area of the rat. J Comp Neurol 411:346–358
41. Soper BD, Weick RF (1980) Hypothalamic and extrahypothalamic mediation of pulsatile discharges of luteinizing hormone in the ovariectomized rat. Endocrinology 106:348–355
42. Gu GB, Simerly RB (1997) Projections of the sexually dimorphic anteroventral periventricular nucleus in the female rat. J Comp Neurol 384:142–164
43. Herbison AE (2008) Estrogen positive feedback to gonadotropin-releasing hormone (GnRH) neurons in the rodent: the case for the rostral periventricular area of the third ventricle (RP3V). Brain Res Rev 57:277–287
44. Simerly RB (1998) Organization and regulation of sexually dimorphic neuroendocrine pathways. Behav Brain Res 92:195–203
45. Blache D, Fabre-Nys CJ, Venier G (1991) Ventromedial hypothalamus as a target for oestradiol action on proceptivity, receptivity and luteinizing hormone surge of the ewe. Brain Res 546:241–249
46. Caraty A, Fabre-Nys C, Delaleu B, Locatelli A, Bruneau G, Karsch FJ, Herbison A (1998) Evidence that the mediobasal hypothalamus is the primary site of action of estradiol in inducing the preovulatory gonadotropin releasing hormone surge in the ewe. Endocrinology 139:1752–1760
47. Cheng G, Coolen LM, Padmanabhan V, Goodman RL, Lehman MN (2010) The kisspeptin/neurokinin B/dynorphin (KNDy) cell population of the arcuate nucleus: sex differences and effects of prenatal testosterone in sheep. Endocrinology 151:301–311
48. Hess DL, Wilkins RH, Moossy J, Chang JL, Plant TM, McCormack JT, Nakai Y, Knobil E (1977) Estrogen-induced gonadotropin surges in decerebrated female rhesus monkeys with medial basal hypothalamic peninsulae. Endocrinology 101:1264–1271
49. Knobil E, Plant TM, Wildt L, Belchetz PE, Marshall G (1980) Control of the rhesus monkey menstrual cycle: permissive role of hypothalamic gonadotropin-releasing hormone. Science 207:1371–1373
50. Krey LC, Butler WR, Knobil E (1975) Surgical disconnection of the medial basal hypothalamus and pituitary function in the rhesus monkey. I. Gonadotropin secretion. Endocrinology 96:1073–1087
51. Norman RL, Resko JA, Spies HG (1976) The anterior hypothalamus: how it affects gonadotropin secretion in the rhesus monkey. Endocrinology 99:59–71
52. Cogen P, Antunes JL, Louis KM, Dyrenfurth I, Ferin M (1980) The effects of anterior hypothalamic disconnection on gonadotropin secretion in the female rhesus monkey. Endocrinology 107:677–683
53. Smith JT, Cunningham MJ, Rissman EF, Clifton DK, Steiner RA (2005) Regulation of Kiss1 gene expression in the brain of the female mouse. Endocrinology 146:3686–3692
54. Navarro VM, Castellano JM, Fernandez-Fernandez R, Barreiro ML, Roa J, Sanchez-Criado JE, Aguilar E, Dieguez C, Pinilla L, Tena-Sempere M (2004) Developmental and hormonally regulated messenger ribonucleic acid expression of KiSS-1 and its putative receptor, GPR54, in rat hypothalamus and potent luteinizing hormone-releasing activity of KiSS-1 peptide. Endocrinology 145:4565–4574
55. Shughrue PJ, Lane MV, Merchenthaler I (1997) Regulation of progesterone receptor messenger ribonucleic acid in the rat medial preoptic nucleus by estrogenic and antiestrogenic compounds: an in situ hybridization study. Endocrinology 138:5476–5484
56. Smith JT, Dungan HM, Stoll EA, Gottsch ML, Braun RE, Eacker SM, Clifton DK, Steiner RA (2005) Differential regulation of KiSS-1 mRNA expression by sex steroids in the brain of the male mouse. Endocrinology 146:2976–2984
57. Clarkson J, Shamas S, Mallinson S, Herbison AE (2012) Gonadal steroid induction of kisspeptin peptide expression in the rostral periventricular area of the third ventricle during postnatal development in the male mouse. J Neuroendocrinol 24(6):907–915

58. Kim J, Semaan SJ, Clifton DK, Steiner RA, Dhamija S, Kauffman AS (2011) Regulation of Kiss1 expression by sex steroids in the amygdala of the rat and mouse. Endocrinology 152:2020–2030
59. Garcia-Galiano D, van Ingen Schenau D, Leon S, Krajnc-Franken MA, Manfredi-Lozano M, Romero-Ruiz A, Navarro VM, Gaytan F, van Noort PI, Pinilla L, Blomenrohr M, Tena-Sempere M (2012) Kisspeptin signaling is indispensable for neurokinin B, but not glutamate, stimulation of gonadotropin secretion in mice. Endocrinology 153:316–328
60. Roseweir AK, Kauffman AS, Smith JT, Guerriero KA, Morgan K, Pielecka-Fortuna J, Pineda R, Gottsch ML, Tena-Sempere M, Moenter SM, Terasawa E, Clarke IJ, Steiner RA, Millar RP (2009) Discovery of potent kisspeptin antagonists delineate physiological mechanisms of gonadotropin regulation. J Neurosci 29:3920–3929
61. Mayer C, Acosta-Martinez M, Dubois SL, Wolfe A, Radovick S, Boehm U, Levine JE (2010) Timing and completion of puberty in female mice depend on estrogen receptor alpha-signaling in kisspeptin neurons. Proc Natl Acad Sci USA 107:22693–22698
62. Ohkura S, Takase K, Matsuyama S, Mogi K, Ichimaru T, Wakabayashi Y, Uenoyama Y, Mori Y, Steiner RA, Tsukamura H, Maeda KI, Okamura H (2009) Gonadotrophin-releasing hormone pulse generator activity in the hypothalamus of the goat. J Neuroendocrinol 21:813–821
63. Goodman RL, Lehman MN, Smith JT, Coolen LM, de Oliveira CV, Jafarzadehshirazi MR, Pereira A, Iqbal J, Caraty A, Ciofi P, Clarke IJ (2007) Kisspeptin neurons in the arcuate nucleus of the ewe express both dynorphin A and neurokinin B. Endocrinology 148:5752–5760
64. Navarro VM, Gottsch ML, Chavkin C, Okamura H, Clifton DK, Steiner RA (2009) Regulation of gonadotropin-releasing hormone secretion by kisspeptin/dynorphin/neurokinin B neurons in the arcuate nucleus of the mouse. J Neurosci 29:11859–11866
65. Wakabayashi Y, Nakada T, Murata K, Ohkura S, Mogi K, Navarro VM, Clifton DK, Mori Y, Tsukamura H, Maeda K, Steiner RA, Okamura H (2010) Neurokinin B and dynorphin a in kisspeptin neurons of the arcuate nucleus participate in generation of periodic oscillation of neural activity driving pulsatile gonadotropin-releasing hormone secretion in the goat. J Neurosci 30:3124–3132
66. Wiegand SJ, Terasawa E, Bridson WE (1978) Persistent estrus and blockade of progesterone-induced LH release follows lesions which do not damage the suprachiasmatic nucleus. Endocrinology 102:1645–1648
67. Clarkson J, d'Anglemont de Tassigny X, Moreno AS, Colledge WH, Herbison AE (2008) Kisspeptin-GPR54 signaling is essential for preovulatory gonadotropin-releasing hormone neuron activation and the luteinizing hormone surge. J Neurosci 28:8691–8697
68. Pineda R, Garcia-Galiano D, Roseweir A, Romero M, Sanchez-Garrido MA, Ruiz-Pino F, Morgan K, Pinilla L, Millar RP, Tena-Sempere M (2010) Critical roles of kisspeptins in female puberty and preovulatory gonadotropin surges as revealed by a novel antagonist. Endocrinology 151:722–730
69. Barbacka-Surowiak G, Surowiak J, Stoklosowa S (2003) The involvement of suprachiasmatic nuclei in the regulation of estrous cycles in rodents. Reprod Biol 3:99–129
70. de la Iglesia HO, Schwartz WJ (2006) Minireview: timely ovulation: circadian regulation of the female hypothalamo-pituitary-gonadal axis. Endocrinology 147:1148–1153
71. Robertson JL, Clifton DK, de la Iglesia HO, Steiner RA, Kauffman AS (2009) Circadian regulation of Kiss1 neurons: implications for timing the preovulatory gonadotropin-releasing hormone/luteinizing hormone surge. Endocrinology 150:3664–3671
72. Williams WP 3rd, Jarjisian SG, Mikkelsen JD, Kriegsfeld LJ (2011) Circadian control of kisspeptin and a gated GnRH response mediate the preovulatory luteinizing hormone surge. Endocrinology 152:595–606
73. Smarr BL, Morris E, de la Iglesia HO (2012) The dorsomedial suprachiasmatic nucleus times circadian expression of Kiss1 and the luteinizing hormone surge. Endocrinology 153:2839–2850
74. Semaan SJ, Murray EK, Poling MC, Dhamija S, Forger NG, Kauffman AS (2010) BAX-dependent and BAX-independent regulation of Kiss1 neuron development in mice. Endocrinology 151:5807–5817
75. Caraty A, Smith JT, Lomet D, Ben Said S, Morrissey A, Cognie J, Doughton B, Baril G, Briant C, Clarke IJ (2007) Kisspeptin synchronizes preovulatory surges in cyclical ewes and causes ovulation in seasonally acyclic ewes. Endocrinology 148:5258–5267

76. Sebert ME, Lomet D, Said SB, Monget P, Briant C, Scaramuzzi RJ, Caraty A (2010) Insights into the mechanism by which kisspeptin stimulates a preovulatory LH surge and ovulation in seasonally acyclic ewes: potential role of estradiol. Domest Anim Endocrinol 38:289–298
77. Clarke IJ (1987) GnRH and ovarian hormone feedback. Oxf Rev Reprod Biol 9:54–95
78. Gottsch ML, Navarro VM, Zhao Z, Glidewell-Kenney C, Weiss J, Jameson JL, Clifton DK, Levine JE, Steiner RA (2009) Regulation of Kiss1 and dynorphin gene expression in the murine brain by classical and nonclassical estrogen receptor pathways. J Neurosci 29:9390–9395
79. Hoffman GE, Le WW, Franceschini I, Caraty A, Advis JP (2011) Expression of fos and in vivo median eminence release of LHRH identifies an active role for preoptic area kisspeptin neurons in synchronized surges of LH and LHRH in the ewe. Endocrinology 152:214–222
80. Tomikawa J, Homma T, Tajima S, Shibata T, Inamoto Y, Takase K, Inoue N, Ohkura S, Uenoyama Y, Maeda K, Tsukamura H (2010) Molecular characterization and estrogen regulation of hypothalamic KISS1 gene in the pig. Biol Reprod 82:313–319
81. Smith JT, Shahab M, Pereira A, Pau KY, Clarke IJ (2010) Hypothalamic expression of KISS1 and gonadotropin inhibitory hormone genes during the menstrual cycle of a non-human primate. Biol Reprod 83:568–577
82. Porteous R, Petersen SL, Yeo SH, Bhattarai JP, Ciofi P, de Tassigny XD, Colledge WH, Caraty A, Herbison AE (2011) Kisspeptin neurons co-express met-enkephalin and galanin in the rostral periventricular region of the female mouse hypothalamus. J Comp Neurol 519:3456–3469
83. Kauffman AS, Navarro VM, Kim J, Clifton D, Steiner RA (2009) Sex differences in the regulation of Kiss1/NKB neurons in Juvenile Mice: implications for the timing of puberty. Am J Physiol Endocrinol Metab 297(5):E1212–E1221
84. Li D, Mitchell D, Luo J, Yi Z, Cho SG, Guo J, Li X, Ning G, Wu X, Liu M (2007) Estrogen regulates KiSS1 gene expression through estrogen receptor alpha and SP protein complexes. Endocrinology 148:4821–4828
85. Glidewell-Kenney C, Hurley LA, Pfaff L, Weiss J, Levine JE, Jameson JL (2007) Nonclassical estrogen receptor alpha signaling mediates negative feedback in the female mouse reproductive axis. Proc Natl Acad Sci USA 104:8173–8177
86. Dhillo WS, Chaudhri OB, Thompson EL, Murphy KG, Patterson M, Ramachandran R, Nijher GK, Amber V, Kokkinos A, Donaldson M, Ghatei MA, Bloom SR (2007) Kisspeptin-54 stimulates gonadotropin release most potently during the preovulatory phase of the menstrual cycle in women. J Clin Endocrinol Metab 92:3958–3966
87. Smith JT, Saleh SN, Clarke IJ (2009) Seasonal and cyclical change in the luteinizing hormone response to kisspeptin in the ewe. Neuroendocrinology 90:283–291
88. Roa J, Vigo E, Castellano JM, Navarro VM, Fernandez-Fernandez R, Casanueva FF, Dieguez C, Aguilar E, Pinilla L, Tena-Sempere M (2006) Hypothalamic expression of KiSS-1 system and gonadotropin-releasing effects of kisspeptin in different reproductive states of the female Rat. Endocrinology 147:2864–2878
89. Roa J, Vigo E, Castellano JM, Gaytan F, Navarro VM, Aguilar E, Dijcks FA, Ederveen AG, Pinilla L, van Noort PI, Tena-Sempere M (2008) Opposite roles of estrogen receptor (ER)-alpha and ERbeta in the modulation of luteinizing hormone responses to kisspeptin in the female rat: implications for the generation of the preovulatory surge. Endocrinology 149:1627–1637
90. Roa J, Vigo E, Castellano JM, Gaytan F, Garcia-Galiano D, Navarro VM, Aguilar E, Dijcks FA, Ederveen AG, Pinilla L, van Noort PI, Tena-Sempere M (2008) Follicle-stimulating hormone responses to kisspeptin in the female rat at the preovulatory period: modulation by estrogen and progesterone receptors. Endocrinology 149:5783–5790
91. Yamada S, Uenoyama Y, Kinoshita M, Iwata K, Takase K, Matsui H, Adachi S, Inoue K, Maeda KI, Tsukamura H (2007) Inhibition of metastin (Kisspeptin-54)-GPR 54 signaling in the arcuate nucleus-median eminence region during lactation in rats. Endocrinology 148:2226–2232
92. Li Q, Roa A, Clarke IJ, Smith JT (2012) Seasonal variation in GnRH response to kisspeptin in sheep: possible kisspeptin regulation of the kisspeptin receptor. Neuroendocrinology 96(3):212–221

Chapter 14
Kisspeptin and GnRH Pulse Generation

Hiroaki Okamura, Hiroko Tsukamura, Satoshi Ohkura, Yoshihisa Uenoyama, Yoshihiro Wakabayashi, and Kei-ichiro Maeda

Abstract The reproductive neuropeptide gonadotropin-releasing hormone (GnRH) has two modes of secretion. Besides the surge mode, which induces ovulation in females, the pulse mode of GnRH release is essential to cause various reproductive events in both sexes, such as spermatogenesis, follicular development, and sex steroid synthesis. Some environmental cues control gonadal activities through modulating GnRH pulse frequency. Researchers have looked for the anatomical location of the mechanism generating GnRH pulses, the GnRH pulse generator, in the brain, because an artificial manipulation of GnRH pulse frequency is of therapeutic importance to stimulate or suppress gonadal activity. Discoveries of kisspeptin and, consequently, KNDy (kisspeptin/neurokinin B/dynorphin) neurons in the hypothalamus have provided a clue to the possible location of the GnRH pulse generator. Our analyses of hypothalamic multiple-unit activity revealed that KNDy neurons located in the hypothalamic arcuate nucleus might play a central role in the generation of GnRH pulses in goats, and perhaps other mammalian species. This chapter further discusses the possible mechanisms for GnRH pulse generation.

H. Okamura • Y. Wakabayashi
Animal Physiology Research Unit, Division of Animal Sciences, National Institute of Agrobiological Sciences, Tsukuba, Ibaraki, Japan
e-mail: hokamu@affrc.go.jp

H. Tsukamura • Y. Uenoyama
Graduate School of Bioagricultural Sciences, Laboratory of Reproductive Science, Nagoya University, Nagoya, Japan

K.-i. Maeda (✉)
Animal Breeding, Veterinary Medical Sciences, The University of Tokyo, Tokyo, Japan
e-mail: akeimaed@mail.ecc.u-tokyo.ac.jp

S. Ohkura
Graduate School of Bioagricultural Sciences, Laboratory of Animal Production, Nagoya University, Nagoya, Japan

A.S. Kauffman and J.T. Smith (eds.), *Kisspeptin Signaling in Reproductive Biology*, Advances in Experimental Medicine and Biology 784, DOI 10.1007/978-1-4614-6199-9_14,

Introduction

There are two modes of gonadotropin-releasing hormone (GnRH) and luteinizing hormone (LH) secretion: one mode is the surge, necessary for ovulation in females, and the other is the pulse, required for the tonic support of reproductive function in both sexes. For example, GnRH pulses are needed to initiate the process of reproductive cycles, such as estrous cycles, in females. Follicular development is stimulated by the increase in frequency of GnRH/LH pulses, resulting in a surge-like secretion of estrogen from the mature follicles. The increased estrogen acts in the brain to cause the GnRH surge to induce ovulation in females. In contrast, males do not generate GnRH surges, and therefore only have the pulse mode of GnRH secretion, to maintain testicular activities such as spermatogenesis and steroidogenesis. Therefore, manipulation of the activity of the GnRH pulse generator is of therapeutic potential in both sexes, and the GnRH pulse generator is a good target for the development of drugs that might control fertility. This chapter focuses on the involvement of kisspeptin, and other related peptides, in the generation of GnRH pulses in mammals.

Discovery of Pulsatile LH Secretion

Pulsatile secretion of LH was first described, in monkeys, in 1970 by Knobil [1]. This was only a few years after the establishment of a radioimmunoassay for LH in the blood [2]. Knobil had noticed that the concentration of LH in the blood fluctuated significantly from assay to assay, or from time to time, in monkeys. He then utilized frequent blood collections in monkeys to determine the cause of these fluctuations. The resultant data exposed a beautiful series of plasma LH concentrations displaying repetitive abrupt increases in LH followed by an exponential decrease, the distinguishing feature of pulses [1].

The discovery of LH pulses changed the concept of hormone actions, because gonadal activity was subsequently shown in rhesus macaques to be controlled by the "frequency" of LH pulses [3–5]. The greater the LH pulse frequency, the greater the resultant gonadal activity. Knobil's experiments elegantly proved that gonadal activity is completely dependent on the pulse frequency of LH release. After the discovery of pulsatile LH secretion in monkeys, reproductive endocrinologists began to reveal the pulsatile nature of LH secretion in various other mammalian species, including rats [6], sheep [7], cows [8], pigs [9], and horses [10], although frequent blood sampling was sometimes difficult in some species under no anesthesia and freely moving conditions. These data reiterated the importance of LH pulse frequency for the regulation of gonadal activities. Consistent across species, more frequent LH pulses are found during the follicular phase, whereas the pulse frequency is lower during luteal phase [8, 11, 12]. In seasonal animals, such as sheep, LH pulses are more frequent during the breeding season and less frequent in the nonbreeding season [13]. Interestingly, the frequency of the pulse is negatively correlated with the size of the body [14].

Discovery of Gonadotropin-Releasing Hormone Pulses and Surges

There is little doubt that the pulsatile nature of pituitary LH secretion is caused by the pulsatile release of GnRH from nerve terminals located in the median eminence, because GnRH is considered to be the single hypothalamic releasing factor stimulating pituitary LH secretion. Initially, however, this was only a belief and not based on solid evidence. The pulsatile nature of GnRH secretion was first seen in 1982 in a landmark study by Clarke and Cummins [15] and later examined in greater detail by Moenter et al. in the early 1990s. Both groups used a skillful technique of portal cannulation in sheep and very frequent portal blood collections (e.g., 30-s intervals!) to demonstrate beautiful GnRH pulses, each of which corresponded to simultaneous LH pulses [16]. The width of GnRH pulses was found to be narrower than LH pulses, suggesting that the half-life of GnRH in the portal blood is much shorter than the half-life of LH in the peripheral circulation [16]. This pioneering work demonstrated the clear relationship between GnRH and LH secretion, and supported the earlier studies by Knobil's group demonstrating that when pulses of GnRH were infused to monkeys bearing hypothalamic lesions and abolished pulsatile LH secretion, LH secretion was restored in a pulsatile fashion, with each LH pulse corresponding beautifully to each experimental GnRH pulse [17]. Additionally, artificial pulsatile infusion of GnRH, with 1-h intervals, stimulated the ovary to produce complete menstrual cycles [3], whereas monkeys exposed to less frequent GnRH pulses showed no sign of ovarian activity [18].

In addition to identifying GnRH pulses in their sheep portal samples, Moenter et al. also observed robust periodic GnRH surges [19]. The discovery of GnRH surges leads to a dramatic turnaround in the theory of the LH surge formation, because researchers had previously believed that a high frequency of LH pulses during the preovulatory period caused a surge. This model held that when the frequency of LH pulses was too high to be effectively cleared from the circulation, the blood LH concentration would not decline and would keep increasing (i.e., a surge) until pulse frequency eventually drops. However, this idea was rejected after the discovery that a huge amount of GnRH is released just prior to LH surges, and the GnRH surge release continues even after the end of LH surges [19]. Currently, researchers believe that the GnRH and LH surges are generated by a mechanism different from that generating GnRH/LH pulses.

Anatomical Location of the GnRH Pulse Generator

The anatomical location of the GnRH pulse-generating mechanism has always been a big puzzle for reproductive endocrinologists. The first work describing the possible location of the center for pulsatile GnRH secretion was conducted by Halasz and Pupp [20], who utilized a micro "Halasz" knife in rats to isolate specific brain regions from the rest of the brain. They found that isolating the mediobasal hypothalamus (MBH), including the pituitary, abolishes ovulation but not follicular development [20]. This was confirmed later by Blake and Sawyer [21], who demonstrated that complete hypothalamic deafferentation spares LH pulses in ovariectomized (OVX) rats. These experiments clearly showed that the brain center generating GnRH/LH pulses was located within the hypothalamic area isolated by the Halasz's knife, namely, the MBH. According to this data, the GnRH pulse generator may not involve GnRH neurons themselves, because very few GnRH cell bodies are located in the MBH of most animal species (with the exception of primates, in which most of GnRH neurons are located in the area [22]). The MBH location of the GnRH pulse generator was also confirmed by fetal MBH transplantation in rats that had brain lesions which abolished GnRH pulses [23]. A type of deafferentation called posterior-anterior deafferentation (PAD), which cuts the anterior part of the arcuate nucleus (ARC) off, abolished pulsatile LH secretion in rats, but the pulse was restored with transplantation of fetal MBH tissues (but not fetal cortical tissues). These findings indicate the presence of a GnRH pulse-generating mechanism in the MBH region.

On the other hand, evidence also suggests that GnRH neurons themselves are equipped with an intrinsic GnRH pulse-generating mechanism. This was first demonstrated in GT-1 cells, which are immortalized by introducing T antigen to the mouse genome to induce GnRH-producing tumor cells. GT-1 cells show periodic excitation, resulting in pulsatile GnRH release into the culture medium [24]. Further evidence came from primary cultures of rhesus monkey GnRH neurons taken from the fetal olfactory placode, the anatomical region where GnRH neurons originate and migrate from to the hypothalamus during development. The idea to obtain a pure population of GnRH neurons from the monkey fetus came from the laboratory of Terasawa and enabled the demonstration of pulsatile activation of GnRH neurons in vitro. These primary GnRH neurons displayed periodic increases in intracellular calcium concentrations [25]. Terasawa's group also found that the periodic increases in intracellular calcium levels in cultured GnRH neurons are synchronized with each other [26]. The authors considered that these calcium increases cause GnRH pulses.

It is evident that GnRH is released in fixed intervals from GnRH neuronal terminals. The synchronized release of GnRH from each nerve terminal appears to require coordinated activation of GnRH neurons from neuronal afferents. There are three mechanistic possibilities for synchronizing GnRH neuronal output. First, GnRH cell bodies make contacts with each other, as evidenced by reports of morphological contacts between GnRH neuronal processes [27]. However, somatosomatic or dendrodendritic contacts between GnRH neurons are quite rare in the POA of rats [27]. Second, the synchronization of GnRH releases from each nerve terminal might be achieved by contact between multiple GnRH terminals in the median eminence, because the median eminence is one of the sites where there is a convergence of various bioactive substances acting to regulate the release GnRH [28]. There might be the third possibility that GnRH cells may all be synchronized by an upstream "clock" that affects all GnRH cells at the same time, resulting in simultaneous

GnRH output from the various GnRH cells. However, there is no experimental evidence yet to support the last possibility.

The discovery of kisspeptin might help to settle the controversy over the location of the GnRH pulse generator and synchronization of GnRH release. However, there are still difficulties we must overcome in order to unravel the mechanism of GnRH pulse generation. In the rest of this chapter, we will discuss the possibility that kisspeptin neurons play a major role in generating GnRH pulses in multiple mammalian species.

MUA Recording of the GnRH Pulse Generator Activity at Close Vicinity of Kisspeptin Neurons in the Arcuate Nucleus

The Knobil laboratory was the first to identify changes in the multiple-unit activity (MUA) corresponding to changes in LH pulses [29]. By recording electrical activity in the MBH, the neural activity of the putative GnRH pulse generator was successfully represented as periodic bursts of MUA (termed MUA volleys) in monkeys [29–35], rats [36–40], and goats [41–46]. Those studies unambiguously demonstrated that the pulsatile discharge of GnRH into the portal vessels is governed by neural substrates in the MBH that fire a high-frequency volley of action potentials. However, none of the aforementioned studies successfully identified a specific neuronal population within the MBH that was responsible for the generation of the MUA volley.

The MUA volley was observed in the MBH in all animals, regardless of the difference in the distribution of GnRH neurons between species; GnRH cells are relatively abundant in the MBH of monkeys [47, 48], moderately so in goats [49], and few, if any, in rats [50, 51]. Moreover, during the LH surge, when the activity of GnRH neurons was extremely enhanced, the basal MUA activity did not change and the MUA volley frequency decreased rather than increased [30, 31, 42, 44]. These findings strongly suggest that the MUA volley originates outside of the GnRH neuronal network. It was proposed that the observed bursts of MUA in the MBH might reflect the pulsatile activation of GnRH fibers as they traverse en passant to the ME; in this case, the GnRH pulse would be triggered by another unidentified group of oscillators. Thus, the neural substrate of the GnRH pulse generator was still to be determined.

When MUA is measured in goats through an electrode targeted to the posterior ARC (which is part of the MBH), in which a number of kisspeptin neurons are concentrated (Fig. 14.1a), rhythmic MUA volleys are found at regular intervals and are temporally associated with LH pulses (Fig. 14.1b) in both gonadectomized males [52] and females [53]. Furthermore, treatment of OVX goats with estradiol (E2) increases the intervolley interval (i.e., decreases the MUA frequency), while the duration of the volley is decreased (Fig. 14.2a–c). The frequency of the MUA volley in goats is also profoundly decreased by progesterone (P) (Fig. 14.2d) [53]. These results are likely to reflect the negative feedback actions of gonadal sex steroids. Because these

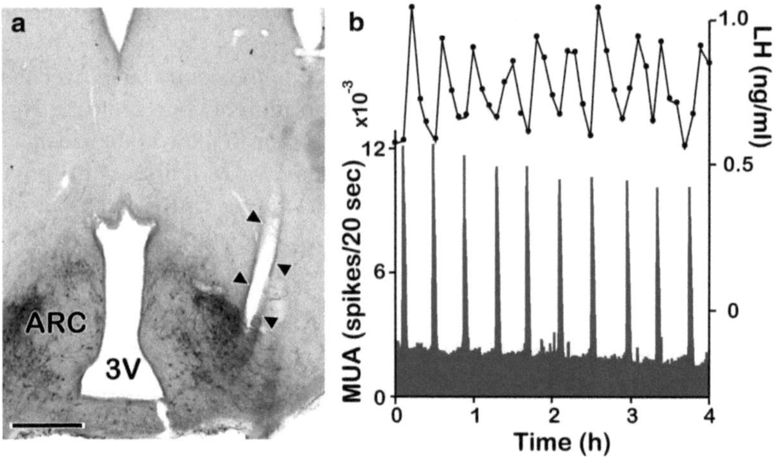

Fig. 14.1 MUA recording at close vicinity of kisspeptin neurons in the ARC. (**a**) A photomicrograph showing the placement of MUA recording electrode in a section immunostained for kisspeptin. *Arrowheads* indicate the area where a trace of a bundle of electrodes is observed. *ARC* arcuate nucleus; *3V* third ventricle. Scale bar: 1 mm. (**b**) Representative profiles of the MUA and plasma LH concentrations in an OVX goat. Panel (**b**) was modified from Wakabayashi Y, et al. Neurokinin B and dynorphin A in kisspeptin neurons of the arcuate nucleus participate in generation of periodic oscillation of neural activity driving pulsatile gonadotropin-releasing hormone secretion in the goat. J Neurosci. 2010 Feb 24;30(8):3124–32. With permission from *Journal of Neuroscience*

results are consistent with those previously demonstrated [30, 35, 42], it is reasonable to conclude that the MUA volley observed at close vicinity of ARC kisspeptin neurons represents the GnRH pulse generator activity. These results lead us to propose a compelling idea that the population of ARC kisspeptin neurons is the intrinsic source of the GnRH pulse generator [53–55]. However, the argument remains circumstantial at this moment. Because the MUA is the summation of the electrical activity of multiple neurons around the electrode, it is still possible that the MUA volley originates from a population of anonymous non-kisspeptin neurons residing in the same vicinity as kisspeptin neurons.

Anatomical Aspects of ARC Kisspeptin Neurons in Relation to the GnRH Pulse Generator

In theory, the GnRH pulse generator should possess several neural characteristics to perform its tasks, including the generation of rhythmic oscillations, electrophysiological synchronization, transmission of the signal of rhythmic oscillation to GnRH neurons, elicitation of a pulsatile GnRH discharge, and processing of the negative feedback action of gonadal steroids. It appears that the functional and anatomical characteristics of ARC kisspeptin neurons meet these requirements.

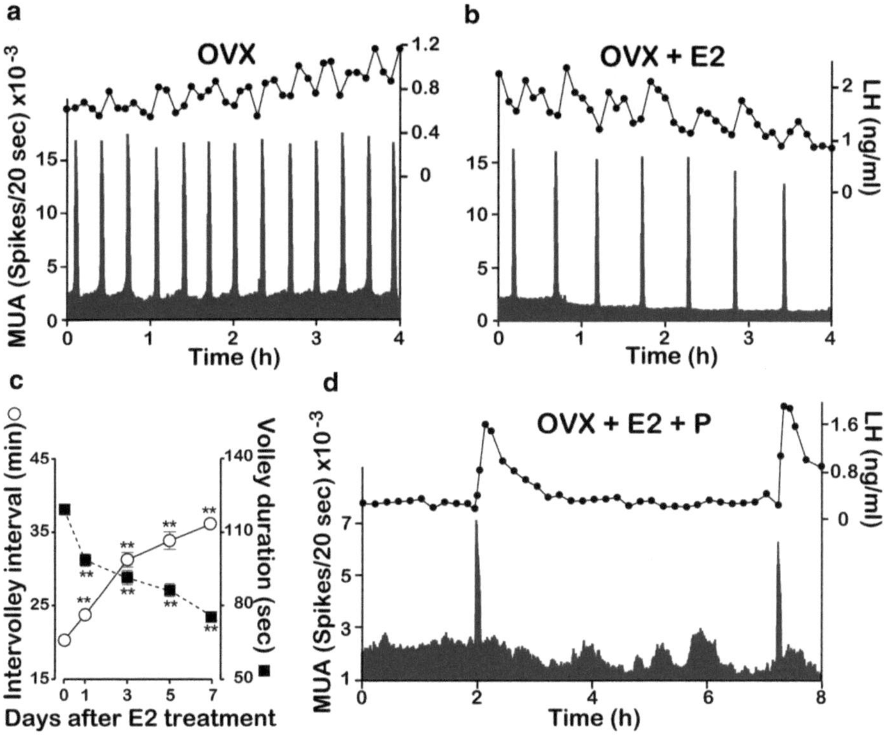

Fig. 14.2 Effects of ovarian steroids on the MUA and LH secretion. (**a**) Representative profiles of the MUA and plasma LH concentrations in an OVX goat. (**b**) Representative profiles of the MUA and plasma LH concentrations in an E2-treated OVX goat. (**c**) Changes in the intervolley interval (*blank circle*) and volley duration (*solid square*) of the MUA volley after the E2 treatment. Data were collected for 6 h (12:00–18:00) in each day, and values are expressed as mean ± SEM in three goats. $^{**}p<0.01$ compared with those on Day 0. (**d**) Representative profiles of the MUA and plasma LH concentrations in an E2 plus P-treated OVX goat. Note that the MUA volley is invariably accompanied by an LH pulse, regardless of the steroidal milieu. Panel (**d**) was reproduced from Wakabayashi Y, et al. Neurokinin B and dynorphin A in kisspeptin neurons of the arcuate nucleus participate in generation of periodic oscillation of neural activity driving pulsatile gonadotropin-releasing hormone secretion in the goat. J Neurosci. 2010 Feb 24;30(8):3124–32. With permission from *Journal of Neuroscience*

Using the ovine model, Goodman et al. [56] were the first to document that kisspeptin neurons in the ARC co-express neurokinin B (NKB) and dynorphin A (Dyn). Since then, the colocalization of kisspeptin with either NKB or Dyn—or both—in ARC neurons was identified in a variety of mammals, including mice [57, 58], rats [59], goats [53], monkeys [60], and humans [61]. Therefore, concomitant expression of these three peptides in single ARC neurons appears to be a common feature across mammalian species. Those neurons, therefore, have been referred to as KNDy (kisspeptin/*N*KB/*Dy*n) neurons [62].

Anatomical evidence indicates that KNDy neurons comprise a neuronal network interconnected by axon (and/or dendritic) collaterals. For example, in the rodent [63] and ovine [63] ARC, NKB/Dyn neurons receive close appositions from fibers containing NKB/Dyn. Dyn neurons in the ARC form synaptic contacts with Dyn fibers [64]. It is therefore not surprising that kisspeptin/NKB and kisspeptin/Dyn neurons are surrounded by their own dense network of fibers [53]. Moreover, an anterograde tracer study in rats revealed that NKB neurons in the ARC are bilaterally interconnected by NKB axons [65]. Importantly, NKB neurons in the ARC contain NKB receptors (NK3R) [63, 66] and ARC *Kiss1* neurons express both NK3R and KOR [57]. These reports suggest that NKB/NK3R and Dyn/KOR signaling pathways might play a role in an auto-feedback loop (or paracrine feedback loop) of KNDy neurons [55, 57, 62, 67, 68]. However, it should be noted that one study recently reported that KNDy neurons in the male mouse do not to express KOR [58], which is inconsistent with this group's earlier report. Other KOR-expressing interneurons mediating Dyn's action might be involved in the auto-feedback loop of KNDy neurons in the ARC.

NKB/NK3R signaling is thought to play a role in stimulating neuronal activity [69], whereas Dyn/KOR (Dyn receptor) signaling is considered to participate in suppressing neuronal activity [70, 71]. By possessing these two opposing signaling mechanisms and forming an anatomical network structure, the population of ARC KNDy neurons seems to possess the required framework for a role as a GnRH pulse generator. For example, reciprocal interactions between NKB/NK3R and Dyn/KOR (or other inhibitory signaling mediating the Dyn action) signaling would make it possible to generate pseudo-pacemaking activities, providing the oscillatory drive of the GnRH pulse generator. The neural network would be suitable for electrophysiological synchronization of individual neurons.

Kisspeptin fibers make extensive associations with GnRH axons in the ME [72–75], and kisspeptin could therefore act as the output of the pulse generator to influence GnRH neurons. Electron microscopy has revealed that kisspeptin axon terminals are in fact in close apposition to GnRH axon terminals [73, 74]. Considering the fact that NKB is contained in KNDy neurons, but not in POA kisspeptin neurons (Fig. 14.3a–c), and that a majority of those kisspeptin fibers in the ME also contain NKB (Fig. 14.3d–f) [73, 75, 76], it is likely that KNDy neurons send, although not exclusively, dense projections to the ME [62, 68] and interact with Kiss1r on GnRH axon terminals. However, the presence of Kiss1r protein on GnRH axon fibers has yet to be demonstrated since there is currently not a good Kiss1r antibody.

It is thought that the GnRH pulse generator is responsive to the negative feedback actions of gonadal steroids [77]. Although there are several populations of neurons that contain sex steroid receptors in the hypothalamus, such as GABA [78], neuropeptide Y [79], substance P [80], somatostatin [81], beta-endorphin [82], or dopamine [82] neurons, KNDy neurons are conspicuous in that virtually all of them express both estrogen receptor alpha [63, 83–86] and progesterone receptor [64, 87] in the female or androgen receptors in the male [88]. This anatomical property further supports the possibility that the KNDy neurons may comprise for the GnRH pulse generator.

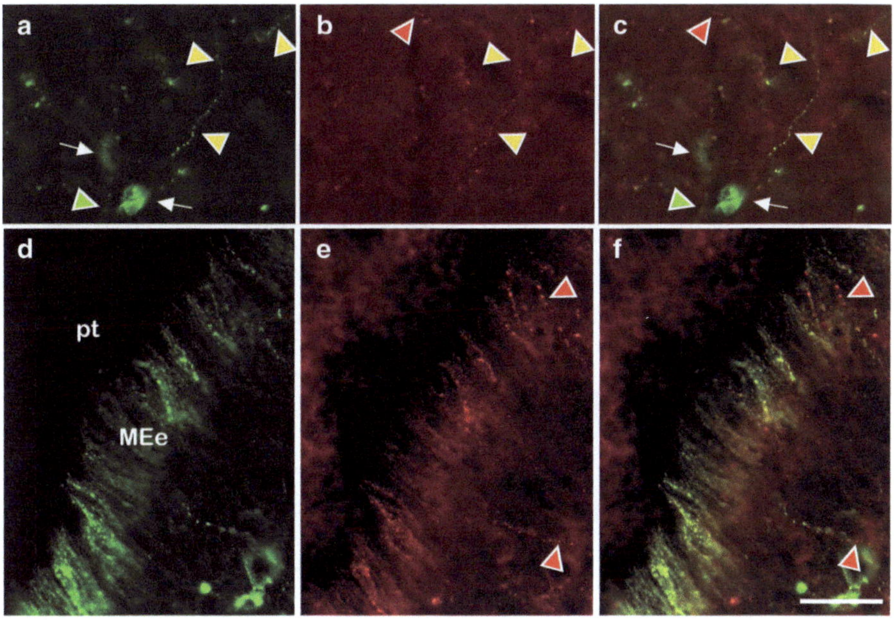

Fig. 14.3 Dual labeling of kisspeptin and NKB in the E2-treated OVX goat. Photomicrographs of sections of the POA (**a**–**c**) or ME (**d**–**f**) immunostained for kisspeptin (**a** and **d**) and NKB (**b** and **e**). (**c**, **f**) are computer-aided merged images of (**a**) and (**b**), or (**d**) and (**e**), respectively. The *arrows* in (**a**) and (**c**) indicate cell bodies containing exclusively kisspeptin immunoreactivity. The *green*, *red*, and *yellow arrowheads* show kisspeptin, NKB, and kisspeptin/NKB positive fibers. Note that a majority of kisspeptin positive fibers contain NKB immunoreactivity (**f**) at the ME. *MEe* the external layer of the ME; *pt* pars tuberalis. Scale bar: 50 μm

Roles of NKB/NK3R, Dyn/KOR, and Kisspeptin/Kiss1r Signaling Pathways in the GnRH Pulse Generation

NKB/NK3R Signaling

The involvement of NKB in the control of GnRH/LH secretion was initially proposed based on morphological changes in NKB neurons in the ARC (the infundibular nucleus in primates) of postmenopausal women and experimental animals [67, 83]. The proposition is strongly supported by the finding that mutations in either Tac3 or Tacr3 (which encode NKB and NK3R, respectively) cause severe gonadotropin deficiency in humans [89, 90] and that Tacr3 null mice show reduced gonadal activities, although they are not completely infertile [91]. Those studies predicted the stimulatory action of NKB on GnRH/LH secretion, but initial reports provided a controversial view indicating that senktide (a selective NK3R agonist) decreased LH secretion in rats [92] and mice [57].

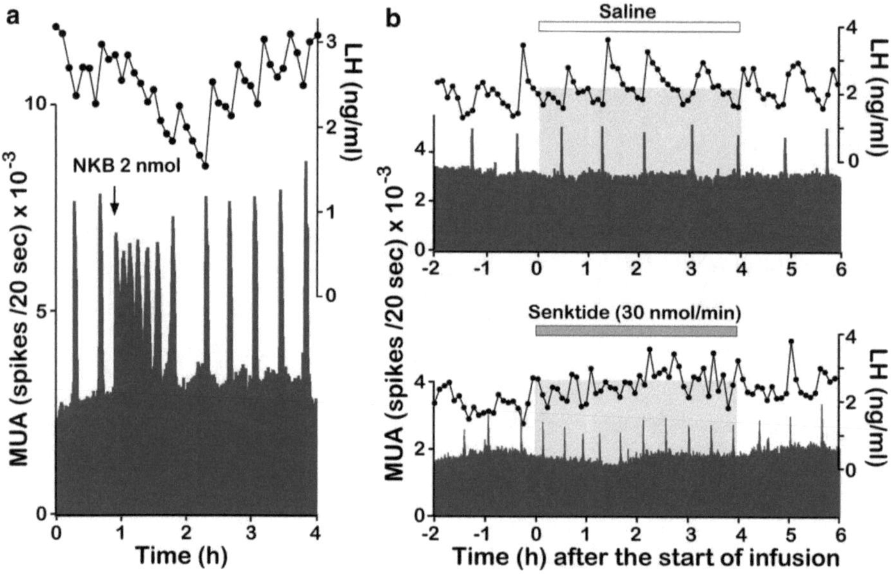

Fig. 14.4 Effects of NK3R agonists on the MUA and LH secretion. (**a**) Representative profiles of the MUA and plasma LH concentrations in an OVX goat that received a bolus icv injection of NKB at an indicated time point. (**b**) Representative profiles of the MUA and plasma LH concentrations in an E2-teated OVX goat received iv infusion of saline (*upper*) or senktide (*lower*) for 4 h. Note that the change in LH concentrations during the senktide infusion is an enhanced pulse frequency (although some pulses are ambiguous) but not increase or decrease in overall concentrations. Panel (**a**) was modified from Wakabayashi Y, et al. Neurokinin B and dynorphin A in kisspeptin neurons of the arcuate nucleus participate in generation of periodic oscillation of neural activity driving pulsatile gonadotropin-releasing hormone secretion in the goat. J Neurosci. 2010 Feb 24;30(8):3124–32. With permission from *Journal of Neuroscience*

Our electrophysiological studies clarified the physiological role of NKB by examining effects of activation or blockade of the NKB/NK3R signaling pathway on the GnRH pulse generator activity in goats, using MUA recordings aimed at KNDy neurons. A bolus intracerebroventricular (icv) administration of NKB immediately induces multiple MUA volleys in the area where KNDy neurons reside, followed by a slight quiescent period before the resumption of spontaneous MUA volleys (Fig. 14.4a) [53]. When senktide was peripherally infused, the intervolley interval of the MUA volley was decreased and was maintained at a relatively constant level throughout the infusion period (Fig. 14.4b) [93]. On the other hand, the blockade of NKB/NK3R signaling by peripheral administration of an NK3R antagonist significantly decreased the occurrence of MUA volleys (Wakabayashi et al., unpublished data). These results suggest that the role of NKB/NK3R signaling is to stimulate the pulse generator activity in the ARC region, which may in fact be the GnRH pulse generator. Because KNDy neurons contain NKB receptors [57, 58, 63, 66], and icv administration of senktide induces cFos in KNDy neurons [59, 94], it is likely that a population of KNDy neurons in the ARC is at least one of the sites of NKB's stimulatory action. Indeed, recent electrophysiological studies using *Kiss1-CreGFP* transgenic mice demonstrated that NKB elicits trains of action potentials in *Kiss1* neurons in the ARC via NK3R [58].

MUA studies in goats also uncovered an important aspect in the pulse-generating mechanism. Since the activation of NK3R by either a bolus administration of NKB or continuous infusion of senktide resulted in intermittent MUA volleys, rather than a single sustained rise in the MUA, it is hypothesized that the stimulatory action of NKB/NK3R signaling on MUA firing is counteracted by some endogenous inhibitory drive, which operates immediately after the induction of the MUA volley and gradually reduces its inhibitory tone thereafter.

Dyn/KOR Signaling

It has been shown that administration of naloxone, a nonselective opioid receptor antagonist, increases the frequency of LH pulses [94] and bursts of the GnRH pulse generator [34, 43]. Moreover, a series of elegant studies in sheep indicated that the inhibitory effect of P on pulsatile GnRH/LH secretion is mediated by endogenous opioid peptides, namely, Dyn [95–97]. In support of this, icv administration of Dyn in goats suppresses the occurrence of the MUA volleys in the ARC region, resulting in a marked increase in the intervolley interval after the treatment (Fig. 14.5a). On the other hand, the blockade of Dyn/KOR signaling by icv administration of nor-binaltorphimine (nor-BNI, a selective KOR antagonist) reduced the intervolley interval and increased the volley duration (Fig. 14.5b, c) [53], indicating that the GnRH pulse generator activity is under a tonic suppression by endogenous Dyn. In vasopressin neurons of the supraoptic nucleus, Dyn/KOR signaling has been suggested to participate in termination of the phasic firing and the release of vasopressin by an autosynaptic loop [98, 99]. With an analogy to vasopressin neurons, it is proposed that Dyn/KOR signaling plays a role in extinguishing the bursts of KNDy neurons in the ARC and regulating the duration of nadir between each bout of bursts.

Kisspeptin/Kiss1r Signaling

Peripheral injection of kisspeptin-10 [39, 54], or central administration of the full-length kisspeptin (Wakabayashi et al., unpublished data), which elicits a robust release of LH, has no effect on either amplitude or frequency of the MUA volley. In a preliminary experiment, we observed in goats that the blockade of kisspeptin/Kiss1r signaling by a continuous activation of Kiss1r resulted in a complete suppression of LH secretion and no detectable LH pulses in plasma, as demonstrated in

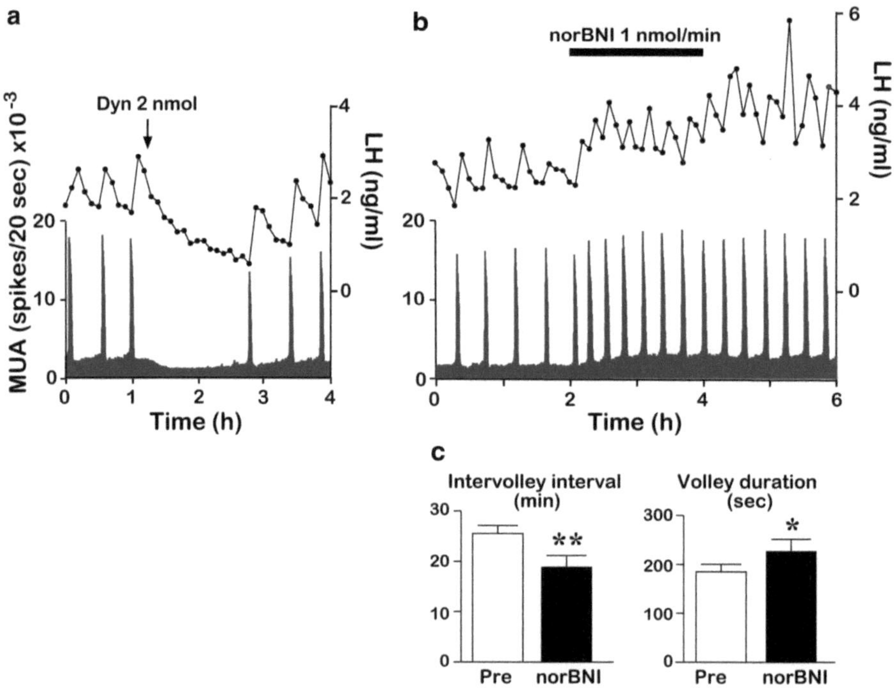

Fig. 14.5 Effects of KOR agonist or antagonist on the MUA and LH secretion in an OVX goat. (**a**) Representative profiles of the MUA and plasma LH concentrations in an OVX goat that received a bolus icv injection of Dyn at the indicated time point. (**b**) Representative profiles of the MUA and plasma LH concentrations in the goat that received icv infusion of KOR antagonist (nor-BNI) for 2 h. (**c**) Changes in the intervolley interval and volley duration before (Pre, *blank bar*) and during (*solid bar*) the nor-BNI infusion periods. Values are expressed as mean ± SEM in four goats. $^{**}p<0.01$, $^{*}p<0.05$ compared with respective Pre values. Reproduced from Wakabayashi Y, et al. Neurokinin B and dynorphin A in kisspeptin neurons of the arcuate nucleus participate in generation of periodic oscillation of neural activity driving pulsatile gonadotropin-releasing hormone secretion in the goat. J Neurosci. 2010 Feb 24;30(8):3124–32. With permission from *Journal of Neuroscience*

other species [100–102], whereas the occurrence of MUA volleys was unchanged [103]. Furthermore, the expression of Kiss1r was not detected in KNDy neurons [75, 104]. These results suggest that kisspeptin/Kiss1r signaling is not involved in the GnRH pulse-generating mechanism per se. However, fibers surrounding KNDy neurons contain not only NKB and Dyn [53, 62, 63, 105] but also kisspeptin [54, 86, 106]. Moreover, treatment with a kisspeptin antagonist into the ARC suppresses pulsatile LH secretion [107]. Therefore, the possibility that kisspeptin may have some functions in the control of GnRH/LH secretion by acting on other cells than KNDy neurons still cannot be ruled out. However, it is also likely that the kisspeptin antagonist treatment did not affect the GnRH pulse generator, but rather diffused to the median eminence where it was able to block kisspeptin stimulation of GnRH fibers, resulting in suppressed LH secretion.

It is very likely that the primary role of kisspeptin/Kiss1r signaling in the GnRH pulse generation mechanism is to transmit volleys of action potentials from the pulse generator to GnRH neurons and regulate pulsatile GnRH secretion at the level of the ME. Several lines of evidence support this notion. First, in monkeys, kisspeptin is secreted into the ME episodically and is temporally associated with pulsatile GnRH secretion [108]. Second, kisspeptin stimulates GnRH release from the ME in vivo [108] and in vitro [74, 75, 109], potentially acting via Kiss1r [109]. Third, administration of a kisspeptin antagonist directly into the ME suppresses pulsatile GnRH release [110].

Interaction of NKB and Kisspeptin Signaling

Human genetic studies [89, 90, 111, 112] indicate that kisspeptin and NKB signaling play pivotal roles in the control of reproduction by facilitating GnRH secretion. In concert, it has been demonstrated in a variety of species that activation of Kiss1r [113] or NK3R [58–60, 114–117] increases LH secretion. Moreover, it has been shown that administration of antagonists for either Kiss1r [75, 110] or NKB receptor [60, 118] suppresses LH secretion. Their similar physiological characteristics and concomitant existence in KNDy neurons suggest an intimate association between kisspeptin and NKB signaling.

Recently it has been demonstrated that the blockade of kisspeptin/Kiss1r signaling by Kiss1r desensitization [115] or in Kiss1r KO mice [116] abrogates the stimulatory action of senktide on LH secretion, whereas the block of NKB/NK3R signaling by NK3R desensitization does not affect the ability of kisspeptin to stimulate LH secretion [115]. We have observed in goats that the blockade of kisspeptin/Kiss1r signaling completely eliminates LH pulses without affecting the MUA volley [103], whereas the occurrences of the MUA volley and LH pulses are concomitantly postponed after the injection of NK3R antagonist (Wakabayashi et al., unpublished data). Furthermore, GnRH neurons possess Kiss1r [75, 104, 119] but not NK3R [58, 66], but see Krajewski et al. [120], and NK3R agonists have no effect on electrophysiological activities of GnRH neurons in vitro [58]. Thus, it is plausible to conclude that NKB/NK3R signaling is upstream from kisspeptin/Kiss1r signaling, and that the activation of NK3R stimulates, via kisspeptin/Kiss1r signaling, a discharge of GnRH, and thus LH [59, 115–117].

We reported that icv administration of NKB induced a distinct MUA volley, with an accompanying LH pulse, in P-treated OVX goats, whereas the association of the MUA volley and LH pulse was ambiguous in some instances in OVX and E2-treated OVX goats, and overall LH secretion was reduced by a high dose (but not a low dose) of NKB [52]. However, with the latter, the initial event after NKB treatment was a discharge of LH, which was followed by a gradual decline of basal LH levels (Fig. 14.4a). In those animals with reduced LH secretion, several MUA volleys that had an extraordinarily shorter intervolley interval were induced, and there was a slight pause before the normal spontaneous MUA volley were reestablished. We assume that this pause resulted in an extended decline of basal LH levels, leading to an apparent reduction in LH secretion. Excessive activation of NK3R might therefore cause dysfunction among the NKB/Dyn-kisspeptin-GnRH-LH cascade, such as a hyperenhancement of the Dyn/KOR signaling tone, before the resumption of normal bursting activities of KNDy neurons. This may, at least in part, be responsible for the inconsistent results of LH responses to pharmacological NK3R agonist treatments [40, 53, 57, 92].

Electrophysiological Properties of the GnRH Pulse Generator

Knobil and colleagues uncovered the single unit components of the MUA underlying the operation of the GnRH pulse generator by cluster analysis in monkeys [33]. The results indicated that the MUA volley is the consequence of coincidental increases in the firing rate of individual cells that are active even during the intervals between volleys, rather than the activation of previously silent cells. Thus, neurons consisting of the GnRH pulse generator appear to have electrophysiological properties for both spontaneous and burst activities. In this context, it is of great interest that recent findings in Kiss1-CreGFP mice [121] and genetically intact guinea pigs [122] show that ARC kisspeptin neurons do possess such electrophysiological properties. Levine [123, 124] has proposed in his model of the GnRH pulse-generating mechanism that the random activity of any neurons within an interconnected network would initiate the process of the pulse-generating activity. It is conceivable that spontaneous activity in ARC kisspeptin neurons plays a role to generate such random activity, though this requires further investigation.

A Putative Mechanism of the GnRH Pulse Generation

Taken all together, we propose, although highly speculative, the following working hypothesis for the mechanism of GnRH pulse generation [55]:

1. KNDy neurons in the ARC send projections to GnRH terminals in the ME, while their collaterals and/or dendrites form a bilateral neural network connecting each other (Fig. 14.6a).

Fig. 14.6 (continued) inhibits the bursting activities. Progesterone enhances the inhibitory tone of Dyn/KOR signaling, which acts to reduce the frequency of the periodic burst. Estrogen attenuates the stimulatory tone of NKB/N3R signaling and the excitability of KNDy neurons, which act to shorten the duration of each burst and to reduce the frequency of the periodic burst, respectively. (**c**) A sustained activation of KNDy neurons by continuous administration of NK3R agonist results in an apparent rise in the random activity, leading to an increase in the frequency of the burst. (**d**) A sustained attenuation of KOR signaling by continuous administration of KOR antagonist also produces an increase in the frequency of the burst. See text for details

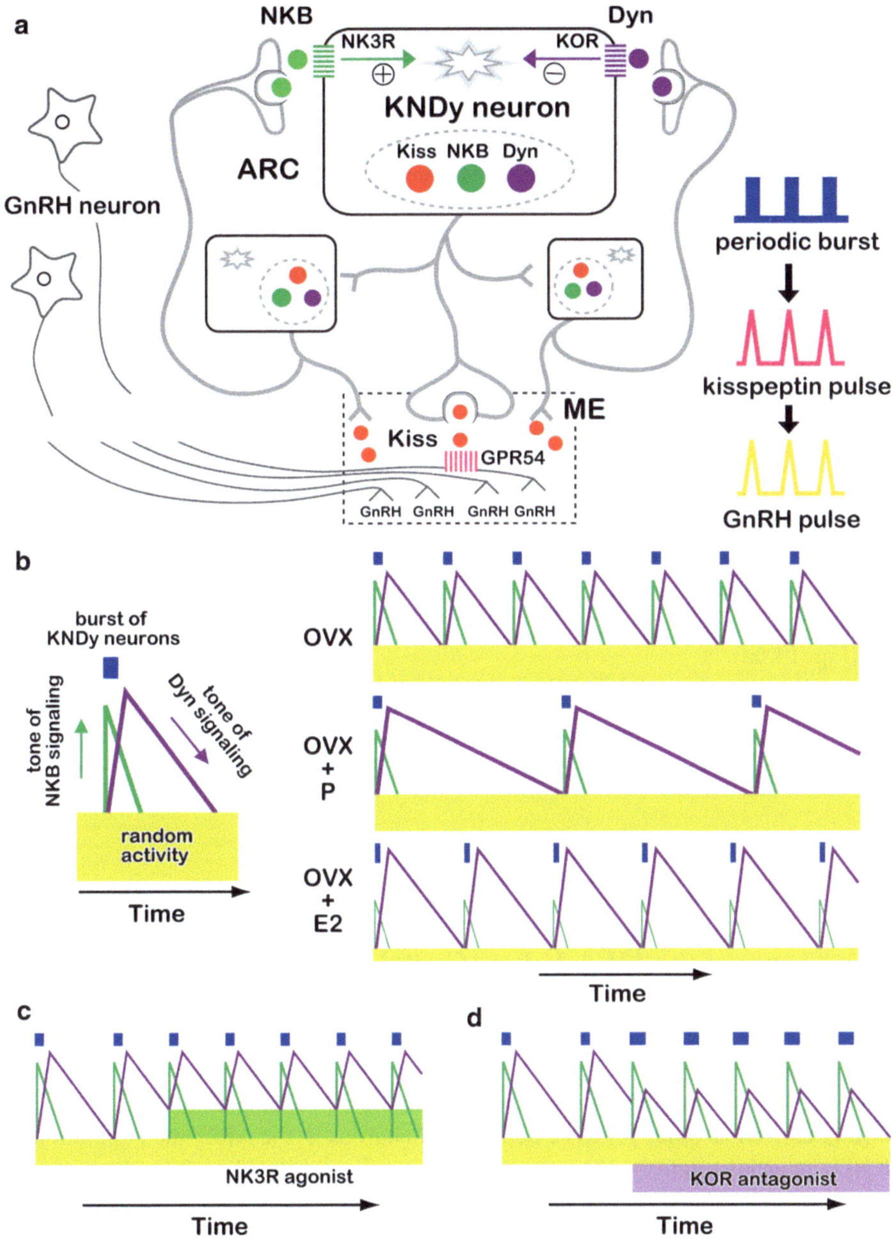

Fig. 14.6 A speculative hypothesis for the role of KNDy neurons in the generation of pulsatile GnRH release. (**a**) A population of KNDy neurons forms a neural network connected by their axon collaterals (and/or dendrites). Through the reciprocal actions of NKB/NK3R and Dyn/KOR signaling in the KNDy neuron network, episodic bursts are periodically generated, each of which, in turn, induces pulsatile discharge of kisspeptin at the ME and hence, pulsatile GnRH release into the portal circulation. (**b**) It is assumed that three components are involved in the generation of the burst: the random activity of any neuron within the network that initiates the burst, NKB/NK3r signaling that evokes synchronized bursting activities in the network, and Dyn/KOR signaling that

2. The random activity of any neuron within the KNDy neuron network would propagate among other neurons in the network through NKB/NK3 signaling to evoke synchronized bursting activities (volleys of action potentials) among KNDy neurons, which may function as a kind of positive feedback mechanism.
3. At the same time, Dyn would also be released by bursting activities in KNDy neurons, and Dyn/KOR signaling is considered to act, with a slight time lag (perhaps caused by differences of secretory mechanism or cellular signal transduction processes between NKB/NK3R and Dyn/KOR signaling), to extinguish these bursts, resulting in the net activity of the KNDy neuronal network to be an episodic oscillation (Fig. 14.6b).
4. It is suggested that Dyn/KOR signaling then imposes a prolonged quiescence, or a refractory period, which lasts until the drive of Dyn/KOR signaling diminishes enough to allow the propagation of random activities again.
5. The reciprocal interaction between the stimulatory tone of NKB/NK3R signaling and the inhibitory tone of Dyn/KOR signaling would generate intermittent oscillations, providing a pseudo-pacemaking activity in the KNDy neuron network (Fig. 14.6b).
6. Each oscillation would induce a pulse of kisspeptin release at the ME, which in turn would trigger a discharge of GnRH through kisspeptin/Kiss1r signaling, producing a pulsatile mode of GnRH secretion into the portal circulation (Fig. 14.6a).

This hypothesis is in accord with that of other research laboratories who have established the KNDy cell model [57, 62, 68] as well as the model proposed by Levine [123, 124] before the discovery of kisspeptin.

Implications Based on the Hypothesis

The Source of the MUA Volley (GnRH Pulse Generator Activity)

The MUA volleys, which represent electrophysiological manifestations of the GnRH pulse generator, can be monitored at the posterior ARC (Fig. 14.1). Although there are several neuronal populations, such as NPY [46], dopamine [82], substance P [80], as well as other yet to be determined neurons in the ARC, the population of KNDy neurons might be the only one that is fully equipped with the prerequisite neural mechanisms to act as the GnRH pulse generator, i.e., generating rhythmic oscillation, synchronizing activities within the population, and transmitting the rhythmic activity to GnRH neurons. Moreover, the negative feedback action of E2 on LH secretion, which is mediated by the GnRH pulse generator [77], is completely diminished by a pharmacological ablation of KNDy neurons [125]. Thus, it is plausible that the population of KNDy neurons is the intrinsic source of the MUA volley observed at the posterior ARC in goats [53, 54] as well as in the MBH of monkeys [29–35], rats [36–40], and goats [41–46].

Putative Mechanisms Underlying the Negative Feedback Actions of Steroid Hormones

Mechanisms of the negative feedback action of gonadal steroids can be, at least in part, explained by the schema shown in Fig. 14.6b. Progesterone is a potent inhibitor of pulsatile GnRH secretion in many species. KNDy neurons contain Dyn and receptors for P [64, 87], and P increases the expression of Dyn [97]. Therefore, it is suggested that P enhances the inhibitory drive of Dyn/KOR signaling, leading to a reduction in the frequency of burst activities in KNDy neurons (Fig. 14.6b). This speculation is in concert with the previous finding that blockade of Dyn/KOR signaling reverses the inhibitory effect of P on pulsatile LH secretion in rats [126] and sheep [127]. It appears that Dyn/KOR signaling may play a critical role in determining the length of the refractory period after the burst in KNDy neurons.

One aspect of E2 negative feedback is a decrease in the amplitude (amount) of LH secretion. The expression of not only NKB [57, 128, 129] but also NK3R [57], in the ARC, is decreased by E2, suggesting that E2 acts to attenuate the stimulatory drive of NKB/NK3 signaling. Figure 14.6b indicates that such E2 action would lead to "thinning" of the burst of KNDy neurons, which might be reflected as a marked decrease in the duration of the MUA volley after E2 treatment (Fig. 14.2c). Given that the release of kisspeptin to GnRH neuronal projections in the ME is mainly under the control of the burst activity of KNDy neurons (Fig. 14.6a), the shortening of the burst of KNDy neurons by E2 would result in a decline in the amount of GnRH released during each pulse. This may represent one aspect of the negative feedback action of E2. Moreover, it has been indicated in many species that E2 also reduces the expression of kisspeptin in the ARC [85, 87, 106, 130], which may also contribute to the decreased amount of GnRH released per pulse.

The other aspect of the E2 action is its negative effect on the frequency of GnRH/LH pulses. It has been shown that E2 also reduces the frequency of the MUA volley and LH pulses in several species, including rats [38], monkeys [30], and goats (Fig. 14.2 [42, 44, 53]), although this action of E2 seems less conspicuous in sheep [131, 132]. Because the inhibitory effect of E2 is much smaller than P (Fig. 14.2), it seems unlikely that Dyn/KOR signaling mediates the E2 action. Instead, other mechanisms may also be involved in the negative feedback action of E2. One possible mechanism is the alteration of neuronal excitability. It is possible that E2 reduces the excitability of KNDy neurons through modifying electrophysiological properties of the cell membrane, as shown in mouse GnRH neurons [133], leading to the attenuation of spontaneous activity of individual neurons. This would decrease, in a stochastic manner, the occurrence of the random activity that initiates the bursting process in the KNDy neuron. Although highly speculative, it is suggested that neuronal mechanisms involving E2 actions in KNDy neurons, such as the excitability for example, are associated with the pathway of the control of GnRH secretion by nutrition, because the inhibitory influence of several nutritional stressors on the GnRH pulse generator is more conspicuous in the presence of E2 than its absence [32, 46, 134, 135].

Putative Mechanism for the Action of Pheromones on the GnRH Pulse Generator

In goats and sheep, exposure of seasonally anestrous females to the male pheromone results in an out-of-seasonal ovulation [136, 137]. Because the initial endocrine event following the reception of the pheromone is the stimulation of pulsatile GnRH/LH secretion, it is suggested that the central target of the pheromone signal is the GnRH pulse generator [45]. We examined whether the KNDy neuronal network was involved in the pheromone action in OVX goats using MUA recording with the electrode aimed at KNDy neurons. Exposure to the male pheromone, between two successive MUA volleys, immediately induced an MUA volley and an accompanying LH pulse [138]. This pheromone effect on the MUA volley and LH secretion was abrogated by the treatment with an NK3R antagonist (Sakamoto et al., unpublished data). Further, the pheromone evoked the MUA volley but not LH pulses when kisspeptin/KOR signaling was blocked (Sakamoto et al., unpublished data). Therefore, it seems conceivable that the action of the male pheromone is indeed mediated by the KNDy neuronal network [139]. Interestingly, the effect of the pheromone was time dependent, i.e., the pheromone was not able to induce the MUA volley immediately after the preceding MUA volley, and the ability of the pheromone in inducing the MUA volley increased towards the occurrence of the next MUA volley [138]. This suggests that pheromone action may be counteracted by the inhibitory tone of Dyn/KOR signaling, which we propose would gradually decrease from the maximum to the basal level during the refractory period (Fig. 14.6b). These pheromone studies also reveal a note of caution that should be taken into account when observing the GnRH/LH response to an experimental stimulation of KNDy neurons. If a stimulus acts at the level of Kiss1r (e.g., kisspeptin), one would be able to expect a consistent result. However, if the stimulus acts at the levels of NK3R (e.g., senktide), it is possible that the GnRH/LH response to the treatment is variable depending on the timing of the treatment between two spontaneously occurring bursts of KNDy neurons.

Perspective on the Application

GnRH neurons are charged with the role of maintaining the ever-present basal levels of circulating gonadotropins for the normal functioning of the gonads. Because continuous exposure of the gonadotrophs to GnRH results in the abolishment of gonadotropin secretion, a pulsatile mode of GnRH discharge is obligatory to produce sustained gonadotropin secretion [3]. In this context, it is of interest that continuous infusion of NKB (Fig. 14.4b) or nor-BNI (Fig. 14.5b) induced frequent MUA volleys rather than a sustained raise in the MUA. Our hypothesis envisages that the frequency of periodic bursts in KNDy neurons can be increased by continuously raising the stimulatory tone of NKB/NK3R signaling by NK3R agonists (Fig. 14.6c) or reducing the inhibitory tone of Dyn/KOR signaling by KOR antagonists (Fig. 14.6d). The preliminary result detailed in this chapter (Fig. 14.2b) partially supports this proposition. There are several occasions in which insufficient LH pulse frequency causes reproductive disorders, such as women with anorexia nervosa [138], exercise amenorrhea [140], or hyperprolactinemia [141]. Our proposed model implies that NKB agonists and KOR antagonists may hold promise as novel therapeutic drugs to accelerate or improve gonadal activities via their ability to enhance the GnRH pulse generator activity.

References

1. Dierschke DJ, Bhattacharya AN, Atkinson LE, Knobil E (1970) Circhoral oscillations of plasma LH levels in the ovariectomized rhesus monkey. Endocrinology 87(5):850–853
2. Midgley AR Jr (1966) Radioimmunoassay: a method for human chorionic gonadotropin and human luteinizing hormone. Endocrinology 79(1):10–18
3. Knobil E, Plant TM, Wildt L, Belchetz PE, Marshall G (1980) Control of the rhesus monkey menstrual cycle: permissive role of hypothalamic gonadotropin-releasing hormone. Science 207(4437):1371–1373
4. Wildt L, Hausler A, Marshall G, Hutchison JS, Plant TM, Belchetz PE, Knobil E (1981) Frequency and amplitude of gonadotropin-releasing hormone stimulation and gonadotropin secretion in the rhesus monkey. Endocrinology 109(2):376–385
5. Pohl CR, Richardson DW, Hutchison JS, Germak JA, Knobil E (1983) Hypophysiotropic signal frequency and the functioning of the pituitary-ovarian system in the rhesus monkey. Endocrinology 112(6):2076–2080
6. Gay VL, Sheth NA (1972) Evidence for a periodic release of LH in castrated male and female rats. Endocrinology 90(1):158–162
7. Butler WR, Malven PV, Willett LB, Bolt DJ (1972) Patterns of pituitary release and cranial output of LH and prolactin in ovariectomized ewes. Endocrinology 91(3):793–801
8. Rahe CH, Owens RE, Fleeger JL, Newton HJ, Harms PG (1980) Pattern of plasma luteinizing hormone in the cyclic cow: dependence upon the period of the cycle. Endocrinology 107(2):498–503
9. Van de Wiel DF, Erkens J, Koops W, Vos E, Van Landeghem AA (1981) Periestrous and midluteal time courses of circulating LH, FSH, prolactin, estradiol-17 beta and progesterone in the domestic pig. Biol Reprod 24(2):223–233
10. Fitzgerald BP, I'Anson H, Loy RG, Legan SJ (1983) Evidence that changes in LH pulse frequency may regulate the seasonal modulation of LH secretion in ovariectomized mares. J Reprod Fertil 69(2):685–692
11. Foster DL, Lemons JA, Jaffe RB, Niswender GD (1975) Sequential patterns of circulating luteinizing hormone and follicle-stimulating hormone in female sheep from early postnatal life through the first estrous cycles. Endocrinology 97(4):985–994
12. Sollenberger MJ, Carlsen EC, Johnson ML, Veldhuis JD, Evans WS (1990) Specific physiological regulation of luteinizing hormone secretory events throughout the human menstrual cycle: new insights into the pulsatile mode of gonadotropin release. J Neuroendocrinol 2(6):845–852
13. Lincoln GA, Short RV (1980) Seasonal breeding: nature's contraceptive. Recent Prog Horm Res 36:1–52
14. Minabe S, Uenoyama Y, Tsukamura H, Maeda K-I (2011) Analysis of pulsatile and surge-like luteinizing hormone secretion with frequent blood sampling in female mice. J Reprod Dev 57(5):660–664

15. Clarke IJ, Cummins JT (1982) The temporal relationship between gonadotropin releasing hormone (GnRH) and luteinizing hormone (LH) secretion in ovariectomized ewes. Endocrinology 111(5):1737–1739
16. Moenter SM, Brand RM, Midgley AR, Karsch FJ (1992) Dynamics of gonadotropin-releasing hormone release during a pulse. Endocrinology 130(1):503–510
17. Plant TM, Krey LC, Moossy J, McCormack JT, Hess DL, Knobil E (1978) The arcuate nucleus and the control of gonadotropin and prolactin secretion in the female rhesus monkey (Macaca mulatta). Endocrinology 102(1):52–62
18. Belchetz PE, Plant TM, Nakai Y, Keogh EJ, Knobil E (1978) Hypophysial responses to continuous and intermittent delivery of hypothalamic gonadotropin-releasing hormone. Science 202(4368):631–633
19. Moenter SM, Brand RC, Karsch FJ (1992) Dynamics of gonadotropin-releasing hormone (GnRH) secretion during the GnRH surge: insights into the mechanism of GnRH surge induction. Endocrinology 130(5):2978–2984
20. Halasz B, Pupp L (1965) Hormone secretion of the anterior pituitary gland after physical interruption of all nervous pathways to the hypophysiotrophic area. Endocrinology 77(3):553–562
21. Blake CA, Sawyer CH (1974) Effects of hypothalamic deafferentation on the pulsatile rhythm in plasma concentrations of luteinizing hormone in ovariectomized rats. Endocrinology 94(3):730–736
22. Silverman AJ, Antunes JL, Ferin M, Zimmerman EA (1977) The distribution of luteinizing hormone-releasing hormone (LHRH) in the hypothalamus of the rhesus monkey. Light microscopic studies using immunoperoxidase technique. Endocrinology 101(1):134–142
23. Ohkura S, Tsukamura H, Maeda K-I (1992) Effects of transplants of fetal mediobasal hypothalamus on luteinizing hormone pulses impaired by hypothalamic deafferentation in adult ovariectomized rats. Neuroendocrinology 55(4):422–426
24. Martinez de la Escalera G, Choi AL, Weiner RI (1992) Generation and synchronization of gonadotropin-releasing hormone (GnRH) pulses: intrinsic properties of the GT1-1 GnRH neuronal cell line. Proc Natl Acad Sci U S A 89(5):1852–1855
25. Terasawa E, Schanhofer WK, Keen KL, Luchansky L (1999) Intracellular Ca2+ oscillations in luteinizing hormone-releasing hormone neurons derived from the embryonic olfactory placode of the rhesus monkey. J Neurosci 19(14):5898–5909
26. Richter TA, Keen KL, Terasawa E (2002) Synchronization of Ca2+ oscillations among primate LHRH neurons and nonneuronal cells in vitro. J Neurophysiol 88(3):1559–1567
27. Witkin JW, Silverman AJ (1985) Synaptology of luteinizing hormone-releasing hormone neurons in rat preoptic area. Peptides 6(2):263–271
28. Halasz B, Kiss J, Molnar J (1989) Regulation of the gonadotropin-releasing hormone (GnRH) neuronal system: morphological aspects. J Steroid Biochem 33(4B):663–668
29. Knobil E (1981) Patterns of hypophysiotropic signals and gonadotropin secretion in the rhesus monkey. Biol Reprod 24(1):44–49
30. Kesner JS, Wilson RC, Kaufman JM, Hotchkiss J, Chen Y, Yamamoto H, Pardo RR, Knobil E (1987) Unexpected responses of the hypothalamic gonadotropin-releasing hormone "pulse generator" to physiological estradiol inputs in the absence of the ovary. Proc Natl Acad Sci U S A 84(23):8745–8749
31. O'Byrne KT, Thalabard JC, Grosser PM, Wilson RC, Williams CL, Chen MD, Ladendorf D, Hotchkiss J, Knobil E (1991) Radiotelemetric monitoring of hypothalamic gonadotropin-releasing hormone pulse generator activity throughout the menstrual cycle of the rhesus monkey. Endocrinology 129(3):1207–1214
32. Chen MD, O'Byrne KT, Chiappini SE, Hotchkiss J, Knobil E (1992) Hypoglycemic 'stress' and gonadotropin-releasing hormone pulse generator activity in the rhesus monkey: role of the ovary. Neuroendocrinology 56(5):666–673
33. Cardenas H, Ordog T, O'Byrne KT, Knobil E (1993) Single unit components of the hypothalamic multiunit electrical activity associated with the central signal generator that directs the pulsatile secretion of gonadotropic hormones. Proc Natl Acad Sci U S A 90(20):9630–9634

34. Grosser PM, O'Byrne KT, Williams CL, Thalabard JC, Hotchkiss J, Knobil E (1993) Effects of naloxone on estrogen-induced changes in hypothalamic gonadotropin-releasing hormone pulse generator activity in the rhesus monkey. Neuroendocrinology 57(1):115–119
35. O'Byrne KT, Knobil E (1993) Electrophysiological approaches to gonadotrophin releasing hormone pulse generator activity in the rhesus monkey. Hum Reprod 8(suppl 2):37–40
36. Kawakami M, Uemura T, Hayashi R (1982) Electrophysiological correlates of pulsatile gonadotropin release in rats. Neuroendocrinology 35(1):63–67
37. Nishihara M, Hiruma H, Kimura F (1991) Interactions between the noradrenergic and opioid peptidergic systems in controlling the electrical activity of luteinizing hormone-releasing hormone pulse generator in ovariectomized rats. Neuroendocrinology 54(4):321–326
38. Nishihara M, Sano A, Kimura F (1994) Cessation of the electrical activity of gonadotropin-releasing hormone pulse generator during the steroid-induced surge of luteinizing hormone in the rat. Neuroendocrinology 59(6):513–519
39. Kinsey-Jones JS, Li XF, Luckman SM, O'Byrne KT (2008) Effects of kisspeptin-10 on the electrophysiological manifestation of gonadotropin-releasing hormone pulse generator activity in the female rat. Endocrinology 149(3):1004–1008
40. Kinsey-Jones JS, Grachev P, Li XF, Lin YS, Milligan SR, Lightman SL, O'Byrne KT (2012) The inhibitory effects of neurokinin B on GnRH pulse generator frequency in the female rat. Endocrinology 153(1):307–315
41. Mori Y, Nishihara M, Tanaka T, Shimizu T, Yamaguchi M, Takeuchi Y, Hoshino K (1991) Chronic recording of electrophysiological manifestation of the hypothalamic gonadotropin-releasing hormone pulse generator activity in the goat. Neuroendocrinology 53(4):392–395
42. Tanaka T, Mori Y, Hoshino K (1992) Hypothalamic GnRH pulse generator activity during the estradiol-induced LH surge in ovariectomized goats. Neuroendocrinology 56(5):641–645
43. Ito K, Tanaka T, Mori Y (1993) Opioid peptidergic control of gonadotropin-releasing hormone pulse generator activity in the ovariectomized goat. Neuroendocrinology 57(4):634–639
44. Tanaka T, Mori Y, Hoshino K (1994) Long-term recording of hypothalamic GnRH pulse generation activity during programmed administration of progesterone and estradiol in the ovariectomized goat. J Reprod Dev 40(3):183–188
45. Hamada T, Nakajima M, Takeuchi Y, Mori Y (1996) Pheromone-induced stimulation of hypothalamic gonadotropin-releasing hormone pulse generator in ovariectomized, estrogen-primed goats. Neuroendocrinology 64(4):313–319
46. Ichimaru T, Mori Y, Okamura H (2001) A possible role of neuropeptide Y as a mediator of undernutrition to the hypothalamic gonadotropin-releasing hormone pulse generator in goats. Endocrinology 142(6):2489–2498
47. Silverman AJ, Antunes JL, Abrams GM, Nilaver G, Thau R, Robinson JA et al (1982) The luteinizing hormone-releasing hormone pathways in rhesus (Macaca mulatta) and pigtailed (Macaca nemestrina) monkeys: new observations on thick, unembedded sections. J Comp Neurol 211(3):309–317
48. Goldsmith PC, Thind KK, Song T, Kim EJ, Boggant JE (1990) Location of the neuroendocrine gonadotropin-releasing hormone neurons in the monkey hypothalamus by retrograde tracing and immunostaining. J Neuroendocrinol 2(2):157–168
49. Hamada T, Shimizu T, Ichikawa M, Mori Y (1992) Immunohistochemical study on gonadotropin-releasing hormone neurons in the Shiba goat brain. J Reprod Dev 38(2):133–142
50. Hoffman GE, Gibbs FP (1982) LHRH pathways in rat brain: 'deafferentation' spares a sub-chiasmatic LHRH projection to the median eminence. Neuroscience 7(8):1979–1993
51. Witkin JW, Paden CM, Silverman AJ (1982) The luteinizing hormone-releasing hormone (LHRH) systems in the rat brain. Neuroendocrinology 35(6):429–438
52. Ohkura S, Uenoyama Y, Yamada S, Homma T, Takase K, Inoue N, Maeda K-I, Tsukamura H (2009) Physiological role of metastin/kisspeptin in regulating gonadotropin-releasing hormone (GnRH) secretion in female rats. Peptides 30(1):49–56

53. Wakabayashi Y, Nakada T, Murata K, Ohkura S, Mogi K, Navarro VM, Clifton DK, Mori Y, Tsukamura H, Maeda K-I, Steiner RA, Okamura H (2010) Neurokinin B and dynorphin A in kisspeptin neurons of the arcuate nucleus participate in generation of periodic oscillation of neural activity driving pulsatile gonadotropin-releasing hormone secretion in the goat. J Neurosci 30(8):3124–3132
54. Ohkura S, Takase K, Matsuyama S, Mogi K, Ichimaru T, Wakabayashi Y, Uenoyama Y, Mori Y, Steiner RA, Tsukamura H, Maeda K-I, Okamura H (2009) Gonadotrophin-releasing hormone pulse generator activity in the hypothalamus of the goat. J Neuroendocrinol 21(10):813–821
55. Maeda K-I, Ohkura S, Uenoyama Y, Wakabayashi Y, Oka Y, Tsukamura H, Okamura H (2010) Neurobiological mechanisms underlying GnRH pulse generation by the hypothalamus. Brain Res 1364:103–115
56. Goodman RL, Lehman MN, Smith JT, Coolen LM, de Oliveira CV, Jafarzadehshirazi MR, Pereira A, Iqbal J, Caraty A, Ciofi P, Clarke IJ (2007) Kisspeptin neurons in the arcuate nucleus of the ewe express both dynorphin A and neurokinin B. Endocrinology 148(12):5752–5760
57. Navarro VM, Gottsch ML, Chavkin C, Okamura H, Clifton DK, Steiner RA (2009) Regulation of gonadotropin-releasing hormone secretion by kisspeptin/dynorphin/neurokinin B neurons in the arcuate nucleus of the mouse. J Neurosci 29(38):11859–11866
58. Navarro VM, Gottsch ML, Wu M, Garcia-Galiano D, Hobbs SJ, Bosch MA, Pinilla L, Clifton DK, Dearth A, Ronnekleiv OK, Braun RE, Palmiter RD, Tena-Sempere M, Alreja M, Steiner RA (2011) Regulation of NKB pathways and their roles in the control of Kiss1 neurons in the arcuate nucleus of the male mouse. Endocrinology 152(11):4265–4275
59. Navarro VM, Castellano JM, McConkey SM, Pineda R, Ruiz-Pino F, Pinilla L, Clifton DK, Tena-Sempere M, Steiner RA (2011) Interactions between kisspeptin and neurokinin B in the control of GnRH secretion in the female rat. Am J Physiol Endocrinol Metab 300(1):E202–E210
60. Ramaswamy S, Seminara SB, Ali B, Ciofi P, Amin NA, Plant TM (2010) Neurokinin B stimulates GnRH release in the male monkey (Macaca mulatta) and is colocalized with kisspeptin in the arcuate nucleus. Endocrinology 151(9):4494–4503
61. Hrabovszky E, Ciofi P, Vida B, Horvath MC, Keller E, Caraty A, Bloom SR, Ghatei MA, Dhillo WS, Liposits Z, Kallo I (2010) The kisspeptin system of the human hypothalamus: sexual dimorphism and relationship with gonadotropin-releasing hormone and neurokinin B neurons. Eur J Neurosci 31(11):1984–1998
62. Lehman MN, Coolen LM, Goodman RL (2010) Minireview: kisspeptin/neurokinin B/dynorphin (KNDy) cells of the arcuate nucleus: a central node in the control of gonadotropin-releasing hormone secretion. Endocrinology 151(8):3479–3489
63. Burke MC, Letts PA, Krajewski SJ, Rance NE (2006) Coexpression of dynorphin and neurokinin B immunoreactivity in the rat hypothalamus: morphologic evidence of interrelated function within the arcuate nucleus. J Comp Neurol 498(5):712–726
64. Foradori CD, Coolen LM, Fitzgerald ME, Skinner DC, Goodman RL, Lehman MN (2002) Colocalization of progesterone receptors in parvicellular dynorphin neurons of the ovine preoptic area and hypothalamus. Endocrinology 143(11):4366–4374
65. Krajewski SJ, Burke MC, Anderson MJ, McMullen NT, Rance NE (2010) Forebrain projections of arcuate neurokinin B neurons demonstrated by anterograde tract-tracing and monosodium glutamate lesions in the rat. Neuroscience 166(2):680–697
66. Amstalden M, Coolen LM, Hemmerle AM, Billings HJ, Connors JM, Goodman RL, Lehman MN (2010) Neurokinin 3 receptor immunoreactivity in the septal region, preoptic area and hypothalamus of the female sheep: colocalisation in neurokinin B cells of the arcuate nucleus but not in gonadotrophin-releasing hormone neurones. J Neuroendocrinol 22(1):1–12
67. Rance NE (2009) Menopause and the human hypothalamus: evidence for the role of kisspeptin/neurokinin B neurons in the regulation of estrogen negative feedback. Peptides 30(1):111–122
68. Rance NE, Krajewski SJ, Smith MA, Cholanian M, Dacks PA (2010) Neurokinin B and the hypothalamic regulation of reproduction. Brain Res 1364:116–128

69. Zhang L, Brass LF, Manning DR (2009) The Gq and G12 families of heterotrimeric G proteins report functional selectivity. Mol Pharmacol 75(1):235–241
70. Kelly MJ, Qiu J, Ronnekleiv OK (2003) Estrogen modulation of G-protein-coupled receptor activation of potassium channels in the central nervous system. Ann N Y Acad Sci 1007:6–16
71. Sadja R, Alagem N, Reuveny E (2003) Gating of GIRK channels: details of an intricate, membrane-delimited signaling complex. Neuron 39(1):9–12
72. Ramaswamy S, Guerriero KA, Gibbs RB, Plant TM (2008) Structural interactions between kisspeptin and GnRH neurons in the mediobasal hypothalamus of the male rhesus monkey (Macaca mulatta) as revealed by double immunofluorescence and confocal microscopy. Endocrinology 149(9):4387–4395
73. Matsuyama S, Ohkura S, Mogi K, Wakabayashi Y, Mori Y, Tsukamura H, Maeda K-I, Ichikawa M, Okamura H (2011) Morphological evidence for direct interaction between kisspeptin and gonadotropin-releasing hormone neurons at the median eminence of the male goat: an immunoelectron microscopic study. Neuroendocrinology 94(4):323–332
74. Uenoyama Y, Inoue N, Pheng V, Homma T, Takase K, Yamada S, Ajiki K, Ichikawa M, Okamura H, Maeda K-I, Tsukamura H (2011) Ultrastructural evidence of kisspeptin-gonadotrophin-releasing hormone (GnRH) interaction in the median eminence of female rats: implication of axo-axonal regulation of GnRH release. J Neuroendocrinol 23(10):863–870
75. Smith JT, Li Q, Yap KS, Shahab M, Roseweir AK, Millar RP, Clarke IJ (2011) Kisspeptin is essential for the full preovulatory LH surge and stimulates GnRH release from the isolated ovine median eminence. Endocrinology 152(3):1001–1012
76. True C, Kirigiti M, Ciofi P, Grove KL, Smith MS (2011) Characterisation of arcuate nucleus kisspeptin/neurokinin B neuronal projections and regulation during lactation in the rat. J Neuroendocrinol 23(1):52–64
77. Karsch FJ, Bittman EL, Foster DL, Goodman RL, Legan SJ, Robinson JE (1984) Neuroendocrine basis of seasonal reproduction. Recent Prog Horm Res 40:185–232
78. Flugge G, Oertel WH, Wuttke W (1986) Evidence for estrogen-receptive GABAergic neurons in the preoptic/anterior hypothalamic area of the rat brain. Neuroendocrinology 43(1):1–5
79. Skinner DC, Herbison AE (1997) Effects of photoperiod on estrogen receptor, tyrosine hydroxylase, neuropeptide Y, and beta-endorphin immunoreactivity in the ewe hypothalamus. Endocrinology 138(6):2585–2595
80. Akesson TR, Micevych PE (1988) Estrogen concentration by substance P-immunoreactive neurons in the medial basal hypothalamus of the female rat. J Neurosci Res 19(4):412–419
81. Herbison AE (1994) Somatostatin-immunoreactive neurones in the hypothalamic ventromedial nucleus possess oestrogen receptors in the male and female rat. J Neuroendocrinol 6(3):323–328
82. Lehman MN, Karsch FJ (1993) Do gonadotropin-releasing hormone, tyrosine hydroxylase-, and beta-endorphin-immunoreactive neurons contain estrogen receptors? A double-label immunocytochemical study in the Suffolk ewe. Endocrinology 133(2):887–895
83. Rance NE, Young WS III (1991) Hypertrophy and increased gene expression of neurons containing neurokinin-B and substance-P messenger ribonucleic acids in the hypothalami of postmenopausal women. Endocrinology 128(5):2239–2247
84. Goubillon ML, Forsdike RA, Robinson JE, Ciofi P, Caraty A, Herbison AE (2000) Identification of neurokinin B-expressing neurons as an highly estrogen-receptive, sexually dimorphic cell group in the ovine arcuate nucleus. Endocrinology 141(11):4218–4225
85. Smith JT, Cunningham MJ, Rissman EF, Clifton DK, Steiner RA (2005) Regulation of Kiss1 gene expression in the brain of the female mouse. Endocrinology 146(9):3686–3692
86. Franceschini I, Lomet D, Cateau M, Delsol G, Tillet Y, Caraty A (2006) Kisspeptin immunoreactive cells of the ovine preoptic area and arcuate nucleus co-express estrogen receptor alpha. Neurosci Lett 401(3):225–230
87. Smith JT, Clay CM, Caraty A, Clarke IJ (2007) KiSS-1 messenger ribonucleic acid expression in the hypothalamus of the ewe is regulated by sex steroids and season. Endocrinology 148(3):1150–1157

88. Smith JT, Dungan HM, Stoll EA, Gottsch ML, Braun RE, Eacker SM et al (2005) Differential regulation of KiSS-1 mRNA expression by sex steroids in the brain of the male mouse. Endocrinology 146(7):2976–2984
89. Topaloglu AK, Reimann F, Guclu M, Yalin AS, Kotan LD, Porter KM, Serin A, Mungan NO, Cook JR, Ozbek MN, Imamoglu S, Akalin NS, Yuksel B, O'Rahilly S, Semple RK (2009) TAC3 and TACR3 mutations in familial hypogonadotropic hypogonadism reveal a key role for Neurokinin B in the central control of reproduction. Nat Genet 41(3):354–358
90. Young J, Bouligand J, Francou B, Raffin-Sanson ML, Gaillez S, Jeanpierre M, Grynberg M, Kamenicky P, Chanson P, Brailly-Tabard S, Guiochon-Mantel A (2010) TAC3 and TACR3 defects cause hypothalamic congenital hypogonadotropic hypogonadism in humans. J Clin Endocrinol Metab 95(5):2287–2295
91. Yang JJ, Caligioni CS, Chan YM, Seminara SB (2012) Uncovering novel reproductive defects in neurokinin B receptor null mice: closing the gap between mice and men. Endocrinology 153(3):1498–1508
92. Sandoval-Guzman T, Rance NE (2004) Central injection of senktide, an NK3 receptor agonist, or neuropeptide Y inhibits LH secretion and induces different patterns of Fos expression in the rat hypothalamus. Brain Res 1026(2):307–312
93. Wakabayashi Y, Yamamura T, Ohkura S, Homma T, Sakamoto K, Mori Y, Okamura H (2012) Senktide, a neurokinin B receptor agonist, stimulates pulsatile LH secretion through a mechanism mediated by the GnRH pulse generator in goats. In: Society for Neuroscience 42nd annual meeting (Neuroscience 2012), New Orleans, 2012, #90.07
94. Sakamoto K, Murata K, Wakabayashi Y, Yayou K, Ohkura S, Takeuchi Y, Mori Y, Okamura H (2012) Central administration of neurokinin B activates kisspeptin/NKB neurons in the arcuate nucleus and stimulates luteinizing hormone secretion in ewes during the non-breeding season. J Reprod Dev. J Reprod Dev 58(6):700–706
95. Whisnant CS, Goodman RL (1988) Effects of an opioid antagonist on pulsatile luteinizing hormone secretion in the ewe vary with changes in steroid negative feedback. Biol Reprod 39(5):1032–1038
96. Whisnant CS, Goodman RL (1994) Effect of anterior hypothalamic deafferentation on the negative feedback of gonadal steroids on luteinizing hormone pulse frequency in the ewe. Domest Anim Endocrinol 11(2):151–159
97. Foradori CD, Goodman RL, Adams VL, Valent M, Lehman MN (2005) Progesterone increases dynorphin A concentrations in cerebrospinal fluid and preprodynorphin messenger ribonucleic acid levels in a subset of dynorphin neurons in the sheep. Endocrinology 146(4):1835–1842
98. Brown CH, Ludwig M, Leng G (1998) kappa-opioid regulation of neuronal activity in the rat supraoptic nucleus in vivo. J Neurosci 18(22):9480–9488
99. Iremonger KJ, Bains JS (2009) Retrograde opioid signaling regulates glutamatergic transmission in the hypothalamus. J Neurosci 29(22):7349–7358
100. Seminara SB, Dipietro MJ, Ramaswamy S, Crowley WF Jr, Plant TM (2006) Continuous human metastin 45–54 infusion desensitizes G protein-coupled receptor 54-induced gonadotropin-releasing hormone release monitored indirectly in the juvenile male Rhesus monkey (Macaca mulatta): a finding with therapeutic implications. Endocrinology 147(5):2122–2126
101. Ramaswamy S, Seminara SB, Pohl CR, DiPietro MJ, Crowley WF Jr, Plant TM (2007) Effect of continuous intravenous administration of human metastin 45-54 on the neuroendocrine activity of the hypothalamic-pituitary-testicular axis in the adult male rhesus monkey (Macaca mulatta). Endocrinology 148(7):3364–3370
102. Roa J, Vigo E, Garcia-Galiano D, Castellano JM, Navarro VM, Pineda R, Dieguez C, Aguilar E, Pinilla L, Tena-Sempere M (2008) Desensitization of gonadotropin responses to kisspeptin in the female rat: analyses of LH and FSH secretion at different developmental and metabolic states. Am J Physiol Endocrinol Metab 294(6):E1088–E1096

103. Wakabayashi Y, Yamamura T, Ohkura S, Tanaka T, Kusaka M, Okamura H (2011) Effects of chronic administration of a metastin analog, TAK-683, on the GnRH pulse generator activity in ovariectomized goats. In: The Endocrine Society 93rd annual meeting & expo (ENDO 2011), Boston, 2011, #P2-264
104. Herbison AE, de Tassigny X, Doran J, Colledge WH (2010) Distribution and postnatal development of Gpr54 gene expression in mouse brain and gonadotropin-releasing hormone neurons. Endocrinology 151(1):312–321
105. Foradori CD, Amstalden M, Goodman RL, Lehman MN (2006) Colocalisation of dynorphin a and neurokinin B immunoreactivity in the arcuate nucleus and median eminence of the sheep. J Neuroendocrinol 18(7):534–541
106. Adachi S, Yamada S, Takatsu Y, Matsui H, Kinoshita M, Takase K, Sugiura H, Ohtaki T, Matsumoto H, Uenoyama Y, Tsukamura H, Inoue K, Maeda K-I (2007) Involvement of anteroventral periventricular metastin/kisspeptin neurons in estrogen positive feedback action on luteinizing hormone release in female rats. J Reprod Dev 53(2):367–378
107. Li XF, Kinsey-Jones JS, Cheng Y, Knox AM, Lin Y, Petrou NA, Roseweir A, Lightman SL, Milligan SR, Millar RP, O'Byrne KT (2009) Kisspeptin signalling in the hypothalamic arcuate nucleus regulates GnRH pulse generator frequency in the rat. PLoS One 4(12):e8334
108. Keen KL, Wegner FH, Bloom SR, Ghatei MA, Terasawa E (2008) An increase in kisspeptin-54 release occurs with the pubertal increase in luteinizing hormone-releasing hormone-1 release in the stalk-median eminence of female rhesus monkeys in vivo. Endocrinology 149(8):4151–4157
109. d'Anglemont de Tassigny X, Fagg LA, Carlton MB, Colledge WH (2008) Kisspeptin can stimulate gonadotropin-releasing hormone (GnRH) release by a direct action at GnRH nerve terminals. Endocrinology 149(8):3926–3932
110. Roseweir AK, Kauffman AS, Smith JT, Guerriero KA, Morgan K, Pielecka-Fortuna J, Pineda R, Gottsch ML, Tena-Sempere M, Moenter SM, Terasawa E, Clarke IJ, Steiner RA, Millar RP (2009) Discovery of potent kisspeptin antagonists delineate physiological mechanisms of gonadotropin regulation. J Neurosci 29(12):3920–3929
111. de Roux N, Genin E, Carel JC, Matsuda F, Chaussain JL, Milgrom E (2003) Hypogonadotropic hypogonadism due to loss of function of the KiSS1-derived peptide receptor GPR54. Proc Natl Acad Sci U S A 100(19):10972–10976
112. Seminara SB, Messager S, Chatzidaki EE, Thresher RR, Acierno JS Jr, Shagoury JK et al (2003) The GPR54 gene as a regulator of puberty. N Engl J Med 349(17):1614–1627
113. Oakley AE, Clifton DK, Steiner RA (2009) Kisspeptin signaling in the brain. Endocr Rev 30(6):713–743
114. Billings HJ, Connors JM, Altman SN, Hileman SM, Holaskova I, Lehman MN, McManus CJ, Nestor CC, Jacobs BH, Goodman RL (2010) Neurokinin B acts via the neurokinin-3 receptor in the retrochiasmatic area to stimulate luteinizing hormone secretion in sheep. Endocrinology 151(8):3836–3846
115. Ramaswamy S, Seminara SB, Plant TM (2011) Evidence from the agonadal juvenile male rhesus monkey (Macaca mulatta) for the view that the action of neurokinin B to trigger gonadotropin-releasing hormone release is upstream from the kisspeptin receptor. Neuroendocrinology 94(3):237–245
116. Garcia-Galiano D, van Ingen Schenau D, Leon S, Krajnc-Franken MA, Manfredi-Lozano M, Romero-Ruiz A, Navarro VM, Gaytan F, van Noort PI, Pinilla L, Blomenrohr M, Tena-Sempere M (2012) Kisspeptin signaling is indispensable for neurokinin B, but not glutamate, stimulation of gonadotropin secretion in mice. Endocrinology 153(1):316–328
117. Nestor CC, Briscoe AM, Davis SM, Valent M, Goodman RL, Hileman SM (2012) Evidence of a role for kisspeptin and neurokinin B in puberty of female sheep. Endocrinology 153(6):2756–2765
118. Noritake K, Matsuoka T, Ohsawa T, Shimomura K, Sanbuissho A, Uenoyama Y, Maeda K-I, Tsukamura H (2011) Involvement of neurokinin receptors in the control of pulsatile luteinizing hormone secretion in rats. J Reprod Dev 57(3):409–415

119. Irwig MS, Fraley GS, Smith JT, Acohido BV, Popa SM, Cunningham MJ, Gottsch ML, Clifton DK, Steiner RA (2004) Kisspeptin activation of gonadotropin releasing hormone neurons and regulation of KiSS-1 mRNA in the male rat. Neuroendocrinology 80(4):264–272
120. Krajewski SJ, Anderson MJ, Iles-Shih L, Chen KJ, Urbanski HF, Rance NE (2005) Morphologic evidence that neurokinin B modulates gonadotropin-releasing hormone secretion via neurokinin 3 receptors in the rat median eminence. J Comp Neurol 489(3):372–386
121. Gottsch ML, Popa SM, Lawhorn JK, Qiu J, Tonsfeldt KJ, Bosch MA, Kelly MJ, Ronnekleiv OK, Sanz E, McKnight GS, Clifton DK, Palmiter RD, Steiner RA (2011) Molecular properties of Kiss1 neurons in the arcuate nucleus of the mouse. Endocrinology 152(11):4298–4309
122. Qiu J, Fang Y, Bosch MA, Ronnekleiv OK, Kelly MJ (2011) Guinea pig kisspeptin neurons are depolarized by leptin via activation of TRPC channels. Endocrinology 152(4):1503–1514
123. Levine JE (1999) Pulsatility in primates—a perspective from the placode. Endocrinology 140(3):1033–1035
124. Levine JE (2000) The hypothalamus as a major integrating center. In: Conn P, Freeman ME (eds) Neuroendocrinology in physiology and medicine. Humana Press, Totowa, NJ, pp 75–93
125. Mittelman-Smith MA, Williams H, Krajewski-Hall SJ, Lai J, Ciofi P, McMullen NT, Rance NE (2012) Arcuate kisspeptin/neurokinin B/dynorphin (KNDy) neurons mediate the estrogen suppression of gonadotropin secretion and body weight. Endocrinology 153(6):2800–2812
126. Gallo RV (1990) Kappa-opioid receptor involvement in the regulation of pulsatile luteinizing hormone release during early pregnancy in the rat. J Neuroendocrinol 2(5):685–691
127. Goodman RL, Coolen LM, Anderson GM, Hardy SL, Valent M, Connors JM, Fitzgerald ME, Lehman MN (2004) Evidence that dynorphin plays a major role in mediating progesterone negative feedback on gonadotropin-releasing hormone neurons in sheep. Endocrinology 145(6):2959–2967
128. Rance NE, Bruce TR (1994) Neurokinin B gene expression is increased in the arcuate nucleus of ovariectomized rats. Neuroendocrinology 60(4):337–345
129. Pillon D, Caraty A, Fabre-Nys C, Bruneau G (2003) Short-term effect of oestradiol on neurokinin B mRNA expression in the infundibular nucleus of ewes. J Neuroendocrinol 15(8):749–753
130. Rometo AM, Krajewski SJ, Voytko ML, Rance NE (2007) Hypertrophy and increased kisspeptin gene expression in the hypothalamic infundibular nucleus of postmenopausal women and ovariectomized monkeys. J Clin Endocrinol Metab 92(7):2744–2750
131. Karsch FJ (1984) The hypothalamus and anterior pituitary gland. In: Austin CR, Short RV (eds) Reproduction in mammals: 3. Hormonal control of reproduction. Cambridge University Press, Cambridge
132. Goodman RL (1994) Neuroendocrine control of the ovine estrous cycle. In: Knobil E, Neill JD (eds) The physiology of reproduction, 2nd edn. Raven Press, New York, pp 659–709
133. Sun J, Chu Z, Moenter SM (2010) Diurnal in vivo and rapid in vitro effects of estradiol on voltage-gated calcium channels in gonadotropin-releasing hormone neurons. J Neurosci 30(11):3912–3923
134. Cagampang FR, Maeda K-I, Tsukamura H, Ohkura S, Ota K (1991) Involvement of ovarian steroids and endogenous opioids in the fasting-induced suppression of pulsatile LH release in ovariectomized rats. J Endocrinol 129(3):321–328
135. Nagatani S, Tsukamura H, Maeda K-I (1994) Estrogen feedback needed at the paraventricular nucleus or A2 to suppress pulsatile luteinizing hormone release in fasting female rats. Endocrinology 135(3):870–875
136. Martin GB, Oldham CM, Cognie Y, Pearce DT (1986) The physiological responses of anovulatory ewes to the introduction of rams—a review. Livest Prod Sci 15:219–247
137. Delgadillo JA, Gelez H, Ungerfeld R, Hawken PA, Martin GB (2009) The 'male effect' in sheep and goats—revisiting the dogmas. Behav Brain Res 200(2):304–314
138. Murata K, Wakabayashi Y, Sakamoto K, Tanaka T, Takeuchi Y, Mori Y, Okamura H (2011) Effects of brief exposure of male pheromone on multiple-unit activity at close proximity to kisspeptin neurons in the goat arcuate nucleus. J Reprod Dev 57(2):197–202

139. Okamura H, Murata K, Sakamoto K, Wakabayashi Y, Ohkura S, Takeuchi Y, Mori Y (2010) Male effect pheromone tickles the gonadotrophin-releasing hormone pulse generator. J Neuroendocrinol 22(7):825–832
140. Veldhuis JD, Evans WS, Demers LM, Thorner MO, Wakat D, Rogol AD (1985) Altered neuroendocrine regulation of gonadotropin secretion in women distance runners. J Clin Endocrinol Metab 61(3):557–563
141. Sauder SE, Frager M, Case GD, Kelch RP, Marshall JC (1984) Abnormal patterns of pulsatile luteinizing hormone secretion in women with hyperprolactinemia and amenorrhea: responses to bromocriptine. J Clin Endocrinol Metab 59(5):941–948

Chapter 15
Interactions Between Kisspeptins and Neurokinin B

Víctor M. Navarro

Abstract Reproductive function is tightly regulated by an intricate network of central and peripheral factors; however, the precise mechanism triggering critical reproductive events, such as puberty onset, remains largely unknown. Recently, the neuropeptides kisspeptin (encoded by *Kiss1*) and neurokinin B (NKB, encoded by *TAC3* in humans and *Tac2* in rodents) have been placed as essential gatekeepers of puberty. Studies in humans and rodents have revealed that loss-of-function mutations in the genes encoding either kisspeptin and NKB or their receptors, Kiss1r and neurokinin 3 receptor (NK3R), lead to impaired sexual maturation and infertility. Kisspeptin, NKB, and dynorphin A are co-expressed in neurons of the arcuate nucleus (ARC), so-called *K*isspeptin/*N*KB/*Dy*n (KNDy) neurons. Importantly, these neurons also co-express NK3R. Compelling evidence suggests a stimulatory role of NKB (or the NK3R agonist, senktide) on LH release in a number of species. This effect is likely mediated by autosynaptic inputs of NKB on KNDy neurons to induce the secretion of gonadotropin-releasing hormone (GnRH) in a kisspeptin-dependent manner, with the coordinated actions of other neuroendocrine factors, such as dynorphin, glutamate, or GABA. Thus, we have proposed a model in which NKB feeds back to the KNDy neuron to shape the pulsatile release of kisspeptin, and hence GnRH, in a mechanism also dependent on the sex steroid level. Additionally, NKB may contribute to the regulation of the reproductive function by metabolic cues. Investigating how NKB and kisspeptin interact to regulate the gonadotropic axis will offer new insights into the control of GnRH release during puberty onset and the maintenance of the reproductive function in adulthood, offering a platform for the understanding and treatment of a number of reproductive disorders.

V.M. Navarro, Ph.D. (✉)
Division of Endocrinology, Diabetes and Hypertension, Brigham and Women's Hospital and Harvard Medical School, Boston, MA 02115, USA
e-mail: vnavarro@partners.org

A.S. Kauffman and J.T. Smith (eds.), *Kisspeptin Signaling in Reproductive Biology*, Advances in Experimental Medicine and Biology 784, DOI 10.1007/978-1-4614-6199-9_15,

Introduction (Why Study Neurokinin B in a Kisspeptin Setting?)

Perpetuation of the species is an essential, but extremely energy costly, endeavor for most animals—especially for mammals [1, 2]. For this reason, it is not surprising that a vast array of neurotransmitters and endocrine factors are devoted to the precise control of the gonadotropic axis and, hence, translate the information of environmental and internal cues into a specific timing and pattern of gonadotropin-releasing hormone (GnRH) release. A key aspect that deserves to be emphasized is the incapability of GnRH neurons to show a direct response to some important modulators of reproductive function, e.g., the negative feedback of sex steroids [3, 4]. In recent years, *Kiss1* neurons have been shown to be—directly or indirectly—receptive to numerous regulatory cues, including sex steroids and metabolic and circadian factors [5]. This has placed kisspeptin in the spotlight to play a major role as a regulator of GnRH release. Yet, it is conceivable that such a critical function for the species as reproduction cannot rely exclusively on a single molecule (kisspeptin) and, hence, a number of essential "fine-tuners" and "fail-safes" might exist to ensure reproductive success. Indeed, in 2009, the endocrine community witnessed the emergence of neurokinin B (NKB) as a critical player in the control of gonadotropin release [6]. In another example of reverse translational research—suitably called "from bedside to benchside"—human genetic studies revealed that patients bearing inactivating mutations in the gene encoding NKB (*TAC3*) or its receptor, neurokinin 3 receptor (NK3R, encoded by *TACR3*), displayed hypogonadotropic hypogonadism and closely resembled the phenotype of patients with loss-of-function mutations in the genes that encode kisspeptin (*KISS1*) and the kisspeptin receptor (*KISS1R*, also known as *GPR54*) [6–13]. Some of these findings have also been partially recapitulated in *Tacr3* null mice [14], indicating that the NKB/NK3R system plays a role in the control of gonadotropin secretion in different species. Altogether, given the clear parallelism in the reproductive phenotype of humans (and mice) suffering from congenital inactivation of the kisspeptin/Kiss1r or the NKB/NK3R systems, it is conceivable that the actions of these two neuroendocrine systems interact to control GnRH release. As a result, significant efforts in the field have been recently devoted to puzzle out this interaction in what may constitute a nodal regulatory center in the control of reproductive function. This chapter intends to offer a concise overview of the latest achievements in the characterization of the reproductive facet of the NKB/NK3R system, with attention paid to the implications for the central mechanisms that govern GnRH release.

The NKB/NK3R System: Structure and Distribution in the Brain

Neurokinin B

NKB belongs to the tachykinin family of peptides that initially included the neuropeptides substance P (SP), neurokinin A (NKA), and NKB and more recently, endokinins and hemokinins [15]. Tachykinins are peptides comprised of 10–11 amino

Fig. 15.1 Schematic representation of the *TAC3* gene depicting the exon encoding NKB

acid residues in length that share a common carboxy-terminal amino acid sequence (Phe-X-Gly-Leu-Met-NH_2), where X corresponds to an aliphatic (NKA and NKB) or an aromatic (SP) residue [16]. The gene-encoding NKB (*TAC3* in higher primates and *Tac2* in rodents) is divided into seven exons, five of which encode the precursor preprotachykinin B [16–18]. Following proteolytic cleavage, this precursor leads to, first, proneurokinin B, and then NKB (initially contained in exon 5) (Fig. 15.1) [16].

The expression of NKB mRNA and protein displays a dispersed distribution in the brain of all studied species to date. In humans, prominent populations of *TAC3-positive* neurons have been identified in the infundibular nucleus, anterior hypothalamic area, septal region, diagonal band of Broca, bed nucleus of the stria terminalis, amygdala, and neocortex [19, 20]. In the rat, *Tac2* expression has been found in the cerebral cortex, hippocampus, amygdaloid complex, bed nucleus of the stria terminalis, ventral pallidum, habenula, olfactory bulb, dorsomedial nucleus, ventromedial nucleus, lateral hypothalamic area (LHA), caudate-putamen, medial preoptic area, arcuate nucleus (ARC), lateral mammillary bodies, superior colliculus, central gray, and dorsal horn of the spinal cord (Fig. 15.2) [21–23]. Of note, mice display a roughly similar distribution of *Tac2* mRNA, although, unlike rats, mice express *Tac2* neither in the hippocampus nor in the nucleus of the lateral olfactory tract [24]. Immunohistochemistry studies depicting the distribution of NKB closely parallel the neuroanatomical mapping of the gene transcript, and importantly, also offer conclusive information on the localization of NKB fibers, thus pointing to potential areas of NKB action [25–31]. Focusing on the population of NKB neurons in the ARC, projections from this nucleus have been described through a combination of tract tracing and double labeling techniques with specific known co-transmitters of NKB in this neuronal site (e.g., dynorphin and kisspeptin) [25, 27, 28, 30–32]. NKB fibers have been found to form a dense network within the ARC and the median eminence. From the ARC of the rat, NKB neurons branch to innervate rostral brain areas, such as the magnocellular and parvocellular nucleus, the anteroventral periventricular nucleus (AVPV), preoptic area, septal nuclei, and the bed nucleus of the stria terminalis [30, 33]. In addition, NKB neurons have been shown to project dorsally to the dorsomedial nucleus, periventricular nucleus, ventromedial nucleus, and LHA, and also may extend caudally to the ventral premammillary nucleus [30].

Neurokinin 3 Receptor

Tachykinins bind a family of G-protein-coupled receptors (GPCRs)—including neurokinin receptor 1 (NK1R), NK2R, and NK3R—to mediate their biological effect. Although interactions of the three initial tachykinins with each of these

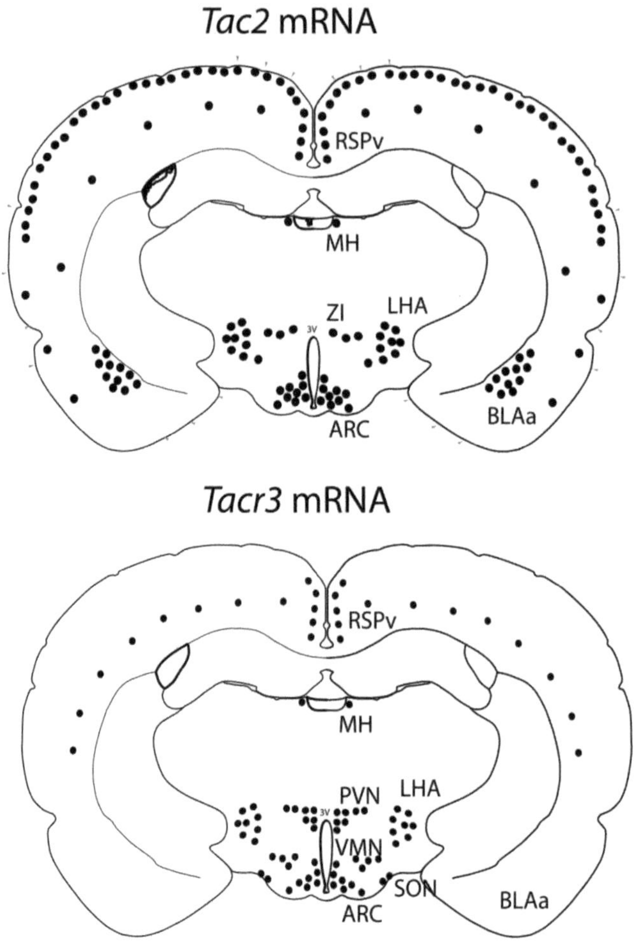

Fig. 15.2 Schematic representation of a coronal section of a rat brain at the arcuate level depicting *Tac2* and *Tacr3* expression. *SON* supraoptic nucleus; *BLAa* basolateral nucleus of the amygdala; *BMAa* basomedial nucleus of the amygdala; *MH* medial habenular nucleus; *LHA* lateral hypothalamus; *ZI* zona incerta; *VMH* ventromedial hypothalamic nucleus; *PVH* paraventricular hypothalamic nucleus; *ARC* arcuate nucleus; *3V* third ventricle

receptors have been described previously, their potency of affinity varies as follows: NK1R, SP > NKA > NKB; NK2R, NKA > NKB > SP; and NK3R, NKB > NKA > SP [34, 35]. Within the context of this review, it is conceivable that this affinity of NKB for other tachykinin receptors, even though marginal, might compromise the interpretation of the studies using NKB itself to decipher its putative roles in reproductive control. Thus, in order to overcome this limitation, the synthetic peptide senktide, a highly selective and potent agonist of NK3R [16, 34], has been systematically used in the majority of the experimental studies.

All three tachykinin receptors are encoded by genes divided into five exons with identical distribution of intronic sequences. NK3R, encoded by *TACR3* in humans and *Tacr3* in rodents, leads to a longer amino acid sequence than NK1R or NK2R, with a slight difference in length between species (465 and 452 residues in humans and rats, respectively) [16].

NK3R is mainly located in the central nervous system (CNS) and the spinal cord [15, 16, 36], although it has been also described in uterus, mesenteric vein, gut neurons, and placenta [37, 38]. Within the CNS of the rat, *Tacr3* mRNA is detected in the olfactory bulb, cortex, amygdala, hippocampus, medial habenula, zona incerta, substantia nigra, ventral tegmental area, interpeduncular nucleus, raphe nuclei, dorsal tegmental nucleus, nucleus of the solitary tract, striatum, dentate gyrus and subiculum, medial septum, diagonal band of Broca, ventral pallidum, globus pallidus, bed nucleus of the stria terminalis, ARC, paraventricular and supraoptic nuclei of the hypothalamus, dorsal and lateral regions of the posterior hypothalamus, premammillary and mammillary nuclei, midbrain central gray, cerebellum, parabrachial nuclei, nucleus of the spinal trigeminal tract, dorsal horn of the spinal cord, and the retina (Fig. 15.2) [22, 29, 39–41]. Of note, *Tacr3* shows a maturational process in the distribution of the expression of the gene along postnatal development. Thus, prepubertal animals show more abundant expression in the amygdala (basolateral amygdalar nucleus and basomedial amygdalar nucleus) and LHA than peripubertal animals [40]. At the protein level, immunohistochemistry studies depicting the distribution of NK3R resemble the brain areas where the *Tacr3* messenger is detected [25, 29, 42–45].

Regulation of the NKB/NK3R System by Sex Steroids

It is well established that gonadal steroids, i.e., estrogens and testosterone, feed back to the CNS to exert a key role in the central control of the reproductive axis by regulating the tonic release of GnRH (negative feedback) and the preovulatory surge-like release of GnRH in females (positive feedback) [46]. In spite of this, neurons that synthesize GnRH cannot respond directly to the homeostatic feedback of sex steroids, i.e., GnRH cells express neither estrogen receptor alpha (ERα) nor androgen receptor (AR) [47, 48] and only a subset express ERβ [49, 50]. However, the action of ERβ in GnRH neurons, while not yet well characterized, does not seem to be critical for, at least, the negative feedback control of sex steroids upon GnRH release although, of note, it may participate in the regulation of the excitability of GnRH neurons [51]. In this context, it is tenable that the actions of sex steroid hormones on the reproductive axis must target upstream modulators of GnRH neurons, such us (probably) NKB neurons in the ARC. Thus, in keeping with this proposed role, this population of NKB neurons may directly sense circulating levels of estradiol (E_2) and testosterone (T) through the expression of ERα [22, 52–56] and AR [57, 58]. A number of studies in adult mammals have demonstrated that, indeed, these ARC neurons exhibit a robust inhibition of *TAC3/Tac2* mRNA and NKB protein levels in the presence of E_2 or T [22, 40, 52–56, 59, 60] and, consequently, the opposite (i.e., stimulation) is true in E_2/T-deprived situations, such as the postmenopausal stage or post-gonadectomy [22, 28, 52, 53, 56, 59, 61–63]. Importantly, not only NKB but also its receptor, NK3R, is subjected to regulation by sex steroids. Thus, *Tacr3* mRNA expression is inhibited in the presence of E_2 [22, 56]. Of note, this is reminiscent of the regulation of *Kiss1* mRNA expression in the ARC by sex steroids [64, 65], which indicates that the Kiss1 and NKB systems likely control GnRH release in the same direction in critical regulatory pathways, such as sex steroid negative feedback, through interactions that yet remain to be fully understood. In this sense, an elegant study published recently by Mittelman-Smith and collaborators demonstrated that ablation of *K*isspeptin/*N*KB/*Dy*n (KNDy) neurons impairs the compensatory rise of LH after gonadectomy, i.e., the removal of steroid negative feedback, thus adding further support to the critical role of these ARC neurons in the control of GnRH release [66].

It is striking, nonetheless, that not all NKB neuronal populations in the hypothalamus behave similarly to circulating levels of estradiol. Thus, estrogens significantly inhibit NKB neurons in the ARC, as mentioned previously, while other populations of NKB neurons may display diametrically different regulation. In this vein, NKB neurons in the LHA exhibit remarkable stimulation of *Tac2* expression in the presence of E_2 [67]. Again, this fact evokes comparisons with the Kiss1 system, recalling the dual regulation by sex steroids that *Kiss1* expression undergoes when comparing ARC vs. AVPV populations [64]. The LHA is known to hold neural centers that control metabolism and, therefore, indirectly, reproductive function [68, 69]; however, whether the NKB/NK3R system in the LHA exercises a role in this control remains to be assessed.

The regulation of the NKB system by circulating levels of sex steroids—at least in the ARC—is not restricted to adult individuals. Studies in mice have depicted a striking sexual dimorphism of prepubertal animals in the sensitivity of *Tac2* expression (in the ARC) to E_2. Thus, in the absence of E_2 after gonadectomy, juvenile female mice respond with the expected compensatory rise of ARC *Tac2* expression [70], which is accompanied by the rise in ARC *Kiss1* mRNA and plasma LH levels, just as adult animals would respond [64]. However, their male counterparts appear indisposed to exhibit similar responses, unlike their adult male equivalents [70]. This fact not only suggests a clear sexual dimorphism in the physiology of the Kiss1/NKB neurons in the ARC prepubertally, but may also hold key aspects for the differential timing in puberty onset (later in males), which also needs to be investigated in more detail. An important phenomenon that may help to understand the contribution of the NKB system to puberty onset may rely in the apparent differential sensitivity to the negative feedback of sex steroids on the NKB system compared to the *Kiss1* system. Whereas both genes (*Kiss1* and *Tac2*) are susceptible to regulation by E_2 prepubertally, *Tac2* expression seems to be less sensitive to the rising levels of gonadal steroids than *Kiss1*. This may account for a sex steroid-independent increase in NKB levels that could, possibly, contribute to the activation of the Kiss1/GnRH axis prepubertally [71].

Finally, it is worth mentioning that the NKB system may be susceptible to the organizing effects of sex steroids during the critical period of sexual differentiation of the brain, i.e., perinatally. Under a physiologic sex steroid environment, the population of NKB neurons in the ARC is larger in females than in males [28, 55, 67, 72]; however, supraphysiological doses of E_2 or T in both sexes during this critical period leads to a significant reduction in the number of NKB neurons in the ARC that persists throughout the animal's life span [55, 67]. Given the exceedingly high hormone doses used in these studies, it remains unclear if normal physiological levels of perinatal sex steroids similarly modulate NKB neuron development.

On the whole, such a modulation of the NKB/NK3R system by sex steroids from early developmental stages to adulthood further unveils a potential role for this system in the central control of the reproductive axis.

Identification of the Hypothalamic KNDy Neuron

The emerging interest to decipher the role of NKB in the central control of reproduction arose, as mentioned above, in 2009 when human genetic studies documented hypogonadotropic hypogonadism in patients bearing loss-of-function mutations in *TAC3* and *TAC3R* genes [6]. Admittedly, however, the potential role of NKB as modulator of GnRH release had been previously recognized in humans, primates, rodents, and sheep [25–27, 29, 32, 52, 53, 55, 57, 58, 60, 61, 63, 73–75]. Nonetheless, the concept of NKB as a regulator of reproductive function has recently advanced to now being considered part of an intricate network of central factors that govern the acquisition and maintenance of reproductive function. In this vein, despite the fact that NKB (transcript and protein) has been described in the hypothalamus for decades, the manner in which NKB interacts with other hypothalamic factors is only starting to be deciphered. Initial studies in sheep and rats documented, by immunohistochemistry, a high degree of co-localization of NKB with the endogenous opioid peptide, dynorphin A (Dyn), in the ARC [25, 27]. These findings paved the way for a seminal study by Goodman and collaborators who showed that NKB/Dyn neurons in the ARC of ewes are also kisspeptin neurons [32]. Thereafter, the co-expression of these three neuropeptides was confirmed in mice, goats, and monkeys at the levels of mRNA by *in situ* hybridization and protein by immunohistochemistry (Fig. 15.3) [31, 56, 76], suggesting high physiological relevance for this phenomenon, which, otherwise, would not have been maintained throughout evolution. This conserved co-expression has spurred the scientific community to rename this population of neurons as KNDy neurons [77]. In addition, of great significance for the role of NKB in reproductive biology is the fact that virtually all KNDy neurons also co-express NK3R [25, 29, 42, 43, 56]. Interestingly, studies in mice and sheep have revealed that the unambiguous co-expression of kisspeptin with NKB, dynorphin, and NK3R is exclusive to the population of *Kiss1* neurons in the ARC, while the other major population of *Kiss1* neurons located in the AVPV

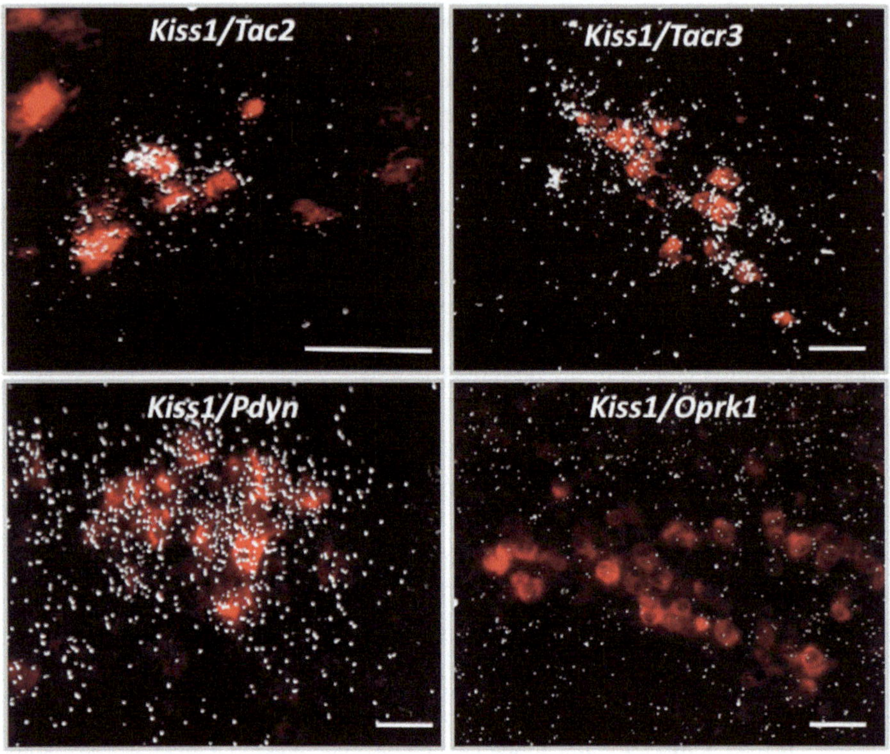

Fig. 15.3 Representative photomicrographs of mouse ARC illustrating the co-expression of *Kiss1* (labeled in *red* with digoxigenin coupled to *vector red*) and NKB (*Tac2*), NK3R (*Tacr3*), dynorphin A (*Pdyn*), and KOR (*Oprk1*) represented by *silver grains*. Scale bars = 50 μm

and the preoptic area (in mice and sheep, respectively) is virtually devoid of these kisspeptin co-transmitters [42, 43, 56]. Consequently, it is reasonable to infer that the role of NKB in the central control of reproductive function must be related to the role of kisspeptin in the ARC, for example, in the negative feedback of sex steroids upon the gonadotropic axis.

An additional aspect in the physiology of NKB neurons that merits special attention is the identification of the neuronal linage that generates the ARC population. Recent work in male mice indicates that while the vast majority of *Kiss1* neurons in the ARC seem to form a homogeneous population and collectively express *Tac2* [56], only approximately half of *Tac2* neurons in the ARC (at least in the male mouse) appear to co-express *Kiss1* [59]. This fact demonstrates a subdivision of this neuronal group with possible functional differences that yet remain to be unfolded. To note, this phenomenon is in keeping with a previous description of two populations of NKB fibers (with and without kisspeptin co-expression) in the median eminence of female rats [78].

Gonadotropin Responses to NKB

The initial characterization of the action of NKB upon gonadotropin release has been controversial. The first studies documented null, or even inhibitory, actions of NKB or the selective NK3R agonist, senktide, on LH release. A study by Sandoval-Guzmán and Rance showed that ovariectomized and estradiol-replaced rats exhibited a substantial decrease in circulating LH levels after senktide treatment [74]. To note, these animals failed to display the expected phenotype subsequent to estradiol replacement, i.e., lowering of the postcastration rise of LH, and, therefore, they should be considered as ovariectomized and sham-replaced animals. In the same vein, the latter rationale could also apply to a recent study by Kinsey-Jones et al. [79]. As a consequence, considering this limitation, the above observations would essentially be in keeping with previous reports documenting inhibitory actions of senktide (or NKB) in gonadectomized mice, rats, and goats [22, 56, 76]. These results, in fact, were clearly contradictory to the initially predicted stimulation of LH release by NKB on the basis of the human studies, in which patients suffering constitutive deficiency in the NKB system exhibited hypogonadotropic hypogonadism [6, 8–10, 13]. Interestingly, this deficiency in humans seemed to be exclusively restricted to LH release, since FSH levels appeared relatively normal [8]. This feature of NKB-null patients suggests a residual low-frequency release of GnRH (NKB-independent), which may result as a consequence of the action of high vs. low pulses of GnRH in the discrimination of LH or FSH release, respectively [80, 81]. Despite the above observations, the overall literature on this topic substantiates a remarkable discrepancy in the response to NKB or senktide in different physiological settings. However, a growing number of studies are offering irrefutable demonstrations of the robust stimulatory action of NKB upon gonadotropin release in a number of species [22, 31, 40, 43, 59, 76, 82, 83]. In this sense, while further investigation is needed to fully understand this paradox (stimulation or inhibition of LH release after senktide treatment), studies in gonadectomized and sham-replaced vs. E_2-replaced animals strongly suggest the need of physiological levels of circulating sex steroids (acting on ERα, most likely, in KNDy neurons) to allow the stimulation of LH release by NKB. This is also in line with the stimulatory action of senktide on LH secretion in the follicular, but not luteal, phase in sheep [43]. Of note, continuous stimulation of NK3R with senktide leads to desensitization of the receptor [83], which might, hypothetically, imply that under the elevated levels of kisspeptin and NKB reached in gonadectomized animals, the additional stimulation of NK3R by senktide could account for the observed decrease of LH release in E_2-deprived animals. Admittedly, a recent publication indicates that, in this scenario, dynorphin may be hyperstimulated by senktide and, consequently, mediate this inhibition [79]. This inhibitory action, however, remains to be carefully explored.

Notwithstanding the important role of NKB upon GnRH release, NKB's characterization initially focused on the female. Intriguingly, subsequent studies in the male are now suggesting a more complex regulation of the NKB system than previously anticipated. In this sense, initial studies in male mice pointed to a likely sexual dimorphism in the response to senktide. Corander et al. first documented the absence of LH release after NKB administration in intact adult male mice [84]; however, shortly after, in a study performed in adult male mice, we reported a positive response to senktide administration, with a robust increase in LH release and also, to a lesser extent, FSH [59]. Although speculative, this discrepancy might be due to different efficacy in the activation of NK3R after administration of senktide [59] vs. NKB itself [84] or, purportedly, to the delivery method, i.e., limitation of NKB to dissolve into solution in saline vehicles [84].

Recently, we have expanded our knowledge of the actions of senktide on LH release in male and female rats along postnatal development. Intact female rats at every assessed developmental stage (i.e., infantile, juvenile, prepubertal, pubertal, and adult in diestrus) display conspicuous stimulation of LH secretion after senktide treatment [22, 67]. Male rats, however, progress from being responsive to senktide prepubertally to becoming irresponsive from puberty onwards, dissociating themselves from adult female rats and male mice [67]. These data unveil a striking sexual dimorphism in the rat as well as important differences in otherwise closely related species, i.e., mice and rats. The precise mechanisms underlying these differences require further investigation in order to clarify the role of NKB in the control of gonadotropin secretion.

Kiss1 Neurons as Primary Targets of NKB Action

As described previously, KNDy neurons express NK3R [25, 26], which supports the possibility of autosynaptic loops within the network of NKB/kisspeptin fibers surrounding KNDy neurons in the ARC [25, 30, 33]. The following evidence supports the contention that NKB acts upstream of, or immediately on, *Kiss1* neurons in the ARC: (a) senktide and NKB strongly depolarize *Kiss1* neurons in the mouse, and this effect is prevented by the NKB antagonist SB222200 [59]; (b) central administration of senktide to gonadectomized and estradiol-replaced ($OVX+E_2$) female rats induces *c-fos* mRNA expression in *Kiss1* neurons [22]; (c) goats treated centrally with NKB exhibit a clear increase in the frequency and amplitude of multiunit activity (MUA) volleys in the ARC, which are invariably mirrored by LH pulses [76], yet the administration of kisspeptin modifies LH pulses without modifying MUA volleys [85].

Kiss1 neurons, however, express a number of other co-transmitters (e.g., dynorphin and glutamate). For this reason, any action of NKB upon GnRH release, evoked by the activation of ARC *Kiss1* neurons, could not be entirely attributed to the release of kisspeptin. For instance, these kisspeptin neurons have been also shown to innervate the tuberoinfundibular dopaminergic (TIDA) neurons in the ARC [86]. Yet, compelling evidence indicates that kisspeptin release is, indeed, necessary for the reproductive role of NKB given the following data: (a) mice bearing nonfunctional kisspeptin receptors (*Kiss1r* KO mice) are irresponsive to the central administration of senktide in terms of LH release [82]; (b) juvenile monkeys showing blunted LH responses due to NK3R desensitization still respond to kisspeptin; however, monkeys subjected to Kiss1r desensitization showed a significant reduction in their response to senktide treatment [31, 83], although, admittedly, they did present a marginal (perhaps kisspeptin-independent) stimulation of LH release.

On the whole, these data strongly suggest a putative action of NKB upon ARC *Kiss1* neurons; however, at the same time, we cannot rule out any additional action of this peptide on other brain areas or even different neuronal populations within the ARC. Along these lines, several studies document the expression of NK3R in GnRH terminals of rats [25, 29] and *Tacr3* mRNA in GnRH neurons in the mouse [87] as well as in the immortalized GT1-7 cell line—a model of differentiated GnRH neurons—[88]; however, a call of caution should be added since immortalized cell lines do not always resemble in vivo models. Despite the latter observations, recent studies document the absence of NK3R immuno-localization and *Tacr3* mRNA in GnRH neurons of sheep [42] and mice [59], respectively, adding further controversy to the role of NKB in the control of GnRH secretion. Indeed, mounting evidence seems to further support the contention that GnRH neurons are devoid—or only present marginal expression—of NK3R and therefore should not be considered as the primary site of action of NKB to regulate GnRH release. For example, GnRH neurons coupled to green fluorescent protein (GnRH-GFP) and subjected to whole-cell recordings do not display signs of activation after senktide treatment in the mouse [59]. This latter study, nonetheless, shows lack of activity at the level of the GnRH cell body, but does not rule out the presence of NK3R at the level of GnRH terminals. In this sense, Corander and colleagues demonstrated that, unlike kisspeptin [89], addition of NKB to hypothalamic explants from male rats (devoid of GnRH cell bodies) did not evoke any effect on GnRH release, which, initially, would preclude any direct action at the level of these GnRH fiber terminals [84].

The Roles of NKB/Kisspeptin Interactions in the Control of Reproductive Function

Generation of GnRH Pulses

Although this book includes a specific chapter devoted to this topic, it is, however, worth highlighting the potential implications for reproductive physiology that interactions of NKB and kisspeptin (and dynorphin) may have for the normal functioning of the gonadotropic axis. There is convincing evidence to conclude that kisspeptin release is pulsatile and that those pulses correlate with GnRH/LH pulses, based on studies in monkeys and goats [76, 90, 91]. Consequently, we have proposed a model whereby NKB from KNDy neurons acts autosynaptically on the NK3R of the same neuron, through recurrent collaterals, as well as on other neighboring KNDy neurons (Fig. 15.4) [56, 76, 92, 93]. This action of NKB would evoke a synchronized release of kisspeptin within the ARC, a fact of critical importance in order

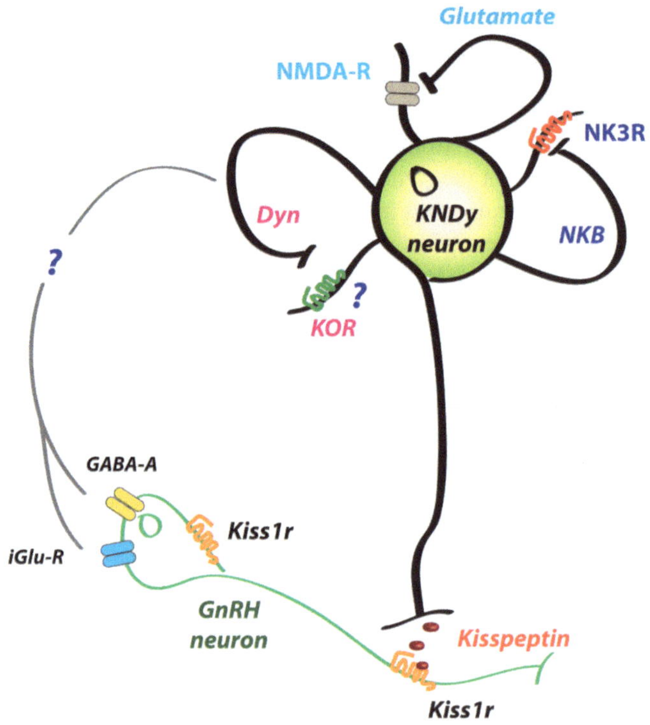

Fig. 15.4 Schematic representation of a KNDy neuron depicting possible autoregulatory loops and interactions with GnRH neurons

to present a unique GnRH pulse after every kisspeptin pulse. This possibility is in keeping with the dense network of NKB fibers surrounding KNDy neurons in the ARC [25, 30, 33]. Next in this model, kisspeptin would reach the kisspeptin receptor at the GnRH terminals to induce GnRH release to the median eminence. Of note, there is no direct evidence of the presence of Kiss1r at the level of GnRH terminals; however, this could be inferred based on the observations that (a) kisspeptin can elicit GnRH secretion from explants of the mediobasal hypothalamus, which contains few, if any, GnRH cell bodies [89, 94–96] and (b) kisspeptin fibers form close appositions at the GnRH fiber terminals [97–99]. At the same time, dynorphin (an endogenous opioid peptide with known inhibitory action upon gonadotropin release [100, 101]), would act mainly on yet unknown intermediate neurons to eventually shut down NKB and kisspeptin release, therefore creating a kisspeptin pulse. Noteworthy, each neuropeptide released from KNDy neurons seems to target a different set of neurons. Thus, (a) kisspeptins act mainly on GnRH neurons, since KNDy neurons themselves in the ARC are devoid of Kiss1 receptor [94] and do not respond to kisspeptin [59]; (b) NKB seems to primarily activate KNDy neurons, since GnRH neurons (at least in mice and sheep) do not express detectable levels of NK3R [43, 59]; (c) GnRH neurons do not express the dynorphin receptor (κ-opioid receptor, KOR) [102, 103], and its expression in KNDy neurons seems to be marginal [56, 59]. These findings allude to intermediate neurons as the primary site of action of dynorphin in the context of kisspeptin release, which, allegedly, would play a substantial role by increasing the turnaround time of the inhibitory signal before acting back on KNDy neurons, therefore allowing for the appearance of a kisspeptin pulse. Of note, as research on the characterization of KNDy neurons advances, new co-transmitters emerge, which, in the near future, may make the descriptive term "KNDy" obsolete. In this vein, glutamate and glutamate receptors, also described in KNDy neurons [104], may constitute additional autosynaptic regulatory loops [82, 104, 105] to "fine-tune" kisspeptin release (Fig. 15.4). It has been speculated that this or a similar mechanism may be involved in the attenuation of the severe reproductive phenotype of *Tacr3* null mice compared to humans described recently by Yang and colleagues [14, 106]. Moreover, recent studies also suggest an action (direct and indirect) of GnRH itself on KNDy neurons to, eventually, modulate GnRH release [107]; however, this contention requires further research. For instance, the presence of the GnRH receptor in KNDy neurons has yet to be demonstrated.

Control of Puberty Onset

GnRH release undergoes a series of developmental changes, progressing from an activated neonatal period, followed by a dormant stage during the infantile and juvenile ages, to puberty onset, which is characterized by the appearance of GnRH pulses with increasing amplitude and frequency. A number of central and peripheral factors have been posed to mediate this awakening of the reproductive axis; however, exactly what triggers puberty onset remains elusive.

In recent years, hypothalamic kisspeptin has become a likely candidate to serve as a gatekeeper of puberty onset. Kisspeptin is the most potent secretagogue of GnRH described to date [5], and compelling evidence shows that *Kiss1* expression and synaptic contacts between Kiss1 neurons in the hypothalamus increase at the time of puberty onset in rodents [92, 108, 109], suggesting that hypothalamic *Kiss1* neurons play an important role during this maturational process [110]. Furthermore, chronic administration of kisspeptin to prepubertal female rats advances puberty onset [111], which is prevented in the presence of a kisspeptin antagonist [112]. In this context, if we assume that NKB has a role in the control of kisspeptin release at the onset of puberty, we could expect that impairments in the NKB/NK3R system would directly translate into disorders in the timing of puberty. Indeed, as mentioned previously, a number of studies have documented impuberism associated with hypogonadotropic hypogonadism (HH) in humans bearing inactivating mutations in the genes encoding NKB and NK3R, which has been partially recapitulated in *Tacr3* knockout mice. To note, *Tac2*/*Tacr3* expression in the hypothalamus of rats (and markedly *Tacr3* in the ARC) increases prior to puberty onset [22, 40], apparently anticipating the increase of kisspeptin immunoreactivity reported in the rostral periventricular area of the hypothalamus in prepubertal female mice [113].

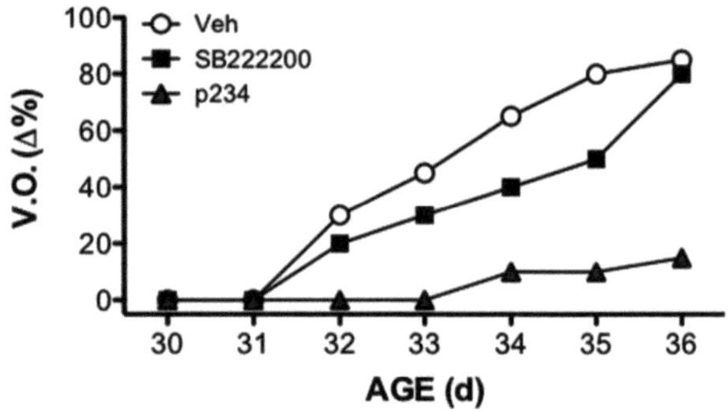

Fig. 15.5 Percentage of vaginal opening (V.O.) in peripubertal female rats treated chronically with central injections of vehicle, the NKB antagonist SB222200 or the kisspeptin antagonist p234. Figure composed based on the original data presented in refs. [40, 112]

Importantly, despite the documented sensitivity of *Tac2* expression to the negative feedback of sex steroids prepubertally in the mouse [70], the major regulatory pathway that drives the stimulation of *Tac2* expression prior to puberty onset seems to be independent and, seemingly, dominates over the rising levels of gonadal steroids that occur during puberty onset [40, 71, 114].

Additionally, in line with a role for NKB in the timing of puberty, senktide induces strong secretory responses of LH in juvenile monkeys [31], prepubertal male and female rats [67] and prepubertal ewes [115], suggesting that the gonadotropic axis is responsive to NKB stimulation before puberty onset. On the whole, these observations, along with recent clinical observations suggesting that the stimulatory action NKB on the gonadotropic axis may be more prominent during early stages of sexual maturation [10], are in line with an eventual stimulatory role of NKB upon *Kiss1* expression during pubertal maturation.

Moreover, studies of chronic blockade of NKB signaling using the NK3R antagonist, SB222200, to prepubertal female rats induce a slight delay in the timing of puberty onset (Fig. 15.5) [40]. The effect of this antagonist to block pubertal progression is, however, not as effective as the chronic administration of a kisspeptin antagonist, which results in a marked suppression of vaginal opening (Fig. 15.5) and gonadal weights—as indirect markers of the rise of circulating sex steroids during puberty onset [112]. In sum, based on the above observations, it is reasonable to deduce a stimulatory role of NKB in the timing of puberty, an effect which may be dependent on kisspeptin signaling.

NKB as a Signaling System of the Metabolic Status

Reproduction is an extremely energy-demanding function and, as such, subjected to regulation by metabolic cues that eventually contribute to the regulation of GnRH release [1, 116]. Mounting data during the last few years suggest that Kiss1 neurons in the ARC play a critical role conveying metabolic information onto the hypothalamic centers that control the attainment and maintenance of reproductive function. For instance, models of metabolic stress, such as acute fasting, impinge a significant restraint on the hypothalamic expression of *Kiss1*/kisspeptins in pubertal animals, which can be reverted by exogenous kisspeptin [117]. Given the fact that the NKB/NK3R system is also present in the metabolic conveyor that KNDy neurons in the ARC constitute, it was reasonable to assume that the action of NKB is also subjected to metabolic regulation during (and after) puberty. In this sense, a recent study documented the sensitivity of this system to metabolic cues [118]. In more detail, a later study in pubertal female rats demonstrates a significant suppression of *Tacr3*, and to a lesser extent *Tac2*, in the ARC after 48-h fasting [40], replicating the previously described expression profile of *Kiss1* in both the ARC and AVPV [117]. Moreover, in this study, LH responses to senktide administration in pubertal (36-d) rats are not only preserved but even augmented in fasting conditions, suggesting a possible sensitization of its stimulatory effects under conditions of negative energy balance—again, resembling previous findings on the gonadotropin-releasing actions of kisspeptins [117, 119]. Additionally, chronic administration of senktide to prepubertal female rats subjected to caloric restriction was sufficient to partially rescue markers of puberty onset, e.g., vaginal opening and LH secretion [40], as previously reported for kisspeptin [117]. On the other hand, situations of exceedingly high caloric intake, such as rats subjected to high fat diet prepubertally, exhibited precocious puberty that correlates with the advancement in the timing of *Kiss1* and *Tac2* expression that, in turn, was associated with the advancement in LH pulsatility [120].

Altogether, the above observations suggest that the NKB system is subjected to modulation by metabolic cues, at least during pubertal progression, probably facilitating the transmission of the energy status of the organism on to the *Kiss1* system, most likely, at the level of the ARC. However, kisspeptin-independent pathways for the stimulatory effects of NKB on the gonadotropic axis at puberty cannot be excluded—which remains to be explored. Noteworthy, recent studies associate KNDy neurons in the ARC with the regulatory (inhibition) effect that estrogens exert on body weight [66]. This finding poses the KNDy neuron as a nodal regulatory center for the integration of reproductive axis and energy balance; however, the mechanisms underlying this effect need to be deciphered.

Conclusion and Future Perspectives

Our knowledge of the neuronal interactions that potentially impinge the accurate functioning of the endocrine system is rapidly evolving. Particularly, in recent years, reproductive neuroendocrinologists are witnessing the appearance of a constellation of central factors that significantly contribute to the modulation of GnRH release. In this context, since 2003, the scientific community has enthusiastically welcomed *Kiss1* neurons as key elements to answer remaining open questions in the physiology of GnRH release. More recently, the NKB system has added a new level of complexity to *Kiss1* neurons. Compelling evidence suggests that NKB plays a critical role in the control of kisspeptin release, at least at the level of the ARC. Not only is NKB able to modulate gonadotropin release through, according to recent studies, its action on *Kiss1* neurons but, also, serves as a conveyor of additional regulatory factors, e.g., developmental and metabolic, to *Kiss1* neurons, thus contributing to the exquisite regulation of kisspeptin release.

Many aspects of the physiology of the NKB/NK3R system in the context of reproduction remain to be fully characterized. For instance, a vivid debate is currently ongoing in the field regarding the putative target/s of NKB, with evidence on both *Kiss1* and GnRH neurons that may only reflect species differences or, perhaps, residual levels of redundancy that, under constitutive absence of one or another, may lead to potential compensation. Additionally, important aspects regarding the sexual differentiation of the response of gonadotropins to NKB (or senktide), as recently observed in the rat, constitute a mystery that demands to be resolved. Indeed, reconciliation of the results in male mice and rats, and a detailed comparison between both sexes, to determine which one better resembles the human phenotype would clearly help the scientific community to establish a working model with translational potential to humans. Overall, elucidating how the brain triggers the neuroendocrine events that lead to the attainment and maintenance of reproductive function will provide the intellectual platform for understanding certain disorders of reproduction, including delayed or precocious puberty—and perhaps guide us toward improved therapies for their treatment.

Acknowledgments This research was supported by the Marie Curie Outgoing International Fellowship within the seventh Framework Programme of the European Union. The author is grateful to Drs. Amy Oakley and Leonor Pinilla for their constructive comments on the manuscript and Drs. Manuel Tena-Sempere, Don Clifton, and Robert Steiner for their contribution to some of the studies highlighted in this chapter.

Conflict of interest: The author reports no conflicts of interest.
Disclosure: The author has nothing to disclose.

References

1. Hill JW, Elmquist JK, Elias CF (2008) Hypothalamic pathways linking energy balance and reproduction. Am J Physiol Endocrinol Metab 294:E827–E832
2. Ojeda SR, Advis JP, Andrews WW (1980) Neuroendocrine control of the onset of puberty in the rat. Fed Proc 39:2365–2371
3. Herbison A (2006) Physiology of the GnRH neuronal network. In: Neil J, Knobil E (eds) Physiology of reproduction. Academic, San Diego, CA, pp 1415–1482
4. Herbison AE, Robinson JE, Skinner DC (1993) Distribution of estrogen receptor-immunoreactive cells in the preoptic area of the ewe: co-localization with glutamic acid decarboxylase but not luteinizing hormone-releasing hormone. Neuroendocrinology 57:751–759
5. Oakley AE, Clifton DK, Steiner RA (2009) Kisspeptin signaling in the brain. Endocr Rev 30:713–743
6. Topaloglu AK, Reimann F, Guclu M, Yalin AS, Kotan LD, Porter KM, Serin A, Mungan NO, Cook JR, Ozbek MN, Imamoglu S, Akalin NS, Yuksel B, O'Rahilly S, Semple RK (2009) TAC3 and TACR3 mutations in familial hypogonadotropic hypogonadism reveal a key role for neurokinin B in the central control of reproduction. Nat Genet 41:354–358

7. de Roux N, Genin E, Carel JC, Matsuda F, Chaussain JL, Milgrom E (2003) Hypogonadotropic hypogonadism due to loss of function of the KiSS1-derived peptide receptor GPR54. Proc Natl Acad Sci U S A 100:10972–10976
8. Francou B, Bouligand J, Voican A, Amazit L, Trabado S, Fagart J, Meduri G, Brailly-Tabard S, Chanson P, Lecomte P, Guiochon-Mantel A, Young J (2011) Normosmic congenital hypogonadotropic hypogonadism due to TAC3/TACR3 mutations: characterization of neuroendocrine phenotypes and novel mutations. PLoS One 6:e25614
9. Fukami M, Maruyama T, Dateki S, Sato N, Yoshimura Y, Ogata T (2010) Hypothalamic dysfunction in a female with isolated hypogonadotropic hypogonadism and compound heterozygous TACR3 mutations and clinical manifestation in her heterozygous mother. Horm Res Paediatr 73:477–481
10. Gianetti E, Tusset C, Noel SD, Au MG, Dwyer AA, Hughes VA, Abreu AP, Carroll J, Trarbach E, Silveira LF, Costa EM, de Mendonca BB, de Castro M, Lofrano A, Hall JE, Bolu E, Ozata M, Quinton R, Amory JK, Stewart SE, Arlt W, Cole TR, Crowley WF, Kaiser UB, Latronico AC, Seminara SB (2010) TAC3/TACR3 mutations reveal preferential activation of gonadotropin-releasing hormone release by neurokinin B in neonatal life followed by reversal in adulthood. J Clin Endocrinol Metab 95:2857–2867
11. Seminara SB, Messager S, Chatzidaki EE, Thresher RR, Acierno JS Jr, Shagoury JK, Bo-Abbas Y, Kuohung W, Schwinof KM, Hendrick AG, Zahn D, Dixon J, Kaiser UB, Slaugenhaupt SA, Gusella JF, O'Rahilly S, Carlton MB, Crowley WF Jr, Aparicio SA, Colledge WH (2003) The GPR54 gene as a regulator of puberty. N Engl J Med 349:1614–1627
12. Topaloglu AK, Tello JA, Kotan LD, Ozbek MN, Yilmaz MB, Erdogan S, Gurbuz F, Temiz F, Millar RP, Yuksel B (2012) Inactivating KISS1 mutation and hypogonadotropic hypogonadism. N Engl J Med 366:629–635
13. Young J, Bouligand J, Francou B, Raffin-Sanson ML, Gaillez S, Jeanpierre M, Grynberg M, Kamenicky P, Chanson P, Brailly-Tabard S, Guiochon-Mantel A (2010) TAC3 and TACR3 defects cause hypothalamic congenital hypogonadotropic hypogonadism in humans. J Clin Endocrinol Metab 95:2287–2295
14. Yang JJ, Caligioni CS, Chan YM, Seminara SB (2012) Uncovering novel reproductive defects in neurokinin B receptor null mice: closing the gap between mice and men. Endocrinology 153(3):1498–1508
15. Page NM (2005) New challenges in the study of the mammalian tachykinins. Peptides 26:1356–1368
16. Almeida TA, Rojo J, Nieto PM, Pinto FM, Hernandez M, Martin JD, Candenas ML (2004) Tachykinins and tachykinin receptors: structure and activity relationships. Curr Med Chem 11:2045–2081
17. Bonner TI, Affolter HU, Young AC, Young WS III (1987) A cDNA encoding the precursor of the rat neuropeptide, neurokinin B. Brain Res 388:243–249
18. Page NM, Morrish DW, Weston-Bell NJ (2009) Differential mRNA splicing and precursor processing of neurokinin B in neuroendocrine tissues. Peptides 30:1508–1513
19. Chawla MK, Gutierrez GM, Young WS III, McMullen NT, Rance NE (1997) Localization of neurons expressing substance P and neurokinin B gene transcripts in the human hypothalamus and basal forebrain. J Comp Neurol 384:429–442
20. Pinto FM, Almeida TA, Hernandez M, Devillier P, Advenier C, Candenas ML (2004) mRNA expression of tachykinins and tachykinin receptors in different human tissues. Eur J Pharmacol 494:233–239
21. Marksteiner J, Sperk G, Krause JE (1992) Distribution of neurons expressing neurokinin B in the rat brain: immunohistochemistry and in situ hybridization. J Comp Neurol 317:341–356
22. Navarro VM, Castellano JM, McConkey SM, Pineda R, Ruiz-Pino F, Pinilla L, Clifton DK, Tena-Sempere M, Steiner RA (2011) Interactions between kisspeptin and neurokinin B in the control of GnRH secretion in the female rat. Am J Physiol Endocrinol Metab 300:E202–E210
23. Warden MK, Young WS III (1988) Distribution of cells containing mRNAs encoding substance P and neurokinin B in the rat central nervous system. J Comp Neurol 272:90–113

24. Duarte CR, Schutz B, Zimmer A (2006) Incongruent pattern of neurokinin B expression in rat and mouse brains. Cell Tissue Res 323:43–51
25. Burke MC, Letts PA, Krajewski SJ, Rance NE (2006) Coexpression of dynorphin and neurokinin B immunoreactivity in the rat hypothalamus: morphologic evidence of interrelated function within the arcuate nucleus. J Comp Neurol 498:712–726
26. Ciofi P, Krause JE, Prins GS, Mazzuca M (1994) Presence of nuclear androgen receptor-like immunoreactivity in neurokinin B-containing neurons of the hypothalamic arcuate nucleus of the adult male rat. Neurosci Lett 182:193–196
27. Foradori CD, Amstalden M, Goodman RL, Lehman MN (2006) Colocalisation of dynorphin a and neurokinin B immunoreactivity in the arcuate nucleus and median eminence of the sheep. J Neuroendocrinol 18:534–541
28. Hrabovszky E, Molnar CS, Sipos M, Vida B, Ciofi P, Borsay BA, Sarkadi L, Herczeg L, Bloom SR, Ghatei MA, Dhillo WS, Kallo I, Liposits Z (2011) Sexual dimorphism of kisspeptin and neurokinin B immunoreactive neurons in the infundibular nucleus of aged men and women. Front Endocrinol 2:80. doi:10.3389/fendo.2011.00080
29. Krajewski SJ, Anderson MJ, Iles-Shih L, Chen KJ, Urbanski HF, Rance NE (2005) Morphologic evidence that neurokinin B modulates gonadotropin-releasing hormone secretion via neurokinin 3 receptors in the rat median eminence. J Comp Neurol 489:372–386
30. Krajewski SJ, Burke MC, Anderson MJ, McMullen NT, Rance NE (2010) Forebrain projections of arcuate neurokinin B neurons demonstrated by anterograde tract-tracing and monosodium glutamate lesions in the rat. Neuroscience 166:680–697
31. Ramaswamy S, Seminara SB, Ali B, Ciofi P, Amin NA, Plant TM (2010) Neurokinin B stimulates GnRH release in the male monkey (Macaca mulatta) and is colocalized with kisspeptin in the arcuate nucleus. Endocrinology 151:4494–4503
32. Goodman RL, Lehman MN, Smith JT, Coolen LM, de Oliveira CV, Jafarzadehshirazi MR, Pereira A, Iqbal J, Caraty A, Ciofi P, Clarke IJ (2007) Kisspeptin neurons in the arcuate nucleus of the ewe express both dynorphin A and neurokinin B. Endocrinology 148:5752–5760
33. Yeo SH, Herbison AE (2011) Projections of arcuate nucleus and rostral periventricular kisspeptin neurons in the adult female mouse brain. Endocrinology 152:2387–2399
34. Mussap CJ, Geraghty DP, Burcher E (1993) Tachykinin receptors: a radioligand binding perspective. J Neurochem 60:1987–2009
35. Seabrook GR, Bowery BJ, Hill RG (1995) Pharmacology of tachykinin receptors on neurones in the ventral tegmental area of rat brain slices. Eur J Pharmacol 273:113–119
36. Pantaleo N, Chadwick W, Park SS, Wang L, Zhou Y, Martin B, Maudsley S (2010) The mammalian tachykinin ligand-receptor system: an emerging target for central neurological disorders. CNS Neurol Disord Drug Targets 9:627–635
37. Valero MS, Fagundes DS, Grasa L, Arruebo MP, Plaza MA, Murillo MD (2011) Contractile effect of tachykinins on rabbit small intestine. Acta Pharmacol Sin 32:487–494
38. Page NM, Bell NJ (2002) The human tachykinin NK1 (short form) and tachykinin NK4 receptor: a reappraisal. Eur J Pharmacol 437:27–30
39. Ding YQ, Shigemoto R, Takada M, Ohishi H, Nakanishi S, Mizuno N (1996) Localization of the neuromedin K receptor (NK3) in the central nervous system of the rat. J Comp Neurol 364:290–310
40. Navarro VM, Ruiz-Pino F, Sanchez-Garrido MA, Garcia-Galiano D, Hobbs SJ, Manfredi-Lozano M, León S, Sangiao-Alvarellos S, Castellano JM, Clifton DK, Pinilla L, Steiner RA, Tena-Sempere M (2012) Role of neurokinin B in the control of female puberty and its modulation by metabolic status. J Neurosci 153(1):316–328
41. Shughrue PJ, Lane MV, Merchenthaler I (1996) In situ hybridization analysis of the distribution of neurokinin-3 mRNA in the rat central nervous system. J Comp Neurol 372:395–414
42. Amstalden M, Coolen LM, Hemmerle AM, Billings HJ, Connors JM, Goodman RL, Lehman MN (2009) Neurokinin 3 receptor immunoreactivity in the septal region, preoptic area and hypothalamus of the female sheep: colocalisation in neurokinin B cells of the arcuate nucleus but not in gonadotropin-releasing hormone neurones. J Neuroendocrinol 22:1–12

43. Billings HJ, Connors JM, Altman SN, Hileman SM, Holaskova I, Lehman MN, McManus CJ, Nestor CC, Jacobs BH, Goodman RL (2010) Neurokinin B acts via the neurokinin-3 receptor in the retrochiasmatic area to stimulate luteinizing hormone secretion in sheep. Endocrinology 151:3836–3846
44. Duque E, Mangas A, Salinas P, Diaz-Cabiale Z, Narvaez JA, Covenas R (2013) Mapping of alpha-neo-endorphin- and neurokinin B-immunoreactivity in the human brainstem. Brain Struct Funct 218(1):131–149
45. Koutcherov Y, Ashwell KW, Paxinos G (2000) The distribution of the neurokinin B receptor in the human and rat hypothalamus. Neuroreport 11:3127–3131
46. Fink G (2000) Neuroendocrine regulation of pituitary function: general principles. In: Conn PM, Freeman ME (eds) Neuroendocrinology in physiology and medicine. Humana Press, Totowa, pp 107–134
47. Herbison AE, Skinner DC, Robinson JE, King IS (1996) Androgen receptor-immunoreactive cells in ram hypothalamus: distribution and co-localization patterns with gonadotropin-releasing hormone, somatostatin and tyrosine hydroxylase. Neuroendocrinology 63:120–131
48. Huang X, Harlan RE (1993) Absence of androgen receptors in LHRH immunoreactive neurons. Brain Res 624:309–311
49. Hrabovszky E, Kallo I, Szlavik N, Keller E, Merchenthaler I, Liposits Z (2007) Gonadotropin-releasing hormone neurons express estrogen receptor-beta. J Clin Endocrinol Metab 92:2827–2830
50. Hrabovszky E, Shughrue PJ, Merchenthaler I, Hajszan T, Carpenter CD, Liposits Z, Petersen SL (2000) Detection of estrogen receptor-beta messenger ribonucleic acid and 125I-estrogen binding sites in luteinizing hormone-releasing hormone neurons of the rat brain. Endocrinology 141:3506–3509
51. Chu Z, Takagi H, Moenter SM (2010) Hyperpolarization-activated currents in gonadotropin-releasing hormone (GnRH) neurons contribute to intrinsic excitability and are regulated by gonadal steroid feedback. J Neurosci 30:13373–13383
52. Danzer SC, Price RO, McMullen NT, Rance NE (1999) Sex steroid modulation of neurokinin B gene expression in the arcuate nucleus of adult male rats. Brain Res Mol Brain Res 66:200–204
53. Dellovade TL, Merchenthaler I (2004) Estrogen regulation of neurokinin B gene expression in the mouse arcuate nucleus is mediated by estrogen receptor alpha. Endocrinology 145:736–742
54. Eghlidi DH, Haley GE, Noriega NC, Kohama SG, Urbanski HF (2010) Influence of age and 17beta-estradiol on kisspeptin, neurokinin B, and prodynorphin gene expression in the arcuate-median eminence of female rhesus macaques. Endocrinology 151:3783–3794
55. Goubillon ML, Forsdike RA, Robinson JE, Ciofi P, Caraty A, Herbison AE (2000) Identification of neurokinin B-expressing neurons as an highly estrogen-receptive, sexually dimorphic cell group in the ovine arcuate nucleus. Endocrinology 141:4218–4225
56. Navarro VM, Gottsch ML, Chavkin C, Okamura H, Clifton DK, Steiner RA (2009) Regulation of gonadotropin-releasing hormone secretion by kisspeptin/dynorphin/neurokinin B neurons in the arcuate nucleus of the mouse. J Neurosci 29:11859–11866
57. Ciofi P, Lapirot OC, Tramu G (2007) An androgen-dependent sexual dimorphism visible at puberty in the rat hypothalamus. Neuroscience 146:630–642
58. Ciofi P, Leroy D, Tramu G (2006) Sexual dimorphism in the organization of the rat hypothalamic infundibular area. Neuroscience 141:1731–1745
59. Navarro VM, Gottsch ML, Wu M, Garcia-Galiano D, Hobbs SJ, Bosch MA, Pinilla L, Clifton DK, Dearth A, Ronnekleiv OK, Braun RE, Palmiter RD, Tena-Sempere M, Alreja M, Steiner RA (2011) Regulation of NKB pathways and their roles in the control of Kiss1 neurons in the arcuate nucleus of the male mouse. Endocrinology 152:4265–4275
60. Pillon D, Caraty A, Fabre-Nys C, Bruneau G (2003) Short-term effect of oestradiol on neurokinin B mRNA expression in the infundibular nucleus of ewes. J Neuroendocrinol 15:749–753

61. Abel TW, Voytko ML, Rance NE (1999) The effects of hormone replacement therapy on hypothalamic neuropeptide gene expression in a primate model of menopause. J Clin Endocrinol Metab 84:2111–2118
62. Rance NE (2009) Menopause and the human hypothalamus: evidence for the role of kisspeptin/neurokinin B neurons in the regulation of estrogen negative feedback. Peptides 30:111–122
63. Rance NE, Bruce TR (1994) Neurokinin B gene expression is increased in the arcuate nucleus of ovariectomized rats. Neuroendocrinology 60:337–345
64. Smith JT, Cunningham MJ, Rissman EF, Clifton DK, Steiner RA (2005) Regulation of Kiss1 gene expression in the brain of the female mouse. Endocrinology 146:3686–3692
65. Smith JT, Dungan HM, Stoll EA, Gottsch ML, Braun RE, Eacker SM, Clifton DK, Steiner RA (2005) Differential regulation of KiSS-1 mRNA expression by sex steroids in the brain of the male mouse. Endocrinology 146:2976–2984
66. Mittelman-Smith MA, Williams H, Krajewski-Hall SJ, Lai J, Ciofi P, McMullen NT, Rance NE (2012) Arcuate kisspeptin/neurokinin B/dynorphin (KNDy) neurons mediate the estrogen suppression of gonadotropin secretion and body weight. Endocrinology 153:2800–2812
67. Ruiz-Pino F, Navarro VM, Bentsen A, García-Galiano D, Sánchez-Garrido MA, Clifton DK, Steiner RA, Pinilla L, Mikkelsen J, Tena-Sempere M (2012) Neurokinin B and the control of the gonadotropic axis in the rat: developmental changes, sexual dimorphism and regulation by gonadal steroids. Endocrinology 153(10):4818–4829
68. Nurmio M, Tena-Sempere M, Toppari J (2010) Orexins and the regulation of the hypothalamic-pituitary-testicular axis. Acta Physiol (Oxf) 198:349–354
69. Williamson-Hughes PS, Grove KL, Smith MS (2005) Melanin concentrating hormone (MCH): a novel neural pathway for regulation of GnRH neurons. Brain Res 1041:117–124
70. Kauffman AS, Navarro VM, Kim J, Clifton DK, Steiner RA (2009) Sex differences in the regulation of Kiss1/NKB neurons in juvenile mice: implications for the timing of puberty. Am J Physiol Endocrinol Metab 297:E1212–E1221
71. Gill JC, Navarro VM, Kwong C, Noel SD, Martin C, Xu S, Clifton DK, Carroll RS, Steiner RA, Kaiser UB (2012) Increased neurokinin B (Tac2) expression in the mouse arcuate nucleus is an early marker of pubertal onset with differential sensitivity to sex steroid negative feedback than Kiss1. Endocrinology 153(10):4883–4893
72. Rance NE, Krajewski SJ, Smith MA, Cholanian M, Dacks PA (2010) Neurokinin B and the hypothalamic regulation of reproduction. Brain Res 1364:116–128
73. Robinson JE, Birch RA, Taylor JA, Foster DL, Padmanabhan V (2002) In utero programming of sexually differentiated gonadotropin releasing hormone (GnRH) secretion. Domest Anim Endocrinol 23:43–52
74. Sandoval-Guzman T, Rance NE (2004) Central injection of senktide, an NK3 receptor agonist, or neuropeptide Y inhibits LH secretion and induces different patterns of Fos expression in the rat hypothalamus. Brain Res 1026:307–312
75. Rance NE, Young WS III (1991) Hypertrophy and increased gene expression of neurons containing neurokinin-B and substance-P messenger ribonucleic acids in the hypothalami of postmenopausal women. Endocrinology 128:2239–2247
76. Wakabayashi Y, Nakada T, Murata K, Ohkura S, Mogi K, Navarro VM, Clifton DK, Mori Y, Tsukamura H, Maeda K, Steiner RA, Okamura H (2010) Neurokinin B and dynorphin A in kisspeptin neurons of the arcuate nucleus participate in generation of periodic oscillation of neural activity driving pulsatile gonadotropin-releasing hormone secretion in the goat. J Neurosci 30:3124–3132
77. Lehman MN, Coolen LM, Goodman RL (2010) Minireview: kisspeptin/neurokinin B/dynorphin (KNDy) cells of the arcuate nucleus: a central node in the control of gonadotropin-releasing hormone secretion. Endocrinology 151:3479–3489
78. True C, Kirigiti M, Ciofi P, Grove KL, Smith MS (2011) Characterisation of arcuate nucleus kisspeptin/neurokinin B neuronal projections and regulation during lactation in the rat. J Neuroendocrinol 23:52–64

79. Kinsey-Jones JS, Grachev P, Li XF, Lin YS, Milligan SR, Lightman SL, O'Byrne KT (2012) The inhibitory effects of neurokinin B on GnRH pulse generator frequency in the female rat. Endocrinology 153(1):307–315
80. Ciccone NA, Xu S, Lacza CT, Carroll RS, Kaiser UB (2010) Frequency-dependent regulation of follicle-stimulating hormone beta by pulsatile gonadotropin-releasing hormone is mediated by functional antagonism of bZIP transcription factors. Mol Cell Biol 30:1028–1040
81. Marshall JC, Dalkin AC, Haisenleder DJ, Paul SJ, Ortolano GA, Kelch RP (1991) Gonadotropin-releasing hormone pulses: regulators of gonadotropin synthesis and ovulatory cycles. Recent Prog Horm Res 47:155–187; discussion 159–188
82. García-Galiano D, León-Tellez S, Manfredi-Lozano M, Romero-Ruiz A, Navarro VM, Pinilla L, Tena-Sempere M (2012) Kisspeptin signaling is indispensable for neurokinin B, but not glutamate, stimulation of gonadotropin secretion in mice. Endocrinology 153(1):316–328
83. Ramaswamy S, Seminara SB, Plant TM (2011) Evidence from the agonadal juvenile male rhesus monkey (Macaca mulatta) for the view that the action of neurokinin B to trigger gonadotropin-releasing hormone release is upstream from the kisspeptin receptor. Neuroendocrinology 94(3):237–245
84. Corander MP, Challis BG, Thompson EL, Jovanovic Z, Loraine Tung YC, Rimmington D, Huhtaniemi IT, Murphy KG, Topaloglu AK, Yeo GS, O'Rahilly S, Dhillo WS, Semple RK, Coll AP (2010) The effects of neurokinin B upon gonadotropin release in male rodents. J Neuroendocrinol 22:181–187
85. Ohkura S, Takase K, Matsuyama S, Mogi K, Ichimaru T, Wakabayashi Y, Uenoyama Y, Mori Y, Steiner RA, Tsukamura H, Maeda KI, Okamura H (2009) Gonadotropin-releasing hormone pulse generator activity in the hypothalamus of the goat. J Neuroendocrinol 21:813–821
86. Sawai N, Iijima N, Takumi K, Matsumoto K, Ozawa H (2012) Immunofluorescent histochemical and ultrastructural studies on the innervation of kisspeptin/neurokinin B neurons to tuberoinfundibular dopaminergic neurons in the arcuate nucleus of rats. Neurosci Res 74(1):10–16
87. Todman MG, Han SK, Herbison AE (2005) Profiling neurotransmitter receptor expression in mouse gonadotropin-releasing hormone neurons using green fluorescent protein-promoter transgenics and microarrays. Neuroscience 132:703–712
88. Glidewell-Kenney CA, Grove AMH, Iyer K, Mellon PL (2010) The neurokinin 3 receptor agonist, senktide, reduces gonadotropin releasing hormone synthesis and secretion in the inmortalized GT1-7 hypothalamic cell model. In: Society for Neuroscience 40th annual meeting, San Diego
89. Castellano JM, Navarro VM, Fernandez-Fernandez R, Castano JP, Malagon MM, Aguilar E, Dieguez C, Magni P, Pinilla L, Tena-Sempere M (2006) Ontogeny and mechanisms of action for the stimulatory effect of kisspeptin on gonadotropin-releasing hormone system of the rat. Mol Cell Endocrinol 257–258:75–83
90. Guerriero KA, Keen KL, Terasawa E (2012) Developmental increase in kisspeptin-54 release in vivo is independent of the pubertal increase in estradiol in female rhesus monkeys (Macaca mulatta). Endocrinology 153(4):1887–1897
91. Keen KL, Wegner FH, Bloom SR, Ghatei MA, Terasawa E (2008) An increase in kisspeptin-54 release occurs with the pubertal increase in luteinizing hormone-releasing hormone-1 release in the stalk-median eminence of female rhesus monkeys in vivo. Endocrinology 149:4151–4157
92. Navarro VM, Tena-Sempere M (2011) Neuroendocrine control by kisspeptins: role in metabolic regulation of fertility. Nat Rev Endocrinol 8(1):40–53
93. Navarro VM (2012) New insights into the control of pulsatile GnRH release: the role of Kiss1/neurokinin B neurons. Front Endocrinol (Lausanne) 3:48
94. d'Anglemont de Tassigny X, Fagg LA, Carlton MB, Colledge WH (2008) Kisspeptin can stimulate gonadotropin-releasing hormone (GnRH) release by a direct action at GnRH nerve terminals. Endocrinology 149:3926–3932

95. Irwig MS, Fraley GS, Smith JT, Acohido BV, Popa SM, Cunningham MJ, Gottsch ML, Clifton DK, Steiner RA (2004) Kisspeptin activation of gonadotropin releasing hormone neurons and regulation of KiSS-1 mRNA in the male rat. Neuroendocrinology 80:264–272
96. Thompson EL, Patterson M, Murphy KG, Smith KL, Dhillo WS, Todd JF, Ghatei MA, Bloom SR (2004) Central and peripheral administration of kisspeptin-10 stimulates the hypothalamic-pituitary-gonadal axis. J Neuroendocrinol 16:850–858
97. Kallo I, Vida B, Deli L, Molnar CS, Hrabovszky E, Caraty A, Ciofi P, Coen CW, Liposits Z (2012) Co-Localisation of kisspeptin with galanin or neurokinin B in afferents to mouse GnRH neurones. J Neuroendocrinol 24:464–476
98. Matsuyama S, Ohkura S, Mogi K, Wakabayashi Y, Mori Y, Tsukamura H, Maeda K, Ichikawa M, Okamura H (2011) Morphological evidence for direct interaction between kisspeptin and gonadotropin-releasing hormone neurons at the median eminence of the male goat: an immunoelectron microscopic study. Neuroendocrinology 94:323–332
99. Uenoyama Y, Inoue N, Pheng V, Homma T, Takase K, Yamada S, Ajiki K, Ichikawa M, Okamura H, Maeda KI, Tsukamura H (2011) Ultrastructural evidence of kisspeptin-gonadotropin-releasing hormone interaction in the median eminence of female rats: implication of axo-axonal regulation of GnRH release. J Neuroendocrinol 23(10):863–870
100. Kinoshita F, Nakai Y, Katakami H, Imura H (1982) Suppressive effect of dynorphin-(1-13) on luteinizing hormone release in conscious castrated rats. Life Sci 30:1915–1919
101. Schulz R, Wilhelm A, Pirke KM, Gramsch C, Herz A (1981) Beta-endorphin and dynorphin control serum luteinizing hormone level in immature female rats. Nature 294:757–759
102. Mitchell V, Prevot V, Jennes L, Aubert JP, Croix D, Beauvillain JC (1997) Presence of mu and kappa opioid receptor mRNAs in galanin but not in GnRH neurons in the female rat. Neuroreport 8:3167–3172
103. Sannella MI, Petersen SL (1997) Dual label in situ hybridization studies provide evidence that luteinizing hormone-releasing hormone neurons do not synthesize messenger ribonucleic acid for mu, kappa, or delta opiate receptors. Endocrinology 138:1667–1672
104. Cravo RM, Margatho LO, Osborne-Lawrence S, Donato J Jr, Atkin S, Bookout AL, Rovinsky S, Frazao R, Lee CE, Gautron L, Zigman JM, Elias CF (2011) Characterization of Kiss1 neurons using transgenic mouse models. Neuroscience 173:37–56
105. Gottsch ML, Popa SM, Lawhorn JK, Qiu J, Tonsfeldt KJ, Bosch MA, Kelly MJ, Ronnekleiv OK, Sanz E, McKnight GS, Clifton DK, Palmiter RD, Steiner RA (2011) Molecular properties of kiss1 neurons in the arcuate nucleus of the mouse. Endocrinology 152:4298–4309
106. Steiner RA, Navarro VM (2012) Tacking toward reconciliation on Tacr3/TACR3 mutations. Endocrinology 153(4):1578–1581
107. Moenter SM, Burger LL, Chu Z (2011) Gonadotropin-releasing hormone (GnRH) regulates the excitability of neurokinin B (NKB) neurons of the mouse arcuate nucleus. Society for Neuroscience Annual Meeting, Washington, DC
108. Clarkson J, Herbison AE (2006) Postnatal development of kisspeptin neurons in mouse hypothalamus; sexual dimorphism and projections to gonadotropin-releasing hormone neurons. Endocrinology 147:5817–5825
109. Navarro VM, Castellano JM, Fernandez-Fernandez R, Barreiro ML, Roa J, Sanchez-Criado JE, Aguilar E, Dieguez C, Pinilla L, Tena-Sempere M (2004) Developmental and hormonally regulated messenger ribonucleic acid expression of KiSS-1 and its putative receptor, GPR54, in rat hypothalamus and potent luteinizing hormone-releasing activity of KiSS-1 peptide. Endocrinology 145:4565–4574
110. Roa J, Navarro VM, Tena-Sempere M (2011) Kisspeptins in reproductive biology: consensus knowledge and recent developments. Biol Reprod 85(4):650–660
111. Navarro VM, Fernandez-Fernandez R, Castellano JM, Roa J, Mayen A, Barreiro ML, Gaytan F, Aguilar E, Pinilla L, Dieguez C, Tena-Sempere M (2004) Advanced vaginal opening and precocious activation of the reproductive axis by KiSS-1 peptide, the endogenous ligand of GPR54. J Physiol 561:379–386

112. Pineda R, Garcia-Galiano D, Roseweir A, Romero M, Sanchez-Garrido MA, Ruiz-Pino F, Morgan K, Pinilla L, Millar RP, Tena-Sempere M (2010) Critical roles of kisspeptins in female puberty and preovulatory gonadotropin surges as revealed by a novel antagonist. Endocrinology 151:722–730
113. Clarkson J, Boon WC, Simpson ER, Herbison AE (2009) Postnatal development of an estradiol-kisspeptin positive feedback mechanism implicated in puberty onset. Endocrinology 150:3214–3220
114. Gill JC, Wang O, Kakar S, Martinelli E, Carroll RS, Kaiser UB (2010) Reproductive hormone-dependent and -independent contributions to developmental changes in kisspeptin in GnRH-deficient hypogonadal mice. PLoS One 5:e11911
115. Nestor CC, Briscoe AM, Davis SM, Valent M, Goodman RL, Hileman SM (2012) Evidence of a role for kisspeptin and neurokinin B in puberty of female sheep. Endocrinology 153:2756–2765
116. Castellano JM, Bentsen AH, Mikkelsen JD, Tena-Sempere M (2010) Kisspeptins: bridging energy homeostasis and reproduction. Brain Res 1364:129–138
117. Castellano JM, Navarro VM, Fernandez-Fernandez R, Nogueiras R, Tovar S, Roa J, Vazquez MJ, Vigo E, Casanueva FF, Aguilar E, Pinilla L, Dieguez C, Tena-Sempere M (2005) Changes in hypothalamic KiSS-1 system and restoration of pubertal activation of the reproductive axis by kisspeptin in undernutrition. Endocrinology 146:3917–3925
118. True C, Kirigiti MA, Kievit P, Grove KL, Smith MS (2011) Leptin is not the critical signal for kisspeptin or luteinising hormone restoration during exit from negative energy balance. J Neuroendocrinol 23:1099–1112
119. Tovar S, Vazquez MJ, Navarro VM, Fernandez-Fernandez R, Castellano JM, Vigo E, Roa J, Casanueva FF, Aguilar E, Pinilla L, Dieguez C, Tena-Sempere M (2006) Effects of single or repeated intravenous administration of kisspeptin upon dynamic LH secretion in conscious male rats. Endocrinology 147:2696–2704
120. Li XF, Lin YS, Kinsey-Jones JS, O'Byrne KT (2012) High-fat diet increases LH pulse frequency and kisspeptin-neurokinin B expression in puberty-advanced female rats. Endocrinology 153(9):4422–4431

Chapter 16
Electrophysiology of Kisspeptin Neurons

Meenakshi Alreja

Abstract Kisspeptin is an important regulator of reproduction. Electrophysiological studies show that kisspeptin neurons of the arcuate nucleus that co-localize neurokinin B and dynorphin (aka KNDy neurons) fire action potentials in a tonic, irregular, or burst firing manner. Gonadectomy dramatically alters the membrane properties of KNDY neurons from male mice and induces somatic hypertrophy. NMDA, leptin, and neurokinin B are potent activators of KNDY neuron electrical activity and GABA inhibits KNDY neurons. The firing pattern of kisspeptin neurons located in the RP3V fluctuates with the estrus cycle and is strongly modulated by glutamate and GABA. Thus, kisspeptin neurons are capable of burst firing, and their activity is modulated by sex steroids and other regulatory factors.

Introduction

Electrophysiological studies provide an excellent tool for dissecting out cellular and molecular mechanisms that control cell firing, especially when in vivo and in vitro approaches are combined with neuropharmacological analysis. With respect to the reproductive system, in vivo approaches in large animals have the distinct advantage of allowing simultaneous monitoring of neuronal firing patterns and gonadotropin release mechanisms. In vitro electrophysiology allows study of intrinsic membrane properties, as well as second messenger and ionic mechanisms.

Studies on the electrophysiological properties of identified kisspeptin neurons lagged until recently, due to a lack of appropriate tools for unequivocal identification

M. Alreja, Ph.D. (✉)
Yale University School of Medicine, CMHC W-215, 34 Park Street,
New Haven, CT 06508, USA
e-mail: meenakshi.alreja@yale.edu

A.S. Kauffman and J.T. Smith (eds.), *Kisspeptin Signaling in Reproductive Biology*, Advances in Experimental Medicine and Biology 784, DOI 10.1007/978-1-4614-6199-9_16,

of recorded neurons. Earlier studies utilized labor-intensive double-labeling studies for identification of recorded neurons. The recent development of transgenic kisspeptin mouse models in which reporter genes are expressed have enabled a direct examination of the electrophysiological and neuropharmacological characteristics of kisspeptin neurons [1, 2].

Although still in their infancy, electrophysiological studies are providing significant insights into how kisspeptin neurons might control reproductive circuits, mediate sex steroid feedback, and link metabolic to reproductive pathways. In this chapter, we describe the advances made in understanding the cellular, molecular, and morphological properties of kisspeptin neurons located in the arcuate nucleus (ARC-KNDy neurons) and in the sexually dimorphic anteroventral periventricular and periventricular preoptic nuclei (RP3V-Kiss1 neurons), primarily using in vitro electrophysiological approaches in guinea pig and mice.

ARC-KNDy neurons co-localize kisspeptin, neurokinin B (NKB), and dynorphin [3–5]. These neurons are equally well represented in adults of both sexes [6] and have been implicated in mediating sex steroid-dependent negative feedback [7]. KNDy neurons also express neurokinin 3 receptor (Tac3r) mRNA [5] and have been suggested to mediate the effects of neurokinin B on reproduction. KNDy cells may also play a role in the metabolic regulation of reproduction, perhaps in mediating the reproductive effects of leptin [8–11].

RP3V-Kiss1 neurons are sparse in males but prevalent in females [6], and may participate in the female-specific preovulatory GnRH/LH surge via sex steroid-mediated positive feedback mechanisms [7, 12–14]. The transcriptional activity of RP3V-Kiss1 neurons has been reported to be dependent on circadian signaling [15, 16], which may relate to the circadian-timed onset of the LH surge. The electrophysiological studies described below have, amongst other goals, aimed to clarify the role of the above two kisspeptin neuronal subpopulations in mediating sex steroid feedback and in linking reproductive circuits to metabolic and circadian pathways.

Kisspeptin Neurons of the Arcuate Nucleus

Membrane Properties and Modulation by Sex Steroids

A study of the intrinsic membrane properties of neurons enables an understanding of the firing patterns and rhythms a cell may be capable of generating. This knowledge is especially critical for the neuroendocrine reproductive system, as pulsatile secretion of GnRH is essential for sustaining normal gonadotropin secretion and reproductive fertility in mammals of both sexes. Accumulating evidence suggests that this pulsatility originates in kisspeptin neurons of the arcuate nucleus, which provide a rhythmic drive to the GnRH neurons that culminates in pulsatile gonadotropin secretion.

The first evidence for pulsatility in ARC-KNDy neurons came from in vivo goat studies, which observed repetitive electrical burst activity in the vicinity of putative ARC-KNDy neurons in the medial basal hypothalamus, but not when electrode placements were distant from where ARC-KNDy neurons would reside [17]. Moreover, the recorded burst activity synchronized with discrete pulses of LH and was reduced by gonadal steroid treatments [4]. Additionally, the burst activity continued unabated following systemic infusions of kisspeptin, which enhance GnRH and LH secretion, suggesting that the bursting neurons are located upstream of GnRH neurons [17].

More recently, in vitro studies on identified kisspeptin neurons, enabled by the development of transgenic mouse models that express GFP selectively in kisspeptin cells, have probed for evidence of pulsatility. Initial studies show that about 50% of ARC-KNDy-GFP neurons are quiescent in vitro, in brain slices of adult ovariectomized female mice, while the remaining neurons fire action potentials in a tonic or an irregular manner [1]. Importantly, ARC-KNDy neurons have the capability to fire action potentials in bursts, as they express the requisite underlying channels that permit burst firing. Thus, kisspeptin neurons in both the guinea pig [18] and the mouse exhibit the hyperpolarization-activated cyclic nucleotide-gated mixed cationic current (H current) and the T-type calcium current. These currents are well known to underlie endogenous burst firing in several CNS neurons. The glutamate agonist, NMDA, also triggers burst firing in KNDy cells (see Sect. Neurotransmitter and Neuropeptide Modulation of ARC-KNDy Neurons).

Similar to the ovariectomized female, our studies show that ARC-KNDy neurons from male mice also exhibit tonic, irregular, and bursting firing patterns (Fig. 16.1a). Quiescent cells can acquire any of the three firing patterns upon depolarization. Currently, it is unclear whether ARC-KNDy cells can switch their firing pattern from tonic to irregular/bursting or vice versa. We have also found that sex steroids exert a dramatic influence on the membrane properties of ARC-KNDy neurons, which is consistent with the presence of sex steroid receptors in these neurons [7, 19]. Thus, ARC-KNDy neurons from male GDX mice are more likely to be quiescent in their resting state in vitro; only 10% of KNDy-GDX neurons fire spontaneously, whereas 40% of KNDY neurons from gonadally intact mice fire spontaneously in brain slices (Fig. 16.1b). The mean resting membrane potential of the quiescent neurons in GDX neurons is similar to that of quiescent neurons in the intact group (Fig. 16.1c). However, KNDY neurons from GDX male mice have lower membrane resistances, about half that of KNDy neurons from intact adult mice (Fig. 16.1d), and are less excitable—requiring double the amounts of current to elicit an action potential (Fig. 16.1e). The gonadectomy-induced changes noted above take several days to develop after castration (*not shown*), and thus, are unlikely to be due to an acute withdrawal of sex steroids in the excised slice preparation. Nevertheless, it would be important to examine *Kiss1* neuron properties under controlled sex steroid levels in brain slices. It should also be stated that the above studies on ARC-KNDy neurons were performed using the coronal slice preparation. Since the membrane properties of a neuron are dependent on its dendritic and axonal arbor, properties of Kiss1 neurons should also be examined in

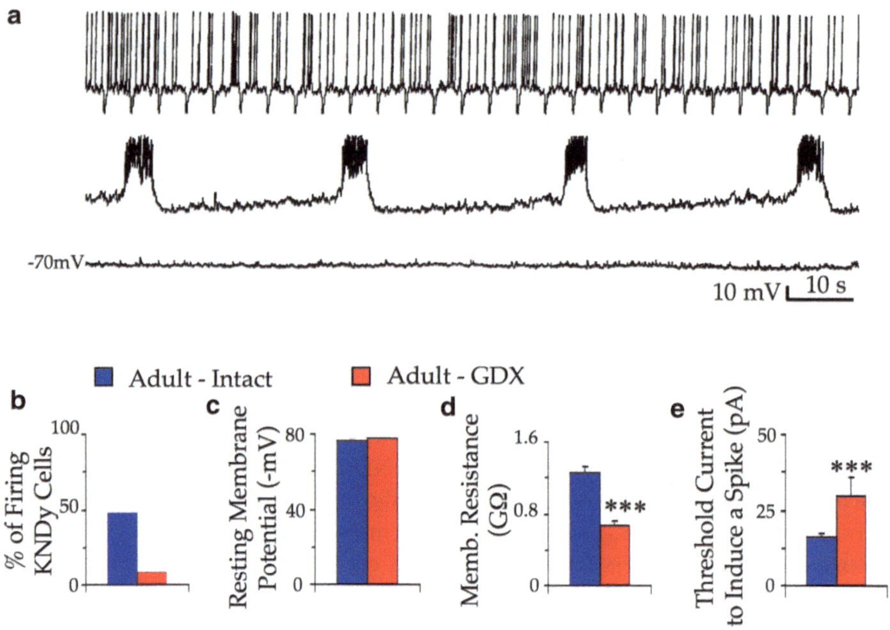

Fig. 16.1 Gonadectomy alters the membrane properties of KNDy-GFP neurons. (**a**) Chart records from 3 KNDy-GFP neurons show that under basal recording conditions, tonic firing, burst firing, and quiescent ARC-KNDy neurons are present in brain slices from male mice. (**b**) Bar chart shows that the fraction of spontaneously firing cells is significantly higher in brain slices prepared from adult gonad-intact mice vs. GDX mice. (**c**). Bar charts show that the resting membrane potential of quiescent ARC-KNDy cells is indistinguishable in the intact and GDX groups. (**d**, **e**) Bar charts show dramatic alterations in membrane resistance and excitability properties of KNDY cells from GDX vs. gonad-intact male mice. The membrane resistance of ARC-KNDy neurons in the GDX group is 50% lower than that of the gonad-intact groups, and KNDy-GDX neurons require significantly larger currents to induce an action potential

slices with different orientations [20]. It should, however, be noted that in morphological studies, long processes were commonly observed in neurons filled in the coronal slice preparation (see Sect. Morphology of ARC-KNDy Neurons and Modulation by Sex Steroids).

The reduced fraction of spontaneously firing cells in the GDX group may seem paradoxical, since gonadectomy upregulates the expression of *Kiss1*, *Tac2*, and *Dyn* in KNDy neurons and indirect evidence would suggest that increased kisspeptin production drives the post-castration rise in luteinizing hormone (LH) [21]. A simple explanation might be that the sex steroid receptors on KNDy neurons drive the transcriptional machinery, but not their firing activity and the subsequent release of contained peptides. Thus, input(s) extrinsic to KNDy neurons in the slice preparation may be required for driving the sustained synchronized firing of KNDy neurons following gonadectomy. These inputs could derive from other ERα-receptor expressing hypothalamic neurons, such as the POMC or the SF-1/VMH neurons; ERα deletion in POMC neurons reduces castration-induced increase in gonadotropin levels at least in the female [22]. Dopaminergic and GHRH neurons located within the arcuate nucleus also express ERα [23, 24]. NPY/AgRP neurons do not express ERα [25]. Multiple extrahypothalamic neurons, such as cholinergic, adrenergic, serotonergic, and/or the dopaminergic neurons, also express sex steroid receptors; further studies will be required to determine whether these inputs modulate the activity of ARC-KNDy neurons.

Post-castration-induced increases in LH secretion could also, theoretically, be sustained by mechanisms not requiring KNDy neurons. For example, cholinergic neurons express sex steroid receptors [26, 27], and cholinergic fibers make close contact with GnRH neurons [28]. Moreover, acetylcholine directly activates GnRH neurons via nicotinic receptors and a nicotinic receptor antagonist, mecamylamine, significantly reduces castration-induced increases in LH secretion [29]. Thus, both KNDy cell-dependent and -independent mechanisms may subserve the castration-induced increase in gonadotropin secretion.

It should be noted that although the reduced firing activity of KNDy neurons in GDX slices may appear paradoxical, the finding that a vast majority of KNDy cells (90%) share the quiescent state in GDX slices also implies that KNDy neurons in the GDX state are primed for synchronized activity upon arrival of an excitatory input. In contrast, only half of the KNDY cells are in the same firing state in slices taken from gonad-intact mice, which may explain the lower amplitude of LH pulses in intact vs. GDX mice.

Morphology of ARC-KNDy Neurons and Modulation by Sex Steroids

Reduced levels of circulating sex steroids in menopausal women and overectomized monkeys have long been known to induce hypertrophy in NKB somata in the arcuate nucleus [30, 31] (which can now be presumed to be KNDy neurons). We have observed a similarly remarkable hypertrophy in ARC-KNDy neurons in male mice following gonadectomy, using neurobiotin-filling in conjunction with whole cell patch-clamp recordings in vitro (Fig. 16.2a, b). ARC-KNDy neurons, that have bipolar or tripolar morphologies with long processes, doubled in soma cross-sectional area following gonadectomy (Fig. 16.2c). This increased cross-sectional area, presumably, underlies the drop in membrane resistance and associated changes in the intrinsic membrane properties of KNDy-GDX neurons described above. Detailed analyses will be required to determine if these somatic changes are also accompanied by changes in dendritic processes. Thus, modulation by sex steroids may be a general property of ARC-KNDy neurons across species, allowing them to convey sex steroid feedback from the gonads [19, 32, 33].

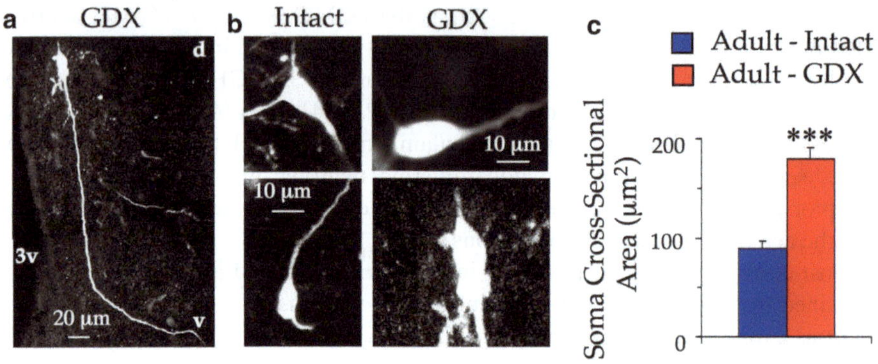

Fig. 16.2 Gonadectomy induces somatic hypertrophy in KNDy-GFP neurons. (**a**) Shows a neurobiotin-filled ARC-KNDy neuron from an adult-GDX male. Note the bipolar shape and the long process. (**b**) Shows neurobiotin-filled ARC-KNDy cell somas from intact and GDX male mice. Note the larger size of the two somas from the GDX group (*right column*). (**c**) Bar chart summarizes the strikingly larger cross-sectional area of KNDY-GDX neurons

Neurotransmitter and Neuropeptide Modulation of ARC-KNDy Neurons

Neuropharmacological studies conducted thus far show that the activity of ARC-KNDy neurons is strongly modulated by amino acid neurotransmitters. The excitatory glutamate receptor agonist, NMDA, reliably induces membrane oscillations in both guinea pig and female mouse KNDy neurons [1, 18]. These oscillations are mediated via a direct action of NMDA on ionotropic receptors in ARC-KNDy cells, as they persist after blockade of synaptic transmission. The inhibitory neurotransmitter, GABA, has an opposing effect—it strongly inhibits ARC-KNDy neurons [1, 18]. Inhibitory inputs to ARC-KNDY cells could potentially provide a feedback circuit for sustaining burst firing in ARC-KNDy neurons. Whether GABA or glutamate antagonists alter the firing activity of ARC-KNDy neurons is not yet known. Future studies will also be required to determine if the effects of NMDA and GABA on ARC-KNDy neurons are age- or sex-dependent.

We have recently shown that the activity of ARC-KNDy neurons is strongly modulated by the neuropeptide, neurokinin B (NKB) [2], that is contained within the ARC-KNDy neurons. In contrast, our preliminary data suggests a lack of effect of kisspeptin on ARC-KNDY neurons (*not shown*). Whether the NKB input to ARC-KNDY neurons originates from ARC-KNDy neurons themselves remains unknown. Brief applications of NKB (15 s) depolarize ARC-KNDY neurons and may elicit volleys of action potentials in a tonic or burst pattern with varied response durations (Fig. 16.3a, b, h). The NKB effect shows little evidence of desensitization on repeated applications (Fig. 16.3c) and is graded in nature (Fig. 16.3i). Surprisingly, the amplitude of the NKB depolarization is strikingly similar in ARC-KNDy neurons from gonad-intact vs. GDX adult male mice (Fig. 16.3d), despite the significantly lower

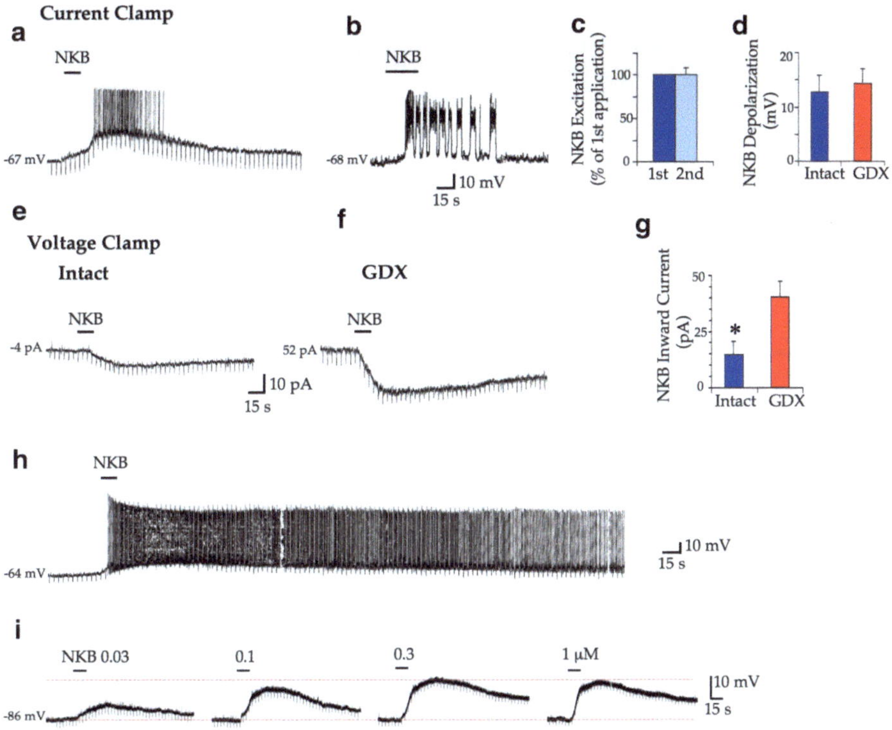

Fig. 16.3 Neurokinin B activates KNDY-GFP neurons from male mice, independent of gonadal status. (**a**) Current-clamp recording from a quiescent KNDy-GFP neuron, recorded from a brain slice obtained from an adult male mouse with intact gonads, shows that NKB (0.1 µM, 15 s) produced a 9 mV depolarization and induced action potentials. The NKB response lasted for 6.5 min in this cell. (**b**) Shows another quiescent neuron in which NKB induced burst firing. (**c, d**) Bar chart shows that the amplitude of the NKB response was reproducible on repeated applications, and statistically similar in the intact vs. GDX groups. (**e–g**) Voltage-clamp recordings and summary bar chart show that the NKB-induced inward current was larger in the KNDY-GDX group. Note the higher membrane conductance in the GDX example (as illustrated by the length of the *vertical lines*) that is reflective of the low membrane resistance. The larger NKB current in GDX explains the lack of difference in the amplitude of the NKB-induced depolarization in the two groups despite the low membrane resistance of GDX neurons. (**h**) Shows a KNDY neuron in which the NKB (0.1 µM, 15 s) response lasted for >20 min. (**i**) Shows that NKB excitation is concentration-dependent. Increasing concentrations of NKB were applied to a quiescent ARC-KNDy neuron. A maximal depolarization of 17 mV was obtained with 0.3 and 1 µM NKB

membrane resistance and reduced excitability of KNDy-GDX neurons described above. Further analysis of NKB effects using voltage-clamp recordings revealed that NKB-induced inward currents in KNDy-GDX neurons are significantly larger when compared to intact-KNDy cells (Fig. 16.3g), thus explaining the similar magnitude of NKB-induced depolarization in the two groups. Presumably, when KNDy-GDX cells hypertrophy, the neurons compensate by inserting additional neurokinin

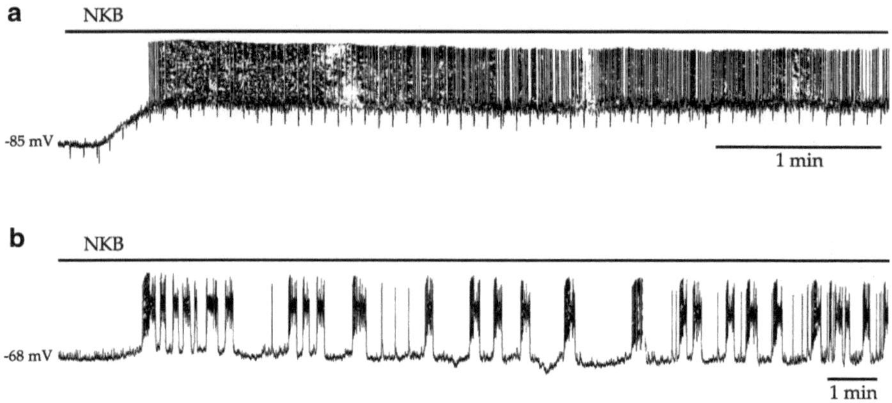

Fig. 16.4 Bath-applied NKB induces persistent tonic or bursting firing activity in KNDy-GFP neurons. (**a**) Bath-applied NKB (15 min) depolarized a quiescent ARC-KNDy neuron and produced sustained tonic firing activity with little evidence of desensitization. (**b**) In another cell, NKB applied for 35 min produced persistent bursting activity without any apparent desensitization

receptors into the cell membrane such that the firing response of the ARC-KNDy cell to NKB is not compromised by gonadectomy in the male. These findings are consistent with the robust increase in gonadotropin secretion observed following NKB infusions in intact male mice and castrated adult male monkeys [2, 34]. However, in female mice, the NKB effect on gonadotropin secretion is modulated by sex steroids [5, 35]; whether this reflects a sex difference remains to be determined and awaits electrophysiology studies on NKB effects in the female under different sex steroid environments.

Consistent with the non-desensitizing nature of the NKB response to repeated applications described above, prolonged bath applications of NKB for 15–35 min produce persistent tonic or burst firing activity (Fig. 16.4a, b). NKB activation of ARC-KNDy neurons is attenuated by the NK3 receptor antagonist, SB222200 (Fig. 16.5a, b), and senktide, an NK3R agonist, activates KNDy neurons (Fig. 16.5c), indicating a role for NK3 receptors.

In vivo, in the ovariectomized goat, intracerebroventricular infusions of NKB induce multiunit volleys in the vicinity of ARC-KNDY cells, and dynorphin, which is also co-localized in KNDy neurons, inhibits multiunit activity. Our in vitro data suggests that the NKB effect may have occurred via an NKB-induced depolarization of ARC-KNDy cells [4]. Further in vitro studies will be required to determine the specific sites of action of dynorphin.

Leptin, an adiposity signal from the fat cells, acts on brain neurons via the leptin receptor (LepRb) to communicate the availability of energy reserves. Leptin also regulates puberty, fertility, and sex behaviors [36–38]; mutations in leptin signaling pathways result in reproductive deficits [39]. Leptin can reverse hypogonadotropic hypogonadism associated with reduced body fat [40–43], and leptin is a potential

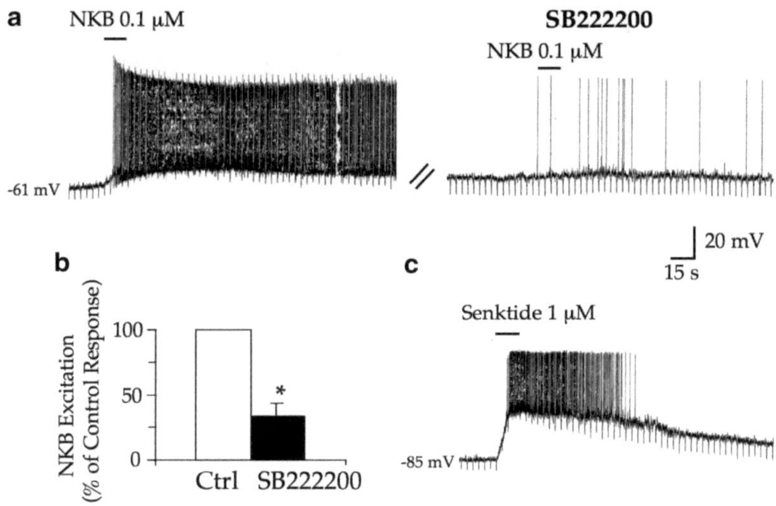

Fig. 16.5 NKB activates KNDy-GFP neurons via NK3 receptors. (**a**) Traces show an NKB-responsive cell from a GDX mouse, in which the NK3R antagonist, SB222200 (3 μM, 15 min), attenuated the response to a subsequent (at 22 min) application of NKB. (**b**) Bar chart summarizes the antagonist effect of SB222200, which reduced the NKB response to less than 35% of control value in the cells tested. (**c**) Shows a cell that was strongly activated by senktide (1 μM, 15 s), a selective NK3R receptor agonist, further confirming the presence of an NK3R-mediated activation in ARC-KNDy neurons

treatment for hypothalamic amenorrhea [44]. However, the mechanism(s) via which leptin regulates reproductive functions remain unclear. Multiple studies have concluded that GnRH neurons do not express LepRb [11, 45, 46] focusing attention on kisspeptin neurons, although a recent study casts doubt on the role of kisspeptin neurons, at least in mediating the effects of leptin on puberty [47]. Manipulation of leptin levels consistently regulates *Kiss1* mRNA message—fasting suppresses *Kiss1* mRNA expression and leptin restores these levels. $Lep^{ob/ob}$ mice also show reduced *Kiss1* mRNA levels in the ARC that are partially restored by leptin treatment [8]; total hypothalamic *Kiss1* expression (which included both the ARC and the AVPV) does not change significantly, signifying dilution of the effect by extra-ARC regions [48]. Leptin also enhances *Kiss1* mRNA levels in conditions associated with reduced energy stores, such as in streptozotocin-induced diabetic rats [49]. Studies detailed below suggest complex mechanisms may underlie the effects of leptin on reproductive functions.

Kisspeptin neurons of RP3V have consistently been shown not to express leptin receptors [50], and the proportion of ARC-KNDy cells expressing leptin receptors is controversial. For example, Smith et al. reported LepRb expression in up to 40% of ARC-KNDy neurons [8], whereas Louis et al. found that LepR co-localizes in only 0–6% of ARC-KNDy neurons in mice and ewes [11]. Backholer et al. reported a high level of leptin receptor expression in ARC-Kiss1 cells of the ewe [10].

Using pSTAT3-immunoreactivity as a marker of a direct response to leptin, Cravo et al. found that 15% of ARC-KNDy neurons in the diestrus female mouse respond directly to leptin [50]. Thus, there is a huge discrepancy in these studies. Caution should be used to interpret results from STAT3 signaling, as it may not be required for leptin's regulation of reproduction [51].

Multiple indirect mechanisms have been evoked to explain leptin's actions on the reproductive axis. Leptin may act indirectly, via the pre-mammillary nucleus [52, 53] and/or via POMC neurons [10, 54]. Louis et al. have demonstrated that hypothalamic neurons located in the striohypothalamic nucleus, preoptic area, and ventral pre-mammillary nucleus may directly innervate GnRH neurons, and Kiss1 neurons may be directly innervated by POMC or other ARC neurons expressing leptin receptors.

In the female guinea pig, direct effects of leptin on arcuate Kiss1 expressing cells have been described in a significant majority of neurons, using the electrophysiological approach. These effects are mediated via activation of a nonselective cationic conductance and are blocked by the TRPC channel blocker, 2-APB, as well as by inhibitors of Jak, PI3K, and PLCY, demonstrating coupling to the JAK-PI3K-PLCY pathway [18]. Further electrophysiological studies will be required to determine if mouse ARC-KNDY neurons also respond directly to leptin. In conclusion, multiple indirect and direct mechanisms may convey the effects of leptin on reproductive parameters.

Kisspeptin Neurons of the RP3V

Intrinsic Properties

The sexually dimorphic, estrogen receptor-α-expressing kisspeptin neurons of the RP3V are considered integral to the control of preovulatory surge mechanisms [12–14]. The membrane properties of RP3V neurons in female mice were first reported in 2010 using labor-intensive double-labeling studies [55]. Similar to KNDy neurons of the arcuate nucleus, RP3V-Kiss1 neurons fire action potentials in irregular, tonic, or burst firing patterns. Published data suggests that a vast majority of RP3V-Kiss1 spontaneously discharge action potentials in brain slices, irrespective of the phase of the estrus cycle [55]. However, the firing rate and the pattern of firing in RP3V-Kiss1 neurons may fluctuate with the phase of the cycle. Thus, RP3V-Kiss1 neurons fire at lower rates in brain slices prepared from diestrus vs. proestrus/estrus mice, and their firing pattern switches from predominantly tonic to irregular on transition from diestrus to proestrus. In general, the spontaneous firing rate of RP3V-Kiss1 neurons ranges from 0.2 to 7.7 Hz in brain slices, and is higher than that of the neighboring non-Kiss1 cells. Future studies will be required to determine if other membrane properties, such as resting membrane potential and membrane resistance, fluctuate with the estrus cycle, and how changes in these properties alter the receptivity of RP3V neurons to afferent inputs.

Neurotransmitter and Neuropeptide Modulation of RP3V Kisspeptin Neurons

Similar to the ARC-KNDy cells, RP3V-Kiss1 neurons are strongly activated by glutamate and inhibited by GABA. Antagonist data also suggests that, at least in vitro, RP3V-Kiss1 neuron activity is not under either an excitatory or inhibitory tone due to endogenously released glutamate or GABA [55].

Since both kisspeptin expression and neuronal activity (measured using c-fos) in RP3V-Kiss1 neurons fluctuate on a circadian basis [16], there is considerable interest in whether the activity of RP3V-Kiss1 neurons is modulated by neuropeptides associated with circadian signaling. Interestingly, the two classical transmitters released from the suprachiasmatic nucleus, vasopressin and vasoactive intestinal peptide (VIP), have thus far been shown to have no direct effect on the activity of RP3V-Kiss1 neurons [55]. If future studies corroborate these findings, indirect mechanisms will need to be investigated to comprehend mechanisms underlying circadian linking to RP3V kisspeptin neurons.

RFRP-3, another neuropeptide implicated in reproduction, has no effect on RP3V-Kiss1 neurons [55]. Similar to ARC-KNDy neurons, kisspeptin also does not affect RP3V-Kiss1 neurons, suggesting lack of autoregulation via released kisspeptin.

Conclusions

Kisspeptin cells of the arcuate nucleus and the RP3V region fire action potentials in a tonic, irregular, or bursting pattern in vitro in brain slices, and are inhibited by GABA and activated by glutamate. The membrane properties of ARC-KNDy neurons are strongly modulated by sex steroids. Similarly, the firing activity and pattern of RP3V-Kiss1 neurons is estrus cycle-dependent, but is not directly regulated by the circadian cycle peptides, VIP, and vasopressin. NKB and NK3R agonists are potent activators of KNDy neurons, providing a mechanistic interpretation of how NKB sustains normal reproductive functions. Leptin activates ARC-KNDy neuron, supporting their role in linking metabolism to reproduction.

References

1. Gottsch ML, Popa SM, Lawhorn JK, Qiu J, Tonsfeldt KJ, Bosch MA et al (2011) Molecular properties of kiss1 neurons in the arcuate nucleus of the mouse. Endocrinology 152(11):4298–4309
2. Navarro VM, Gottsch ML, Wu M, García-Galiano D, Hobbs SJ, Bosch MA et al (2011) Regulation of NKB pathways and their roles in the control of kiss1 neurons in the arcuate nucleus of the male mouse. Endocrinology 152(11):4265–4275
3. Goodman RL, Lehman MN, Smith JT, Coolen LM, de Oliveira CV, Jafarzadehshirazi MR et al (2007) Kisspeptin neurons in the arcuate nucleus of the ewe express both dynorphin A and neurokinin B. Endocrinology 148(12):5752–5760

4. Wakabayashi Y, Nakada T, Murata K, Ohkura S, Mogi K, Navarro VM et al (2010) Neurokinin B and dynorphin a in kisspeptin neurons of the arcuate nucleus participate in generation of periodic oscillation of neural activity driving pulsatile gonadotropin-releasing hormone secretion in the goat. J Neurosci 30(8):3124–3132
5. Navarro VM, Gottsch ML, Chavkin C, Okamura H, Clifton DK, Steiner RA (2009) Regulation of gonadotropin-releasing hormone secretion by kisspeptin/dynorphin/neurokinin B neurons in the arcuate nucleus of the mouse. J Neurosci 29(38):11859–11866
6. Clarkson J, Herbison AE (2006) Postnatal development of kisspeptin neurons in mouse hypothalamus; sexual dimorphism and projections to gonadotropin-releasing hormone neurons. Endocrinology 147(12):5817–5825
7. Smith JT, Cunningham MJ, Rissman EF, Clifton DK, Steiner RA (2005) Regulation of kiss1 gene expression in the brain of the female mouse. Endocrinology 146(9):3686–3692
8. Smith JT, Acohido BV, Clifton DK, Steiner RA (2006) Kiss-1 neurones are direct targets for leptin in the ob/ob mouse. J Neuroendocrinol 18(4):298–303
9. Tena-Sempere M (2006) Kiss-1 and reproduction: focus on its role in the metabolic regulation of fertility. Neuroendocrinology 83(5–6):275–281
10. Backholer K, Smith JT, Rao A, Pereira A, Iqbal J, Ogawa S et al (2010) Kisspeptin cells in the ewe brain respond to leptin and communicate with neuropeptide Y and proopiomelanocortin cells. Endocrinology 151(5):2233–22343
11. Louis GW, Greenwald-Yarnell M, Phillips R, Coolen LM, Lehman MN, Myers MG (2011) Molecular mapping of the neural pathways linking leptin to the neuroendocrine reproductive axis. Endocrinology 152(6):2302–2310
12. Smith JT, Popa SM, Clifton DK, Hoffman GE, Steiner RA (2006) Kiss1 neurons in the forebrain as central processors for generating the preovulatory luteinizing hormone surge. J Neurosci 26(25):6687–6694
13. Adachi S, Yamada S, Takatsu Y, Matsui H, Kinoshita M, Takase K et al (2007) Involvement of anteroventral periventricular metastin/kisspeptin neurons in estrogen positive feedback action on luteinizing hormone release in female rats. J Reprod Dev 53(2):367–378
14. Clarkson J, d'Anglemont de Tassigny X, Moreno AS, Colledge WH, Herbison AE (2008) Kisspeptin-Gpr54 signaling is essential for preovulatory gonadotropin-releasing hormone neuron activation and the luteinizing hormone surge. J Neurosci 28(35):8691–8697
15. Khan AR, Kauffman AS (2012) The role of kisspeptin and RFRP-3 neurons in the circadian-timed preovulatory luteinizing hormone surge. J Neuroendocrinol 24(1):131–143
16. Robertson JL, Clifton DK, de la Iglesia HO, Steiner RA, Kauffman AS (2009) Circadian regulation of kiss1 neurons: implications for timing the preovulatory gnrh/LH surge. Endocrinology 150(8):3664–3671
17. Ohkura S, Takase K, Matsuyama S, Mogi K, Ichimaru T, Wakabayashi Y et al (2009) Gonadotropin-releasing hormone pulse generator activity in the hypothalamus of the goat. J Neuroendocrinol 21(10):813–821
18. Qiu J, Fang Y, Bosch MA, Rønnekleiv OK, Kelly MJ (2011) Guinea pig kisspeptin neurons are depolarized by leptin via activation of TRPC channels. Endocrinology 152(4):1503–1514
19. Smith JT, Dungan HM, Stoll EA, Gottsch ML, Braun RE, Eacker SM et al (2005) Differential regulation of kiss-1 mrna expression by sex steroids in the brain of the male mouse. Endocrinology 146(7):2976–2984
20. Constantin S, Piet R, Iremonger K, HwaYeo S, Clarkson J, Porteous R, Herbison AE (2012) Gnrh neuron firing and response to GABA *in vitro* depend on acute brain slice thickness and orientation. Endocrinology 153(8):3758–3769
21. Oakley AE, Clifton DK, Steiner RA (2009) Kisspeptin signaling in the brain. Endocr Rev 30(6):713–743
22. Xu Y, Nedungadi TP, Zhu L, Sobhani N, Irani BG, Davis KE et al (2011) Distinct hypothalamic neurons mediate estrogenic effects on energy homeostasis and reproduction. Cell Metab 14(4):453–465
23. Addison ML, Rissman EF (2012) Sexual dimorphism of growth hormone in the hypothalamus: regulation by estradiol. Endocrinology 153(4):1898–1907

24. Mitchell V, Loyens A, Spergel DJ, Flactif M, Poulain P, Tramu G, Beauvillain JC (2003) A confocal microscopic study of gonadotropin-releasing hormone (gnrh) neuron inputs to dopaminergic neurons containing estrogen receptor alpha in the arcuate nucleus of gnrh-green fluorescent protein transgenic mice. Neuroendocrinology 77(3):198–207
25. Olofsson LE, Pierce AA, Xu AW (2009) Functional requirement of agrp and NPY neurons in ovarian cycle-dependent regulation of food intake. Proc Natl Acad Sci USA 106(37):15932–15937
26. Vanderhorst VG, Gustafsson JA, Ulfhake B (2005) Estrogen receptor-alpha and -beta immunoreactive neurons in the brainstem and spinal cord of male and female mice: relationships to monoaminergic, cholinergic, and spinal projection systems. J Comp Neurol 488(2):152–179
27. McEwen B (2002) Estrogen actions throughout the brain. Recent Prog Horm Res 57:357–384
28. Turi GF, Liposits Z, Hrabovszky E (2008) Cholinergic afferents to gonadotropin-releasing hormone neurons of the rat. Neurochem Int 52(4–5):723–728
29. Wu M, Kauffman AS, Alreja M (2012) Acetylcholine activates gnrh neurons and sustains gonadotropin secretion via low-affinity nicotinic receptors in mice. Endo Soc. Abstr 2012:SUN-687
30. Rometo AM, Krajewski SJ, Voytko ML, Rance NE (2007) Hypertrophy and increased kisspeptin gene expression in the hypothalamic infundibular nucleus of postmenopausal women and ovariectomized monkeys. J Clin Endocrinol Metab 92(7):2744–2750
31. Rance NE, Young WS (1991) Hypertrophy and increased gene expression of neurons containing neurokinin-b and substance-p messenger ribonucleic acids in the hypothalami of postmenopausal women. Endocrinology 128(5):2239–2247
32. Popa SM, Clifton DK, Steiner RA (2008) The role of kisspeptins and GPR54 in the neuroendocrine regulation of reproduction. Annu Rev Physiol 70:213–238
33. Danzer SC, Price RO, McMullen NT, Rance NE (1999) Sex steroid modulation of neurokinin B gene expression in the arcuate nucleus of adult male rats. Brain Res Mol Brain Res 66(1–2):200–204
34. Ramaswamy S, Seminara SB, Ali B, Ciofi P, Amin NA, Plant TM (2010) Neurokinin B stimulates gnrh release in the male monkey (macaca mulatta) and is colocalized with kisspeptin in the arcuate nucleus. Endocrinology 151(9):4494–4503
35. Navarro VM, Castellano JM, McConkey SM, Pineda R, Ruiz-Pino F, Pinilla L et al (2011) Interactions between kisspeptin and neurokinin B in the control of gnrh secretion in the female rat. Am J Physiol Endocrinol Metab 300(1):E202–E210
36. Krasnow SM, Steiner RA (2006) Physiological mechanisms integrating metabolism and reproduction. In: Neil JD (ed) Knobil and Neill's physiology of reproduction, 3rd edn. Elsevier, Philadelphia, pp 2553–2696
37. Hill JW, Elmquist JK, Elias CF (2008) Hypothalamic pathways linking energy balance and reproduction. Am J Physiol Endocrinol Metab 294(5):E827–E832
38. Farooqi IS (2002) Leptin and the onset of puberty: insights from rodent and human genetics. Semin Reprod Med 20(2):139–144
39. Montez JM, Soukas A, Asilmaz E, Fayzikhodjaeva G, Fantuzzi G, Friedman JM (2005) Acute leptin deficiency, leptin resistance, and the physiologic response to leptin withdrawal. Proc Natl Acad Sci USA 102(7):2537–2542
40. Ahima RS, Prabakaran D, Mantzoros C, Qu D, Lowell B, Maratos-Flier E, Flier JS (1996) Role of leptin in the neuroendocrine response to fasting. Nature 382(6588):250–252
41. Chehab FF, Lim ME, Lu R (1996) Correction of the sterility defect in homozygous obese female mice by treatment with the human recombinant leptin. Nat Genet 12(3):318–320
42. Barash IA, Cheung CC, Weigle DS, Ren H, Kabigting EB, Kuijper JL et al (1996) Leptin is a metabolic signal to the reproductive system. Endocrinology 137(7):3144–3147
43. Nagatani S, Guthikonda P, Thompson RC, Tsukamura H, Maeda KI, Foster DL (1998) Evidence for gnrh regulation by leptin: leptin administration prevents reduced pulsatile LH secretion during fasting. Neuroendocrinology 67(6):370–376
44. Chou SH, Chamberland JP, Liu X, Matarese G, Gao C, Stefanakis R et al (2011) Leptin is an effective treatment for hypothalamic amenorrhea. Proc Natl Acad Sci USA 108(16):6585–6590

45. Finn PD, Cunningham MJ, Pau KY, Spies HG, Clifton DK, Steiner RA (1998) The stimulatory effect of leptin on the neuroendocrine reproductive axis of the monkey. Endocrinology 139(11):4652–4662
46. Quennell JH, Mulligan AC, Tups A, Liu X, Phipps SJ, Kemp CJ et al (2009) Leptin indirectly regulates gnrh neuronal function. Endocrinology 150(6):2805–2812
47. Donato J, Cravo RM, Frazão R, Gautron L, Scott MM, Lachey J et al (2011) Leptin's effect on puberty in mice is relayed by the ventral premammillary nucleus and does not require signaling in kiss1 neurons. J Clin Invest 121(1):355–368
48. Luque RM, Kineman RD, Tena-Sempere M (2007) Regulation of hypothalamic expression of kiss-1 and GPR54 genes by metabolic factors: analyses using mouse models and a cell line. Endocrinology 148(10):4601–4611
49. Castellano JM, Navarro VM, Fernández-Fernández R, Roa J, Vigo E, Pineda R et al (2006) Expression of hypothalamic kiss-1 system and rescue of defective gonadotropic responses by kisspeptin in streptozotocin-induced diabetic male rats. Diabetes 55(9):2602–2610
50. Cravo RM, Margatho LO, Osborne-Lawrence S, Donato J, Atkin S, Bookout AL et al (2011) Characterization of kiss1 neurons using transgenic mouse models. Neuroscience 173:37–56
51. Bates SH, Stearns WH, Dundon TA, Schubert M, Tso AW, Wang Y et al (2003) STAT3 signalling is required for leptin regulation of energy balance but not reproduction. Nature 421(6925):856–859
52. Donato J, Silva RJ, Sita LV, Lee S, Lee C, Lacchini S et al (2009) The ventral premammillary nucleus links fasting-induced changes in leptin levels and coordinated luteinizing hormone secretion. J Neurosci 29(16):5240–5250
53. Donato J, Cravo RM, Frazão R, Elias CF (2011) Hypothalamic sites of leptin action linking metabolism and reproduction. Neuroendocrinology 93(1):9–18
54. Backholer K, Bowden M, Gamber K, Bjørbaek C, Iqbal J, Clarke IJ (2010) Melanocortins mimic the effects of leptin to restore reproductive function in lean hypogonadotropic ewes. Neuroendocrinology 91(1):27–40
55. Ducret E, Gaidamaka G, Herbison AE (2010) Electrical and morphological characteristics of anteroventral periventricular nucleus kisspeptin and other neurons in the female mouse. Endocrinology 151(5):2223–2232

Chapter 17
Metabolic Regulation of Kisspeptin

Juan Manuel Castellano and Manuel Tena-Sempere

Abstract Body energy balance and metabolic signals are important modulators of puberty and reproductive function, so that perturbations of metabolism and energy reserves (ranging from persistent energy insufficiency to morbid obesity) are frequently linked to reproductive disorders. The mechanisms for the tight association between body metabolic state and reproduction are multifaceted, and likely involve numerous peripheral hormones and central transmitters. In recent years, a prominent role of kisspeptins in the central pathways responsible for conveying metabolic information into the brain centers responsible for reproductive control, and specifically GnRH neurons, has been proposed on the basis of a wealth of expression and functional data. In this chapter, we will summarize such evidence, with special attention to the potential (direct and/or indirect) interaction of leptin and kisspeptin pathways. In addition, other potential metabolic modulators of kisspeptin signaling, as well as some of the putative molecular mechanisms for the metabolic regulation of *Kiss1* will be briefly reviewed. Conflictive data, including those questioning an essential role of *Kiss1* neurons in mediating leptin effects on the reproductive axis, will be also discussed. All in all, we aim to provide an integral and balanced view of the physiological relevance and potential mechanisms for the metabolic control of the kisspeptin system, as important pathway for the integral regulation of energy balance, puberty onset, and fertility.

J.M. Castellano • M. Tena-Sempere, M.D., Ph.D. (✉)
Department of Cell Biology, Physiology and Immunology, University of Cordoba,
Avda. Mendndez Pidal s/n, 14004 Cordoba, Spain
e-mail: b72caroj@uco.cs; filtesem@uco.es

A.S. Kauffman and J.T. Smith (eds.), *Kisspeptin Signaling in Reproductive Biology*, Advances in Experimental Medicine and Biology 784, DOI 10.1007/978-1-4614-6199-9_17,

Introduction: Reproductive Function and Energy Balance Are Closely Linked

Reproduction is essential for the survival and perpetuation of any given species; yet, reproductive maturation and fertility are dispensable at the individual level and, as such, various conditions that perturb organism homeostasis frequently result in reproductive impairment. Regulation of reproductive function is highly sophisticated and requires complex regulatory networks, which impinge upon the so-called hypothalamic–pituitary–gonadal (HPG) axis [1, 2]. Among the different regulators of this neurohormonal axis, it is well known that reproductive function is sensitive to the magnitude of energy reserves and the modulatory actions of diverse nutritional and metabolic factors [3, 4]. Indeed, this close link between fatness and fertility was suggested from ancient times based on intuitive knowledge, as evidenced by the fact that fertility symbols were represented as obese female figures. Nonetheless, it was not until 1960s and 1970s when this contention acquired a scientific basis thanks to the pioneering work of Kennedy in rodents and, later, Frisch in humans [5–7]. This combination of experimental and clinical data set the ground for the formulation of the so-called "critical fat mass hypothesis" that pointed out the need to reach a certain threshold of body (fat) mass in order to attain complete pubertal development and an appropriate reproductive function in adulthood. This phenomenon is especially relevant in the female, where acquisition of sufficient fuel stores is indispensable in order to successfully face the significant amount of energy needed for pregnancy and lactation [8]. Nonetheless, in both sexes, reproductive maturation and function are sensitive to conditions of metabolic stress [9]. All in all, this phenomenon illustrates the tight association, and possibly joint control, of the neuroendocrine networks governing energy balance and reproductive function (Fig. 17.1) [3, 4].

Our knowledge of the neurohormonal pathways responsible for the metabolic control of puberty onset and gonadotropic function has considerably enlarged during the last two decades. Among the numerous endocrine regulators involved in the integral control of reproduction and metabolism, the adipocyte hormone, leptin, has been recognized as an essential neuroendocrine integrator linking the magnitude of body fat stores and different endocrine axes, including the reproductive system [3, 4]. In terms of body energy homeostasis, leptin is secreted by the white adipose tissue in proportion to the amount of body energy stores and acts as anorexigenic and thermogenic factor at the hypothalamic level, thus contributing to dynamically adjust energy requirements, fat reserves, and food intake [8]. It is worth noting that, in terms of reproductive control, leptin seems to play a key role as permissive, rather than trigger, signal for puberty onset and fertility [10, 11], so that threshold leptin levels are mandatory, but not sufficient per se, to attain a normal pubertal development and to maintain reproductive function in adulthood. Yet, recent data suggest that leptin might not be as critical as originally thought in restoring normal gonadotropic function during exit from negative energy balance condition in rodents and sheep [12, 13]; a phenomenon which is yet to be fully validated. As an

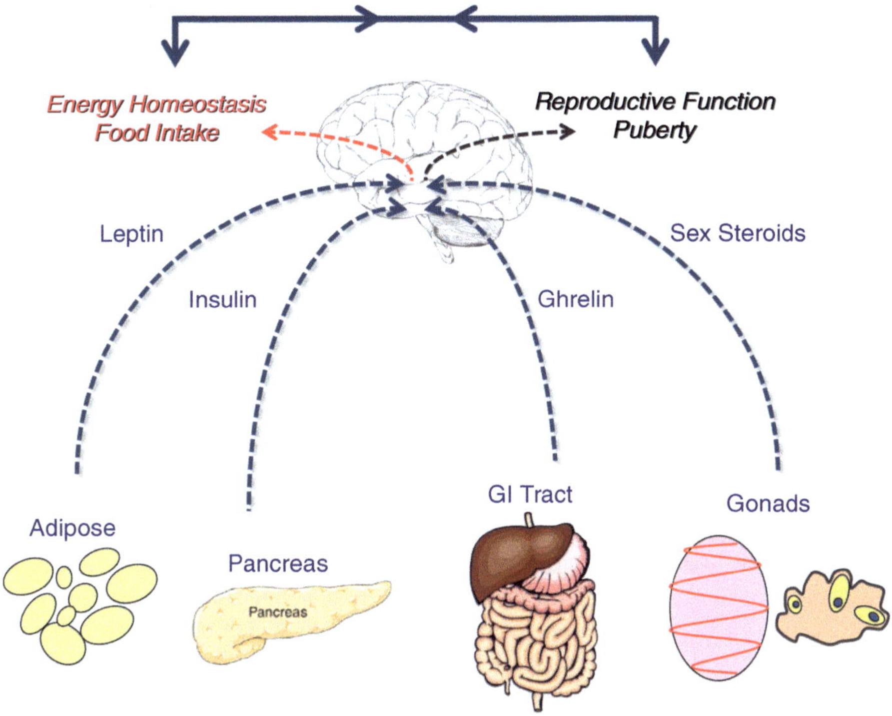

Fig. 17.1 Schematic diagram of various hormonal signals, originated from various peripheral tissues, involved in the integral control of food intake, energy balance, and reproductive function (including puberty). Paradigmatic examples of hormones with proven roles in the integral control of food intake and the reproductive axis are presented, including factors from adipose tissue (leptin: inhibitory signal for food intake; stimulatory/permissive signal for puberty and reproduction), pancreas (insulin; same profile as leptin in terms of actions on food intake and reproduction), gastrointestinal tract (ghrelin: stimulatory signal for food intake; inhibitory signal for the reproductive axis), and the gonads (sex steroids, such as estrogens, which are food-intake suppressing signals and conduct both negative and positive feedback effects on the gonadotropic axis). The central mechanisms whereby these peripheral metabolic signals affect reproductive function remain to be fully characterized

additional call of caution, it is possible that the set of metabolic signals, and their physiological relevance, in the regulation of puberty (the period of full activation of the HPG axis) vs. fertility (maintenance of the capacity to reproduce during adulthood) might be different.

Compelling evidence suggests that the metabolic control of the HPG axis, and specifically leptin-mediated actions, are conducted mainly at central levels, where leptin is thought to modulate the activity of hypothalamic gonadotropin-releasing hormone (GnRH) neurons [14]. The effects of leptin, however, seem to be conducted indirectly, as GnRH neurons do not physiologically express leptin receptors [15], suggesting the participation of intermediate pathways that would convey the modulatory actions of the adipocyte hormone. As summarized in the following sections, strong evidence points out that *Kiss1* neurons are sensitive to changes in body energy status and respond to leptin with increased *Kiss1* expression; therefore, they would be well suited to participate in the afferent networks responsible for the metabolic regulation of GnRH neurons. Yet, whether leptin effects are conducted directly or indirectly on *Kiss1* neurons is presently under considerable debate and investigation.

Metabolic Control of the *Kiss1* System: Evidence from Expression Studies

As described elsewhere in this textbook, during the last few years, *Kiss1* neurons in the basal forebrain have been recognized as master elements of the reproductive brain, which operate as key conduits for transmitting the regulatory actions of numerous modulators of the HPG axis, from gonadal steroids to photoperiodic stimuli [16–18]. In order to demonstrate if metabolic and nutritional cues, with known impacts on the reproductive axis, operate via *Kiss1* neurons for their modulation of the GnRH system, initial expression and functional analyses were conducted to experimentally test two key issues: (a) whether the hypothalamic *Kiss1* system, as evaluated by changes in *Kiss1* mRNA expression, is altered in conditions of metabolic distress known to perturb puberty and/or gonadotropin secretion; and (b) whether exogenous administration of kisspeptin, as a means to replace its defective endogenous levels, may rescue reproductive deficits seen in those conditions [19].

Studies conducted initially were designed as to provide *proof of principle* for the above contentions and hence were mainly focused in models of persistent or substantial negative energy balance, which are linked to different degrees of reproductive dysfunction. Thus, studies in pubertal male and females rats under acute fasting documented a significant decrease in the hypothalamic expression of *Kiss1* mRNA that was associated with lowering of serum LH levels [20]. Similar observations were later obtained in adult female rats [21, 22] and adult male mice, where time-course analyses of the effects on hypothalamic expression of *Kiss1* gene revealed a rapid drop in its mRNA levels as soon as 12-h after beginning of food-restriction [23]. Of note, suppression of the hypothalamic expression of *Kiss1* has been also reported in other models of metabolic stress coupled to negative energy balance and impaired reproductive function, such as uncontrolled experimental (streptozotocin-induced) diabetes, where a marked reduction in the hypothalamic *Kiss1* mRNA expression, coupled to a substantial suppression of circulating LH and sex steroid levels, has been described in male and female rats [24, 25]. These findings add further strength to the contention that energy insufficiency inhibits hypothalamic expression of *Kiss1* gene; a phenomenon that seems to be relevant for the suppression of the reproductive axis in conditions of sustained energy deficit.

Although the initial expression studies unveiled the sensitivity of the hypothalamic *Kiss1* system to changes in body energy stores and metabolic cues, these did not thoroughly address the neuroanatomical location of the reported alterations of *Kiss1* expression within the hypothalamus. However, this is a relevant aspect, given the different roles and regulatory features of *Kiss1* neurons in the arcuate nucleus (ARC) and anteroventral periventricular area (AVPV). Notwithstanding, studies in leptin-deficient (ob/ob) male mice showed back in 2006 a significant reduction of *Kiss1* mRNA levels at the ARC in this model of congenital absence of leptin [26]. Likewise, a severe decrease of *Kiss1* mRNA expression in the ARC was observed in chronically underfed rats during the pubertal transition [27]. On the other hand, adult OVX female rats replaced with estrogen and submitted to short-term fasting displayed reduced *Kiss1* mRNA levels at the AVPV [28]. Moreover, 50% caloric restriction resulted in decreased *Kiss1* mRNA levels in the ARC and AVPV in rats [12], a phenomenon that was also observed during lactation, another condition of negative energy balance [29]. Finally, sheep under dietary restriction experienced a robust suppression of *Kiss1* mRNA levels at the ARC and preoptic area [30]. Altogether, these evidences suggest that *Kiss1* neurons in both the ARC and rostral/AVPV may be targets for metabolic regulation. It is yet to be defined whether metabolic-induced changes in these two populations of *Kiss1* neurons are similar in magnitude or whether they depend on sex, prevailing metabolic status and/or stage of development.

As related issue, most of the studies conducted so far on the metabolic regulation of the *Kiss1* system have addressed changes in its hypothalamic mRNA levels following various forms of metabolic stress. To what extend these mRNA changes translate into alterations of kisspeptin content, or even kisspeptin release, at these hypothalamic sites remains scarcely evaluated. Notwithstanding, it has been documented that 48-h fasting in pubertal female rats induces a significant suppression of kisspeptin-immunoreactivity (IR) and the number of kisspeptin-positive neurons in the ARC [31]. Admittedly, however, no obvious decrease in kisspeptin-IR was detectable in adult female rats after a similar period of food deprivation. The fact that, in that particular study, animals were adult and gonadal-intact (and thus subjected to compensatory changes in endogenous sex steroid milieu), as well as the features of the immunohistochemical analyses that did not allow for proper detection of discrete quantitative changes in kisspeptin protein expression but rather gross changes in the numbers of kisspeptin-positive cells, may explain this discrepancy [31]. Of note, more protracted conditions of negative energy balance, such as lactation, have been reported to cause a decrease in kisspeptin-IR in the ARC [29]. Indeed, the ARC population of *Kiss1* neurons is also sensitive to other forms of metabolic stress, as evidenced by the decrease in the number of kisspeptin-IR detected in adult male rats submitted to an acute inflammatory challenge by administration of bacterial LPS [32].

Finally, expression analyses targeting the hypothalamic *Kiss1* system have been also conducted in conditions of obesity; yet, rather limited data have been produced so far on the effects of situations of persistent overweight on the hypothalamic expression of *Kiss1*. These analyses, however, are particularly interesting, given the rising prevalence of obesity and its potential reproductive comorbidities, which are likely to include perturbed pubertal timing and sub-fertility [33–36]. To our knowledge, the first evidence suggesting (subtle) alterations of the hypothalamic *Kiss1* system in conditions of energy excess came from our studies in long-term diet-induced obese mice that, despite very low levels of circulating testosterone, displayed roughly preserved levels of hypothalamic *Kiss1* mRNA. Considering that severely decreased testosterone levels should have brought about an increase in the hypothalamic expression of *Kiss1* mRNA [37], the above observations were interpreted as evidence for some degree of impairment of the *Kiss1* system, and particularly of its responses to key regulators, such as sex steroids, in obesity [23]. More recently, persistent obesity in DBA/2J mice (a strain that is prone to obesity-induced infertility) has been shown to evoke a significant reduction of *Kiss1* mRNA levels in the ARC and AVPV, as well as in the number of immunoreactive *Kiss1* neurons in the AVPV [38]. On the other hand, studies in male rats transiently exposed to high fat diet have shown increases of *Kiss1* mRNA levels at the hypothalamus [21]. In the same vein, our findings in peripubertal female rats subjected to early postnatal overfeeding suggested a putative correlation between early-onset overweight, advanced vaginal opening (as index of pubertal maturation) and increased *Kiss1* mRNA levels in the hypothalamus, as well as possibly increased numbers of kisspeptin fibers in the AVPV [39]. Yet, a recent report has failed to detect significant changes in *Kiss1* mRNA levels in the ARC or AVPV at the time of vaginal opening in female rats submitted to postnatal overnutrition [40]. In any event, the above observations in various models of increased body weight/obesity in rodents illustrate that, depending on the degree and duration of obesity, stimulatory or inhibitory responses in terms of *Kiss1* expression and gonadotropic function might be observed. Admittedly, however, the mechanisms for *Kiss1* alterations in conditions of obesity are likely different from those of metabolic distress associated to energy insufficiency. Likewise, the potential impact of the adaptative/pathological changes frequently linked to obesity on the hypothalamic *Kiss1* system has not been explored. On note, obesity (and type-2 diabetes) have been suggested to affect the hypothalamic *Kiss1* system in humans [41]; yet, this hypothesis needs to be experimentally validated.

Metabolic Control of the *Kiss1* System: Evidence from Functional Analyses

In spite of the wealth of expression data, demonstration of any causal link between the perturbations of the hypothalamic *Kiss1* system and alterations of the reproductive axis in conditions of metabolic distress would require additional functional analyses. With this aim, a number of (mainly pharmacological) *proof-of-principle* studies have been conducted in order to evaluate whether prevention of the expected drop of kisspeptin levels in situations of energy insufficiency, by means of its exogenous administration, would be sufficient to ameliorate or normalize different reproductive parameters in those situations. These studies, in rather extreme metabolic conditions, have nonetheless illustrated that by enhancing the endogenous kisspeptin tone, without any other neuroendocrine manipulation, several indices of pubertal/reproductive failure can be rescued, thus reinforcing the pathophysiological relevance of *Kiss1* alterations in such situations. As a call of caution, studies in this area have used protocols of central administration of high doses of the peptide, which might have caused an elevation of the endogenous kisspeptin tone over physiological levels.

The first evidence in this front came from our studies on the effects of repeated icv injections of kisspeptin-10 to immature female rats, submitted to chronic subnutrition during the pubertal transition; a manipulation that was sufficient to rescue vaginal opening (in a significant proportion of cases) and to induce potent gonadotropic and estrogenic responses, despite the prevailing suppression of circulating levels of gonadotropins due to malnutrition [20]. In the same vein, repeated administration of kisspeptin-10 to hypogonadotropic diabetic male rats ameliorated testicular and prostate weights, and normalized circulating LH and testosterone levels [24]. Indeed, central injection of kisspeptin-10 to uncontrolled diabetic male and female rats was able to reverse its hypogonadotropic state [24, 25]. Of note, in both models of metabolic distress (subnutrition and diabetes) rescue of the reproductive indices was achieved in spite of the lack of direct effects of chronic kisspeptin treatment on body weights or any improvement of other metabolic parameters.

The functionality of the *Kiss1* system in different situations of metabolic stress has been also evaluated by monitoring acute gonadotropin responses to kisspeptin-10 administration. Thus, gonadotropin responses to acute injection of kisspeptin in fasted male and female rats were not only preserved but even enhanced, despite the drop of gonadotropins levels due to food deprivation [20, 42]. Moreover, the duration of LH responses to kisspeptin-10 was protracted in female rats subjected to chronic subnutrition [43], suggesting that conditions of persistent energy deficit modify the patterns of desensitization to continuous/repeated kisspeptin stimulation. The above evidence may suggest a state of hyper-responsiveness to kisspeptin in situations of negative energy balance, a phenomenon also observed in terms of GnRH secretion by hypothalamic explants from fasted rats [20]. While this phenomenon might be related with sensitization at the receptor level (*see below*), the possibility that such enhanced responses to kisspeptin may derive from a greater accumulation of GnRH and/or gonadotropin stores due to the prevailing state of negative energy balance cannot be excluded either.

One possible explanation for such hyper-responsiveness is that conditions of negative energy balance evoke a decrease of the endogenous kisspeptin tone, which would induce a compensatory state of augmented responsiveness to the neuropeptide. This phenomenon might involve an increase in the expression of *Gpr54* (i.e., *Kiss1r*), as protracted fasting in pubertal rats elicited an increase in its hypothalamic mRNA levels [20]. Of note, however, shorter protocols of fasting in adult mice have been shown to reduce *Gpr54* mRNA expression [23], suggesting a complex dynamics in the changes of kisspeptin release and receptor sensitivity during the course of conditions of persistent negative energy balance. As further illustration of such a complexity, it has been reported that, contrary to rodents, fasting decreases the net responsiveness of the HPG axis to kisspeptin stimulation in the monkey [44]. Several factors, including differences in terms of species, age and duration of fasting, may account for some the above discrepancies. In any event, it seems clear that metabolic stress does not only alter *Kiss1*/kisspeptin expression at the hypothalamus, but also causes functional alterations in terms of GnRH/gonadotropin responsiveness to the stimulatory effects of kisspeptins in different species.

Leptin Signaling and the Hypothalamic *Kiss1* System

The signals responsible for the metabolic control of the hypothalamic *Kiss1* system have been actively investigated. For obvious reasons, because of its paramount physiological relevance in the metabolic gating of puberty and fertility, the first efforts in this front aimed to evaluate whether leptin might operate as major regulator of *Kiss1* expression, as surrogate marker of kisspeptin signaling activity. As mentioned earlier in this chapter, by the time of identification of the reproductive dimension of kisspeptin, it had been thoroughly documented that leptin is an important signal of energy sufficiency for the HPG axis [4]. Moreover, (indirect) evidence for its capacity to stimulate/facilitate GnRH neuron firing at the hypothalamus had been presented at that time [3, 4, 8, 45]. However, the apparent absence of functional leptin receptors in GnRH neurons, as conclusively demonstrated recently [15], suggested the potential involvement of afferent pathways (sensitive to leptin actions) for conveying its effects onto GnRH neurons.

Initial studies provided significant and convincing evidence for a putative role of *Kiss1* neurons in such afferent pathways (see Fig. 17.2). Yet, it must be stressed that most of these initial studies involved rather extreme in vivo models of leptin deficiency, as well as some in vitro experiments using immortalized cell lines. Moreover, the use of rather high doses of leptin (as proof-of-principle approach) might question the physiological relevance of some of the reported stimulatory effects of leptin on the *Kiss1* system under certain conditions, such as the exit from situations of negative energy balance [12]. While we believe that these features do not preclude the usefulness of most of the experimental approaches reported so far, they bring a call of caution when directly extrapolating these observations to "real" regulatory networks and physiological conditions.

The first study to demonstrate direct actions of leptin on *Kiss1* neurons was carried out in gonadectomized ob/ob mice. The results of this work showed (a) a significant suppression of *Kiss1* mRNA levels at the ARC in the absence of leptin; (b) the restoration of *Kiss1* mRNA expression by leptin administration; and (c) the expression of functional leptin receptors in a significant proportion (>40%) of Kiss1 neurons in the ARC [26]. It is important to note that in that study, the animals were gonadectomized in order to prevent fluctuations in endogenous testosterone levels as potential confounding factor for analysis of *Kiss1* mRNA levels [26]. The fact that, at least a fraction of *Kiss1* neurons do express the gene encoding leptin receptors has been recently confirmed in ARC *Kiss1* neurons [46]. In the latter study, the analyses were carried out in transgenic mice that allowed GFP tagging and subsequent isolation of *Kiss1* neurons in vivo [46]. However, according to that study, only a modest subset (<15%) of *Kiss1* neurons displayed conventional pSTAT3 responses to leptin treatment, whereas *Kiss1* neurons in the AVPV did not apparently express the leptin receptor gene [46].

As additional proof of the sensitivity of the hypothalamic *Kiss1* system to the regulatory actions of leptin, central infusion of leptin (at doses previously used in

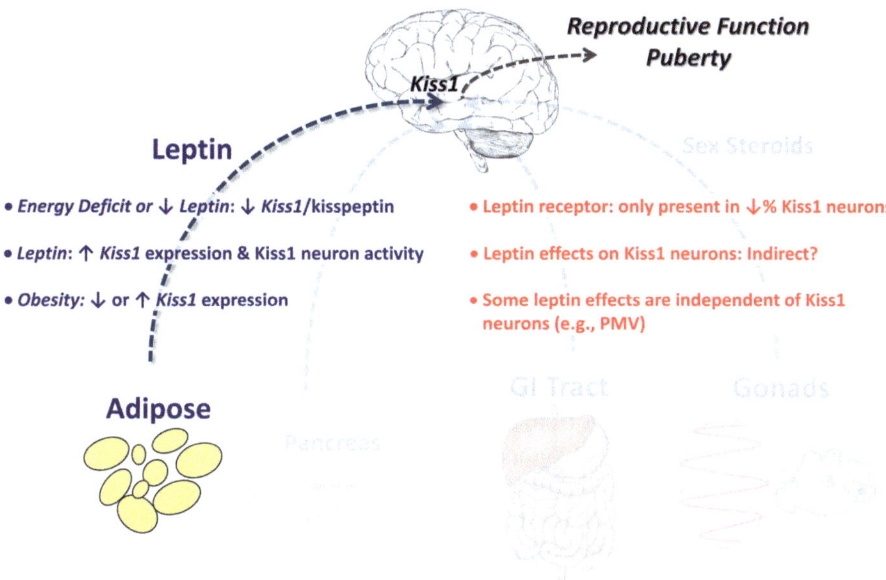

Fig. 17.2 Schematic diagram of the potential actions of leptin in the metabolic gating of puberty and reproduction, and its potential link with the hypothalamic *Kiss1* system. Evidences for a regulatory role of energy balance and leptin on hypothalamic *Kiss1*/kisspeptin expression are summarized in *blue*. Recent data suggesting indirect effects of leptin on *Kiss1* neurons, and *Kiss1*-independent pathways for leptin effects on the reproductive axis, are highlighted in *red*

several metabolic studies), but not insulin, to male rats with uncontrolled diabetes, caused by administration of streptozotocin, has been shown to normalize hypothalamic *Kiss1* gene expression in this model of profound metabolic distress, which is characterized, among other alterations, by a state of hypogonadotropic hypogonadism, linked to marked hypoleptinemia and hypoinsulinemia [24]. Key reproductive parameters, including LH and testosterone secretion, were ameliorated in diabetic male rats by central leptin treatment [24]. In good agreement with those observations, leptin was able to increase *Kiss1* gene expression in the murine hypothalamic cell line, N6 [23], as well as in primary cultures of human fetal GnRH-secreting neuroblasts [47]. In the same vein, leptin has been recently reported to induce the depolarization of *Kiss1* neurons at the ARC via activation of TRPC channels in the guinea pig [48]; a species where ARC *Kiss1* neurons seem to express functional leptin receptors [48]. Additionally, studies in the sheep have documented the expression of the leptin receptor gene in both ARC and POA *Kiss1* neurons, and leptin has been shown to increase *Kiss1* mRNA expression in those hypothalamic nuclei [30]. As a whole, these findings support the existence of a leptin–kisspeptin–GnRH pathway, whereby leptin actions could be signaled onto GnRH neurons via modulation of kisspeptin afferents, thereby allowing proper maturation and function of the HPG axis in conditions of energy/leptin abundance.

Leptin Actions on Hypothalamic *Kiss1* Neurons: Direct, Indirect, or Independent?

Despite the wealth of data suggesting direct regulatory actions of leptin on *Kiss1* neurons (*see previous section*), this contention has been recently challenged by two elegant studies in rodents, which support the possibility that leptin may act indirectly on *Kiss1* neurons. First, a study by Donato and coworkers has shown that the onset of puberty is roughly preserved in female mice selectively and congenitally lacking leptin receptors in *Kiss1*-expressing cells, as generated by the use of Cre-loxP approaches. The fact that *Kiss1* expression is widely spread not only within the brain, but also in other peripheral tissues at early stages of maturation, may compromise the specificity of this phenotype, as it might not be solely caused by selective elimination of leptin receptors from postnatal *Kiss1* neurons in the hypothalamus [49]. Likewise, the development of compensatory mechanisms in this transgenic mouse of congenital elimination of leptin receptors cannot be dismissed.

Notwithstanding, more recent studies have added further strength to the possibility of indirect effects of leptin on *Kiss1* (or GnRH) neurons. Thus, using a combination of expression (immunohistochemical) analyses in WT and transgenic mouse lines, as well as neuroanatomical studies in sheep, it has been documented that only a marginal subset of *Kiss1* neurons in the ARC appear to express leptin receptors, with virtually absent canonical pSTAT3 responses to leptin administration in this neuronal population [50]. Yet, it is stressed here that, in this study, the number of kisspeptin-positive cells in the ARC was remarkably low; a feature which might have underestimated this population and hence the degree of co-expression of leptin receptors. Of note, the latter study also identified uncharacterized populations leptin receptor-expressing cells in the close vicinity of *Kiss1* neurons in the ARC and AVPV [50], whose role as transmitters of leptin effects onto the *Kiss1* pathways is yet to be elucidated.

As a whole, these pieces of evidence strongly suggest that a (substantial) fraction of the effects of leptin on *Kiss1*/kisspeptin expression in the hypothalamus might be transmitted via intermediate pathways (see Fig. 17.2). However, some aspects of this phenomenon are yet to be fully clarified. For instance, the apparent discrepancy between RNA (detectable) and protein (nearly null levels) expression, in terms of leptin receptor, in *Kiss1* neurons needs to be explained, but it may have a methodological basis, including differences in the thresholds of detection of immunohistochemical and molecular assays, and/or might be caused by the expression of alternative isoforms of leptin receptors, whose functional relevance in the hypothalamus has been previously described in other settings [51]. In addition, species differences cannot be excluded either; thus, while studies in mice have pointed out that only a rather modest subset of *Kiss1* neurons express the leptin receptor gene [46], analyses in sheep suggest that the majority of *Kiss1* neurons in the ARC and POA do express it [30], while in the guinea pig, 36% of ARC *Kiss1* neurons in the guinea pig showed detectable leptin receptor gene expression [48]. Finally, differences in the end-points used to monitor leptin effects (e.g., pSTAT3 vs. electrical responses) should be also considered when comparing data from different studies.

Regardless of the predominant (direct and/or indirect) mode of action of leptin on *Kiss1* neurons, the influence of the adipocyte hormone on the hypothalamic *Kiss1* system might not only have physiological relevance, but also pathophysiological implications. Indeed, marked changes in leptin levels have been related to diverse reproductive disorders, including anorexia nervosa, extreme physical exercise, or low body weight [52, 53]. In those conditions, endogenous kisspeptin levels would be expected to be low and kisspeptin analogues might have also potential therapeutic implications, given the preserved, if not augmented, responses to kisspeptins in situations of energy deficiency, at least in rodents [20, 42]. In addition, it has been reported that leptin is able to restore ovarian cyclicity in women with hypothalamic amenorrhea linked to conditions of energy deficit [53, 54]. In this scenario, it is tenable (although yet to be proven) that exogenous kisspeptin might conduct similar positive effects in affected humans, in good agreement with recent data of acute LH responses to kisspeptin in women with hypothalamic amenorrhea [55], and previous results in rodent models of subnutrition in puberty [20].

As final note in this section, it is stressed that the above evidence does not preclude that part of leptin effects on the GnRH system might be conducted in a kisspeptin-independent manner. For instance, leptin receptor-expressing cells have detected in the ventral premammillary nucleus (PMV). This area is a target for leptin effects and, despite the lack of *Kiss1* neurons at this site, seems to play an essential role in conveying leptin effects onto GnRH neurons and, hence, the reproductive axis [49, 56]. Admittedly, however, it remains possible that PMV circuits and those involving *Kiss1* neurons, in other hypothalamic nuclei, may interplay/cooperate in the central control of puberty and fertility by leptin and/or other metabolic cues.

Molecular Mediators and Developmental Aspects of Leptin/Metabolic Control of the *Kiss1* System

In spite of the ongoing debate on whether leptin acts, in some cases, directly on *Kiss1* neurons or modulate afferents to these neurons, putative molecular mediators for such direct and/or indirect effects on the *Kiss1* system have been exposed in recent years. As example, the cellular energy sensor, mammalian target of rapamycin (mTOR), which has been shown to operate as cellular metabolic gauge [57–61], and mediator of some of the effects of leptin in the central control of food intake [62, 63], is likely involved in conveying part of the regulatory effects of leptin on the reproductive brain, by virtue of its ability to regulate *Kiss1* expression.

Persistent blockade of central mTOR signaling resulted in delayed puberty and reduced LH levels; yet, gonadotropin responsiveness to key activators of the gonadotropic axis, such as kisspeptins, was preserved following mTOR blockade, suggesting specificity for these inhibitory effects. In the same vein, central inhibition of mTOR signaling prevented the permissive/stimulatory effects of leptin on puberty onset, and caused a substantial suppression of *Kiss1* mRNA levels in key hypothalamic centers, such as the ARC and (to a lesser extent) the AVPV. Conversely, activation of mTOR by central injection of l-leucine partially reversed the state of hypogonadotropism induced by chronic subnutrition; a condition of low leptin levels [27]. All in all, these observations are compatible with the existence of a leptin–mTOR–kisspeptin pathway that may play a physiological role in the metabolic control of puberty onset and fertility. Admittedly, however, the cellular basis for such pathway, and whether mTOR operates as modulator within *Kiss1* cells and/or in afferent neuronal pathways, is yet to be clarified and warrants further investigation.

Another putative molecular modulator of *Kiss1* gene expression is the cAMP responsive element-binding protein-1 (Creb1)-regulated transcription coactivator-1 (Crct1; also termed TORC1, which stands for *Transducer of Regulated Creb*). It was reported in 2008 that Crtc1 KO mice display not only obese and hyperphagic phenotypes, but are also infertile [64]. The underlying mechanisms for such a combined alteration of energy homeostasis and reproduction seem to involve the disturbance (due to the lack of Crtc1) of the ability of leptin to stimulate the expression of the genes encoding CART (cocaine- and amphetamine-regulated transcript, which operate as anorexigenic neuropeptide controlling food intake) and kisspeptins. On the latter, leptin has been proven to dephosphorylate (and activate) Crtc1, which in turn stimulates the recruitment of Crct1 to *Kiss1* gene promoter. In addition, dephosphorylation of Crtc1 enhanced *Kiss1* gene expression in GT1-7 cells, and Crtc1 over-expression increased *Kiss1* promoter activity [64]. However, an independent report was unable to replicate the consequences of functional inactivation of Crtc1 on mouse fertility [65]. This discordant observation, together with novel expression and functional genomic data suggesting a nondirect mode of action of leptin on *Kiss1* neurons, bring some doubts on the actual physiological role of Crtc1 in mediating leptin effects on the HPG axis in vivo.

In addition to putative molecular mediators, the developmental effects of metabolic signals in the pubertal maturation of the hypothalamic *Kiss1* system have been recently explored. As described elsewhere in this book, developing *Kiss1* neurons are sensitive to the organizing actions of other key regulators of the HPG axis, such as sex steroids [37]. Considering the increasing prevalence of early-onset metabolic disorders, including childhood obesity that may affect the timing of puberty, and the proven sensitivity of the hypothalamic *Kiss1* system to conditions of metabolic distress, specific analysis of the *early origins* of alterations of this system linked to metabolic distress are warranted. In this context, a recent study from our group using models of postnatal nutritional challenge, by means of manipulation of litter size, strongly suggested that early changes in body weight and energy homeostasis might be also an important regulator of later *Kiss1*/kisspeptin expression at puberty. Thus, female rats bred in large litters (as model of early underfeeding) were leaner and displayed delayed vaginal opening, despite they being allowed to eat ad libitum from weaning onwards. These animals had also low serum leptin levels and reduced *Kiss1* mRNA levels and kisspeptin-positive neurons at the hypothalamus during puberty [39]; roughly similar findings have been recently reported in female mice subjected to postnatal undernutrition [66]. In good agreement, gestational undernutrition has been shown to decrease hypothalamic *Kiss1* expression and to delay puberty onset in female rats [67]. Conversely, female rats bred in small litters (as model of early postnatal overfeeding) were heavier and showed earlier entry into puberty, when they also displayed elevated serum leptin concentrations and increased *Kiss1* mRNA levels at the hypothalamus, which was associated with a trend for a higher number of kisspeptin-positive fibers in the AVPV [39]. Whether these alterations stem from nutritional/leptin changes at early postnatal periods and/or during puberty is yet to be defined. Of note, a very recent study of early nutritional manipulation, roughly similar to our previous study, confirmed changes in the age of puberty onset in both male (delayed in the postnatally underfed group) and female rats (advanced in the postnatally overfed group), but failed to demonstrate any significant change in hypothalamic *Kiss1* mRNA expression among the different experimental models [40]. It is possible that differences in the timing of tissue sampling for expression analysis (fixed d-36 in our study vs. age of external sign of puberty) and/or the setting up of the litters (100% females in our study vs. 50% females–50% males) might partially explain the above discrepancies. In addition, relevant reproductive/metabolic parameters, such as gonadotropin and leptin levels at puberty, were not evaluated in the latter study; parameters that may help to clarify these discordances. Finally, a recent study suggested that the hypothalamic *Kiss1* system is particularly sensitive to inhibition by acute fasting during juvenile, rather than the infantile period [68]. All in all, the influence of different forms of metabolic stress on the early organization and development of *Kiss1* circuits, and the potential relevance of such phenomenon on the timing of puberty, warrants further investigation.

Other Putative Metabolic Regulators of the Hypothalamic *Kiss1* System

In addition to leptin, other *metabolic* or *nutritional* cues may participate also in the modulation of kisspeptin signaling, although to date this possibility has been addressed only fragmentarily and further characterization of the whole set of metabolic signals involved in the control of the *Kiss1* system is eagerly awaited. Among potential candidates, the gut-derived hormone, ghrelin, has been suggested to inhibit the *Kiss1* system in rat hypothalamus. Ghrelin is secreted by specific endocrine (X/A) cells of the gastric mucosa and operates as circulating orexigenic molecule, with effects that are opposite to those of leptin, thus functioning as signal for energy insufficiency [9]. In keeping with such a role, we, as well as others, have documented that ghrelin is a negative modifier of puberty onset and/or gonadotropin secretion in a variety of species (including rodents, sheep, monkey, and human), acting mainly at central levels [9, 69]. Recently, ghrelin has been shown to inhibit hypothalamic *Kiss1* mRNA expression in female rats, and this phenomenon, which needs to be confirmed, may contribute to the suppression of the HPG axis induced by elevated levels of ghrelin (see Fig. 17.3). The putative molecular mechanisms and mode of action of ghrelin on *Kiss1* neurons remain totally unexplored.

Other potential metabolic regulators of the *Kiss1* system are NPY and melanocortins, although the amount of experimental data supporting this possibility is also scarce. NPY, which operates as potent orexigenic signal in the brain [70, 71], may

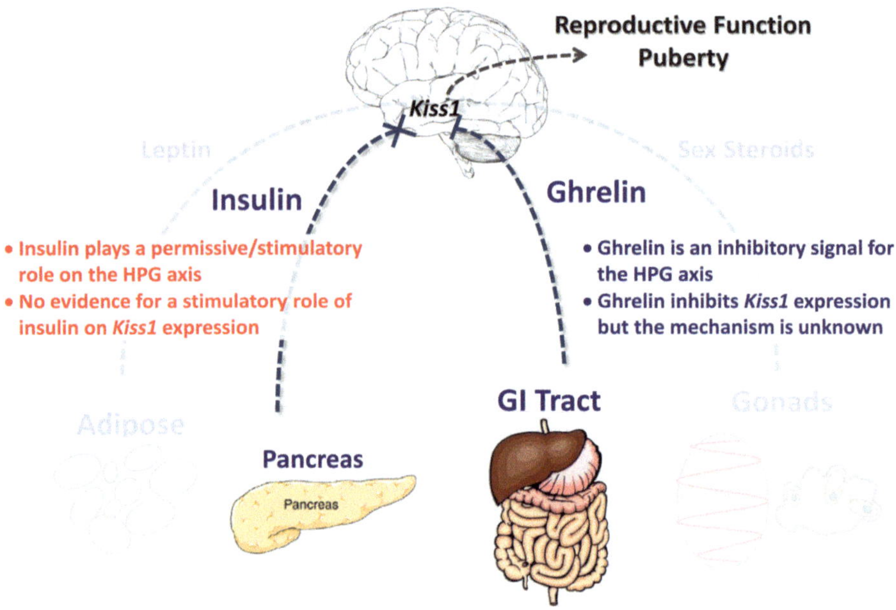

Fig. 17.3 Schematic diagram of the potential actions of the pancreas-derived hormone, insulin, and the gut-born hormone, ghrelin, in the metabolic control of reproduction. While insulin exerts stimulatory/permissive effects on the reproductive axis, experimental data from different species supports a predominant inhibitory role of ghrelin the control of puberty and reproduction. Evidence supporting (or not) the potential link of these hormones with the hypothalamic *Kiss1* system is itemized

function as stimulatory factor for *Kiss1* expression, since *Kiss1* mRNA levels are decreased in the hypothalamus of NPY KO mice, whereas NPY enhanced *Kiss1* mRNA expression in the hypothalamic cell line, N6 [23]. Such stimulatory effect is somewhat counterintuitive, given the proven increase of NPY expression following conditions of negative energy balance, in which both *Kiss1* and the gonadotropic axis appear to be suppressed, and the documented inhibitory effects of NPY on GnRH secretion [72]. Yet, it is also known that at certain physiological and experimental conditions, NPY may drive a stimulatory signal to the GnRH system [73]; conditions where *Kiss1* pathways might operate as putative effector. In any event, since leptin has been shown to suppress the expression of NPY at specific neuronal populations in the ARC [74], NPY is not likely to operate as a mediator for the stimulatory effects of leptin on *Kiss1* neurons, but rather to act as independent modulator.

In addition, melanocortins, which are anorectic neuropeptide products of proopio-melanocortin (POMC) neurons in the ARC, have been recently suggested to stimulate hypothalamic *Kiss1* gene expression at the preoptic area in the sheep [75]. In contrast, the orexigenic neuropeptide, melanin-concentrating hormone (MCH), which is prominently expressed in the lateral hypothalamus, suppressed kisspeptin-induced stimulation of GnRH neurons [76]. Yet, to our knowledge, the effects of MCH on *Kiss1* expression have not been reported to date. Finally, insulin does not appear to significantly contribute to the regulation of *Kiss1* expression, as suggested by studies in models of diabetic male rats centrally infused with insulin, which displayed persistently reduced *Kiss1* mRNA levels despite insulin administration (see Fig. 17.3). In addition, insulin failed to stimulate *Kiss1* expression in the hypothalamic cell line, N6, in vitro [23, 24].

Kisspeptins and the Control of Food Intake and Energy Balance: Physiological Relevance?

Recent evidence suggests that, in addition to their primary roles in the control of the reproductive axis, kisspeptins might also participate in the central networks controlling food intake and energy balance. Indeed, bidirectional interactions between the systems governing energy homeostasis and the HPG axis have been reported for a large number of neuropeptides and hormones [4]. Accordingly, initial studies addressing the roles of kisspeptins in the metabolic control of reproduction evaluated also the potential impact of kisspeptins on food intake and body weight. Yet, those analyses failed to demonstrate any significant change in the patterns of food intake or the hypothalamic expression of relevant genes in the control of energy balance, such as NPY, POMC, Agouti-related peptide (AgRP), and CART, after central administration of kisspeptin [20, 77]. These results questioned the physiological contribution of kisspeptin signaling in the central pathways controlling energy homeostasis.

This contention, however, has been challenged by recent electrophysiological analyses that have documented the ability of kisspeptin-10 to activate the population of anorexigenic POMC neurons, and to inhibit the orexigenic NPY neurons in the ARC in rodents [78]. As complementary finding, central injection of kisspeptin have been shown to suppress nocturnal food intake in mice [79]. As a whole, the above data would suggest that, under specific conditions, kisspeptins may operate also as feeding suppressing signals, probably via activation ARC POMC neurons, which do express Gpr54. While this possibility is appealing, further experimental analyses, including the assessment of the specific roles of Gpr54 signaling in POMC neurons, are needed in order to confirm or refute this possibility, and to explain discrepancies between previous and recent literature.

As final note, it is stressed here that estrogens, which are proven regulators of central *Kiss1* expression [37], are potent food-intake suppressor signals acting in different hypothalamic nuclei, so that ovariectomy causes a rapid increase in body weight in rodents. Very recently, using a toxic ablation approach, Rance and collaborators evaluated the consequences of elimination of *Kiss1* neurons in the ARC, which co-express the neuropeptide, neurokinin-B, and its receptor, NK3R, on the effects of estrogen upon the negative feedback control of gonadotropin secretion [80]. Curiously enough, in that study ablation of ARC *Kiss1* neurons abolished the effects of gonadectomy (increase) and estrogen (reduction) on body weight. This observation would suggest that the integrity of ARC kisspeptin pathways is necessary for the conduction of the anorexigenic action of estradiol. These findings, however, seem difficult to reconcile with the facts that estrogen inhibits *Kiss1* expression in the ARC, and that kisspeptins (as is the case for estrogens) are thought to conduct a primary anorexigenic action in the hypothalamus [79]. The possibility that the above phenomenon may stem from elimination of other co-transmitters of ARC *Kiss1* neurons, such as dynorphin, warrants further investigation.

Concluding Remarks

While the influence of body weight on puberty and fertility has been intuitively known for ages, the neuroendocrine mechanisms underlying the metabolic control of the reproductive axis have begun to be exposed only in the past decades, in which the discovery of leptin and, more recently, kisspeptins, are considered important milestones. Indeed, in the last few years we have learnt that *Kiss1* neurons can *sense* extreme metabolic conditions and are likely to participate in the transmission of such metabolic information to GnRH neurons. Among the peripheral signals involved in the control of *Kiss1* neurons, compelling evidence from proof-of-principle studies, addressing the effects of either the lack of leptin or its supplementation, strongly suggest that, in keeping with its key roles in energy homeostasis and body weight control, leptin participates in the metabolic regulation of the hypothalamic *Kiss1* system. By virtue of leptin ability to stimulate *Kiss1* expression, it is likely that it allows the proper functioning of the reproductive axis in conditions of energy sufficiency.

The reality, however, is probably not so simple and, certainly, the metabolic control of the *Kiss1* system likely goes beyond the regulatory actions of leptin, just as much as the metabolic control of puberty and fertility is not all about kisspeptin regulation. For instance, very recently, we have become aware that *Kiss1* neurons might not be direct targets of leptin, which in turn may operate (largely) via indirect mechanisms. But, is this the case in all species, and if so, what is the reason for the apparent expression of leptin receptors in *Kiss1* neurons? Similarly, we have learnt that some of the positive effects of leptin on *Kiss1* expression are detectable only at rather high (probably pharmacological) levels, but this does not preclude a physiological role of this adipose hormone in the metabolic control of reproductive onset and fertility. On the other hand, although the field of the metabolic control of the *Kiss1* system has been largely dominated by leptin studies, mounting, but as yet limited, evidence has pointed out that other metabolic signals besides leptin can also participate in the control of *Kiss1* neurons. If so, what is the basis and physiological relevance of such actions? These are but few examples of "uncertainties" and open questions in the field of the metabolic control of kisspeptins that warrant specific investigation and are likely to draw considerable attention in the near future.

Acknowledgments The authors are indebted with the members of the research team at the Physiology Section of the University of Cordoba, who actively participated in the generation of experimental data discussed herein. The work from the authors' laboratory summarized in this article was supported by grants BFU 2005-07446, BFU 2008-00984, and BFU2011-25021 (Ministerio de Economía and Competitividad, Spain; covered in part by EU-FEDER funds), grants from Instituto de Salud Carlos III (Red de Centros RCMN C03/08 and Project PI042082; Ministerio de Sanidad, Spain), and EU research contracts EDEN QLK4-CT-2002-00603 and DEER FP7-ENV-2007-1. CIBER is an initiative of Instituto de Salud Carlos III (Ministerio de Sanidad, Spain).

References

1. Schwartz NB (2000) Neuroendocrine regulation of reproductive cyclicity. In: Conn PM, Freeman ME (eds) Neuroendocrinology in physiology and medicine, vol. Humana, Totowa, NJ, pp 135–146
2. Tena-Sempere M, Huhtaniemi I (2003) Gonadotropins and gonadotropin receptors. In: Fauser BCJM (ed) Reproductive medicine—molecular, cellular and genetic fundamentals, vol. Partenon Publishing, New York, pp 225–244
3. Fernandez-Fernandez R, Martini AC, Navarro VM, Castellano JM, Dieguez C, Aguilar E, Pinilla L, Tena-Sempere M (2006) Novel signals for the integration of energy balance and reproduction. Mol Cell Endocrinol 254–255:127–132
4. Tena-Sempere M (2007) Roles of ghrelin and leptin in the control of reproductive function. Neuroendocrinology 86:229–241
5. Kennedy GC, Mitra J (1963) Body weight and food intake as initiating factors for puberty in the rat. J Physiol 166:408–418
6. Frisch RE, Revelle R (1970) Height and weight at menarche and a hypothesis of critical body weights and adolescent events. Science 169:397–399
7. Frisch RE, McArthur JW (1974) Menstrual cycles: fatness as a determinant of minimum weight for height necessary for their maintenance or onset. Science 185:949–951
8. Casanueva FF, Dieguez C (1999) Neuroendocrine regulation and actions of leptin. Front Neuroendocrinol 20:317–363
9. Tena-Sempere M (2008) Ghrelin as a pleotrophic modulator of gonadal function and reproduction. Nat Clin Pract Endocrinol Metab 4:666–674
10. Castellano JM, Roa J, Luque RM, Dieguez C, Aguilar E, Pinilla L, Tena-Sempere M (2009) KiSS-1/kisspeptins and the metabolic control of reproduction: physiologic roles and putative physiopathological implications. Peptides 30:139–145
11. Roa J, Garcia-Galiano D, Castellano JM, Gaytan F, Pinilla L, Tena-Sempere M (2010) Metabolic control of puberty onset: new players, new mechanisms. Mol Cell Endocrinol 324:87–94
12. True C, Kirigiti MA, Kievit P, Grove KL, Susan Smith M (2011) Leptin is not the critical signal for kisspeptin or luteinising hormone restoration during exit from negative energy balance. J Neuroendocrinol 23:1099–1112
13. Szymanski LA, Schneider JE, Friedman MI, Ji H, Kurose Y, Blache D, Rao A, Dunshea FR, Clarke IJ (2007) Changes in insulin, glucose and ketone bodies, but not leptin or body fat content precede restoration of luteinising hormone secretion in ewes. J Neuroendocrinol 19:449–460
14. Cunningham MJ, Clifton DK, Steiner RA (1999) Leptin's actions on the reproductive axis: perspectives and mechanisms. Biol Reprod 60:216–222
15. Quennell JH, Mulligan AC, Tups A, Liu X, Phipps SJ, Kemp CJ, Herbison AE, Grattan DR, Anderson GM (2009) Leptin indirectly regulates gonadotropin-releasing hormone neuronal function. Endocrinology 150:2805–2812
16. Oakley AE, Clifton DK, Steiner RA (2009) Kisspeptin signaling in the brain. Endocr Rev 30:713–743
17. Roa J, Aguilar E, Dieguez C, Pinilla L, Tena-Sempere M (2008) New frontiers in kisspeptin/GPR54 physiology as fundamental gatekeepers of reproductive function. Front Neuroendocrinol 29:48–69

18. Simonneaux V, Ansel L, Revel FG, Klosen P, Pevet P, Mikkelsen JD (2009) Kisspeptin and the seasonal control of reproduction in hamsters. Peptides 30:146–153
19. Tena-Sempere M (2006) KiSS-1 and reproduction: focus on its role in the metabolic regulation of fertility. Neuroendocrinology 83:275–281
20. Castellano JM, Navarro VM, Fernandez-Fernandez R, Nogueiras R, Tovar S, Roa J, Vazquez MJ, Vigo E, Casanueva FF, Aguilar E, Pinilla L, Dieguez C, Tena-Sempere M (2005) Changes in hypothalamic KiSS-1 system and restoration of pubertal activation of the reproductive axis by kisspeptin in undernutrition. Endocrinology 146:3917–3925
21. Brown RE, Imran SA, Ur E, Wilkinson M (2008) KiSS-1 mRNA in adipose tissue is regulated by sex hormones and food intake. Mol Cell Endocrinol 281:64–72
22. Matsuzaki T, Iwasa T, Kinouchi R, Yoshida S, Murakami M, Gereltsetseg G, Yamamoto S, Kuwahara A, Yasui T, Irahara M (2011) Fasting reduces the kiss1 mRNA levels in the caudal hypothalamus of gonadally intact adult female rats. Endocr J 58:1003–1012
23. Luque RM, Kineman RD, Tena-Sempere M (2007) Regulation of hypothalamic expression of KiSS-1 and GPR54 genes by metabolic factors: analyses using mouse models and a cell line. Endocrinology 148:4601–4611
24. Castellano JM, Navarro VM, Fernandez-Fernandez R, Roa J, Vigo E, Pineda R, Dieguez C, Aguilar E, Pinilla L, Tena-Sempere M (2006) Expression of hypothalamic KiSS-1 system and rescue of defective gonadotropic responses by kisspeptin in streptozotocin-induced diabetic male rats. Diabetes 55:2602–2610
25. Castellano JM, Navarro VM, Roa J, Pineda R, Sanchez-Garrido MA, Garcia-Galiano D, Vigo E, Dieguez C, Aguilar E, Pinilla L, Tena-Sempere M (2009) Alterations in hypothalamic KiSS-1 system in experimental diabetes: early changes and functional consequences. Endocrinology 150:784–794
26. Smith JT, Acohido BV, Clifton DK, Steiner RA (2006) KiSS-1 neurones are direct targets for leptin in the ob/ob mouse. J Neuroendocrinol 18:298–303
27. Roa J, Garcia-Galiano D, Varela L, Sanchez-Garrido MA, Pineda R, Castellano JM, Ruiz-Pino F, Romero M, Aguilar E, Lopez M, Gaytan F, Dieguez C, Pinilla L, Tena-Sempere M (2009) The mammalian target of rapamycin as novel central regulator of puberty onset via modulation of hypothalamic Kiss1 system. Endocrinology 150:5016–5026
28. Kalamatianos T, Grimshaw SE, Poorun R, Hahn JD, Coen CW (2008) Fasting reduces KiSS-1 expression in the anteroventral periventricular nucleus (AVPV): effects of fasting on the expression of KiSS-1 and neuropeptide Y in the AVPV or arcuate nucleus of female rats. J Neuroendocrinol 20:1089–1097
29. True C, Kirigiti M, Ciofi P, Grove KL, Smith MS (2011) Characterisation of arcuate nucleus kisspeptin/neurokinin B neuronal projections and regulation during lactation in the rat. J Neuroendocrinol 23:52–64
30. Backholer K, Smith JT, Rao A, Pereira A, Iqbal J, Ogawa S, Li Q, Clarke IJ (2010) Kisspeptin cells in the ewe brain respond to leptin and communicate with neuropeptide Y and proopiomelanocortin cells. Endocrinology 151:2233–2243
31. Castellano JM, Bentsen AH, Mikkelsen JD, Tena-Sempere M (2010) Kisspeptins: bridging energy homeostasis and reproduction. Brain Res 1364:129–138
32. Castellano JM, Bentsen AH, Romero M, Pineda R, Ruiz-Pino F, Garcia-Galiano D, Sanchez-Garrido MA, Pinilla L, Mikkelsen JD, Tena-Sempere M (2010) Acute inflammation reduces kisspeptin immunoreactivity at the arcuate nucleus and decreases responsiveness to kisspeptin independently of its anorectic effects. Am J Physiol Endocrinol Metab 299:E54–E61
33. Rachon D, Teede H (2010) Ovarian function and obesity—interrelationship, impact on women's reproductive lifespan and treatment options. Mol Cell Endocrinol 316:172–179
34. Biro FM, Wien M (2010) Childhood obesity and adult morbidities. Am J Clin Nutr 91:1499S–1505S
35. Ahmed ML, Ong KK, Dunger DB (2009) Childhood obesity and the timing of puberty. Trends Endocrinol Metab 20:237–242
36. Loret de Mola JR (2009) Obesity and its relationship to infertility in men and women. Obstet Gynecol Clin North Am 36:333–346, ix

37. Garcia-Galiano D, Pinilla L, Tena-Sempere M (2012) Sex steroids and the control of the Kiss1 system: developmental roles and major regulatory actions. J Neuroendocrinol 24:22–33
38. Quennell JH, Howell CS, Roa J, Augustine RA, Grattan DR, Anderson GM (2011) Leptin deficiency and diet-induced obesity reduce hypothalamic kisspeptin expression in mice. Endocrinology 152:1541–1550
39. Castellano JM, Bentsen AH, Sanchez-Garrido MA, Ruiz-Pino F, Romero M, Garcia-Galiano D, Aguilar E, Pinilla L, Dieguez C, Mikkelsen JD, Tena-Sempere M (2011) Early metabolic programming of puberty onset: impact of changes in postnatal feeding and rearing conditions on the timing of puberty and development of the hypothalamic kisspeptin system. Endocrinology 152:3396–3408
40. Smith JT, Spencer SJ (2012) Preweaning over- and underfeeding alters onset of puberty in the rat without affecting kisspeptin. Biol Reprod 86(145):141–148
41. George JT, Millar RP, Anderson RA (2010) Hypothesis: kisspeptin mediates male hypogonadism in obesity and type 2 diabetes. Neuroendocrinology 91:302–307
42. Tovar S, Vazquez MJ, Navarro VM, Fernandez-Fernandez R, Castellano JM, Vigo E, Roa J, Casanueva FF, Aguilar E, Pinilla L, Dieguez C, Tena-Sempere M (2006) Effects of single or repeated intravenous administration of kisspeptin upon dynamic LH secretion in conscious male rats. Endocrinology 147:2696–2704
43. Roa J, Vigo E, Garcia-Galiano D, Castellano JM, Navarro VM, Pineda R, Dieguez C, Aguilar E, Pinilla L, Tena-Sempere M (2008) Desensitization of gonadotropin responses to kisspeptin in the female rat: analyses of LH and FSH secretion at different developmental and metabolic states. Am J Physiol Endocrinol Metab 294:E1088–E1096
44. Wahab F, Aziz F, Irfan S, Zaman WU, Shahab M (2008) Short-term fasting attenuates the response of the HPG axis to kisspeptin challenge in the adult male rhesus monkey (Macaca mulatta). Life Sci 83:633–637
45. Hill JW, Elmquist JK, Elias CF (2008) Hypothalamic pathways linking energy balance and reproduction. Am J Physiol Endocrinol Metab 294:E827–E832
46. Cravo RM, Margatho LO, Osborne-Lawrence S, Donato J Jr, Atkin S, Bookout AL, Rovinsky S, Frazao R, Lee CE, Gautron L, Zigman JM, Elias CF (2011) Characterization of Kiss1 neurons using transgenic mouse models. Neuroscience 173:37–56
47. Morelli A, Marini M, Mancina R, Luconi M, Vignozzi L, Fibbi B, Filippi S, Pezzatini A, Forti G, Vannelli GB, Maggi M (2008) Sex steroids and leptin regulate the "first Kiss" (KiSS 1/G-protein-coupled receptor 54 system) in human gonadotropin-releasing-hormone-secreting neuroblasts. J Sex Med 5:1097–1113
48. Qiu J, Fang Y, Bosch MA, Ronnekleiv OK, Kelly MJ (2011) Guinea pig kisspeptin neurons are depolarized by leptin via activation of TRPC channels. Endocrinology 152:1503–1514
49. Donato J Jr, Cravo RM, Frazao R, Gautron L, Scott MM, Lachey J, Castro IA, Margatho LO, Lee S, Lee C, Richardson JA, Friedman J, Chua S Jr, Coppari R, Zigman JM, Elmquist JK, Elias CF (2011) Leptin's effect on puberty in mice is relayed by the ventral premammillary nucleus and does not require signaling in Kiss1 neurons. J Clin Invest 121:355–368
50. Louis GW, Greenwald-Yarnell M, Phillips R, Coolen LM, Lehman MN, Myers MG Jr (2011) Molecular mapping of the neural pathways linking leptin to the neuroendocrine reproductive axis. Endocrinology 152:2302–2310
51. Tena-Sempere M, Barreiro ML (2002) Leptin in male reproduction: the testis paradigm. Mol Cell Endocrinol 188:9–13
52. Welt CK (2007) Will leptin become the treatment of choice for functional hypothalamic amenorrhea? Nat Clin Pract Endocrinol Metab 3:556–557
53. Welt CK, Chan JL, Bullen J, Murphy R, Smith P, DePaoli AM, Karalis A, Mantzoros CS (2004) Recombinant human leptin in women with hypothalamic amenorrhea. N Engl J Med 351:987–997
54. Chou SH, Chamberland JP, Liu X, Matarese G, Gao C, Stefanakis R, Brinkoetter MT, Gong H, Arampatzi K, Mantzoros CS (2011) Leptin is an effective treatment for hypothalamic amenorrhea. Proc Natl Acad Sci U S A 108:6585–6590

55. Jayasena CN, Nijher GM, Chaudhri OB, Murphy KG, Ranger A, Lim A, Patel D, Mehta A, Todd C, Ramachandran R, Salem V, Stamp GW, Donaldson M, Ghatei MA, Bloom SR, Dhillo WS (2009) Subcutaneous injection of kisspeptin-54 acutely stimulates gonadotropin secretion in women with hypothalamic amenorrhea, but chronic administration causes tachyphylaxis. J Clin Endocrinol Metab 94:4315–4323
56. Donato J Jr, Silva RJ, Sita LV, Lee S, Lee C, Lacchini S, Bittencourt JC, Franci CR, Canteras NS, Elias CF (2009) The ventral premammillary nucleus links fasting-induced changes in leptin levels and coordinated luteinizing hormone secretion. J Neurosci 29:5240–5250
57. Wullschleger S, Loewith R, Hall MN (2006) TOR signaling in growth and metabolism. Cell 124:471–484
58. Chiang GG, Abraham RT (2007) Targeting the mTOR signaling network in cancer. Trends Mol Med 13:433–442
59. Tsang CK, Qi H, Liu LF, Zheng XF (2007) Targeting mammalian target of rapamycin (mTOR) for health and diseases. Drug Discov Today 12:112–124
60. Martin DE, Hall MN (2005) The expanding TOR signaling network. Curr Opin Cell Biol 17:158–166
61. Schmelzle T, Hall MN (2000) TOR, a central controller of cell growth. Cell 103:253–262
62. Cota D, Proulx K, Smith KA, Kozma SC, Thomas G, Woods SC, Seeley RJ (2006) Hypothalamic mTOR signaling regulates food intake. Science 312:927–930
63. Woods SC, Seeley RJ, Cota D (2008) Regulation of food intake through hypothalamic signaling networks involving mTOR. Annu Rev Nutr 28:295–311
64. Altarejos JY, Goebel N, Conkright MD, Inoue H, Xie J, Arias CM, Sawchenko PE, Montminy M (2008) The Creb1 coactivator Crtc1 is required for energy balance and fertility. Nat Med 14:1112–1117
65. Breuillaud L, Halfon O, Magistretti PJ, Pralong FP, Cardinaux JR (2009) Mouse fertility is not dependent on the CREB coactivator Crtc1. Nat Med 15:989–990; author reply 991
66. Caron E, Ciofi P, Prevot V, Bouret SG (2012) Alteration in neonatal nutrition causes perturbations in hypothalamic neural circuits controlling reproductive function. J Neurosci 32:11486–11494
67. Iwasa T, Matsuzaki T, Murakami M, Fujisawa S, Kinouchi R, Gereltsetseg G, Kuwahara A, Yasui T, Irahara M (2010) Effects of intrauterine undernutrition on hypothalamic Kiss1 expression and the timing of puberty in female rats. J Physiol 588:821–829
68. Iwasa T, Matsuzaki T, Murakami M, Kinouchi R, Gereltsetseg G, Fujisawa S, Kuwahara A, Yasui T, Irahara M (2010) Sensitivities of mRNA expression levels of Kiss1 and its receptor, Kiss1r, to nutritional status are changed during the developmental period in female rats. J Endocrinol 207:195–202
69. Fernandez-Fernandez R, Navarro VM, Barreiro ML, Vigo EM, Tovar S, Sirotkin AV, Casanueva FF, Aguilar E, Dieguez C, Pinilla L, Tena-Sempere M (2005) Effects of chronic hyperghrelinemia on puberty onset and pregnancy outcome in the rat. Endocrinology 146:3018–3025
70. Kalra SP, Crowley WR (1992) Neuropeptide Y: a novel neuroendocrine peptide in the control of pituitary hormone secretion, and its relation to luteinizing hormone. Front Neuroendocrinol 13:1–46
71. Pralong FP (2010) Insulin and NPY pathways and the control of GnRH function and puberty onset. Mol Cell Endocrinol 324:82–86
72. Xu J, Kirigiti MA, Cowley MA, Grove KL, Smith MS (2009) Suppression of basal spontaneous gonadotropin-releasing hormone neuronal activity during lactation: role of inhibitory effects of neuropeptide Y. Endocrinology 150:333–340
73. Kalra SP, Kalra PS (2004) NPY and cohorts in regulating appetite, obesity and metabolic syndrome: beneficial effects of gene therapy. Neuropeptides 38:201–211
74. Schwartz MW, Seeley RJ, Campfield LA, Burn P, Baskin DG (1996) Identification of targets of leptin action in rat hypothalamus. J Clin Invest 98:1101–1106
75. Backholer K, Smith J, Clarke IJ (2009) Melanocortins may stimulate reproduction by activating orexin neurons in the dorsomedial hypothalamus and kisspeptin neurons in the preoptic area of the ewe. Endocrinology 150:5488–5497

76. Wu M, Dumalska I, Morozova E, van den Pol A, Alreja M (2009) Melanin-concentrating hormone directly inhibits GnRH neurons and blocks kisspeptin activation, linking energy balance to reproduction. Proc Natl Acad Sci U S A 106:17217–17222
77. Thompson EL, Patterson M, Murphy KG, Smith KL, Dhillo WS, Todd JF, Ghatei MA, Bloom SR (2004) Central and peripheral administration of kisspeptin-10 stimulates the hypothalamic-pituitary-gonadal axis. J Neuroendocrinol 16:850–858
78. Fu LY, van den Pol AN (2010) Kisspeptin directly excites anorexigenic proopiomelanocortin neurons but inhibits orexigenic neuropeptide Y cells by an indirect synaptic mechanism. J Neurosci 30:10205–10219
79. Stengel A, Wang L, Goebel-Stengel M, Tache Y (2011) Centrally injected kisspeptin reduces food intake by increasing meal intervals in mice. Neuroreport 22:253–257
80. Mittelman-Smith MA, Williams H, Krajewski-Hall SJ, Lai J, Ciofi P, McMullen NT, Rance NE (2012) Arcuate kisspeptin/neurokinin B/dynorphin (KNDy) neurons mediate the estrogen suppression of gonadotropin secretion and body weight. Endocrinology 153:2800–2812

Chapter 18
Circadian Regulation of Kisspeptin in Female Reproductive Functioning

Lance J. Kriegsfeld

Abstract Female reproductive functioning requires the precise temporal organization of numerous neuroendocrine events by a master circadian brain clock located in the suprachiasmatic nucleus. Across species, including humans, disruptions to circadian timing result in pronounced deficits in ovulation and fecundity. The present chapter provides an overview of the circadian control of female reproduction, underscoring the significance of kisspeptin as a key locus of integration for circadian and steroidal signaling necessary for the initiation of ovulation.

Introduction

The circadian timing system universally coordinates central and peripheral physiology, providing temporal structure to homeostatic regulation and ensuring that physiological processes are maintained within optimal operating limits given changing demands over the course of day and night. Because hormones enter the general circulation, these signaling molecules have widespread influence over physiology and behavior, and represent ideal communicators of timing information. As a result, it is important to understand the means by which circadian rhythms in endocrine secretions are generated and the functional consequences of this temporal communication for downstream target systems.

Converging lines of evidence indicate a critical role for circadian timing in successful female reproduction across mammalian species, including humans. Women with irregular work or sleep cycles, for example, exhibit reduced fertility [1] and an

L.J. Kriegsfeld, Ph.D. (✉)
Department of Psychology and Helen Wills Neuroscience Institute, Neurobiology Laboratory, University of California, 3210 Tolman Hall, 1650, Berkeley, CA 94720-1650, USA
e-mail: Kriegsfeld@berkeley.edu

A.S. Kauffman and J.T. Smith (eds.), *Kisspeptin Signaling in Reproductive Biology*, Advances in Experimental Medicine and Biology 784, DOI 10.1007/978-1-4614-6199-9_18,

increased spontaneous abortion rate [2], suggesting the importance of circadian functioning in ovulation and pregnancy maintenance. Likewise, the luteinizing hormone (LH) surge that initiates ovulation occurs in early morning in women [3] and diurnal rodents [4], but in early evening in nocturnal rodents (reviewed in [5]). In rodents, disruptions of the master circadian clock in the brain, its neural output, or the genes regulating cellular clock function lead to pronounced abnormalities in ovulation and fecundity [6–8]. Given the necessity of proper hormonal timing in female reproductive health, and the experimental tractability of the reproductive axis, the study of female reproductive functioning represents an ideal model system for understanding circadian endocrine control. As is apparent throughout this book, kisspeptin is essential for reproductive functioning and, not surprisingly, represents a critical node in the network of circadian regulation underlying female reproductive health. The present chapter provides a broad overview of circadian control of female reproduction, underscoring the significance of kisspeptin signaling within this framework.

The Circadian Timing System

Four decades ago, two studies provided strong evidence that the suprachiasmatic nucleus (SCN) of the hypothalamus was the locus of the master circadian pacemaker in mammals [9, 10]. In these initial studies, electrolytic lesions of the SCN abolished rhythms in locomotor and drinking behavior and adrenal glucocorticoids, suggesting that either the circadian clock is localized to the SCN, that the SCN is part of a larger network responsible for circadian rhythm generation, or that fibers of passage required for circadian functioning traveled through this neural locus. Numerous converging lines of evidence since these initial investigations, from a host of laboratories, have confirmed the role of the SCN as a master pacemaker. For example, transplants of donor SCN tissue into the brains of arrhythmic, SCN-lesioned hosts restore circadian rhythmicity in behavior [11, 12]. Importantly, rhythms are restored with the period of the donor SCN, indicating that the transplanted tissue does not act by restoring host-brain function but that the “clock” is contained in the transplanted tissue. Furthermore, circadian rhythms in neural firing rate persist in isolated SCN tissue maintained in culture [13], demonstrating that input from extra-SCN brain sites is not necessary for circadian rhythm generation. Although circadian rhythms are endogenously generated, in order to be adaptive for an organism, these rhythms must be synchronized to the external environment. This entrainment is accomplished via direct neural projections from intrinsically photosensitive retinal ganglion cells to the circadian clock in the SCN [14–19].

At the cellular level, circadian rhythms are generated by 24-h autoregulatory transcriptional/translational feedback loops consisting of “clock” genes and their protein products (Fig. 18.1) [20–23]. In mammals, the feedback loop begins in the cell nucleus where CLOCK and BMAL1 proteins heterodimerize and drive the transcription of the Period (*Per1* and *Per2*) and Cryptochrome (*Cry1* and *Cry2*) genes

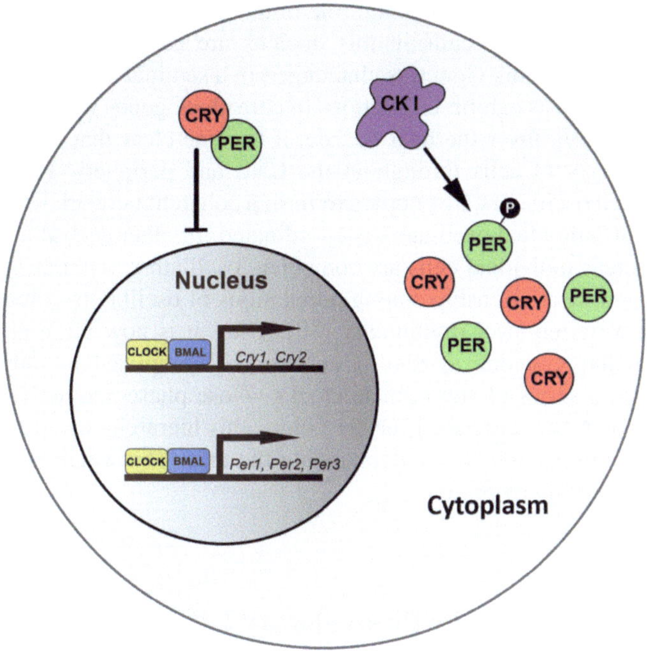

Fig. 18.1 A simplified model of the intracellular mechanisms responsible for mammalian circadian rhythm generation. The process begins when CLOCK and BMAL1 proteins dimerize to drive the transcription of the Per (*Per1* and *Per2*) and Cry (*Cry1* and *Cry2*) genes. In turn, Per and Cry are translocated to the cytoplasm and translated into their respective proteins. Throughout the day, PER and CRY proteins rise within the cell cytoplasm. When levels of PER and CRY reach a threshold, they form heterodimers, feed back to the cell nucleus, and negatively regulate CLOCK:BMAL1-mediated transcription of their own genes. This feedback loop takes approximately 24 h, thereby leading to an intracellular circadian rhythm. See text for additional details

by binding to the E-box (CACGTG) domain on their gene promoters. Once translated, PER and CRY proteins build in the cytoplasm of the cell over the course of the day, and inevitably form hetero- and homodimers that feed back to the cell nucleus to inhibit CLOCK:BMAL1-mediated transcription. The timing of nuclear entry is balanced by regulatory kinases that phosphorylate the PER and CRY proteins, leading to their degradation [21, 24, 25]. Two other promoter elements, DBP/E4BP4 binding elements (D boxes) and REV-ERBα/ROR binding elements (RREs) [26], also participate in cellular clock function. REV-ERBα, an orphan nuclear receptor, negatively regulates the activity of the CLOCK:BMAL1. The same mechanism controlling *Per* and *Cry* gene transcription also controls transcription of REV-ERBα. Similarly, the transcription factor DPB is positively regulated by the CLOCK:BMAL1 complex [27] and acts as an important output mechanism, driving rhythmic transcription of other output genes via a PAR basic leucine zipper (PAR bZIP) [28].

Historically, given the fundamental role of the SCN in maintaining rhythmicity, it was believed that only cells in this nucleus are capable of intrinsic rhythm generation. However, this view was challenged in a seminal study establishing that cultured rat fibroblasts exhibit oscillations in core clock genes when presented with a serum shock [29]. Over the next decade, it became clear that oscillations are a pervasive property of cells throughout the CNS and periphery. However, in the SCN, individual oscillators are coupled to form a coherent network where the phase relationship of individual oscillators is coordinated. In other central and peripheral systems, whereas individual cells are competent oscillators, without SCN communication, the phase relationship among populations of oscillators is lost, leading to an abolition of tissue-level rhythmicity [30]. Thus, it is now more appropriate to conceptualize the "circadian system" as an assembly comprised not only of a master clock, but also a series of subordinate clocks whose phase and activity is coordinated by the SCN. As described further below, this hierarchy of circadian control has important implications for understanding the means by which kisspeptin initiates the LH surge and ovulation.

Circadian Control of the Preovulatory LH Surge and Ovulation

Discovery of Ovulatory Circadian Control

The neural mechanisms regulating ovulation are under circadian control in many spontaneously ovulating species, ensuring that the timing of maximal fertility is concomitant with the period of highest sexual motivation [31, 32]. Superimposed upon this circadian control is a dependence of the reproductive cycle on estradiol to ensure proper maturation of the oocyte at the time of ovulation. The notion that rodent ovulation is under circadian control was first proposed based on data from a classic study by Everett and Sawyer [33]. In this seminal paper, Everett and Sawyer sought to determine if ovulation required "neurogenic" activation of the reproductive axis. To explore this possibility, rats were injected with barbiturate to inhibit neural communication at the time of ovulation. Through a series of studies, they determined that a single barbiturate injection delayed ovulation by 24 h, despite the fact that the impact of the drug was short-lived. Based on this observation, the authors proposed that a 24 h neural signal initiates ovulation.

Over two decades following the work of Everett and Sawyer, Zucker and colleagues performed an elegant series of studies providing further evidence for a 24 h clock in ovulation and associated sexual motivation in Syrian hamsters (*Mesocricetus auratus*). When held in a light:dark (LD) cycle, ovulation and the onset of behavioral receptivity occur precisely every 96 h in this species [34]. This rhythm in reproductive activity is endogenously generated and persists in constant conditions with a period four times their free-running circadian period. Exposing hamsters housed in constant darkness (DD) to deuterium oxide (a treatment that results in a lengthening of the free-running period), resulted in a lengthening of the ovulatory cycle to precisely four times the period of an individual's circadian rhythm, establishing that the ovulatory cycle and locomotor rhythms are governed by a comparable endogenous timing system [35]. Because estrous and activity onset were coupled temporally, it was suggested that the LH surge and locomotor activity are controlled by either a single, endogenous oscillator or a coupled, multioscillator system that regulates the rhythms of each process independently [35]. The former hypothesis postulated that the reproductive axis "tracks" four circadian cycles and ovulation occurs after the count is complete.

Converging lines of evidence over the next three decades established that both of these hypotheses are partially correct. We now know that the SCN provides a daily, stimulatory signal to the reproductive axis each day of the estrous cycle, closely preceding the active phase, in most spontaneously ovulating rodents [5, 36, 37], indicating that a single clock subserves both processes. However, this signal is only effective at stimulating the GnRH system to produce the LH surge in the presence of estradiol concentrations above a critical threshold. Prior to the day of proestrus, the developing ovarian follicles secrete insufficient estradiol to fulfill this criterion. The nature of the daily stimulatory signal from the SCN can be unmasked by implanting animals with estradiol capsules that result in proestrus concentrations of this hormone; in this case, daily LH surges occur [37–39]. Regarding the second hypothesis suggesting a multioscillator organization, although distinct clocks do not underlie locomotor rhythms and estrus, a hierarchical clock structure exists in which the SCN acts as the master pacemaker coordinating rhythmicity in subordinate oscillator systems of the reproductive axis, an arrangement discussed further below.

Sex Steroid and Circadian Integration in Ovulatory Control

Through negative-feedback effects of sex steroids, LH is maintained at low concentrations throughout most of the ovulatory cycle, paradoxically, high concentrations of estradiol are required for the SCN to trigger ovulation (i.e., positive feedback) [5, 36, 37, 40]. The site(s) of integration for positive and negative-feedback effects of estradiol with the circadian timing system are complex and not fully understood. As described below, current evidence indicates that one important site of integration for the positive feedback effects of estradiol with circadian signaling is the kisspeptin network in the anteroventral periventricular nucleus (AVPV).

Circadian Neurochemical Communication and Ovulation

Despite the fact that the SCN is crucial for both locomotor behavior and the initiation of ovulation, each process is likely mediated by distinct communication modalities. In arrhythmic, SCN-lesioned hamsters, fetal SCN transplants restore locomotor, but not endocrine, rhythms [41–43]. Restoration of locomotor and other behavioral rhythms occurs in the absence of neural outgrowth from the grafted SCN, suggesting that intact neural connections are required for endocrine rhythms, whereas behavioral rhythms can be supported by a diffusible signal. Perhaps the best evidence indicating that the SCN communicates via neural connections to the GnRH system to initiate the LH surge comes from studies using hamsters with "split" activity rhythms [44]. When housed in constant light, some hamsters exhibit a splitting in behavior, with two daily activity bouts separated by 12 h, each reflecting an antiphase oscillation of the left and right sides of the bilateral SCN [44–46]. Under these circumstances, ovariectomized (OVX) hamsters treated with estradiol exhibit two LH surges in a 24-h period, each phase-locked to an individual activity bout [47]. Because the SCN communicates principally ipsilaterally [48–51], the authors hypothesized that, if a neural output signal from the SCN initiates the GnRH/LH surge, then one hemispheric set of GnRH neurons should be activated, ipsilateral to the activated SCN, with each locomotor activity bout. Conversely, if controlled by a diffusible signal, then the GnRH system should be activated concurrently on both sides of the brain, twice daily, 12 h apart [52]. Under these split conditions, the former possibility manifested, confirming the importance of neural SCN communication to the GnRH system in ovulatory control. As described below, more recent studies by this group and others establish that SCN communication to kisspeptin cells in the AVPV underlies, in part, this neural signaling cascade initiating the GnRH/LH surge [53–56].

Direct SCN Signaling to the GnRH System

The SCN communicates both monosynaptically and multisynaptically to the GnRH system. Whereas much progress has been made in understanding both modes of communication, most studies to date have investigated the contribution of individual pathways in isolation, making it difficult to ascertain the significance of interactions between these pathways. One SCN neuropeptide that has received considerable attention as a modulator of ovulation is vasoactive intestinal polypeptide (VIP). Neurons synthesizing VIP are located in the retinorecipient, ventrolateral SCN "core" [57–59], and project monsynaptically to GnRH neurons [49, 60] that express the VIP receptor $VPAC_2$ [61]. SCN-derived VIP input to GnRH neurons is sexually dimorphic, with female rats exhibiting higher VIPergic innervation [60], suggesting a critical role in female reproductive functioning. Likewise, the number of VIP-GnRH contacts increases during the post-pubertal transition to reproductive competence [62], a time when estradiol positive feedback is first established. VIP-innervated GnRH neurons exhibit lower activation levels in middle-aged female rats, suggesting that the degradation of this signaling cascade participates in the transition to reproductive senescence in female rodents [63]. In reproductively competent females, GnRH neurons receiving VIPergic input preferentially express the neural activation marker, FOS, at the time of the LH surge [64]. In vivo antisense antagonism of VIP production in the SCN abolishes GnRH/FOS activation in female rats, providing further support for the necessity of VIP output in surge generation [65, 66]. Finally, blocking the $VPAC_2$ receptor attenuates GnRH neuronal cell firing during the afternoon surge in female, estradiol-treated mice [67]. Together, these lines of evidence suggest that direct VIP projections from the SCN to the GnRH system positively drive the GnRH/LH surge. The potential means by which this pathway synergizes with multisynaptic projections from the SCN to the reproductive axis is discussed further below.

Indirect SCN Signaling to the GnRH System

Historically, it was thought that estradiol positive feedback occurred through the convergence of estrogenic and circadian signaling at the level of GnRH neurons. The observation that GnRH neurons do not express estrogen receptor α (ERα), the estrogen receptor subtype mediating the positive feedback effects of estradiol [68–70], motivated the search for additional neural loci at which stimulatory circadian and estrogenic signals converge. The AVPV emerged as a likely site as neurons in this brain region send monosynaptic projections to GnRH cells, express FOS coincident with the LH surge, and lesions of the AVPV eliminate estrous cyclicity in both intact and OVX, estradiol-treated rats [8, 71–74]. Moreover, the SCN sends pronounced monosynaptic projections to cells in the AVPV that express ERα [68, 75–77], leading to the search for the SCN neurochemical cell phenotypes sending projections to the AVPV.

Vasopressinergic (AVPergic) cells in the dorsomedial SCN target ERα—expressing cells in the AVPV [50, 75, 76, 78–80], and AVP injections produce surge-like LH levels in SCN-lesioned, OVX, estradiol-treated rats [81]. Likewise, cells in this brain region express the vasopressin receptor, $V1_a$ [82, 83]. Antiestrogens targeting the AVPV inhibit the LH surge in OVX, estradiol-treated rats [84], confirming the importance of estrogen signaling in ovulation. Vasopressin gene transcription in the SCN is directly controlled by the molecular clockwork at the cellular level [85–87] and is released in a circadian manner [88], with a peak coinciding with the onset of the LH surge [89, 90]. AVP release is synchronous with GnRH secretion in cocultures of medial preoptic area (mPOA) and SCN brains slices [91]. By contrast, central AVP receptor antagonists attenuate the LH surge in proestrous rats [92]. Finally, the inability of *clock* mutant mice to generate an LH surge is associated with diminished *AVP* mRNA expression in the SCN, a phenotype that can be restored via central injections of AVP, further linking this peptide to the circadian control of ovulation [6].

In rodents, *Kiss1* mRNA expressing cells in the AVPV and arcuate (ARC) nuclei exhibit robust ERα labeling [93–97]. The effects of estradiol on kisspeptin activity, however, varies by nucleus, with ovariectomy decreasing *Kiss1* mRNA in the AVPV and increasing *Kiss1* expression in the ARC, pointing to a role for kisspeptin in estradiol positive and negative feedback, respectively [94, 98]. Exogenous kisspeptin administration potently induces LH release as well as upregulates FOS expression in GnRH neurons [93, 99–102]. One recent study used genetic ablation strategies to explore the importance of kisspeptin in puberty and adult reproductive functioning. Genetic ablation of kisspeptin cells or cells expressing GPR54 throughout development does not impact female puberty onset or fertility in adult animals, although loss of GPR54-expressing cells results in blunted LH, reduced ovarian weights, and irregular estrous cycling [103]. In adult animals, acute ablation of kisspeptin neurons markedly disrupts fertility and estrous cyclicity, whereas removal of ~93% of GPR54-expressing GnRH cells results in more mild reductions in LH, fertility, and estrous cycling [103]. Together, these findings suggest that kisspeptin signaling is required for adult female reproductive functioning and compensatory mechanisms can overcome the necessity for kisspeptin when this gene is inactivated throughout development. However, given that as few as three GnRH neurons are sufficient to support activity of the HPG axis [104, 105], it is unclear whether GPR54-expressing GnRH neurons can be dispensed with in LH surge control if a few GPR54-expressing cell remain following the ablation.

The observation across rodent species that the SCN projects to the AVPV, and that this brain region is essential for production of the LH surge, combined with the knowledge that AVPV kisspeptin cells respond positively to estradiol, made these cells an attractive target of exploration in the initiation of GnRH/LH surge. An early study showing that *Kiss1* cells in the AVPV express FOS at the time of the LH surge in naturally cycling, OVX, estradiol-treated rats [106, 107], provided strong support for this possibility. To explore the circadian control of this activation pattern, Kauffman and colleagues maintained mice in constant conditions and examined *Kiss1* mRNA and the percentage of *Kiss1* cells expressing the *c-fos* gene (Fig. 18.2). The maintenance of a circadian pattern in the absence of environmental cues would suggest endogenous rhythmic control through a circadian mechanism rather than a rhythm driven by environmental time cues. The authors established that the circadian pattern of *Kiss1* expression and the percentage of *Kiss1/c-fos* cells persist in constant darkness [56], with peak expression of both measures coordinated with the LH surge, suggesting endogenous circadian regulation of this cell population rather than reliance on external temporal cues. This daily pattern of *Kiss1* expression, and *Kiss1* cells expressing *c-fos*, is abolished by ovariectomy, and reinstated following steady-state estradiol replacement [56], indicating a permissive role for estradiol in the circadian control of *Kiss1* in this species.

These results point to either an endogenous, self-sustained rhythm in kisspeptin neurons, circadian control through upstream projections from the SCN, or a combination of both mechanisms of control. We examined these possibilities in Syrian hamsters. Consistent with findings in mice, FOS expression in kisspeptin immunoreactive (ir) cells expressed a daily rhythm in OVX, estradiol-treated hamsters, with peak coexpression concomitant with the timing of the LH surge. In contrast to results observed in mice, ovariectomy results in a blunted rhythm of kisspeptin-FOS coexpression, but not its abolition [54]. These latter findings point to potential species differences in the role that estrogen plays in AVPV kisspeptin regulation and/or posttranscriptional modification of the *Kiss1* gene, resulting in differences in cells

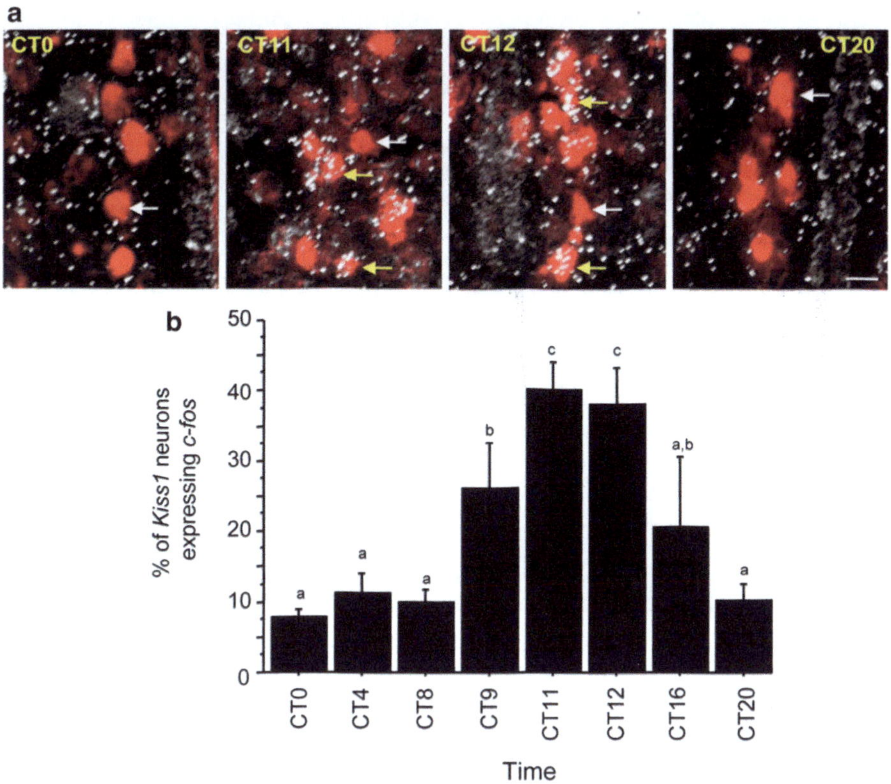

Fig. 18.2 (**a**) Representative photomicrographs of Kiss1 mRNA and c-fos mRNA coexpression in the AVPV of OVX, E_2-treated female mice housed in constant conditions and killed at different times throughout the circadian day. Kiss1-containing neurons were visualized with Vector Red substrate, and c-fos mRNA was marked by the presence of silver grains. *White arrows* denote example Kiss1 cells lacking c-fos; *yellow arrows* denote example Kiss1 cells coexpressing c-fos. (**b**) Mean (±SEM) percentage of Kiss1 mRNA-containing neurons in the AVPV that coexpress c-fos in OVX, E_2-treated female mice killed at one of eight times throughout the circadian day. There was a significant effect of time ($P<0.01$) with increased coexpression of Kiss1 and c-fos in the late afternoon/early evening. Values with different letters differ significantly from each other. $n=4$–6 animals per group. From Robertson JL, Clifton DK, de la Iglesia HO, Steiner RA, Kauffman AS. Circadian regulation of Kiss1 neurons: implications for timing the preovulatory gonadotropin-releasing hormone/luteinizing hormone surge. Endocrinology. 2009;150(8):3664–71. Reprinted with permission from The Endocrine Society

visualized with mRNA versus protein analyses. To determine whether the SCN projects to kisspeptin cells to mediate these observed rhythms, we examined projections from VIPergic and AVPergic SCN cells, given the role of these neuropeptides in positively driving the LH surge. We found that AVPergic SCN cells project directly to a majority of kisspeptin-ir cells, whereas VIPergic SCN cells did not (Fig. 18.3). In mice, AVPergic projections to AVPV kisspeptin cells have also been identified, with synapses confirmed at the electron microscopy level, suggesting

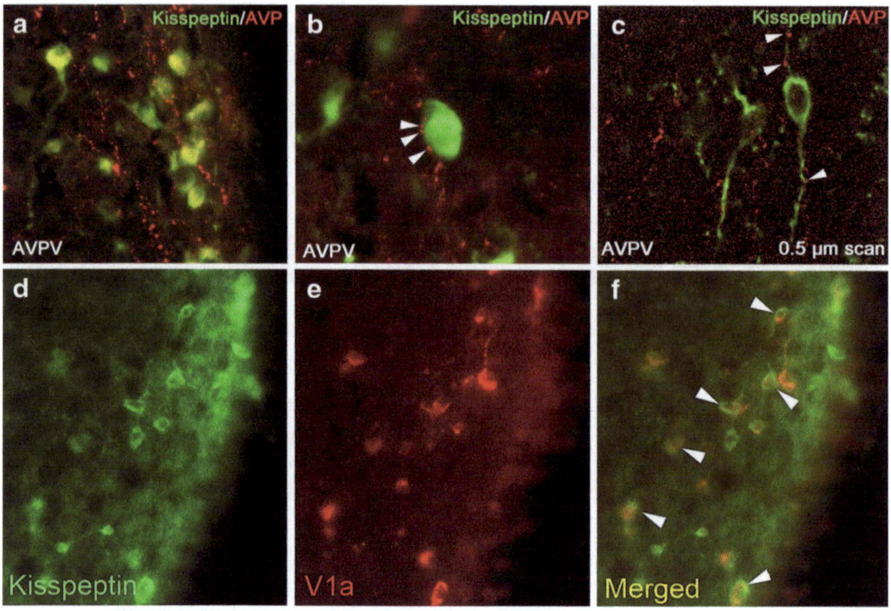

Fig. 18.3 Kisspeptin-ir cells in the hamster AVPV receive SCN-derived fiber contacts expressing AVP-ir. (**a**) Low-power photomicrographs of AVP-ir in the AVPV, in which kisspeptin cell bodies receive extensive AVP-ir fiber contacts. (**b**) High-power photomicrograph showing several presumptive AVP-ir terminal boutons on a kisspeptin-ir cell body at the light level. (**c**) Confocal image (0.5 μm scan taken at ×400) confirming AVP-ir contacts upon kisspeptin-ir cell body and processes. In (**b**, **c**), *arrows* are indicative of close contacts. (**d**–**f**) Kisspeptin cells in the AVPV express the V1a receptor. (**d**) Low-power photomicrographs of kisspeptin-ir cells in the AVPV, (**e**) V1a-ir cells in the AVPV, and (**f**) the merged image showing overlap between kisspeptin-ir and V1a-ir. From Williams WP, 3rd, Jarjisian SG, Mikkelsen JD, Kriegsfeld LJ. Circadian control of kisspeptin and a gated GnRH response mediate the preovulatory luteinizing hormone surge. Endocrinology. 2011;152(2):595–606. Epub 2010/12/31. Reprinted with permission from The Endocrine Society

a common mechanism of circadian control across species [108]. Additionally, these authors established that estrogen increases the percent of GnRH cells with AVPergic terminal appositions, providing an additional mechanism by which estrogen integrates with circadian signaling to stimulate kisspeptin neurons at the time of ovulation [108].

More recently a creative series of studies by de la Iglesia and colleagues further established the specific means by which the SCN regulates *Kiss1* expression and the LH surge [53]. First, to determine if the SCN communicates ipsilaterally to *Kiss1-expressing* neurons in the AVPV, and to assess the significance of this communication for GnRH cell activation at the time of the surge, the authors lesioned one half of this nucleus and examined AVPV *Kiss1* expression and FOS expression in GnRH cells in female rats (Fig. 18.4). Both *Kiss1* expression and the percentage of GnRH cells expressing FOS were reduced on the side of the brain ipsilateral to the lesion,

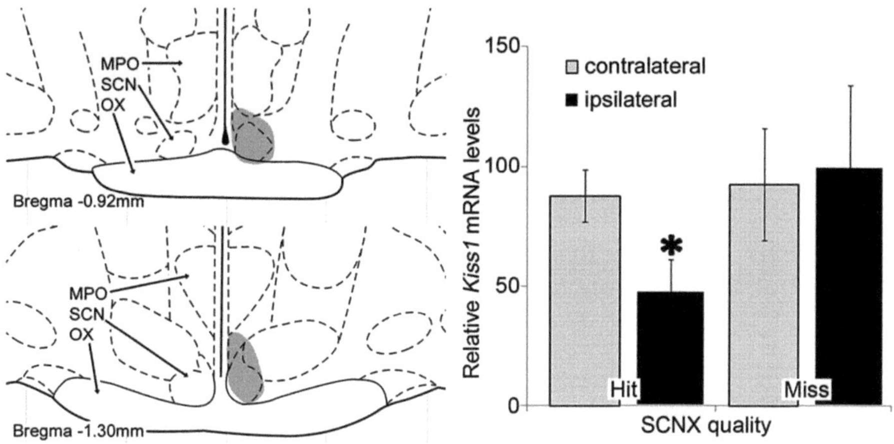

Fig. 18.4 Ipsilateral SCN projections gate the daily activation of GnRH cells and the AVPV Kiss1 expression rhythms in OVX+E_2 rats. (*Left*) Schematic coronal section showing the lesion location at two rostrocaudal levels from an animal that received a successful unilateral SCN lesion (*shaded*). MPO, SCN, and optic chiasm (OX) are indicated. (*Right*) Kiss1 expression is significantly lower on the ipsilateral than on the contralateral side of the lesion in animals with successful unilateral SCN lesions but not in animals in which the unilateral lesion missed the SCN. Bars, mean ± SEM. *Asterisk*, statistically different ($n=5$ successful lesions, paired Student t test, $P=0.016$; $n=7$ missed lesions, paired Student t test, $P=0.33$). From Horvath TL, Cela V, van der Beek EM. Gender-specific apposition between vasoactive intestinal peptide-containing axons and gonadotrophin-releasing hormone-producing neurons in the rat. Brain Res. 1998;795(1–2):277–281, with permission from Elsevier Limited. And from Smarr BL, Morris E, de la Iglesia HO. The Dorsomedial Suprachiasmatic Nucleus Times Circadian Expression of Kiss1 and the Luteinizing Hormone Surge. Endocrinology. 2012. Epub 2012/03/29, with permission from The Endocrine Society

suggesting that the SCN communicates ipsilaterally to AVPV kisspeptin cells and that this stimulation is necessary for activation of the GnRH system. By using a lighting regimen previously established by this group to desynchronize rhythms in the dorsomedial and ventrolateral SCN [109], the authors show that the dorsomedial SCN oscillates independently of the LD cycle and that both AVPV *Kiss1* expression and the LH surge are coupled to oscillations in this SCN subregion (Fig. 18.5) [53]. Together with previous data supporting a role for AVPergic circadian control of AVPV kisspeptin neurons [54–56, 110], these data provide strong evidence for an important role of this indirect SCN-kisspeptin circuit in initiating the GnRH/LH surge to stimulate ovulation.

Previous findings indicate that that administration of AVP can only induce the preovulatory LH surge within a narrow time window [111], suggesting a gated mechanism of control at SCN target loci. The gating of SCN information flow may be controlled within kisspeptin cells in the AVPV, at the level of GnRH neurons, or a combination of both mechanisms. To select among these possibilities, we examined whether (1) kisspeptin cells within the AVPV respond in a time-dependent

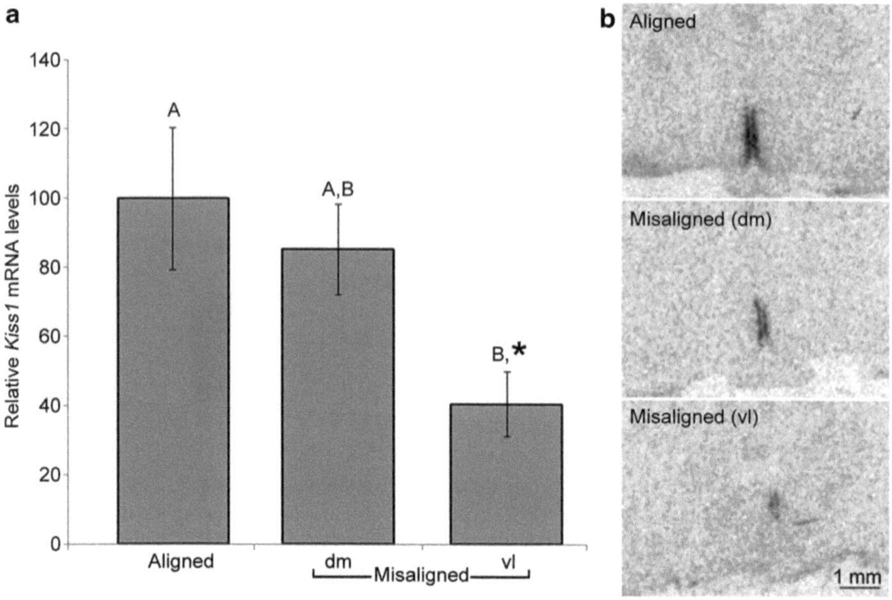

Fig. 18.5 Circadian Kiss1 expression is coupled to an oscillator within the dorsomedial SCN. (**a**) Kiss1 expression on misaligned days 1 h before dmSCN-associated activity onset, but not 1 h before the vlSCN-associated locomotor activity onset, is similar to that in aligned animals before the locomotor activity onset [one way ANOVA ($P=0.05$)]. Bars, mean ± SEM; groups not sharing the same capital letter are statistically significant according to Tukey post hoc comparisons ($P=0.05$). *Asterisk*, significantly different from aligned and misaligned taken together, planned comparison, $P=0.02$. (**b**) Representative Kiss1 ISH autoradiographic films from animals killed on aligned days just before locomotor activity onset and on misaligned days just before dmSCN-associated activity onset (dm) or just before the vlSCN-associated activity onset (vl). From Smarr BL, Morris E, de la Iglesia HO. The Dorsomedial Suprachiasmatic Nucleus Times Circadian Expression of Kiss1 and the Luteinizing Hormone Surge. Endocrinology. 2012. Epub 2012/03/29. Reprinted with permission from The Endocrine Society

manner to AVP stimulation and (2) if GnRH neurons display time-dependent sensitivity to kisspeptin signaling. If time-dependent sensitivity is controlled at the level of the AVPV, then one would expect kisspeptin cells to exhibit daily changes in sensitivity to AVP stimulation and contain an endogenous time-keeping mechanism. Alternatively, if the gating of control occurs within GnRH cells, then one would expect the GnRH system to display daily sensitivity in response to both AVP and kisspeptin administration. Our findings indicate that the kisspeptin system responds indiscriminately to AVP administration, regardless of time of day, whereas the GnRH system is only sensitive to kisspeptin stimulation at the time that the surge would normally occur (Fig. 18.6) [54]. These results further support the notion that kisspeptin cells do not keep circadian time but, instead, that their activity is driven by AVPergic SCN cells. However, these findings point to an important role for autonomous circadian oscillators in GnRH cells underlying time-dependent

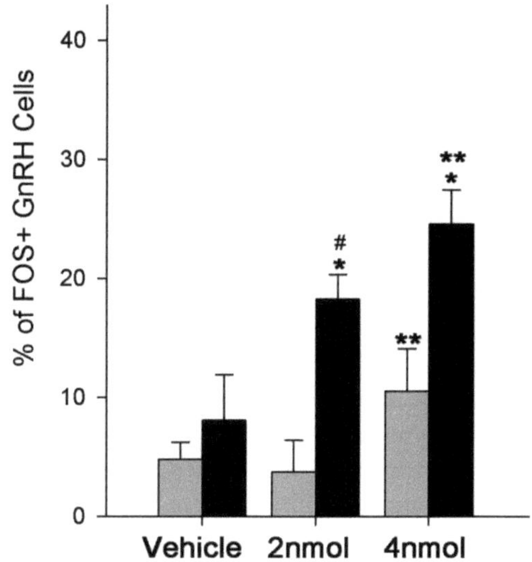

Fig. 18.6 The activation of GnRH after kisspeptin administration is time dependent. The percentage of GnRH-ir cells expressing FOS after vehicle, 2 nmol, or 4 nmol kisspeptin administration at zietgerber time (ZT) 1 or ZT 11 in OVX hamsters. Mean (±SEM) percentage of POA GnRH-ir cells expressing FOS after kisspeptin administration at ZT 1 or ZT 11. From Williams WP, 3rd, Jarjisian SG, Mikkelsen JD, Kriegsfeld LJ. Circadian control of kisspeptin and a gated GnRH response mediate the preovulatory luteinizing hormone surge. Endocrinology. 2011;152(2):595–606. Epub 2010/12/31. Reprinted with permission from The Endocrine Society

sensitivity to upstream signaling. Alternatively, it is possible that the master clock in the SCN communicates timing information to GnRH cells that do not maintain the capacity for endogenous rhythmicity. Given that SCN-derived VIPergic cells project monosynaptically to GnRH neurons, this cell phenotype represents an ideal candidate to communicate such timing information. Finally, a combination of both mechanisms may underlie such daily changes in GnRH cell sensitivity, with VIPergic SCN communication synchronizing independent GnRH cellular oscillators to coordinate the timing in their responsiveness to upstream signaling.

Clock Genes and Kisspeptin Control of Ovulation

The same clock genes that drive circadian rhythms at the cellular level in the SCN are also expressed throughout other brain regions and peripheral tissues. GnRH cells express clock genes, both in vitro [112, 113] and in vivo [114, 115]. Mice with a mutation in the essential circadian clock gene, *Clock*, display abnormal estrous cycles and abnormal LH surge induction in response AVP administration [6], suggesting an important role for clock genes (potentially in GnRH cells) in normal estrous cycling and ovulation. Studies in which the rescue of *Clock* is restricted to the SCN of these mice would help to clarify whether circadian disruption at the level of the SCN or GnRH neurons in responsible for the deficits observed in these animals. Consistent with our findings in vivo, immortalized GnRH neurons (i.e., GT1-7 cells) exhibit ~24 h changes in sensitivity to kisspeptin and VIP signaling [113]. These findings suggest that an endogenous timing mechanism in GnRH cells gates daily changes in responsiveness to upstream, stimulatory neurochemicals, at least under these culture conditions. More recently, Chappell and colleagues found that estrogen-treated GT1-7 cells exhibit a rhythm in *GPR54* expression that is abolished in the absence of estrogen [116]. These results are intriguing given the absence of ERα in this cell population, suggesting participation of ERβ in this rhythm of *GRP54* transcription. Together with our findings in vivo [54], these findings indicate that GnRH cells maintains the ability to oscillate independently, potentially as a mechanism gating responsiveness to signals initiating ovulation. Whether or not SCN communication to the GnRH system is necessary to maintain phase coherence among this population of cells represents an important question for future enquiry.

Circadian Control of Estrogen Negative Feedback

Most work to date on the circadian control of ovulation has focused on estrogen positive feedback. Few studies have explored the neurochemical substrates and target loci at which circadian and estrogenic signals converge to maintain LH at low concentrations throughout the majority of the estrous cycle (i.e., negative feedback), and the mechanisms that suppress negative feedback at the time of the LH surge. Because estrogen suppresses kisspeptin activity in the Arc [94, 98], this population of cells is thought to be an important component of estradiol negative feedback. Whether the SCN projects to kisspeptin cells in the Arc has not been explored and represents an important potential mechanism by which suppression of positive drive to the reproductive axis might be coordinated by the circadian system prior to the LH surge.

We explored whether a neuropeptide in the same family as kisspeptin, gonadotropin-inhibitory hormone (GnIH; also known as RFamide-related peptide-3), participates in circadian-controlled estradiol negative feedback. In 2000, Tsutsui and colleagues isolated a dodecapeptide from Japanese quail brain that dose-dependently inhibits gonadotropin release from cultured quail pituitary that they named GnIH [117]. GnIH acts as a pronounced negative regulator of HPG axis activity in all species investigated, including hamsters, mice, rats, cattle, sheep, nonhuman primates and humans (reviewed in [118–121]). Recently, one manuscript showed that GnIH leads to increased LH secretion in male hamsters, suggesting that it is important to consider sex and physiological state when investigating the impact of this neuropeptide [122]. In rodents, GnIH cell bodies are localized to the dorsomedial hypothalamus (DMH) with projections forming close appositions to GnRH cells [123, 124], suggesting the potential for direct, neural regulation of the GnRH network. Additionally, recent work in mice and Siberian hamsters indicates that a marked number of GnRH neurons express GnIH receptor [125, 126]. Whether or not endogenous GnIH acts on the pituitary, in addition to GnRH cells, in mammalian species requires further investigation to clarify discrepant findings across studies and species [127–131].

We first examined the possibility that GnIH participates in mediating the negative-feedback effects of estradiol. Treatment of female rats with GnIH results in marked inhibition of GnRH neuronal activity at the time of the LH surge, providing support for this possibility [132]. In Syrian hamsters, we found that GnIH-ir cells express ERα and respond to acute estradiol treatment with increased FOS expression, suggesting activation by gonadal steroids [123]. Contrasting results were observed in one recent report, with treatment of mice with estrogen for 4 days leading to a decrease in GnIH mRNA expression [133]. The discrepancy in the impact of estrogen in mice and hamster may result from the timing at which the brains were collected for analysis and the dose/duration of estrogen treatment. Both of these possibilities represent interesting areas of investigation that will help to clarify the specific role of this neuropeptide in female reproductive axis regulation.

We next explored whether GnIH neurons might be a locus of integration for steroidal and circadian signals, providing a mechanism to coordinate the removal of estradiol negative feedback with SCN-mediated stimulation of the GnRH/LH surge. First, we examined the pattern of GnIH cellular activity, uncovering a daily pattern with trough activity at the time of the LH surge, suggesting the removal of negative feedback at this time [127]. Additionally, using anterograde tract tracing, we found that the SCN projects to a large proportion of GnIH cells, providing a mechanism for timing removal of negative drive on the GnRH system (Fig. 18.7). Finally, by exploiting the "splitting" phenomenon seen in hamsters house in LL described previously, we found that activation of the GnIH system is asymmetrical. Importantly, this asymmetry is opposite to that seen for the GnRH system, suggesting that the SCN concomitantly activates ipsilateral GnRH cells at the same time as removing the suppressive influence of GnIH on the same side of the brain (Fig. 18.7) [127]. Whether or not removal of negative feedback is required for the LH surge requires further empirical investigation. A recent series of studies by Pineda et al. using a selective antagonist (RF9) of GnIH provides converging evidence for a role of GnIH in the LH surge, with injections of RF9 increasing LH throughout the ovulatory cycle [134]. These findings are consistent with a role for GnIH in maintaining low LH concentration throughout the ovulatory cycle. Together, these findings and those for kisspeptin, point to an important role for these neuropeptides in the integration of positive and negative effects of estradiol with circadian signaling in the generation of the GnRH/LH surge.

Circadian Implications for Reproductive Aging

Age-related decline in reproductive axis function is common across species, with initiation of this decline typically occurring midway through life. In female rodents, this waning in reproductive function occurs, at least in part, from a reduction in the ability of the SCN to stimulate the neural circuits underlying ovulatory functioning. In middle-aged hamsters, for example, the peak level of LH is delayed and exhibits

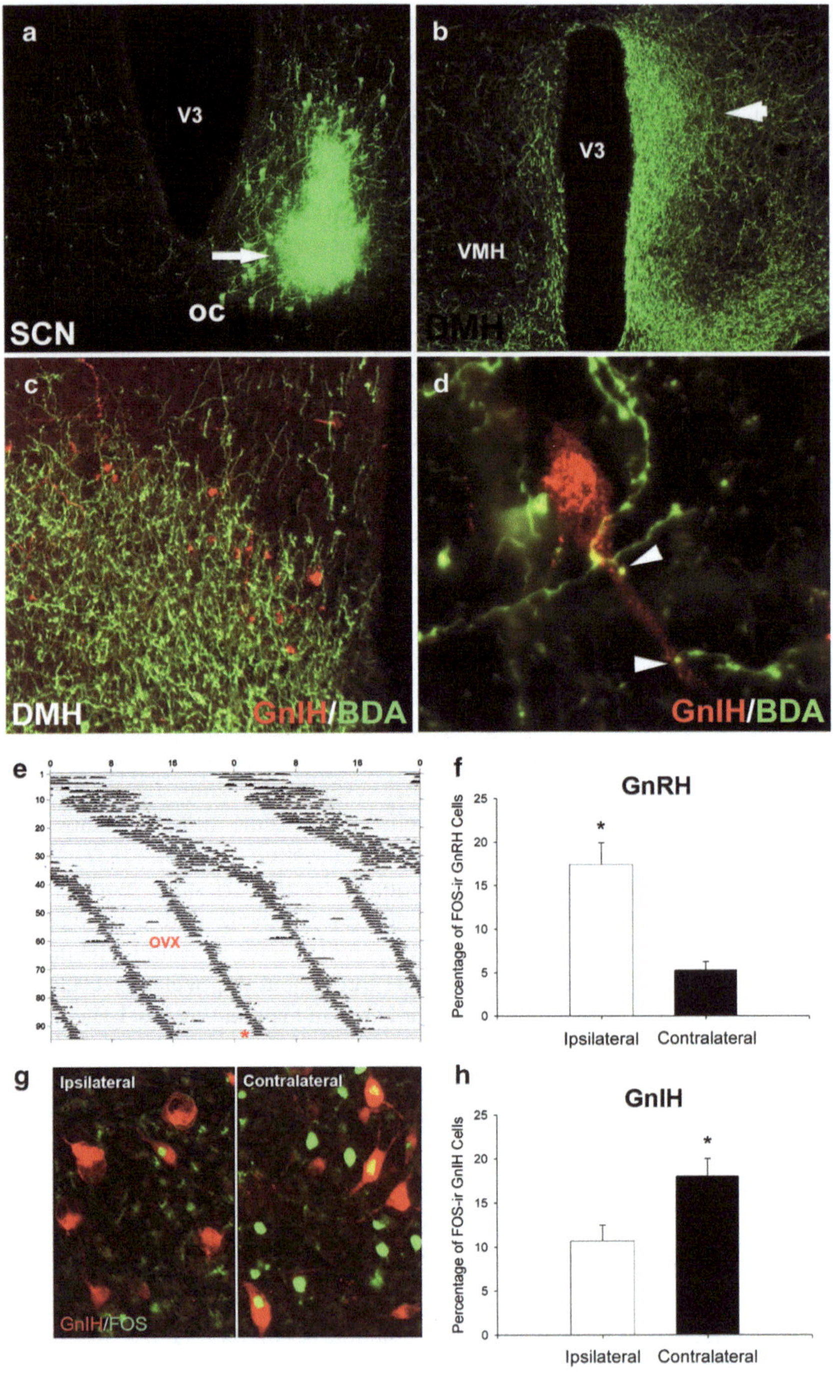
a
V3
oc
SCN
b
V3
VMH
DMH
c
DMH
GnIH/BDA
d
GnIH/BDA
e
OVX
f
GnRH
Percentage of FOS-ir GnRH Cells
Ipsilateral
Contralateral
g
Ipsilateral
Contralateral
GnIH/FOS
h
GnIH
Percentage of FOS-ir GnIH Cells
Ipsilateral
Contralateral

a reduced amplitude relative to young, reproductively healthy animals [135]. These changes are associated with deficits in GnRH cell activation. In regularly cycling young females, 34–40% of GnRH neurons express immediate early genes on the afternoon of proestrus compared to 9–14% in middle-aged animals; the total number of GnRH-immunoreactive neurons does not differ between the two groups [136]. This finding suggests that the SCN may not provide adequate stimulation to the GnRH system in aged animals. In support of this possibility, VIP mRNA, but not AVP mRNA, becomes arrhythmic in the SCN of middle-aged female hamsters [137], and suppression of SCN VIP in young female hamsters results in accelerated deficits in GnRH activation and an LH surge that mimics that of an aged population [65, 66]. Together, these findings suggest that age-related deficits in the neuroendocrine mechanisms mediating ovulation may result from loss of function at the level of the SCN.

Whether or not alterations in circadian signaling to the kisspeptin system contribute to reductions in reproductive competence with age has not been directly explored. However, two recent studies provide suggestive evidence for this possibility [138, 139]. In middle-aged female rats, *Kiss1* mRNA and protein are reduced at the time of estradiol positive feedback relative to young animals, whereas kisspeptin injections into the preoptic area rescue LH release and surge amplitude [138, 139]. These findings are consistent with the notion that reduced circadian drive to kisspeptin neurons participates in aging-associated reductions in LH surge function. Alternatively, it is possible that deficits in estradiol signaling to AVPV kisspeptin cells with advancing age account for reductions in estradiol-induced kisspeptin activity. Whether either or both of these possibilities account for ovulatory decline with advancing age requires further exploration. Nonetheless, given these recent findings, it is noteworthy that kisspeptin not only represents a key neuropeptide initiating the onset of reproduction, but may also be a critical player in the transition to reproductive senescence, further pointing to a prominent role for this neuropeptide in maintaining reproductive competence.

Fig. 18.7 SCN fibers project to GnIH-ir cells in the DMH. (**a**) Example injection site from an injection of biotinylated dextran amine (BDA) that filled the ventrolateral aspect of the SCN; (**b**) low-power photomicrograph indicating terminal fibers from the SCN project to the DMH, principally ipsilaterally. (**c, d**) Examples of SCN projections in close apposition to GnIH-ir cells in the DMH at the light level ((**c**), low power; (**d**), high power). (**e–h**) Lateralization of GnRH and GnIH activation in split hamsters. The pattern of SCN activation is lateralized in animals exhibiting two daily bouts of activity. GnIH cellular activity is lateralized and opposite to that of GnRH, indicating SCN-mediated removal of GnIH inhibition at the time of the LH surge. (**e**) Actogram of wheel-running activity in estradiol-implanted, OVX hamsters kept in constant light conditions (LL) from Day 1 onwards (days indicated by *y* axis, hours by *x* axis). Split hamsters were killed (*asterisk*) 1 h before the onset of one of the two activity bouts. (**g**) Photomicrograph of FOS activation in GnIH cells of split hamsters, showing ipsilaterally reduced GnIH activation when SCN and GnRH activation are high. Mean (±SEM) percentage of FOS-ir (**f**) GnRH and (**h**) GnIH cells in split hamsters. *Asterisk*, significantly different from the opposite hemisphere's activational state, $P<0.05$. From Gibson EM, Humber SA, Jain S, Williams WP, 3rd, Zhao S, Bentley GE, et al. Alterations in RFamide-related peptide expression are coordinated with the preovulatory luteinizing hormone surge. Endocrinology. 2008;149(10):4958–69. Epub 2008/06/21. Reprinted with permission from The Endocrine Society

Conclusions and Considerations

Given the importance of circadian timing in female reproductive functioning and health, and the key role played by kisspeptin in the circadian network driving ovulation, it is critical that a full understanding of the specific mechanisms and neurochemicals pathways driving the timed production and secretion of this neuropeptide be garnered. The present overview points to the SCN as a central clockwork driving the coordinated activity of a host of positive and negative regulators of the reproductive axis (Fig. 18.8). Future studies in which circadian signaling to relevant nodes downstream of the SCN, as well as circadian timing in these targets loci,

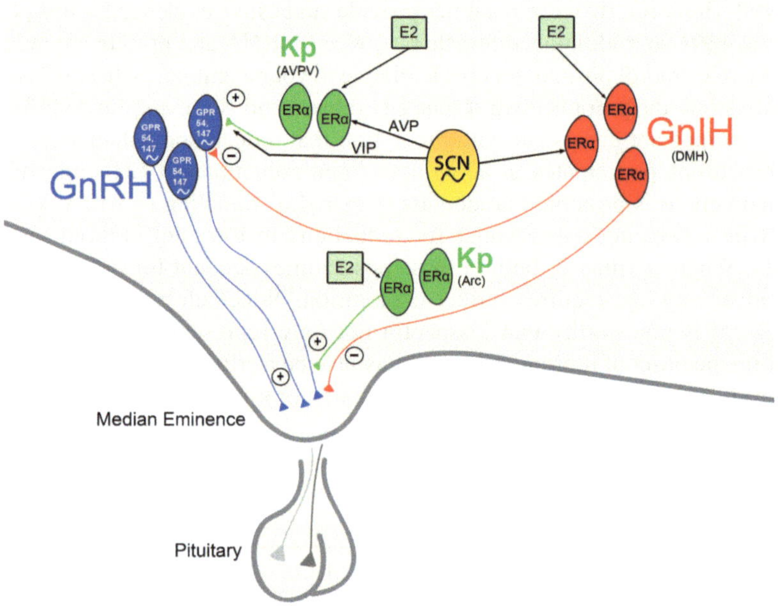

Fig. 18.8 Proposed model of circadian initiation of the preovulatory LH surge in spontaneously ovulating rodents by major positive and negative regulators of GnRH neuronal activity. Kisspeptin cells in the AVPV are active at the time of the LH surge. Whereas estrogen-responsive kisspeptin cells have not been definitively shown to project specifically to GnRH neurons, the emergence and sexual dimorphism of kisspeptin cells and fibers that project to GnRH cell bodies provide compelling evidence for the direct connection between these two neural phenotypes. Connections between the GnIH and GnRH systems indicate a putative role for GnIH in modulating the negative-feedback effects of estrogen with SCN communication allowing for removal of negative feedback on the reproductive axis during the time of the LH surge. Kisspeptin cells in the ARC likely serve to modify GnRH output at the level of the terminal. See text for additional details. Figure as originally published in Williams WP 3rd, Kriegsfeld LJ. Circadian control of neuroendocrine circuits regulating female reproductive function. Front Endocrinol (Lausanne). 2012;3:60. Epub 2012 May 21 doi: 10.3389/fendo.2012.00060

is manipulated through pharmacology and molecular/cellular biology will help to further establish, and clarify, the specific role played by this circadian timing hierarchy in female reproduction. For example, does GnRH exhibit time-dependent specificity to all modulatory neurochemicals, only kisspeptin, or only positive regulators of GnRH activity? Likewise, in addition to estrogen signaling, what other positive and negative regulators of the reproductive axis converge on kisspeptin cells to integrate with circadian signaling and precisely time the GnRH/LH surge? Recent evidence in mice suggests that at least GnIH acts directly on a subset of kisspeptin cells [125], providing a mechanism for kisspeptin cell inhibition prior to initiation of ovulation and sexual motivation. Additionally, it is unclear why the SCN projects both directly to the GnRH and kisspeptin systems to positively drive GnRH secretion when integration at a single locus is more parsimonious. Finally, most studies to date investigate GnRH secretagogues (e.g., GABA, dynorphin, glutamate) in isolation or pairs. Future studies applying multivariate connectomic, proteomic, and genomic analyses to understand the complex interactions among these systems, and their integration with circadian control, will be necessary to gain a full understanding of the complexity underlying normal female reproduction and apply this knowledge to disorders of puberty and infertility.

References

1. Ahlborg G Jr, Axelsson G, Bodin L (1996) Shift work, nitrous oxide exposure and subfertility among Swedish midwives. Int J Epidemiol 25(4):783–790
2. Nurminen T (1998) Shift work and reproductive health. Scand J Work Environ Health 24(suppl 3):28–34
3. Kerdelhue B, Brown S, Lenoir V, Queenan JT Jr, Jones GS, Scholler R et al (2002) Timing of initiation of the preovulatory luteinizing hormone surge and its relationship with the circadian cortisol rhythm in the human. Neuroendocrinology 75(3):158–163
4. Mahoney MM, Sisk C, Ross HE, Smale L (2004) Circadian regulation of gonadotropin-releasing hormone neurons and the preovulatory surge in luteinizing hormone in the diurnal rodent, Arvicanthis niloticus, and in a nocturnal rodent, Rattus norvegicus. Biol Reprod 70(4):1049–1054
5. Christian CA, Moenter SM (2010) The neurobiology of preovulatory and estradiol-induced gonadotropin-releasing hormone surges. Endocr Rev 31(4):544–577
6. Miller BH, Olson SL, Turek FW, Levine JE, Horton TH, Takahashi JS (2004) Circadian clock mutation disrupts estrous cyclicity and maintenance of pregnancy. Curr Biol 14(15):1367–1373
7. Nunez AA, Stephan FK (1977) The effects of hypothalamic knife cuts on drinking rhythms and the estrus cycle of the rat. Behav Biol 20(2):224–234
8. Wiegand SJ, Terasawa E (1982) Discrete lesions reveal functional heterogeneity of suprachiasmatic structures in regulation of gonadotropin secretion in the female rat. Neuroendocrinology 34(6):395–404
9. Stephan FK, Zucker I (1972) Circadian rhythms in drinking behavior and locomotor activity of rats are eliminated by hypothalamic lesions. Proc Natl Acad Sci U S A 69(6):1583–1586
10. Moore RY, Eichler VB (1972) Loss of a circadian adrenal corticosterone rhythm following suprachiasmatic lesions in the rat. Brain Res 42(1):201–206
11. Ralph MR, Foster RG, Davis FC, Menaker M (1990) Transplanted suprachiasmatic nucleus determines circadian period. Science (New York, NY) 247(4945):975–978

12. Lehman MN, Silver R, Gladstone WR, Kahn RM, Gibson M, Bittman EL (1987) Circadian rhythmicity restored by neural transplant. Immunocytochemical characterization of the graft and its integration with the host brain. J Neurosci 7(6):1626–1638
13. Green DJ, Gillette R (1982) Circadian rhythm of firing rate recorded from single cells in the rat suprachiasmatic brain slice. Brain Res 245(1):198–200
14. Morin LP, Allen CN (2006) The circadian visual system, 2005. Brain Res Rev 51(1):1–60
15. Hattar S, Liao HW, Takao M, Berson DM, Yau KW (2002) Melanopsin-containing retinal ganglion cells: architecture, projections, and intrinsic photosensitivity. Science (New York, NY) 295(5557):1065–1070
16. Provencio I, Rollag MD, Castrucci AM (2002) Photoreceptive net in the mammalian retina. This mesh of cells may explain how some blind mice can still tell day from night. Nature 415(6871):493
17. Panda S, Sato TK, Castrucci AM, Rollag MD, DeGrip WJ, Hogenesch JB et al (2002) Melanopsin (Opn4) requirement for normal light-induced circadian phase shifting. Science (New York, NY) 298(5601):2213–2216
18. Berson DM, Dunn FA, Takao M (2002) Phototransduction by retinal ganglion cells that set the circadian clock. Science (New York, NY) 295(5557):1070–1073
19. Ruby NF, Brennan TJ, Xie X, Cao V, Franken P, Heller HC et al (2002) Role of melanopsin in circadian responses to light. Science (New York, NY) 298(5601):2211–2213
20. Maywood ES, O'Neill JS, Chesham JE, Hastings MH (2007) Minireview: the circadian clockwork of the suprachiasmatic nuclei—analysis of a cellular oscillator that drives endocrine rhythms. Endocrinology 148(12):5624–5634
21. Mohawk JA, Takahashi JS (2011) Cell autonomy and synchrony of suprachiasmatic nucleus circadian oscillators. Trends Neurosci 34(7):349–358
22. Reppert SM, Weaver DR (2002) Coordination of circadian timing in mammals. Nature 418(6901):935–941
23. Chen R, Schirmer A, Lee Y, Lee H, Kumar V, Yoo SH et al (2009) Rhythmic PER abundance defines a critical nodal point for negative feedback within the circadian clock mechanism. Mol Cell 36(3):417–430
24. Lowrey PL, Shimomura K, Antoch MP, Yamazaki S, Zemenides PD, Ralph MR et al (2000) Positional syntenic cloning and functional characterization of the mammalian circadian mutation tau. Science (New York, NY) 288(5465):483–492
25. Wang GQ, Du YZ, Tong J (2007) Daily oscillation and photoresponses of clock gene, Clock, and clock-associated gene, arylalkylamine N-acetyltransferase gene transcriptions in the rat pineal gland. Chronobiol Int 24(1):9–20
26. Ueda HR, Hayashi S, Chen W, Sano M, Machida M, Shigeyoshi Y et al (2005) System-level identification of transcriptional circuits underlying mammalian circadian clocks. Nat Genet 37(2):187–192
27. Ripperger JA, Schibler U (2006) Rhythmic CLOCK-BMAL1 binding to multiple E-box motifs drives circadian Dbp transcription and chromatin transitions. Nat Genet 38(3):369–374
28. Lavery DJ, Lopez-Molina L, Margueron R, Fleury-Olela F, Conquet F, Schibler U et al (1999) Circadian expression of the steroid 15 alpha-hydroxylase (Cyp2a4) and coumarin 7-hydroxylase (Cyp2a5) genes in mouse liver is regulated by the PAR leucine zipper transcription factor DBP. Mol Cell Biol 19(10):6488–6499
29. Balsalobre A, Damiola F, Schibler U (1998) A serum shock induces circadian gene expression in mammalian tissue culture cells. Cell 93(6):929–937
30. Welsh DK, Yoo SH, Liu AC, Takahashi JS, Kay SA (2004) Bioluminescence imaging of individual fibroblasts reveals persistent, independently phased circadian rhythms of clock gene expression. Curr Biol 14(24):2289–2295
31. Nequin LG, Alvarez J, Schwartz NB (1975) Steroid control of gonadotropin release. J Steroid Biochem 6(6):1007–1012
32. Sarkar DK, Chiappa SA, Fink G, Sherwood NM (1976) Gonadotropin-releasing hormone surge in pro-oestrous rats. Nature 264(5585):461–463

33. Everett JW, Sawyer CH (1950) A 24-hour periodicity in the "LH-release apparatus" of female rats, disclosed by barbiturate sedation. Endocrinology 47(3):198–218
34. Alleva JJ, Waleski MV, Alleva FR (1971) A biological clock controlling the estrous cycle of the hamster. Endocrinology 88(6):1368–1379
35. Fitzgerald K, Zucker I (1976) Circadian organization of the estrous cycle of the golden hamster. Proc Natl Acad Sci U S A 73(8):2923–2927
36. Kriegsfeld LJ, Silver R (2006) The regulation of neuroendocrine function: timing is everything. Horm Behav 49(5):557–574
37. Legan SJ, Karsch FJ (1975) A daily signal for the LH surge in the rat. Endocrinology 96(1):57–62
38. Legan SJ, Coon GA, Karsch FJ (1975) Role of estrogen as initiator of daily LH surges in the ovariectomized rat. Endocrinology 96(1):50–56
39. Christian CA, Mobley JL, Moenter SM (2005) Diurnal and estradiol-dependent changes in gonadotropin-releasing hormone neuron firing activity. Proc Natl Acad Sci U S A 102(43):15682–15687
40. de la Iglesia HO, Schwartz WJ (2006) Minireview: timely ovulation: circadian regulation of the female hypothalamo-pituitary-gonadal axis. Endocrinology 147(3):1148–1153
41. Silver R, Lehman MN, Gibson M, Gladstone WR, Bittman EL (1990) Dispersed cell suspensions of fetal SCN restore circadian rhythmicity in SCN-lesioned adult hamsters. Brain Res 525(1):45–58
42. Meyer-Bernstein EL, Jetton AE, Matsumoto SI, Markuns JF, Lehman MN, Bittman EL (1999) Effects of suprachiasmatic transplants on circadian rhythms of neuroendocrine function in golden hamsters. Endocrinology 140(1):207–218
43. Silver R, LeSauter J, Tresco PA, Lehman MN (1996) A diffusible coupling signal from the transplanted suprachiasmatic nucleus controlling circadian locomotor rhythms. Nature 382(6594):810–813
44. de la Iglesia HO, Meyer J, Carpino A Jr, Schwartz WJ (2000) Antiphase oscillation of the left and right suprachiasmatic nuclei. Science (New York, NY) 290(5492):799–801
45. Yan L, Foley NC, Bobula JM, Kriegsfeld LJ, Silver R (2005) Two antiphase oscillations occur in each suprachiasmatic nucleus of behaviorally split hamsters. J Neurosci 25(39):9017–9026
46. Tavakoli-Nezhad M, Schwartz WJ (2005) c-Fos expression in the brains of behaviorally "split" hamsters in constant light: calling attention to a dorsolateral region of the suprachiasmatic nucleus and the medial division of the lateral habenula. J Biol Rhythms 20(5):419–429
47. Swann JM, Turek FW (1985) Multiple circadian oscillators regulate the timing of behavioral and endocrine rhythms in female golden hamsters. Science (New York, NY) 228(4701):898–900
48. Kriegsfeld LJ, Leak RK, Yackulic CB, LeSauter J, Silver R (2004) Organization of suprachiasmatic nucleus projections in Syrian hamsters (Mesocricetus auratus): an anterograde and retrograde analysis. J Comp Neurol 468(3):361–379
49. Van der Beek EM, Horvath TL, Wiegant VM, Van den Hurk R, Buijs RM (1997) Evidence for a direct neuronal pathway from the suprachiasmatic nucleus to the gonadotropin-releasing hormone system: combined tracing and light and electron microscopic immunocytochemical studies. J Comp Neurol 384(4):569–579
50. DeVries GJ, Buijs RM, Van Leeuwen FW, Caffe AR, Swaab DF (1985) The vasopressinergic innervation of the brain in normal and castrated rats. J Comp Neurol 233(2):236–254
51. Watts AG, Swanson LW (1987) Efferent projections of the suprachiasmatic nucleus: II. Studies using retrograde transport of fluorescent dyes and simultaneous peptide immunohistochemistry in the rat. J Comp Neurol 258(2):230–252
52. de la Iglesia HO, Meyer J, Schwartz WJ (2003) Lateralization of circadian pacemaker output: activation of left- and right-sided luteinizing hormone-releasing hormone neurons involves a neural rather than a humoral pathway. J Neurosci 23(19):7412–7414
53. Smarr BL, Morris E, de la Iglesia HO (2012) The dorsomedial suprachiasmatic nucleus times circadian expression of Kiss1 and the luteinizing hormone surge. Endocrinology 153(6): 2839–2850

54. Williams WP III, Jarjisian SG, Mikkelsen JD, Kriegsfeld LJ (2011) Circadian control of kisspeptin and a gated GnRH response mediate the preovulatory luteinizing hormone surge. Endocrinology 152(2):595–606
55. Vida B, Deli L, Hrabovszky E, Kalamatianos T, Caraty A, Coen CW et al (2010) Evidence for suprachiasmatic vasopressin neurones innervating kisspeptin neurones in the rostral periventricular area of the mouse brain: regulation by oestrogen. J Neuroendocrinol 22(9): 1032–1039
56. Robertson JL, Clifton DK, de la Iglesia HO, Steiner RA, Kauffman AS (2009) Circadian regulation of Kiss1 neurons: implications for timing the preovulatory gonadotropin-releasing hormone/luteinizing hormone surge. Endocrinology 150(8):3664–3671
57. Moore RY, Speh JC, Leak RK (2002) Suprachiasmatic nucleus organization. Cell Tissue Res 309(1):89–98
58. Ibata Y, Takahashi Y, Okamura H, Kawakami F, Terubayashi H, Kubo T et al (1989) Vasoactive intestinal peptide (VIP)-like immunoreactive neurons located in the rat suprachiasmatic nucleus receive a direct retinal projection. Neurosci Lett 97(1–2):1–5
59. Tanaka M, Ichitani Y, Okamura H, Tanaka Y, Ibata Y (1993) The direct retinal projection to VIP neuronal elements in the rat SCN. Brain Res Bull 31(6):637–640
60. Horvath TL, Cela V, van der Beek EM (1998) Gender-specific apposition between vasoactive intestinal peptide-containing axons and gonadotrophin-releasing hormone-producing neurons in the rat. Brain Res 795(1–2):277–281
61. Smith MJ, Jiennes L, Wise PM (2000) Localization of the VIP2 receptor protein on GnRH neurons in the female rat. Endocrinology 141(11):4317–4320
62. Kriegsfeld LJ, Silver R, Gore AC, Crews D (2002) Vasoactive intestinal polypeptide contacts on gonadotropin-releasing hormone neurones increase following puberty in female rats. J Neuroendocrinol 14(9):685–690
63. Krajnak K, Rosewell KL, Wise PM (2001) Fos-induction in gonadotropin-releasing hormone neurons receiving vasoactive intestinal polypeptide innervation is reduced in middle-aged female rats. Biol Reprod 64(4):1160–1164
64. van der Beek EM, van Oudheusden HJ, Buijs RM, van der Donk HA, van den Hurk R, Wiegant VM (1994) Preferential induction of c-fos immunoreactivity in vasoactive intestinal polypeptide-innervated gonadotropin-releasing hormone neurons during a steroid-induced luteinizing hormone surge in the female rat. Endocrinology 134(6):2636–2644
65. Gerhold LM, Rosewell KL, Wise PM (2005) Suppression of vasoactive intestinal polypeptide in the suprachiasmatic nucleus leads to aging-like alterations in cAMP rhythms and activation of gonadotropin-releasing hormone neurons. J Neurosci 25(1):62–67
66. Harney JP, Scarbrough K, Rosewell KL, Wise PM (1996) In vivo antisense antagonism of vasoactive intestinal peptide in the suprachiasmatic nuclei causes aging-like changes in the estradiol-induced luteinizing hormone and prolactin surges. Endocrinology 137(9):3696–3701
67. Christian CA, Moenter SM (2008) Vasoactive intestinal polypeptide can excite gonadotropin-releasing hormone neurons in a manner dependent on estradiol and gated by time of day. Endocrinology 149(6):3130–3136
68. Herbison AE, Theodosis DT (1992) Localization of oestrogen receptors in preoptic neurons containing neurotensin but not tyrosine hydroxylase, cholecystokinin or luteinizing hormone-releasing hormone in the male and female rat. Neuroscience 50(2):283–298
69. Dorling AA, Todman MG, Korach KS, Herbison AE (2003) Critical role for estrogen receptor alpha in negative feedback regulation of gonadotropin-releasing hormone mRNA expression in the female mouse. Neuroendocrinology 78(4):204–209
70. Wintermantel TM, Campbell RE, Porteous R, Bock D, Grone HJ, Todman MG et al (2006) Definition of estrogen receptor pathway critical for estrogen positive feedback to gonadotropin-releasing hormone neurons and fertility. Neuron 52(2):271–280
71. Wiegand SJ, Terasawa E, Bridson WE, Goy RW (1980) Effects of discrete lesions of preoptic and suprachiasmatic structures in the female rat. Alterations in the feedback regulation of gonadotropin secretion. Neuroendocrinology 31(2):147–157

72. Ronnekleiv OK, Kelly MJ (1988) Plasma prolactin and luteinizing hormone profiles during the estrous cycle of the female rat: effects of surgically induced persistent estrus. Neuroendocrinology 47(2):133–141
73. Gu GB, Simerly RB (1997) Projections of the sexually dimorphic anteroventral periventricular nucleus in the female rat. J Comp Neurol 384(1):142–164
74. Le WW, Berghorn KA, Rassnick S, Hoffman GE (1999) Periventricular preoptic area neurons coactivated with luteinizing hormone (LH)-releasing hormone (LHRH) neurons at the time of the LH surge are LHRH afferents. Endocrinology 140(1):510–519
75. Watson RE Jr, Langub MC Jr, Engle MG, Maley BE (1995) Estrogen-receptive neurons in the anteroventral periventricular nucleus are synaptic targets of the suprachiasmatic nucleus and peri-suprachiasmatic region. Brain Res 689(2):254–264
76. de la Iglesia HO, Blaustein JD, Bittman EL (1995) The suprachiasmatic area in the female hamster projects to neurons containing estrogen receptors and GnRH. Neuroreport 6(13):1715–1722
77. Shughrue PJ, Lane MV, Merchenthaler I (1997) Comparative distribution of estrogen receptor-alpha and -beta mRNA in the rat central nervous system. J Comp Neurol 388(4):507–525
78. Leak RK, Moore RY (2001) Topographic organization of suprachiasmatic nucleus projection neurons. J Comp Neurol 433(3):312–334
79. Hoorneman EM, Buijs RM (1982) Vasopressin fiber pathways in the rat brain following suprachiasmatic nucleus lesioning. Brain Res 243(2):235–241
80. Leak RK, Moore RY (2001) Topographic organization of suprachiasmatic nucleus projection neurons. J Comp Neurol 433(3):312–334
81. Palm IF, Van Der Beek EM, Wiegant VM, Buijs RM, Kalsbeek A (1999) Vasopressin induces a luteinizing hormone surge in ovariectomized, estradiol-treated rats with lesions of the suprachiasmatic nucleus. Neuroscience 93(2):659–666
82. Ostrowski NL, Lolait SJ, Young WS III (1994) Cellular localization of vasopressin V1a receptor messenger ribonucleic acid in adult male rat brain, pineal, and brain vasculature. Endocrinology 135(4):1511–1528
83. Funabashi T, Shinohara K, Mitsushima D, Kimura F (2000) Estrogen increases arginine-vasopressin V1a receptor mRNA in the preoptic area of young but not of middle-aged female rats. Neurosci Lett 285(3):205–208
84. Petersen SL, Barraclough CA (1989) Suppression of spontaneous LH surges in estrogen-treated ovariectomized rats by microimplants of antiestrogens into the preoptic brain. Brain Res 484(1–2):279–289
85. Grace CO, Fink G, Quinn JP (1999) Characterization of potential regulatory elements within the rat arginine vasopressin proximal promoter. Neuropeptides 33(1):81–90
86. Munoz E, Brewer M, Baler R (2002) Circadian transcription. Thinking outside the E-box. J Biol Chem 277(39):36009–36017
87. Jin X, Shearman LP, Weaver DR, Zylka MJ, de Vries GJ, Reppert SM (1999) A molecular mechanism regulating rhythmic output from the suprachiasmatic circadian clock. Cell 96(1):57–68
88. Shinohara K, Honma S, Katsuno Y, Abe H, Honma K (1994) Circadian rhythms in the release of vasoactive intestinal polypeptide and arginine-vasopressin in organotypic slice culture of rat suprachiasmatic nucleus. Neurosci Lett 170(1):183–186
89. Schwartz WJ, Coleman RJ, Reppert SM (1983) A daily vasopressin rhythm in rat cerebrospinal fluid. Brain Res 263(1):105–112
90. Kalsbeek A, Buijs RM, Engelmann M, Wotjak CT, Landgraf R (1995) In vivo measurement of a diurnal variation in vasopressin release in the rat suprachiasmatic nucleus. Brain Res 682(1–2):75–82
91. Funabashi T, Shinohara K, Mitsushima D, Kimura F (2000) Gonadotropin-releasing hormone exhibits circadian rhythm in phase with arginine-vasopressin in co-cultures of the female rat preoptic area and suprachiasmatic nucleus. J Neuroendocrinol 12(6):521–528

92. Funabashi T, Aiba S, Sano A, Shinohara K, Kimura F (1999) Intracerebroventricular injection of arginine-vasopressin V1 receptor antagonist attenuates the surge of luteinizing hormone and prolactin secretion in proestrous rats. Neurosci Lett 260(1):37–40
93. Gottsch ML, Cunningham MJ, Smith JT, Popa SM, Acohido BV, Crowley WF et al (2004) A role for kisspeptins in the regulation of gonadotropin secretion in the mouse. Endocrinology 145(9):4073–4077
94. Smith JT, Dungan HM, Stoll EA, Gottsch ML, Braun RE, Eacker SM et al (2005) Differential regulation of KiSS-1 mRNA expression by sex steroids in the brain of the male mouse. Endocrinology 146(7):2976–2984
95. Clarkson J, Herbison AE (2006) Postnatal development of kisspeptin neurons in mouse hypothalamus; sexual dimorphism and projections to gonadotropin-releasing hormone neurons. Endocrinology 147(12):5817–5825
96. Revel FG, Saboureau M, Masson-Pevet M, Pevet P, Mikkelsen JD, Simonneaux V (2006) Kisspeptin mediates the photoperiodic control of reproduction in hamsters. Curr Biol 16(17):1730–1735
97. Greives TJ, Mason AO, Scotti MA, Levine J, Ketterson ED, Kriegsfeld LJ et al (2007) Environmental control of kisspeptin: implications for seasonal reproduction. Endocrinology 148(3):1158–1166
98. Smith JT, Cunningham MJ, Rissman EF, Clifton DK, Steiner RA (2005) Regulation of Kiss1 gene expression in the brain of the female mouse. Endocrinology 146(9):3686–3692
99. Irwig MS, Fraley GS, Smith JT, Acohido BV, Popa SM, Cunningham MJ et al (2004) Kisspeptin activation of gonadotropin releasing hormone neurons and regulation of KiSS-1 mRNA in the male rat. Neuroendocrinology 80(4):264–272
100. Matsui H, Takatsu Y, Kumano S, Matsumoto H, Ohtaki T (2004) Peripheral administration of metastin induces marked gonadotropin release and ovulation in the rat. Biochem Biophys Res Commun 320(2):383–388
101. Navarro VM, Castellano JM, Fernandez-Fernandez R, Tovar S, Roa J, Mayen A et al (2005) Characterization of the potent luteinizing hormone-releasing activity of KiSS-1 peptide, the natural ligand of GPR54. Endocrinology 146(1):156–163
102. Navarro VM, Castellano JM, Fernandez-Fernandez R, Tovar S, Roa J, Mayen A et al (2005) Effects of KiSS-1 peptide, the natural ligand of GPR54, on follicle-stimulating hormone secretion in the rat. Endocrinology 146(4):1689–1697
103. Mayer C, Boehm U (2011) Female reproductive maturation in the absence of kisspeptin/GPR54 signaling. Nat Neurosci 14(6):704–710
104. Silverman AJ, Zimmerman EA, Gibson MJ, Perlow MJ, Charlton HM, Kokoris GJ et al (1985) Implantation of normal fetal preoptic area into hypogonadal mutant mice: temporal relationships of the growth of gonadotropin-releasing hormone neurons and the development of the pituitary/testicular axis. Neuroscience 16(1):69–84
105. Gibson MJ, Kokoris GJ, Silverman AJ (1988) Positive feedback in hypogonadal female mice with preoptic area brain transplants. Neuroendocrinology 48(2):112–119
106. Smith JT, Popa SM, Clifton DK, Hoffman GE, Steiner RA (2006) Kiss1 neurons in the forebrain as central processors for generating the preovulatory luteinizing hormone surge. J Neurosci 26(25):6687–6694
107. Adachi S, Yamada S, Takatsu Y, Matsui H, Kinoshita M, Takase K et al (2007) Involvement of anteroventral periventricular metastin/kisspeptin neurons in estrogen positive feedback action on luteinizing hormone release in female rats. J Reprod Dev 53(2):367–378
108. Vida B, Deli L, Hrabovszky E, Kalamatianos T, Caraty A, Coen CW et al (2010) Evidence for suprachiasmatic vasopressin neurons innervating kisspeptin neurons in the rostral periventricular area of the mouse brain: regulation by oestrogen. J Neuroendocrinol 22(9):1032–1039
109. de la Iglesia HO, Cambras T, Schwartz WJ, Diez-Noguera A (2004) Forced desynchronization of dual circadian oscillators within the rat suprachiasmatic nucleus. Curr Biol 14(9):796–800
110. Khan AR, Kauffman AS (2011) The role of kisspeptin and RFRP-3 neurons in the circadian-timed preovulatory luteinizing hormone surge. J Neuroendocrinol 24(1):131–143

111. Palm IF, van der Beek EM, Wiegant VM, Buijs RM, Kalsbeek A (2001) The stimulatory effect of vasopressin on the luteinizing hormone surge in ovariectomized, estradiol-treated rats is time-dependent. Brain Res 901(1–2):109–116
112. Chappell PE, White RS, Mellon PL (2003) Circadian gene expression regulates pulsatile gonadotropin-releasing hormone (GnRH) secretory patterns in the hypothalamic GnRH-secreting GT1-7 cell line. J Neurosci 23(35):11202–11213
113. Zhao S, Kriegsfeld LJ (2009) Daily changes in GT1-7 cell sensitivity to GnRH secretagogues that trigger ovulation. Neuroendocrinology 89(4):448–457
114. Hickok JR, Tischkau SA (2010) In vivo circadian rhythms in gonadotropin-releasing hormone neurons. Neuroendocrinology 91(1):110–120
115. Olcese J, Domagalski R, Bednorz A, Weaver DR, Urbanski HF, Reuss S et al (2003) Expression and regulation of mPer1 in immortalized GnRH neurons. Neuroreport 14(4):613–618
116. Tonsfeldt KJ, Goodall CP, Latham KL, Chappell PE (2011) Oestrogen induces rhythmic expression of the Kisspeptin-1 receptor GPR54 in hypothalamic gonadotrophin-releasing hormone-secreting GT1-7 cells. J Neuroendocrinol 23(9):823–830
117. Tsutsui K, Saigoh E, Ukena K, Teranishi H, Fujisawa Y, Kikuchi M et al (2000) A novel avian hypothalamic peptide inhibiting gonadotropin release. Biochem Biophys Res Commun 275(2):661–667
118. Bentley GE, Tsutsui K, Kriegsfeld LJ (2010) Recent studies of gonadotropin-inhibitory hormone (GnIH) in the mammalian hypothalamus, pituitary and gonads. Brain Res 1364:62–71
119. Kriegsfeld LJ, Gibson EM, Williams WP III, Zhao S, Mason AO, Bentley GE et al (2010) The roles of RFamide-related peptide-3 in mammalian reproductive function and behaviour. J Neuroendocrinol 22(7):692–700
120. Tsutsui K, Bentley GE, Kriegsfeld LJ, Osugi T, Seong JY, Vaudry H (2010) Discovery and evolutionary history of gonadotrophin-inhibitory hormone and kisspeptin: new key neuropeptides controlling reproduction. J Neuroendocrinol 22(7):716–727
121. Smith JT, Clarke IJ (2010) Gonadotropin inhibitory hormone function in mammals. Trends Endocrinol Metab 21(4):255–260
122. Ancel C, Bentsen AH, Sebert ME, Tena-Sempere M, Mikkelsen JD, Simonneaux V (2012) Stimulatory effect of RFRP-3 on the gonadotrophic axis in the male Syrian hamster: the exception proves the rule. Endocrinology 153(3):1352–1363
123. Kriegsfeld LJ, Mei DF, Bentley GE, Ubuka T, Mason AO, Inoue K et al (2006) Identification and characterization of a gonadotropin-inhibitory system in the brains of mammals. Proc Natl Acad Sci U S A 103(7):2410–2415
124. Johnson MA, Tsutsui K, Fraley GS (2007) Rat RFamide-related peptide-3 stimulates GH secretion, inhibits LH secretion, and has variable effects on sex behavior in the adult male rat. Horm Behav 51(1):171–180
125. Rizwan MZ, Poling MC, Corr M, Cornes PA, Augustine RA, Quennell JH et al (2012) RFamide-related peptide-3 receptor gene expression in GnRH and kisspeptin neurons and GnRH-dependent mechanism of action. Endocrinology 153(8):3770–3779
126. Ubuka T, Inoue K, Fukuda Y, Mizuno T, Ukena K, Kriegsfeld LJ et al (2012) Identification, expression, and physiological functions of Siberian hamster gonadotropin-inhibitory hormone. Endocrinology 153(1):373–385
127. Gibson EM, Humber SA, Jain S, Williams WP III, Zhao S, Bentley GE et al (2008) Alterations in RFamide-related peptide expression are coordinated with the preovulatory luteinizing hormone surge. Endocrinology 149(10):4958–4969
128. Hinuma S, Shintani Y, Fukusumi S, Iijima N, Matsumoto Y, Hosoya M et al (2000) New neuropeptides containing carboxy-terminal RFamide and their receptor in mammals. Nat Cell Biol 2(10):703–708
129. Murakami M, Matsuzaki T, Iwasa T, Yasui T, Irahara M, Osugi T et al (2008) Hypophysiotropic role of RFamide-related peptide-3 in the inhibition of LH secretion in female rats. J Endocrinol 199(1):105–112

130. Rizwan MZ, Porteous R, Herbison AE, Anderson GM (2009) Cells expressing RFamide-related peptide-1/3, the mammalian gonadotropin-inhibitory hormone orthologs, are not hypophysiotropic neuroendocrine neurons in the rat. Endocrinology 150(3):1413–1420
131. Yin H, Ukena K, Ubuka T, Tsutsui K (2005) A novel G protein-coupled receptor for gonadotropin-inhibitory hormone in the Japanese quail (Coturnix japonica): identification, expression and binding activity. J Endocrinol 184(1):257–266
132. Anderson GM, Relf HL, Rizwan MZ, Evans JJ (2009) Central and peripheral effects of RFamide-related peptide-3 on luteinizing hormone and prolactin secretion in rats. Endocrinology 150(4):1834–1840
133. Molnar CS, Kallo I, Liposits Z, Hrabovszky E (2011) Estradiol down-regulates RF-amide-related peptide (RFRP) expression in the mouse hypothalamus. Endocrinology 152(4):1684–1690
134. Pineda R, Garcia-Galiano D, Sanchez-Garrido MA, Romero M, Ruiz-Pino F, Aguilar E et al (2010) Characterization of the potent gonadotropin-releasing activity of RF9, a selective antagonist of RF-amide-related peptides and neuropeptide FF receptors: physiological and pharmacological implications. Endocrinology 151(4):1902–1913
135. Wise PM (1982) Alterations in proestrous LH, FSH, and prolactin surges in middle-aged rats. Proc Soc Exp Biol Med 169(3):348–354
136. Lloyd JM, Hoffman GE, Wise PM (1994) Decline in immediate early gene expression in gonadotropin-releasing hormone neurons during proestrus in regularly cycling, middle-aged rats. Endocrinology 134(4):1800–1805
137. Krajnak K, Kashon ML, Rosewell KL, Wise PM (1998) Aging alters the rhythmic expression of vasoactive intestinal polypeptide mRNA but not arginine vasopressin mRNA in the suprachiasmatic nuclei of female rats. J Neurosci 18(12):4767–4774
138. Lederman MA, Lebesgue D, Gonzalez VV, Shu J, Merhi ZO, Etgen AM et al (2010) Age-related LH surge dysfunction correlates with reduced responsiveness of hypothalamic anteroventral periventricular nucleus kisspeptin neurons to estradiol positive feedback in middle-aged rats. Neuropharmacology 58(1):314–320
139. Neal-Perry G, Lebesgue D, Lederman M, Shu J, Zeevalk GD, Etgen AM (2009) The excitatory peptide kisspeptin restores the luteinizing hormone surge and modulates amino acid neurotransmission in the medial preoptic area of middle-aged rats. Endocrinology 150(8): 3699–3708

Chapter 19
Kisspeptin and Seasonality of Reproduction

Iain J. Clarke and Alain Caraty

Abstract Wild and domesticated species display seasonality in reproductive function, controlled predominantly by photoperiod. Seasonal alterations in breeding status are caused by changes in the secretion of gonadotropin-releasing hormone (GnRH) that are mediated by upstream neuronal afferents that regulate the GnRH cells. In particular, kisspeptin appears to play a major role in seasonality of reproduction, transducing the feedback effect of gonadal steroids as well as having an independent (nonsteroid dependent) circannual rhythm. A substantial body of data on this issue has been obtained from studies in sheep and hamsters and this is reviewed here in detail. Kisspeptin function is upregulated during the breeding season in sheep, stimulating reproductive function, but contradictory data are found in Siberian and Syrian hamsters. The relative quiescence of kisspeptin cells in the nonbreeding season can be counteracted by administration of the peptide, leading to activation of reproductive function. Although there is a major role for melatonin in the transduction of photoperiod to the reproductive system, kisspeptin cells do not appear to express the melatonin receptor, so the means by which seasonality changes the level of kisspeptin activity remains unknown.

I.J. Clarke (✉)
Department of Physiology, School of Biomedical Sciences, Monash University,
PO Box 13F, Clayton, VIC 3800, Australia
e-mail: iain.clarke@med.monash.edu.au

A. Caraty
Unité Mixte de Recherche, Physiologie de la Reproduction et des Comportements,
Université de Tours/Haras Nationaux, Nouzilly, France

A.S. Kauffman and J.T. Smith (eds.), *Kisspeptin Signaling in Reproductive Biology*,
Advances in Experimental Medicine and Biology 784, DOI 10.1007/978-1-4614-6199-9_19,

Introduction

Most subhuman mammals, both domesticated and wild, display seasonality in reproductive function that is controlled by day length [1]. In the broadest sense, animals are either long-day breeders or short-day breeders, such that the reproductive axis is activated by increasing day length or decreasing day length, respectively. Thus, sheep are short-day breeders whereas hamsters are long-day breeders. In the simplest terms, day length is "measured" by the perception of light through the eye, transmitting a signal to the suprachiasmatic nucleus (SCN), which sends indirect neuronal signals to the pineal gland, which secretes melatonin exclusively during the hours of darkness (reviewed in refs. [2, 3]). Thus, the duration of melatonin secretion, which accurately reflects the duration of darkness (and hence, the photoperiod), is the key endogenous signal which induces seasonal changes in reproductive status. The daily rhythm in CLOCK gene expression within the SCN is a fundamental driver of this system of detection of day length, leading to a distinct seasonality of reproduction that is well characterized in hamsters [4] and sheep [5, 6].

This review focuses on reproductive function, and in this respect, it is important to note that seasonal change in reproductive function is characterized, in sheep at least, by a change in the secretion of gonadotropin-releasing hormone (GnRH) [7]. A nonsteroidal component of the mechanism that causes a fundamental change in the frequency of luteinizing hormone (LH) pulses was shown in ovariectomized (OVX) ewes [8]. Earlier work showed that there is an estrogen-dependent mechanism that also underlies the transition between breeding and nonbreeding seasons, such that estrogen has a much greater negative feedback effect on the GnRH-gonadotropin axis in the nonbreeding season [9]. The sheep is an ideal model for the collection of hypophyseal portal blood, allowing for the measurement of hypothalamic secretions [10], and this model was used to show that estrogen acts within the brain to cause this negative feedback effect [11]. During the nonbreeding season, estrogen reduces pulse frequency of GnRH secretion, whereas this does not occur in the breeding season, at least in sheep [12]. Herein, we review the evidence that this seasonally regulated estrogen feedback is mediated through kisspeptin neurons.

Although feedback control of GnRH secretion by gonadal steroids is fundamental to reproductive function [11, 13], GnRH cells do not express estrogen receptor-α (ER-α) [14], but do express ERβ [15]. Because ERα is the predominant mediator of the feedback effects of estrogen [16], other ERα-expressing cells in the brain promulgate the feedback effects of estrogen to the GnRH axis. Although various types of neurons have been shown to express the relevant sex steroid receptors, a major conduit for steroid feedback to GnRH cells remained elusive for many years [17]. The discovery that kisspeptin and its cognate receptor are essential for normal reproduction [18, 19] was a landmark in our understanding of reproductive control, because kisspeptin cells are highly responsive to gonadal steroid regulation and appear to transmit feedback signals to GnRH cells. While this does not exclude a role for other sex steroid receptive neurons, kisspeptin cells appear to be of primary importance in the control of GnRH secretion. Accordingly, the role of kisspeptin in seasonal reproduction is of little surprise. This review focuses on how kisspeptin cells alter in function during seasonal breeding cycles, with particular reference to work in sheep and hamsters.

Location and Projections of Kisspeptin Neurons

The major populations of kisspeptin cells in the mammalian brain are found in the arcuate nucleus (ARC) of the hypothalamus and the preoptic area (POA). Regarding the latter population, slight species variation is observed. In the two species under consideration here, with respect to photoperiodic control of reproduction, sheep display a population of cells in the lateral POA [20], whereas the hamster has a rostrally located population in the anteroventral periventricular nucleus (AVPV) [21].

Of particular relevance are findings from anterograde and retrograde neuronal tracing studies between the ARC and the POA of the ovine brain, which indicate that there is very limited direct input to GnRH cell bodies from the former region [22, 23]. Virtually none of the ARC kisspeptin cells of the sheep brain are retrogradely labeled when Fluorogold injections are placed in the basal POA, where most GnRH cell bodies are found [24]. Although we see no ARC-to-GnRH pathways with conventional tracing methods, it remains possible that the ARC cells contact GnRH dendrites, which are not readily seen with conventional methodology. Certainly, in the rodent (mouse) brain, there is at least some input to GnRH cells in the POA from kisspeptin cells in the ARC [25], but no similar information is available for hamsters. It appears most likely that any influence that is exerted by the ARC kisspeptin cells on GnRH cell bodies is via at least one interneuron, in sheep at least.

Another possible means by which ARC kisspeptin cells may stimulate GnRH secretion is at the level of the median eminence. Significant projections of kisspeptin cells are to the external (neurosecretory zone) of the median eminence [26], designated ARC cells by virtue of their co-expression of neurokinin B (NKB); only the ARC kisspeptin cells coproduce this peptide [27]. Kisspeptin may act at this neuroanatomical level to regulate GnRH secretion, as indicated by in vitro studies with isolated mouse mediobasal hypothalamus [28] or sheep median eminence [26]. Although we observe kisspeptin fibers in the neurosecretory zone of the ovine median eminence, and Kiss1r (the kisspeptin receptor) is expressed in the pituitary gland [29], only low levels of the kisspeptin peptide are detectable in portal blood [29]. Thus, kisspeptin does not appear to affect LH or FSH release by direct action on pituitary gonadotropes [29]. Accordingly, any involvement that kisspeptin cells have with respect to season is due to action within the brain to regulate GnRH cells.

In sheep, kisspeptin cells in the lateral POA are retrogradely labeled when Fluorogold is injected into the basal POA of the sheep brain [24], which explains the close appositions between GnRH cell bodies and varicose fibers containing kisspeptin [30]. GnRH cells express kisspeptin receptors, allowing for direct action of kisspeptin from the POA cells [31]. In addition, it has been shown, in the mouse brain at least, that kisspeptin cells communicate to neurons producing nitric oxide in the POA, and this is another indirect means by which kisspeptin signals from the ARC might be relayed to GnRH cells [32].

Kisspeptin and the Feedback Effects of Estrogen

Kisspeptin cells in the ARC are involved in transmission of both negative and positive feedback signals to GnRH cells in the sheep, which is different to the case in the rodent, where positive feedback involves the AVPV population of kisspeptin cells and negative feedback is effected through the ARC cells [33]. In the ewe, the fact that the positive feedback signal is initiated in the ARC, not the POA, was indicated by estrogen implant studies, coupled with measures of GnRH secretion [34]. The kisspeptin cells are those ARC neurons most likely to be involved; these cells show an upregulation in gene expression and peptide production in the late follicular phase of the estrous cycle [35, 36] and robust c-Fos labeling (as an index of neuronal activation) in response to estrogen injection [36]. Regarding the latter, this is a signal event that causes a time-delayed surge in GnRH and LH secretion [37], which causes ovulation. In addition to this *initiation* event in the ARC that culminates in GnRH/LH surge secretion, the POA kisspeptin cells are activated at the time of the surge, *facilitating* the positive feedback event [38]. This latter observation is interesting since only 50 % of the POA kisspeptin neurons are seen to co-express ERα in the ewe [20].

A fundamental question is how it is possible for one set of neurons, viz. ARC kisspeptin neurons, to mediate *both* negative and positive feedback effects of estrogen in the sheep. One way to come to terms with this is to consider the positive feedback event to be transient, whereas negative feedback is operative on a continuous basis. Thus, when ovariectomized ewes are challenged with a single, acute intravenous injection of estradiol-17β, the ARC kisspeptin cells are activated (c-Fos labeling) within 1 h, and this leads to a surge in GnRH/LH secretion (positive feedback) [36]. On the other hand, when ovariectomized ewes receive constant-release estradiol-17β implants, for a period of 2 weeks, the same population of ARC kisspeptin cells shows reduced function (negative feedback) (Fig. 19.1) [36].

It should also be appreciated that GnRH neurons are also controlled by a wide variety of cells in the brainstem, mid-brain, hypothalamus, and forebrain [39–42]. Convergent pathways from different regions of the brain may regulate GnRH cells through polysynaptic pathways. The fact that estrogen feedback may involve cells other than kisspeptin cells is indicated in a large number of studies, but not reviewed here. An example, however, of one such pathway is the estrogen-receptive A1 noradrenergic neurons of the brainstem, which project to the bed nucleus of the stria terminalis (BNST) and POA, and may provide direct input to GnRH cells [40, 43]. These noradrenergic cells appear to be involved in estrogen feedback, since there is evidence of noradrenaline release in the POA at the time of the estrogen-induced LH surge in the ewe [44], as in other species, such as the rat [45]. These other circuits should be kept in mind when considering control of reproduction by kisspeptin, if for no other reason than to take account of the multifaceted regulation of

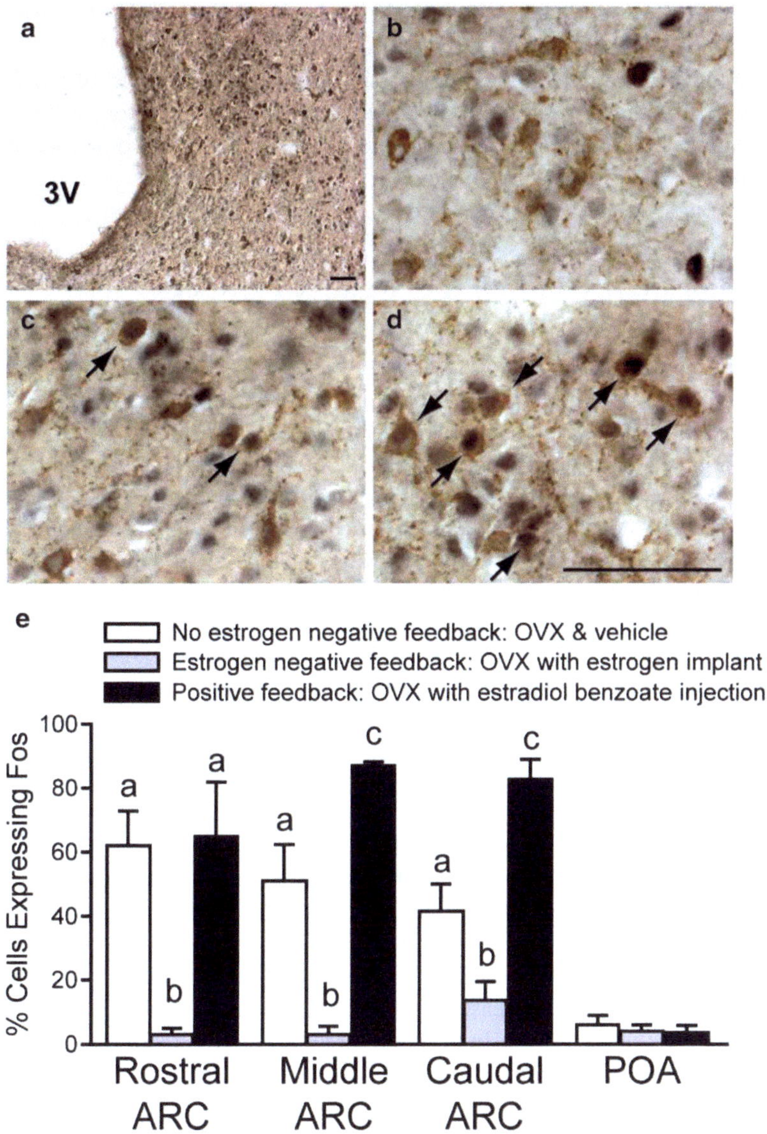

Fig. 19.1 Activation of ARC kisspeptin cells by a surge-inducing stimulus, and reduction in activity in the same cell group with chronic estrogen treatment. Ovariectomized ewes received 50 μg estradiol-17β as a single i.v. injection (acute treatment) 1 h before brain collection, or estradiol-17β implants for 2 weeks (chronic treatment). Panels (**a**–**d**) show kisspeptin (*brown*) and Fos (*black*) labeling by immunohistochemistry (*scale bars* 50 μm). (**a**) Is a low power image, showing the location of the kisspeptin cells in the ARC, (**b**–**d**) show Fos/kisspeptin labeling in animals given either an estrogen implant (**b**), no treatment (**c**), or an i.v. injection (**d**). The percent kisspeptin cells that had c-FOS labeling is shown in (**e**), with labeling in the POA given also, for comparison. The values are means ± standard error and groups with different letters are significantly ($P<0.05$) different. *3V* third ventricle. Note that acute treatment increases c-FOS labeling, whereas chronic treatment reduces c-FOS labeling, of kisspeptin cells; this shows that the same cells can respond to *both* negative and positive feedback effects of estrogen in opposite ways. *Scale bars* are 50 μm. From Smith JT, Li Q, Pereira A, Clarke IJ. Kisspeptin neurons in the ovine arcuate nucleus and preoptic area are involved in the preovulatory luteinizing hormone surge. Endocrinology. 2009 Dec;150(12):5530–8. Reprinted with permission from The Endocrine Society

reproduction by various brain systems, subserving the influence of diet, stress, mood, etc. For example, forebrain glutamatergic cells and inhibitory cells utilizing gamma amino butyric acid (GABA) as a neurotransmitter are also important in the regulation of GnRH secretion [41, 46–48].

Kisspeptin Cells Transmit the Negative Feedback Effect of Sex Steroids to GnRH Cells

As mentioned above, GnRH cells do not express estrogen ERα, but it is clear that sex steroids regulate the secretion of GnRH. Unlike GnRH cells, ARC kisspeptin cells express ERα and progesterone receptors [20]. Also, in ewes, upregulation of expression of the kisspeptin gene, *Kiss1*, is seen in the ARC following ovariectomy [49], and increased numbers of kisspeptin cells are also seen by immunohistochemistry [50]. Thus, estrogen exerts a tonic restraint on ARC kisspeptin cells in ewes [36]. Virtually all kisspeptin cells in the ovine ARC co-express dynorphin (DYN) and NKB [27]. This has led to the naming of these cells as KNDy (K-kisspeptin; N-neurokinin B; Dy-dynorphin) cells [51]. In sheep, there is good evidence that DYN plays a role in mediating the negative feedback effects of progesterone [52, 53]. NKB most likely stimulates the reproductive axis in sheep, since intracerebroventricular injection of the agonist senktide increases plasma LH levels, and implant studies suggest that this is due to action on the cognate receptors for NKB in the retrochiasmatic area [54]. Another possible role of NKB is positive autoregulation of kisspeptin cells, since the ARC kisspeptin cells express NKB receptors [55], and there is evidence that NKB is required for the function of kisspeptin cells [56]. There is little currently known about how NKB expression in kisspeptin cells relates to seasonality.

Effects of castration and sex steroid replacement on kisspeptin cells have also been studied in hamsters. Castration of male Syrian hamsters during long days (i.e., the breeding season) increased *Kiss1* gene expression in the ARC and reduced *Kiss1* expression in the AVPV. In females, under long days, ovariectomy increased *Kiss1-expressing* cell numbers in the ARC, with no change in the AVPV [57]. This indicates a clear sex difference in the AVPV in the response to photoperiod, in this species at least. Increased *Kiss1* gene expression in the ARC following gonadectomy is consistent with the general finding across species that negative feedback of gonadal hormones is exerted in this region. Under short days (i.e., the nonbreeding season for hamsters), *Kiss1*-expressing cells in the ARC of castrate males were not changed by chronic testosterone treatment, but there was an increase in *Kiss1* cell number in the AVPV. In female hamsters, chronic treatment with estrogen during short days reduced *Kiss1* expression in the ARC and increased expression in the AVPV [57]. Again, sex differences were observed, particularly at the level of ARC, and this should be taken into account when attempting to decipher the seasonal changes that occur in the hypothalamus of the hamster.

In another study, effects of castration, photoperiod, and testosterone replacement on kisspeptin-ir cell numbers were measured in the brains of male Siberian hamsters [58].

In non-operated animals, kisspeptin-ir cell number in the AVPV was reduced in short days, compared to long days, but cell numbers in the ARC were higher in the long-day photoperiod, consistent with an earlier observation from this group [21]. There was no significant effect of castration of male Siberian hamsters on kisspeptin levels in either nucleus, in either photoperiod; this is clearly different to the result obtained in Syrian hamsters [57]. Testosterone treatment increased kisspeptin-ir cell number in the AVPV of castrated Siberian hamster males, an effect which was much greater under short days (in which kisspeptin cell numbers are very low to begin with) [58]. This result, at least under short-day photoperiod, is consistent with that seen in the Syrian hamster (see above), and the result for the AVPV is similar to that seen in the mouse [59]. On the other hand, testosterone reduces *Kiss1* expression in the ARC of the castrated mouse [59], but does not appear to do so in either Syrian or Siberian hamsters, indicating key species differences.

It has been noted that changes in kisspeptin gene expression or peptide levels in the ARC could relate to changes in metabolic regulation with season [21]. Another point of consideration is that the AVPV is considered to be the site of positive feedback effects of estrogen on GnRH/LH secretion in mice and rats (and hamsters), whereas the ARC is the seat of negative feedback control via kisspeptin cells [33]. Accordingly, it is interesting that there are seasonal changes in the AVPV of the hamster in both sexes. These data are complex and further studies are required before the seasonal control of reproduction in the two different hamster species is fully understood.

Kisspeptin and Seasonal Reproduction

As indicated in the Introduction, an increase in the negative feedback effect of estrogen on GnRH/LH secretion is a fundamental cause of seasonal anestrus in the ewe [9, 11]. In normal Soay ewes, expression of *Kiss1* in the ARC fell when the animals were transitioned from a photoperiod of 8 h light and 16 h dark to longer day lengths [60], consistent with the short-day breeding pattern of this species. Other studies have shown that the number of *Kiss1*-expressing and kisspeptin-immunoreactive (-ir) cells in the ARC is greater in the brains of ovariectomized ewes at the time they would normally (if gonad int.ct) be breeding. This effect was seen irrespective of whether the animals were treated with long-term estrogen implants [30, 49], but the inhibitory effects of long-term estrogen treatment on *Kiss1* mRNA and kisspeptin expression in the ARC (indicative of negative feedback) are greater during the nonbreeding season [30]. These data suggest the seasonal change in sensitivity to estrogen, which is a major mechanism for seasonal reproduction (see above), is affected, at least in part, by changing responsiveness of the kisspeptin cells to estrogen. The subcellular mechanisms that underlie this fundamental change in responsiveness to estrogen have not been elucidated, and it is not known if the level of ERα expression in ARC kisspeptin cells changes with season. This seasonal effect on *Kiss1* was seen in the ARC, but not the POA, supporting the notion that the negative feedback effect is promulgated by this subset of kisspeptin cells. In another study [61], however, a significant increase in immunoreactive kisspeptin cells was seen in both the ARC and the POA of ewes transferred to from long-day (16 h light:8 h dark) to short-day (8 h light:16 h dark) photoperiod (replicating the shift to the breeding season). In this latter study, the sheep were ovariectomized and given estrogen implants, and the increase in kisspeptin-ir neuron number was seen in the same time frame as the release from estrogen negative feedback. This adds further weight to the assertion that the seasonality of reproduction in the ewe involves a change in the responsiveness of kisspeptin cells to estrogen.

In addition to seasonal changes in kisspeptin gene expression and synthesis in the ovine brain, there is also greater kisspeptin input to the GnRH neurons in the breeding than nonbreeding season [30]. This most likely originates from the POA kisspeptin cells (see above). Thus, at least in sheep, both the level of kisspeptin expression and the level of kisspeptin input to GnRH neurons are higher during the breeding season, while the negative feedback effects of estrogen on kisspeptin are lower at this time of the year. This further asserts the case for a major role of kisspeptin in the seasonality of reproduction.

As the seasonal change in kisspeptin expression in ewes is replicated by manipulation of photoperiod, it appears this change in the diurnal pattern of the light/dark cycle may be the primary stimulus governing kisspeptin change. Work in hamsters supports this notion, although there are differences between Syrian and Siberian hamsters that complicate interpretation in these species. Transfer from long days (16 h light:8 h dark) to short days (8 h light:16 h dark) reduced the number of cells expressing *Kiss1* in male and female Syrian hamsters, in both the AVPV and the ARC, as measured via in situ hybridization [55, 62–64]. These data are consistent with those obtained in the sheep [49], but different to those seen in the Siberian hamster, in which immunohistochemistry was used to measure an increase in kisspeptin-ir cell number in the ARC under short-day photoperiod [21]. Interestingly, Siberian hamsters that did not respond to short days (with a reduction in reproductive function) showed no increase in kisspeptin cell numbers, which remained similar to those seen under long days [21]. Although these papers indicate a critical species difference between Syrian and Siberian hamsters, it should be noted that one study [21] counted kisspeptin-ir cells, whereas the other [57] counted *Kiss1-expressing* cells, and thus, different techniques were used. However, it has been recognized for some time that Syrian and Siberian hamsters may differ in the mechanisms that underlie seasonality of reproductive function [65]. These differences may, at least in part, be due to the differences in the neural sites of action of melatonin. Thus, in Siberian hamsters, lesion of the SCN prevents the inhibitory effect of exogenous melatonin infusions [66], but this is not the case in Syrian hamsters [67].

Role of the A14/A15 Dopaminergic Nucleus

In sheep, dopaminergic neurons located in the A15 region of the forebrain appear to be involved in the seasonality of breeding (reviewed in refs. [68, 69]). These dopaminergic cells act in some way to reduce GnRH pulse frequency during the nonbreeding season [70, 71]. Although these A15 dopaminergic cells are activated by estrogen during the nonbreeding season, they do not express ERα [72]. Estrogen regulation may be via afferents to these cells, such as glutamatergic cells of the POA [73]. Neither do the A14/A15 dopaminergic cells project directly to GnRH cell bodies [74], so the question arises as to how there might be any involvement in the regulation of reproduction. The answer may be found in anterograde tracing experiments, which show projections from the A14/A15 region to the caudal ARC and the median eminence [69]; it is possible that these dopaminergic cells regulate kisspeptin cells at this level, but further work is required before this is established as a bona fide pathway. A working model was provided by Goodman et al. for the sheep [69], which is that estrogen feedback to the ventromedial hypothalamus and the retrochiasmatic nucleus is relayed to the A15 dopaminergic elements that then project to kisspeptin neurons in the ARC to regulate kisspeptin control of the GnRH neurons. The mechanism by which melatonin, and hence, photoperiod, might regulate this pathway is not known.

Role of Melatonin in the Photoperiodic Control of Kisspeptin Neurons

Evidence for seasonal changes in kisspeptin cell activity being controlled by melatonin has been provided by work in hamsters, because the decline in *Kiss1* expression that is seen in Syrian hamsters under short-days is lost if the animals are pinealectomized [63]. This outcome was recapitulated in a later study, such that the number of *Kiss1*-expressing cells in the ARC of male and female Syrian hamsters held in short days was increased following pinealectomy (i.e., removal of endogenous melatonin), with no change in the AVPV population [57]. Furthermore, pinealectomy of hamsters in long days prevents *Kiss1* downregulation upon subsequent transfer to short-day photoperiod [63]. Because it is accepted that melatonin acts at the level of the mediobasal hypothalamus [75, 76] in sheep at least, this observation is consistent with some mechanism at this level—perhaps the ARC—to control reproductive function. Moreover, these Syrian hamster data suggest that *Kiss1* expression in the AVPV and ARC is downregulated in short days via different mechanisms: in the ARC, short days inhibit *Kiss1* via a direct melatonin effect on the hypothalamus, whereas in the AVPV, the short-day decrease in *Kiss1* expression appears to be secondary to the melatonin-driven reduction of sex steroid feedback levels.

Under long days (active reproductive photoperiod), male Syrian hamsters treated with melatonin for 8 weeks showed a reduction in both paired testis weight and plasma testosterone levels [57]. This was accompanied by a fall in the number of *Kiss1*-expressing cells in both the AVPV and the ARC, consistent with the effect of simulated transfer from long-day to short-day photoperiod. In the absence of testes, melatonin caused a reduction in cell number in the ARC within 1 week (with no change in AVPV cell number), leading to the suggestion that testosterone may potentiate the effect of melatonin [57]. This work in Syrian hamsters supports the notion that melatonin drives the photoperiodic control of reproduction, through a mechanism upstream of the kisspeptin cells. It has not been established that kisspeptin cells express melatonin receptors in this species, but kisspeptin cells do not express melatonin receptor 1A (the signaling form of the receptor) in the ovine brain [77]. Because the time frame of the action of melatonin (several days or even weeks) compared to the immediate effects of kisspeptin on GnRH secretion, it seems most likely that some other cell type transduces the melatonin signal to the ARC kisspeptin cells. Recent evidence for a circadian-based molecular mechanism within the pars tuberalis of the pituitary couples the variations of the melatonin signal with the variations of thyroid hormones within the hypothalamus [78]. Given that most of the kisspeptin cells of the ARC express the thyroid hormone receptor [79], it is possible that there may be some involvement of melatonin to modulate thyroid hormone secretion, which, in turn, impacts upon kisspeptin function, to govern seasonal changes in breeding patterns.

Kisspeptin and Gonadotropin Inhibitory Hormone as Key Reciprocal Regulators in Seasonal Breeding

Gonadotropin inhibitory hormone (GnIH), also called RFRP-3, is a neuropeptide produced in cells of the dorsomedial nucleus/paraventricular nucleus of the hypothalamus in mammals. GnIH was originally isolated from the quail brain [80] and is associated with seasonality in birds [81]. In the ewe, GnIH gene expression and protein production are higher during the nonbreeding season than in the breeding season [30], and terminal projections from GnIH cells to GnRH neurons increase during the nonbreeding season [30]. Using an antagonist (RF9) to the GnIH receptor, Caraty et al. [82] were able to demonstrate release of LH, indicating that endogenous GnIH must act to restrain GnRH/LH secretion. The response to the antagonist was higher in the nonbreeding season than the breeding season, perhaps for the same reason that responses to kisspeptin are greater at this time (see above). These recent findings suggest that kisspeptin and GnIH act in concert to control seasonal cycles of breeding in mammals such as sheep. It should be noted however, that this antagonist may act through the putative GnIH receptor (also known as NPFF1) or NPFF2 receptors, having cardiovascular effects [83], and may also interact with other, as yet unidentified receptors. In the sheep, GnIH exerts dual effects on the GnRH neurons, as well as on pituitary gonadotropes. The effect on the gonadotropes is by means of projections of the GnIH neurons to the external zone of the median eminence [84] and secretion into the hypophyseal portal blood [85]. Portal blood levels of GnIH are higher in the anestrous season, indicating a role in the suppression of pituitary function in this reproductive state. There is virtually no projection of GnIH neurons to the external zone of the median eminence in the hamster [86], so pituitary action may not be relevant in these animals.

In hamsters, the level of GnIH expression has also been found to respond to season, but with results different to those seen in sheep. In particular, studies in Syrian and Siberian hamsters surprisingly showed increased levels of GnIH expression and numbers of immunoreactive cells under long-day (stimulatory) photoperiod [87], with no change being observed in Wistar rats (which are not seasonal breeders). Castration had no effect on GnIH gene expression during long days, and testosterone treatment had no effect in short days, leading to the conclusion that GnIH is not regulated by this gonadal steroid. Pinealectomy prevented the reduction in GnIH gene expression under short days, and melatonin treatment of long-day animals led to a reduction in GnIH gene expression within 3 weeks [87]. Most recently, another study in Siberian hamsters confirmed the results of the earlier work, confirming the changes in GnIH expression with long- and short-day photoperiod and regulation by melatonin, with corroborative immunohistochemical data [86]. In addition, these authors showed that GnIH fiber input to GnRH cells was reduced under short-day photoperiod. Importantly, it was shown that intracerebroventricular infusion of GnIH inhibited plasma LH levels in animals under long-day photoperiod but stimulated LH levels when animals were under short-day photoperiod. This led the authors to speculate that GnIH is inhibitory when LH levels are high and stimulatory when LH levels are low. These data implicate GnIH in "seasonality" in hamsters, but the higher levels of activity of GnIH neurons under long-day photoperiod (the breeding condition) indicate that it is unlikely that changes in GnIH are of major importance in the reproductive state that is controlled by photoperiod. However, GnIH certainly has significant roles in the regulation of appetite [84, 88] and stress [89], and the former is controlled by photoperiod in seasonally breeding mammals.

In summary, in sheep at least, there is good indication that seasonality is due to reciprocal changes in the influence of kisspeptin and GnIH on GnRH secretion, with the added effect of an increase in GnIH action on pituitary gonadotropes in anestrus; the same is not true for hamsters.

Response to Kisspeptin in Breeding and Nonbreeding "Seasons"

In sheep, the response to kisspeptin is greater in the nonbreeding season than in the breeding season [90], which appears to be due to a higher level of kisspeptin receptor expression in GnRH neurons and a higher GnRH response to kisspeptin in the nonbreeding season [31]. In a consistent manner, a higher response to kisspeptin, in terms of LH secretion, is seen in Siberian hamsters under short-day (nonbreeding) photoperiod compared to long-day (breeding) photoperiod [91]. The higher response to kisspeptin in nonbreeding animals may be a reflection of the lower GnRH/LH pulse frequency at this time, allowing for greater build up of releasable pools of GnRH and LH respectively.

Induction of Reproductive Function with Kisspeptin During Photoperiodic Quiescence

The recognition that kisspeptin cells are major regulatory elements of the gonadotropic axis (for review, see ref. [92]), together with the above-mentioned observations, strongly indicates that acyclicity in ewes during the nonbreeding season is associated with reduced kisspeptin function in the ARC. In order to determine

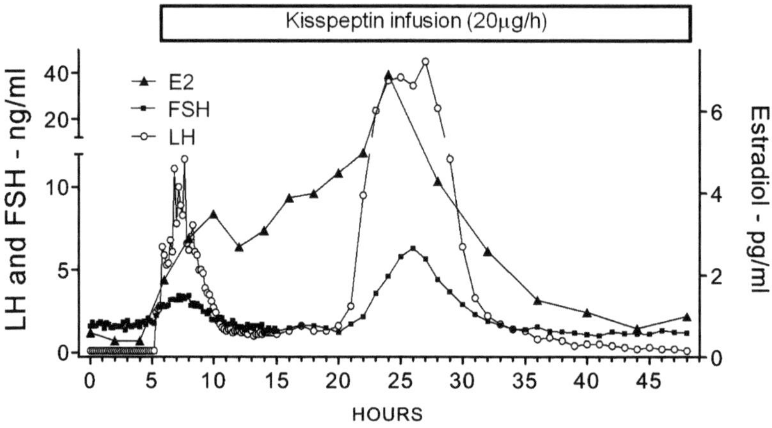

Fig. 19.2 Effect of an intravenous infusion of kisspeptin-10 on LH, FSH, and estradiol secretion patterns in a representative acyclic anestrous ewe. Administration of kisspeptin led to an immediate increase in plasma concentrations of LH (*open circles*) and FSH (*closed squares*) for a 4–5 h period. This was followed by a reduction in the concentrations of the two gonadotropins, but LH concentrations remained above baseline until a preovulatory surge of the two gonadotropins was eventually observed. Plasma estradiol concentrations (*black triangles*) were also increased by kisspeptin infusion and remained elevated until the occurrence of a gonadotropin surge. Redrawn from ref. [95]

whether kisspeptin treatment reactivates the gonadotropic axis and causes ovulation, we conducted studies in two breeds of sheep raised in both Northern and Southern hemispheres. Intravenous infusion of a low dose of kisspeptin-10 was given to acyclic anoestrus ewes over 48 h [93]. This treatment stimulated gonadotropin secretion, induced LH surges, and caused ovulation in more than 80 % of the anestrous animals. Analysis of the hormonal changes induced by this chronic kisspeptin infusion is illustrated in Fig. 19.2, indicating a sharp, albeit transient, increase in serum gonadotropin levels that was maintained for 4–5 h. Thereafter, while FSH returned to basal levels, LH secretion stayed at follicular phase levels (1–2 ng/mL) up until a well-synchronized LH surge occurred some 16–22 h later. Thus, in terms of mechanism, kisspeptin infusion induced a sequence of events similar to a normal follicular phase, which culminated in the positive feedback of estrogen in these animals with a quiescent hypothalamo-pituitary axis. This included an initial rapid stimulation of GnRH secretion, leading to the stimulation of LH secretion to follicular phase levels, which subsequently activated the ovaries and ultimately led to the initiation of the positive feedback circuits within the brain and the induction of the preovulatory LH surge.

An interesting question is why stimulation by kisspeptin can lead to a sustained level of LH secretion and not be overcome by the negative feedback effect of estrogen, which is pronounced in the anestrous season. At least one likely explanation is that this mechanism involves regulation of kisspeptin cells of the ARC and, since the treatment with exogenous kisspeptin overrides this, it allows a sustained GnRH secretion. Another interesting finding of these studies was that constant infusion of kisspeptin is efficacious in stimulating gonadotropin secretions, but intermittent intravenous injection was not [93]. Other work in primates has suggested that constant infusion of high kisspeptin amounts causes downregulation of the kisspeptin receptor [19]. In peripubertal female rats, circulating LH concentrations remain elevated after 7 days of continuous cerebral infusion of kisspeptin, an effect which is not observed in cyclic rats [94]. Therefore, it is very likely that, within a given range of concentrations, which might vary according to the species and developmental stage, continuous infusion of kisspeptin can stimulate GnRH, and hence gonadotropin secretion, for extended periods of time. This latter observation is especially relevant from a practical perspective, as it might lead to new treatments to control fertility. Indeed, it was demonstrated that 24 h of constant infusion of kisspeptin in acyclic anestrous ewes was the minimal duration required to induce the secretion of estrogen [95] to a level known to cause positive feedback [96]. For this to be practicable, however, we need to produce kisspeptin agonists with higher potency than the natural molecules, but there is only one report of such a compound with slightly higher activity than kisspeptin-10 [97]. An alternative strategy to address the problem could be to increase the half-life of the kisspeptin molecule in plasma, rather than its intrinsic potency. This strategy has already been applied successfully in the case of GnRH agonist molecules used in the treatment of various forms of cancers [98]. Generation of long-lasting agonists of kisspeptin could provide sustained stimulation of the GnRH/gonadotropic axis.

In essence, the demonstration that i.v. administration of kisspeptin can stimulate ovulation in seasonally anestrous ewes and in prepubertal ewes [99] offers a means of controlling reproduction in farm animals. For practical use, however, the development of new pharmacological tools will nevertheless prove essential to devise new strategies and make progress towards practical exploitation.

As with the issues relating to the role of kisspeptin and response to photoperiod in hamsters, our current understanding is somewhat unclear because of differences between the two species. Chronic central administration of kisspeptin-10 for 4 weeks to male Syrian hamsters on short days was shown to increase testis weight and plasma testosterone to levels seen under long-day photoperiod [63]. However, a paper by Greives et al. [58] indicated that various treatment paradigms of peripheral administration of kisspeptin-10 (infusion by osmotic mini-pump or daily i.p. injection) failed to activate reproductive function in male Siberian hamsters on short-day photoperiod. Neither did daily kisspeptin-10 injections prevent reproductive regression during transition from long days to short days. At face value, this indicates that male Siberian hamsters are markedly different from female sheep, in which kisspeptin infusion activates the reproductive axis in the nonbreeding season. On the other hand, continuous treatment of Syrian hamsters with kisspeptin-54 delivered by osmotic mini-pumps did not stimulate the reproductive axis in short-day (inhibitory) photoperiod, but intermittent treatment (two injections each day) did so [100]. These authors suggested that failure of continuous delivery could have been partly due to degradation of kisspeptin in the mini-pumps, an inappropriate dose, and/or downregulation of the kisspeptin receptors after chronic treatment. Another point of difference between the two studies was that Ansel et al. [100] used kisspeptin-54, whereas Greives et al. [58] used kisspeptin-10. Nevertheless, there is little difference in the potency of the two forms, even though kisspeptin-54 shows a slightly delayed effect [101].

The above findings raise the question of why does continuous peripheral kisspeptin treatment of sexually quiescent Syrian hamsters fail to activate the gonads, whereas continuous treatment of ewes does so Furthermore, one might ask why intermittent kisspeptin treatment of anestrous ewes was not effective, whereas constant infusion was so [93]. The answers are not yet known, but may be because the acute response in the anestrous ewe is a rise in LH secretion, sufficient to cause a rise in plasma estrogen levels, and it is the latter that then leads to a positive feedback response and an ovulatory LH surge. In this respect, nuances in the response to kisspeptin, in relation to species and the type of response being sought (male vs. female), need careful consideration.

Concluding Remarks

Kisspeptin cells appear to be integrally involved in the transitions into and out of the breeding seasons. Generally, there is an increase in the activity of ARC kisspeptin cells in the breeding season and a reduction in the nonbreeding season. Because these cells are estrogen responsive, they are able to transduce the enhanced negative feedback effect of estrogen that suppresses GnRH secretion in the nonbreeding season. Consistent with the findings in terms of cellular function, some kisspeptin treatments allow maintenance of reproductive function in hamsters housed in short-day (inhibitory) photoperiod and cause ovulation in anestrous ewes.

Data from hamsters and sheep strongly suggest that the seasonal changes in kisspeptin function are due to photoperiod, and more specifically, the pattern of melatonin secretion, but a means by which melatonin acts to cause these changes is a fundamental issue that requires some resolution before we can fully understand mechanisms of seasonal breeding.

In sheep at least, there appears to be a reciprocal change in the function of GnIH cells and kisspeptin cells that are complementary, such that reduced activity of kisspeptin and increased function of GnIH both suppress reproductive function in the anestrous season. This does not seem to be the case in Syrian or Siberian hamsters, where GnIH expression is surprisingly decreased during inhibitory (short day) photoperiod. Even though kisspeptin cells play a pivotal role in the reproductive response to photoperiod, a greater level of understanding of the afferents to kisspeptin cells is required before we can determine how seasonality of reproduction is actually controlled by melatonin.

References

1. Robinson TJ (1959) The oestrus cycle of the ewe and doe. In: Cole HH, Cupps PT (eds) Reproduction in domestic animals. Academic, New York, pp 292–333
2. Boden MJ, Kennaway DJ (2006) Circadian rhythms and reproduction [review]. Reproduction 132(3):379–392
3. Kennaway DJ (2005) The role of circadian rhythmicity in reproduction [review]. Hum Reprod Update 11(1):91–101
4. Reiter RJ (1980) The pineal and its hormones in the control of reproduction in mammals [review]. Endocr Rev 1(2):109–131
5. Lincoln GA, Andersson H, Loudon A (2003) Clock genes in calendar cells as the basis of annual timekeeping in mammals—a unifying hypothesis [review]. J Endocrinol 179(1):1–13
6. Karsch FJ, Bittman EL, Foster DL, Goodman RL, Legan SJ, Robinson JE (1984) Neuroendocrine basis of seasonal reproduction [review]. Recent Prog Horm Res 40:185–232
7. Barrell GK, Moenter SM, Caraty A, Karsch FJ (1992) Seasonal changes of gonadotropin-releasing hormone secretion in the ewe. Biol Reprod 46(6):1130–1135
8. Robinson JE, Radford HM, Karsch FJ (1985) Seasonal changes in pulsatile luteinizing hormone (LH) secretion in the ewe: relationship of frequency of LH pulses to day length and response to estradiol negative feedback. Biol Reprod 33(2):324–334
9. Legan SJ, Karsch FJ, Foster DL (1977) The endocrine control of seasonal reproductive function in the ewe: a marked change in response to the negative feedback action of estradiol on luteinizing hormone secretion. Endocrinology 101(3):818–824
10. Clarke IJ, Cummins JT (1982) The temporal relationship between gonadotropin releasing hormone (GnRH) and luteinizing hormone (LH) secretion in ovariectomized ewes. Endocrinology 111(5):1737–1739
11. Karsch FJ, Cummins JT, Thomas GB, Clarke IJ (1987) Steroid feedback inhibition of pulsatile secretion of gonadotropin-releasing hormone in the ewe. Biol Reprod 36(5):1207–1218
12. Karsch FJ, Dahl GE, Evans NP, Manning JM, Mayfield KP, Moenter SM et al (1993) Seasonal changes in gonadotropin-releasing hormone secretion in the ewe: alteration in response to the negative feedback action of estradiol. Biol Reprod 49(6):1377–1383
13. Clarke IJ (1993) Variable patterns of gonadotropin-releasing hormone secretion during the estrogen-induced luteinizing hormone surge in ovariectomized ewes. Endocrinology 133(4):1624–1632
14. Herbison AE (1998) Multimodal influence of estrogen upon gonadotropin-releasing hormone neurons. Endocr Rev 19(3):302–330
15. Herbison AE, Theodosis DT (1992) Localization of oestrogen receptors in preoptic neurons containing neurotensin but not tyrosine hydroxylase, cholecystokinin or luteinizing hormone-releasing hormone in the male and female rat. Neuroscience 50(2):283–298
16. Dorling AA, Todman MG, Korach KS, Herbison AE (2003) Critical role for estrogen receptor alpha in negative feedback regulation of gonadotropin-releasing hormone mRNA expression in the female mouse. Neuroendocrinology 78(4):204–209
17. Tilbrook AJ, Turner AI, Clarke IJ (2002) Stress and reproduction: central mechanisms and sex differences in non-rodent species. Stress 5(2):83–100
18. de Roux N, Genin E, Carel JC, Matsuda F, Chaussain JL, Milgrom E (2003) Hypogonadotropic hypogonadism due to loss of function of the KiSS1-derived peptide receptor GPR54. Proc Natl Acad Sci U S A 100(19):10972–10976
19. Seminara SB, Dipietro MJ, Ramaswamy S, Crowley WF Jr, Plant TM (2006) Continuous human metastin 45–54 infusion desensitizes G protein-coupled receptor 54-induced gonadotropin-releasing hormone release monitored indirectly in the juvenile male Rhesus monkey (Macaca mulatta): a finding with therapeutic implications. Endocrinology 147(5):2122–2126
20. Franceschini I, Lomet D, Cateau M, Delsol G, Tillet Y, Caraty A (2006) Kisspeptin immunoreactive cells of the ovine preoptic area and arcuate nucleus co-express estrogen receptor alpha. Neurosci Lett 401(3):225–230

21. Greives TJ, Mason AO, Scotti MA, Levine J, Ketterson ED, Kriegsfeld LJ et al (2007) Environmental control of kisspeptin: implications for seasonal reproduction. Endocrinology 148(3):1158–1166
22. Backholer K, Smith JT, Rao A, Pereira A, Iqbal J, Ogawa S et al (2010) Kisspeptin cells in the ewe brain respond to leptin and communicate with neuropeptide Y and proopiomelanocortin cells. Endocrinology 151(5):2233–2243
23. Pompolo S, Rawson JA, Clarke IJ (2001) Projections from the arcuate/ventromedial region of the hypothalamus to the preoptic area and bed nucleus of stria terminalis in the brain of the ewe; lack of direct input to gonadotropin-releasing hormone neurons. Brain Res 904(1):1–12
24. Backholer K, Smith J, Clarke IJ (2009) Melanocortins may stimulate reproduction by activating orexin neurons in the dorsomedial hypothalamus and kisspeptin neurons in the preoptic area of the ewe. Endocrinology 150(12):5488–5497
25. Kallo I, Vida B, Deli L, Molnar CS, Hrabovszky E, Caraty A et al (2012) Co-localisation of kisspeptin with galanin or neurokinin B in afferents to mouse GnRH neurones. J Neuroendocrinol 24(3):464–476
26. Smith JT, Li Q, Yap KS, Shahab M, Roseweir AK, Millar RP et al (2011) Kisspeptin is essential for the full preovulatory LH surge and stimulates GnRH release from the isolated ovine median eminence. Endocrinology 152(3):1001–1012
27. Goodman RL, Lehman MN, Smith JT, Coolen LM, de Oliveira CV, Jafarzadehshirazi MR et al (2007) Kisspeptin neurons in the arcuate nucleus of the ewe express both dynorphin A and neurokinin B. Endocrinology 148(12):5752–5760
28. d'Anglemont de Tassigny X, Fagg LA, Carlton MB, Colledge WH (2008) Kisspeptin can stimulate gonadotropin-releasing hormone (GnRH) release by a direct action at GnRH nerve terminals. Endocrinology 149(8):3926–3932
29. Smith JT, Rao A, Pereira A, Caraty A, Millar RP, Clarke IJ (2008) Kisspeptin is present in ovine hypophysial portal blood but does not increase during the preovulatory luteinizing hormone surge: evidence that gonadotropes are not direct targets of kisspeptin in vivo. Endocrinology 149(4):1951–1959
30. Smith JT, Coolen LM, Kriegsfeld LJ, Sari IP, Jaafarzadehshirazi MR, Maltby M et al (2008) Variation in kisspeptin and RFamide-related peptide (RFRP) expression and terminal connections to gonadotropin-releasing hormone neurons in the brain: a novel medium for seasonal breeding in the sheep. Endocrinology 149(11):5770–5782
31. Li Q, Roa A, Clarke IJ, Smith JT (2012) Seasonal variation in the gonadotropin-releasing hormone response to kisspeptin in sheep: possible kisspeptin regulation of the kisspeptin receptor. Neuroendocrinology 96(3):212–221
32. Hanchate NK, Parkash J, Bellefontaine N, Mazur D, Colledge WH, d'Anglemont de Tassigny X et al (2012) Kisspeptin-GPR54 signaling in mouse NO-synthesizing neurons participates in the hypothalamic control of ovulation. J Neurosci 32(3):932–945
33. Oakley AE, Clifton DK, Steiner RA (2009) Kisspeptin signaling in the brain [review]. Endocr Rev 30(6):713–743
34. Caraty A, Fabre-Nys C, Delaleu B, Locatelli A, Bruneau G, Karsch FJ et al (1998) Evidence that the mediobasal hypothalamus is the primary site of action of estradiol in inducing the preovulatory gonadotropin releasing hormone surge in the ewe. Endocrinology 139(4):1752–1760
35. Estrada KM, Clay CM, Pompolo S, Smith JT, Clarke IJ (2006) Elevated KiSS-1 expression in the arcuate nucleus prior to the cyclic preovulatory gonadotrophin-releasing hormone/lutenising hormone surge in the ewe suggests a stimulatory role for kisspeptin in oestrogen-positive feedback. J Neuroendocrinol 18(10):806–809
36. Smith JT, Li Q, Pereira A, Clarke IJ (2009) Kisspeptin neurons in the ovine arcuate nucleus and preoptic area are involved in the preovulatory luteinizing hormone surge. Endocrinology 150(12):5530–5538
37. Clarke IJ (1995) The preovulatory LH surge A case of a neuroendocrine switch. Trends Endocrinol Metab 6(7):241–247
38. Hoffman GE, Le WW, Franceschini I, Caraty A, Advis JP (2011) Expression of fos and in vivo median eminence release of LHRH identifies an active role for preoptic area

kisspeptin neurons in synchronized surges of LH and LHRH in the ewe. Endocrinology 152(1):214–222
39. Iqbal J, Pompolo S, Sakurai T, Clarke IJ (2001) Evidence that orexin-containing neurones provide direct input to gonadotropin-releasing hormone neurones in the ovine hypothalamus. J Neuroendocrinol 13(12):1033–1041
40. Pompolo S, Ischenko O, Pereira A, Iqbal J, Clarke IJ (2005) Evidence that projections from the bed nucleus of the stria terminalis and from the lateral and medial regions of the preoptic area provide input to gonadotropin releasing hormone (GNRH) neurons in the female sheep brain. Neuroscience 132(2):421–436
41. Pompolo S, Pereira A, Kaneko T, Clarke IJ (2003) Seasonal changes in the inputs to gonadotropin-releasing hormone neurones in the ewe brain: an assessment by conventional fluorescence and confocal microscopy. J Neuroendocrinol 15(5):538–545
42. Pompolo S, Pereira A, Scott CJ, Fujiyma F, Clarke IJ (2003) Evidence for estrogenic regulation of gonadotropin-releasing hormone neurons by glutamatergic neurons in the ewe brain: an immunohistochemical study using an antibody against vesicular glutamate transporter-2. J Comp Neurol 465(1):136–144
43. Pereira A, Rawson J, Jakubowska A, Clarke IJ (2010) Estradiol-17beta-responsive A1 and A2 noradrenergic cells of the brain stem project to the bed nucleus of the stria terminalis in the ewe brain: a possible route for regulation of gonadotropin releasing hormone cells. Neuroscience 165(3):758–773
44. Clarke IJ, Scott CJ, Pereira A, Rawson J (1999) Levels of dopamine beta hydroxylase immunoreactivity in the preoptic hypothalamus of the ovariectomised ewe following injection of oestrogen: evidence for increased noradrenaline release around the time of the oestrogen-induced surge in luteinizing hormone. J Neuroendocrinol 11(7):503–512
45. Mohankumar PS, Thyagarajan S, Quadri SK (1994) Correlations of catecholamine release in the medial preoptic area with proestrous surges of luteinizing hormone and prolactin: effects of aging. Endocrinology 135(1):119–126
46. Constantin S, Jasoni CL, Wadas B, Herbison AE (2010) Gamma-aminobutyric acid and glutamate differentially regulate intracellular calcium concentrations in mouse gonadotropin-releasing hormone neurons. Endocrinology 151(1):262–270
47. Herbison AE (1997) Estrogen regulation of GABA transmission in rat preoptic area. Brain Res Bull 44(4):321–326
48. Kuehl-Kovarik MC, Pouliot WA, Halterman GL, Handa RJ, Dudek FE, Partin KM (2002) Episodic bursting activity and response to excitatory amino acids in acutely dissociated gonadotropin-releasing hormone neurons genetically targeted with green fluorescent protein. J Neurosci 22(6):2313–2322
49. Smith JT, Clay CM, Caraty A, Clarke IJ (2007) KiSS-1 messenger ribonucleic acid expression in the hypothalamus of the ewe is regulated by sex steroids and season. Endocrinology 148(3):1150–1157
50. Pompolo S, Pereira A, Estrada KM, Clarke IJ (2006) Colocalization of kisspeptin and gonadotropin-releasing hormone in the ovine brain. Endocrinology 147(2):804–810
51. Cheng G, Coolen LM, Padmanabhan V, Goodman RL, Lehman MN (2010) The kisspeptin/neurokinin B/dynorphin (KNDy) cell population of the arcuate nucleus: sex differences and effects of prenatal testosterone in sheep. Endocrinology 151(1):301–311
52. Foradori CD, Goodman RL, Adams VL, Valent M, Lehman MN (2005) Progesterone increases dynorphin a concentrations in cerebrospinal fluid and preprodynorphin messenger ribonucleic acid levels in a subset of dynorphin neurons in the sheep. Endocrinology 146(4):1835–1842
53. Goodman RL, Coolen LM, Anderson GM, Hardy SL, Valent M, Connors JM et al (2004) Evidence that dynorphin plays a major role in mediating progesterone negative feedback on gonadotropin-releasing hormone neurons in sheep. Endocrinology 145(6):2959–2967
54. Billings HJ, Connors JM, Altman SN, Hileman SM, Holaskova I, Lehman MN et al (2010) Neurokinin B acts via the neurokinin-3 receptor in the retrochiasmatic area to stimulate luteinizing hormone secretion in sheep. Endocrinology 151(8):3836–3846

55. Amstalden M, Coolen LM, Hemmerle AM, Billings HJ, Connors JM, Goodman RL et al (2010) Neurokinin 3 receptor immunoreactivity in the septal region, preoptic area and hypothalamus of the female sheep: colocalisation in neurokinin B cells of the arcuate nucleus but not in gonadotrophin-releasing hormone neurones. J Neuroendocrinol 22(1):1–12
56. Young J, George JT, Tello JA, Francou B, Bouligand J, Guiochon-Mantel A et al (2012) Kisspeptin restores pulsatile LH secretion in patients with neurokinin B signaling deficiencies: physiological, pathophysiological and therapeutic implications. Neuroendocrinology [ePub ahead of print]
57. Ansel L, Bolborea M, Bentsen AH, Klosen P, Mikkelsen JD, Simonneaux V (2010) Differential regulation of kiss1 expression by melatonin and gonadal hormones in male and female Syrian hamsters. J Biol Rhythms 25(2):81–91
58. Greives TJ, Kriegsfeld LJ, Demas GE (2008) Exogenous kisspeptin does not alter photoperiod-induced gonadal regression in Siberian hamsters (Phodopus sungorus). Gen Comp Endocrinol 156(3):552–558
59. Smith JT, Dungan HM, Stoll EA, Gottsch ML, Braun RE, Eacker SM et al (2005) Differential regulation of KiSS-1 mRNA expression by sex steroids in the brain of the male mouse. Endocrinology 146(7):2976–2984
60. Wagner GC, Johnston JD, Clarke IJ, Lincoln GA, Hazlerigg DG (2008) Redefining the limits of day length responsiveness in a seasonal mammal. Endocrinology 149(1):32–39
61. Chalivoix S, Bagnolini A, Caraty A, Cognie J, Malpaux B, Dufourny L (2010) Effects of photoperiod on kisspeptin neuronal populations of the ewe diencephalon in connection with reproductive function. J Neuroendocrinol 22(2):110–118
62. Revel FG, Ansel L, Klosen P, Saboureau M, Pevet P, Mikkelsen JD et al (2007) Kisspeptin: a key link to seasonal breeding [review]. Rev Endocr Metab Disord 8(1):57–65
63. Revel FG, Saboureau M, Masson-Pevet M, Pevet P, Mikkelsen JD, Simonneaux V (2006) Kisspeptin mediates the photoperiodic control of reproduction in hamsters. Curr Biol 16(17):1730–1735
64. Simonneaux V, Ansel L, Revel FG, Klosen P, Pevet P, Mikkelsen JD (2009) Kisspeptin and the seasonal control of reproduction in hamsters. Peptides 30(1):146–153
65. Bartness TJ, Powers JB, Hastings MH, Bittman EL, Goldman BD (1993) The timed infusion paradigm for melatonin delivery: what has it taught us about the melatonin signal, its reception, and the photoperiodic control of seasonal responses? [review]. J Pineal Res 15(4):161–190
66. Bartness TJ, Goldman BD, Bittman EL (1991) SCN lesions block responses to systemic melatonin infusions in Siberian hamsters. Am J Physiol 260(1 pt 2):R102–R112
67. Bittman EL, Crandell RG, Lehman MN (1989) Influences of the paraventricular and suprachiasmatic nuclei and olfactory bulbs on melatonin responses in the golden hamster. Biol Reprod 40(1):118–126
68. Thiery JC, Gayrard V, Le Corre S, Viguie C, Martin GB, Chemineau P et al (1995) Dopaminergic control of LH secretion by the A15 nucleus in anoestrous ewes. J Reprod Fertil Suppl 49:285–296
69. Goodman RL, Jansen HT, Billings HJ, Coolen LM, Lehman MN (2010) Neural systems mediating seasonal breeding in the ewe. J Neuroendocrinol 22(7):674–681
70. Havern RL, Whisnant CS, Goodman RL (1994) Dopaminergic structures in the ovine hypothalamus mediating estradiol negative feedback in anestrous ewes. Endocrinology 134(4):1905–1914
71. Meyer SL, Goodman RL (1985) Neurotransmitters involved in mediating the steroid-dependent suppression of pulsatile luteinizing hormone secretion in anestrous ewes: effects of receptor antagonists. Endocrinology 116(5):2054–2061
72. Lehman MN, Durham DM, Jansen HT, Adrian B, Goodman RL (1996) Dopaminergic A14/A15 neurons are activated during estradiol negative feedback in anestrous, but not breeding season, ewes. Endocrinology 137(10):4443–4450
73. Singh SR, Hileman SM, Connors JM, McManus CJ, Coolen LM, Lehman MN et al (2009) Estradiol negative feedback regulation by glutamatergic afferents to A15 dopaminergic neurons: variation with season. Endocrinology 150(10):4663–4671

74. Tillet Y, Thibault J (1989) Catecholamine-containing neurons in the sheep brainstem and diencephalon: immunohistochemical study with tyrosine hydroxylase (TH) and dopamine-beta-hydroxylase (DBH) antibodies. J Comp Neurol 290(1):69–104
75. Malpaux B, Daveau A, Maurice F, Gayrard V, Thiery JC (1993) Short-day effects of melatonin on luteinizing hormone secretion in the ewe: evidence for central sites of action in the mediobasal hypothalamus. Biol Reprod 48(4):752–760
76. Malpaux B, Daveau A, Maurice-Mandon F, Duarte G, Chemineau P (1998) Evidence that melatonin acts in the premammillary hypothalamic area to control reproduction in the ewe: presence of binding sites and stimulation of luteinizing hormone secretion by in situ microimplant delivery. Endocrinology 139(4):1508–1516
77. Li Q, Rao A, Pereira A, Clarke IJ, Smith JT (2011) Kisspeptin cells in the ovine arcuate nucleus express prolactin receptor but not melatonin receptor. J Neuroendocrinol 23(10):871–882
78. Dardente H (2012) Melatonin-dependent timing of seasonal reproduction by the pars tuberalis: pivotal roles for long daylengths and thyroid hormones. J Neuroendocrinol 24(2):249–266
79. Dufourny L, Gennetay D, Beltramo M (2011) Kisspeptin neurons contain thyroid hormone receptor alpha in the ovine hypothalamus. In: Proceedings of the Society for Neuroscience annual meeting, Washington, 2011, Poster 712.12
80. Tsutsui K, Saigoh E, Ukena K, Teranishi H, Fujisawa Y, Kikuchi M et al (2000) A novel avian hypothalamic peptide inhibiting gonadotropin release. Biochem Biophys Res Commun 275(2):661–667
81. Bentley GE, Perfito N, Ukena K, Tsutsui K, Wingfield JC (2003) Gonadotropin-inhibitory peptide in song sparrows (Melospiza melodia) in different reproductive conditions, and in house sparrows (Passer domesticus) relative to chicken-gonadotropin-releasing hormone. J Neuroendocrinol 15(8):794–802
82. Caraty A, Blomenrohr M, Vogel GM, Lomet D, Briant C, Beltramo M (2012) RF9 powerfully stimulates gonadotrophin secretion in the ewe: evidence for a seasonal threshold of sensitivity. J Neuroendocrinol 24(5):725–736
83. Simonin F, Schmitt M, Laulin JP, Laboureyras E, Jhamandas JH, MacTavish D et al (2006) RF9, a potent and selective neuropeptide FF receptor antagonist, prevents opioid-induced tolerance associated with hyperalgesia. Proc Natl Acad Sci U S A 103(2):466–471
84. Clarke IJ, Sari IP, Qi Y, Smith JT, Parkington HC, Ubuka T et al (2008) Potent action of RFamide-related peptide-3 on pituitary gonadotropes indicative of a hypophysiotropic role in the negative regulation of gonadotropin secretion. Endocrinology 149(11):5811–5821
85. Smith JT, Ross Young I, Veldhuis JD, Clarke IJ (2012) Gonadotropin-inhibitory hormone (GnIH) secretion into the ovine hypophyseal portal system. Endocrinology 153(7):3368–3375
86. Ubuka T, Inoue K, Fukuda Y, Mizuno T, Ukena K, Kriegsfeld LJ et al (2012) Identification, expression, and physiological functions of Siberian hamster gonadotropin-inhibitory hormone. Endocrinology 153(1):373–385
87. Revel FG, Saboureau M, Pevet P, Simonneaux V, Mikkelsen JD (2008) RFamide-related peptide gene is a melatonin-driven photoperiodic gene. Endocrinology 149(3):902–912
88. Johnson MA, Tsutsui K, Fraley GS (2007) Rat RFamide-related peptide-3 stimulates GH secretion, inhibits LH secretion, and has variable effects on sex behavior in the adult male rat. Horm Behav 51(1):171–180
89. Kirby ED, Geraghty AC, Ubuka T, Bentley GE, Kaufer D (2009) Stress increases putative gonadotropin inhibitory hormone and decreases luteinizing hormone in male rats. Proc Natl Acad Sci U S A 106(27):11324–11329
90. Smith JT, Saleh SN, Clarke IJ (2009) Seasonal and cyclical change in the luteinizing hormone response to kisspeptin in the ewe. Neuroendocrinology 90(3):283–291
91. Greives TJ, Long KL, Burns CM, Demas GE (2011) Response to exogenous kisspeptin varies according to sex and reproductive condition in Siberian hamsters (Phodopus sungorus). Gen Comp Endocrinol 170(1):172–179
92. Roa J, Navarro VM, Tena-Sempere M (2011) Kisspeptins in reproductive biology: consensus knowledge and recent developments. Biol Reprod 85(4):650–660

93. Caraty A, Smith JT, Lomet D, Ben Said S, Morrissey A, Cognie J et al (2007) Kisspeptin synchronizes preovulatory surges in cyclical ewes and causes ovulation in seasonally acyclic ewes. Endocrinology 148(11):5258–5267
94. Roa J, Vigo E, Garcia-Galiano D, Castellano JM, Navarro VM, Pineda R et al (2008) Desensitization of gonadotropin responses to kisspeptin in the female rat: analyses of LH and FSH secretion at different developmental and metabolic states. Am J Physiol Endocrinol Metab 294(6):E1088–E1096
95. Sebert ME, Lomet D, Said SB, Monget P, Briant C, Scaramuzzi RJ et al (2010) Insights into the mechanism by which kisspeptin stimulates a preovulatory LH surge and ovulation in seasonally acyclic ewes: potential role of estradiol. Domest Anim Endocrinol 38(4):289–298
96. Ben Said S, Lomet D, Chesneau D, Lardic L, Canepa S, Guillaume D et al (2007) Differential estradiol requirement for the induction of estrus behavior and the luteinizing hormone surge in two breeds of sheep. Biol Reprod 76(4):673–680
97. Curtis AE, Cooke JH, Baxter JE, Parkinson JR, Bataveljic A, Ghatei MA et al (2010) A kisspeptin-10 analog with greater in vivo bioactivity than kisspeptin-10. Am J Physiol Endocrinol Metab 298(2):E296–E303
98. Whelan P (2010) Triptorelin embonate: a 6-month formulation for prostate cancer [review]. Expert Opin Pharmacother 11(17):2929–2932
99. Redmond JS, Macedo GG, Velez IC, Caraty A, Williams GL, Amstalden M (2011) Kisspeptin activates the hypothalamic-adenohypophyseal-gonadal axis in prepubertal ewe lambs. Reproduction 141(4):541–548
100. Ansel L, Bentsen AH, Ancel C, Bolborea M, Klosen P, Mikkelsen JD et al (2011) Peripheral kisspeptin reverses short photoperiod-induced gonadal regression in Syrian hamsters by promoting GNRH release. Reproduction 142(3):417–425
101. Mikkelsen JD, Bentsen AH, Ansel L, Simonneaux V, Juul A (2009) Comparison of the effects of peripherally administered kisspeptins. Regul Pept 152(1–3):95–100

Chapter 20
Stress Regulation of Kisspeptin in the Modulation of Reproductive Function

Pasha Grachev, Xiao Feng Li, and Kevin O'Byrne

Abstract Stressful stimuli abound in modern society and have shaped evolution through altering reproductive development, behavior, and physiology. The recent identification of kisspeptin as an important component of the hypothalamic regulatory circuits involved in reproductive homeostasis sparked a great deal of research interest that subsequently implicated kisspeptin signaling in the relay of metabolic, environmental, and physiological cues to the hypothalamo–pituitary–gonadal axis. However, although it is widely recognized that exposure to stress profoundly impacts on reproductive function, the roles of kisspeptin within the complex mechanisms underlying stress regulation of reproduction remain poorly understood. We and others have recently demonstrated that a variety of experimental stress paradigms downregulate the expression of kisspeptin ligand and receptor within the reproductive brain. Coincidently, these stressors also inhibit gonadotropin secretion and delay pubertal onset—processes that rely on kisspeptin signaling. However, a modest literature is inconsistent with an exclusively suppressive influence of stress on the reproductive axis and suggests that complicated neural interactions and signaling mechanisms translate the stress response into reproductive perturbations. The purpose of this chapter is to review the evidence for a novel role of kisspeptin signaling in the modulation of reproductive function by stress and to broaden the understanding of this timely phenomenon.

P. Grachev • X.F. Li
Division of Women's Health, School of Medicine,
King's College London, London, UK
e-mail: pasha.grachev@kcl.ac.uk

K. O'Byrne (✉)
Division of Women's Health, King's College London,
2.92 W Hodgkin Building, Guy's Campus, London SE1 1UL, UK
e-mail: kevin.obyrne@kcl.ac.uk

A.S. Kauffman and J.T. Smith (eds.), *Kisspeptin Signaling in Reproductive Biology*, Advances in Experimental Medicine and Biology 784, DOI 10.1007/978-1-4614-6199-9_20,

Introduction

Kisspeptin has been widely recognized as a key regulator of the hypothalamo–pituitary–gonadal (HPG) axis, and thus reproductive development and function. The roles of kisspeptin in the control of sexual differentiation, puberty, surge and pulsatile modes of gonadotropin secretion, gonadal steroid hormone feedback, pregnancy, lactation, and reproductive senescence have been well studied in a wide variety of clinical and animal models, as well as in vitro. However, the discovery of kisspeptin was associated with research on metastasis suppression mechanisms in malignant melanoma cells [1], and kisspeptin has been implicated in several peripheral, as well as central, processes, both physiological and pathological, that have little direct relevance to reproduction, including hippocampal seizures [2], cardiovascular vasoconstriction [3], renal fluid homeostasis [4], and pancreatic glucose-induced insulin secretion [5].

Although, by far, the most significant populations of kisspeptin-expressing neurons are found in the arcuate nucleus (ARC) and preoptic area (including the anteroventral periventricular (AVPV) and preoptic periventricular (PeN) nuclei in rodents) of the hypothalamus, both chiefly involved in regulation of reproductive functions, kisspeptin fibers are abundant in the paraventricular nucleus (PVN) of the hypothalamus [6], where perikarya of neurons that secrete corticotropin-releasing factor (CRF, a key neuroendocrine mediator of the stress response) and the stress-associated peptide hormones, arginine vasopressin (AVP) and oxytocin, are located [7]. Furthermore, numerous kisspeptin neurons project to limbic structures (namely the bed nucleus of the stria terminalis (BNST) and the medial nucleus of the amygdala (MeA)) [6], which innervate the PVN via gamma-aminobutyric acid (GABA)ergic inputs [8]. Therefore mammalian neuroanatomy potentially permits an intimate interaction between the HPG and stress axes, within which kisspeptin signaling may play a key role.

Experimental interventions commonly used so simulate a range of stressful situations and evoke a physiological stress response are referred to as stress paradigms. Administration of the *Escherichia coli* endotoxin, lipopolysaccharide (LPS), elicits reproducible acute inflammation that serves as a paradigm of immunological stress. Restraint of experimental animals is a well-defined acute psychological stress model. Social isolation has been used to replicate the effects of psychosocial stress, and is considered a mild stressor. Social status or rank reflects a more complex and chronic psychosocial stressor. The paradigms of metabolic stress are comparatively more diverse: nutritional disbalance, short-term fasting, chronic malnutrition, insulin-induced hypoglycemia and exercise are all considered to exert physiological stress on individuals, although perhaps through different mechanisms. These have been extensively reviewed in the context of pubertal onset [9, 10] and reproductive function in adults [11]. Numerous other stress paradigms are routinely employed in research, and many share overlapping effects on the stress response and reproductive processes; a comprehensive review of each is beyond the scope of this chapter. Therefore we focused our discussion on the restraint, LPS, insulin-induced hypoglycemia and social isolation experimental stress paradigms.

Puberty, Stress and Kisspeptin

One major facet of reproductive research that concerns the effects of both kisspeptin and stress is puberty. Characterized as the transition from juvenile to adult state of reproductive development, growth, and adrenal maturation, pubertal timing in mammals (including man) is controlled by a multiplicity of complex interactions between genetic and environmental factors, with the latter providing fine tuning to maximize reproductive potential to fit the prevailing or predicted environment. There is unequivocal evidence that chronic stress exposure suppresses the activity of the HPG axis and delays puberty. However, less well recognized is the advancement of puberty by environmental factors, such as psychosocial stress within the family domain, including absence of father, as well as parental and mother–daughter conflict [12], with striking parallels in an animal model of weak parent–offspring bonding [13]. The age at which girls are reaching puberty has been trending downwards in recent decades, coincident with the increasing prevalence of overweight and obesity; this is in keeping with evidence that overnutrition advances puberty [14] and body fatness and a rapid elevation in BMI are predictors of earlier onset of puberty. However, recent studies reveal that this downward trend is evident irrespective of BMI, suggesting that other factors are involved [14, 15]. Animal models of delayed puberty (e.g., intrauterine growth retardation [16, 17] or neonatal exposure to LPS [18]) or advanced puberty (e.g., low maternal care [13] or high-fat diet [19]) also fail to show a clear correlation between body weight, body fat, or its central signaling biomarker, leptin, and the timing of puberty.

Kisspeptin/Kiss1r signaling is a prerequisite for physiological development of the reproductive system, since inactivating mutations in *KISS1* [20] or *KISS1R* [21, 22] result in pubertal failure and hypogonadotropic hypogonadism in humans. *Kiss1*- and *Kiss1r*-knockout mouse models largely produce a similar reproductive phenotype [23–25], although *Kiss1*-knockdown results in a less severe phenotype [26]. However, a recent report challenged the dogma that kisspeptin signaling is required for puberty and adult fertility in mice [27]. Nevertheless, levels of hypothalamic *Kiss1* and *Kiss1r* expression peak at puberty (Fig. 20.1) [18, 28, 29], and kisspeptin administration stimulates precocious puberty in rats [30]. Furthermore, polymorphisms in the *KISS1* gene [31] and activating mutations in *KISS1R* [32] are associated with central precocious puberty in humans. A variety of well-defined acute and chronic stress paradigms and experimental parameters of pubertal development have facilitated in vivo research aiming to elucidate the effects of stress on reproductive maturation. However, to this end, few studies have directly addressed the involvement of kisspeptin in the alteration of pubertal dynamics in response to stress.

Exposure to an immune challenge, such as LPS, in the early neonatal period has been shown to have profound and long-lasting effects on the stress response throughout later life in the rat, evident through increased CRF gene expression in the PVN of the hypothalamus and an increase in the pulse frequency and amplitude of corticosterone (CORT) release [33]. We have shown that neonatal exposure to

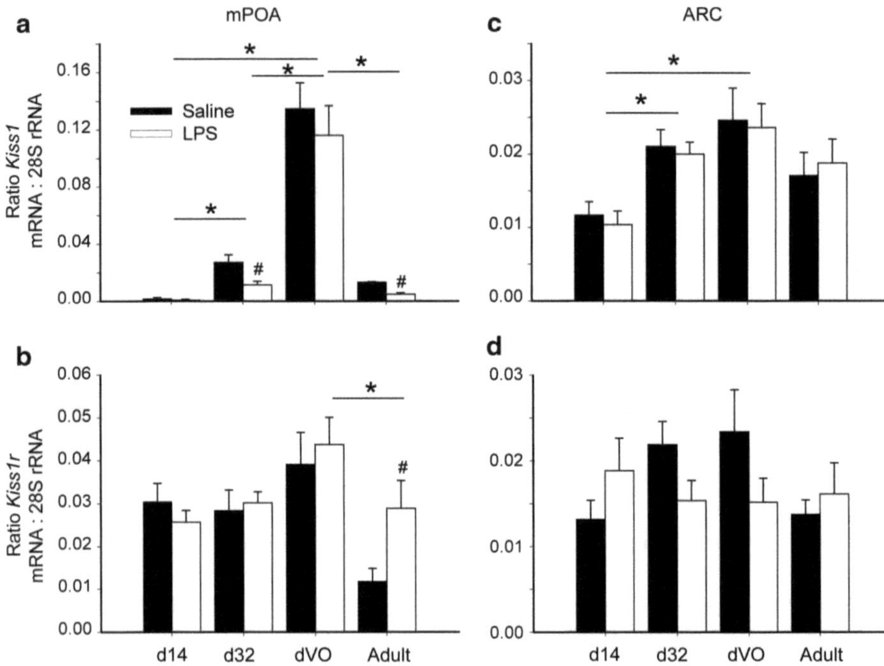

Fig. 20.1 Effects of neonatal lipopolysaccharide (LPS, 50 μg/kg ip) or saline control given at postnatal day (pnd) 3–5 on kisspeptin (*Kiss1*) mRNA expression in the medial preoptic area (mPOA) (**a**) and arcuate nucleus (ARC) (**c**) and on kisspeptin receptor (*Kiss1r*) mRNA expression in the mPOA (**b**) and ARC (**d**) in female rats at pnd 14, pnd 32, (the day of vaginal opening (dVO)), and at 11 weeks of age (Adult). *Kiss1* and *Kiss1r* mRNA levels were measured in brain micropunch samples from the mPOA or ARC using real-time reverse transcriptase–polymerase chain reaction. Quantification for *Kiss1*, *Kiss1r*, and 28S rRNA was carried out on all samples; the values are expressed as a ratio of *Kiss1* mRNA and 28S rRNA, or *Kiss1r* mRNA and 28S rRNA (mean ± SEM). $^{*}P<0.05$ vs. the respective treatment group at different time points; $^{\#}P<0.05$ vs. saline control at same time point; $n=5$–9 per group (from Knox AM, Li XF, Kinsey-Jones JS, Wilkinson ES, Wu XQ, Cheng YS, et al. Neonatal lipopolysaccharide exposure delays puberty and alters hypothalamic Kiss1 and Kiss1r mRNA expression in the female rat. J Neuroendocrinol. 2009 Aug;21(8):683–9. Reprinted with permission from John Wiley & Sons)

LPS delays puberty and disrupts estrous cyclicity, concordant with downregulation of *Kiss1*, but not *Kiss1r*, mRNA expression in the medial preoptic area (mPOA) (Fig. 20.1) [18, 34]. This decrease in *Kiss1* expression could provide a mechanism for the observed delay of puberty. By contrast, the lack of effect of neonatal LPS treatment on *Kiss1* or *Kiss1r* expression in the ARC would indicate that kisspeptin/Kiss1r signaling in this brain region is not an obvious contributing factor to pubertal delay. Interestingly, the postponement of LPS treatment from postnatal days 3 and 5 to postnatal days 7 and 9 failed to delay puberty, suggesting there is a discrete developmental time window that is sensitive to immunological challenge [18].

A variety of stressors that suppress GnRH pulse generator frequency, including LPS, downregulate hypothalamic *Kiss1* and *Kiss1r* expression in adult rats [35, 36]. Although the neural inputs to the mPOA and ARC that mediate the stress-induced suppression of *Kiss1* and *Kiss1r* expression within these loci remain to be determined, CRF is a prime candidate, not only because of its pivotal role in stress-induced suppression of the GnRH pulse generator [37–41], but also with respect to data showing that intracerebroventricular (icv) administration of CRF profoundly decreased *Kiss1* and *Kiss1r* mRNA levels in both the mPOA and ARC [35]. Moreover, we have shown that icv administration of CRF or CRF antagonist delays or advances puberty, respectively, without altering the circulating levels of CORT, suggestive of CRF regulation of kisspeptin at puberty in the female rat [42]. Indeed, the delay in puberty consequent to administration of CRF was associated with a marked reduction in *Kiss1* mRNA expression in the mPOA, but not the ARC [42]. Interestingly, we have shown a reduction in *CRF*, CRF receptor type 1 (*CRF-R1*), and CRF receptor type 2 (*CRF-R2*) mRNA expression in the PVN, the core regulatory component of the HPA axis, across pubertal transition in female rats (Fig. 20.2) [42]. Since bilateral lesions of the PVN do not alter the response of the HPG axis to stress in adult rats [43], the site and mechanism of action of the endogenous CRF tone that plays a critical role in the timing of puberty remain elusive. In the Japanese quail, genetic selection for high or low HPA axis stress responsivity delays or advances puberty, respectively [44]. A decrease in *CRF* and *CRF-R1* expression was observed also in the mPOA, with no change in the ARC, across the pubertal transition in the rat (Fig. 20.2) [42]. However, possible interaction between CRF and kisspeptin signaling systems in the mPOA remains to be examined.

The importance of the limbic brain, in particular the amygdaloid complex, in the control of reproductive function and stress responsivity is well recognized [8, 45]. Overexpression of CRF in the central nucleus of the amygdala (CeA) disrupts estrous cyclicity and reduces GnRH expression [46]. Further, there is upregulation of CRF in the CeA with delayed puberty in neonatally LPS-treated female rats (Li XF and O'Byrne KT; unpublished observation). The MeA is also a major regulator of the HPG and HPA axes [47], and there are extensive reciprocal connections between the MeA and CeA [48, 49]. Stimulation and ablation studies in prepubertal rats have revealed a critical role for the MeA in the timing of puberty, with lesions advancing puberty [50] and stimulation delaying puberty [51]. Moreover, we have shown that intra-MeA administration of CRF delays puberty in female rats (Li XF and O'Byrne KT; unpublished observation). Although extensive projections from the MeA to the mPOA [49] provide an anatomical substrate for a potential interaction with kisspeptin signaling systems that may mediate this inhibitory effect on pubertal timing, their neurochemical phenotype and modus operandi remain unknown.

Recent studies have shown that kisspeptin neurons project to many brain loc., including the PVN, which are not considered components of the HPG axis [6]. Although kisspeptin administration did not alter glucocorticoid secretion in vivo in monkeys [52] and rats [53], it did decrease *CRF* mRNA expression in a PVN neural cell line, thus raising the possibility of a novel mechanism for the regulation of the

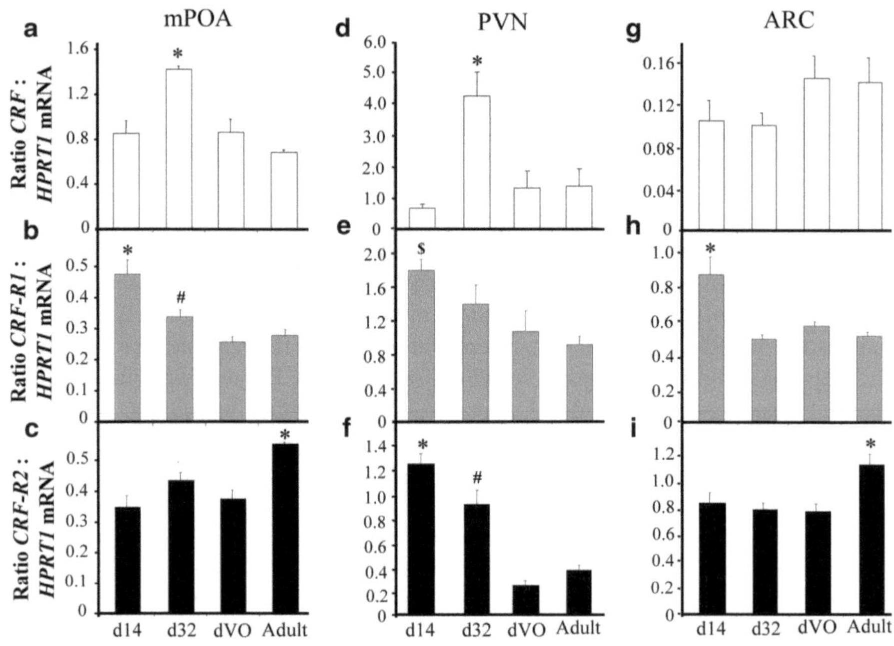

Fig. 20.2 Developmental changes in CRF, CRF-R1, and CRF-R2 across puberty in the medial preoptic area (mPOA) (**a**–**c**), hypothalamic paraventricular nucleus (PVN) (**d**–**f** respectively), and arcuate nucleus (ARC) (**g**–**i** respectively). Time points include postnatal day 14 (d14), 32 (d32), day of vaginal opening (dVO: 40.9±0.9 days, mean±SEM), and adult (postnatal day 77). *CRF*, *CRF-R1,* and *CRF-R2* mRNA levels were measured in brain micropunch samples from the mPOA, PVN, and ARC using a real-time reverse transcriptase–polymerase chain reaction. Note the peak in *CRF* mRNA levels in the mPOA and PVN in the late prepubertal phase that has receded by the time of puberty and the gradual decline in *CRF-R1* expression in the mPOA across the pubertal transition. Quantification for *CRF*, *CRF-R1*, *CRF-R2,* and *HPRT1* mRNA was carried out on all samples and the values are expressed as a ratio of *CRF*, *CRF-R1,* or *CRF-R2* to *HPRT1* mRNA (mean±SEM). $^{*}P<0.05$ vs. other time points; $^{\#}P<0.05$ vs. dVO; $^{\$}P<0.05$ vs. adult; $n=6$–10 per group (from Kinsey-Jones JS, Li XF, Knox AM, Lin YS, Milligan SR, Lightman SL, et al. Corticotrophin-releasing factor alters the timing of puberty in the female rat. J Neuroendocrinol. 2010 Feb;22(2):102-9. Reprinted with permission from John Wiley & Sons)

HPA axis by kisspeptin, as proposed by Rao and colleagues [53]. Indeed, the inverse relationship between PVN CRF and mPOA *Kiss1* expression across the pubertal transition in the female rat (Figs. 20.1 and 20.2) is striking, though causality is unknown. In conclusion, a variety of stressors impact on pubertal timing, potentially by interfering with hypothalamic kisspeptin/Kiss1r signaling, though further research is required to elucidate the neural mechanisms involved in this interaction. The dynamics of pubertal development constitute a powerful correlate of HPG axis function vulnerable to stress, and the wealth of evidence for the involvement of kisspeptin makes the developing reproductive system a useful tool for the study of interactions between stress and kisspeptin signaling.

The GnRH Pulse Generator, Stress, and Kisspeptin

The dynamics of the pulsatile secretion of GnRH are believed to be controlled by a neural construct, probably resident within the mediobasal hypothalamus (MBH), termed the "GnRH pulse generator" [54]. The discovery of kisspeptin as a powerful and indispensible GnRH secretagogue, and the ensuing surge of academic discoveries linked to kisspeptin signaling, provide ample grounds on which to base the hypothesis that ARC kisspeptin neurons that coexpress neurokinin B (NKB), dynorphin A (Dyn), and the alpha subtype of the estrogen receptor (ERα) are the substrate of the GnRH pulse generator. These so-called KNDy neurons generate stimulatory and inhibitory signals to downstream mediators of reproductive function, provide estradiol-sensitive feedback to GnRH neurons, and seem to relay a wide variety of physiological and environmental stimuli to the HPG axis. Furthermore, the hypophysial GnRH and systemic LH pulses, as well as multiple electrophysiological manifestations of the GnRH pulse generator, serve as components of what can be described as a detailed bioassay for the many extrinsic stimuli that influence the HPG axis. For this reason, the study of the GnRH pulse generator is an indispensible tool for addressing the interactions of the stress system and the HPG axis.

The regulation of kisspeptin signaling by stress neural networks is evident in adult as well as peripubertal animals, despite a decline in ARC and mPOA kisspeptin expression following puberty [18, 55]. In adult female rats treated with LPS, levels of *Kiss1* and *Kiss1r* mRNAs were decreased in the ARC and mPOA, in association with a suppression of GnRH pulse generator frequency (Figs. 20.3 and 20.4) [35]. Administration of kisspeptin reversed the reduction in serum LH levels associated with acute inflammation, and the LPS-induced suppression of LH secretion and *Kiss1* mRNA expression are blocked by the anti-inflammatory drug, indomethacin [56]. Adult male rats treated with LPS have less abundant kisspeptin-immunoreactive cell bodies in the ARC than saline-injected controls, and the LPS-induced reduction in LH and testosterone (T) secretion in this animal model is independent of the anorexic effects of inflammation [57].

Adult female rats treated neonatally with LPS show decreased *Kiss1* but elevated *Kiss1r* mRNA expression in the mPOA (Fig. 20.1) [18]. The same repercussions are observed following acute or chronic CORT administration in adult female rats but with identical changes in *Kiss1/Kiss1r* expression extending to the ARC (Fig. 20.5) [35]. Furthermore, such early life stress exposure programs the stress response in adulthood by inducing chronic hypercorticosteronemia [33] and upregulating the mPOA expression of *CRF-R1* mRNA in response to acute homotypic stress exposure [58]. Such programming evidently sensitizes the animals to stress exposure in later life, with augmented stress-induced CORT secretion [33], disrupted estrous cyclicity [34], and suppression of the GnRH pulse generator [58] in rats treated neonatally with LPS than in saline-treated controls. On the contrary, neonatal LPS challenge in male rat pups attenuated the response to homotypic stress challenge in adulthood by restricting the increase in hypothalamic and testicular expression of mRNA encoding the proinflammatory cytokines TNF-α and IL-1β in response to LPS injection in

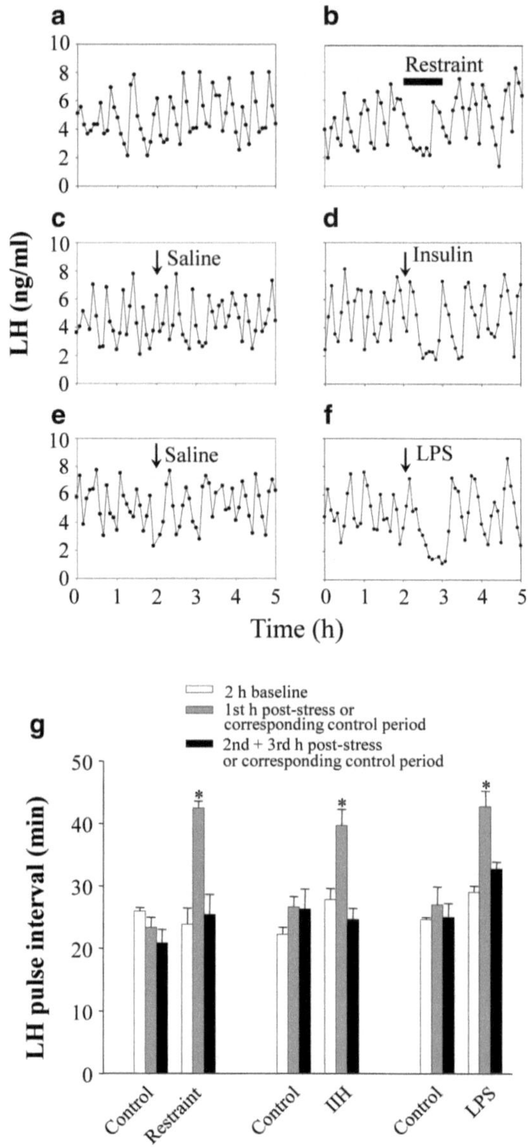

Fig. 20.3 Effects of restraint, insulin-induced hypoglycemia (IIH), and lipopolysaccharide (LPS) stress on pulsatile LH secretion in ovariectomized estradiol (E_2)-replaced rats. Representative examples showing (**a**) normal LH pulses in a rat in the absence of restraint stress, (**b**) restraint (1 h) stress-induced interruption of LH pluses, (**c**) normal LH pulses in a rat receiving an intravenous injection of vehicle (0.3 mL saline), (**d**) the inhibitory effects of insulin-induced hypoglycemic (insulin, 0.25 IU/kg in 0.3 mL saline i.v.) on LH pulses, (**e**) normal LH pulses in a rat receiving an intravenous injection of vehicle (0.3 mL saline) for the immunological stress, and (**f**) the inhibitory effects of LPS (0.5 μg/kg in 0.3 mL saline i.v.) on LH pulses. (**g**) Summaries showing the effects the three different stress paradigms on pulsatile LH secretion. *$P<0.05$ vs. 2-h baseline control period within the same treatment group; $n=8$ per group (from Kinsey-Jones JS, Li XF, Knox AM, Lin YS, Milligan SR, Lightman SL, et al. Corticotrophin-releasing factor alters the timing of puberty in the female rat. J Neuroendocrinol. 2010 Feb;22(2):102-9. Reprinted with permission from John Wiley & Sons)

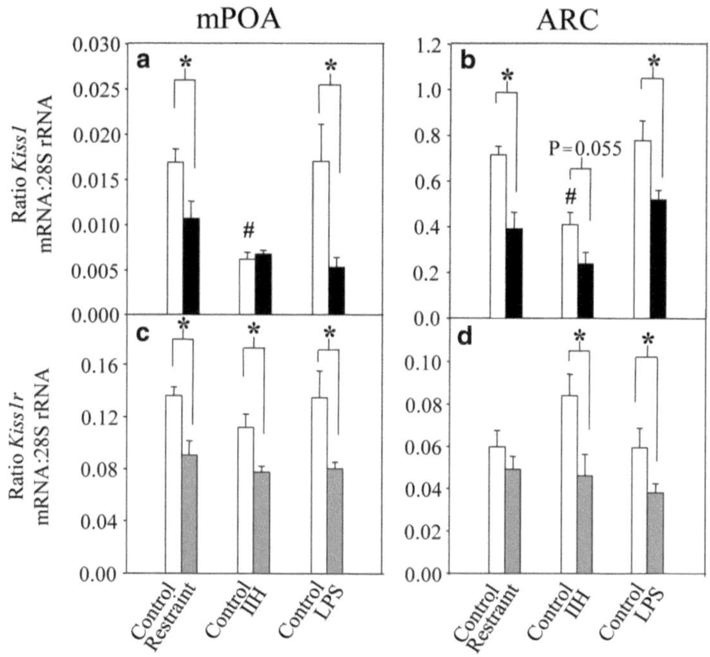

Fig. 20.4 Effects of restraint, insulin-induced hypoglycemia (IIH), and lipopolysaccharide (LPS) stress on kisspeptin (*Kiss1*) mRNA expression in the medial preoptic area (mPOA) (**a**) and arcuate nucleus (ARC) (**b**), and on kisspeptin receptor (*Kiss1r*) mRNA expression in the mPOA (**c**) and ARC (**d**) in ovariectomized estradiol (E_2)-replaced rats. *Kiss1* and *Kiss1r* mRNA levels were measured in brain micropunch samples from the mPOA or ARC using real-time RT-PCR. Quantification for *Kiss1*, *Kiss1r*, and 28S rRNA was carried out on all samples and the values are expressed as a ratio of *Kiss1* mRNA and 28S rRNA, or *Kiss1r* mRNA and 28S rRNA (mean ± SEM). $^*P<0.05$ vs. respective control. $^{\#}P<0.05$ vs. restraint or LPS controls; $n=8$ per group (from Kinsey-Jones JS, Li XF, Knox AM, Wilkinson ES, Zhu XL, Chaudhary AA, et al. Down-regulation of hypothalamic kisspeptin and its receptor, Kiss1r, mRNA expression is associated with stress-induced suppression of luteinising hormone secretion in the female rat. J Neuroendocrinol. 2009 Jan;21(1):20-9. Reprinted with permission from John Wiley & Sons)

adulthood, and thus preventing inflammatory suppression of LH secretion [59]. Indeed, the HPA axis is differentially modulated by the male and female gonadal steroids: hypothalamic activation of ERα augments the diurnal and stress-induced rises in circulating CORT, thus impairing glucocorticoid negative feedback on the HPA axis [60], while T enhances this feedback mechanism by increasing the availability of bioactive CORT, thereby decreasing circulating glucocorticoid levels [61]. In summary, neonatal LPS challenge modulates kisspeptin/Kiss1r signaling in adults and differentially programs responsiveness to stress later in adulthood in males and females.

Intravenous administration of insulin dramatically decreases plasma glucose, inducing acute hypoglycemia—a well-established paradigm of metabolic stress. Such stress rapidly elevates serum cortisol and suppresses both the endocrine and the electrophysiological correlates of the GnRH pulse generator in ovariectomized

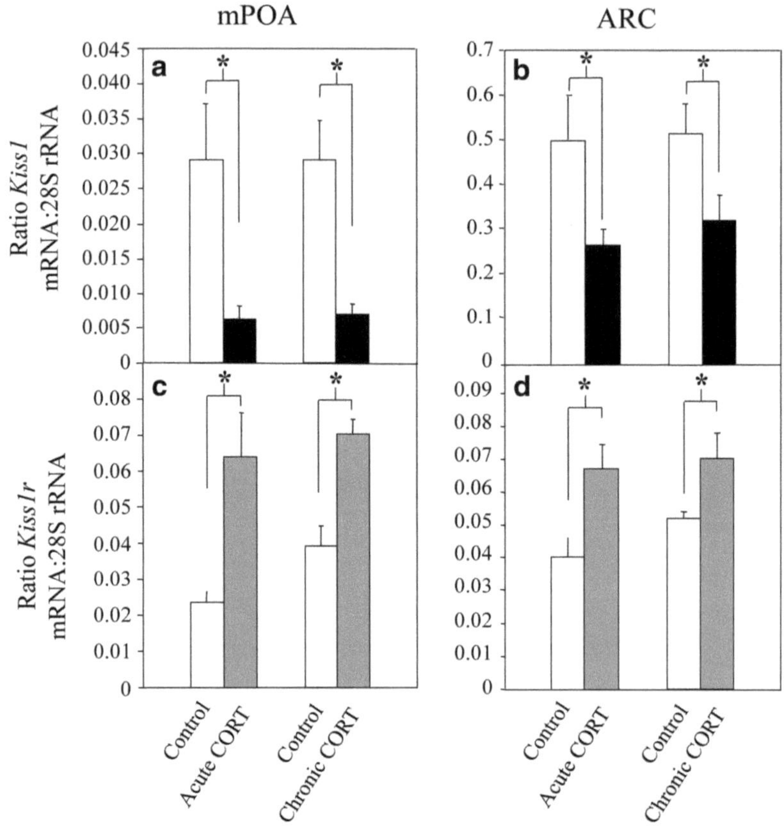

Fig. 20.5 Effects of acute (2 mg/kg, s.c. injection) or chronic (2×200 mg 21-day release s.c. pellets) CORT administration on kisspeptin (*Kiss1*) mRNA expression in the medial preoptic area (mPOA) (**a**) and arcuate nucleus (ARC) (**b**), and on *Kiss1r* mRNA expression in the mPOA (**c**) and ARC (**d**) in ovariectomized estradiol (E_2)-replaced rats. *Kiss1* and *Kiss1r* mRNA levels were measured in brain micropunch samples from the mPOA or ARC using real-time RT-PCR. Quantification for *Kiss1*, *Kiss1r*, and 28S rRNA was carried out on all samples and the values are expressed as a ratio of *Kiss1* mRNA and 28S rRNA, or *Kiss1r* mRNA and 28S rRNA (mean ± SEM). $*P<0.05$ vs. respective controls; $n=6–8$ per group (from Kinsey-Jones JS, Li XF, Knox AM, Wilkinson ES, Zhu XL, Chaudhary AA, et al. Down-regulation of hypothalamic kisspeptin and its receptor, Kiss1r, mRNA expression is associated with stress-induced suppression of luteinising hormone secretion in the female rat. J Neuroendocrinol. 2009 Jan;21(1):20-9. Reprinted with permission from John Wiley & Sons)

rhesus monkeys [62]. The inhibitory effect of insulin-induced hypoglycemia on pulsatile secretion of LH in adult female monkey and rats is CRF-dependent [38, 62], and involves CRF-R2, but not CRF-R1, at least in the rat [40]. Like other stressors, such as restraint (psychological) and LPS (immunological), hypoglycemia down-regulates *Kiss1* mRNA expression in the ARC and *Kiss1r* mRNA expression in the ARC and mPOA (Figs. 20.3 and 20.4) [35]. However, since kisspeptin expression is sensitive to metabolic cues [10], overnight fasting alone was sufficient to decrease basal levels of the mPOA *Kiss1* message, an effect that was not amplified by insulin administration (Fig. 20.4) [35]. In fact, prolonged (72 h) fasting suppresses ARC kisspeptin expression and LH secretion, and prolongs the estrous cycle in gonadal intact adult female rats [63]. In summary, the suppression of the ARC kisspeptin expression by acute metabolic deprivation appears responsible for the inhibition of pulsatile LH secretion and the subsequent prolongation of the estrous cycle, and may be mediated by CRF-R2.

Psychological stress is often harder to define than other forms of stress, although restraint stress applied to experimental animals has long been employed as a model for a range of emotional and behavioral disturbances and psychiatric disorders. Clinically relevant psychological stressors stimulate the HPA axis and the sympathetic nervous system, as does acute restraint. Furthermore, this stress paradigm effectively suppresses the HPG axis, as is evident from disrupted LH pulsatility and downregulation of ARC *Kiss1* and mPOA *Kiss1* and *Kiss1r* mRNA expression following restraint in adult female rats (Figs. 20.3 and 20.4) [35]. However, unlike other stressors, restraint has no effect on ARC expression of *Kiss1r* [35]. The restraint-induced suppression of the LH pulse in adult female rats is blocked by direct injections of a nonselective CRF antagonist into the locus ceruleus; however, such treatment did not affect the hypoglycemic stress-induced suppression of LH secretion [64]. icv administration of either type-specific (CRF-R1 or CRF-R2) CRF-R antagonists also blocked the inhibitory effects of restraint stress on pulsatile LH secretion in this animal model [39, 40]. Since the ARC and mPOA expression of *Kiss1* and *Kiss1r* mRNA is robustly downregulated by icv administration of CRF [35], it is logical to hypothesize that the effects of restraint stress on LH secretion are downstream of the CRF-induced suppression of kisspeptin/Kiss1r signaling.

Isolation has been widely accepted as a mild social stressor. In female mice the stress of chronic isolation increases anxiety [65], undermines the regularity of estrous cyclicity, and downregulates mPOA *Kiss1r* expression [66]. Adult rat social isolates display a dysregulated CORT response, with acute stress (predator odor) markedly exacerbating the rise in CORT compared to grouped animals—an effect that is enhanced with age [67]. Individual reactivity or vulnerability to stress can reflect differences in dominance status among social animals. Variation in basal glucocorticoid levels has been identified to be dependent upon social rank in numerous species. Dominant female squirrel monkeys [68], cynomolgus monkeys [69], and baboons [70] demonstrate reduced cortisol levels compared with subordinates in the same social group, an effect that is also observed in the teleost fish (*Haplochromis burtoni*) [71], the naked mole rat (*Heterocephalus glaber*) [72], pig [73], and horse [74]. Similarly, subordinate mice [75] and rats [76] display higher basal CORT release than their dominant counterpart. In response to stress, however, subordinate rats displayed a significantly reduced, or even completely absent, CORT response as compared to their dominant counterparts [76]. This outcome is not uncommon in other species, including primates (e.g., olive baboon) [77], although this phenotype can be markedly modified subject to particular stylistic traits of social behavior in primate species [77]. Thus, subordination may result in changes in both the basal and stress-induced activity of the HPA axis. Generally, these facets have been interpreted as reflecting the chronic stress of social subordinance.

Social status is important in determining reproductive function in addition to stress reactivity in a wide variety of species. The common marmoset is a well-studied primate model of socially mediated infertility, in which subordinate females display impaired hypothalamic GnRH secretion and anovulation [78], and receipt of aggression from female conspecifics reduces LH pulse frequency [79]. The role of kisspeptin signaling in socially induced infertility in the common marmoset or other primates has not been reported. However, subordination in other species, including the naked mole rat, which exhibit extreme socially induced suppression of the HPG axis [80] displays downregulated kisspeptin in the AVPV [81]. Similarly, in the teleost fish, *Astatotilapia burtoni*, reduced *Kiss1r* expression was observed in whole brain of reproductively suppressed subordinate males [82].

Naturally, there are many caveats in extrapolating these data to humans. Nevertheless, it is of note that about 35% of women presenting with secondary amenorrhea are diagnosed with functional hypothalamic amenorrhea. In these women, effective stress management results in reversal of the extant low LH pulse frequency and hypercortisolism, and restoration of fertility in the vast majority [83]. The cynomolgus monkey has proved an exquisite primate model of functional hypothalamic amenorrhea, in which CRF-R1 antagonism restores normal GnRH pulse generator frequency suppressed by a combination of mild psychosocial and metabolic stressors that simulate the stressor indices of functional hypothalamic amenorrhea in women [84, 85]. Although the role of kisspeptin signaling has not been examined in this primate model, intermittent kisspeptin administration stimulates gonadotropin release in women with hypothalamic amenorrhea [86]. Furthermore, continuous kisspeptin infusion restores pulsatile LH secretion in men with hypogonadotropic hypogonadism resulting from loss-of-function mutations in NKB ligand and receptor [87].

A vast body of literature supports the notion that the GnRH pulse generator, as well as other reproductive networks within which kisspeptin signaling plays a key role, is exquisitely sensitive to changes within the environment, either as a direct consequence of perceived stressful stimuli or through autonomic suppression of nonessential physiological processes. Kisspeptin has widely been dubbed a central gatekeeper of reproductive function; however, such a moniker does not adequately befit the suitor in the context of stress suppression of the HPG axis. Expression of Kisspeptin and/or its receptor is downregulated by a variety of stressors and mediators of the stress response, "surrendering" gracelessly at the mere sight of "intruders" and thus allowing the instillation of a suppressive "regime." It is perhaps this submissive nature of the kisspeptin/Kiss1r system that allows stressful stimuli to deregulate reproductive processes and, in some individuals, lets this develop into reproductive pathologies. To consolidate the grounds for an interaction between kisspeptin signaling and the stress response, we proceed to reviewing recent data that suggest potential neural mechanisms for this interaction.

Mechanisms Integrating Stress Networks and Kisspeptin Signaling

Despite many similarities in which various stressors activate the HPA axis and/or the sympathetic nervous system, as well as inhibit the HPG axis, a number of disparities are also evident from current literature. First, the central mediators of the stress response are differentially implicated in the suppression of the GnRH pulse generator. Consistent with a CRF-dependent mechanism of stress suppression of the HPG axis are the findings that, in the female rat, (1) icv administration of CRF [38] or a selective CRF-R2 agonist [39] prolongs the LH pulse interval; (2) CRF-R2 antagonism blocks the suppression of pulsatile LH secretion induced by restraint [39], insulin-induced hypoglycemia, and LPS [40]; and (3) the inhibitory effect of restraint stress on LH secretion is blocked by a selective CRF-R1 antagonist [40]. However, blockade of CRF-R1 did not affect the hypoglycemia- and LPS-induced suppression of the GnRH pulse generator [40]. Furthermore, while CRF, insulin-induced hypoglycemia, and LPS downregulated the expression of both kisspeptin and its receptor in the ARC and mPOA, restraint did not affect the ARC expression of *Kiss1r* [35]. These findings point at the notion that different stressors employ differential neural mechanisms for the activation of the stress response and suppression of the reproductive axis.

Second, while various stress paradigms, as well as icv CRF administration, potently downregulate *Kiss1* and *Kiss1r* mRNA expression in the ARC and mPOA, concomitant with suppression of pulsatile LH secretion, both acute and chronic administration of physiological CORT downregulate *Kiss1*, but upregulate *Kiss1r* expression, with no net effect on LH pulses [35]. Indeed, acute stress-induced CORT release has no effect on pulsatile LH secretion in the monkey [88]. Further, treatment with metyrapone, an inhibitor of adrenal steroidogenesis that prevented the CRF-induced rise in CORT levels, did not reverse the inhibitory effects of CRF on GnRH pulse generator frequency in the rhesus monkey [89]. The decrease in the ARC expression of *Kiss1* mRNA following CORT administration has also been shown in male mice, and CORT was hypothesized to influence kisspeptin neurons directly, in light of the expression of glucocorticoid receptors (GR) by these cells [36], despite previous reports that gonadal GR mediate the stress-induced CORT feedback on the HPG axis [90]. In mice, however, CORT did decrease LH levels, albeit at a pharmacological dose [36]. Kisspeptin, on the other hand, affects neither CORT nor adrenocorticotropin hormone (ACTH) levels under basal or restraint stress conditions, in adult male rats [53]. It is apparent, therefore, that central and peripheral processes of the HPA axis modulate hypothalamic kisspeptin signaling in different ways. It is important to note that components of the stress axis do not act in isolation in vivo, and hence multiple parameters of the stress response need to be considered when extrapolating conclusions regarding the stress feedback to the kisspeptin/Kiss1r system.

Third, different neuronal populations appear to mediate the suppressive effects of different types of stress. In addition to the aforementioned involvement of CRF-R1 and CRF-R2 in the suppression of pulsatile LH secretion under various stress paradigms, neurons projecting from the locus ceruleus [64], the BNST [41], and the amygdala [45], as well as several neuropeptides, including AVP, Dyn, and calcitonin gene-related peptide (CGRP), along with the chiefly inhibitory amino acid neurotransmitter γ-aminobutyric acid (GABA), have all been shown to play key roles in the stress-induced suppression of reproductive function. These mechanisms are further discussed below. Finally, while stress is commonly known to confer an energetic advantage to an organism at the expense of its nonessential bodily functions, e.g., reproduction, in some scenarios, stressful stimuli may serve to enhance the functions of the HPG axis and, in particular, induce premature LH surges and ovulation [91, 92]. These observations hint at a complex network of stress-reactive neurons spanning numerous regions of the brain, without a clearly identified central mediator of the integration of stress stimuli with reproductive function, in the context of kisspeptin signaling.

To date, various limbic structures have been differentially implicated in the modulation of the effects of different stressors on the HPA axis [8]. We have recently shown evidence for the involvement of the MeA and CeA in psychogenic and immunological stress-induced suppression of the GnRH pulse generator, respectively [45]. Adult female rats in which the MeA was chemically lessoned showed an attenuated restraint-induced, but *not* hypoglycemia- or LPS-induced, decrease in LH pulse frequency [45]. The extent of the LPS-induced suppression of pulsatile LH secretion was strongly reduced in CeA-lesioned animals vs. intact controls, though the CeA lesions had *no* effect on the LH pulse following restraint or hypoglycemic stress [45]. These data suggest that these two amygdaloid loci are not involved in mediating the hypoglycemic stress-induced suppression of the LH pulse. Indeed, we have previously shown that removal of the area postrema of the caudal brainstem prevents insulin-induced glucoprivic suppression of pulsatile LH secretion [93].

CRF neurons of the BNST connect the PVN with the amygdala, hippocampus, and prefrontal cortex [94]. Further, non-CRF interneurons project to mPOA GnRH perikarya [95]. When stimulated electrochemically, lateral BNST neurons prevent ovulation by abrogating the preovulatory surge of LH, while medial BNST neurons advanced the LH surge in proestrus rats [96]. Administration of CRF directly into the dorsolateral BNST of adult female rats dose dependently suppresses pulsatile LH secretion and activates GABA neurons in the mPOA [41]. Furthermore, intra-BNST injection of a selective CRF-R2 antagonist blocked the restraint-induced, but not IIH-induced, suppression of pulsatile LH secretion [41].

The brainstem noradrenergic locus ceruleus is innervated by CRF neurons projecting from CeA [97], BNST [98], and PVN [99]. Bilateral electrolytic lesions of the locus ceruleus disrupt both the pulsatile [100] and surge [101] modes of LH release. Intra-cerulear administration of CRF to adult OVX rats dose dependently suppresses pulsatile LH secretion and activates GABA neurons in the mPOA [64]. These effects are amplified by estradiol [64]. Furthermore, intra-cerulear injection of a nonselective CRF receptor antagonist blocked the restraint-induced, but not glucoprivic, suppression of pulsatile LH secretion [64].

Although CRF signaling appears indispensible in mediating stress-induced suppression of the HPG axis, other neuropeptide and amino acid signaling molecules have also been implicated in the inhibition of GnRH neurons. AVP and CRF, both expressed by parvocellular neurons of the PVN, act synergistically to stimulate ACTH secretion from the anterior pituitary, although different mechanisms are involved in regulating AVP and CRF expression in response to acute and chronic stress exposure [102]. AVP neurons of the suprachiasmatic nucleus (SCN) of the hypothalamus innervate the kisspeptin neuron population in the AVPV, and estradiol has been shown to increase the number of these monosynaptic connections [103]. Moreover, icv administration of kisspeptin increases plasma AVP levels in rats [104], although without altering the firing rate of AVP/oxytocin neurons of the supraoptic nucleus of the hypothalamus [105]. This could suggest that kisspeptin is involved in a feedback mechanism bridging the stress response with reproductive function.

CGRP neurons project from the caudal aspect of the organum vasculosum of the lamina terminalis (OVLT), via the AVPV, to the medial preoptic nucleus [106], and form direct contacts with CRF neurons of the CeA [107, 108]. Central administration of CGRP induces c-fos expression in the CeA [109], mPOA, and PVN [110]. Moreover, CGRP-treated rats demonstrate elevated circulating CORT levels [110, 111] and GnRH pulse generator inhibition [110, 112], effects that are dependent on CRF-R1 signaling [109]. Administered intra-mPOA, CGRP dose dependently suppressed LH pulse frequency [112], while intra-PVN administration of CGRP stimulates ACTH and CORT release [111]. A GnRH cell line expresses CGRP receptors, and CGRP dose dependently suppresses GnRH release in vitro [113]. These data strongly implicate hypothalamic CGRP signaling in the inhibition of the GnRH pulse generator in response to HPA axis activation.

The endogenous opioid peptide, Dyn, is coexpressed with kisspeptin and NKB in ARC KNDy neurons that project to GnRH neurons [114, 115]. Central administration of a Dyn analog, selective for the kappa-subtype of the opioid receptor (KOR), suppresses pulsatile LH secretion in female rats [116], and KOR antagonism blocks the CGRP- and NKB-mediated suppression of pulsatile LH secretion [116, 117]. Because endogenous opioid peptides play an important role in stress-induced suppression of the reproductive axis, and since KNDy neurons express KOR, Dyn might inhibit the GnRH pulse generator through autocrine inhibition of the kisspeptin tone.

RF-amide-related peptide type-3 (RFRP-3), a member of the same RF-amine superfamily as kisspeptin and a mammalian ortholog of avian gonadotropin-inhibitory hormone (GnIH), is inhibitory to both GnRH neuron firing [118, 119] and LH secretion [120, 121], while RFRP receptor antagonists are stimulatory to LH secretion [122, 123]. In addition to the widespread distribution of RFRP-3 neuron fibers in the hypothalamus, with abundant close appositions to GnRH cells [124], these fibers are also present in the hypothalamic PVN and limbic areas, including the BNST and MeA [120, 125]. Taken together, these data provide an anatomical substrate for the hypothesis that stress-induced suppression of the HPG axis is mediated, in part, by RFRP-3.

Indeed, there are numerous PVN CRF neurons with close RFRP-3 fiber appositions [126] and there is abundant expression of *GPR147* mRNA, encoding the putative RFRP receptor, in the PVN [127]. Further, central administration of RFRP-3 evokes c-fos expression in the PVN, increases ACTH release, and induces anxiety in the rat [127]. It has also been suggested that RFRP-3 plays a key role in stress-induced suppression of gonadotropin secretion, since an increase in hypothalamic *RFRP* mRNA expression was associated with reduced LH secretion in response to immobilization stress in the rat [128]. In addition, hypothalamic RFRP-3 neurons express CRF-R1 and GR [128], and CRF per se increased *GPR147* mRNA expression, concomitant with reduced *GNRH1* mRNA levels in the GnRH-expressing N39 cells in vitro [129]. Moreover, RFRP-3 blocks kisspeptin-induced activation of GnRH neurons [119]. It is conceivable that functional interactions between the excitatory kisspeptin and inhibitory GnIH signaling pathways fine-tune the response of the GnRH neural system to stressors, thus minimizing reproductive dysfunction.

GABAergic transmission underlies a significant proportion of inhibitory stimuli in the brain, including those at GnRH neurons [130, 131]. Activity of GABA neurons in the mPOA is stimulated by central CRF administration [41, 64], as well as by a range of stress paradigms [45, 132–134]. We have recently shown mPOA GABA receptors ($GABA_AR$ and $GABA_BR$) to be differentially involved in mediating the suppressive effects of immunological and psychological stress on LH secretion, respectively [135]. Furthermore, both $GABA_AR$ and $GABA_BR$ antagonists block the inhibition of the LH pulse by CRF [135]. Therefore, the CRF neurons projecting to the mPOA seem to mediate the GABA receptor-dependent effects of stress on the GnRH neural system. Since AVPV kisspeptin neurons are predominantly GABAergic [136] and $GABA_AR$ antagonism potently induces kisspeptin secretion in the stalk-median eminence of prepubertal monkeys [137], the balance between stimulatory kisspeptin and inhibitory GABA inputs appears to govern marked changes in the pattern of GnRH neurosecretion. Indeed, the pubertal increase in GnRH release is preceded by a reduction in GABAergic inhibition, and disinhibition of GnRH secretion through $GABA_AR$ antagonism induces precocious puberty in juvenile monkeys [138]. Thus, it is probable that stress-induced suppression of GnRH secretion is secondary to the inhibition of kisspeptin and/or GnRH neurons by GABA.

Conclusion

The complexity of the neural networks and signaling mechanisms involved in the translation of sensory stimuli into a stress response, complicated further by peripheral feedback mechanisms (CORT and ACTH), as well as evidence for the convergence of multiple stimulatory and inhibitory neuronal populations, makes a reductionist approach to the interpretation of the interactions between stress and reproduction suboptimal. Direct neural pathways are not a prerequisite for stress suppression of kisspeptin signaling. Synaptic transmission through interneurons [139], and/or paracrine volume transmission of neuropeptidergic signals [140], may also be involved in the regulation of GnRH neurosecretion by signaling systems associated with the stress response. Hypotheses arising through anatomical observations must be consolidated by functional data; however, intracellular events should also be taken into consideration. While considerable effort has been made to untangle the individual neural pathways that may contribute to the stress suppression of reproductive function, revealing the true complexity of the networks in question, a meaningful model of the complex interaction between stress and reproduction is warranted. Perhaps future mathematical modeling will aid the assembly of the components of the stress and reproductive systems brought to light to date and, ultimately, provide a better understanding of the neural connectivity and signaling mechanisms involved. In summary, kisspeptin signaling is a novel component of the stress neurocircuitry implicated in the perturbation of reproductive development and function under stress conditions.

References

1. Lee JH, Miele ME, Hicks DJ, Phillips KK, Trent JM, Weissman BE et al (1996) KiSS-1, a novel human malignant melanoma metastasis-suppressor gene. J Natl Cancer Inst 88(23):1731–1737
2. Arai AC (2009) The role of kisspeptin and GPR54 in the hippocampus. Peptides 30(1):16–25
3. Mead EJ, Maguire JJ, Kuc RE, Davenport AP (2007) Kisspeptins are novel potent vasoconstrictors in humans, with a discrete localization of their receptor, G protein-coupled receptor 54, to atherosclerosis-prone vessels. Endocrinology 148(1):140–147
4. Nakamura Y, Aoki S, Xing Y, Sasano H, Rainey WE (2007) Metastin stimulates aldosterone synthesis in human adrenal cells. Reprod Sci 14(8):836–845
5. Bowe JE, King AJ, Kinsey-Jones JS, Foot VL, Li XF, O'Byrne KT et al (2009) Kisspeptin stimulation of insulin secretion: mechanisms of action in mouse islets and rats. Diabetologia 52(5):855–862
6. Clarkson J, d'Anglemont de Tassigny X, Colledge WH, Caraty A, Herbison AE (2009) Distribution of kisspeptin neurones in the adult female mouse brain. J Neuroendocrinol 21(8):673–682
7. Landgraf R, Neumann ID (2004) Vasopressin and oxytocin release within the brain: a dynamic concept of multiple and variable modes of neuropeptide communication. Front Neuroendocrinol 25(3–4):150–176
8. Herman JP, Ostrander MM, Mueller NK, Figueiredo H (2005) Limbic system mechanisms of stress regulation: hypothalamo-pituitary-adrenocortical axis. Prog Neuropsychopharmacol Biol Psychiatry 29(8):1201–1213
9. Cameron JL (1991) Metabolic cues for the onset of puberty. Horm Res 36(3–4):97–103
10. Castellano JM, Bentsen AH, Mikkelsen JD, Tena-Sempere M (2010) Kisspeptins: bridging energy homeostasis and reproduction. Brain Res 1364:129–138
11. Aubert ML, Pierroz DD, Gruaz NM, d'Alleves V, Vuagnat BA, Pralong FP et al (1998) Metabolic control of sexual function and growth: role of neuropeptide Y and leptin. Mol Cell Endocrinol 140(1–2):107–113
12. Deardorff J, Ekwaru JP, Kushi LH, Ellis BJ, Greenspan LC, Mirabedi A et al (2011) Father absence, body mass index, and pubertal timing in girls: differential effects by family income and ethnicity. J Adolesc Health 48(5):441–447
13. Cameron NM, Champagne FA, Parent C, Fish EW, Ozaki-Kuroda K, Meaney MJ (2005) The programming of individual differences in defensive responses and reproductive strategies in the rat through variations in maternal care. Neurosci Biobehav Rev 29(4–5):843–865
14. Aksglaede L, Juul A, Olsen LW, Sorensen TI (2009) Age at puberty and the emerging obesity epidemic. PLoS One 4(12):e8450

15. Morris DH, Jones ME, Schoemaker MJ, Ashworth A, Swerdlow AJ (2010) Determinants of age at menarche in the UK: analyses from the Breakthrough Generations Study. Br J Cancer 103(11):1760–1764
16. Engelbregt MJ, van Weissenbruch MM, Popp-Snijders C, Lips P, Delemarre-van de Waal HA (2001) Body mass index, body composition, and leptin at onset of puberty in male and female rats after intrauterine growth retardation and after early postnatal food restriction. Pediatr Res 50(4):474–478
17. Iwasa T, Matsuzaki T, Murakami M, Fujisawa S, Kinouchi R, Gereltsetseg G et al (2010) Effects of intrauterine undernutrition on hypothalamic Kiss1 expression and the timing of puberty in female rats. J Physiol 588(Pt 5):821–829
18. Knox AM, Li XF, Kinsey-Jones JS, Wilkinson ES, Wu XQ, Cheng YS et al (2009) Neonatal lipopolysaccharide exposure delays puberty and alters hypothalamic Kiss1 and Kiss1r mRNA expression in the female rat. J Neuroendocrinol 21(8):683–689
19. Boukouvalas G, Gerozissis K, Kitraki E (2010) Fat feeding of rats during pubertal growth leads to neuroendocrine alterations in adulthood. Cell Mol Neurobiol 30(1):91–99
20. Topaloglu AK, Tello JA, Kotan LD, Ozbek MN, Yilmaz MB, Erdogan S et al (2012) Inactivating KISS1 mutation and hypogonadotropic hypogonadism. N Engl J Med 366(7):629–635
21. Seminara SB, Messager S, Chatzidaki EE, Thresher RR, Acierno JS Jr, Shagoury JK et al (2003) The GPR54 gene as a regulator of puberty. N Engl J Med 349(17):1614–1627
22. de Roux N, Genin E, Carel JC, Matsuda F, Chaussain JL, Milgrom E (2003) Hypogonadotropic hypogonadism due to loss of function of the KiSS1-derived peptide receptor GPR54. Proc Natl Acad Sci U S A 100(19):10972–10976
23. Funes S, Hedrick JA, Vassileva G, Markowitz L, Abbondanzo S, Golovko A et al (2003) The KiSS-1 receptor GPR54 is essential for the development of the murine reproductive system. Biochem Biophys Res Commun 312(4):1357–1363
24. d'Anglemont de Tassigny X, Fagg LA, Dixon JP, Day K, Leitch HG, Hendrick AG et al (2007) Hypogonadotropic hypogonadism in mice lacking a functional Kiss1 gene. Proc Natl Acad Sci U S A 104(25):10714–10719
25. Chan YM, Broder-Fingert S, Wong KM, Seminara SB (2009) Kisspeptin/Gpr54-independent gonadotrophin-releasing hormone activity in Kiss1 and Gpr54 mutant mice. J Neuroendocrinol 21(12):1015–1023
26. Lapatto R, Pallais JC, Zhang D, Chan YM, Mahan A, Cerrato F et al (2007) Kiss1-/- mice exhibit more variable hypogonadism than Gpr54-/- mice. Endocrinology 148(10):4927–4936
27. Mayer C, Boehm U (2011) Female reproductive maturation in the absence of kisspeptin/GPR54 signaling. Nat Neurosci 14(6):704–710
28. Navarro VM, Castellano JM, Fernandez-Fernandez R, Barreiro ML, Roa J, Sanchez-Criado JE et al (2004) Developmental and hormonally regulated messenger ribonucleic acid expression of KiSS-1 and its putative receptor, GPR54, in rat hypothalamus and potent luteinizing hormone-releasing activity of KiSS-1 peptide. Endocrinology 145(10):4565–4574
29. Shahab M, Mastronardi C, Seminara SB, Crowley WF, Ojeda SR, Plant TM (2005) Increased hypothalamic GPR54 signaling: a potential mechanism for initiation of puberty in primates. Proc Natl Acad Sci U S A 102(6):2129–2134
30. Navarro VM, Fernandez-Fernandez R, Castellano JM, Roa J, Mayen A, Barreiro ML et al (2004) Advanced vaginal opening and precocious activation of the reproductive axis by KiSS-1 peptide, the endogenous ligand of GPR54. J Physiol 561(Pt 2):379–386
31. Luan X, Zhou Y, Wang W, Yu H, Li P, Gan X et al (2007) Association study of the polymorphisms in the KISS1 gene with central precocious puberty in Chinese girls. Eur J Endocrinol 157(1):113–118
32. Teles MG, Bianco SD, Brito VN, Trarbach EB, Kuohung W, Xu S et al (2008) A GPR54-activating mutation in a patient with central precocious puberty. N Engl J Med 358(7):709–715
33. Shanks N, Windle RJ, Perks PA, Harbuz MS, Jessop DS, Ingram CD et al (2000) Early-life exposure to endotoxin alters hypothalamic-pituitary-adrenal function and predisposition to inflammation. Proc Natl Acad Sci U S A 97(10):5645–5650

34. Wu XQ, Li XF, Ye B, Popat N, Milligan SR, Lightman SL et al (2011) Neonatal programming by immunological challenge: effects on ovarian function in the adult rat. Reproduction 141(2):241–248
35. Kinsey-Jones JS, Li XF, Knox AM, Wilkinson ES, Zhu XL, Chaudhary AA et al (2009) Down-regulation of hypothalamic kisspeptin and its receptor, Kiss1r, mRNA expression is associated with stress-induced suppression of luteinising hormone secretion in the female rat. J Neuroendocrinol 21(1):20–29
36. Wang O, Lanjuin A, Basko J, Dulac C, Majzoub J (2011) Glucocorticoids inhibit kisspeptin neurons during stress-induced reproductive inhibition. Endocr Rev 32(03_MeetingAbstracts):P3–P141
37. Rivier C, Vale W (1984) Influence of corticotropin-releasing factor on reproductive functions in the rat. Endocrinology 114(3):914–921
38. Cates PS, Li XF, O'Byrne KT (2004) The influence of 17beta-oestradiol on corticotrophin-releasing hormone induced suppression of luteinising hormone pulses and the role of CRH in hypoglycaemic stress-induced suppression of pulsatile LH secretion in the female rat. Stress 7(2):113–118
39. Li XF, Bowe JE, Lightman SL, O'Byrne KT (2005) Role of corticotropin-releasing factor receptor-2 in stress-induced suppression of pulsatile luteinizing hormone secretion in the rat. Endocrinology 146(1):318–322
40. Li XF, Bowe JE, Kinsey-Jones JS, Brain SD, Lightman SL, O'Byrne KT (2006) Differential role of corticotrophin-releasing factor receptor types 1 and 2 in stress-induced suppression of pulsatile luteinising hormone secretion in the female rat. J Neuroendocrinol 18(8):602–610
41. Li XF, Lin YS, Kinsey-Jones JS, Milligan SR, Lightman SL, O'Byrne KT (2011) The role of the bed nucleus of the stria terminalis in stress-induced inhibition of pulsatile luteinising hormone secretion in the female rat. J Neuroendocrinol 23(1):3–11
42. Kinsey-Jones JS, Li XF, Knox AM, Lin YS, Milligan SR, Lightman SL et al (2010) Corticotrophin-releasing factor alters the timing of puberty in the female rat. J Neuroendocrinol 22(2):102–109
43. Rivest S, Rivier C (1991) Influence of the paraventricular nucleus of the hypothalamus in the alteration of neuroendocrine functions induced by intermittent footshock or interleukin. Endocrinology 129(4):2049–2057
44. Satterlee DG, Marin RH, Jonest RB (2002) Selection of Japanese quail for reduced adreno-cortical responsiveness accelerates puberty in males. Poult Sci 81(7):1071–1076
45. Lin Y, Li X, Lupi M, Kinsey-Jones JS, Shao B, Lightman SL et al (2011) The role of the medial and central amygdala in stress-induced suppression of pulsatile LH secretion in female rats. Endocrinology 152(2):545–555
46. Keen-Rhinehart E, Michopoulos V, Toufexis DJ, Martin EI, Nair H, Ressler KJ et al (2009) Continuous expression of corticotropin-releasing factor in the central nucleus of the amygdala emulates the dysregulation of the stress and reproductive axes. Mol Psychiatry 14(1):37–50
47. Solomon MB, Jones K, Packard BA, Herman JP (2010) The medial amygdala modulates body weight but not neuroendocrine responses to chronic stress. J Neuroendocrinol 22(1):13–23
48. Jolkkonen E, Pitkanen A (1998) Intrinsic connections of the rat amygdaloid complex: projections originating in the central nucleus. J Comp Neurol 395(1):53–72
49. Usunoff KG, Schmitt O, Itzev DE, Haas SJ, Lazarov NE, Rolfs A et al (2009) Efferent projections of the anterior and posterodorsal regions of the medial nucleus of the amygdala in the mouse. Cells Tissues Organs 190(5):256–285
50. Relkin R (1971) Relative efficacy of pinealectomy, hypothalamic and amygdaloid lesions in advancing puberty. Endocrinology 88(2):415–418
51. Bar-Sela M, Critchlow V (1966) Delayed puberty following electrical stimulation of amygdala in female rats. Am J Physiol 211(5):1103–1107
52. Ramaswamy S, Gibbs RB, Plant TM (2009) Studies of the localisation of kisspeptin within the pituitary of the rhesus monkey (Macaca mulatta) and the effect of kisspeptin on the release of non-gonadotropic pituitary hormones. J Neuroendocrinol 21(10):795–804

53. Rao YS, Mott NN, Pak TR (2011) Effects of kisspeptin on parameters of the HPA axis. Endocrine 39(3):220–228
54. Knobil E (1988) The hypothalamic gonadotrophic hormone releasing hormone (GnRH) pulse generator in the rhesus monkey and its neuroendocrine control. Hum Reprod 3(1):29–31
55. Bentsen AH, Ansel L, Simonneaux V, Tena-Sempere M, Juul A, Mikkelsen JD (2010) Maturation of kisspeptinergic neurons coincides with puberty onset in male rats. Peptides 31(2):275–283
56. Iwasa T, Matsuzaki T, Murakami M, Shimizu F, Kuwahara A, Yasui T et al (2008) Decreased expression of kisspeptin mediates acute immune/inflammatory stress-induced suppression of gonadotropin secretion in female rat. J Endocrinol Invest 31(7):656–659
57. Castellano JM, Bentsen AH, Romero M, Pineda R, Ruiz-Pino F, Garcia-Galiano D et al (2010) Acute inflammation reduces kisspeptin immunoreactivity at the arcuate nucleus and decreases responsiveness to kisspeptin independently of its anorectic effects. Am J Physiol Endocrinol Metab 299(1):E54–E61
58. Li XF, Kinsey-Jones JS, Knox AM, Wu XQ, Tahsinsoy D, Brain SD et al (2007) Neonatal lipopolysaccharide exposure exacerbates stress-induced suppression of luteinizing hormone pulse frequency in adulthood. Endocrinology 148(12):5984–5990
59. Iwasa T, Matsuzaki T, Murakami M, Kinouchi R, Ogata R, Kuwahara A et al (2009) Neonatal lipopolysaccharide exposure attenuates the homotypic stress-induced suppression of LH secretion in adulthood in male rat. Int J Dev Neurosci 27(4):345–349
60. Weiser MJ, Handa RJ (2009) Estrogen impairs glucocorticoid dependent negative feedback on the hypothalamic-pituitary-adrenal axis via estrogen receptor alpha within the hypothalamus. Neuroscience 159(2):883–895
61. Viau V, Meaney MJ (2004) Testosterone-dependent variations in plasma and intrapituitary corticosteroid binding globulin and stress hypothalamic-pituitary-adrenal activity in the male rat. J Endocrinol 181(2):223–231
62. Chen MD, Ordog T, O'Byrne KT, Goldsmith JR, Connaughton MA, Knobil E (1996) The insulin hypoglycemia-induced inhibition of gonadotropin-releasing hormone pulse generator activity in the rhesus monkey: roles of vasopressin and corticotropin-releasing factor. Endocrinology 137(5):2012–2021
63. Matsuzaki T, Iwasa T, Kinouchi R, Yoshida S, Murakami M, Gereltsetseg G et al (2011) Fasting reduces the kiss1 mRNA levels in the caudal hypothalamus of gonadally intact adult female rats. Endocr J 58(11):1003–1012
64. Mitchell JC, Li XF, Breen L, Thalabard JC, O'Byrne KT (2005) The role of the locus coeruleus in corticotropin-releasing hormone and stress-induced suppression of pulsatile luteinizing hormone secretion in the female rat. Endocrinology 146(1):323–331
65. Palanza P, Gioiosa L, Parmigiani S (2001) Social stress in mice: gender differences and effects of estrous cycle and social dominance. Physiol Behav 73(3):411–420
66. Hasen NS, O'Leary KA, Auger AP, Schuler LA (2010) Social isolation reduces mammary development, tumor incidence, and expression of epigenetic regulators in wild-type and p53-heterozygotic mice. Cancer Prev Res (Phila) 3(5):620–629
67. Hermes GL, Delgado B, Tretiakova M, Cavigelli SA, Krausz T, Conzen SD et al (2009) Social isolation dysregulates endocrine and behavioral stress while increasing malignant burden of spontaneous mammary tumors. Proc Natl Acad Sci U S A 106(52):22393–22398
68. Gonzalez CA, Coe CL, Levine S (1982) Cortisol responses under different housing conditions in female squirrel monkeys. Psychoneuroendocrinology 7(2–3):209–216
69. Shively CA (1998) Social subordination stress, behavior, and central monoaminergic function in female cynomolgus monkeys. Biol Psychiatry 44(9):882–891
70. Sapolsky RM, Alberts SC, Altmann J (1997) Hypercortisolism associated with social subordinance or social isolation among wild baboons. Arch Gen Psychiatry 54(12):1137–1143
71. Fox HE, White SA, Kao MH, Fernald RD (1997) Stress and dominance in a social fish. J Neurosci 17(16):6463–6469

72. Clarke FM, Faulkes CG (1997) Dominance and queen succession in captive colonies of the eusocial naked mole-rat, Heterocephalus glaber. Proc Biol Sci 264(1384):993–1000
73. Fernandez X, Meunier-Salaun MC, Mormede P (1994) Agonistic behavior, plasma stress hormones, and metabolites in response to dyadic encounters in domestic pigs: interrelationships and effect of dominance status. Physiol Behav 56(5):841–847
74. Alexander SL, Irvine CH (1998) The effect of social stress on adrenal axis activity in horses: the importance of monitoring corticosteroid-binding globulin capacity. J Endocrinol 157(3):425–432
75. Fano E, Sanchez-Martin JR, Arregi A, Castro B, Alonso A, Brain P et al (2001) Social stress paradigms in male mice: Variations in behavior, stress and immunology. Physiol Behav 73(1–2):165–173
76. Blanchard DC, Spencer RL, Weiss SM, Blanchard RJ, McEwen B, Sakai RR (1995) Visible burrow system as a model of chronic social stress: behavioral and neuroendocrine correlates. Psychoneuroendocrinology 20(2):117–134
77. Virgin CE Jr, Sapolsky RM (1997) Styles of male social behavior and their endocrine correlates among low-ranking baboons. Am J Primatol 42(1):25–39
78. Abbott DH, Hodges JK, George LM (1988) Social status controls LH secretion and ovulation in female marmoset monkeys (Callithrix jacchus). J Endocrinol 117(3):329–339
79. O'Byrne KT, Lunn SF, Dixson AF (1989) Naloxone reversal of stress-induced suppression of LH release in the common marmoset. Physiol Behav 45(5):1077–1080
80. Faulkes CG, Abbott DH, Jarvis JU, Sherriff FE (1990) LH responses of female naked mole-rats, Heterocephalus glaber, to single and multiple doses of exogenous GnRH. J Reprod Fertil 89(1):317–323
81. Zhou S, Holmes MM, Forger NG, Goldman BD, Lovern MB, Caraty A et al (2010) Release from socially-induced reproductive suppression in eusocial naked mole-rats (Heterocephalus glaber) is marked by increased kisspeptin-immunoreactive cell bodies in the hypothalamic anteroventral periventricular nucleus. Society for Neuroscience, San Diego, CA, Contract No.: 594.16/HHH29
82. Grone BP, Maruska KP, Korzan WJ, Fernald RD (2010) Social status regulates kisspeptin receptor mRNA in the brain of Astatotilapia burtoni. Gen Comp Endocrinol 169(1):98–107
83. Berga SL, Marcus MD, Loucks TL, Hlastala S, Ringham R, Krohn MA (2003) Recovery of ovarian activity in women with functional hypothalamic amenorrhea who were treated with cognitive behavior therapy. Fertil Steril 80(4):976–981
84. Williams NI, Berga SL, Cameron JL (2007) Synergism between psychosocial and metabolic stressors: impact on reproductive function in cynomolgus monkeys. Am J Physiol Endocrinol Metab 293(1):E270–E276
85. Herod SM, Pohl CR, Cameron JL (2011) Treatment with a CRH-R1 antagonist prevents stress-induced suppression of the central neural drive to the reproductive axis in female macaques. Am J Physiol Endocrinol Metab 300(1):E19–E27
86. Jayasena CN, Nijher GM, Chaudhri OB, Murphy KG, Ranger A, Lim A et al (2009) Subcutaneous injection of kisspeptin-54 acutely stimulates gonadotropin secretion in women with hypothalamic amenorrhea, but chronic administration causes tachyphylaxis. J Clin Endocrinol Metab 94(11):4315–4323
87. Young J, George JT, Tello JA, Francou B, Bouligand J, Guiochon-Mantel A et al (2012) Kisspeptin Restores Pulsatile LH Secretion in Patients with Neurokinin B Signaling Deficiencies: Physiological. Pathophysiological and Therapeutic Implications, Neuroendocrinology
88. Chen MD, O'Byrne KT, Chiappini SE, Hotchkiss J, Knobil E (1992) Hypoglycemic 'stress' and gonadotropin-releasing hormone pulse generator activity in the rhesus monkey: role of the ovary. Neuroendocrinology 56(5):666–673
89. Williams CL, Nishihara M, Thalabard JC, Grosser PM, Hotchkiss J, Knobil E (1990) Corticotropin-releasing factor and gonadotropin-releasing hormone pulse generator activity in the rhesus monkey. Electrophysiological studies. Neuroendocrinology 52(2):133–137
90. Dong Q, Salva A, Sottas CM, Niu E, Holmes M, Hardy MP (2004) Rapid glucocorticoid mediation of suppressed testosterone biosynthesis in male mice subjected to immobilization stress. J Androl 25(6):973–981

91. Puder JJ, Freda PU, Goland RS, Ferin M, Wardlaw SL (2000) Stimulatory effects of stress on gonadotropin secretion in estrogen-treated women. J Clin Endocrinol Metab 85(6):2184–2188
92. Tarin JJ, Hamatani T, Cano A (2010) Acute stress may induce ovulation in women. Reprod Biol Endocrinol 8:53
93. Cates PS, O'Byrne KT (2000) The area postrema mediates insulin hypoglycaemia-induced suppression of pulsatile LH secretion in the female rat. Brain Res 853(1):151–155
94. Spencer SJ, Buller KM, Day TA (2005) Medial prefrontal cortex control of the paraventricular hypothalamic nucleus response to psychological stress: possible role of the bed nucleus of the stria terminalis. J Comp Neurol 481(4):363–376
95. Hahn JD, Kalamatianos T, Coen CW (2003) Studies on the neuroanatomical basis for stress-induced oestrogen-potentiated suppression of reproductive function: evidence against direct corticotropin-releasing hormone projections to the vicinity of luteinizing hormone-releasing hormone cell bodies in female rats. J Neuroendocrinol 15(8):732–742
96. Beltramino C, Taleisnik S (1980) Dual action of electrochemical stimulation of the bed nucleus of the stria terminalis on the release of LH. Neuroendocrinology 30(4):238–242
97. Van Bockstaele EJ, Colago EE, Valentino RJ (1998) Amygdaloid corticotropin-releasing factor targets locus coeruleus dendrites: substrate for the co-ordination of emotional and cognitive limbs of the stress response. J Neuroendocrinol 10(10):743–757
98. Van Bockstaele EJ, Peoples J, Valentino RJ (1999) A.E. Bennett Research Award. Anatomic basis for differential regulation of the rostrolateral peri-locus coeruleus region by limbic afferents. Biol Psychiatry 46(10):1352–1363
99. Curtis AL, Bello NT, Connolly KR, Valentino RJ (2002) Corticotropin-releasing factor neurones of the central nucleus of the amygdala mediate locus coeruleus activation by cardiovascular stress. J Neuroendocrinol 14(8):667–682
100. Anselmo-Franci JA, Rocha-Barros VM, Franci CR, McCann SM (1999) Locus ceruleus lesions block pulsatile LH release in ovariectomized rats. Brain Res 833(1):86–92
101. Helena CV, Franci CR, Anselmo-Franci JA (2002) Luteinizing hormone and luteinizing hormone-releasing hormone secretion is under locus coeruleus control in female rats. Brain Res 955(1–2):245–252
102. Ma XM, Aguilera G (1999) Transcriptional responses of the vasopressin and corticotropin-releasing hormone genes to acute and repeated intraperitoneal hypertonic saline injection in rats. Brain Res Mol Brain Res 68(1–2):129–140
103. Vida B, Deli L, Hrabovszky E, Kalamatianos T, Caraty A, Coen CW et al (2010) Evidence for suprachiasmatic vasopressin neurones innervating kisspeptin neurones in the rostral periventricular area of the mouse brain: regulation by oestrogen. J Neuroendocrinol 22(9):1032–1039
104. Ten SC, Gu SY, Niu YF, An XF, Yan M, He M (2010) Central administration of kisspeptin-10 inhibits water and sodium excretion of anesthetized male rats and the involvement of arginine vasopressin. Endocr Res 35(3):128–136
105. Scott V, Brown CH (2011) Kisspeptin activation of supraoptic nucleus neurons in vivo. Endocrinology 152(10):3862–3870
106. Herbison AE (1992) Identification of a sexually dimorphic neural population immunoreactive for calcitonin gene-related peptide (CGRP) in the rat medial preoptic area. Brain Res 591(2):289–295
107. Harrigan EA, Magnuson DJ, Thunstedt GM, Gray TS (1994) Corticotropin releasing factor neurons are innervated by calcitonin gene-related peptide terminals in the rat central amygdaloid nucleus. Brain Res Bull 33(5):529–534
108. Dobolyi A, Irwin S, Makara G, Usdin TB, Palkovits M (2005) Calcitonin gene-related peptide-containing pathways in the rat forebrain. J Comp Neurol 489(1):92–119
109. Bowe JE, Li XF, Kinsey-Jones JS, Brain SD, Lightman SL, O'Byrne KT (2008) The role of corticotrophin-releasing hormone receptors in the calcitonin gene-related peptide-induced suppression of pulsatile luteinising hormone secretion in the female rat. Stress 11(4):312–319
110. Li XF, Bowe JE, Mitchell JC, Brain SD, Lightman SL, O'Byrne KT (2004) Stress-induced suppression of the gonadotropin-releasing hormone pulse generator in the female rat: a novel neural action for calcitonin gene-related peptide. Endocrinology 145(4):1556–1563

111. Dhillo WS, Small CJ, Jethwa PH, Russell SH, Gardiner JV, Bewick GA et al (2003) Paraventricular nucleus administration of calcitonin gene-related peptide inhibits food intake and stimulates the hypothalamo-pituitary-adrenal axis. Endocrinology 144(4):1420–1425
112. Li XF, Kinsey-Jones JS, Bowe JE, Wilkinson ES, Brain SD, Lightman SL et al (2009) A role for the medial preoptic area in CGRP-induced suppression of pulsatile LH secretion in the female rat. Stress 12(3):259–267
113. Kinsey-Jones JS, Li XF, Bowe JE, Brain SD, Lightman SL, O'Byrne KT (2005) Effect of calcitonin gene-related peptide on gonadotrophin-releasing hormone mRNA expression in GT1-7 cells. J Neuroendocrinol 17(9):541–544
114. Krajewski SJ, Anderson MJ, Iles-Shih L, Chen KJ, Urbanski HF, Rance NE (2005) Morphologic evidence that neurokinin B modulates gonadotropin-releasing hormone secretion via neurokinin 3 receptors in the rat median eminence. J Comp Neurol 489(3):372–386
115. Krajewski SJ, Burke MC, Anderson MJ, McMullen NT, Rance NE (2010) Forebrain projections of arcuate neurokinin B neurons demonstrated by anterograde tract-tracing and monosodium glutamate lesions in the rat. Neuroscience 166(2):680–697
116. Bowe JE, Li XF, Kinsey-Jones JS, Paterson S, Brain SD, Lightman SL et al (2005) Calcitonin gene-related peptide-induced suppression of luteinizing hormone pulses in the rat: the role of endogenous opioid peptides. J Physiol 566(Pt 3):921–928
117. Kinsey-Jones JS, Grachev P, Li XF, Lin YS, Milligan SR, Lightman SL et al (2012) The inhibitory effects of neurokinin B on GnRH pulse generator frequency in the female rat. Endocrinology 153(1):307–315
118. Ducret E, Anderson GM, Herbison AE (2009) RFamide-related peptide-3, a mammalian gonadotropin-inhibitory hormone ortholog, regulates gonadotropin-releasing hormone neuron firing in the mouse. Endocrinology 150(6):2799–2804
119. Wu M, Dumalska I, Morozova E, van den Pol AN, Alreja M (2009) Gonadotropin inhibitory hormone inhibits basal forebrain vGluT2-gonadotropin-releasing hormone neurons via a direct postsynaptic mechanism. J Physiol 587(Pt 7):1401–1411
120. Johnson MA, Tsutsui K, Fraley GS (2007) Rat RFamide-related peptide-3 stimulates GH secretion, inhibits LH secretion, and has variable effects on sex behavior in the adult male rat. Horm Behav 51(1):171–180
121. Clarke IJ, Qi Y, Puspita Sari I, Smith JT (2009) Evidence that RF-amide related peptides are inhibitors of reproduction in mammals. Front Neuroendocrinol 30(3):371–378
122. Pineda R, Garcia-Galiano D, Sanchez-Garrido MA, Romero M, Ruiz-Pino F, Aguilar E et al (2010) Characterization of the inhibitory roles of RFRP3, the mammalian ortholog of GnIH, in the control of gonadotropin secretion in the rat: in vivo and in vitro studies. Am J Physiol Endocrinol Metab 299(1):E39–E46
123. Caraty A, Blomenrohr M, Vogel GM, Lomet D, Briant C, Beltramo M (2012) RF9 powerfully stimulates gonadotrophin secretion in the ewe: Evidence for a seasonal threshold of sensitivity. J Neuroendocrinol 24(5):725–736
124. Clarke IJ (2011) Control of GnRH secretion: one step back. Front Neuroendocrinol 32(3):367–375
125. Yano T, Iijima N, Kakihara K, Hinuma S, Tanaka M, Ibata Y (2003) Localization and neuronal response of RFamide related peptides in the rat central nervous system. Brain Res 982(2):156–167
126. Qi Y, Oldfield BJ, Clarke IJ (2009) Projections of RFamide-related peptide-3 neurones in the ovine hypothalamus, with special reference to regions regulating energy balance and reproduction. J Neuroendocrinol 21(8):690–697
127. Kaewwongse M, Takayanagi Y, Onaka T (2011) Effects of RFamide-related peptide (RFRP)-1 and RFRP-3 on oxytocin release and anxiety-related behaviour in rats. J Neuroendocrinol 23(1):20–27
128. Kirby ED, Geraghty AC, Ubuka T, Bentley GE, Kaufer D (2009) Stress increases putative gonadotropin inhibitory hormone and decreases luteinizing hormone in male rats. Proc Natl Acad Sci U S A 106(27):11324–11329

129. Kageyama K, Hasegawa G, Akimoto K, Yamagata S, Tamasawa N, Suda T (2012) Differential regulation of gonadotropin-releasing hormone by corticotropin-releasing factor family peptides in hypothalamic N39 cells. Peptides 33(1):149–155
130. Clarkson J, Herbison AE (2006) Development of GABA and glutamate signaling at the GnRH neuron in relation to puberty. Mol Cell Endocrinol 254–255:32–38
131. Sliwowska JH, Billings HJ, Goodman RL, Lehman MN (2006) Immunocytochemical colocalization of GABA-B receptor subunits in gonadotropin-releasing hormone neurons of the sheep. Neuroscience 141(1):311–319
132. Akema T, He D, Sugiyama H (2005) Lipopolysaccharide increases gamma-aminobutyric acid synthesis in medial preoptic neurones in association with inhibition of steroid-induced luteinising hormone surge in female rats. J Neuroendocrinol 17(10):672–678
133. Sirivelu MP, Burnett R, Shin AC, Kim C, MohanKumar PS, MohanKumar SM (2009) Interaction between GABA and norepinephrine in interleukin-1beta-induced suppression of the luteinizing hormone surge. Brain Res 1248:107–114
134. Bowers G, Cullinan WE, Herman JP (1998) Region-specific regulation of glutamic acid decarboxylase (GAD) mRNA expression in central stress circuits. J Neurosci 18(15):5938–5947
135. Lin YS, Li XF, Shao B, Hu MH, Goundry AL, Jeyaram A et al (2012) The role of GABAergic signalling in stress-induced suppression of gonadotrophin-releasing hormone pulse generator frequency in female rats. J Neuroendocrinol 24(3):477–488
136. Cravo RM, Margatho LO, Osborne-Lawrence S, Donato J Jr, Atkin S, Bookout AL et al (2011) Characterization of Kiss1 neurons using transgenic mouse models. Neuroscience 173:37–56
137. Terasawa E, Kurian JR, Guerriero KA, Kenealy BP, Hutz ED, Keen KL (2010) Recent discoveries on the control of gonadotrophin-releasing hormone neurones in nonhuman primates. J Neuroendocrinol 22(7):630–638
138. Keen KL, Burich AJ, Mitsushima D, Kasuya E, Terasawa E (1999) Effects of pulsatile infusion of the GABA(A) receptor blocker bicuculline on the onset of puberty in female rhesus monkeys. Endocrinology 140(11):5257–5266
139. Neal-Perry G, Lebesgue D, Lederman M, Shu J, Zeevalk GD, Etgen AM (2009) The excitatory peptide kisspeptin restores the luteinizing hormone surge and modulates amino acid neurotransmission in the medial preoptic area of middle-aged rats. Endocrinology 150(8):3699–3708
140. Leng G, Ludwig M (2006) Jacques Benoit Lecture. Information processing in the hypothalamus: peptides and analogue computation. J Neuroendocrinol 18(6):379–392

Chapter 21
Effects of Environmental Endocrine Disruptors and Phytoestrogens on the Kisspeptin System

Heather B. Patisaul

Abstract Sex steroid hormones, most notably estradiol, play a pivotal role in the sex-specific organization and function of the kisspeptin system. Endocrine-disrupting compounds are anthropogenic or naturally occurring compounds that interact with steroid hormone signaling. Thus, these compounds have the potential to disrupt the sexually dimorphic ontogeny and function of kisspeptin signaling pathways, resulting in adverse effects on neuroendocrine physiology. This chapter reviews the small but growing body of evidence for endocrine disruption of the kisspeptin system by the exogenous estrogenic compounds bisphenol A, polychlorinated biphenyl mixtures, and the phytoestrogen genistein. Disruption is region, sex, and compound specific, and associated with shifts in the timing of pubertal onset, irregular estrous cycles, and altered sociosexual behavior. These effects highlight that disruption of kisspeptin signaling pathways could have wide ranging effects across multiple organ systems, and potentially underlies a suite of adverse human health trends including precocious female puberty, idiopathic infertility, and metabolic syndrome.

Introduction

Environmental endocrine disruptors (EDCs) have garnered considerable attention in recent years, partly because of their omnipresence, but also because the endocrine disruption hypothesis provides a plausible explanation for the rapid prevalence of numerous neuroendocrine disorders. For example, in the United States and other Western countries, the age of female pubertal onset is undeniably advancing [1–5]. Although theories regarding the underlying etiology of this trend abound, the specific confluence

H.B. Patisaul (✉)
Department of Biology, North Carolina State University, Raleigh, NC 27695, USA
e-mail: Heather_Patisaul@ncsu.edu

A.S. Kauffman and J.T. Smith (eds.), *Kisspeptin Signaling in Reproductive Biology*, Advances in Experimental Medicine and Biology 784, DOI 10.1007/978-1-4614-6199-9_21,

of external forces driving this acceleration of female maturation remains largely unidentified. If "puberty begins with a kiss [6]," is it conceivable that perturbation of kisspeptin signaling by environmental hormone mimics might underlie this and other adverse health trends? Disruption of the kisspeptin system is an appealing hypothesis because aberrant organization or function of this system could manifest as a wide range of seemingly disparate, but related, disorders including altered pubertal timing, fertility problems, and metabolic disorder. Although this area of kisspeptin research is relatively new and limited in scope, it is evident that the ontogeny and function of kisspeptin (and related RF-amide) pathways, particularly the sexually dimorphic aspects, are vulnerable to endocrine disruption. Because kisspeptin signaling pathways likely evolved to help coordinate reproductive status with salient environmental cues, this may be an unfortunate liability of a system honed to optimize reproductive fitness. Thus, understanding how kisspeptin signaling pathways are perturbed by EDCs could have a profound impact on the fields of evolutionary ecology and toxicology by demonstrating that disruption within one key neuroendocrine system could ultimately underlie a suite of adverse human health trends including early female puberty, unexplained infertility [7–9], obesity [10, 11], and related neuroendocrine disorders.

The Endocrine Disruption Hypothesis

The term "endocrine disruptor" was coined at a small working group convened at the Wingspread Conference Center in 1991 to discuss what common mechanism(s) might underlie a suite of concerning effects observed in wildlife [12, 13] and if these phenomena were predictive of adverse human health outcomes. Effects included thinning eggshells in birds, abnormal gonadal and genital morphology in alligators, altered sex ratios in turtle clutches, and a rapidly increasing prevalence of intersex in fish. Controversial since it was first voiced, the endocrine disruption hypothesis posits that environmental chemicals have the capacity to interfere with the endocrine system and confer disease in wildlife and humans. Within less than a decade, numerous international agencies assembled their own meetings to weigh in on the issue and craft a common definition. The United States Environmental Protection Agency (EPA) currently defines an EDC as, "an exogenous chemical substance or mixture that alters the structure or function(s) of the endocrine system and causes adverse effects at the level of the organism, its progeny, populations or subpopulations of organisms based on scientific principles, data, weight-of-evidence, and the precautionary principle" (EDSTAC Final Report; available at http://www.epa.gov/endo/pubs/edspoverview/finalrpt.htm). Although it remains a highly controversial topic, in 2009, the Endocrine Society issued a statement supporting the hypothesis and concluding that experimental and epidemiological studies have sufficiently converged with human clinical observations "to implicate EDCs as a significant concern to public health." In June of 2012, The Endocrine Society followed up on this document and published its own definition of an EDC, which is nearly identical to the EPA's but eliminates the requirement that the effects be "adverse."

To date, the EPA has identified hundreds of compounds that meet this definition and thousands of others are suspected of having similar properties (for more information visit http://www.epa.gov/endocrine/Project.html) [14, 15]. These compounds are contained in a wide array of consumer products, including cosmetics and other personal care items, pesticides, plastics, building materials, food containers, medical equipment, epoxy resins, paper products, furniture, electronics, and as "inert" ingredients in pharmaceuticals. They also contaminate our air, water, food supply, and bodies, including those of the unborn [16]. Not all EDCs are anthropogenic, however. Numerous plant-derived compounds, most notably the phytoestrogens, also meet the defining criteria of an EDC. Understanding how these naturally occurring compounds impact neuroendocrine systems, such as the kisspeptin system, will help elucidate the evolved mechanisms by which organisms are sensitive to synthetic compounds with similar structural and chemical properties. Growing awareness of how ubiquitous EDC exposure has become has generated substantial, and growing, public concern regarding the potential long-term risks they pose for wildlife and human health. Systematic evaluation of this risk requires uncovering the mechanisms by which these compounds act within our bodies to confer disease.

Endocrine Disruption or Environmental Sensing?

The idea that environmental chemicals can interact with mammalian physiology is not a radical notion. All living things evolved in a soup of chemicals, and interspecies chemical warfare is a defining element, if not a critical driver, of evolutionary history. For example, plants and animals generate potent neurotoxins, such as venoms and poisons, to protect against predation or enhance their own deadly prowess. Cytochrome P450 enzymes (CYP), required for steroid hormone synthesis, appear to have evolved as a defense against botanical poisons and then repurposed [17]. Moreover, plant alkaloids have historically been, and continue to be, our single greatest source of medicinal therapeutics. Some of the most well known are caffeine, cocaine, nicotine, strychnine, morphine, quinine, and mescaline, a powerful hallucinogen. Plant-derived compounds can also be endocrine disrupting [18]. Phytoestrogens have proven to be an important environmental cue and endocrine-active compound for numerous species, including rodents [19], birds [20], cheetahs [21], and grazers like cattle and sheep and the southern white rhinoceros [22–25]. There are several classes of phytoestrogens, all of which structurally resemble mammalian estrogens. Like bisphenol A (BPA) and other synthetic EDCs, phytoestrogens have historically been thought to act primarily through estrogen receptors [26], but they are also tyrosine-kinase inhibitors [27, 28] and modulate DNA methylation and chromatin configuration [29]. Genistein (GEN) is a member of the most well-known class, the isoflavones, which are most abundant in soybeans and soy-based foods. Others, such as the coumestans, are prevalent in alfalfa, clover, and other pasture legumes, including the ones sheep and other herbivores typically graze on. Phytoestrogens play an important role in plant defense [30], including the recruitment of nitrogen-fixing bacteria [31] and conferring resistance to fungi [32]. Thus plants produce them when under stress including disease, drought, and extreme temperatures. As such, at high levels they are a signal of poor environmental conditions and thus convey important information about when it might be an optimal time to invest in reproduction. Not surprisingly, a high level of phytoestrogen intake has been shown to suppress ovulation and fertility, even in humans [19].

Sheep raised on legume-rich pastures develop a well-characterized suite of reproductive pathologies resulting in reduced conception rates and embryonic loss [9–11]. Recognized since the 1940s, this syndrome is called "clover disease" and can be ameliorated by rearing the animals on subterranean clover cultivars that produce fewer phytoestrogens. Prolonged exposure, however, can ultimately result in permanent infertility, even in adult ewes [25]. Other features of clover disease include estrous cycle irregularity, pyometra, endometrial hyperplasia, leiomyoma of the cervix and uterus, and cystic ovaries. These effects highlight that exposure to inappropriate levels, or during particularly vulnerable windows of development, can result in deleterious effects. Human reproductive physiology is also responsive to phytoestrogens. Menstrual cycle irregularities have been reported in humans consuming a soy-rich diet [16] and use of soy-based infant formula has been associated with a higher risk of menstrual cycle disorders and uterine fibroids in young women [33, 34]. Anthropogenic EDCs likely exploit these evolved systems for sensing external cues of environmental quality, resulting in similar pathologies. Investigation of the specific mechanisms by which synthetic and naturally occurring EDCs act to affect vertebrate physiology has now begun to hone in on the kisspeptin system because it is rapidly becoming clear that it is an essential driver of reproductive maturation and function across species.

Exploration of how EDCs perturb the organization and function of the kisspeptin system has proven to be a useful approach for the kisspeptin field in general because it yields both critical information about the underlying mechanisms contributing to EDC pathology, but also how this system responds to ecologically relevant environmental cues across the life span. Because a core element of EDC research is the timing-specific consequences of exposure, to date, most of the work on the kisspeptin system within this field has focused on the long-term, sex-specific, impacts of perinatal exposure. This research has already begun to identify species, age, and sex-specific sensitivities to endogenous hormones and EDCs within kisspeptin signaling pathways across different life stages. As the field grows, it will ultimately yield a wealth of information about how species differences in the kisspeptin system may have evolved, and a better understanding of how this system responds to ecologically relevant environmental cues across the life span.

Kisspeptin System Endocrine Disruptors: Bisphenol A, Polychlorinated Biphenyls, and Genistein

Although they are now recognized to have multiple modes of action within the endocrine system, EDCs are thought to act, primarily, by interfering with steroid hormone signaling. The subclass of EDCs that has been most comprehensively studied is the group which acts on nuclear estrogen receptors (ERs) (for detailed reviews, see [16, 35–37]). This group contains many well-known EDCs, including some (i.e., dichlorodiphenyltrichloroethane (DDT), diethylstilbesterol (DES), and the PCBs (polychlorinated biphenyls)) which are no longer used in the USA because of their well-documented impacts on wildlife and human health. These compounds remain a health concern, however, because they are still present in the environment and in our bodies. Once bound to a nuclear ER (ERα; Esr1 or ERβ; Esr2), each EDC induces a unique conformational change which ultimately determines the activity of the complex (agonist or antagonist).

One feature of ERα and ERβ which allows for such a diverse range of EDCs to impact their function is a relatively large and promiscuous ligand-binding pocket. Although they are highly selective for estradiol (E2), other endogenous and exogenous ligands have been shown to bind with relatively high affinity. ERβ, for example, is activated by E2 in the sub-nanomolar range, but the binding affinity for the androgen metabolite 5alpha-androstane-3beta,17beta-diol (3β-diol) is similar, suggesting 3β-diol can also act as a potent endogenous ligand [26, 38]. Structurally, 3β-diol and estrogenic EDCs share several properties with E2, most importantly phenolic rings with hydroxyls that can readily undergo hydrogen binding with the amino acids lining the interior of the ligand-binding pocket [39, 40]. In general, the relative binding affinities of environmental EDCs are at least 1,000-fold lower than E2, with the exception of some phytoestrogens (100-fold lower than E2) [26, 41]. Their transcriptional activity is also typically orders of magnitude lower than for E2. Thus, historically, estrogenic EDCs have been considered "weak" estrogens. Although it has rapidly become apparent that this definition is too narrow and that their activity is more complex, context specific, and dose dependent (for a detailed review, see [16]), it is their estrogenic properties that make them capable of interacting with the kisspeptin system. Exposures during development are hypothesized to be particularly adverse because disrupted sex-specific organization of kisspeptin pathways could confer lifelong consequences including altered pubertal timing, infertility, and metabolic disorders. To date, only three compounds have been studied for their capacity to interfere with kisspeptin signaling pathways: BPA, PCB mixtures, and the phytoestrogen GEN (Fig. 21.1).

BPA was initially developed as a synthetic estrogen [42] and entered commercial production in the 1950s. It is now a high-volume production chemical incorporated in numerous products from which it readily leaches, including polycarbonate plastics, the epoxy resins that line the interior of food cans, dental sealants, thermal paper receipts, and plastic water pipes [43–47]. Human exposure is nearly ubiquitous, with urinary levels higher in children than adults [48]. Although long considered weakly estrogenic, the specific mechanisms by which BPA interacts with molecular and cellular targets within the hypothalamus and elsewhere are not yet clearly established [49]. Classically, BPA is thought to disrupt nuclear ER activity, but it has also been shown to have rapid actions via membrane ERs [49, 50].

PCBs are organochlorides of which there are 209 congeners with different degrees of chlorination. Although their physical and chemical properties vary across congeners, in general, their low flammability, high thermal conductivity, and exceptional solubility in organic solvents and oils made them suitable for a wide range of applications. They were used in a variety of combinations as coolant fluids in electric motors,

Compound	Structure	Description
Estradiol	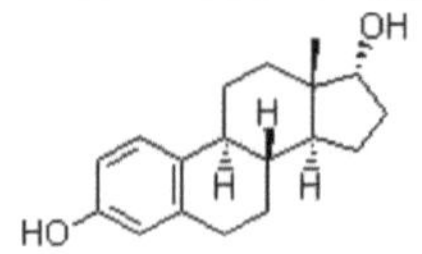	Endogenous Estrogen
BPA	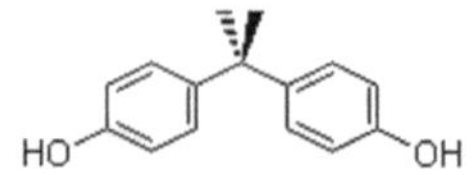	Component of plastics/epoxy resins
GEN	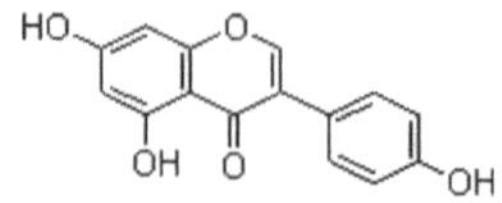	Isoflavone Phytoestrogen
PCBs	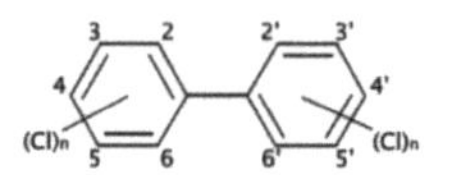	Coolant, plasticizer (no longer used but persists in the environment)

Fig. 21.1 Structure of 17β-estradiol and three common EDCs. Generally, the presence of a phenolic ring indicates the potential for estrogenic activity either by classical or rapid signaling pathways

transformers, and capacitors, but also as plasticizers and stabilizers in paints and cement, fire retardants, hydraulic fluid, adhesives, pesticide mixtures, and sealants. Structurally, there are two classes of PCBs, a coplanar group which makes them chemically similar to dioxins and a noncoplanar group which is less toxic. PCBs were almost always sold as complex mixtures, each of which differed considerably in terms of toxicity. Many are known to have estrogenic, antiestrogenic, or antiandrogenic properties [51]. In general, the desirable physical properties of the PCBs which made them so useful, including their exceptional physical and chemical stability, unfortunately confer resistance to decomposition. In 1979, they were classified as persistent organic pollutants by the EPA and subsequently banned in the USA. Because of their persistence, they remain prevalent in the environment, and tend to bioaccumulate up the food chain and within lipid-rich tissues. Thus, body burdens remain high in some populations [16, 52], and human exposure is ubiquitous and ongoing.

GEN is an isoflavone phytoestrogen found in soy and other legumes, and has been extensively studied for both its beneficial and endocrine-disrupting properties [19, 53]. Intake of GEN is increasing primarily because people are choosing to consume more soy-rich foods (such as soy milk, tofu, and tempeh) to take advantage of their purported health benefits. Increased consumption has been associated with reduced risk of hormone-dependent cancer (breast and prostate), cardiovascular disease, and osteoporosis [19]. Importantly, in Western populations that do not traditionally consume an abundance of soy-rich foods, intake may be higher than perceived because soy is increasingly being incorporated into a wide array (upwards of 60%) of processed food products, such as cereals, crackers, processed meats, bread, and granola bars to increase their nutritional value [54]. Dietary supplements containing high levels of GEN and other isoflavonoid phytoestrogens are also widely available [55]. Exposure

Table 21.1 Comparison of E2, BPA, PCB, and GEN levels in newborns, infants, and adults

	E2 ng/mL	GEN ng/mL	BPA ng/mL	PCBs ng/mL	References
Plasma, adult Western woman (across the cycle)	0.25–5.4	1–2	0.3–5	0.1–14.3	[59–65]
Amniotic fluid, Western		0.4–1.7			[66, 67]
Cord blood, Western			1–3	0.1–7.2	[62, 64]
Plasma, adult Japanese woman		7.2–83	1.4–2		[58, 68–70]
Cord blood, Japan		19.4–45	2.2		[58, 68, 70, 71]
Breast milk		8–13.5	1.3	5.4–63.1	[64, 72, 73]
Plasma, breast-fed infant	<0.04–0.08	2–4.7			[56, 74]
Plasma, infant-fed bovine formula	<0.04–0.08	9.4			[56]
Plasma, infant-fed soy formula	<0.04–0.08	684–757			[56, 75]

GEN levels are typically higher than BPA or PCB levels for all age groups and in all fluids listed
Compiled values represent the range of those previously reported and do not take into account methodological differences (including controlling for lipid levels for the PCB values listed), sample sizes, or differences between steady state and peak levels. Thus, values presented should be considered representative. Circulating E2 levels were obtained from references [76–78] and the UK General Practice Notebook (http://www.gpnotebook.co.uk/simplepage.cfm?ID=570818627&linkID=24801&cook=yes) (Table adapted from [19])

to GEN and other phytoestrogens is typically much higher than for most synthetic EDCs, even among individuals that do not consume a legume-rich diet (Table 21.1). Thus, it is important to consider the potential long-term impacts of exposure, particularly during development. Concern regarding the use of soy-based infant formula is emerging because epidemiology studies have associated its use with elevated risk of menstrual irregularities and uterine fibroids [33, 34]. Infants maintained on these formulas consume as much as 6–9 mg/kg of isoflavone phytoestrogens per day, which is 6–11-fold higher than a typical adult exposure [56]. GEN exposure can also occur in utero through placental transfer or after birth via lactation [57, 58].

Mechanisms of Kisspeptin System Endocrine Disruption

The detailed neuroanatomy, sexual differentiation, and steroid hormone regulation of the kisspeptin system are provided in other chapters, and thus will not be repeated here. Key aspects that make this system vulnerable to endocrine disruption, including how kisspeptin is influenced by perinatal estrogens, are worth highlighting (Fig. 21.2). In rodents, two primary hypothalamic populations of kisspeptin neurons have been characterized: one in a region comprising the anteroventral periventricular (AVPV) nucleus and the medial aspects of the rostral periventricular nucleus (AVPV/PeN, an area sometimes referred to as the rostral periventricular area of the third ventricle (RP3V)), and a second in the arcuate (ARC) nucleus. Additionally, a third, smaller population was subsequently identified in the medial amygdala (MeA) [79, 80]. There is now evidence that

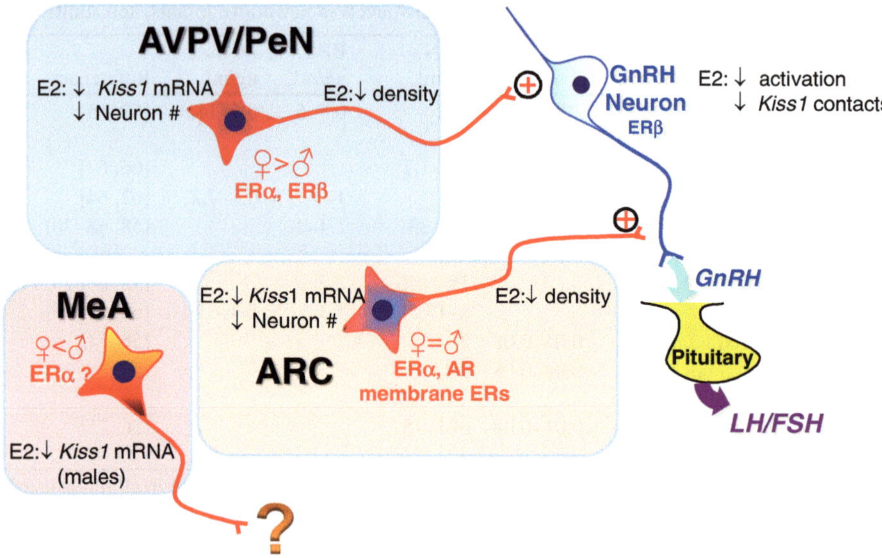

Fig. 21.2 Influence of perinatal estrogen (E2) exposure on the three primary populations of Kp neurons and their efferent projections in female rodents. In general, early life exposure to estrogens masculinizes the kisspeptin system including *Kiss1* mRNA expression, neuron number, density of efferent projections, and putative synapses on GnRH neurons. The MeA population is an exception in that neonatal E2 administration has the opposite effect and acts to decrease male *Kiss1* expression. For each population, sex differences in neuron number are indicated, as well as their co-occurrence with ER subtypes. All three have now been shown to be impacted by EDCs as detailed in Table 21.2 (Figure adapted from a slide generously provided by Alexander Kauffman)

all three of these populations are sensitive to EDCs, particularly during the neonatal critical period. *Kiss1* expression is also found in numerous peripheral tissues important for reproduction and energy balance including the gonads, placenta, adipose tissue, pancreas, and pituitary [81–84], but to date, nothing is known about how EDCs might influence *Kiss1* expression or kisspeptin activity in these tissues. EDC exposure is hypothesized to contribute to diseases, such as metabolic syndrome and compromised fertility, involving these *Kiss1*-expressing organ systems. Thus, it is appealing to speculate that endocrine disruption of the peripheral kisspeptin system may play an important, but as yet unappreciated, role in the etiology of these disorders.

Within the hypothalamus, the ontogeny and function of kisspeptin signaling pathways are profoundly influenced by gonadal steroid hormones. Importantly, early life exposure to estrogens sexually differentiates this system, resulting in distinct anatomical and functional differences between males and females (Table 21.2). For example, in rodents, females have more AVPV/PeN *Kiss1* neurons, fibrous projections, putative contacts on GnRH neurons, and *Kiss1* mRNA expression than males, a sex difference that emerges around the second week of life [85–88]. In the adult rodent ARC, *Kiss1* expression and neuron number appear to be equivalent in both sexes [88, 89], although the density of *Kiss1* fibers in the ARC is appreciably greater in females (the functional significance of which remains poorly understood). Intriguingly, unlike in the AVPV/PeN, *Kiss1* mRNA levels in the ARC are robust

Table 21.2 Summary of perinatal EDC effects on the rat kisspeptin (Kp) system

	AVPV/PeN		ARC		MeA		Physiological outcomes		
	Females	Males	Females	Males	Females	Males	Females	Males	References
E2, EE, or EB	Fewer Kp cells Lower *Kiss1* expression Fewer Kp fibers Fewer Kp contacts on GnRH neurons Elevated *Kiss1r* at birth	Reduced *Kiss1r* at birth	Fewer Kp cells Lower *Kiss1* expression Fewer Kp fibers	None observed	None observed	Lower *Kiss1* expression	Early vaginal opening Anestrus Impaired steroid positive feedback		[96, 101, 102, 111]
BPA	No effect on Kp fiber density Fewer *Kiss1* cells preweaning	More Kp neurons	Fewer Kp fibers	None observed	None observed	Lower *Kiss1* expression	Early vaginal opening Irregular estrus	Decreased basal LH levels Feminization of steroid positive feedback Elevated anxiety	[91, 96, 101, 103, 104]
PCBs	Elevated *Kiss1r* at birth Fewer Kp fibers	Reduced *Kiss1r* at birth No observed effects in adulthood					Impaired steroid positive feedback	Delayed puberty Lower serum progesterone levels Lower serum androgen levels (some congeners)	[111] [112]
GEN	Fewer Kp fibers	None observed	No impact on Kp fibers	None observed			Early vaginal opening Anestrus Impaired steroid positive feedback		[88, 119]

In general, perinatal EDCs have a masculinizing effect on the female kisspeptin system, particularly in the AVPV/PeN. Relatively less is known about EDC effects on the male kisspeptin system. To date, data is only available from rats, highlighting the need for future work in greater breadth of species and sexually dimorphic at birth (higher in females) and then equivalent by postnatal day (PND) 19 and thereafter [87, 90]. The functional significance of this transient early life sex difference remains to be determined. Exogenous administration of estrogens, or aromatizable androgens, during the neonatal critical period produces the male phenotype [91, 92], demonstrating that this is a particularly sensitive window of vulnerability to endocrine disruption and that organizational changes are likely to be permanent.

Although a small subset of AVPV/PeN kisspeptin neurons co-express ERβ in females [93, 94], suggesting that either ER subtype could contribute to the estrogen-dependent sex-specific ontogeny and function of the kisspeptin system, it is generally accepted that ERα is the primary ER isoform mediating estrogen activity on kisspeptin neurons. This is particularly clear in the AVPV/PeN. Estradiol administration successfully stimulates AVPV/PeN kisspeptin neurons in ERβ knockout mice but not in ERα knockout mice [93]. Additionally, a role for ERα in the masculinization of the AVPV/PeN kisspeptin system is suggested by the observation that neonatal administration of the ERα-selective agonist PPT, but not the ERβ-selective agonist DPN, results in abrogated AVPV/PeN *Kiss1* mRNA levels [95] and immunoreactive (–ir) fibers [96] in female rats. Thus, EDCs with the capacity to interact with ERα, or its expression, presumably have the greatest potential to disrupt the sex-specific ontogeny and function of the AVPV/PeN aspect of the kisspeptin system. In the ARC, steroid action is more complex and may be conferred by other routes including the androgen receptor and "nonclassical" estrogen signaling pathways [97]. For example, ARC *Kiss1* expression is suppressed by neonatal administration of E2, but not ligands selective for ERα or ERβ, suggesting that the E2 effect must be mediated via a "nonclassical" pathway (Fig. 21.3) [95]. These regional differences in E2 signaling within the kisspeptin system suggest that endocrine disruptors could have region-specific effects, and that compounds capable of interacting with membrane ERs may specifically modulate the ARC population of kisspeptin neurons.

Comparatively less is known about the MeA population of kisspeptin neurons. It is sexually dimorphic in mature rodents, with higher levels of *Kiss1* in males (Table 21.2) [79]. Similar to the AVPV/PeN population, *Kiss1* expression in the MeA appears to be exclusively driven by ERs because estradiol, but not the non-aromatizable androgen DHT, modulates expression [79]. The functional role of this population remains to be elucidated, although it is hypothesized to be involved in coordinating responses to pheromonal and other olfactory cues related to social and reproductive behavior.

BPA Effects on the Rodent Kisspeptin System

Females

Exposure to BPA has been shown to impact female reproductive physiology in numerous rodent models (reviewed in [16]). Effects include accelerated female puberty, irregular estrous cycles, and premature anestrous. These effects could result from disruption anywhere within the hypothalamic pituitary gonadal (HPG) axis, from the

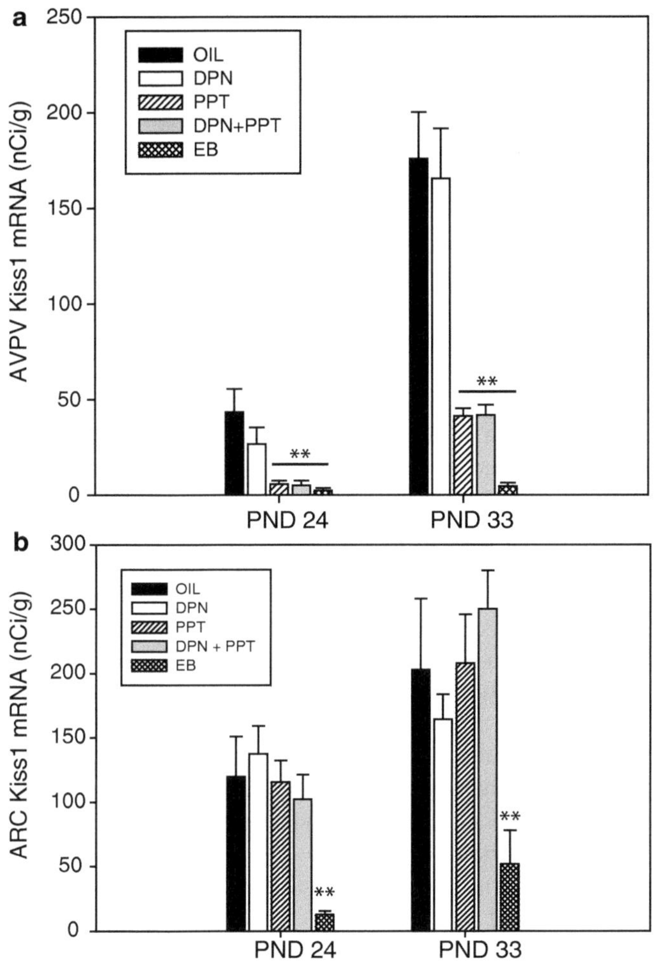

Fig. 21.3 Impact of neonatal exposure to agonists selective for ERα or ERβ on hypothalamic *Kiss1* expression in the pubertal female rat. Neonatal administration of estradiol benzoate (EB) abrogated *Kiss1* expression in both the AVPV/PeN and the ARC. The ERα agonist PPT masculinized AVPV/PeN expression levels, but had no impact on *Kiss1* expression in the ARC. Neonatal agonism of ERβ by DPN had no effect on *Kiss1* levels in either region. Interestingly, coadministration of the ER agonists did not recapitulate the effect of EB in the ARC suggesting that a nonclassical mechanism of estrogen signaling confers the effect of EB in this hypothalamic region. These data suggest that developmental exposure to EDCs could have region-specific effects on *Kiss1* neurons, especially if they have differential activity on ERα and ERβ (Modified from Patisaul HB, Losa-Ward SM, Todd KL, McCaffrey KA, Mickens JA. Influence of ERbeta selective agonism during the neonatal period on the sexual differentiation of the rat hypothalamic-pituitary-gonadal (HPG) axis. Biol Sex Differ. 2012 Jan 19;3(1):2. Copyright ©2012 Patisaul et al.; licensee BioMed Central Ltd)

ovary to the hypothalamus. Reproductive maturation and function is coordinated by the release of gonadotropins [98, 99]; thus, altered GnRH activity could underlie the suite of reproductive effects induced by BPA exposure. Because kisspeptin neurons are considered critical "gatekeepers" of GnRH activity, disruption of this sex-specific function could be a central mechanism by which exposure to an estrogenic endocrine disruptor, such as BPA, induces a suite of reproductive effects.

The first indication that the kisspeptin system could be vulnerable to endocrine disruption was reported in 2009. Subcutaneous injection of 100 or 500 μg BPA between PNDs 1 and 5 resulted in reduced hypothalamic *Kiss1* levels on PND 30 in rats of both sexes [91]. In males, this effect persisted through PND 75 and was accompanied by a persistent decrease in basal LH levels, suggesting that steroid negative feedback on gonadotropin secretion was impaired. This effect implicates the ARC as the site of disruption, a possibility that could not be delineated from this initial study because *Kiss1* expression was quantified in whole hypothalamus. Thus, it remained unclear which population of *Kiss1* neurons BPA impacted: AVPV/PeN, ARC, or both. This study used relatively high doses of BPA which are not relevant when considering the potential impacts of human exposure. Importantly, however, it established the "proof of principle" that the kisspeptin system could be impacted by EDCs. It also identified the neonatal critical period as a vulnerable exposure window, and demonstrated that both sexes are at risk.

A subsequent series of papers provided the first piece of evidence that BPA effects on the kisspeptin system could be region and sex specific. Newborn rats were subcutaneously injected with BPA (50 μg/kg or 50 mg/kg) daily for the first four days of life. Estradiol benzoate (EB) was used as a positive control because it is well established that estrogen administration during this critical period masculinizes the HPG axis. Consistent with what had been published previously [100], females exposed to the lower dose displayed early vaginal opening, a hallmark of puberty in the rat, and females in both exposure groups developed irregular estrous cycles [101]. The animals were then ovariectomized and sequentially administered EB and progesterone over 48 h to stimulate a gonadotropin surge [102]. GnRH activation, assessed by quantifying the co-localization of GnRH and FOS, in both BPA exposure groups was as robust as in the unexposed controls, suggesting that BPA did not masculinize this aspect of the HPG axis. Accordingly, the density of kisspeptin-ir fibers in the AVPV/PeN was not significantly impacted by BPA exposure but significantly lower in females masculinized by EB [96].

By contrast, in the ARC, the plexus of kisspeptin-ir fibers was less dense in the 50 mg/kg BPA exposed females, an observation consistent with impaired steroid negative feedback and may account for the observed estrous cycle irregularities. A potential limitation of these studies, however, is the use of kisspeptin-ir fibers as a marker of kisspeptin content. This was done because the soma do not readily label in the rat, particularly in the AVPV/PeN. A subsequent study by this group showed that the magnitude of the sex difference in kisspeptin-ir fibers closely mirrors that of *Kiss1* mRNA in the AVPV/PeN [88]. By contrast, although the sex difference in the density of ARC kisspeptin-ir fibers is pronounced by weaning, there is no sex difference in the number of kisspeptin-ir soma or *Kiss1* expression [88, 89]. An incomplete understanding of the functional relevance of the sex difference in ARC fiber density makes the impact of EDCs on this aspect of the kisspeptin system difficult to interpret. One possibility is that elevated fiber labeling in females results from increased kisspeptin storage or transport. Conversely, kisspeptin secretion could be greater in males thereby resulting in lower levels of immunoreactivity. That the density of the ARC kisspeptin-ir fiber plexus is clearly influenced by sex steroids, resulting in reduced levels in males and estrogen-exposed females, also raises questions as to their point of origin. If the fibers arise from the AVPV/PeN population, then this sex difference in ARC fiber density is consistent with the sex difference in neuron number and *Kiss1* expression found in the AVPV/PeN.

Males

Data regarding the vulnerability of the male kisspeptin system to BPA is limited and conflicting. Subcutaneous injection of 50 μg/kg BPA over the first 4 days of life did not affect kisspeptin-ir fiber density in either the AVPV/PeN or the ARC of adult male rats [96]. A follow-up study by the same group employing a similar exposure paradigm but with an additional dose (50 μg/kg or 50 mg/kg BPA) found that on postnatal day (PND) 10, when *Kiss1* expression is just beginning to become apparent in the AVPV/PeN, expression was unaltered by BPA compared to unexposed control males [103]. In contrast, females exposed to either dose had fewer *Kiss1*-expressing neurons compared to unexposed females. ARC *Kiss1* levels were unchanged in either sex on PND 4 or 10.

Exposure spanning a longer duration or encompassing a different developmental critical period may be required to modify male kisspeptin signaling pathways [104]. Males exposed to 2 μg/kg BPA from the tenth day of gestation through the first week after birth by injection to the dam had significantly more AVPV/PeN kisspeptin perikarya than unexposed males in peripuberty and adulthood. Moreover, these males displayed an E2-induced LH surge comparable in magnitude to females, indicating that BPA inhibited the defeminization of the neuroendocrine circuitry required to mediate steroid positive feedback. This was accompanied by a significant reduction in GnRH neuron numbers, an observation which is difficult to reconcile in light of the other data. An important caveat of this study is that the axonal transport blocker colchicine was administered prior to sacrifice to enhance visualization of kisspeptin soma, which is not readily immunolabeled in the rat. Although colchicine is often used for this technical purpose, it can also alter the synthesis of the compound of interest [105], making it a potential confounder for a toxicological study.

Emerging evidence indicates that *Kiss1* expression in the male MeA is also vulnerable to BPA exposure. In rats, exposure to low levels of BPA through drinking water across gestation and lactation (via the dam) and peripuberty (via direct consumption) abrogates *Kiss1* expression in the pubertal male MeA, but not the female MeA. Decreased MeA *Kiss1* expression was accompanied by higher levels of anxiety and the downregulation of other genes associated with sociosexual behavior including ERβ (Patisaul et al., in review). The downregulation of *Kiss1* by BPA was recapitulated by Ethinyl estradiol (EE) exposure, suggesting it is an estrogenic effect, an outcome which is unusual because estrogen typically masculinizes, rather than demasculinizes the rodent brain. Importantly, serum BPA levels in this study were within the range reported for humans (Table 21.1; [43]), suggesting that the results may have human relevance. These data provide the first indication of what the functional role of this Kp population might be and the first piece of evidence that they are sensitive to low levels of environmental estrogens.

Overall, there is sparse but clear evidence of BPA effects on the ontogeny and function of kisspeptin signaling pathways in both sexes (Table 21.2). What remains to be determined is if effects occur at levels considered relevant to humans and in species other than rodents. Human plasma BPA levels are generally in the range of 2–4 ng/mL [43] and exposure is primarily oral, so this route of administration is considered more relevant when assessing the potential for human risk. It will also be important to explore effects in non-rodent species because, as discussed at length in the preceding chapters, there are substantive species differences in the location and function of kisspeptin neurons, most notably in the preoptic region [106].

PCB Effects on the Rodent Kisspeptin System

Females

In rodents, perinatal PCB exposure induces a suite of reproductive effects comparable to BPA. Human relevant exposure levels have sex-dependent effects on reproductive physiology, hypothalamic sexual differentiation, and neuroendocrine gene and protein expression in adulthood [107, 108]. Two studies, both conducted by the same group, have shown that PCBs can interfere with the ontogeny of the kisspeptin system. Because PCBs are present in the environment and the body as mixtures, the impacts of two different PCB mixtures were examined. The first, Aroclor 1221, was a popular commercial mixture composed of lightly chlorinated PCB congeners with a half-life on the order of days. The second consisted of a reconstituted PCB mixture containing three heavily chlorinated PCB congeners (PCB138, PCB153, and PCB180) with half-lives on the order of decades and known to bioaccumulate in the environment and human tissues [109, 110]. Exposure on only two days of late gestation (gestational days 16 and 18) to 1 mg/kg of either mixture resulted in significantly elevated preoptic levels of the kisspeptin receptor (*Kiss1r,* previously identified as GPR54) in females but reduced levels in males, by the day of birth. This effect is likely estrogenic because it was recapitulated by gestational EB exposure in both sexes [111].

A cohort of offspring were subsequently reared through young adulthood and irregular estrous cycles were observed in both PCB exposed groups [112]. The ovaries appeared healthy, suggesting an overall disruption within the neural aspects of the HPG axis. The density of kisspeptin-ir fibers in the AVPV/PeN was significantly reduced by both PCB mixtures and to the same degree as EB, indicating masculinization. Correspondingly, the percentage of GnRH neurons co-expressing FOS on the afternoon of proestrus was as low in the PCB exposed groups as the EB exposed group, an observation indicative of the failure to generate a preovulatory GnRH surge via steroid positive feedback. Corroborating indices of masculinization included male-typical AVPV volume in both PCB exposed groups, and male-like levels of ERα-ir neurons in the A1221 exposed group. Intriguingly, luteinizing hormone (LH) levels were significantly attenuated in the A1221 exposed group, but not the PCB mixture group, despite significantly impaired GnRH activation in the latter group. It remains to be determined what the compensatory mechanism for this, resulting in normal LH levels, might be.

Males

In contrast to females, effects of gestational PCB exposure were relatively mild in adult males. AVPV/PeN kisspeptin-ir was unaltered as were AVPV ERα cell numbers. Serum testosterone levels were significantly depreciated in the group exposed to the PCB mixture but not EB or A1221. Serum progesterone levels, however, were significantly lower in both PCB groups but unaltered by EB, suggesting an alternative mechanism of action for this endpoint. Notably, male puberty was delayed by PCB exposure, an effect which may indicate disruption of ARC kisspeptin signaling pathways. This possibility remains to be examined.

Importantly, the dose of Aroclor used for these two studies has previously been shown to produce reproductive effects in female rats [113, 114], and the dose of PCB congeners approximates the typical body burden of human infants [107, 115]. Because these exposure levels employed for these studies are considered to be human relevant [113, 114, 116], these results raise the possibility that the human kisspeptin system may be vulnerable to disruption by exposure to PCBs and other EDCs at current exposure levels (Table 21.2).

Influence of Genistein and other Phytochemicals on the Kisspeptin System

Considering that kisspeptin neurons are likely a critical component of a complex neural network tuned by evolutionary forces to coordinate reproductive activity in response to a milieu of environmental cues [117, 118], it is arguably expected that this system should be sensitive to phytoestrogens. Research spanning six decades has established that phytoestrogens interfere with the organizational role of endogenous estrogen in the developing mammalian reproductive system [19]. Very limited data, however, regarding the impact of phytoestrogens on kisspeptin signaling networks has been generated thus far. These studies have primarily used approaches similar to those described for BPA and the PCBs, which focus on the impacts of acute exposure over discrete developmental periods.

Similar to what has been reported for the PCBs, developmental exposure to GEN, an isoflavone found in soy and other legumes, has a masculinizing influence on the rat female kisspeptin system (Table 21.2). Early life exposure to GEN, like BPA and the PCBs, advances female puberty and produces estrous cycle irregularities [119], implying disrupted ontogeny of the HPG axis. In female rats, neonatal exposure to 10 mg/kg GEN, but not 1 mg/kg, resulted in a lower density of AVPV/PeN kisspeptin-ir fibers across the pubertal transition [88], an effect which persisted into adulthood and is indicative of masculinization (Fig. 21.4) [119]. In the ARC, kisspeptin-ir fiber density was unaltered by GEN but significantly lessened by developmental EB exposure [88, 119]. Early life exposure to GEN also resulted in impaired GnRH activation in ovariectomized, hormone replaced females [119], an effect which is consistent with masculinization of AVPV/PeN kisspeptin signaling pathways. Interestingly, the phytoestrogen metabolite equol did not confer a similar suite of effects on the female kisspeptin system when administered at the same dose [119], despite being considered a more potent estrogen agonist than GEN [120]. Moreover, no significant impacts of postnatal GEN exposure were found on adult male kisspeptin-ir levels [96]. The collective effects of developmental GEN exposure are similar to those produced by PCBs and emphasize the sex-specific vulnerability of the kisspeptin system to other estrogenic EDCs.

From Endocrine Disruption to the Fetal Basis of Adult Disease

Since its conception in 1991, the EDC field has rapidly gained a foothold in clinical endocrinology, and this evolution has generated a provocative new idea: diseases of adulthood may begin in the womb. This "Fetal Basis of Adult Disease" hypothesis is an extension of the "Barker hypothesis" which predicted that individuals conceived during times of stress and nutrient deprivation will develop a "thrifty" phenotype that efficiently extracts and retains nutrients [121]. Disease results when this phenotype is mismatched and the individual lives in conditions where nutrients are abundant and thus becomes obese and prone to diabetes, cardiovascular disease, and hypertension. It is now hypothesized that early life exposure to chemicals also contributes to disease states that emerge later in life, particularly reproductive and metabolic disorders [122]. The impacts of smoking and alcohol on fetal development have long been recognized and give credence to this hypothesis, but the long-term impacts of seemingly more innocuous chemicals, such as those that leach from food containers, coat our electronic devices, and escape from furniture into house dust where we then eat and breathe them, remain unclear.

As exemplified by the phytoestrogens, sensitivity to estrogenic chemicals would be adaptive in the right context but also capable of conferring disease. Current evidence supports the hypothesis that the kisspeptin system is a crucial component of a complex neuroendocrine system, which senses environmental signals and responds to optimize critical developmental stages including puberty, mate seeking, and pregnancy. For example, despite decades of research to unlock the "black box" that is mammalian puberty, the confluence of signals required to initiate the precise moment of reproductive maturation remains elusive but clearly requires the integration of numerous environmental cues including dietary quality, stress, body composition, rank, and the stability of the social group. It is also evident, particularly in long-lived species, such as humans and other apes, that external forces can shift the timing of pubertal onset in either direction. In primates, for example, environmental stress generally induces early puberty in females but delays it in males. Thus, age at puberty does not appear to be "predetermined" at birth, but rather established through complex and adaptive gene by environment interactions. Determining the optimal time to undergo reproductive maturation and begin to compete for mates is critical to maximize fitness. Females that mature too slowly risk shortening their overall reproductive life span (and thus their maximum number of pregnancies), while males that mature too quickly risk losing out to stronger, larger males in the competitive struggle for mates. Thus evolution has likely honed an exquisitely sensitive neural system, of which the kisspeptin system is a critical component, to sense the quality and composition of the environment throughout development and subsequently pace the tempo of reproductive maturation and function.

Uncovering the specific mechanisms underlying this phenomenon is paramount to understanding the factors potentially contributing to the advancement of puberty in girls [1–5] and to test the hypothesis that EDC exposure may exacerbate this phenomenon. Thus, ongoing work exploring how EDCs alter the sex-specific maturation of the kisspeptin system and its governance of pubertal onset will yield critical information about the ontogeny of the kisspeptin system, and also how it is sensitive to environmental cues. The hypothesis that disruption of the kisspeptin system during critical windows of development may shift age at puberty is plausible given that the hypothalamic kisspeptin system was first recognized by the discovery that humans with a mutated form of *KISS1R* are hypogonadal and fail to undergo puberty [123, 124]. This phenotype was recapitulated in mice where either *Kiss1* or *Kiss1r* is knocked out [125]. Chronic administration of kisspeptin during the perinatal period accelerates pubertal onset in rats [126] suggesting that premature release may underlie EDC-induced advanced puberty. Kisspeptin initiates puberty by stimulating GnRH neurons (a detailed description of which is provided in other chapters), upon which *Kiss1r* is constitutively expressed [127]. In the rodent AVPV/PeN, *Kiss1* expression and the density of kisspeptin-ir neurons and fibers increase as puberty approaches and is accompanied by a concomitant increase in the percentage of GnRH neurons with kisspeptin-ir appositions [85–88]. Thus, one potential mechanism by which EDC exposure could advance puberty is by enhancing the postweaning rise in AVPV/PeN levels and their synaptic contacts on GnRH neurons. The limited data generated to date do not support this hypothesis, however. Neonatal exposure to GEN or PCBs attenuates AVPV/PeN *Kiss1* fiber density and their contacts on GnRH neurons [128] during the pubertal transition, an effect that is indicative of masculinization rather than accelerated maturation (Fig. 21.4). An alternative hypothesis is that the ARC population of kisspeptin neurons is primarily responsible for conferring the onset of puberty. Although there is limited data supporting this idea

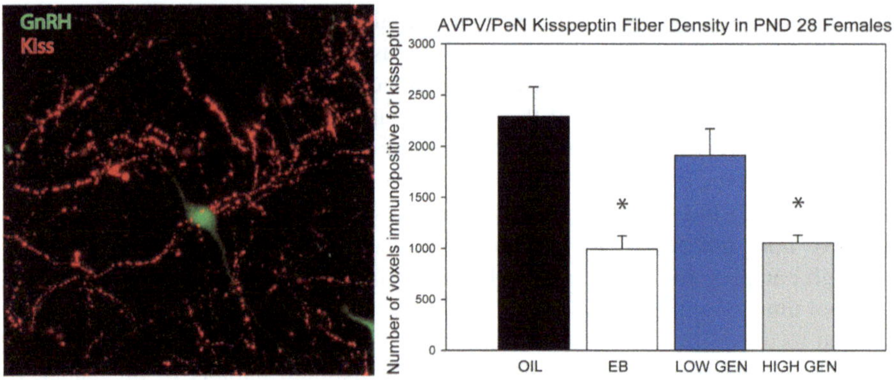

Fig. 21.4 Neonatal exposure to GEN reduces the density of AVPV/PeN fibers immunoreactive for kisspeptin at puberty (PND 28) in female rats. Depicted is a single GnRH neuron (*green*) surrounded by and receiving efferent projections from kisspeptin neurons (*red*). This plexus is significantly denser in females than males, a sex difference which becomes more robust across the pubertal transition. Neonatal exposure to a high (10 mg/kg) but not a low (1 mg/kg) dose of the isoflavone phytoestrogen GEN advances vaginal opening. Early female puberty is accompanied by a marked and significant reduction in the density of kisspeptin fibers in the AVPV/PeN that is recapitulated by estradiol benzoate (EB) exposure and suggestive of masculinization

[129], it remains to be definitively established. Finally, there is the possibility that disruption of other RFamides, such as gonadotropin-inhibiting hormone (GnIH and the human homologs RFRP-1 and RFRP-3 [130]), may also contribute to the steady advancement of female puberty. Data testing this hypothesis is only just beginning to emerge [128].

Future Directions

There is clear and compounding evidence that kisspeptin signaling pathways are vulnerable to endocrine disruption, particularly when exposure occurs in development. By extension, work in this area has also revealed how sensitive this system is to endogenous hormones and the long-term consequences of perturbation by hormone administration in early life. Thus, the endocrine disruption field contributes translational but also fundamental knowledge about the steroid hormone-dependent organization and function of the kisspeptin system across the life span, making it a truly interdisciplinary area of study. As this research progresses, it will advance understanding of the mechanisms by which EDCs act to confer disease, and also how hormones shape the sex-specific development of the kisspeptin system across the life span. Importantly, it will provide insights into the critical developmental windows during which kisspeptin neurons are sensitive to gonadal hormones and the compounds that mimic them, and also the mechanisms by which sex differences in the kisspeptin system emerge.

Numerous questions remain, however, making this a prime area for future study. For example, all of the work in this area to date has focused on perinatal exposure because of the "Fetal Basis of Adult Disease" concept. Thus, nothing is known about the effects of EDC exposure, or hormone administration, during other life stages such as puberty or adulthood. Impacts on the peripheral kisspeptin system also remain unaddressed and will be a particularly critical area of future work. Two noteworthy regions are the pancreas and the gonads because they are prime targets for EDCs, and the physiological role of *Kiss1* and its receptor in these organs remains to be characterized. Moreover, all of the studies to date have been performed in rodents, so it has not yet been established if other species are vulnerable, although the effects of GEN implicate sensitivity in numerous other species, including humans. Work in sheep and other species for which there is considerable knowledge about the neuroanatomical structure and function of the kisspeptin system would be tremendously beneficial. This is particularly critical when considering if any of the EDC effects on the kisspeptin system observed to date are predictive of similar effects in humans. This has become a pressing question because effects in rodents have now been documented at EDC exposure levels considered human relevant. Confirmation of EDC effects in other species would help clarify the possibility that the human kisspeptin system may be liable to disruption and the mechanisms by which EDCs impact human health.

In general, the possibility that EDCs can alter kisspeptin signaling pathways is an appealing hypothesis because it would be profoundly explanatory for a suite of adverse human health effects, both reproductive and metabolic, that currently defy explanation. Understanding the mechanisms by which EDCs alter the ontogeny and function of kisspeptin pathways further elucidates the consequences of endocrine disruption, but also contributes fundamental neuroanatomical data that advances our understanding of the evolution and sex-specific responsivity of this complex system to the environment. It is a rapidly growing field with far-reaching implications for toxicology, endocrinology, and evolutionary biology.

References

1. Herman-Giddens ME, Slora EJ, Wasserman RC, Bourdony CJ, Bhapkar MV, Koch GG et al (1997) Secondary sexual characteristics and menses in young girls seen in office practice: a study from the pediatric research in office settings network. Pediatrics 99(4):505–512
2. Partsch CJ, Sippell WG (2001) Pathogenesis and epidemiology of precocious puberty. Effects of exogenous oestrogens. Hum Reprod Update 7(3):292–302
3. Aksglaede L, Sorensen K, Petersen JH, Skakkebaek NE, Juul A (2009) Recent decline in age at breast development: the Copenhagen Puberty Study. Pediatrics 123(5):e932–e939
4. Proos LA, Hofvander Y, Tuvemo T (1991) Menarcheal age and growth pattern of Indian girls adopted in Sweden. I. Menarcheal age. Acta Paediatr Scand 80(8–9):852–858
5. Parent AS, Teilmann G, Juul A, Skakkebaek NE, Toppari J, Bourguignon JP (2003) The timing of normal puberty and the age limits of sexual precocity: variations around the world, secular trends, and changes after migration. Endocr Rev 24(5):668–693

6. Shahab M, Mastronardi C, Seminara SB, Crowley WF, Ojeda SR, Plant TM (2005) Increased hypothalamic GPR54 signaling: a potential mechanism for initiation of puberty in primates. Proc Natl Acad Sci U S A 102(6):2129–2134
7. Brannian J, Hansen K (2006) Assisted reproductive technologies in South Dakota: the first ten years. S D Med 59(7):291–293
8. Nyboe Andersen A, Erb K (2006) Register data on assisted reproductive technology (ART) in Europe including a detailed description of ART in Denmark. Int J Androl 29(1):12–16
9. Frey KA, Patel KS (2004) Initial evaluation and management of infertility by the primary care physician. Mayo Clin Proc 79(11):1439–43; quiz 43
10. Decherf S, Demeneix BA (2011) The obesogen hypothesis: a shift of focus from the periphery to the hypothalamus. J Toxicol Environ Health B Crit Rev 14(5–7):423–448
11. Castellano JM, Bentsen AH, Mikkelsen JD, Tena-Sempere M (2010) Kisspeptins: bridging energy homeostasis and reproduction. Brain Res 1364:129–138
12. Hotchkiss AK, Rider CV, Blystone CR, Wilson VS, Hartig PC, Ankley GT et al (2008) Fifteen years after "Wingspread"—environmental endocrine disrupters and human and wildlife health: where we are today and where we need to go. Toxicol Sci 105(2):235–259
13. Colborn T, Clement C (1992) Chemically-induced alterations in sexual and functional development: the wildlife/human connection. Princeton Scientific, Princeton, NJ
14. Toppari J, Larsen JC, Christiansen P, Giwercman A, Grandjean P, Guillette LJ Jr et al (1996) Male reproductive health and environmental xenoestrogens. Environ Health Perspect 104(Suppl 4):741–803
15. Crisp TM, Clegg ED, Cooper RL, Wood WP, Anderson DG, Baetcke KP et al (1998) Environmental endocrine disruption: an effects assessment and analysis. Environ Health Perspect 106(Suppl 1):11–56
16. Vandenberg LN, Colborn T, Hayes TB, Heindel JJ, Jacobs DR Jr, Lee DH et al (2012) Hormones and endocrine-disrupting chemicals: low-dose effects and nonmonotonic dose responses. Endocr Rev 33(3):378–455
17. Gonzalez FJ, Nebert DW (1990) Evolution of the P450 gene superfamily: animal-plant 'warfare', molecular drive and human genetic differences in drug oxidation. Trends Genet 6(6):182–186
18. Whitten PL, Naftolin F (1991) Dietary plant estrogens: a biologically active background for estrogen action. In: Hochberg RB, Naftolin F (eds) The new biology of steroid hormones. Raven, New York, pp 155–167
19. Patisaul HB, Jefferson W (2010) The pros and cons of phytoestrogens. Front Neuroendocrinol 31(4):400–419
20. Leopold A, Erwin M, Oh J, Browning B (1976) Phytoestrogens: adverse effects on reproduction in California quail. Science 191(4222):98–100
21. Setchell K, Gosselin S, Welsh M, Johnston J, Balisteri W, Kramer L et al (1987) Dietary estrogens-a probable cause of infertility and liver disease in captive cheetahs. Gastroenterology 93:225–233
22. Adams NR (1995) Detection of the effects of phytoestrogens on sheep and cattle. J Anim Sci 73:1509–1515
23. Adams NR (1995) Organizational and activational effects of phytoestrogens on the reproductive tract of the ewe. Proc Soc Exp Biol Med 208(1):87–91
24. Bennetts HW, Underwood EJ, Shier FL (1946) A specific breeding problem of sheep on subterranean clover pastures in Western Australia. Aust Vet J 22:2
25. Tubbs C, Hartig P, Cardon M, Varga N, Milnes M (2012) Activation of Southern White Rhinoceros (Ceratotherium simum simum) estrogen receptors by phytoestrogens: potential role in the reproductive failure of captive-born females? Endocrinology 153(3):1444–1452
26. Kuiper GG, Lemmen JG, Carlsson B, Corton JC, Safe SH, van der Saag PT et al (1998) Interaction of estrogenic chemicals and phytoestrogens with estrogen receptor beta. Endocrinology 139(10):4252–4263
27. Dixon RA, Ferreira D (2002) Genistein. Phytochemistry 60(3):205–211

28. Messina MJ, Persky V, Setchell KD, Barnes S (1994) Soy intake and cancer risk: a review of the in vitro and in vivo data. Nutr Cancer 21(2):113–131
29. Zhang Y, Chen H (2011) Genistein, an epigenome modifier during cancer prevention. Epigenetics 6(7):888–891
30. Gang DR, Kasahara H, Xia ZQ, Van der Mijnsbrugge K, Bauw G, Boerjan W et al (1999) Evolution of plant defense mechanisms. Relationships of phenylcoumaran benzylic ether reductases to pinoresinol-lariciresinol and isoflavone reductases. J Biol Chem 274(11):7516–7527
31. Bladergroen MR, Spaink HP (1998) Genes and signal molecules involved in the rhizobia-leguminoseae symbiosis. Curr Opin Plant Biol 1(4):353–359
32. Kessmann H, Edwards R, Geno PW, Dixon RA (1990) Stress responses in Alfalfa (Medicago sativa L.): V. Constitutive and elicitor-induced accumulation of isoflavonoid conjugates in cell suspension cultures. Plant Physiol 94(1):227–232
33. D'Aloisio AA, Baird DD, DeRoo LA, Sandler DP (2010) Association of intrauterine and early-life exposures with diagnosis of uterine leiomyomata by 35 years of age in the Sister Study. Environ Health Perspect 118(3):375–381
34. Strom BL, Schinnar R, Ziegler EE, Barnhart KT, Sammel MD, Macones GA et al (2001) Exposure to soy-based formula in infancy and endocrinological and reproductive outcomes in young adulthood. JAMA 286(7):807–814
35. Frye C, Bo E, Calamandrei G, Calza L, Dessi-Fulgheri F, Fernandez M et al (2012) Endocrine disrupters: a review of some sources, effects, and mechanisms of actions on behavior and neuroendocrine systems. J Neuroendocrinol 24(1):144–159
36. Swedenborg E, Pongratz I, Gustafsson JA (2010) Endocrine disruptors targeting ERbeta function. Int J Androl 33(2):288–297
37. Gore AC (2008) Developmental programming and endocrine disruptor effects on reproductive neuroendocrine systems. Front Neuroendocrinol 29(3):358–374
38. Handa RJ, Weiser MJ, Zuloaga DG (2009) A role for the androgen metabolite, 5alpha-androstane-3beta,17beta-diol, in modulating oestrogen receptor beta-mediated regulation of hormonal stress reactivity. J Neuroendocrinol 21(4):351–358
39. Blair RM, Fang H, Branham WS, Hass BS, Dial SL, Moland CL et al (2000) The estrogen receptor relative binding affinities of 188 natural and xenochemicals: structural diversity of ligands. Toxicol Sci 54(1):138–153
40. Shanle EK, Xu W (2011) Endocrine disrupting chemicals targeting estrogen receptor signaling: identification and mechanisms of action. Chem Res Toxicol 24(1):6–19
41. Barkhem T, Carlsson B, Nilsson Y, Enmark E, Gustafsson J, Nilsson S (1998) Differential response of estrogen receptor alpha and estrogen receptor beta to partial estrogen agonists/antagonists. Mol Pharmacol 54(1):105–112
42. Dodds EC, Lawson W (1936) Synthetic estrogenic agents without the phenanthrene nucleus. Nature 137:996
43. Vandenberg LN, Hauser R, Marcus M, Olea N, Welshons WV (2007) Human exposure to bisphenol A (BPA). Reprod Toxicol 24(2):139–177
44. Biedermann S, Tschudin P, Grob K (2010) Transfer of bisphenol A from thermal printer paper to the skin. Anal Bioanal Chem 398(1):571–576
45. Cooper JE, Kendig EL, Belcher SM (2011) Assessment of bisphenol A released from reusable plastic, aluminium and stainless steel water bottles. Chemosphere 85(6):943–947
46. Liao C, Kannan K (2011) Widespread occurrence of bisphenol A in paper an. paper products: implications for human exposure. Environ Sci Technol 45(21):9372–9379
47. Geens T, Goeyens L, Covaci A (2011) Are potential sources for human exposure to bisphenol-A overlooked? Int J Hyg Environ Health 214(5):339–347
48. Calafat AM, Ye X, Wong LY, Reidy JA, Needham LL (2008) Exposure of the U.S. population to bisphenol A and 4-tertiary-octylphenol: 2003–2004. Environ Health Perspect 116(1):39–44
49. vom Saal FS, Akingbemi BT, Belcher SM, Birnbaum LS, Crain DA, Eriksen M et al (2007) Chapel Hill bisphenol A expert panel consensus statement: integration of mechanisms, effects in animals and potential to impact human health at current levels of exposure. Reprod Toxicol 24(2):131–138

50. Belcher SM, Chen Y, Yan S, Wang HS (2012) Rapid estrogen receptor-mediated mechanisms determine the sexually dimorphic sensitivity of ventricular myocytes to 17beta-estradiol and the environmental endocrine disruptor bisphenol a. Endocrinology 153(2):712–720
51. Bonefeld-Jorgensen EC, Andersen HR, Rasmussen TH, Vinggaard AM (2001) Effect of highly bioaccumulated polychlorinated biphenyl congeners on estrogen and androgen receptor activity. Toxicology 158(3):141–153
52. Stein J, Schettler T, Wallinga D, Valenti M (2002) In harm's way: toxic threats to child development. J Dev Behav Pediatr 23(1 Suppl):S13–S22
53. Adlercreutz H (1995) Phytoestrogens: epidemiology and a possible role in cancer protection. Environ Health Perspect 130(Suppl 7):103–112
54. UK-Committee-on-Toxicity (2003) Phytoestrogens and health. Committee on Toxicity of Chemicals in Food, Consumer Products and the Environment, London
55. Setchell KD, Brown NM, Desai P, Zimmer-Nechemias L, Wolfe BE, Brashear WT et al (2001) Bioavailability of pure isoflavones in healthy humans and analysis of commercial soy isoflavone supplements. J Nutr 131(4 Suppl):1362S–1375S
56. Setchell KDR, Zimmer-Nechemias L, Cai J, Heubi JE (1997) Exposure of infants to phyto-oestrogens from soy-based infant formula. Lancet 350:23–27
57. Setchell KD, Zimmer-Nechemias L, Cai J, Heubi JE (1998) Isoflavone content of infant formulas and the metabolic fate of these phytoestrogens in early life. Am J Clin Nutr 68:1453S
58. Todaka E, Sakurai K, Fukata H, Miyagawa H, Uzuki M, Omori M et al (2005) Fetal exposure to phytoestrogens—the difference in phytoestrogen status between mother and fetus. Environ Res 99(2):195–203
59. Valentin-Blasini L, Blount BC, Caudill SP, Needham LL (2003) Urinary and serum concentrations of seven phytoestrogens in a human reference population subset. J Expo Anal Environ Epidemiol 13(4):276–282
60. Adlercreutz H, Fotsis T, Watanabe S, Lampe J, Wahala K, Makela T et al (1994) Determination of lignans and isoflavonoids in plasma by isotope dilution gas chromatography–mass spectrometry. Cancer Detect Prev 18(4):259–271
61. Padmanabhan V, Siefert K, Ransom S, Johnson T, Pinkerton J, Anderson L et al (2008) Maternal bisphenol-A levels at delivery: a looming problem? J Perinatol 28(4):258–263
62. Schonfelder G, Wittfoht W, Hopp H, Talsness CE, Paul M, Chahoud I (2002) Parent bisphenol A accumulation in the human maternal-fetal-placental unit. Environ Health Perspect 110(11):A703–A707
63. Peeters PH, Slimani N, van der Schouw YT, Grace PB, Navarro C, Tjonneland A et al (2007) Variations in plasma phytoestrogen concentrations in European adults. J Nutr 137(5):1294–1300
64. Jacobson JL, Fein GG, Jacobson SW, Schwartz PM, Dowler JK (1984) The transfer of polychlorinated biphenyls (PCBs) and polybrominated biphenyls (PBBs) across the human placenta and into maternal milk. Am J Public Health 74(4):378–379
65. Longnecker MP, Klebanoff MA, Brock JW, Zhou H (2001) Polychlorinated biphenyl serum levels in pregnant subjects with diabetes. Diabetes Care 24(6):1099–1101
66. Engel SM, Levy B, Liu Z, Kaplan D, Wolff MS (2006) Xenobiotic phenols in early pregnancy amniotic fluid. Reprod Toxicol 21(1):110–112
67. Foster WG, Chan S, Platt L, Hughes CL Jr (2002) Detection of phytoestrogens in samples of second trimester human amniotic fluid. Toxicol Lett 129(3):199–205
68. Nagata C, Iwasa S, Shiraki M, Ueno T, Uchiyama S, Urata K et al (2006) Associations among maternal soy intake, isoflavone levels in urine and blood samples, and maternal and umbilical hormone concentrations (Japan). Cancer Causes Control 17(9):1107–1113
69. Arai Y, Uehara M, Sato Y, Kimira M, Eboshida A, Adlercreutz H et al (2000) Comparison of isoflavones among dietary intake, plasma concentration and urinary excretion for accurate estimation of phytoestrogen intake. J Epidemiol 10:127–135
70. Ikezuki Y, Tsutsumi O, Takai Y, Kamei Y, Taketani Y (2002) Determination of bisphenol A concentrations in human biological fluids reveals significant early prenatal exposure. Hum Reprod 17(11):2839–2841

71. Adlercreutz H, Yamada T, Wahala K, Watanabe S (1999) Maternal and neonatal phytoestrogens in Japanese women during birth. Am J Obstet Gynecol 180(3 Pt 1):737–743
72. Ye X, Kuklenyik Z, Needham LL, Calafat AM (2006) Measuring environmental phenols and chlorinated organic chemicals in breast milk using automated on-line column-switching-high performance liquid chromatography-isotope dilution tandem mass spectrometry. J Chromatogr B Analyt Technol Biomed Life Sci 831(1–2):110–115
73. Franke AA, Custer LJ (1996) Daidzein and genistein concentrations in human milk after soy consumption. Clin Chem 42(6):955–964
74. Franke AA, Halm BM, Custer LJ, Tatsumura Y, Hebshi S (2006) Isoflavones in breastfed infants after mothers consume soy. Am J Clin Nutr 84(2):406–413
75. Cao Y, Calafat AM, Doerge DR, Umbach DM, Bernbaum JC, Twaddle NC et al (2009) Isoflavones in urine, saliva, and blood of infants: data from a pilot study on the estrogenic activity of soy formula. J Expo Sci Environ Epidemiol 19(2):223–234
76. Winter JSD, Hughes IA, Reyes FI, Faiman C (1976) Pituitary-gonadal relations in infancy: patterns of serum gonadal steroid concentrations in man from birth to two years of age. J Clin Endocrinol Metabol 42:679–686
77. Farquharson RG, Klopper AI (1984) A study of maternal, retroplacental and umbilical cord estradiol levels in term infants delivered by caesarean section. Eur J Obstet Gynecol Reprod Biol 16(5):315–320
78. Stricker R, Eberhart R, Chevailler MC, Quinn FA, Bischof P (2006) Establishment of detailed reference values for luteinizing hormone, follicle stimulating hormone, estradiol, and progesterone during different phases of the menstrual cycle on the Abbott ARCHITECT analyzer. Clin Chem Lab Med 44(7):883–887
79. Kim J, Semaan SJ, Clifton DK, Steiner RA, Dhamija S, Kauffman AS (2011) Regulation of Kiss1 expression by sex steroids in the amygdala of the rat and mouse. Endocrinology 152(5):2020–2030
80. Wilson C (2011) Neuroendocrinology: Kiss1 expressed in the amygdala in rodents. Nat Rev Endocrinol 7(6):313
81. Terao Y, Kumano S, Takatsu Y, Hattori M, Nishimura A, Ohtaki T et al (2004) Expression of KiSS-1, a metastasis suppressor gene, in trophoblast giant cells of the rat placenta. Biochim Biophys Acta 1678(2–3):102–110
82. Ohtaki T, Shintani Y, Honda S, Matsumoto H, Hori A, Kanehashi K et al (2001) Metastasis suppressor gene KiSS-1 encodes peptide ligand of a G-protein-coupled receptor. Nature 411(6837):613–617
83. Kotani M, Detheux M, Vandenbogaerde A, Communi D, Vanderwinden JM, Le Poul E et al (2001) The metastasis suppressor gene KiSS-1 encodes kisspeptins, the natural ligands of the orphan G protein-coupled receptor GPR54. J Biol Chem 276(37):34631–34636
84. Castellano JM, Gaytan M, Roa J, Vigo E, Navarro VM, Bellido C et al (2006) Expression of KiSS-1 in rat ovary: putative local regulator of ovulation? Endocrinology 147(10):4852–4862
85. Clarkson J, Boon WC, Simpson ER, Herbison AE (2009) Postnatal development of an estradiol-kisspeptin positive feedback mechanism implicated in puberty onset. Endocrinology 150(7):3214–3220
86. Clarkson J, Herbison AE (2006) Postnatal development of kisspeptin neurons in mouse hypothalamus; sexual dimorphism and projections to gonadotropin-releasing hormone neurons. Endocrinology 147(12):5817–5825
87. Cao J, Patisaul HB (2011) Sexually dimorphic expression of hypothalamic estrogen receptors alpha and beta and kiss1 in neonatal male and female rats. J Comp Neurol 519(15):2954–2977
88. Losa SM, Todd KL, Sullivan AW, Cao J, Mickens JA, Patisaul HB (2010) Neonatal exposure to genistein adversely impacts the ontogeny of hypothalamic kisspeptin signaling pathways and ovarian development in the peripubertal female rat. Reprod Toxicol 31(3):280–289
89. Kauffman AS (2009) Sexual differentiation and the Kiss1 system: hormonal and developmental considerations. Peptides 30(1):83–93
90. Poling MC, Kauffman AS (2012) Sexually dimorphic testosterone secretion in prenatal and neonatal mice is independent of kisspeptin-Kiss1r and GnRH signaling. Endocrinology 153(2):782–793

91. Navarro VM, Sanchez-Garrido MA, Castellano JM, Roa J, Garcia-Galiano D, Pineda R et al (2009) Persistent impairment of hypothalamic KiSS-1 system after exposures to estrogenic compounds at critical periods of brain sex differentiation. Endocrinology 150(5):2359–2367
92. Kauffman AS, Gottsch ML, Roa J, Byquist AC, Crown A, Clifton DK et al (2007) Sexual differentiation of Kiss1 gene expression in the brain of the rat. Endocrinology 148(4):1774–1783
93. Smith JT, Cunningham MJ, Rissman EF, Clifton DK, Steiner RA (2005) Regulation of Kiss1 gene expression in the brain of the female mouse. Endocrinology 146(9):3686–3692
94. Smith JT, Popa SM, Clifton DK, Hoffman GE, Steiner RA (2006) Kiss1 neurons in the forebrain as central processors for generating the preovulatory luteinizing hormone surge. J Neurosci 26(25):6687–6694
95. Patisaul HB, Losa-Ward SM, Todd KL, McCaffrey KA, Mickens JA (2012) Influence of ERbeta selective agonism during the neonatal period on the sexual differentiation of the rat hypothalamic-pituitary-gonadal (HPG) axis. Biol Sex Differ 3(1):2
96. Patisaul HB, Todd KL, Mickens JA, Adewale HB (2009) Impact of neonatal exposure to the ERalpha agonist PPT, bisphenol-A or phytoestrogens on hypothalamic kisspeptin fiber density in male and female rats. Neurotoxicology 30(3):350–357
97. Gottsch ML, Navarro VM, Zhao Z, Glidewell-Kenney C, Weiss J, Jameson JL et al (2009) Regulation of Kiss1 and dynorphin gene expression in the murine brain by classical and nonclassical estrogen receptor pathways. J Neurosci 29(29):9390–9395
98. Gorski RA, Mennin SP, Kubo K (1975) The neural and hormonal bases of the reproductive cycle of the rat. Adv Exp Med Biol 54:115–153
99. Elkind-Hirsch K, King JC, Gerall AA, Arimura AA (1981) The luteinizing hormone-releasing hormone (LHRH) system in normal and estrogenized neonatal rats. Brain Res Bull 7(6):645–654
100. Howdeshell KL, Hotchkiss AK, Thayer KA, Vandenbergh JG, vom Saal FS (1999) Exposure to bisphenol A advances puberty. Nature 401(6755):763–764
101. Adewale HB, Jefferson WN, Newbold RR, Patisaul HB (2009) Neonatal bisphenol-A exposure alters rat reproductive development and ovarian morphology without impairing activation of gonadotropin releasing hormone neurons. Biol Reprod 81(4):690–699
102. Birke LI (1984) Effects of estradiol and progesterone on scent-marking behavior of female rats. Horm Behav 18(1):95–98
103. Cao J, Mickens JA, McCaffrey KA, Leyrer SM, Patisaul HB (2012) Neonatal bisphenol A exposure alters sexually dimorphic gene expression in the postnatal rat hypothalamus. Neurotoxicology 33(1):23–36
104. Bai Y, Chang F, Zhou R, Jin PP, Matsumoto H, Sokabe M et al (2011) Increase of anteroventral periventricular kisspeptin neurons and generation of E2-induced LH-surge system in male rats exposed perinatally to environmental dose of bisphenol-A. Endocrinology 152(4):1562–1571
105. Cortes R, Ceccatelli S, Schalling M, Hokfelt T (1990) Differential effects of intracerebroventricular colchicine administration on the expression of mRNAs for neuropeptides and neurotransmitter enzymes, with special emphasis on galanin: an in situ hybridization study. Synapse 6(4):369–391
106. Plant TM, Ramaswamy S (2009) Kisspeptin and the regulation of the hypothalamic-pituitary-gonadal axis in the rhesus monkey (Macaca mulatta). Peptides 30(1):67–75
107. Lackmann GM (2002) Polychlorinated biphenyls and hexachlorobenzene in full-term neonates. Reference values updated. Biol Neonate 81:82–85
108. Lackmann GM, Schaller KH, Angerer J (2004) Organochlorine compounds in breast-fed vs. bottle-fed infants: preliminary results at six weeks of age. Sci Total Environ 329(1–3):289–293
109. Bentzen TW, Muir DC, Amstrup SC, O'Hara TM (2008) Organohalogen concentrations in blood and adipose tissue of Southern Beaufort Sea polar bears. Sci Total Environ 406(1–2):352–367
110. Gladen BC, Doucet J, Hansen LG (2003) Assessing human polychlorinated biphenyl contamination for epidemiologic studies: lessons from patterns of congener concentrations in Canadians in 1992. Environ Health Perspect 111(4):437–443
111. Dickerson SM, Cunningham SL, Gore AC (2011) Prenatal PCBs disrupt early neuroendocrine development of the rat hypothalamus. Toxicol Appl Pharmacol 252(1):36–46

112. Dickerson SM, Cunningham SL, Patisaul HB, Woller MJ, Gore AC (2011) Endocrine disruption of brain sexual differentiation by developmental PCB exposure. Endocrinology 152(2):581–594
113. Steinberg RM, Walker DM, Juenger TE, Woller MJ, Gore AC (2008) Effects of perinatal polychlorinated biphenyls on adult female rat reproduction: development, reproductive physiology, and second generational effects. Biol Reprod 78(6):1091–1101
114. Steinberg RM, Juenger TE, Gore AC (2007) The effects of prenatal PCBs on adult female paced mating reproductive behaviors in rats. Horm Behav 51(3):364–372
115. Lanting CI, Huisman M, Muskiet FA, van der Paauw CG, Essed CE, Boersma ER (1998) Polychlorinated biphenyls in adipose tissue, liver, and brain from nine stillborns of varying gestational ages. Pediatr Res 44(2):222–225
116. Takagi Y, Aburada S, Hashimoto K, Kitaura T (1986) Transfer and distribution of accumulated (14C)polychlorinated biphenyls from maternal to fetal and suckling rats. Arch Environ Contam Toxicol 15(6):709–715
117. Kriegsfeld LJ, Gibson EM, Williams WP 3rd, Zhao S, Mason AO, Bentley GE et al (2010) The roles of RFamide-related peptide-3 in mammalian reproductive function and behaviour. J Neuroendocrinol 22(7):692–700
118. Tsutsui K, Bentley GE, Kriegsfeld LJ, Osugi T, Seong JY, Vaudry H (2010) Discovery and evolutionary history of gonadotrophin-inhibitory hormone and kisspeptin: new key neuropeptides controlling reproduction. J Neuroendocrinol 22(7):716–727
119. Bateman HL, Patisaul HB (2008) Disrupted female reproductive physiology following neonatal exposure to phytoestrogens or estrogen specific ligands is associated with decreased GnRH activation and kisspeptin fiber density in the hypothalamus. Neurotoxicology 29(6):988–997
120. Setchell KD, Clerici C (2010) Equol: pharmacokinetics and biological actions. J Nutr 140(7):1363S–1368S
121. Barker DJ (1997) Maternal nutrition, fetal nutrition, and disease in later life. Nutrition 13(9):807–813
122. Heindel J, Lawler C (2006) Role of exposure to environmental chemicals in developmental origins of health and disease. In: Gluckman P, Hanson M (eds) Developmental origins of health and disease. Cambridge University Press, Cambridge
123. de Roux N, Genin E, Carel JC, Matsuda F, Chaussain JL, Milgrom E (2003) Hypogonadotropic hypogonadism due to loss of function of the KiSS1-derived peptide receptor GPR54. Proc Natl Acad Sci U S A 100(19):10972–10976
124. Seminara SB, Messager S, Chatzidaki EE, Thresher RR, Acierno JS Jr, Shagoury JK et al (2003) The GPR54 gene as a regulator of puberty. N Engl J Med 349(17):1614–1627
125. Clarkson J, d'Anglemont de Tassigny X, Moreno AS, Colledge WH, Herbison AE (2008) Kisspeptin-GPR54 signaling is essential for preovulatory gonadotropin-releasing hormone neuron activation and the luteinizing hormone surge. J Neurosci 28(35):8691–8697
126. Navarro VM, Castellano JM, Fernandez-Fernandez R, Barreiro ML, Roa J, Sanchez-Criado JE et al (2004) Developmental and hormonally regulated messenger ribonucleic acid expression of KiSS-1 and its putative receptor, GPR54, in rat hypothalamus and potent luteinizing hormone-releasing activity of KiSS-1 peptide. Endocrinology 145(10):4565–4574
127. Messager S, Chatzidaki EE, Ma D, Hendrick AG, Zahn D, Dixon J et al (2005) Kisspeptin directly stimulates gonadotropin-releasing hormone release via G protein-coupled receptor 54. Proc Natl Acad Sci U S A 102(5):1761–1766
128. Losa-Ward SM, Todd KL, McCaffrey KA, Tsutsui K, Patisaul HB (2012) Disrupted organization of RFamide pathways in the hypothalamus is associated with advanced puberty in female rats neonatally exposed to bisphenol A. Biol Reprod 87(2):28
129. Kauffman AS, Navarro VM, Kim J, Clifton D, Steiner RA (2009) Sex differences in the regulation of Kiss1/NKB neurons in juvenile mice: implications for the timing of puberty. Am J Physiol Endocrinol Metab 297(5):E1212–E1221
130. Ubuka T, Morgan K, Pawson AJ, Osugi T, Chowdhury VS, Minakata H et al (2009) Identification of human GnIH homologs, RFRP-1 and RFRP-3, and the cognate receptor, GPR147 in the human hypothalamic pituitary axis. PLoS One 4(12):e8400

Chapter 22
Model Systems for Studying Kisspeptin Signalling: Mice and Cells

William H. Colledge, Joanne Doran, and Hua Mei

Abstract Kisspeptins are a family of overlapping neuropeptides, encoded by the *Kiss1* gene, that are required for activation and maintenance of the mammalian reproductive axis. Kisspeptins act within the hypothalamus to stimulate release of gonadotrophic releasing hormone and activation of the pituitary-gonadal axis. Robust model systems are required to dissect the regulatory mechanisms that control *Kiss1* neuronal activity and to examine the molecular consequences of kisspeptin signalling. While studies in normal animals have been important in this, transgenic mice with targeted mutations affecting the kisspeptin signalling pathway have played a significant role in extending our understanding of kisspeptin physiology. Knock-out mice recapitulate the reproductive defects associated with mutations in humans and provide an experimentally tractable model system to interrogate regulatory feedback mechanisms. In addition, transgenic mice with cell-specific expression of modulator proteins such as the CRE recombinase or fluorescent reporter proteins such as GFP allow more sophisticated analyses such as cell or gene ablation or electrophysiological profiling. At a less complex level, immortalized cell lines have been useful for studying the role of kisspeptin in cell migration and metastasis and examining the intracellular signalling events associated with kisspeptin signalling.

W.H. Colledge, B.Sc., Ph.D. (✉)
Department of Physiology, Development and Neuroscience, University of Cambridge, Downing Street, Cambridge, CB2 3EG, UK
e-mail: whc23@cam.ac.uk

J. Doran
Takeda Cambridge, Cambridge, UK

H. Mei
Department of Opthalmology, Cornea Division, University of California, Los Angeles, CA, USA

A.S. Kauffman and J.T. Smith (eds.), *Kisspeptin Signaling in Reproductive Biology*, Advances in Experimental Medicine and Biology 784, DOI 10.1007/978-1-4614-6199-9_22,

Introduction

Mammalian fertility is regulated by neuroendocrine signals in the hypothalamus that integrate hormonal, metabolic and environmental cues with activation of the reproductive axis. Gonadotropin releasing hormone (GnRH) neurons are vital for the normal function of the reproductive axis as they release GnRH into the hypophyseal portal system to stimulate gonadotrophic hormone secretion from the anterior pituitary gland. The gonadotrophic hormones, luteinizing hormone (LH) and follicle-stimulating hormone (FSH), act on the gonads to increase sex steroid production required for gametogenesis, ovulation and normal fertility.

Kisspeptin and its G-protein coupled receptor (KISS1R or GPR54) are required for central activation of the hypothalamic pituitary-gonadal axis at puberty and regulation of reproductive function in many species, including mammals [1, 2]. Kisspeptins are produced by the *Kiss1* gene, which encodes a 145 or 138 amino-acid precursor protein in humans, depending upon a frameshift polymorphism, that is cleaved into overlapping C-terminal amidated peptides (KP54, KP14, KP13 and KP10) with similar binding affinities for the kisspeptin receptor [3–5].

Kisspeptins are expressed predominantly in two discrete neuronal populations in the rodent hypothalamus, the anterioventral periventricular nucleus (AVPV) and the arcuate nucleus (ARC) [6]. Kisspeptin neurons make close contact with GnRH neurons [7], which express the kisspeptin receptor [8, 9], and kisspeptin is a potent stimulator of GnRH release [10–12]. The kisspeptin neurons in the AVPV region are sexually dimorphic, with a greater number in female rodents [13, 14]. Kisspeptin expression is increased in AVPV neurons in response to rising oestrogen levels [15], which stimulate the GnRH/LH surge required for ovulation [16, 17]. In contrast, kisspeptin expression in ARC neurons is negatively regulated by oestrogen and testosterone [15, 18]. This differential regulation of kisspeptin expression by sex steroids indicates a clear biological difference between AVPV and ARC *Kiss1* neurons. It is generally accepted that the AVPV *Kiss1* neurons are involved in generating the ovulatory LH surge, while the ARC *Kiss1* neurons may act as a GnRH pulse generator [19]. This pacemaker activity of ARC *Kiss1* neurons is currently thought to be controlled by an interplay between neurokinin B (NKB) and dynorphin, two neuropeptides that are produced by *Kiss1* neurons in the ARC, but not in the AVPV region [20, 21].

Transgenic Mice Generation

Mice with specific alterations to their genome are a powerful tool to study the function of a protein in the context of the whole animal [22, 23]. Understanding the complex neuronal circuitry and the neuropeptides and neurotransmitters that control the reproductive axis requires studying an intact system. There is conservation of function in the reproductive and neuroendocrine systems between mice and other mammalian species which allows gene function to be assessed in a whole animal environment and the data extrapolated to other species, including humans. Manipulation of embryonic stem (ES) cells allows precise genetic modifications to be introduced into mice [24]. These modifications range from precise genetic alterations such as deletions, point mutations and reporter/modulator gene tagging to more extensive mutations such as conditional alleles, chromosomal deletions and translocations [25].

Transgenic Mice with Global Disruption of *Kiss1r*

Five transgenic mouse lines have been generated with global disruption of the *Kiss1r* gene (Fig. 22.1 and Table 22.1). The *Kiss1r*^{tm1Coll} line has a 702 bp deletion encompassing the final 92 bp of exon 1, the whole of intron 1 (509 bp) and the first 101 bp of exon 2 [26]. The *Kiss1r*^{tm1Gstn} line has a 52 bp deletion within exon 2 [27], while the *Kiss1r*$^{tm1.1On}$ line has a deletion of all five coding exons after a CRE-mediated recombination event [28]. These three transgenic lines all contain a *LacZ* reporter gene driven from the *Kiss1r* promoter and can be used to map the expression pattern of the kisspeptin receptor in the mouse brain [9]. The *Kiss1r*^{tm1Rla} line has complete deletion of exon 2 [29], while the *Kiss1r*^{Gt1Stei} line has a proviral insertion in intron 2 with no loss of *Kiss1r* coding sequence [30, 31]. *Kiss1r* expression is disrupted by a splice acceptor within the retrovirus to trap splicing from exon 2 and two polyadenylation signal sequences to terminate transcription. In situ hybridization could not detect any *Kiss1r* transcripts in the mutant mice, but as the coding exons are still intact it is at least theoretically possible that there might be a very low level of residual expression (i.e. a hypomorphic mutation). The retrovirus also encodes the reverse tetracycline-dependent transactivator protein (rtTA), which should be expressed in cells where the *Kiss1r* gene is transcriptionally active. The *Gfp* gene in the provirus will, however, not specifically mark neurons expressing the kisspeptin receptor, as it is driven by its own promoter. In the absence of an antibody with good specificity for the kisspeptin receptor, confirmation that these mice have null alleles has relied on molecular means such as RT-PCR or in situ hybridization.

Transgenic Mice with Global Disruption of *Kiss1*

Three lines of transgenic mice have been described with global disruption of the *Kiss1* gene. In the *Kiss1*^{tm1Coll} line the whole of the *Kiss1* coding region has been deleted and replaced with an IRES LacZ sequence [32]. Similarly, in the *Kis1*$^{tm2(cre/EGFP)Coll}$ line, the coding exons have been replaced with a *Cre/Egfp* fusion gene from the pCAG-Cre:GFP vector from Connie Cepko's laboratory [33, WHC, unpublished]. In the *Kiss1*^{tm1Rla} line, the first *Kiss1* coding exon that contains the translational initiation codon has been deleted [29]. Absence of kisspeptin expression in these mutant mice has been confirmed by RT-PCR and by immunohistochemistry using a well-characterized and specific anti-kisspeptin antibody [34].

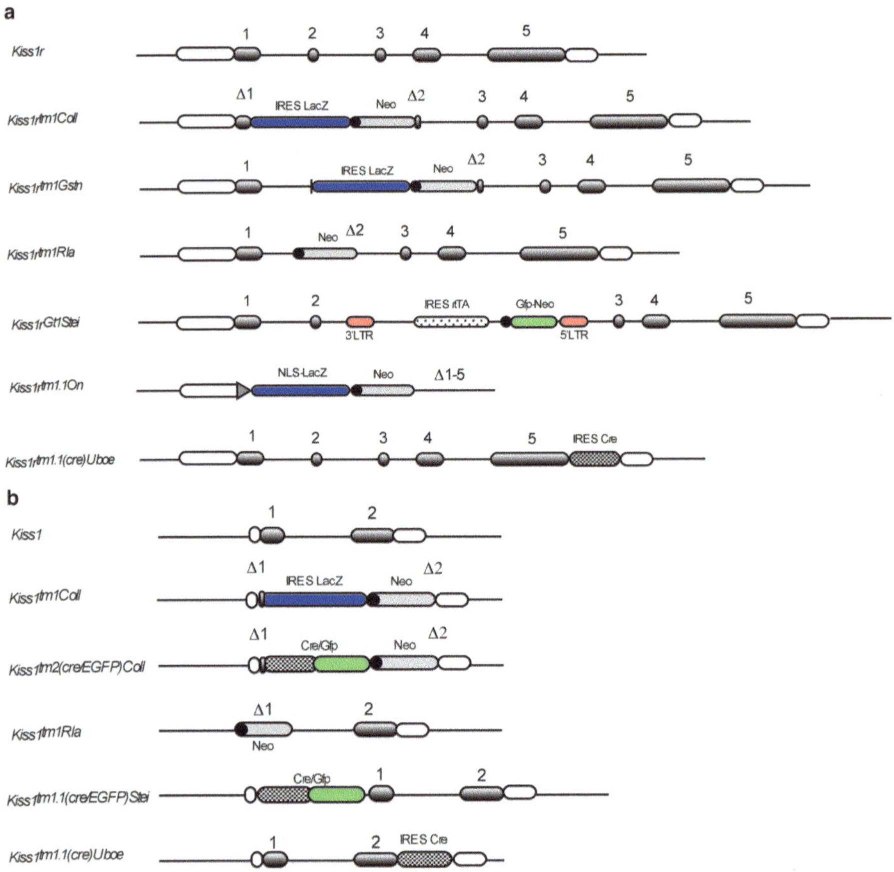

Fig. 22.1 Genomic arrangement of *Kiss1r* and *Kiss1* targetted loci. The nomenclature for these transgenic lines is based on the guidelines established by the International Committee on Standardized Genetic Nomenclature for mice (http://www.informatics.jax.org/mgihome/nomen/index.shtml) and follows that used on the Mouse Genome Informatics website (http://www.informatics.jax.org/). Lab Codes are: Coll (Colledge), Gstn (Gustafson), Rla (Lapatto), Stei (Steiner), On (Organon), Ubeo (Boehm). *Kiss1r* and *Kiss1* coding regions are numbered and non-coding exons are not shaded. Promoters are *shaded black*. Deleted exons are indicated by a *delta* symbol. *IRES*, internal ribosome entry sequence; *neo*, promoter driven neomycin resistance gene; *Cre*, cre recombinase gene; *Gfp*, enhanced green fluorescent protein gene; *LacZ*, β-galactosidase gene; *NLS*, nuclear localization signal sequence; *rtTA*, reverse tetracyclin-responsive transactivator protein; *LTR*, proviral long terminal repeat

Phenotype of *Kiss1r* and *Kiss1* Global Knock-Out Mice

Analysis of the phenotypes of the *Kiss1r* and *Kiss1* mutant mice has shown the importance of kisspeptin signalling in the initiation and maintenance of the reproductive axis [26, 27]. In general, the phenotypes of all the mutant mice lines are

Table 22.1 Summary of transgenic lines

Mouse line	Type of mutation	Major phenotype	Comments	References
Genetic modification of Kiss1r/Gpr54 locus				
Kiss1r^{tm1Coll}	Deletion of parts of exon 1 and 2	Hypogonadotrophic Hypogonadism	Null mutation *Kiss1r* expression tagged with *LacZ*	[26]
Kiss1r^{tm1Gstn}	Deletion of part of exon 2	Hypogonadotrophic Hypogonadism	Null mutation *Kiss1r* expression tagged with *LacZ*	[27]
Kiss1r^{tm1Rla}	Complete deletion of exon 2	Hypogonadotrophic Hypogonadism	Null mutation No reporter gene present	[29]
Kiss1r^{Gt1Stei}	Retroviral insertion intron 2. No loss of coding sequence	Hypogonadotrophic Hypogonadism	May be a hypomorphic mutation Females can be induced to show a modest pre-ovulatory LH surge May retain low level expression of *Kiss1r*	[30, 31]
Kiss1r$^{tm1.1On}$	Deletion of all coding exons after a CRE-mediated recombination step	Hypogonadotrophic Hypogonadism	Null mutation *Kiss1r* expression tagged with NLS-*LacZ*	[28]
Kiss1r$^{tm1.1(cre)Uboe}$	IRES-CRE inserted just downstream of coding sequence in exon 5	Fertile	Neutral mutation. *Kiss1r* still expressed *Kiss1r* expression tagged with CRE	[39]
Genetic modification of Kiss1 locus				
Kiss1^{tm1Coll}	Complete deletion of exon 1 and 2	Hypogonadotrophic Hypogonadism	Null mutation *Kiss1* expression tagged with *LacZ*	[32]
Kiss1^{tm1Rla}	Deletion of exon 1	Hypogonadotrophic Hypogonadism	Null mutation	[29]

(continued)

Table 22.1 (continued)

Mouse line	Type of mutation	Major phenotype	Comments	References
Kiss1$^{tm2(cre/EGFP)Coll}$	Complete deletion of exon 1 and 2	Hypogonadotrophic	Null mutation	Unpublished
		Hypogonadism	Kiss1 expression tagged with CRE-EGFP fusion protein	
Kiss1$^{tm1.1(cre/EGFP)Stei}$	Insertion of CRE/EGFP fusion gene just before exon 1 coding sequence	Males fertile	May retain low level expression of *Kiss1*	[42]
		Females have reduced fertility	*Kiss1* expression tagged with CRE-EGFP fusion protein	
Kiss1$^{tm1.1(cre)Uboe}$	Insertion of IRES-CRE just after *Kiss1* coding sequence	Fertile	Neutral mutation	[43]
			Kiss1 expression tagged with CRE protein	

Lab codes are: Coll (Colledge), Gstn (Gustafson), Rla (Lapatto), Stei (Steiner), On (Organon), Ubeo (Boehm)
AVPV anteroventral periventricular; *tm* targetted mutation; *Gt* gene trap mutation

broadly similar and these have been reviewed previously [22, 23]. Small differences in phenotype are probably due to experimental variation between laboratories. In summary, mice with global disruption of *Kiss1r* or *Kiss1* are sterile and do not progress through puberty. Mutant males have small testes, disruption of spermatogenesis and fail to develop secondary sex glands. Mutant females are acyclic, have thread-like uteri and small ovaries with no Graafian follicle or corpus luteum formation. These effects are caused by low sex steroid and gonadotrophic hormone levels (hypogonadotrophism). The primary defect in the mutant mice is failure of GnRH secretion from the median eminence [12]. In addition, the *Kiss1r* and *Kiss1* mutant mice cannot activate kisspeptin or GnRH neurons as judged by *c-fos* induction or produce the pre-ovulatory LH surge after appropriate hormonal stimulation [17], although it has also been reported that the $Kiss1r^{Gt1Stei}$ mice can be induced to show a modest LH surge [30]. Most of the data, however, suggest that kisspeptin signalling is also required for normal oestrous cycling and ovulation.

While the phenotypes of the global knock-out mice are generally similar, there are some subtle differences between *Kiss1r* and *Kiss1* mice, which may be physiologically informative. The *Kiss1* mutant mice have a slightly less severe reproductive deficit than the *Kiss1r* mice. This is most noticeable in males, where the testes, epididymides and seminal vesicle weights are significantly larger in the *Kiss1* mutant mice compared with the *Kiss1r* mutants and higher numbers of spermatozoa are found in the vas deferens (Fig. 22.2a–d) [29, 35, 36]. In addition, the *Kiss1* mutant mice show more advanced stages of spermatogenesis than the *Kiss1r* mutants, and this is particularly noticeable when the animals are not exposed to dietary phytoestrogens, which can actually stimulate spermatogenesis in these mutant mice [35]. These phenotypic differences are probably related to the slightly higher levels of both circulating and intratesticular testosterone in the *Kiss1* mice compared with the *Kiss1r* mice. These data suggest that the *Kiss1* mutant mice may retain some residual signalling through the kisspeptin receptor, which cannot occur in the *Kiss1r* mutant mice. Consistent with this hypothesis is the finding that the kisspeptin receptor has around a 5% basal signalling activity in the absence of ligand binding when expressed in HEK293 cells [37].

There is also evidence that both the *Kiss1* and *Kiss1r* mutant mice can release low levels of GnRH from the hypothalamus, which may contribute to their partial sexual maturation, although they remain infertile. Female mutant mice of either genotype show periods of vaginal keratinization typically found during oestrous although ovulations do not occur [36]. Treatment of mutant females with acyline, a GnRH antagonist, prevented these periods of keratinization and reduced uterine weights and LH levels in the blood. In male mice, acyline treatment reduced testicular weights and LH levels in both *Kiss1r* and *Kiss1* mutants. The ligand-independent basal activity of the kisspeptin receptor may explain these observations in the *Kiss1* mutant mice, but the effects in the *Kiss1r* mutant mice suggest another pathway for GnRH release. This is unlikely to be mediated by neurokinin B, as it has been shown in the $Kiss1r^{tm1.1On}$ mutant mice that this neuropeptide requires the kisspeptin receptor to stimulate LH release [28]. In contrast, NMDA has been shown to stimulate GnRH/LH release in *Kiss1r* mutant mice [28, 38] hence the residual GnRH release may be mediated by ionotropic glutamate neurotransmission.

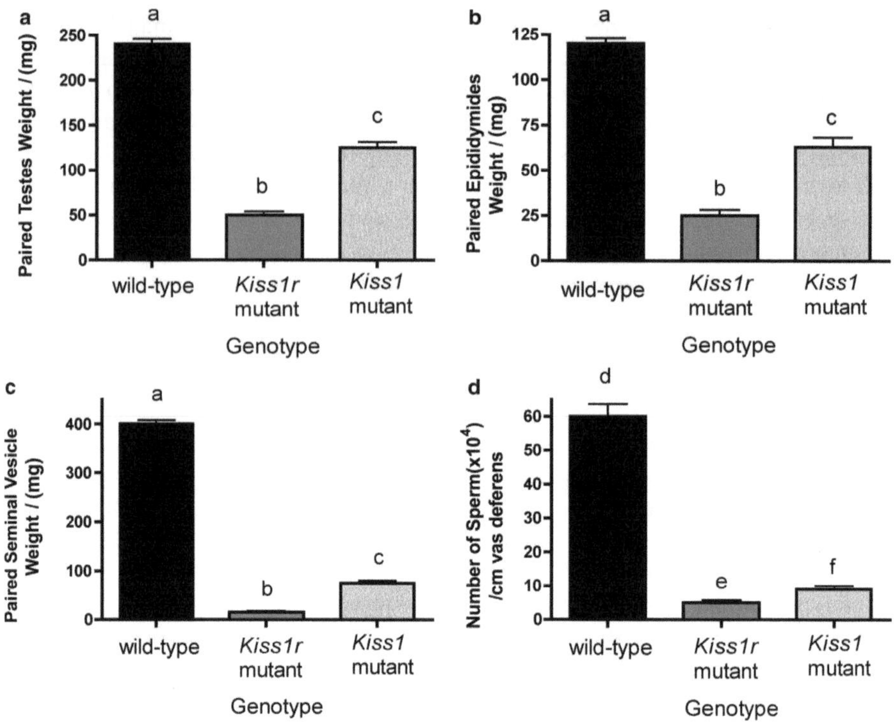

Fig. 22.2 Differences in the extent of hypogonadism between *Kiss1r* and *Kiss1* mutant male mice. Samples were collected from 7-month-old mice with at least 11 mice in each data set. *a* vs. *b* vs. *c* $p<0.05$. *d* vs. *e* vs. *f* $p<0.01$ (one way ANOVA followed by Tukey-Kramer multiple conversion test)

Transgenic Mice with Tagged *Kiss1r* and *Kiss1* Alleles

Recently, several transgenic mice have been generated that have functionally intact *Kiss1r* and *Kiss1* alleles that express modulator proteins that can be used to facilitate a number of downstream studies. The *Kiss1r*$^{tm1.1(cre)Uboe}$ line has an IRES-Cre transgene located just after the translational termination codon in exon 5 of the *Kiss1r* gene [39]. This insertion does not disrupt expression of the kisspeptin receptor and the mice are fertile. Confirmation that CRE expression was restricted to neuronal cells expressing the kisspeptin receptor was obtained by crossing the *Kiss1r*$^{tm1.1(cre)Uboe}$ mice with a reporter strain in which expression of τGFP is dependent upon a CRE-mediated recombination event [40]. Hypothalamic sections through the medial preoptic area showed that 99% of the GnRH neurons were labelled with τGFP consistent with the reported expression of the kisspeptin receptor in these neurons [8, 9, 41].

The *Kiss1*$^{tm1.1(cre/EGFP)Stei}$ line contains a *Cre/Gfp* fusion gene inserted within the first exon of the *Kiss1* gene just upstream of the ATG translational initiation codon [42]. The pattern of GFP expression in these mice recapitulates the expression profile of the *Kiss1* gene, with prominent expression in the ARC and AVPV regions and appropriate regulation by estradiol; although some expression has also been found in cortical neurons (see below). Surprisingly, homozygous mutant mice of both sexes do not show severe fertility defects. The mutant males appear to be nearly normal, with small testes, but they can impregnate females and have normal testosterone levels, whereas the females have markedly impaired fertility, but relatively normal organ weights (Robert Steiner, personal communication). In situ hybridization showed a reduction in the level of *Kiss1* transcripts, so this transgenic mouse model probably carries a hypomorphic mutation rather than a null allele. It is noteworthy that there is a moderately conserved splice acceptor sequence (AGGAGACTGTAGAC) located 21 bases upstream from the ATG codon which may allow some transcripts to splice out the *Cre/Gfp* transgene and allow low levels of kisspeptin protein to be made. Since kisspeptins are very potent stimulators of GnRH secretion, low levels may be sufficient to maintain the reproductive axis.

The *Kiss1*$^{tm1.1(cre)Uboe}$ mice have an IRES-Cre transgene inserted just after the *Kiss1* coding region in exon 2 (Fig. 22.1) [43]. Appropriate expression of the *Cre* transgene in ARC and AVPV *Kiss1* neurons was confirmed by breeding the mice with a line expressing a CRE-activated fluorescent reporter and 97% of *Kiss1* neurons were fluorescently labelled.

In addition to the various transgenic mouse lines that have been generated by gene targeting, several Kiss-Cre lines have also been made by pronuclear microinjection of bacterial artificial chromosome (BAC) constructs. One of these BAC vectors contained a DNA fragment with 109 kb of genomic sequence upstream of a *Cre* transgene inserted at the *Kiss1* translational initiation codon [44]. Three founder lines were generated (J2–3, J2–4 and J2–6), and to validate the expression profile of these lines, the mice were bred with CRE-dependent reporter mice (GFP or β-galactosidase). The *Tg(Kiss1-Cre)J2-4Cfe* line showed the most appropriate expression of the *Cre* transgene with GFP or β-galactosidase expression being found in the AVPV and the ARC regions of the hypothalamus and 93% of these neurons were identified as co-expressing *Kiss1* mRNA by in situ hybridization. The *Tg(Kiss1-Cre)J2-3Cfe* line showed expression in the ARC, but had very little expression in the AVPV region. The reason for the more limited expression pattern found in the *Tg(Kiss1-Cre)J2-3Cfe* mice is not known, but probably results from rearrangement of the *Kiss1* promoter during the BAC integration event and loss of important regulatory elements. Expression of the *Tg(Kiss1-Cre)J2-4Cfe* transgene was also found in the cerebral cortex, which is not a site where *Kiss1* expression has been found in adult mice. This ectopic expression could arise from inappropriate *Kiss1* promoter activity as a result of the site of the transgene integration. Alternatively, the expression in the cortical neurons could indicate that these cells expressed *Kiss1* at some point in their development while in the adult this expression is no longer found. Since the *Kiss1*-Cre transgene was visualized by CRE-mediated activation of a reporter gene, any cell that even transiently expressed *Kiss1* will be permanently marked. This phenomenon has also been observed in the *Kiss1*$^{tm1.1(cre/EGFP)Stei}$ mice, when expression was visualized using a CRE-inducible *LacZ* gene and β-galactosidase positive neurons were found extending outside the ARC region [42] including the cortex (Robert Steiner, personal communication). Thus, it seems likely that some neurons in the cortex express *Kiss1* during development and this expression is switched off in the adult.

Conditional Gene Ablation

One of the limitations of complete disruption of a gene is that the corresponding protein is absent throughout development, which may produce pleiotropic changes in gene expression and result in compensatory developmental changes. To overcome this problem, conditional mutant alleles are usually generated where expression of the gene can be ablated in a tissue-specific or temporal way. This is normally achieved by creating a targeted allele where a coding exon is flanked by sequences that can be induced to undergo recombination to delete the intervening exon. This is usually achieved by CRE-mediated recombination between LoxP sites or FLP-mediated recombination between FRT sites.

Conditional gene ablation has been used to study the role of *Kiss1* neurons in mediating the effects of leptin on puberty in mice. Leptin is produced by white adipose tissue and provides a signal to the hypothalamus about peripheral energy reserves [45]. Mice lacking leptin (*ob/ob*) or the leptin receptor (*db/db*) develop obesity and are sterile. Leptin administration advances the onset of puberty in rodents [46–48] and restores fertility to *ob/ob* mice [49]. To examine how leptin mediates its action on the reproductive axis, mice lacking the leptin receptor (*LepR*) in *Kiss1* neurons were generated by breeding *Tg(Kiss1-Cre)J2-4Cfe* mice with mice carrying a LoxP-modified *LepR* allele [50]. Loss of functional leptin receptors in *Kiss1* neurons was confirmed by treating fasted mice with leptin and examining brain sections for leptin-induced phosphorylation of the STAT3 protein. Quantitation of pSTAT3 immunoreactivity showed a reduction of 90–95% in the number of leptin-responsive *Kiss1* neurons [50]. Mice of either sex showed no problems with the timing of puberty or fertility suggesting that leptin signalling in *Kiss1* neurons is not required for these processes. Moreover, the authors also showed that LepR neurons in the ventral premammillary nucleus seem to be more important in mediating the action of leptin on the reproductive axis. One caveat to these data, however, is that there are still a small number of *Kiss1* neurons (5–10%) that can still respond to leptin signalling in the LepR deleted mice, and it is not known if this would be sufficient to maintain functional inputs into GnRH neurons.

Genetic Ablation of GnRH Neurons

The $Kiss1r^{tm1.1(cre)Uboe}$ mice have been used to genetically ablate GnRH neurons by crossing with a CRE-inducible Diphtheria toxin A transgene [39]. The Diphtheria toxin A inhibits protein synthesis by reducing the activity of elongation factor-2 (EF-2) causing cell death. The number of GnRH neurons in these transgenic mice was reduced to around 9% and the GnRH neurons that remained showed no evidence of expression of the kisspeptin receptor. Thus, these mice lack the downstream pathways for kisspeptin signalling, but surprisingly the female mice showed normal sexual maturation and fertility even though they had significantly smaller ovaries. These data suggest that a small number of GnRH neurons (around 10% of normal) that do not express the kisspeptin receptor are capable of sustaining the reproductive axis.

Genetic Ablation of *Kiss1* Neurons

Unexpectedly, genetic ablation of *Kiss1* neurons by CRE-induced expression of a Diphtheria toxin A protein did not cause sterility in female mice [39]. In contrast, when the *Kiss1* neurons were ablated in adult females at post-natal day 20 (P20) by cell-specific expression of a Diphtheria toxin receptor and toxin delivery, the females were sterile. The authors suggest that this may be due to developmental compensation for loss of the *Kiss1* neurons before P20. One hypothesis is that absence of a signalling molecule from *Kiss1* neurons during development may alter the physiology of GnRH neurons and allow them to respond to other factors that can stimulate GnRH release. The identity of this signalling molecule is not known, nor the compensatory changes that might be induced in the GnRH neurons. These could include induction of new receptor molecules by the GnRH neurons or increased dendritic branching to alter afferent inputs.

One important caveat to these neuronal ablation studies is that a small number of neurons survive and these may still exert physiological effects. Ablation of the GnRH neurons leaves a residual population of around 10%, but it is known that as little as 12% of the GnRH neuron population can initiate puberty in mice, although higher numbers of these neurons are required for full fertility [51]. While these residual neurons do not seem to express the kisspeptin receptor, this was established by lack of activation of a Rosa-YFP transgene [52], which is known to give low expression levels in neurons and may have been below the level of detection. Similarly, there is a residual *Kiss1* neuronal population in the AVPV nucleus of around 3%. The minimum number of *Kiss1* neurons that are required for fertility is not known, but as kisspeptins are extremely potent neuropeptides, relatively few may be sufficient.

Transgenic Mice with Fluorescently—Marked GnRH Neurons

A number of transgenic mouse lines have been developed with GnRH neurons marked with a green fluorescent protein ([53], for review, see [54]) and these have been used to examine the electrophysiological effects of kisspeptins on GnRH neurons. Kisspeptins act on the majority of GnRH neurons to initiate a sustained depolarization event and increase the action potential firing rate [55–57]. Pharmacological studies and an assessment of the current–voltage (I–V) relationships indicate that

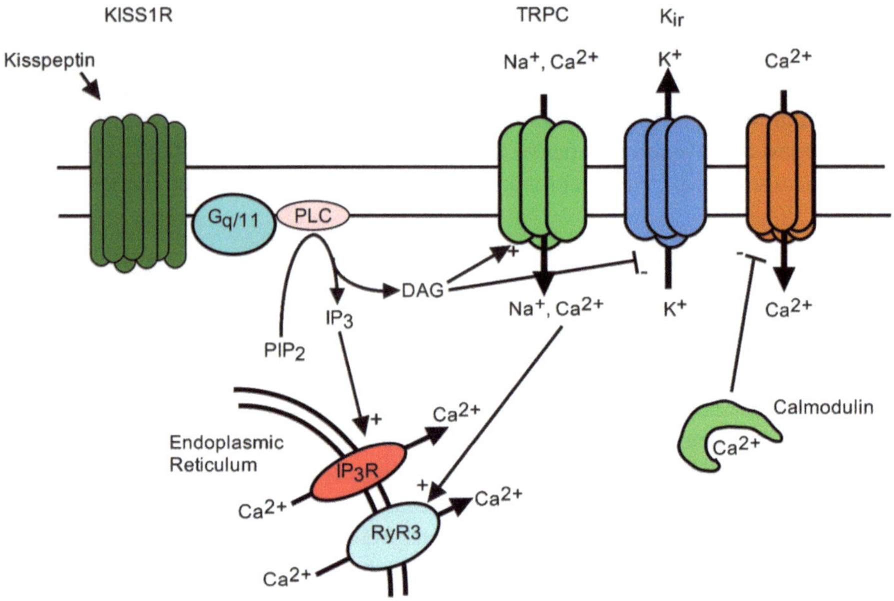

Fig. 22.3 Intracellular signalling events in GnRH neurons after kisspeptin stimulation. Kisspeptin binding to its receptor activates the G-protein $G_{q/11}$ to stimulate phospholipase C (PLC) activity and cleavage of PIP_2 into IP_3 and diacylglycerol (DAG). DAG stimulates canonical transient receptor potential channels (TRPC) and inhibits voltage-gated potassium channels (K_v) to cause membrane depolarization and action potential firing. Intracellular calcium levels are increased by release of Ca^{2+} from endoplasmic reticulum stores (mediated by IP_3R and RyR3) and also by Ca^{2+} influx through TRPC channels. Additional Ca^{2+} may also enter the cell after membrane depolarization by voltage-gated calcium channels (VGCC) which are subsequently inactivated by calmodulin. *PIP_2* phosphatidylinositol 4,5-bisphosphate; *IP_3* inositol triphosphate; *IP_3R* IP_3 receptor; *RyR3* ryanodine receptor 3

depolarization is caused by inhibition of voltage-gated potassium channels (K_v) [57, 58] and activation of canonical transient receptor potential channels (TRPC) [56, 59]. Indeed, individual GnRH neurons have been shown by RT-PCR to express several TRPC channel subunits [59]. GnRH neurons also show a rise in intracellular Ca^{2+} after kisspeptin stimulation [56]. This rise is due not only to inositol-(1,4,5)-triphosphate (IP_3)-mediated Ca^{2+} release from the endoplasmic reticulum, but also to Ca^{2+} entry through TRPC channels and Ca^{2+}-induced Ca^{2+} release via RyR3 (Fig. 22.3) [60]. Kisspeptin-induced membrane depolarization may also activate voltage-gated calcium channels (VGCC) to allow additional Ca^{2+} entry. The VGCC channels would be subsequently closed by a Ca^{2+}/calmodulin-dependent feedback loop.

A small number of GnRH neurons (around 2%) show intrinsic oscillations in membrane potential often accompanied by neuronal firing [61]. Kisspeptin causes these neurons to oscillate closer to the depolarization threshold, effectively making these neurons more sensitive to additional depolarization signals and increasing their chance of firing. If this sub-population of GnRH neurons act as pacemakers to co-ordinate the firing rate of the non-oscillating GnRH neurons, then this could provide a mechanism by which kisspeptin can regulate GnRH pulsatility.

Transgenic Mice with Fluorescently Marked *Kiss1* Neurons

Transgenic mice in which *Kiss1* neurons are marked with a fluorescent protein such as GFP allow unequivocal identification of these neurons in brain slice preparations to interrogate their function and responses to neuromodulators. Patch clamp recordings of GFP-labelled neurons in the ARC of female mice have shown that most *Kiss1* neurons express both a hyperpolarization-induced cation current (h-current) and a T-type Ca^{2+} current, which may produce intrinsic pacemaker activity in *Kiss1* neurons [42]. Expression of the relevant genes (HCN1-4 and $Ca_v3.1$) to produce these pacemaker channels was confirmed by RT-PCR analysis from individual *Kiss1* neurons. *Kiss1* neurons also respond to NMDA or glutamate by increasing their firing rates and to GABA by inhibiting firing rates [42]. In male mice, the majority of the *Kiss1* neurons in the ARC are depolarized by NKB and show an increased firing rate [21]. This observation is consistent with the hypothesis that NKB stimulates GnRH release via activation of *Kiss1* neurons in the ARC.

Cell Lines

Immortalized cell lines provide a less complex model system to study kisspeptin signalling pathways and how these influence cell biology and behaviour. Cell lines have the advantage that they represent a homogeneous cell population without the complex neuronal inputs found in vivo, greatly simplifying the experimental analyses. It is also easier to control experimental variables using cell lines rather than whole animals.

Cell lines were initially used to establish the intracellular signalling pathways associated with kisspeptin binding to its receptor. Originally, immortalized cell lines transfected with *Kiss1r* were used for these studies (CHO, HEK293, NIH 3T3 and Cos7 cells), but similar results have been found in cell lines with endogenous kisspeptin receptor expression (for review, see [62]). Kisspeptin receptor activation was found to couple through $G_{q/11}$ to activate phospholipase C and increase intracellular IP_3 and Ca^{2+} levels [3, 4, 63]. Kisspeptin was also found to activate mitogen-activated-protein-kinase (MAPK) pathways with increased phosphorylation of ERK1/2 and p38MAPK [3]. Activation of the MAPK pathways is thought to occur by Src-dependent transactivation of the EGF receptor [64, 65].

The *Kiss1* gene was originally identified as a suppressor of metastasis [66] and cell lines have been used to study the mechanism by which kisspeptins inhibit cell migration. Migratory inhibition may be caused by increased cell adhesion to the extracellular matrix at focal adhesion points or increased cell–cell binding at adherens junctions. Kisspeptin increases the phosphorylation of focal adhesion kinase

(FAK) and paxillin proteins, which would increase cell adhesion. Migratory inhibition of the trophoblast-derived cell line HTR8SVneo occurs through an ERK1/2-p90rsk pathway to inhibit GSK3β activity and increase β-catenin levels in the cytoplasm [64]. β-Catenin can bind to transmembrane cadherins to promote cell–cell binding [67]. In addition, inhibition of migration of the human ovarian cancer cell line SKOV3 can be overcome by phorbol ester stimulation of protein kinase Cα [68], suggesting that continuous kisspeptin signalling could reduce cell migration by inhibiting PKCα activity. Disruption of PKCα signalling is associated with increased cell adhesion in several different systems [69–71]. In addition to promoting cell adhesion, kisspeptins may decrease cell invasion by reducing the expression of proteases such as MMP-2 [72, 73]. Kisspeptins also inhibit the intracellular signalling cascade of the pro-metastatic G-protein coupled receptor CXCR4 [74].

Immortalized *GnRH* Neurons

Since GnRH neurons are the principal target for kisspeptin, immortalized cell lines derived from these neurons are useful for studying intracellular signalling pathways after kisspeptin stimulation. One of the most widely used GnRH cell lines is GT1-7, which was derived from a hypothalamic tumour in a female mouse induced by transgenic expression of the SV40 large T-antigen expressed from the GnRH promoter [75]. Similarly, immortalized GnRH neuronal cell lines (GN and NLT series sub-clones) have been derived from an olfactory bulb tumour induced in male mice [76, 77] by large T-antigen expression. Since the GN and NLT lines were derived from GnRH neurons during their normal migration from the olfactory bulb, they are generally considered to represent immature GnRH neurons.

To control for the developmental stage at which GnRH neurons are immortalized, Wolfe et al. [78] generated transgenic mice in which a doxycycline-regulated T-antigen transgene was used to select for cell lines (GRT cells) from adult mouse hypothalamic explants. Salvi et al. [79] derived several GnRH cell lines (Gnv clones) by immortalizing cells from isolated hypothalami by lentiviral delivery of a tetracycline—inducible *v-myc* oncogene. An advantage of these approaches is that the immortalized GnRH cell lines are selected in culture, which should limit the acquisition of mutations associated with tumour formation in vivo.

GT1-7 cells retain markers of GnRH neurons including expression of the kisspeptin receptor [80–83] and the ability to secrete GnRH in an autonomous pulsatile manner [84, 85]. It should also be noted, however, that the GT1-7 cell line also shows gene expression not normally found in GnRH neurons including the oestrogen receptor alpha (ERa) and *Kiss1* [80, 83, 86]. Kisspeptin stimulation of GT1-7 cells causes an increase in intracellular Ca^{2+} [87], ERK1/2 phosphorylation and GnRH secretion [80, 82], similar to the responses observed in GnRH neurons.

GT1-7 cells have also been used to study the mechanisms of transcriptional regulation of the *Kiss1* gene. Transient transfection of luciferase reporter plasmids with different lengths of the *Kiss1* promoter has shown that oestrogen can stimulate promoter activity [88]. Chromatin immunoprecipitation (CHIP) analysis showed that ERα and the transcription factor SP1 form a complex in vivo to mediate this oestrogen activation. Interestingly, while SP1 acted as a transcriptional activator, the SP3 protein functioned as a transcriptional repressor [88]. This could provide a mechanism to explain the differential regulation of *Kiss1* transcription by oestrogen between AVPV and ARC *Kiss1* neurons. If *Kiss1* neurons in the AVPV region expressed high levels of SP1, then *Kiss1* expression would be stimulated by oestrogen. Conversely, if *Kiss1* neurons in the ARC expressed high levels of SP3, then *Kiss1* expression would be repressed by oestrogen.

Other transcription factors have been shown to be expressed in ARC *Kiss1* neurons in the rat, including TTF1 [89], YY1, EAP1 and CUX1, and these have been shown to be recruited to the *Kiss1* promoter in HeLa cells by CHIP assays [86]. GT1-7 cells have been used to examine the effects of these transcription factors on *Kiss1* promoter activity. EAP1 and CUX1 isoform p110 were found to repress transcription from the *Kiss1* promoter, while YY1 and TTF1 had no effect in GT1-7 cells [86].

Cell Lines Expressing Kiss1

The laboratory of Denise Belsham has generated a large number of immortalized cell lines from hypothalamic neurons of mice [90] and rats [91]. Cells were immortalized by retroviral transduction of SV40 large T-antigen to primary cultures of dissociated cells from embryonic hypothalamic tissues. Clonal cell lines were isolated for further characterization by PCR to identify their expression profile of neuroendocrine markers. Each cell line had a unique signature of receptor and neuropeptide expression. Only one out of six mouse lines was reported to express *Kiss1* (mHypoE-36/1), but all six mouse lines expressed *Kiss1r* [92]. It is perhaps surprising that the mHypoE-36/1 cell line expresses both *Kiss1* and *Kiss1r* since *Kiss1r* expression has not been reported in *Kiss1* neurons in vivo [9]. The majority (25 out of 32) of the rat cell lines (rHypoE series) expressed *Kiss1* [91]. It is not clear why the proportion of *Kiss1* expressing cell lines was higher in the rat than the mouse; perhaps the rat *Kiss1* neurons are more mitotically active at this embryonic stage than in the mouse and consequently more susceptible to viral transduction. To our knowledge, these *Kiss1* expressing cell lines have not been used yet to study the physiology of kisspeptin neurons.

Limitations of Cell Line Models

While the use of immortalized cell lines has some advantages over whole animal studies, there are a number of limitations and caveats to this reductionist approach. Cell lines are no longer in the context of their normal environment, lacking input from other neurons or glial cells. They are not subjected to the normal regulatory

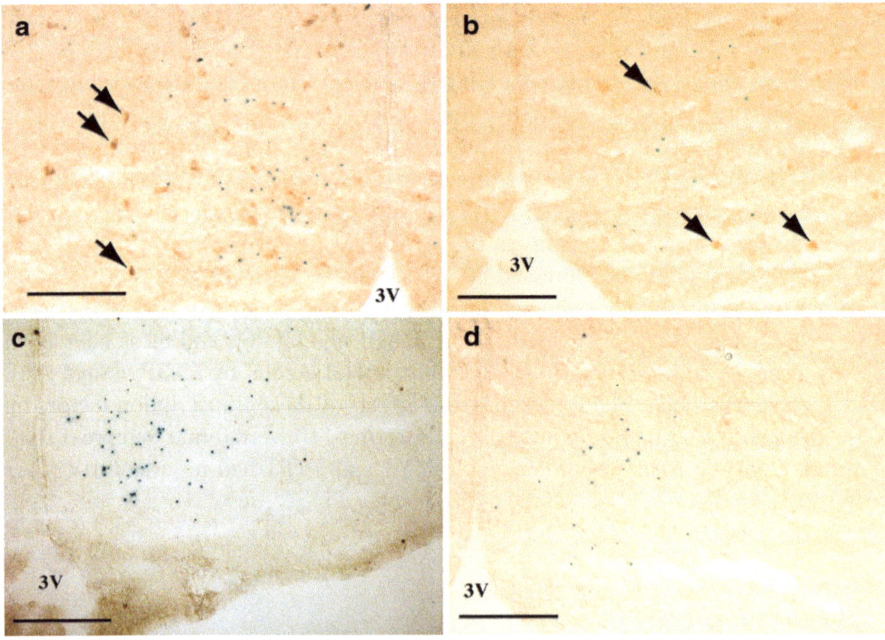

Fig. 22.4 Absence of POMC expression in *Kiss1* neurons. POMC neurons (*arrowed*) were visualized by immunohistochemistry using an anti-beta-endorphin antibody. *Kiss1* neurons were visualized by β-galactosidase activity (*blue dots*). No co-localization of POMC in *Kiss1* neurons was observed in the arcuate nucleus of the hypothalamus of male heterozygous $Kiss1_{tm1Coll}$ mice. (**a**, **c**) 2 weeks old; (**b**, **d**) 1 month old. (**a**, **b**) with 1:5,000 dilution of anti-beta-endorphin antibody (Peninsula Laboratories Inc.). (**c**, **d**) No primary antibody. *3V* third ventricle. Scale bars = 500 μm

inputs that might modulate their physiological responses to neurotransmitters or neuropeptides. Most of the immortalized cell lines are derived from embryonic neurons since retroviral delivery of SV40 T-antigen requires proliferating cells and does not work in post-mitotic adult neurons. This raises the question of how accurately immortalized cells model adult neurons. In addition, cell immortalization is associated with a number of genetic and epigenetic changes, which may alter the gene expression profile of the cell compared with its progenitor. For example, the rHypoE-7 and rHypoE-8 cell lines co-express *Kiss1* and *Pomc* [91], but co-expression has not been found in vivo [93] and we have failed to detect POMC co-expression in the $Kiss1^{tmColl}$ mice (see Fig. 22.4). Moreover, RT-PCR analysis of mRNA isolated from adult Kiss1-GFP neurons has also failed to detect *Pomc* expression [42] (Table 22.2).

Future Directions

Undoubtedly, transgenic mice will continue to play a significant role in understanding the key molecules that regulate the mammalian reproductive axis. Mouse mutagenesis programmes, such as those being undertaken by the International Knock-out

Table 22.2 Cell lines with kisspeptin expression

Cell Line	Origin	Detection method	References	Comments
LNCaP	Human prostatic carcinomas	RIA	[98]	Highest kisspeptin secretion from DU145 cell line
PC3				Mainly Kp54 secreted
DU145				
MIN6	Mouse insuloma	RT-PCR	[99]	Kp has been reported to increase insulin secretion from mouse Islets of Langerhans in culture
Jeg3	Human placental choriocarcinoma	RT-PCR	WHC, unpublished	
MCF-7	Human breast adenocarcinoma	RT-PCR	[100]	High *Kiss1* expression correlates with high expression of AP-2a transcription factor
rHypoE-7 rHypoE-8	Embryonic rat hypothalamus	RT-PCR	[91]	Immortalized by SV40 T-antigen
mHypo-36/1	Embryonic mouse hypothalamus	RT-PCR	[90, 92]	Immortalized by SV40 T-antigen

Mouse Consortium (IKMC, http://www.knockoutmouse.org/), will generate mice with reproductive defects and thereby identify novel genes involved in this process. The remit of the IKMC is to mutate all protein coding genes in mice through gene targeting and gene trapping in C57Bl/6 ES cells [94, 95]. Some of these gene targeting vectors produce a mutant allele that can be used to generate mice with a *LacZ* reporter—tagged allele, a LoxP conditional allele or a null allele [96].

Transgenic mice may also be used to examine the way in which *Kiss1* neurons integrate within the neuronal circuitry of the hypothalamus and how their function is co-ordinated with different physiological states. A fundamental step in understanding how *Kiss1* neurons are regulated by upstream signals requires an accurate definition of their neuronal connections. One approach would be to use a genetically modified pseudorabies virus that requires a CRE-mediated recombination event to generate a replication competent virus [97]. Delivery of this virus into Kiss-CRE transgenic mice will activate the virus specifically in *Kiss1* neurons and allow retrograde spread into synaptically connected neurons, which can be visualized by GFP. Using this approach it should be possible to trace primary, secondary and tertiary inputs into *Kiss1* neurons to map the neuronal circuitry.

Model systems have been essential in understanding the role of kisspeptin signalling in regulating the mammalian reproductive axis. Studies using cell lines provide important clues about regulatory pathways, which can be confirmed in vivo using transgenic mice. Transgenic mice will continue to be used to answer fundamental questions such as the interplay between NKB, dynorphin and other neuromodulators in regulating *Kiss1* neuronal activity and GnRH pulsatility, identification of upstream regulators of *Kiss1* and how *Kiss1* neurons are regulated by afferent inputs from other regions of the brain.

Acknowledgments We would like to thank Takeda Cambridge for their continued support, Ann Silver for proof reading the review and the animal house staff for excellent husbandry. This work was supported by a grant from the BBSRC (BB/F01936X/1) and the Ford Physiology Fund (WHC).

References

1. Oakley AE, Clifton DK, Steiner RA (2009) Kisspeptin signaling in the brain. Endocr Rev 30(6):713–743
2. d'Anglemont de Tassigny X, Colledge WH (2010) The role of kisspeptin signaling in reproduction. Physiology (Bethesda) 25(4):207–217
3. Kotani M, Detheux M, Vandenbogaerde A, Communi D, Vanderwinden JM, Le Poul E et al (2001) The metastasis suppressor gene KiSS-1 encodes kisspeptins, the natural ligands of the orphan G protein-coupled receptor GPR54. J Biol Chem 276(37):34631–34636
4. Muir AI, Chamberlain L, Elshourbagy NA, Michalovich D, Moore DJ, Calamari A et al (2001) AXOR12, a novel human G protein-coupled receptor, activated by the peptide KiSS-1. J Biol Chem 276(31):28969–28975
5. Ohtaki T, Shintani Y, Honda S, Matsumoto H, Hori A, Kanehashi K et al (2001) Metastasis suppressor gene KiSS-1 encodes peptide ligand of a G-protein-coupled receptor. Nature 411(6837):613–617
6. Mikkelsen JD, Simonneaux V (2009) The neuroanatomy of the kisspeptin system in the mammalian brain. Peptides 30(1):26–33
7. Clarkson J, Herbison AE (2006) Postnatal development of kisspeptin neurons in mouse hypothalamus; sexual dimorphism and projections to gonadotropin-releasing hormone neurons. Endocrinology 147(12):5817–5825
8. Messager S, Chatzidaki EE, Ma D, Hendrick AG, Zahn D, Dixon J et al (2005) Kisspeptin directly stimulates gonadotropin-releasing hormone release via G protein-coupled receptor 54. Proc Natl Acad Sci U S A 102(5):1761–1766
9. Herbison AE, de Tassigny X, Doran J, Colledge WH (2010) Distribution and postnatal development of Gpr54 gene expression in mouse brain and gonadotropin-releasing hormone neurons. Endocrinology 151(1):312–321
10. Gottsch ML, Cunningham MJ, Smith JT, Popa SM, Acohido BV, Crowley WF et al (2004) A role for kisspeptins in the regulation of gonadotropin secretion in the mouse. Endocrinology 145(9):4073–4077
11. Navarro VM, Castellano JM, Fernandez-Fernandez R, Barreiro ML, Roa J, Sanchez-Criado JE et al (2004) Developmental and hormonally regulated messenger ribonucleic acid expression of KiSS-1 and its putative receptor, GPR54, in rat hypothalamus and potent luteinizing hormone-releasing activity of KiSS-1 peptide. Endocrinology 145(10):4565–4574
12. d'Anglemont de Tassigny X, Fagg LA, Carlton MB, Colledge WH (2008) Kisspeptin can stimulate gonadotropin-releasing hormone (GnRH) release by a direct action at GnRH nerve terminals. Endocrinology 149(8):3926–3932
13. Clarkson J, d'Anglemont de Tassigny X, Colledge WH, Caraty A, Herbison AE (2009) Distribution of kisspeptin neurones in the adult female mouse brain. J Neuroendocrinol 21(8):673–682
14. Kauffman AS, Gottsch ML, Roa J, Byquist AC, Crown A, Clifton DK et al (2007) Sexual differentiation of Kiss1 gene expression in the brain of the rat. Endocrinology 148(4): 1774–1783
15. Smith JT, Cunningham MJ, Rissman EF, Clifton DK, Steiner RA (2005) Regulation of Kiss1 gene expression in the brain of the female mouse. Endocrinology 146(9):3686–3692
16. Adachi S, Yamada S, Takatsu Y, Matsui H, Kinoshita M, Takase K et al (2007) Involvement of anteroventral periventricular metastin/kisspeptin neurons in estrogen positive feedback action on luteinizing hormone release in female rats. J Reprod Dev 53(2):367–378

17. Clarkson J, d'Anglemont de Tassigny X, Moreno AS, Colledge WH, Herbison AE (2008) Kisspeptin-GPR54 signaling is essential for preovulatory gonadotropin-releasing hormone neuron activation and the luteinizing hormone surge. J Neurosci 28(35):8691–8697
18. Smith JT, Dungan HM, Stoll EA, Gottsch ML, Braun RE, Eacker SM et al (2005) Differential regulation of KiSS-1 mRNA expression by sex steroids in the brain of the male mouse. Endocrinology 146(7):2976–2984
19. Navarro VM (2012) New insights into the control of pulsatile GnRH release: the role of kiss1/neurokinin B neurons. Front Endocrinol (Lausanne) 3:48
20. Lehman MN, Coolen LM, Goodman RL (2010) Minireview: kisspeptin/neurokinin B/dynorphin (KNDy) cells of the arcuate nucleus: a central node in the control of gonadotropin-releasing hormone secretion. Endocrinology 151(8):3479–3489
21. Navarro VM, Gottsch ML, Wu M, Garcia-Galiano D, Hobbs SJ, Bosch MA et al (2011) Regulation of NKB pathways and their roles in the control of Kiss1 neurons in the arcuate nucleus of the male mouse. Endocrinology 152(11):4265–4275
22. Colledge WH (2009) Transgenic mouse models to study Gpr54/kisspeptin physiology. Peptides 30(1):34–41
23. Colledge WH, Mei H, d'Anglemont de Tassigny X (2010) Mouse models to study the central regulation of puberty. Mol Cell Endocrinol 324(1–2):12–20
24. Capecchi MR (2005) Gene targeting in mice: functional analysis of the mammalian genome for the twenty-first century. Nat Rev Genet 6(6):507–512
25. Glaser S, Anastassiadis K, Stewart AF (2005) Current issues in mouse genome engineering. Nat Genet 37(11):1187–1193
26. Seminara SB, Messager S, Chatzidaki EE, Thresher RR, Acierno JS Jr, Shagoury JK et al (2003) The GPR54 gene as a regulator of puberty. N Engl J Med 349(17):1614–1627
27. Funes S, Hedrick JA, Vassileva G, Markowitz L, Abbondanzo S, Golovko A et al (2003) The KiSS-1 receptor GPR54 is essential for the development of the murine reproductive system. Biochem Biophys Res Commun 312(4):1357–1363
28. Garcia-Galiano D, van Ingen Schenau D, Leon S, Krajnc-Franken MA, Manfredi-Lozano M, Romero-Ruiz A et al (2012) Kisspeptin signaling is indispensable for neurokinin B, but not glutamate, stimulation of gonadotropin secretion in mice. Endocrinology 153(1):316–328
29. Lapatto R, Pallais JC, Zhang D, Chan YM, Mahan A, Cerrato F et al (2007) Kiss1-/- mice exhibit more variable hypogonadism than Gpr54-/- mice. Endocrinology 148(10): 4927–4936
30. Dungan HM, Gottsch ML, Zeng H, Gragerov A, Bergmann JE, Vassilatis DK et al (2007) The role of kisspeptin-GPR54 signaling in the tonic regulation and surge release of gonadotropin-releasing hormone/luteinizing hormone. J Neurosci 27(44):12088–12095
31. Kauffman AS, Park JH, McPhie-Lalmansingh AA, Gottsch ML, Bodo C, Hohmann JG et al (2007) The kisspeptin receptor GPR54 is required for sexual differentiation of the brain and behavior. J Neurosci 27(33):8826–8835
32. d'Anglemont de Tassigny X, Fagg LA, Dixon JP, Day K, Leitch HG, Hendrick AG et al (2007) Hypogonadotropic hypogonadism in mice lacking a functional Kiss1 gene. Proc Natl Acad Sci U S A 104(25):10714–10719
33. Matsuda T, Cepko CL (2007) Controlled expression of transgenes introduced by in vivo electroporation. Proc Natl Acad Sci U S A 104(3):1027–1032
34. Franceschini I, Lomet D, Cateau M, Delsol G, Tillet Y, Caraty A (2006) Kisspeptin immunoreactive cells of the ovine preoptic area and arcuate nucleus co-express estrogen receptor alpha. Neurosci Lett 401(3):225–230
35. Mei H, Walters C, Carter R, Colledge WH (2011) Gpr54-/- mice show more pronounced defects in spermatogenesis than Kiss1-/- mice and improved spermatogenesis with age when exposed to dietary phytoestrogens. Reproduction 141(3):357–366
36. Chan YM, Broder-Fingert S, Wong KM, Seminara SB (2009) Kisspeptin/Gpr54-independent gonadotrophin-releasing hormone activity in Kiss1 and Gpr54 mutant mice. J Neuroendocrinol 21(12):1015–1023

37. Pampillo M, Camuso N, Taylor JE, Szereszewski JM, Ahow MR, Zajac M et al (2009) Regulation of GPR54 signaling by GRK2 and {beta}-arrestin. Mol Endocrinol 23(12): 2060–2074
38. d'Anglemont de Tassigny X, Ackroyd KJ, Chatzidaki EE, Colledge WH (2010) Kisspeptin signaling is required for peripheral but not central stimulation of gonadotropin-releasing hormone neurons by NMDA. J Neurosci 30(25):8581–8590
39. Mayer C, Boehm U (2011) Female reproductive maturation in the absence of kisspeptin/ GPR54 signaling. Nat Neurosci 14(6):704–710
40. Wen S, Gotze IN, Mai O, Schauer C, Leinders-Zufall T, Boehm U (2011) Genetic identification of GnRH receptor neurons: a new model for studying neural circuits underlying reproductive physiology in the mouse brain. Endocrinology 152(4):1515–1526
41. Irwig MS, Fraley GS, Smith JT, Acohido BV, Popa SM, Cunningham MJ et al (2004) Kisspeptin activation of gonadotropin releasing hormone neurons and regulation of KiSS-1 mRNA in the male rat. Neuroendocrinology 80(4):264–272
42. Gottsch ML, Popa SM, Lawhorn JK, Qiu J, Tonsfeldt KJ, Bosch MA et al (2011) Molecular properties of Kiss1 neurons in the arcuate nucleus of the mouse. Endocrinology 152(11):4298–4309
43. Mayer C, Acosta-Martinez M, Dubois SL, Wolfe A, Radovick S, Boehm U et al (2010) Timing and completion of puberty in female mice depend on estrogen receptor alpha-signaling in kisspeptin neurons. Proc Natl Acad Sci U S A 107(52):22693–22698
44. Cravo RM, Margatho LO, Osborne-Lawrence S, Donato J Jr, Atkin S, Bookout AL et al (2011) Characterization of Kiss1 neurons using transgenic mouse models. Neuroscience 173:37–56
45. Ahima RS (2005) Central actions of adipocyte hormones. Trends Endocrinol Metab 16(7):307–313
46. Ahima RS, Dushay J, Flier SN, Prabakaran D, Flier JS (1997) Leptin accelerates the onset of puberty in normal female mice. J Clin Invest 99(3):391–395
47. Carro E, Pinilla L, Seoane LM, Considine RV, Aguilar E, Casanueva FF et al (1997) Influence of endogenous leptin tone on the estrous cycle and luteinizing hormone pulsatility in female rats. Neuroendocrinology 66(6):375–377
48. Chehab FF, Lim ME, Lu R (1996) Correction of the sterility defect in homozygous obese female mice by treatment with the human recombinant leptin. Nat Genet 12(3):318–320
49. Mounzih K, Lu R, Chehab FF (1997) Leptin treatment rescues the sterility of genetically obese ob/ob males. Endocrinology 138(3):1190–1193
50. Donato J Jr, Cravo RM, Frazao R, Gautron L, Scott MM, Lachey J et al (2011) Leptin's effect on puberty in mice is relayed by the ventral premammillary nucleus and does not require signaling in Kiss1 neurons. J Clin Invest 121(1):355–368
51. Herbison AE, Porteous R, Pape JR, Mora JM, Hurst PR (2008) Gonadotropin-releasing hormone neuron requirements for puberty, ovulation, and fertility. Endocrinology 149(2): 597–604
52. Srinivas S, Watanabe T, Lin CS, William CM, Tanabe Y, Jessell TM et al (2001) Cre reporter strains produced by targeted insertion of EYFP and ECFP into the ROSA26 locus. BMC Dev Biol 1:4
53. Suter KJ, Song WJ, Sampson TL, Wuarin JP, Saunders JT, Dudek FE et al (2000) Genetic targeting of green fluorescent protein to gonadotropin-releasing hormone neurons: characterization of whole-cell electrophysiological properties and morphology. Endocrinology 141(1):412–419
54. Spergel DJ, Kruth U, Shimshek DR, Sprengel R, Seeburg PH (2001) Using reporter genes to label selected neuronal populations in transgenic mice for gene promoter, anatomical, and physiological studies. Prog Neurobiol 63(6):673–686
55. Han SK, Gottsch ML, Lee KJ, Popa SM, Smith JT, Jakawich SK et al (2005) Activation of gonadotropin-releasing hormone neurons by kisspeptin as a neuroendocrine switch for the onset of puberty. J Neurosci 25(49):11349–11356

56. Liu X, Lee K, Herbison AE (2008) Kisspeptin excites gonadotropin-releasing hormone neurons through a phospholipase C/calcium-dependent pathway regulating multiple ion channels. Endocrinology 149(9):4605–4614
57. Pielecka-Fortuna J, Chu Z, Moenter SM (2008) Kisspeptin acts directly and indirectly to increase gonadotropin-releasing hormone neuron activity and its effects are modulated by estradiol. Endocrinology 149(4):1979–1986
58. Pielecka-Fortuna J, Defazio RA, Moenter SM (2011) Voltage-gated potassium currents are targets of diurnal changes in estradiol feedback regulation and kisspeptin action on gonadotropin-releasing hormone neurons in mice. Biol Reprod 85(5):987–995
59. Zhang C, Roepke TA, Kelly MJ, Ronnekleiv OK (2008) Kisspeptin depolarizes gonadotropin-releasing hormone neurons through activation of TRPC-like cationic channels. J Neurosci 28(17):4423–4434
60. Zhang XB, Spergel DJ (2012) Kisspeptin inhibits high-voltage activated ca2+ channels in GnRH neurons via multiple ca(2+) influx and release pathways. Neuroendocrinology 96(1):68–80
61. Chu Z, Tomaiuolo M, Bertram R, Moenter SM (2012) Two types of burst firing in gonadotropin-releasing hormone (GnRH) neurones. J Neuroendocrinol 24(7):1065–1077
62. Castano JP, Martinez-Fuentes AJ, Gutierrez-Pascual E, Vaudry H, Tena-Sempere M, Malagon MM (2009) Intracellular signaling pathways activated by kisspeptins through GPR54: do multiple signals underlie function diversity? Peptides 30(1):10–15
63. Stafford LJ, Xia C, Ma W, Cai Y, Liu M (2002) Identification and characterization of mouse metastasis-suppressor KiSS1 and its G-protein-coupled receptor. Cancer Res 62(19): 5399–5404
64. Roseweir AK, Katz AA, Millar RP (2012) Kisspeptin-10 inhibits cell migration in vitro via a receptor-GSK3 beta-FAK feedback loop in HTR8SVneo cells. Placenta 33(5):408–415
65. Zajac M, Law J, Cvetkovic DD, Pampillo M, McColl L, Pape C et al (2011) GPR54 (KISS1R) transactivates EGFR to promote breast cancer cell invasiveness. PLoS One 6(6):e21599
66. Lee JH, Miele ME, Hicks DJ, Phillips KK, Trent JM, Weissman BE et al (1996) KiSS-1, a novel human malignant melanoma metastasis-suppressor gene. J Natl Cancer Inst 88(23):1731–1737
67. Daugherty RL, Gottardi CJ (2007) Phospho-regulation of Beta-catenin adhesion and signaling functions. Physiology (Bethesda) 22:303–309
68. Jiang Y, Berk M, Singh LS, Tan H, Yin L, Powell CT et al (2005) KiSS1 suppresses metastasis in human ovarian cancer via inhibition of protein kinase C alpha. Clin Exp Metastasis 22(5):369–376
69. Harrington EO, Loffler J, Nelson PR, Kent KC, Simons M, Ware JA (1997) Enhancement of migration by protein kinase Calpha and inhibition of proliferation and cell cycle progression by protein kinase Cdelta in capillary endothelial cells. J Biol Chem 272(11):7390–7397
70. Thomason HA, Cooper NH, Ansell DM, Chiu M, Merrit AJ, Hardman MJ et al (2012) Direct evidence that PKCalpha positively regulates wound re-epithelialisation: correlation with changes in desmosomal adhesiveness. J Pathol 227(3):346–356
71. O'Neill AK, Gallegos LL, Justilien V, Garcia EL, Leitges M, Fields AP et al (2011) Protein kinase Calpha promotes cell migration through a PDZ-dependent interaction with its novel substrate discs large homolog 1 (DLG1). J Biol Chem 286(50):43559–43568
72. Bilban M, Ghaffari-Tabrizi N, Hintermann E, Bauer S, Molzer S, Zoratti C et al (2004) Kisspeptin-10, a KiSS-1/metastin-derived decapeptide, is a physiological invasion inhibitor of primary human trophoblasts. J Cell Sci 117(pt 8):1319–1328
73. Yoshioka K, Ohno Y, Horiguchi Y, Ozu C, Namiki K, Tachibana M (2008) Effects of a KiSS-1 peptide, a metastasis suppressor gene, on the invasive ability of renal cell carcinoma cells through a modulation of a matrix metalloproteinase 2 expression. Life Sci 83(9–10):332–338
74. Navenot JM, Wang Z, Chopin M, Fujii N, Peiper SC (2005) Kisspeptin-10-induced signaling of GPR54 negatively regulates chemotactic responses mediated by CXCR4: a potential mechanism for the metastasis suppressor activity of kisspeptins. Cancer Res 65(22): 10450–10456

75. Mellon PL, Windle JJ, Goldsmith PC, Padula CA, Roberts JL, Weiner RI (1990) Immortalization of hypothalamic GnRH neurons by genetically targeted tumorigenesis. Neuron 5(1):1–10
76. Radovick S, Wray S, Lee E, Nicols DK, Nakayama Y, Weintraub BD et al (1991) Migratory arrest of gonadotropin-releasing hormone neurons in transgenic mice. Proc Natl Acad Sci U S A 88(8):3402–3406
77. Zhen S, Dunn IC, Wray S, Liu Y, Chappell PE, Levine JE et al (1997) An alternative gonadotropin-releasing hormone (GnRH) RNA splicing product found in cultured GnRH neurons and mouse hypothalamus. J Biol Chem 272(19):12620–12625
78. Wolfe A, Ng Y, Divall SA, Singh SP, Radovick S (2008) Development of an immortalised, post-pubertal gonadotrophin-releasing hormone neuronal cell line. J Neuroendocrinol 20(9):1029–1037
79. Salvi R, Castillo E, Voirol MJ, Glauser M, Rey JP, Gaillard RC et al (2006) Gonadotropin-releasing hormone-expressing neurons immortalized conditionally are activated by insulin: implication of the mitogen-activated protein kinase pathway. Endocrinology 147(2):816–826
80. Quaynor S, Hu L, Leung PK, Feng H, Mores N, Krsmanovic LZ et al (2007) Expression of a functional GPR54-Kisspeptin autoregulatory system in hypothalamic GnRH neurons. Mol Endocrinol 21(12):3062–3070
81. Tonsfeldt KJ, Goodall CP, Latham KL, Chappell PE (2011) Oestrogen induces rhythmic expression of the kisspeptin-1 receptor GPR54 in hypothalamic gonadotrophin-releasing hormone-secreting GT1-7 cells. J Neuroendocrinol 23(9):823–830
82. Novaira HJ, Ng Y, Wolfe A, Radovick S (2009) Kisspeptin increases GnRH mRNA expression and secretion in GnRH secreting neuronal cell lines. Mol Cell Endocrinol 311(1–2):126–134
83. Jacobi JS, Martin C, Nava G, Jeziorski MC, Clapp C, Martinez de la Escalera G (2007) 17-Beta-estradiol directly regulates the expression of adrenergic receptors and kisspeptin/GPR54 system in GT1-7 GnRH neurons. Neuroendocrinology 86(4):260–269
84. Martinez de la Escalera G, Choi AL, Weiner RI (1992) Generation and synchronization of gonadotropin-releasing hormone (GnRH) pulses: intrinsic properties of the GT1-1 GnRH neuronal cell line. Proc Natl Acad Sci U S A 89(5):1852–1855
85. Krsmanovic LZ, Stojilkovic SS, Merelli F, Dufour SM, Virmani MA, Catt KJ (1992) Calcium signaling and episodic secretion of gonadotropin-releasing hormone in hypothalamic neurons. Proc Natl Acad Sci U S A 89(18):8462–8466
86. Mueller JK, Dietzel A, Lomniczi A, Loche A, Tefs K, Kiess W et al (2011) Transcriptional regulation of the human KiSS1 gene. Mol Cell Endocrinol 342(1–2):8–19
87. Ozcan M, Alcin E, Ayar A, Yilmaz B, Sandal S, Kelestimur H (2011) Kisspeptin-10 elicits triphasic cytosolic calcium responses in immortalized GT1-7 GnRH neurones. Neurosci Lett 492(1):55–58
88. Li D, Mitchell D, Luo J, Yi Z, Cho SG, Guo J et al (2007) Estrogen regulates KiSS1 gene expression through estrogen receptor alpha and SP protein complexes. Endocrinology 148(10):4821–4828
89. Mastronardi C, Smiley GG, Raber J, Kusakabe T, Kawaguchi A, Matagne V et al (2006) Deletion of the Ttf1 gene in differentiated neurons disrupts female reproduction without impairing basal ganglia function. J Neurosci 26(51):13167–13179
90. Belsham DD, Cai F, Cui H, Smukler SR, Salapatek AM, Shkreta L (2004) Generation of a phenotypic array of hypothalamic neuronal cell models to study complex neuroendocrine disorders. Endocrinology 145(1):393–400
91. Gingerich S, Wang X, Lee PK, Dhillon SS, Chalmers JA, Koletar MM et al (2009) The generation of an array of clonal, immortalized cell models from the rat hypothalamus: analysis of melatonin effects on kisspeptin and gonadotropin-inhibitory hormone neurons. Neuroscience 162(4):1134–1140
92. Mayer CM, Fick LJ, Gingerich S, Belsham DD (2009) Hypothalamic cell lines to investigate neuroendocrine control mechanisms. Front Neuroendocrinol 30(3):405–423

93. Goodman RL, Lehman MN, Smith JT, Coolen LM, de Oliveira CV, Jafarzadehshirazi MR et al (2007) Kisspeptin neurons in the arcuate nucleus of the ewe express both dynorphin A and neurokinin B. Endocrinology 148(12):5752–5760
94. Skarnes WC, Rosen B, West AP, Koutsourakis M, Bushell W, Iyer V et al (2011) A conditional knockout resource for the genome-wide study of mouse gene function. Nature 474(7351):337–342
95. Ringwald M, Iyer V, Mason JC, Stone KR, Tadepally HD, Kadin JA et al (2011) The IKMC web portal: a central point of entry to data and resources from the International Knockout Mouse Consortium. Nucleic Acids Res 39(Database issue):D849–D855
96. Testa G, Schaft J, van der Hoeven F, Glaser S, Anastassiadis K, Zhang Y et al (2004) A reliable lacZ expression reporter cassette for multipurpose, knockout-first alleles. Genesis 38(3):151–158
97. DeFalco J, Tomishima M, Liu H, Zhao C, Cai X, Marth JD et al (2001) Virus-assisted mapping of neural inputs to a feeding center in the hypothalamus. Science 291(5513): 2608–2613
98. Curtis AE, Murphy KG, Chaudhri OB, Ramachandran R, Young AM, Waxman J et al (2010) Kisspeptin is released from human prostate cancer cell lines but plasma kisspeptin is not elevated in patients with prostate cancer. Oncol Rep 23(6):1729–1734
99. Hauge-Evans AC, Richardson CC, Milne HM, Christie MR, Persaud SJ, Jones PM (2006) A role for kisspeptin in islet function. Diabetologia 49(9):2131–2135
100. Mitchell DC, Abdelrahim M, Weng J, Stafford LJ, Safe S, Bar-Eli M et al (2006) Regulation of KiSS-1 metastasis suppressor gene expression in breast cancer cells by direct interaction of transcription factors activator protein-2alpha and specificity protein-1. J Biol Chem 281(1):51–58

Index

A.S. Kauffman and J.T. Smith (eds.), *Kisspeptin Signaling in Reproductive Biology*, Advances in Experimental Medicine and Biology 784, DOI 10.1007/978-1-4614-6199-9,

H

FSC
www.fsc.org
MIX
Papier aus verantwortungsvollen Quellen
Paper from responsible sources
FSC® C105338

If you have any concerns about our products,
you can contact us on
ProductSafety@springernature.com

In case Publisher is established outside the EU,
the EU authorized representative is:
Springer Nature Customer Service Center GmbH
Europaplatz 3, 69115 Heidelberg, Germany

Printed by Libri Plureos GmbH
in Hamburg, Germany